Lecture Notes in Computer Science 16143

Founding Editors

Gerhard Goos
Juris Hartmanis

Editorial Board Members

Elisa Bertino, *Purdue University, West Lafayette, IN, USA*
Wen Gao, *Peking University, Beijing, China*
Bernhard Steffen ⓘ, *TU Dortmund University, Dortmund, Germany*
Moti Yung ⓘ, *Columbia University, New York, NY, USA*

The series Lecture Notes in Computer Science (LNCS), including its subseries Lecture Notes in Artificial Intelligence (LNAI) and Lecture Notes in Bioinformatics (LNBI), has established itself as a medium for the publication of new developments in computer science and information technology research, teaching, and education.

LNCS enjoys close cooperation with the computer science R & D community, the series counts many renowned academics among its volume editors and paper authors, and collaborates with prestigious societies. Its mission is to serve this international community by providing an invaluable service, mainly focused on the publication of conference and workshop proceedings and postproceedings. LNCS commenced publication in 1973.

Pavithra Prabhakar · Andrea Vandin
Editors

Quantitative Evaluation of Systems and Formal Modeling and Analysis of Timed Systems

Second International Joint Conference, QEST+FORMATS 2025
Aarhus, Denmark, August 26–28, 2025
Proceedings

Editors
Pavithra Prabhakar
Kansas State University
Manhattan, KS, USA

Andrea Vandin
Sant'Anna School of Advanced Studies
Pisa, Italy

ISSN 0302-9743   ISSN 1611-3349 (electronic)
Lecture Notes in Computer Science
ISBN 978-3-032-05791-4   ISBN 978-3-032-05792-1 (eBook)
https://doi.org/10.1007/978-3-032-05792-1

© The Editor(s) (if applicable) and The Author(s), under exclusive license
to Springer Nature Switzerland AG 2026

This work is subject to copyright. All rights are solely and exclusively licensed by the Publisher, whether the whole or part of the material is concerned, specifically the rights of translation, reprinting, reuse of illustrations, recitation, broadcasting, reproduction on microfilms or in any other physical way, and transmission or information storage and retrieval, electronic adaptation, computer software, or by similar or dissimilar methodology now known or hereafter developed.
The use of general descriptive names, registered names, trademarks, service marks, etc. in this publication does not imply, even in the absence of a specific statement, that such names are exempt from the relevant protective laws and regulations and therefore free for general use.
The publisher, the authors and the editors are safe to assume that the advice and information in this book are believed to be true and accurate at the date of publication. Neither the publisher nor the authors or the editors give a warranty, expressed or implied, with respect to the material contained herein or for any errors or omissions that may have been made. The publisher remains neutral with regard to jurisdictional claims in published maps and institutional affiliations.

This Springer imprint is published by the registered company Springer Nature Switzerland AG
The registered company address is: Gewerbestrasse 11, 6330 Cham, Switzerland

If disposing of this product, please recycle the paper.

Preface

This book proudly presents the proceedings of the joint event for the 22nd International Conference on Quantitative Evaluation of SysTems (QEST 2025) and the 23rd International Conference on Formal Modeling and Analysis of Timed Systems (FORMATS 2025). The two flagship conferences are pursuing a strategic process of joining forces and sparking a new joint conference for quantitative modeling, analysis, and verification. This year, for the second time, the two conferences formed the QEST-FORMATS joint conference, which was held on 26–28 August 2025 in Aarhus, Denmark, as part of the CONFEST 2025 umbrella conference.

This edition of QEST and FORMATS continued the strategic merging of the two conferences started last year after both had run in excess of 21 editions. The major activities of this year for continuing implementing the partnership were (a) forming a joint program committee (PC) as a selection of previous members of the two PCs and of the joint one from last year, and inviting new members, (b) having a joint call for papers, (c) having an integrated review process and artifact evaluation adapted to the focus areas of the two conferences, (d) having an integrated PC and review process for artifact evaluation, and (e) having an integrated program.

As a result of maintaining a standard of high-quality publications, reviews, talks, speakers, and discussions despite strong competition, we were delighted to present a strong and diverse QEST-FORMATS 2025 program selected by an expert PC. The 37 PC members, with help from their external reviewers, decided to accept 26 papers out of 57 submissions. Each submission received at least 3 single-blind reviews. All accepted papers were very strong, and most decisions were unanimous. This year, we observed an increase in the number of submissions, reflecting strong support for merging of the two conferences by the community. Artifact evaluation is a well-established part of both QEST and FORMATS. QEST-FORMATS 2025 included an artifact evaluation (AE), carried out by 26 AE Committee members (AEC). Three AEC members reviewed each artifact in terms of consistency and reproducibility of results presented in the paper, completeness, future-proofness, documentation and ease of use, and availability in an online repository with a DOI. Based on these reviews, and following the EAPLS guidelines for artifact badging, every artifact was awarded up to two badges (Available | Functional), to be displayed on the front page of the corresponding paper. From 31 artifacts reviewed, 30 received the Available badge, and 25 the Functional badge.

As usual, the purpose of the joint conference was to foster progress in the quantitative evaluation and verification of systems, to promote the study of fundamental and practical aspects of systems with quantitative nature (such as probability, timing, or cost), and to bring together researchers from different disciplines who share interests in the modeling, design, and analysis of computational systems. The particular focus was to stay on the cutting edge of research by welcoming contributions that tap into data-driven and machine-learning systems, case studies, and tool papers.

A great highlight was having three excellent invited speakers, Christoph Matheja, University of Oldenburg, Germany, and DTU Technical University of Denmark; Alessandro Abate, University of Oxford, UK; Lu Feng, University of Virginia, USA. All three are known worldwide for their strong research on formal and quantitative aspects of systems.

Finally, we have to thank several people sincerely. First, we want to thank all authors who submitted their work to QEST-FORMATS 2025. Likewise, we thank all PC members and all external reviewers for their hard work and timely reviews. The process of evaluating artifacts went smoothly, thanks to Carlos E. Budde and Ratan Lal and their excellent work in chairing the artifact evaluation. All the reviewing and evaluation work was supported by EasyChair, which made our lives much easier. Special thanks go to Marco Paolieri for serving as the QEST+FORMATS publicity chair. We also thank the local organizing committee of CONFEST and both QEST and FORMATS steering committees. The joint event was a memorable occasion, and we want to thank everybody who contributed in any way.

August 2025

Pavithra Prabhakar
Andrea Vandin

Organization

Program Committee Chairs

Pavithra Prabhakar	Kansas State University, USA
Andrea Vandin	Sant'Anna School of Advanced Studies, Italy & Technical University of Denmark, Denmark

Publicity Chair

Marco Paolieri   University of Southern California, USA

Artifact Evaluation Chairs

Carlos E. Budde	Technical University of Denmark, Denmark
Ratan Lal	Northwest Missouri State University, USA

QEST Steering Committee

Erika Ábrahám	RWTH Aachen University, Germany
Benoît Barbot	Université Paris-Est Créteil, France
Ezio Bartocci	TU Wien, Austria
Jane Hillston	University of Edinburgh, UK
Nils Jansen	Ruhr University Bochum, Germany
Diwakar Krishnamurthy	Calgary University, Canada
Andrea Marin	Università Ca' Foscari Venezia, Italy
Guillermo Alberto Perez	University of Antwerp, Belgium
Anne Remke	University of Münster, Germany
Enrico Vicario (Chair)	University of Florence, Italy

FORMATS Steering Committee

Rajeev Alur	University of Pennsylvania, USA
Eugene Asarin	Université Paris Cité, France
Martin Fränzle (Chair)	University of Oldenburg, Germany

Joost-Pieter Katoen — RWTH Aachen University, Germany
Kim G. Larsen — Aalborg University, Denmark
Oded Maler (Founding Chair, 1957–2018) — University of Grenoble, France
Pavithra Prabhakar — Kansas State University, USA
Mariëlle Stoelinga — TU Twente, Netherlands
Wang Yi — Uppsala University, Sweden

Program Web Co-chairs

Marco Paolieri — University of Southern California, USA
Andrea Vandin — Sant'Anna School of Advanced Studies, Pisa & Technical University of Denmark, Denmark

Program Committee

Alessandro Abate — University of Oxford, UK
Ezio Bartocci — TU Wien, Austria
Andrea Burattin — Technical University of Denmark, Denmark
Valentina Castiglioni — Eindhoven University of Technology, Netherlands
Pedro R. D'Argenio — Universidad Nacional de Córdoba – CONICET, Argentina
Thao Dang — CNRS/VERIMAG, France
Taylor T. Johnson — Vanderbilt University, USA
Yusuke Kawamoto — National Institute of Advanced Industrial Science and Technology, Japan
S. Krishna — IIT Bombay, India
Diwakar Krishnamurthy — University of Calgary, Canada
Arnd Hartmanns — University of Twente, Netherlands
Jane Hillston — University of Edinburgh, UK
Engel Lefaucheux — Inria Nancy, Loria, Université de Lorraine, France
Michele Loreti — University of Camerino, Italy
Jun Liu — University of Waterloo, Canada
Andrea Marin — Università Ca' Foscari Venezia, Italy
Dejan Nickovic — Austrian Institute of Technology AIT, Austria
Marco Paolieri — University of Southern California, USA
Loïc Paulevé — CNRS/LaBRI, Bordeaux, France
Tuan Phung-Duc — University of Tsukuba, Japan
Anne Remke — WWU Münster, Germany
Camilo Rocha — Pontificia Universidad Javeriana Cali, Colombia

César Sánchez — IMDEA Software Institute, Spain
Ocan Sankur — Mitsubishi Electric R&D Centre Europe, UK
Ramin Sadre — Université catholique de Louvain, Belgium
Sadegh Soudjani — Max Planck Institute for Software Systems, Germany
Jeremy Sproston — University of Turin, Italy
Max Tschaikowski — Aalborg University, Denmark
Ashutosh Trivedi — University of Colorado Boulder, USA
Hoang Dung Tran — University of Nebraska-Lincoln, USA
Benny Van Houdt — University of Antwerp, Belgium
Mahesh Viswanathan — University of Illinois Urbana-Champaign, USA
Masaki Waga — Kyoto University, Japan
Verena Wolf — Saarland University, Germany
Tichakorn Wongpiromsarn — Iowa State University, USA
Bai Xue — Chinese Academy of Sciences, China
Paolo Zuliani — Università di Roma La Sapienza, Italy

Artifact Evaluation Committee

Adityo Anggraito — Ca'Foscari University, Italy
Chris Johannsen — Iowa State University, USA
Diletta Olliaro — Ca'Foscari University, Italy
Dylan Marinho — Sorbonne University, France
Ernesto Casablanca — Newcastle University, UK
Eshita Zaman — Utah Valley University, USA
Filip Smola — University of Edinburgh, UK
Gaëtan Staquet — Inria, France
Gokul Hariharan — Intel, USA
Kentaro Kobayashi — Tsukuba University & AIST, Japan
Leonardo Picchiami — Università di Roma La Sapienza, Italy
Marco Esposito — Universitè Paris-Est Creteil, France
Marco Lewis — Inria, France
Mark van Wijk — University of Twente, Netherlands
Maurice Laveaux — Eindhoven University of Technology, Netherlands
Mikael Bisgaard Dahlsen-Jensen — Aarhus University, Denmark
Minjian Zhang — University of Illinois Urbana-Champaign, USA
Mohammad Mamduhi — University of Birmingham, UK
Nabarun Deka — University of Illinois Urbana-Champaign, USA
Nausheen Mohammed — University of Illinois Urbana-Champaign, USA
Nicolaj Ø. Jensen — Aalborg Universitet, Denmark
Samuel Sasaki — Vanderbilt University, USA

Simone Pernice   University of Turin, Italy
Sungwoo Choi   University of Nebraska-Lincoln, USA
Thom Badings   University of Oxford, UK
Yuntao Li   University of Nebraska-Lincoln, USA

Contents

A Hyperlogic for Strategies in Stochastic Games 1
Lina Gerlach, Christof Löding, and Erika Ábrahám

Time-Sensitive Importance Splitting 21
Gabriel Dengler, Carlos E. Budde, Laura Carnevali, and Arnd Hartmanns

Active Learning of Mealy Machines with Timers 42
Véronique Bruyère, Bharat Garhewal, Guillermo A. Pérez, Gaëtan Staquet, and Frits W. Vaandrager

Formal Control for Uncertain Systems via Contract-Based Probabilistic Surrogates 62
Oliver Schön, Sofie Haesaert, and Sadegh Soudjani

Statistical Model Checking Beyond Means: Quantiles, CVaR, and the DKW Inequality 83
Carlos E. Budde, Arnd Hartmanns, Tobias Meggendorfer, Maximilian Weininger, and Patrick Wienhöft

Signal Sampling and Optimisation Under Symbolic Timed Automata Constraints 95
Benoît Barbot, Nicolas Basset, Thao Dang, Alexandre Donzé, Marco Esposito, and Dejan Nickovic

Programming and Reasoning in Partially Observable Probabilistic Environments 115
Tobias Gürtler and Benjamin Lucien Kaminski

PyDSMC: Statistical Model Checking for Neural Agents Using the Gymnasium Interface 134
Timo P. Gros, Arnd Hartmanns, Ivo Hoese, Joshua Meyer, Nicola J. Müller, and Verena Wolf

Minimal Per-Flow Backlog Bounds at an Aggregate FIFO Server Under Piecewise-Linear Arrival Curves 157
Lukas Wildberger, Anja Hamscher, and Jens B. Schmitt

Learning Mealy Machines with Sparse Observation Tables 176
Wolffhardt Schwabe, Paul Kogel, and Sabine Glesner

What Are the Odds? Improving Statistical Model Checking of Markov
Decision Processes 195
  Tobias Meggendorfer, Maximilian Weininger, and Patrick Wienhöft

A Product-Form Model for Systems with Aging Objects and Similarities 219
  Andrea Marin, Diletta Olliaro, Sabina Rossi, and Daniel Menasché

Tightening the Frontier of Decidability for Decisiveness 237
  Gaspard Fougea, Serge Haddad, Lina Ye, Shreyas Jain, and Alain Finkel

Fuzzy Fault Trees: the Fast and the Formal 256
  *Thi Kim Nhung Dang, Benedikt Peterseim,
  Milan Lopuhaä-Zwakenberg, and Mariëlle Stoelinga*

Positive Almost-Sure Termination of Polynomial Random Walks 275
  Lorenz Winkler and Laura Kovács

Noninterference Analysis of Deterministically Timed Reversible Systems 293
  Andrea Esposito, Alessandro Aldini, and Marco Bernardo

Controller Synthesis for Parametric Timed Games 314
  *Mikael Bisgaard Dahlsen-Jensen, Baptiste Fievet, Laure Petrucci,
  and Jaco van de Pol*

Symbolic Reduction for Formal Synthesis of Global Lyapunov Functions 333
  Jun Liu and Maxwell Fitzsimmons

Using Communication to Bound Clock Drift in Local-Timed Negotiations 353
  *Abhinav Garg, Madhavan Mukund, Adwitee Roy, B. Srivathsan,
  and Gautham Viswanathan*

Numerical Errors in Quantitative System Analysis With Decision Diagrams 371
  *Sebastiaan Brand, Arend-Jan Quist, Richard M. K. van Dijk,
  and Alfons Laarman*

Modeling Uncertainty: From Simulink to Stochastic Hybrid Automata 389
  *Pauline Blohm, Felix Schulz, Lisa Willemsen, Anne Remke,
  and Paula Herber*

Statistical Bayesian Inference for Stochastic Process Discovery 409
  Pierre Cry, Paolo Ballarini, András Horváth, and Pascale Le Gall

Conservation Analysis and Discrete Probabilistic Approximations
for Parameter Estimation of Biochemical Networks 429
   *Olivier Bouët-Willaumez, Adrien Le Coënt, Benoît Barbot,
and Nihal Pekergin*

Formal Approximations of the Transient Distributions of the M/G/1
Workload Process 449
   Fabian Michel and Markus Siegle

On Choice of Loss Functions for Neural Control Barrier Certificates 468
   Alireza Nadali, Ashutosh Trivedi, and Majid Zamani

Computing the Congestion Phases of Dynamical Systems with Priorities
and Application to Emergency Departments 487
   Xavier Allamigeon, Pascal Capetillo, and Stéphane Gaubert

Author Index 507

A Hyperlogic for Strategies in Stochastic Games

Lina Gerlach[✉], Christof Löding, and Erika Ábrahám

RWTH Aachen University, Aachen, Germany
{gerlach,loeding,abraham}@cs.rwth-aachen.de

Abstract. We propose a probabilistic hyperlogic called HyperSt^2 that can express hyperproperties of strategies in turn-based stochastic games. To the best of our knowledge, HyperSt^2 is the first hyperlogic for stochastic games. HyperSt^2 can relate probabilities of several independent executions of strategies in a stochastic game. For example, in HyperSt^2 it is natural to formalize optimality, i.e., to express that some strategy is better than all other strategies, or to express the existence of Nash equilibria. We investigate the expressivity of HyperSt^2 by comparing it to existing logics for stochastic games, as well as existing hyperlogics. Though the model-checking problem for HyperSt^2 is in general undecidable, we show that it becomes decidable for bounded memory and is in EXPTIME and PSPACE-hard over memoryless deterministic strategies, and we identify a fragment for which the model-checking problem is PSPACE-complete.

1 Introduction

Hyperproperties, formally defined as properties of sets of traces [13], can relate different executions of a system. Hyperlogics like HyperLTL and HyperCTL* [12] have been proposed to express hyperproperties of discrete nondeterministic systems. In the presence of random components in the system behavior, *probabilistic hyperproperties* can express probabilistic relations between several independent executions of a probabilistic, potentially nondeterministic, system. Probabilistic hyperlogics for *Markov decision processes* (*MDPs*) or *discrete-time Markov chains* (*DTMCs*) have been proposed in [1,2,15,22].

Since MDPs can be viewed as single-player stochastic games, it is natural to extend this setting to multiple players (also called *agents*). In *multi-player stochastic games*, several agents may influence the choice of which action is taken, and each action is associated with a probabilistic distribution over successor states. In the class of turn-based games, at every step exactly one of the agents can make a choice, while in concurrent games, every agent may make a choice at every step.

On the one hand, temporal logics like rPATL [20,21] and SGL [4] allow to specify properties of stochastic games, but not at the level of hyperproperties. On the other hand, temporal logics like SL [9] or ATL* [3] have been lifted to hyperlogics, HyperSL [7] and HyperATL* [6], to express hyperproperties of concurrent non-stochastic games, but without considering probabilistic behavior. However, to the best of our knowledge, there does not exist any probabilistic hyperlogic for stochastic games yet.

In this paper, we propose the logic HyperSt^2 for **Hyper**properties of **St**rategies in turn-based **St**ochastic games. HyperSt^2 can relate strategies and compare induced probabilities in the resulting probabilistic computation trees. For example, we can relate several independent executions of a strategy for some player against different strategies of other players. It is thus natural to formalize in HyperSt^2 a notion of "optimality" for strategies, as illustrated by the following example.

Example 1. Consider a two-player turn-based stochastic game with a unique initial state labeled with *init*, where agent 1 wants to reach some set of states T (labeled t). We wish to express that agent 1 has a strategy that is optimal in the sense that for any fixed behavior of agent 2, it achieves the highest winning probability under all of its other strategies. The below HyperSt^2 formula compares two plays of the game, starting in states $\hat{s}$ and $\hat{s}'$, respectively, ("$\forall \hat{s} \forall \hat{s}'$") that are required to be initial ("$init_{\hat{s}} \wedge init_{\hat{s}'}$"). The formula states that there exists a strategy for agent 1 in the play from $\hat{s}$ ("$\langle\!\langle 1 \rangle\!\rangle_{\hat{s}}$"), such that under any strategy of agent 1 in the play from $\hat{s}'$ ("$\|1\|_{\hat{s}'}$"), and any strategy that agent 2 follows in both plays ("$\|2\|_{\hat{s},\hat{s}'}$"), the probability to reach T in the first play ("$\mathbb{P}(\Diamond t_{\hat{s}})$") is at least as high as in the second play ("$\geq \mathbb{P}(\Diamond t_{\hat{s}'})$").

$$\forall \hat{s} \, \forall \hat{s}' \, \langle\!\langle 1 \rangle\!\rangle_{\hat{s}} \, \|1\|_{\hat{s}'} \, \|2\|_{\hat{s},\hat{s}'} \, (init_{\hat{s}} \wedge init_{\hat{s}'}) \Rightarrow \mathbb{P}(\Diamond t_{\hat{s}}) \geq \mathbb{P}(\Diamond t_{\hat{s}'})$$

As another example, in HyperSt^2 we can also express the existence of a Nash equilibrium, i.e., the existence of a strategy for each agent such that all strategies together are optimal in the sense that no agent can improve their outcome by unilaterally deviating from it. In Sect. 3, we give more examples of relevant properties that can be specified in HyperSt^2.

Related Work. Alternating-time Temporal Logic (ATL*) and its fragment ATL [3] are temporal logics for concurrent non-stochastic games, extending CTL* and CTL [17] to multiple agents. Game logic (GL) [3] generalizes ATL* by separating path and strategy quantification. GL subsumes ATL*. Strategy Logic (SL) [9] allows to explicitly quantify over strategies of turn-based non-stochastic games. SL subsumes ATL* and GL on turn-based non-stochastic games.

Stochastic Game Logic (SGL) [4] is a probabilistic variant of ATL* for turn-based stochastic games, and employs deterministic Rabin automata to reason about the probability that some ω-regular objective is satisfied. Another temporal logic for turn-based stochastic games is rPATL [10]. It allows to specify probabilistic or reward-based zero-sum objectives by combining ATL [3], PCTL [19] and the reward operator. rPATL has been extended to concurrent stochastic games [20], and to non-zero-sum objectives [21], namely subgame perfect social welfare optimal Nash equilibria.

HyperATL* [6] is a hyperlogic for strategies in concurrent non-stochastic games combining ATL* and HyperCTL* [12]. HyperATL* is subsumed by HyperSL [7], which is an extension of SL and is also defined for concurrent non-stochastic games. Like SL, HyperSL allows explicit first-order quantification over strategies. Strategy variables are not associated with a specific agent when they are quantified, and may be assigned to several different agents in a strategy profile. The outcomes of strategy profiles can then be bound to named path variables and compared.

HyperPCTL [16] and PHL [15] are probabilistic hyperlogics for MDPs, which can be viewed as single-player stochastic games. HyperPCTL over MDPs extends PCTL with quantification over states and schedulers. PHL extends HyperCTL* with the probability operator and quantification over schedulers. HyperPCTL* [22] is defined only over DTMCs and extends PCTL* [19] with quantification over paths.

Contributions. In this work, we propose the logic HyperSt^2 to express hyperproperties for strategies in turn-based stochastic games. HyperSt^2 combines elements of HyperPCTL [16] and SGL [4]. Like HyperPCTL, HyperSt^2 extends PCTL by quantification over game executions (through quantification over the game's starting states) and quantification over strategies. However, HyperSt^2 allows a more flexible strategy quantification structure than HyperPCTL, adapted to the setting of stochastic games. HyperSt^2 restricts quantification over states to the beginning of the formula, but, like in SGL, quantification over strategies may be nested and thus may also occur inside probability operators. The relationship between state and strategy variables mirrors the HyperPCTL setting: Several state variables may be associated with the same strategy, but at every point in time, each state variable is associated with at most one strategy for every agent.

We investigate the expressiveness of HyperSt^2 by comparing it to temporal logics over turn-based stochastic games, as well as hyperlogics for stochastic systems or non-stochastic games. There is a close relationship between HyperSt^2 and logics for games: HyperSt^2 embeds the fragment of rPATL consisting of those formulas that do not contain reward-based objectives or Nash equilibria nested inside strategy quantifiers, and subsumes the fragment of HyperSL where each strategy variable is associated with a unique agent, thus also subsuming HyperATL*.

While the HyperSt^2 model-checking problem over memoryful, probabilistic strategies is undecidable, we show that the model-checking problem over strategies with bounded memory is decidable, and in particular that it is PSPACE-hard and in EXPTIME over memoryless deterministic strategies. For a fixed number of state quantifiers, we even prove the model-checking problem over memoryless deterministic strategies to be PSPACE-complete.

Organization. We recall preliminary concepts in Sect. 2, before we present in Sect. 3 the syntax and semantics of HyperSt^2 and illustrate its expressiveness with several examples. In Sects. 4 and 5 we investigate the relationship between HyperSt^2 and different logics for stochastic games, as well as hyperlogics. We present our results on the complexity of the HyperSt^2 model-checking problem in Sect. 6. We conclude in Sect. 7 and give an outlook on future work. For details on the syntax and semantics of the compared logics, model-checking algorithms, and proofs we refer to the extended version [18].

2 Preliminaries

Let $\mathbb{N}$ denote the natural numbers (including 0) and $\mathbb{R}$ the reals. For a set S, and some subset $R \subseteq S$, let $\overline{R} = S \setminus R$ denote the complement of R in S.

Definition 1. *A turn-based (stochastic) game (TSG) is a tuple* $\mathcal{G} = (Ags, S, Act, \mathbf{P}, AP, L)$ *where (1) Ags is a finite set of agents, (2) S is a countable set of states partitioned into $S = \bigcup_{g \in Ags} S_g$ with $S_g \cap S_{g'} = \emptyset$ for all $g \neq g'$, (3) Act is a finite set of actions, (4) $\mathbf{P}: S \times Act \times S \to [0,1]$ is a transition probability function such that for all $s \in S$ the set $Act(s) = \{a \in Act \mid \sum_{s' \in S} \mathbf{P}(s, a, s') = 1\}$ of enabled actions is non-empty and $\sum_{s' \in S} \mathbf{P}(s, a, s') = 0$ for all $a \in \overline{Act(s)}$, (5) AP is a set of atomic propositions, and (6) $L: S \to 2^{AP}$ is a labeling function.*

In the following, we always assume that $Ags = \{1, \ldots, k\}$ for some $k \in \mathbb{N}$ and use $S_A = \bigcup_{g \in A} S_g$ for $A \subseteq Ags$. We call a game *non-stochastic* if all transition probabilities are 0 or 1. An *(infinite) path* of a turn-based game $\mathcal{G}$ is a sequence of states $\pi = s_0 s_1 s_2 \ldots \in S^\omega$ such that for all $i \geq 0$ there exists $a \in Act$ with $\mathbf{P}(s_i, a, s_{i+1}) > 0$. A *strategy* resolves the choices of one or several of the agents.

Definition 2. *Let $\mathcal{G} = (Ags, S, Act, \mathbf{P}, AP, L)$ be a TSG and $A \subseteq Ags$ a set of agents. An A-strategy for $\mathcal{G}$ is a tuple $\alpha = (Q, init, mode, act)$ where (1) Q is a countable non-empty set of modes, (2) $init: S \to Q$ selects a starting mode for each state, (3) $mode: Q \times S \to Q$ is a mode transition function, and (4) $act: Q \times S_A \times Act \to [0,1]$ is a probabilistic action selection function with $\sum_{a \in Act(s)} act(q, s, a) = 1$ and $\sum_{a \in \overline{Act(s)}} act(q, s, a) = 0$ for all $q \in Q$, $s \in S_A$.*

Let $Str_A^{\mathcal{G}}$ denote the set of all A-strategies for $\mathcal{G}$; we might omit the index $\mathcal{G}$ if clear from the context. We call an Ags-strategy $\alpha \in Str_{Ags}^{\mathcal{G}}$ a *joint* strategy. A strategy is *finite-memory* if Q is finite, *k-memory* if $|Q| \leq k$ for $k \in \mathbb{N}$, *memoryless* if Q is a singleton, and *deterministic* if $act(q, s, a) \in \{0, 1\}$ for all $(q, s, a) \in Q \times S_A \times Act$. If a strategy is memoryless, we often neglect its mode.

For an A-strategy α, and some $B \subset A$ (or $i \in A$), we use $\alpha|_B$ (or $\alpha|_i$) to denote the strategy α in which the action selection is restricted to the agents B (or agent i). For $A, B \subseteq Ags$, and $\alpha \in Str_A^{\mathcal{G}}$, $\beta \in Str_B^{\mathcal{G}}$, we let $\alpha \leftarrow \beta$ be the $(A \cup B)$-strategy obtained by taking the product of α and β, in which the action selection for the agents in B is done according to β, and for the remaining agents in A according to α, as formalized in the definition below. If A and B are disjoint, then $\alpha \leftarrow \beta = \beta \leftarrow \alpha$, and we also use $\alpha \oplus \beta$ to denote this strategy.

Definition 3. Let $\mathcal{G} = (Ags, S, Act, \mathbf{P}, \mathsf{AP}, L)$ be a TSG, and $\alpha \in Str_A^{\mathcal{G}}, \beta \in Str_B^{\mathcal{G}}$ strategies for sets of agents $A, B \subseteq Ags$. The *update* of α with β is the $(A \cup B)$-strategy $\alpha \leftarrow \beta = (Q^{\alpha \leftarrow \beta}, init^{\alpha \leftarrow \beta}, mode^{\alpha \leftarrow \beta}, act^{\alpha \leftarrow \beta})$ where

- $Q^{\alpha \leftarrow \beta} = Q^{\alpha} \times Q^{\beta}$,
- $init^{\alpha \leftarrow \beta}(s) = (init^{\alpha}(s), init^{\beta}(s))$ for $s \in S$,
- $mode^{\alpha \leftarrow \beta}((q^{\alpha}, q^{\beta}), s) = (mode^{\alpha}(q^{\alpha}, s), mode^{\beta}(q^{\beta}, s))$ for $q^{\alpha} \in Q^{\alpha}, q^{\beta} \in Q^{\beta}$, $s \in S$, and
- $act^{\alpha \leftarrow \beta}((q^{\alpha}, q^{\beta}), s) = \begin{cases} act^{\beta}(q^{\beta}, s) & \text{if } s \in S_B \\ act^{\alpha}(q^{\alpha}, s) & \text{else} \end{cases}$

for $q^{\alpha} \in Q^{\alpha}, q^{\beta} \in Q^{\beta}, s \in S_{A \cup B}$.

Applying a joint strategy to a turn-based game induces a *discrete-time Markov chain*, i.e., a stochastic game with a single agent and without nondeterminism.

Definition 4. A *discrete-time Markov chain* (DTMC) *is a tuple* $\mathcal{D} = (S, \mathbf{P}, \mathsf{AP}, L)$ *where (1)* S *is a finite set of* states, *(2)* $\mathbf{P} \colon S \times S \to [0, 1]$ *is a transition probability function with* $\sum_{s' \in S} \mathbf{P}(s, s') = 1$ *for all* $s \in S$, *(3)* AP *is a set of atomic propositions, and (4)* $L \colon S \to 2^{\mathsf{AP}}$ *is a labeling function.*

Definition 5. *For a TSG* $\mathcal{G} = (Ags, S, Act, \mathbf{P}, \mathsf{AP}, L)$ *and a joint strategy* $\alpha = (Q, init, mode, act) \in Str_{Ags}^{\mathcal{G}}$ *for* $\mathcal{G}$, *we define the* DTMC *induced by* α *in* $\mathcal{G}$ *as* $\mathcal{G}^{\alpha} = (S^{\alpha}, \mathbf{P}^{\alpha}, \mathsf{AP}^{\alpha}, L^{\alpha})$ *with the following components:*

- $S^{\alpha} = S \times Q$,
- $\mathbf{P}^{\alpha}((s, q), (s', q')) = \sum_{a \in Act(s)} \mathbf{P}(s, a, s') \cdot act(s, q, a)$ if $q' = mode(q, s)$ and 0 otherwise, for $s, s' \in S$, and $q, q' \in Q$,
- $\mathsf{AP}^{\alpha} = \mathsf{AP}$, and
- $L^{\alpha}(s, q) = L(s)$ for $s \in S$ and $q \in Q$.

Definition 6. *For* $n \in \mathbb{N}$ *and DTMCs* $\mathcal{D}_1, \ldots, \mathcal{D}_n$ *with* $\mathcal{D}_i = (S_i, \mathbf{P}_i, \mathsf{AP}_i, L_i)$ *for* $i = 1, \ldots, n$, *we define their* Composition $\mathcal{D}_1 \times \ldots \times \mathcal{D}_n$ *as the DTMC* $\mathcal{D} = (S, \mathbf{P}, \mathsf{AP}, L)$ *with (1)* $S = S_1 \times \ldots \times S_n$, *(2)* $\mathbf{P}((s_1, \ldots, s_n), (s'_1, \ldots, s'_n)) = \prod_{i=1}^{n} \mathbf{P}_i(s_i, s'_i)$, *(3)* $\mathsf{AP} = \bigcup_{i=1}^{n} \{p_i \mid p \in \mathsf{AP}_i\}$, *and (4)* $L(s_1, \ldots, s_n) = \bigcup_{i=1}^{n} \{p_i \mid p \in L_i(s_i)\}$ *for* $s_i, s'_i \in S_i$ *for* $i = 1, \ldots, n$.

An *(infinite) path* of a DTMC $\mathcal{D} = (S, \mathbf{P}, \mathsf{AP}, L)$ is a sequence of states $\pi = s_0 s_1 s_2 \ldots \in S^{\omega}$ with $\mathbf{P}(s_i, s_{i+1}) > 0$ for all $i \geq 0$. Let $Paths^{\mathcal{D}}$ denote the set of all paths of $\mathcal{D}$, and $Paths_s^{\mathcal{D}}$ those starting in $s \in S$. For a path $\pi = s_0 s_1 s_2 \ldots$ and $j \in \mathbb{N}$ we use $\pi[j]$ to denote s_j and $\pi[0, j]$ to denote the finite prefix $s_0 \ldots s_j$. A finite path is a finite prefix of an infinite path π.

For $s \in S$, let $\Pr_s^{\mathcal{D}}$ denote the probability measure associated with $\mathcal{D}$ and s. An *objective* for a turn-based game $\mathcal{G}$ is a measurable function $X\colon \textit{Paths}^{\mathcal{G}} \to \{0,1\}$. Let $\Pr_s^{\mathcal{G},\alpha}(X)$ denote the probability of satisfying an objective X over all paths in $\mathcal{G}$ starting in state s under the joint strategy α, i.e., $\Pr_{(s,\textit{init}(s))}^{\mathcal{G}^\alpha}(\{\pi \in \textit{Paths}_{(s,\textit{init}(s))}^{\mathcal{G}^\alpha} \mid X(\pi) = 1\})$. We refer to [5] for a detailed discussion of probability measures. An *objective profile* for two opposing coalitions $C, \overline{C} \subseteq \textit{Ags}$ is a pair of objectives $(X^C, X^{\overline{C}})$. A *Nash equilibrium* for an objective profile is a joint strategy such that neither coalition can improve their expected outcome by unilaterally deviating from this strategy.

Definition 7. *Let* $\mathcal{G} = (\textit{Ags}, S, \textit{Act}, \mathbf{P}, \textsf{AP}, L)$ *be a TSG,* $s \in S$, $C \subseteq \textit{Ags}$, *and* $X^C, X^{\overline{C}}\colon \textit{Paths}^{\mathcal{G}} \to \{0,1\}$. *A joint strategy* $\alpha^* \in \textit{Str}_{\textit{Ags}}^{\mathcal{G}}$ *is a* Nash equilibrium (NE) *for* $(X^C, X^{\overline{C}})$ *from* s *if*

- *for all* $\gamma \in \textit{Str}_C^{\mathcal{G}}$ *we have* $\Pr_s^{\mathcal{G},\alpha^*}(X^C) \geq \Pr_s^{\mathcal{G},\alpha}(X^C)$ *where* $\alpha := \alpha^* \leftarrow \gamma$, *and*
- *for all* $\gamma \in \textit{Str}_{\overline{C}}^{\mathcal{G}}$ *we have* $\Pr_s^{\mathcal{G},\alpha^*}(X^{\overline{C}}) \geq \Pr_s^{\mathcal{G},\alpha}(X^{\overline{C}})$ *where* $\alpha := \alpha^* \leftarrow \gamma$.

3 The Hyperlogic HyperSt^2

We propose the hyperlogic HyperSt^2 for strategies in turn-based stochastic games that combines elements of HyperPCTL and SGL. The general structure of the logic is very similar to HyperPCTL. However, being designed for one-player games only, HyperPCTL requires to first quantify over schedulers and then over states, and quantification is restricted to the beginning of a formula. For multi-player games, a more flexible quantification structure seems more appropriate. Concretely, HyperSt^2 first quantifies over states and then over strategies, and does not restrict strategy quantification to the beginning of the formula but instead allows nested strategy quantification, like SGL.

3.1 Syntax

Our goal is to reason about several independent plays of a game, which can be viewed as probabilistic computation trees from some state under some joint strategy. HyperSt^2 enables this by first quantifying over state variables in order to fix the root states of the computation trees, and then fixing a joint strategy for each play by quantifying over strategies and associating them with the state variables. Thus, each state variable is associated with a play of the game.

Let $\hat{S}$ be an infinite set of state variables, $\textit{Ags}$ some set of agents, and $\textsf{AP}$ a set of atomic propositions. HyperSt^2 formulas are defined by the following grammar:

state-quant.:	φ^{sta}	$::= \exists \hat{s}\, \varphi^{sta} \mid \forall \hat{s}\, \varphi^{sta} \mid \varphi^{str}$
strategy-quant.:	φ^{str}	$::= \langle\!\langle A \rangle\!\rangle_{\hat{R}}\, \varphi^{str} \mid \|A\|_{\hat{R}}\, \varphi^{str} \mid \varphi^{nq}$
non-quant.:	φ^{nq}	$::= \top \mid \mathsf{p}_{\hat{s}} \mid \varphi^{nq} \wedge \varphi^{nq} \mid \neg \varphi^{nq} \mid \varphi^{pr} \sim \varphi^{pr} \mid \hat{s} = \hat{s}' \mid \varphi^{str}$
probability expr.:	φ^{pr}	$::= \mathbb{P}(\varphi^{path}) \mid f(\varphi^{pr}, \ldots, \varphi^{pr})$
path formula:	φ^{path}	$::= \bigcirc \varphi^{nq} \mid \varphi^{nq}\, \mathsf{U}\, \varphi^{nq}$

Definition 3. Let $\mathcal{G} = (Ags, S, Act, \mathbf{P}, AP, L)$ be a TSG, and $\alpha \in Str_A^{\mathcal{G}}$, $\beta \in Str_B^{\mathcal{G}}$ strategies for sets of agents $A, B \subseteq Ags$. The update of α with β is the $(A \cup B)$-strategy $\alpha \leftarrow \beta = (Q^{\alpha \leftarrow \beta}, init^{\alpha \leftarrow \beta}, mode^{\alpha \leftarrow \beta}, act^{\alpha \leftarrow \beta})$ where

- $Q^{\alpha \leftarrow \beta} = Q^\alpha \times Q^\beta$,
- $init^{\alpha \leftarrow \beta}(s) = (init^\alpha(s), init^\beta(s))$ for $s \in S$,
- $mode^{\alpha \leftarrow \beta}((q^\alpha, q^\beta), s) = (mode^\alpha(q^\alpha, s), mode^\beta(q^\beta, s))$ for $q^\alpha \in Q^\alpha$, $q^\beta \in Q^\beta$, $s \in S$, and
- $act^{\alpha \leftarrow \beta}((q^\alpha, q^\beta), s) = \begin{cases} act^\beta(q^\beta, s) & \text{if } s \in S_B \\ act^\alpha(q^\alpha, s) & \text{else} \end{cases}$

for $q^\alpha \in Q^\alpha$, $q^\beta \in Q^\beta$, $s \in S_{A \cup B}$.

Applying a joint strategy to a turn-based game induces a *discrete-time Markov chain*, i.e., a stochastic game with a single agent and without nondeterminism.

Definition 4. A discrete-time Markov chain (DTMC) *is a tuple* $\mathcal{D} = (S, \mathbf{P}, AP, L)$ *where (1)* S *is a finite set of* states, *(2)* $\mathbf{P} \colon S \times S \to [0,1]$ *is a transition probability function with* $\sum_{s' \in S} \mathbf{P}(s, s') = 1$ *for all* $s \in S$, *(3)* AP *is a set of* atomic propositions, *and (4)* $L \colon S \to 2^{AP}$ *is a* labeling function.

Definition 5. *For a TSG* $\mathcal{G} = (Ags, S, Act, \mathbf{P}, AP, L)$ *and a joint strategy* $\alpha = (Q, init, mode, act) \in Str_{Ags}^{\mathcal{G}}$ *for* $\mathcal{G}$, *we define the* DTMC induced by α in $\mathcal{G}$ *as* $\mathcal{G}^\alpha = (S^\alpha, \mathbf{P}^\alpha, AP^\alpha, L^\alpha)$ *with the following components:*

- $S^\alpha = S \times Q$,
- $\mathbf{P}^\alpha((s,q), (s',q')) = \sum_{a \in Act(s)} \mathbf{P}(s, a, s') \cdot act(s, q, a)$ *if* $q' = mode(q, s)$ *and 0 otherwise, for* $s, s' \in S$, *and* $q, q' \in Q$,
- $AP^\alpha = AP$, *and*
- $L^\alpha(s, q) = L(s)$ *for* $s \in S$ *and* $q \in Q$.

Definition 6. *For* $n \in \mathbb{N}$ *and DTMCs* $\mathcal{D}_1, \ldots, \mathcal{D}_n$ *with* $\mathcal{D}_i = (S_i, \mathbf{P}_i, AP_i, L_i)$ *for* $i = 1, \ldots, n$, *we define their* Composition $\mathcal{D}_1 \times \ldots \times \mathcal{D}_n$ *as the DTMC* $\mathcal{D} = (S, \mathbf{P}, AP, L)$ *with (1)* $S = S_1 \times \ldots \times S_n$, *(2)* $\mathbf{P}((s_1, \ldots, s_n), (s_1', \ldots, s_n')) = \prod_{i=1}^n \mathbf{P}_i(s_i, s_i')$, *(3)* $AP = \bigcup_{i=1}^n \{p_i \mid p \in AP_i\}$, *and (4)* $L(s_1, \ldots, s_n) = \bigcup_{i=1}^n \{p_i \mid p \in L_i(s_i)\}$ *for* $s_i, s_i' \in S_i$ *for* $i = 1, \ldots, n$.

An *(infinite) path* of a DTMC $\mathcal{D} = (S, \mathbf{P}, AP, L)$ is a sequence of states $\pi = s_0 s_1 s_2 \ldots \in S^\omega$ with $\mathbf{P}(s_i, s_{i+1}) > 0$ for all $i \geq 0$. Let $Paths^{\mathcal{D}}$ denote the set of all paths of $\mathcal{D}$, and $Paths_s^{\mathcal{D}}$ those starting in $s \in S$. For a path $\pi = s_0 s_1 s_2 \ldots$ and $j \in \mathbb{N}$ we use $\pi[j]$ to denote s_j and $\pi[0, j]$ to denote the finite prefix $s_0 \ldots s_j$. A finite path is a finite prefix of an infinite path π.

For $s \in S$, let $\Pr_s^{\mathcal{D}}$ denote the probability measure associated with $\mathcal{D}$ and s. An *objective* for a turn-based game $\mathcal{G}$ is a measurable function $X \colon \textit{Paths}^{\mathcal{G}} \to \{0,1\}$. Let $\Pr_s^{\mathcal{G},\alpha}(X)$ denote the probability of satisfying an objective X over all paths in $\mathcal{G}$ starting in state s under the joint strategy α, i.e., $\Pr_{(s,init(s))}^{\mathcal{G}^\alpha}(\{\pi \in \textit{Paths}_{(s,init(s))}^{\mathcal{G}^\alpha} \mid X(\pi) = 1\})$. We refer to [5] for a detailed discussion of probability measures. An *objective profile* for two opposing coalitions $C, \overline{C} \subseteq \textit{Ags}$ is a pair of objectives $(X^C, X^{\overline{C}})$. A *Nash equilibrium* for an objective profile is a joint strategy such that neither coalition can improve their expected outcome by unilaterally deviating from this strategy.

Definition 7. *Let $\mathcal{G} = (\textit{Ags}, S, \textit{Act}, \mathbf{P}, \mathsf{AP}, L)$ be a TSG, $s \in S$, $C \subseteq \textit{Ags}$, and $X^C, X^{\overline{C}} \colon \textit{Paths}^{\mathcal{G}} \to \{0,1\}$. A joint strategy $\alpha^* \in \textit{Str}_{\textit{Ags}}^{\mathcal{G}}$ is a* Nash equilibrium *(NE) for $(X^C, X^{\overline{C}})$ from s if*

- *for all $\gamma \in \textit{Str}_{C}^{\mathcal{G}}$ we have $\Pr_s^{\mathcal{G},\alpha^*}(X^C) \geq \Pr_s^{\mathcal{G},\alpha}(X^C)$ where $\alpha := \alpha^* \leftarrow \gamma$, and*
- *for all $\gamma \in \textit{Str}_{\overline{C}}^{\mathcal{G}}$ we have $\Pr_s^{\mathcal{G},\alpha^*}(X^{\overline{C}}) \geq \Pr_s^{\mathcal{G},\alpha}(X^{\overline{C}})$ where $\alpha := \alpha^* \leftarrow \gamma$.*

3 The Hyperlogic HyperSt²

We propose the hyperlogic HyperSt² for strategies in turn-based stochastic games that combines elements of HyperPCTL and SGL. The general structure of the logic is very similar to HyperPCTL. However, being designed for one-player games only, HyperPCTL requires to first quantify over schedulers and then over states, and quantification is restricted to the beginning of a formula. For multi-player games, a more flexible quantification structure seems more appropriate. Concretely, HyperSt² first quantifies over states and then over strategies, and does not restrict strategy quantification to the beginning of the formula but instead allows nested strategy quantification, like SGL.

3.1 Syntax

Our goal is to reason about several independent plays of a game, which can be viewed as probabilistic computation trees from some state under some joint strategy. HyperSt² enables this by first quantifying over state variables in order to fix the root states of the computation trees, and then fixing a joint strategy for each play by quantifying over strategies and associating them with the state variables. Thus, each state variable is associated with a play of the game.

Let $\hat{S}$ be an infinite set of state variables, *Ags* some set of agents, and AP a set of atomic propositions. HyperSt² formulas are defined by the following grammar:

$$
\begin{array}{ll}
\text{state-quant.:} & \varphi^{sta} ::= \exists \hat{s}\, \varphi^{sta} \mid \forall \hat{s}\, \varphi^{sta} \mid \varphi^{str} \\
\text{strategy-quant.:} & \varphi^{str} ::= \langle\!\langle A \rangle\!\rangle_{\hat{R}}\, \varphi^{str} \mid \|A\|_{\hat{R}}\, \varphi^{str} \mid \varphi^{nq} \\
\text{non-quant.:} & \varphi^{nq} ::= \top \mid \mathsf{p}_{\hat{s}} \mid \varphi^{nq} \wedge \varphi^{nq} \mid \neg \varphi^{nq} \mid \varphi^{pr} \sim \varphi^{pr} \mid \hat{s} = \hat{s}' \mid \varphi^{str} \\
\text{probability expr.:} & \varphi^{pr} ::= \mathbb{P}(\varphi^{path}) \mid f(\varphi^{pr}, \ldots, \varphi^{pr}) \\
\text{path formula:} & \varphi^{path} ::= \bigcirc\, \varphi^{nq} \mid \varphi^{nq}\, \mathsf{U}\, \varphi^{nq}
\end{array}
$$

where $\hat{s}, \hat{s}' \in \hat{S}$, $A \subseteq Ags$, $\hat{R} \subseteq \hat{S}$, $\mathsf{p} \in \mathsf{AP}$, $\sim \in \{\leq, <, =, >, \geq\}$, and $f\colon [0,1]^j \to \mathbb{R}$ is a j-ary standard arithmetic operation like addition, subtraction, multiplication, and constants. We sometimes simplify notation by writing, e.g., $\langle\!\langle 1,2\rangle\!\rangle_{\hat{s}}\, \varphi$ instead of $\langle\!\langle 1,2\rangle\!\rangle_{\{\hat{s}\}}\, \varphi$. We use standard syntactic sugar like $\varphi_1 \vee \varphi_2 := \neg(\neg\varphi_1 \wedge \neg\varphi_2)$, $\varphi_1 \Rightarrow \varphi_2 := \neg\varphi_1 \vee \varphi_2$, $\mathbb{P}(\Diamond\varphi) := \mathbb{P}(\top \mathsf{U}\, \varphi)$, $\mathbb{P}(\Box\varphi) := 1 - \mathbb{P}(\Diamond\neg\varphi)$. Extending HyperSt² by adding direct nesting of temporal operators would be straightforward, but would increase the model-checking complexity, analogously to the increased model-checking complexity of PCTL* compared to PCTL.

A state-quantified HyperSt² formula φ^{sta} is *well-formed* if (C1) each occurrence of $\mathsf{p}_{\hat{s}}$ in φ^{sta} is in the scope of a quantifier for $\hat{s}$, (C2) any quantifier $\langle\!\langle A\rangle\!\rangle_{\{\hat{s}_1,\dots,\hat{s}_m\}}$ is in the scope of quantifiers for $\hat{s}_1, \dots, \hat{s}_m$, and (C3) each probability expression $\mathbb{P}(\varphi^{path})$ is in the scope of strategy quantifiers for all agents for all state variables[1].

The following examples illustrate different kinds of properties expressible in HyperSt². Whenever we refer to some set of states T, we assume that the states contained in that set are labeled with a corresponding lower-case atomic proposition t.

Example 2. The HyperSt² formula $\forall \hat{s}\, \langle\!\langle 1\rangle\!\rangle_{\hat{s}}\, \|2\|_{\hat{s}}\, \mathbb{P}(\Diamond t_{\hat{s}}) = 1$ expresses that agent 1 has an almost-sure winning strategy to reach a set of target states T against agent 2. This example does not utilize the full power of HyperSt² as a hyperlogic, since it only employs a single state quantifier.

Example 3. Assume agent 1 wants to reach one of two target sets R and T, and prefers R over T. Reaching R is a possible way for agent 2 to fulfill its objective but there might be other options for agent 2. The following formula expresses that the two agents can collaborate for reaching R almost surely in such a way that (1) the strategy of agent 1 maximizes the probability of reaching R among all strategies for agent 1 ("$\|1\|_{\hat{s}'}$") and against every strategy of agent 2 ("$\|2\|_{\hat{s},\hat{s}'}$"), and (2) agent 1 can play safe in the sense that it can ensure to almost surely reach a state that is either in R, or from which agent 1 has a strategy to almost surely reach T, no matter how agent 2 behaves.

$$\forall \hat{s}\, \forall \hat{s}'\, (init_{\hat{s}} \wedge init_{\hat{s}'}) \Rightarrow \langle\!\langle 1,2\rangle\!\rangle_{\hat{s},\hat{s}'} \Big[\mathbb{P}(\Diamond r_{\hat{s}}) = 1 \wedge$$
$$\underbrace{\|1\|_{\hat{s}'}\, \|2\|_{\hat{s},\hat{s}'}\, \mathbb{P}(\Diamond r_{\hat{s}}) \geq \mathbb{P}(\Diamond r_{\hat{s}'})}_{(1)} \wedge \underbrace{\|2\|_{\hat{s}}\, \mathbb{P}\big[\Diamond\, \big(r_{\hat{s}} \vee \langle\!\langle 1\rangle\!\rangle_{\hat{s}}\, \|2\|_{\hat{s}}\, \mathbb{P}(\Diamond t_{\hat{s}}) = 1\big)\big] = 1}_{(2)} \Big]$$

Example 4. Consider a turn-based *non-stochastic* game with two agents where the agents have the same objectives as in Example 3. The following formula expresses that the two agents can collaborate for reaching R almost surely, and at the same time agent 1 can almost-surely ensure to almost-surely reach T in case agent 2 deviates from their joint strategy:

---

[1] It would be sufficient to require this for all agents for all state variables *occurring in* φ^{path}. We make this stronger requirement to simplify notation.

$\forall \hat{s} \, \forall \hat{s}'. \, (\mathit{init}_{\hat{s}} \wedge \mathit{init}_{\hat{s}'}) \Rightarrow \langle\!\langle 1,2 \rangle\!\rangle_{\hat{s},\hat{s}'}$
$$\Big(\mathbb{P}(\Diamond r_{\hat{s}}) = 1 \wedge \|2\|_{\hat{s}'} \, \mathbb{P}\big[\Box \, \big(\neg (\hat{s} = \hat{s}') \Rightarrow \langle\!\langle 1 \rangle\!\rangle_{\hat{s}'} \, \mathbb{P}(\Diamond t_{\hat{s}'}) = 1 \big) \big] = 1 \Big).$$

Example 5. Assume agents 1 and 2 want to reach goals T^1 and T^2, respectively. The following HyperSt² formula expresses that there exists a Nash equilibrium from some fixed initial state labeled with *init*, i.e., a joint strategy such that, if only one of the agents changes its strategy, then the probability of reaching its respective goal will not increase.

$$\forall \hat{s} \, \forall \hat{s}'. \, (\mathit{init}_{\hat{s}} \wedge \mathit{init}_{\hat{s}'}) \Rightarrow \langle\!\langle 1,2 \rangle\!\rangle_{\hat{s},\hat{s}'} \, \big(\|1\|_{\hat{s}'} \, \mathbb{P}(\Diamond t_{\hat{s}}^1) \geq \mathbb{P}(\Diamond t_{\hat{s}'}^1) \big)$$
$$\wedge \, \big(\|2\|_{\hat{s}'} \, \mathbb{P}(\Diamond t_{\hat{s}}^2) \geq \mathbb{P}(\Diamond t_{\hat{s}'}^2) \big)$$

3.2 Semantics

Without loss of generality, let us now assume that all well-formed HyperSt² formulas are of the form $\mathbb{Q}\hat{s}_1 \ldots \mathbb{Q}\hat{s}_n. \, \psi$ for some $n \in \mathbb{N}$ and some strategy-quantified HyperSt² formula ψ. HyperSt² formulas are evaluated over a context $\Gamma = (\mathcal{G}, \mathit{map}_{\mathit{Str}}, \boldsymbol{s})$ consisting of

- a turn-based game $\mathcal{G} = (\mathit{Ags}, S, \mathit{Act}, \mathbf{P}, \mathsf{AP}, L)$,
- a partial mapping $\mathit{map}_{\mathit{Str}} \colon \hat{S} \times \mathit{Ags} \rightharpoonup \bigcup_{g \in \mathit{Ags}} \mathit{Str}_{\{g\}}^{\mathcal{G}}$ assigning strategies to combinations of state variables and agents s.t. $\mathit{map}_{\mathit{Str}}(\hat{s}, g) \in \mathit{Str}_{\{g\}}^{\mathcal{G}}$, and
- a partial mapping $\boldsymbol{s} \colon \hat{S} \rightharpoonup S$ assigning states to state variables.

Let $\mathcal{G}$ be a turn-based game, $\mathit{map}_{\mathit{Str}} \colon \hat{S} \times \mathit{Ags} \rightharpoonup \bigcup_{g \in \mathit{Ags}} \mathit{Str}_{\{g\}}^{\mathcal{G}}$ a strategy mapping, $\boldsymbol{s} \colon \hat{S} \rightharpoonup S$ a state mapping, and $\Gamma = (\mathcal{G}, \mathit{map}_{\mathit{Str}}, \boldsymbol{s})$. $\mathcal{G}$ satisfies a well-formed HyperSt² formula φ, written $\mathcal{G} \models \varphi$, iff $(\mathcal{G}, \{\}, \{\}) \models \varphi$. We define the semantics of well-formed HyperSt² formulas by structural induction. State variable quantification is evaluated by quantifying over all states of the game structure. A strategy quantifier $\langle\!\langle A \rangle\!\rangle_{\hat{R}}$ or $\|A\|_{\hat{R}}$ is evaluated by quantifying over all A-strategies and assigning the quantified strategy to all state variables $\hat{s} \in \hat{R}$. For each state variable $\hat{s}$, the current assignments of the mappings thus encode a play of the game as the computation tree from $\boldsymbol{s}(\hat{s})$ under $\bigoplus_{g \in \mathit{Ags}} \mathit{map}_{\mathit{Str}}(\hat{s}, g)$. Formally, the semantics of state- and strategy-quantified formulas are as follows.

$\Gamma \models \mathbb{Q}\hat{s}_i \, \varphi^{sta} \quad \Leftrightarrow \quad \mathbb{Q} s_i \in S. \, (\mathcal{G}, \mathit{map}_{\mathit{Str}}, \boldsymbol{s}[\hat{s}_i \mapsto s_i]) \models \varphi^{sta}$
$\Gamma \models \langle\!\langle A \rangle\!\rangle_{\hat{R}} \, \varphi^{str} \quad \Leftrightarrow \quad \exists \alpha \in \mathit{Str}_A^{\mathcal{G}}. \, (\mathcal{G}, \mathit{map}_{\mathit{Str}}[(\hat{s}, g) \mapsto \alpha|_g \text{ for } g \in A, \hat{s} \in \hat{R}], \boldsymbol{s}) \models \varphi^{str}$
$\Gamma \models \|A\|_{\hat{R}} \, \varphi^{str} \quad \Leftrightarrow \quad \forall \alpha \in \mathit{Str}_A^{\mathcal{G}}. \, (\mathcal{G}, \mathit{map}_{\mathit{Str}}[(\hat{s}, g) \mapsto \alpha|_g \text{ for } g \in A, \hat{s} \in \hat{R}], \boldsymbol{s}) \models \varphi^{str}$

where $\mathbb{Q} \in \{\exists, \forall\}$, $i \in \{1, \ldots, n\}$, $A \subseteq \mathit{Ags}$, $\hat{R} \subseteq \{\hat{s}_1, \ldots, \hat{s}_n\}$.

The semantics of non-quantified formulas is defined as follows.

$$\begin{aligned}
&\Gamma \models \top \\
&\Gamma \models \mathsf{p}_{\hat{s}_i} &&\Leftrightarrow \mathsf{p} \in L(s(\hat{s}_i)) \\
&\Gamma \models \varphi_1^{nq} \wedge \varphi_2^{nq} &&\Leftrightarrow \Gamma \models \varphi_1^{nq} \text{ and } \Gamma \models \varphi_2^{nq} \\
&\Gamma \models \neg \varphi^{nq} &&\Leftrightarrow \Gamma \not\models \varphi^{nq} \\
&\Gamma \models \varphi_1^{pr} \sim \varphi_2^{pr} &&\Leftrightarrow [\![\varphi_1^{pr}]\!]_\Gamma \sim [\![\varphi_2^{pr}]\!]_\Gamma \\
&\Gamma \models \hat{s}_i = \hat{s}_j &&\Leftrightarrow s(\hat{s}_i) = s(\hat{s}_j)
\end{aligned}$$

where $i, j \in \{1, \ldots, n\}$, and $\mathsf{p} \in \mathsf{AP}$.

In order to evaluate a probability expression, we build the composition of the DTMCs induced by the strategies fixed in map_{Str}, and evaluate path probabilities in the composed DTMC as usual in PCTL. Recall that conditions (C1)–(C3) assure that whenever we evaluate well-formed $\mathsf{HyperSt}^2$ formulas, probability expressions are evaluated in a context satisfying $Dom(s) = \{\hat{s}_1, \ldots, \hat{s}_n\}$ and $Dom(map_{Str}) = \{\hat{s}_1, \ldots, \hat{s}_n\} \times Ags$. Formally, we let $\alpha_i = (Q_i, init_i, mode_i, act_i) := \bigoplus_{g \in Ags} map_{Str}(\hat{s}_i, g)$ for $i \in \{1, \ldots, n\}$, and $\mathcal{D} := \mathcal{G}^{\alpha_1} \times \ldots \times \mathcal{G}^{\alpha_n}$ with state set $S^\mathcal{D}$, and define the semantics of probability expressions as follows.

$$\begin{aligned}
[\![\mathbb{P}(\varphi^{path})]\!]_\Gamma &= \Pr^\mathcal{D}_r(\pi \in Paths_r^\mathcal{D} \mid \Gamma, \pi \models \varphi^{path}) \\
[\![f(\varphi_1^{pr}, \ldots, \varphi_\ell^{pr})]\!]_\Gamma &= f([\![\varphi_1^{pr}]\!]_\Gamma, \ldots, [\![\varphi_\ell^{pr}]\!]_\Gamma)
\end{aligned}$$

where $\ell \in \mathbb{N}$ and $r := ((s(\hat{s}_1), init_1(s(\hat{s}_1))), \ldots, (s(\hat{s}_n), init_n(s(\hat{s}_n)))) \in S^\mathcal{D}$ is the state corresponding to the current state variable assignment s in $\mathcal{D}$.

Temporal operators intuitively allow to make steps in the quantified plays, moving to some sub-tree in each computation tree. Formally, the temporal operators are evaluated over paths in the composed induced DTMC $\mathcal{D}$. In order to evaluate a subformula at some position on a composed path, we update the state mapping and shift the strategy mapping such that it remembers the history of the game that was played until that position: For $\pi \in Paths^\mathcal{D}$ and $j \in \mathbb{N}$, we use $map_{Str}^{\pi[0,j]}$ for the update of map_{Str} by $\pi[0, j]$ defined as follows. For $i \in \{1, \ldots, n\}$ and $g \in Ags$, we let $(s_i, q_i) := (\pi[j])_i \in S^{\mathcal{G}^{\alpha_i}}$ and $map_{Str}^{\pi[0,j]}(\hat{s}_i, g) := \alpha_i^{\pi[0,j]} := (Q_i, init_i^{q_i}, mode_i, act_i)$ where $\alpha_i = (Q_i, init_i, mode_i, act_i)$ as above and $init_i^{q_i}(s) = q_i$ for all $s \in S$. The semantics of path formulas is then defined as follows, where we use $\pi[j]_S$ to denote the updated state variable mapping s' with $s'(\hat{s}_i) = ((\pi[j])_i)_S$, the projection of $(\pi[j])_i$ to the S-component, for $i = 1, \ldots, n$.

$$\begin{aligned}
&\Gamma, \pi \models \bigcirc \varphi^{nq} \Leftrightarrow (\mathcal{G}, map_{Str}^{\pi[0,1]}, \pi[1]_S) \models \varphi^{nq} \\
&\Gamma, \pi \models \varphi_1^{nq} \, \mathsf{U} \, \varphi_2^{nq} \Leftrightarrow \exists j_2 \geq 0. \, ((\mathcal{G}, map_{Str}^{\pi[0,j_2]}, \pi[j_2]_S) \models \varphi_2^{nq} \wedge \\
&\qquad\qquad\qquad\qquad\qquad \forall 0 \leq j_1 < j_2. \, (\mathcal{G}, map_{Str}^{\pi[0,j_1]}, \pi[j_1]_S) \models \varphi_1^{nq})
\end{aligned}$$

The semantics of the probability operator is well-defined as the set $\{\pi \in Paths_r^\mathcal{D} \mid \Gamma, \pi \models \varphi^{path}\}$ is measurable with analogous reasoning to PCTL.

4 Relationship to Logics for Stochastic Games

Stochastic game logic (SGL) and probabilistic alternating-time temporal logic (PATL) are temporal logics for stochastic games. Both are evaluated over a fixed initial state s_{init} of a game structure, while HyperSt^2 is evaluated over structures without dedicated initial states. Therefore, we can only mimic SGL and PATL in HyperSt^2 if we can assume that the given state s_{init} has a unique label, e.g., *init*. We can then specify that we are only interested in executions where some state variable is initialized with a state labeled with *init*. In fact, if we access the label *init* for this purpose only, we do not increase the observability of the initial state compared to SGL and PATL. The extended version contains the full syntax and semantics of SGL and rPATL. An overview of the relationships between HyperSt^2, SGL and rPATL is presented in Fig. 1a.

4.1 PATL and rPATL

PATL and PATL* [11] are PCTL-style (resp. PCTL*-style) extensions of ATL [3]. They allow to specify zero-sum properties of turn-based and concurrent stochastic games. PATL can be directly embedded into HyperSt^2.[2] Since HyperSt^2 does not allow arbitrary nesting of temporal operators, simulating PATL* in HyperSt^2 is not possible. rPATL and rPATL* [10,20] extend PATL and PATL* with the ability to reason about expected reward. In [21], rPATL is further extended with the possibility to specify *social welfare subgame perfect ϵ-Nash equilibria (SW-SP-ϵ-NE)*. An rPATL formula of the form $\langle\langle C:\overline{C}\rangle\rangle_{max\sim x}$ ($P[\psi^1] + P[\psi^2]$) expresses that the agents have a joint strategy α^* such that (1) the sum of the probabilities of satisfying ψ^1 and ψ^2 relates to x as specified, and (2) α^* is a SW-SP-ϵ-NE, where ψ^1 is the objective for C and ψ^2 is the objective for $\overline{C}$. A *social welfare* Nash equilibrium maximizes the sum of the probability of satisfying the objectives. A Nash equilibrium is considered to be *subgame perfect* if it is a Nash equilibrium in all subgames, i.e., at each state of the game. Let PATL$_{NE}$ be the fragment of rPATL with SW-SP-ϵ-NE but without reward objectives or bounded Until.

In Example 5 we have seen that HyperSt^2 can express Nash equilibria. In order to express SW-SP-ϵ-NE, we need to compare different strategies from the same state. In HyperSt^2, a state variable cannot be associated with several different joint strategies at the same time. However, we can model the desired behavior by quantifying over several state variables and specifying that we are only interested in cases where the state variables are assigned the same state, using "$\hat{s} = \hat{s}'''$". Further, the definition of SW-SP-ϵ-NE first quantifies over strategies and then over subgames, i.e., over states, while in HyperSt^2 we quantify first over states and then over strategies. However, if there is no nested strategy quantification, moving the state quantification to the front results in an equivalent formulation.

---

[2] Technically, PATL allows bounded Until but HyperSt^2 does not. Extending HyperSt^2 with bounded Until would be straightforward and not increase model-checking complexity.

Theorem 1. For $\mathsf{PATL_{NE}}$ formulas ψ^1, ψ^2 without strategy quantifiers (including Nash equilibria), it holds that there exists a $\mathsf{HyperSt}^2$ formula φ' such that

$$\mathcal{G}, s \models_{\mathsf{rPATL}} \langle\!\langle C:\overline{C} \rangle\!\rangle_{max \sim x} (\mathbb{P}[\psi^1] + \mathbb{P}[\psi^2]) \;\Leftrightarrow\; \mathcal{G}' \models_{\mathsf{HyperSt}^2} \varphi'$$

for all turn-based stochastic games $\mathcal{G}$ and states $s \in S$, where $\mathcal{G}'$ is the same as $\mathcal{G}$ but s is additionally labeled with a fresh atomic proposition init.

Proof (Sketch). We construct

$$\varphi' := \forall \hat{s}_0 \forall \hat{s}'_0 \; \forall \hat{s}_1 \forall \hat{s}'_1 \; \exists \hat{s}_2 \exists \hat{s}'_2. \; (\mathit{init}_{\hat{s}_0} \wedge \mathit{init}_{\hat{s}'_0} \wedge \hat{s}_1 = \hat{s}'_1) \Rightarrow (\hat{s}_2 = \hat{s}'_2 \wedge \varphi_{SWSP\epsilon})$$

$$\varphi_{SWSP\epsilon} := \langle\!\langle Ags \rangle\!\rangle_{\hat{s}_0, \hat{s}_1, \hat{s}'_1} \left[\mathbb{P}(\psi^1_{\hat{s}_0}) + \mathbb{P}(\psi^2_{\hat{s}_0}) \sim x \wedge \varphi_{SP\epsilon}(\hat{s}_1, \hat{s}'_1) \right.$$

$$\left. \wedge \|Ags\|_{\hat{s}'_0, \hat{s}_2, \hat{s}'_2} \; (\varphi_{SP\epsilon}(\hat{s}_2, \hat{s}'_2) \Rightarrow \varphi_{SW}(\hat{s}_0, \hat{s}'_0)) \right]$$

$$\varphi_{SP\epsilon}(\hat{s}_i, \hat{s}'_i) := \left(\|C\|_{\hat{s}'_i} \; \mathbb{P}(\psi^1_{\hat{s}_i}) \geq \mathbb{P}(\psi^1_{\hat{s}'_i}) - \epsilon \right) \wedge \left(\|\overline{C}\|_{\hat{s}'_i} \; \mathbb{P}(\psi^2_{\hat{s}_i}) \geq \mathbb{P}(\psi^2_{\hat{s}'_i}) - \epsilon \right)$$

$$\varphi_{SW}(\hat{s}_0, \hat{s}'_0) := \sum_{i=1}^{2} \mathbb{P}(\psi^i_{\hat{s}_0}) \geq \sum_{i=1}^{2} \mathbb{P}(\psi^i_{\hat{s}'_0})$$

where $\psi^i_{\hat{s}}$ corresponds to ψ^i with all atomic propositions indexed by state variable $\hat{s}$. Intuitively, $\varphi_{SWSP\epsilon}$ asks for a joint strategy α^* such that (1) the overall profit under α^* from s relates to x as specified, (2) α^* is a subgame perfect ϵ-NE, and (3) α^* results in a higher overall profit than any other subgame perfect ϵ-NE α from s (social welfare). We associate state variables $\hat{s}_0, \hat{s}_1, \hat{s}'_1$ with α^* and state variables $\hat{s}'_0, \hat{s}_2, \hat{s}'_2$ with α. We use $\hat{s}_0, \hat{s}'_0$ to compare the overall profit from s under α^* and α. We require $\hat{s}_i$ and $\hat{s}'_i$ to be instantiated with the same state for $i = 1, 2$, and use $\hat{s}_1, \hat{s}'_1$ to check whether α^* is a subgame perfect ϵ-NE, and $\hat{s}_2, \hat{s}'_2$ to do the same for α. The size of φ' is linear in the size of the $\mathsf{PATL_{NE}}$ formula $\langle\!\langle C:\overline{C} \rangle\!\rangle_{max \sim x} (\mathbb{P}[\psi^1] + \mathbb{P}[\psi^2])$. For the full proof, see the extended version. □

This idea can be extended to translate the fragment $\mathsf{PATL_{1\text{-}NE}}$ of $\mathsf{PATL_{NE}}$ to $\mathsf{HyperSt}^2$, consisting of those $\mathsf{PATL_{NE}}$ formulas where Nash equilibria are not nested inside strategy quantifiers (including Nash equilibria). Translating Boolean operators, atomic propositions, strategy quantifiers and probability operators is straightforward, and Nash equilibria can be translated as shown in Theorem 1, using fresh state variables for every Nash equilibrium. However, it is not obvious how to express formulas with Nash equilibria nested inside strategy quantifiers or Nash equilibria in $\mathsf{HyperSt}^2$, since then state quantifiers are nested inside probability operators and cannot be pulled to the front, see the extended version for details. Hence, it is unclear whether $\mathsf{PATL_{NE}}$ can be fully simulated in $\mathsf{HyperSt}^2$.

Open Problem. Does $\mathsf{HyperSt}^2$ subsume $\mathsf{PATL_{NE}}$ over turn-based stochastic games?

4.2 Stochastic Game Logic

SGL [4] is a probabilistic variant of ATL for turn-based stochastic games that allows to specify ω-regular conditions on paths via *deterministic Rabin automata (DRA)*.

HyperSt^2 cannot fully subsume SGL because HyperSt^2 employs LTL-style path conditions, which are strictly less expressive than DRA.

When evaluating a probability operator, SGL implicitly quantifies over all memoryful probabilistic strategies for all agents for whom a strategy has not been fixed yet. In HyperSt^2, we enforce explicit strategy quantification for all agents in order to have a more transparent logic and simpler semantics. In SGL, if we restrict the strategies for the explicitly quantified agents to be, e.g., memoryless deterministic, the remaining agents are still allowed memory and randomization, while in HyperSt^2, we can only restrict the strategy class for all agents.

Let SGL$_{\text{LTL}}$ be the fragment of SGL where we restrict path conditions to Boolean and temporal operators and assume explicit strategy quantification. There still is a crucial difference between SGL$_{\text{LTL}}$ and HyperSt^2 in the evaluation of temporal operators: In HyperSt^2, when we evaluate a temporal operator on a path under some strategy mapping, we evaluate the subformula(s) from a position on the path under the *shifted* strategy mapping that remembers the history of the play until that position. In SGL, however, all subformula are evaluated under the *same* strategy again, i.e., we do not remember the play so far. We refer to the extended version for details and an example.

Conjecture 1. SGL$_{\text{LTL}}$ cannot be embedded in HyperSt^2 over TSGs.

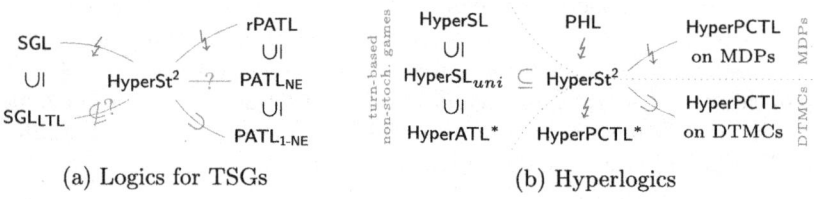

Fig. 1. Overview of the relationship between HyperSt^2 and other logics. ↯: Logics syntactically incompatible. ?: Open problem/depicted relation is a conjecture.

5 Relationship to Hyperlogics

In this section, we investigate the relationship between HyperSt^2 and different hyperlogics for stochastic systems or non-stochastic games, namely HyperPCTL, PHL, and HyperPCTL* as well as HyperSL and HyperATL*. An overview of the relationships between the logics is given in Fig. 1b.

5.1 Hyperlogics for Stochastic Systems

The probabilistic hyperlogics HyperPCTL, PHL and HyperPCTL* take different approaches for extending temporal logics to hyperlogics on stochastic systems. Apart from syntactical differences, they also make different assumptions about initial states: PHL is defined for systems with initial state distributions and HyperPCTL* for systems with a unique initial state, and both quantify only over paths from initial states.[3] In contrast, HyperPCTL—like HyperSt^2—is defined for systems without specified initial states and allows to quantify over all states. A technical difference between the logics is thus that HyperPCTL and HyperSt^2 can reason about unreachable states while HyperPCTL* and PHL cannot.

We note that we decided to base HyperSt^2 on HyperPCTL, and the relationship between PHL, HyperPCTL, and HyperPCTL* is not fully understood yet.

HyperPCTL for DTMCs [2] extends PCTL by explicit quantification over states and allows to express probabilistic relations between several executions of a DTMC. HyperPCTL includes the bounded Until operator, which we decided to leave out in HyperSt^2 for the sake of simplicity. It would be straightforward to extend HyperSt^2 with the bounded Until operator, and this would not increase the model-checking complexity. HyperSt^2 subsumes HyperPCTL without bounded Until on DTMCs.

Proposition 1. *For a HyperPCTL formula $\varphi := Q_1 \hat{s}_1 \ldots Q_n \hat{s}_n . \varphi^{nq}$ with n state quantifiers that does not contain bounded Until, it holds that $\mathcal{D} \models_{\text{HyperPCTL}} \varphi$ iff $\mathcal{G}_\mathcal{D} \models_{\text{HyperSt}^2} Q_1 \hat{s}_1 \ldots Q_n \hat{s}_n . \|1\|_{\hat{s}_1,\ldots,\hat{s}_n} . \varphi^{nq}$ for all DTMCs $\mathcal{D}$ where $\mathcal{G}_\mathcal{D}$ corresponds to $\mathcal{D}$ interpreted as a turn-based game $\mathcal{G}_\mathcal{D}$ with $Ags = \{1\}$.*

HyperPCTL was extended to MDPs [16] by adding quantification over schedulers. HyperSt^2 cannot subsume HyperPCTL on MDPs, since HyperPCTL quantifies first over strategies (also called schedulers in the MDP-setting) and then over states, while HyperSt^2 quantifies first over states and then over strategies.[4] Intuitively, in a HyperSt^2 formula of the form $\forall \hat{s}_1 \forall \hat{s}_2 . \langle\!\langle 1 \rangle\!\rangle_{\hat{s}_1} \langle\!\langle 1 \rangle\!\rangle_{\hat{s}_2} \varphi^{nq}$ we can pick the schedulers depending on the instantiations of *both* state variables, but in a HyperPCTL formula of the form $\exists \hat{\sigma}_1 \exists \hat{\sigma}_2 . \forall \hat{s}_1(\hat{\sigma}_1) \forall \hat{s}_2(\hat{\sigma}_2) . \varphi^{nq}$ we have to find schedulers that work for all possible instantiations of the state variables. Specifically, the behavior of the scheduler assigned to $\hat{\sigma}_1$ cannot depend on the initial state assignment of $\hat{s}_2$.

Conjecture 2. HyperSt^2 does not subsume HyperPCTL on MDPs and vice versa.

---

[3] Note that an MDP with a fixed initial state is a special case of one with an initial distribution, and that an MDP with an initial distribution can be transformed into one with a fresh initial state whose outgoing transitions match the initial distribution.

[4] As mentioned in Sect. 3, while HyperSt^2 is based on HyperPCTL, a more flexible strategy quantification structure, including in particular also nested strategy quantification, seem more appropriate for games, and therefore the quantification order is switched compared to HyperPCTL.

If HyperSt² first quantified over strategies and then over states, then HyperSt² would strictly subsume HyperPCTL (without bounded Until) over MDPs.

For only a *single* state and scheduler variable, the quantification order does not matter over general schedulers: Intuitively, if for all states s there exists a scheduler σ_s such that φ^{nq} holds, we can construct a scheduler satisfying φ^{nq} from every state by choosing the appropriate σ_s depending on the initial assignment of $\hat{s}$. Over memoryless deterministic schedulers, however, the quantification order does matter since we cannot change behavior depending on where we start.

PHL [15] allows to define probabilistic hyperproperties over MDPs with an initial state distribution. PHL formulas start with scheduler quantification[5], followed by a Boolean combination of (1) comparison of probabilistic expressions over LTL marked with scheduler variables, and (2) HyperCTL* formulas. These two types of subformula cannot be mixed. In the first case, paths start at any of the initial states. In the second case, they start at the current state of the path variable that was quantified last (or initially at an initial state). HyperSt² allows to restrict state quantification to initial states but does not (natively) allow to start a new experiment at the current state of an existing experiment.

Neither case is syntactically compatible with HyperSt²: (1) HyperSt² allows nesting probability operators but not direct nesting of temporal operators, while PHL allows the latter but not the former, and (2) HyperCTL* subsumes CTL* while HyperSt² subsumes PCTL, and CTL* and PCTL are incomparable over MDPs.

Conjecture 3. HyperSt² does not subsume PHL on MDPs and vice versa.

HyperPCTL* [22] is defined for DTMCs with a specified initial state. HyperPCTL* extends PCTL* and thus allows direct nesting of temporal operators, which is not allowed in HyperSt². HyperPCTL* can be seen as a probabilistic extension of HyperLTL [12], replacing the explicit path quantifiers from HyperLTL by probability operators annotated with a number of path variables that indicate how to draw the random paths over which the probability operator is evaluated. Paths are drawn from the initial state of the DTMC or the initial state of the path currently assigned to some path variable. While a probability operator is evaluated only over the specified paths, one may still refer to paths drawn outside of the current probability operator but the evolution of such previously drawn paths is 'set in stone'. In contrast, HyperSt² does not support starting new experiments in some subformula since state quantification is restricted to the beginning of the formula. A probability operator is evaluated over all paths starting from the current state assignments, i.e., all paths are drawn anew and no path is 'set in stone' like in HyperPCTL*. These differences illustrate that HyperPCTL* and HyperSt² use very different approaches and thus are hard to compare.

Conjecture 4. HyperPCTL* does not subsume HyperSt² on MDPs and vice versa.

---

[5] To be precise, every formula can be brought into PNF, i.e., transformed such that scheduler quantification only occurs in a prefix.

5.2 Hyperlogics for Non-stochastic Games

HyperSL [7] is an extension of strategy logic [9] that allows to specify strategic hyperproperties for concurrent non-stochastic games over history-dependent deterministic strategies. HyperSL state formulas allow to explicitly quantify over named strategies, combine the quantified strategy variables into strategy profiles, bind the plays resulting from these strategy profiles to path variables, and then evaluate a path formula with respect to these path variables. Path formulas consist of Boolean and temporal operators, atomic propositions indexed by a path variable, and nested state formulas indexed by a path variable. A nested state formula is evaluated from the current state of the indexing path variable. The detailed syntax and semantics are given in the extended version.

HyperSL and HyperSt^2 can combine strategies in similar ways: Several experiments (corresponding to path variables in HyperSL and state variables in HyperSt^2) can share a strategy and strategy quantification can be nested without overwriting previous quantification. However, in HyperSL, all currently fixed strategies and paths are 'forgotten' when nesting strategy quantification in path formulas (by nesting a state formula inside a path formula). In HyperSt^2, on the other hand, nested strategy quantification only overwrites the current strategy for the specified agents and state variables.

HyperSL is defined for concurrent games that have the same actions available for all agents at each state and allows to assign the same strategy variable to different agents. For example, a HyperSL formula of the form $\exists \hat{y}.\ \psi[\hat{\pi} : (\hat{y}, \hat{y})]$ for a concurrent two-player game asks for a strategy $\hat{y}$ such that ψ holds if both agents follow $\hat{y}$. However, in turn-based games, only one agent can make a move at every state and thus sharing strategies between agents does not seem meaningful here. In order to compare the expressivity of HyperSL and HyperSt^2 over turn-based non-stochastic games, it is hence reasonable to only consider the fragment of HyperSL consisting of all formulas where every strategy variable is only assigned to a unique agent. We call this fragment HyperSL$_{uni}$. We will show that each HyperSL$_{uni}$ formula can be transformed into a HyperSt^2 formula. The translation of path formulas is straightforward apart from the translation of a state formula φ nested inside a path formula $\psi = (\varphi)_{\hat{\pi}}$. This nesting starts new experiments (i.e., introduces new path variables) at the current state of another experiment (namely at the first state of the path currently assigned to $\hat{\pi}$). The HyperSt^2 syntax does not allow to start a new experiment inside a non-quantified formula (state quantification is restricted to the prefix of a formula). However, we can simulate the behavior of HyperSL by starting all experiments initially and enforcing that experiments occurring in a nested state formula follow the same strategies as the current indexing experiment until the nested formula is reached.

For example, consider the following HyperSL formula, which serves purely illustrative purposes and does not necessarily express a meaningful property,

$$\varphi_{ex} := \exists \hat{y}_1, \hat{y}_2, \hat{y}_3 \left(\mathsf{p}_{\hat{\pi}_1} \,\mathsf{U}\, \underbrace{\left[\exists \hat{y}'_1, \hat{y}'_2 \, \psi'[\hat{\pi}_3 : (\hat{y}'_1, \hat{y}'_2)] \right]_{\hat{\pi}_2}}_{\psi''} \right) [\hat{\pi}_1 : (\hat{y}_1, \hat{y}_2), \hat{\pi}_2 : (\hat{y}_1, \hat{y}_3)]$$

$$\psi' := \Diamond \left(\mathsf{q}_{\hat{\pi}_3} \wedge \left[\exists \hat{y}''_1, \hat{y}''_2 \, (\widehat{\mathsf{p}_{\hat{\pi}_4}}) [\hat{\pi}_4 : (\hat{y}''_1, \hat{y}''_2)] \right]_{\hat{\pi}_3} \right)$$

It first starts experiments (paths) $\hat{\pi}_1$ and $\hat{\pi}_2$ at the initial state using strategy profiles $(\hat{y}_1, \hat{y}_2)$ and $(\hat{y}_1, \hat{y}_3)$, respectively. On the right-hand side of the Until, it then starts $\hat{\pi}_3$ at the current state of $\hat{\pi}_2$ and then $\hat{\pi}_4$ at the current state of $\hat{\pi}_3$, as illustrated on the right. We can mimic this in HyperSt^2 by starting four experiments $\hat{s}_1, \ldots, \hat{s}_4$ from the initial state and specifying that (1) $\hat{s}_3$ and $\hat{s}_4$ initially follow the same strategies as $\hat{s}_2$, (2) $\hat{s}_4$ mimics $\hat{s}_3$ once we step inside ψ', and (3) $\hat{s}_4$ follows its 'own' strategy profile inside ψ'':

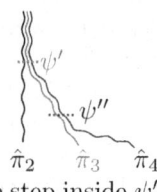

$$\forall \hat{s}_1, \hat{s}_2, \hat{s}_3, \hat{s}_4. \, (\bigwedge_{i=1}^{4} \mathit{init}_{\hat{s}_i}) \Rightarrow T(\varphi_{ex}) \quad \text{where}$$
$$T(\varphi_{ex}) := \langle\!\langle 1 \rangle\!\rangle_{\hat{s}_1,\hat{s}_2,\hat{s}_3,\hat{s}_4} \, \langle\!\langle 2 \rangle\!\rangle_{\hat{s}_1} \, \langle\!\langle 2 \rangle\!\rangle_{\hat{s}_2,\hat{s}_3,\hat{s}_4} \, \mathbb{P}\!\left(\mathsf{p}_{\hat{s}_1} \,\mathsf{U}\, \langle\!\langle 1 \rangle\!\rangle_{\hat{s}_3,\hat{s}_4} \, \langle\!\langle 2 \rangle\!\rangle_{\hat{s}_3,\hat{s}_4} \, T(\psi') \right) = 1$$
$$T(\psi') := \mathbb{P}\!\left(\Diamond \, (\mathsf{q}_{\hat{s}_3} \wedge \langle\!\langle 1 \rangle\!\rangle_{\hat{s}_4} \, \langle\!\langle 2 \rangle\!\rangle_{\hat{s}_4} \, \mathsf{p}_{\hat{s}_4}) \right) = 1$$

Since HyperSL is defined for deterministic memoryful strategies, we also restrict to this strategy class for HyperSt^2. We interpret a non-stochastic game as a stochastic game with transition probabilities 0 or 1. Since HyperSL is defined for games with an initial state s_{init}, but HyperSt^2 is evaluated over structures without initial states, we assume s_{init} to be labeled with $init$ (see also Sect. 4).

Theorem 2. *For every* HyperSL$_{uni}$ *state formula φ, there exists a HyperSt^2 formula φ', such that*

$$\mathcal{G}, s_{init} \models_{\mathsf{HyperSL}} \varphi \quad \Leftrightarrow \quad \mathcal{G}' \models_{\mathsf{HyperSt}^2} \varphi'$$

over memoryful deterministic strategies, for all turn-based non-stochastic games $\mathcal{G}$ and states s_{init} of $\mathcal{G}$, where $\mathcal{G}'$ is $\mathcal{G}$ interpreted as a stochastic game with transition probabilities 0 and 1, where additionally s_{init} is labeled with a fresh atomic proposition init.

Proof (Sketch). We translate a HyperSL$_{uni}$ formula φ with path variables $\hat{\pi}_1, \ldots, \hat{\pi}_l$ to a HyperSt^2 formula of the form $\forall \hat{s}_1 \ldots \forall \hat{s}_l. \, (\bigwedge_{r=1}^{l} \mathit{init}_{\hat{s}_r}) \Rightarrow T(\varphi)$, where state variable $\hat{s}_r$ captures $\hat{\pi}_r$ for $r = 1, \ldots, l$. The size of our translation $T(\varphi)$ is polynomial in the size of φ. Since each strategy variable is associated with exactly one agent by assumption, each strategy variable quantifier in HyperSL$_{uni}$ can be translated to a strategy quantifier in HyperSt^2. The challenge is determining which state variables should be associated with the strategy quantifier since new path variables start at the current state of a specified reference variable, yielding a tree of dependencies between the path variables. In order to mimic this behavior in HyperSt^2, we track these dependencies by determining the current *reference variable* of each path variable in the context of each subformula, i.e., its 'ancestor' in the dependency tree on the level corresponding to the subformula. For example, in the context of φ_{ex}, the reference variables of $\hat{\pi}_4$ with respect to ψ' and φ_{ex} are $\hat{\pi}_3$ and $\hat{\pi}_2$, respectively. In order to translate a strategy quantifier $\exists \hat{y}. \psi$, we thus collect all path variables $\hat{\pi}$ binding $\hat{y}$ and additionally all path variables that have such a variable $\hat{\pi}$ as a reference variable for this subformula. For example, $\hat{y}_3$ in φ_{ex} is bound by $\hat{\pi}_2$, and $\hat{\pi}_2$ is the reference variable of $\hat{\pi}_3$ and $\hat{\pi}_4$ with respect to φ_{ex}. We refer to the extended version for the full proof. □

HyperATL* [6] is a strategic hyperlogic for concurrent non-stochastic games that combines ATL* and HyperCTL*. HyperATL* is subsumed by HyperSL. In particular, existential strategy quantification $\langle\!\langle A \rangle\!\rangle \pi. \varphi$ in HyperATL* can be translated to HyperSL by existentially quantifying over fresh strategy variables for agents in A, and universally quantifying over fresh strategy variables for agents in $\overline{A}$. This translation does not require agents to share strategies. Thus, the resulting HyperSL formula can be translated to HyperSt^2, and HyperSt^2 subsumes HyperATL*.

6 Model-Checking

Since HyperSt^2 subsumes PCTL and can reason about the existence of strategies, we can reduce the strategy-synthesis problem for single-player stochastic games with PCTL objectives [8] to HyperSt^2 model-checking, yielding the following result.

Theorem 3. *The model-checking problem for HyperSt^2 over memoryful probabilistic strategies is undecidable.*

Over bounded-memory strategies we can decide the model-checking problem by encoding it in *non-linear real arithmetic* [14] as shown in the extended version.

Theorem 4. *For $k \in \mathbb{N}$, the model-checking problem for HyperSt^2 over k-memory probabilistic strategies is decidable.*

Theorem 5. *The model-checking problem for HyperSt^2 over memoryless deterministic strategies is in EXPTIME and PSPACE-hard.*

Proof (Sketch). EXPTIME-membership: In the extended version, we give a brute-force EXPTIME decision procedure that resolves state and strategy quantification by testing all possible states or (memoryless deterministic) strategies, respectively. The number of state and strategy assignments that have to be checked is bounded exponentially in the size of the input. In particular, probability expressions/temporal operators do not pose a challenge, but can be handled in the usual way for PCTL on DTMCs by solving a linear equation system with at most $|S|^n$ equations where n is the number of state variables in the formula.

PSPACE-hardness: Model-checking HyperPCTL on DTMCs is PSPACE-hard [2]. HyperSt^2 subsumes HyperPCTL on DTMCs (Proposition 1); over DTMCs the strategy class does not play any role. □

It remains an open question whether tighter upper and lower bounds exist.

Open Problem. Is the model-checking problem for HyperSt^2 over memoryless deterministic strategies in PSPACE? Is it EXPTIME-hard?

For a fixed number of state quantifiers, however, we can give tight bounds.

Theorem 6. *For $n \in \mathbb{N}$, the model-checking problem for HyperSt^2 formulas with n state quantifiers is PSPACE-complete over memoryless deterministic strategies.*

Proof (Sketch). We adapt the PSPACE-completeness proof for SGL over memoryless deterministic strategies [4]. We show that PSPACE-hardness holds already for $n = 1$. For membership, we inductively define the type of a HyperSt^2 formula as a class of the polynomial time hierarchy and prove that an upper bound for the complexity of the model-checking problem for a class of formulas of the same type is given by this type. For the full proof, see the extended version. □

7 Conclusion

We have proposed a hyperlogic for strategies in turn-based stochastic games that allows to express probabilistic relations between several executions of a game, where an execution is a probabilistic computation tree resulting from fixed strategies for the agents, rooted at a state of the game. To the best of our knowledge, HyperSt^2 is the first hyperlogic for stochastic games. We illustrated that HyperSt^2 can express interesting properties of stochastic games, like the existence of optimal strategies or Nash equilibria. HyperSt^2 subsumes a fragment of rPATL over turn-based stochastic games, and a fragment of HyperSL that in turn subsumes HyperATL* over turn-based non-stochastic games. We have established that the HyperSt^2 model-checking problem is undecidable over general strategies but decidable for bounded memory. Over memoryless deterministic strategies, the model-checking problem is in EXPTIME and PSPACE-hard, but PSPACE-complete if we fix the number of state quantifiers. In future work, it would be interesting to extend HyperSt^2 with the ability to reason about (expected) rewards, and to lift the logic to *concurrent* stochastic games.

Acknowledgments. Lina Gerlach is supported by the DFG RTG 2236/2 *UnRAVeL*.

Disclosure of Interests. The authors have no competing interests to declare that are relevant to the content of this article.

References

1. Ábrahám, E., Bartocci, E., Bonakdarpour, B., Dobe, O.: Probabilistic hyperproperties with nondeterminism. In: Hung, D.V., Sokolsky, O. (eds.) Automated Technology for Verification and Analysis. ATVA 2020. LNCS, vol. 12302, pp. 518–534. Springer, Cham (2020). https://doi.org/10.1007/978-3-030-59152-6_29
2. Ábrahám, E., Bonakdarpour, B.: HyperPCTL: a temporal logic for probabilistic hyperproperties. In: McIver, A., Horvath, A. (eds.) Quantitative Evaluation of Systems. QEST 2018. LNCS, vol. 11024, pp. 20–35. Springer, Cham (2018). https://doi.org/10.1007/978-3-319-99154-2_2
3. Alur, R., Henzinger, T.A., Kupferman, O.: Alternating-time temporal logic. J. ACM **49**(5), 672–713 (2002)
4. Baier, C., Brázdil, T., Größer, M., Kucera, A.: Stochastic game logic. Acta Informatica **49**(4), 203–224 (2012)
5. Baier, C., Katoen, J.P.: Principles of Model Checking. MIT Press, Cambridge (2008)
6. Beutner, R., Finkbeiner, B.: HyperATL*: a logic for hyperproperties in multi-agent systems. Log. Methods Comput. Sci. **19**(2) (2023)
7. Beutner, R., Finkbeiner, B.: Hyper Strategy Logic (March 2024)
8. Brázdil, T., Brozek, V., Forejt, V., Kucera, A.: Stochastic games with branching-time winning objectives. In: 21th IEEE Symposium on Logic in Computer Science (LICS 2006), 12–15 August 2006, Seattle, WA, USA, Proceedings, pp. 349–358. IEEE Computer Society (2006)
9. Chatterjee, K., Henzinger, T.A., Piterman, N.: Strategy logic. In: Caires, L., Vasconcelos, V.T. (eds.) CONCUR 2007 – Concurrency Theory. CONCUR 2007. LNCS, vol. 4703, pp. 59–73. Springer, Berlin, Heidelberg (2007). https://doi.org/10.1007/978-3-540-74407-8_5
10. Chen, T., Forejt, V., Kwiatkowska, M., Parker, D., Simaitis, A.: Automatic verification of competitive stochastic systems. Form. Methods Syst. Des. **43**(1), 61–92 (2013)
11. Chen, T., Lu, J.: Probabilistic alternating-time temporal logic and model checking algorithm. In: Lei, J. (ed.) Fourth International Conference on Fuzzy Systems and Knowledge Discovery, FSKD 2007, 24–27 August 2007, Haikou, Hainan, China, Proceedings, Volume 2, pp. 35–39. IEEE Computer Society (2007)
12. Clarkson, M.R., Finkbeiner, B., Koleini, M., Micinski, K.K., Rabe, M.N., Sánchez, C.: Temporal logics for hyperproperties. In: Abadi, M., Kremer, S. (eds.) Principles of Security and Trust. POST 2014. LNCS, vol. 8414, pp. 265–284. Springer, Berlin, Heidelberg (2014). https://doi.org/10.1007/978-3-642-54792-8_15
13. Clarkson, M.R., Schneider, F.B.: Hyperproperties. J. Comput. Secur. **18**(6), 1157–1210 (2010)
14. Collins, G.E.: Quantifier elimination for real closed fields by cylindrical algebraic decompostion. In: Brakhage, H. (eds.) Automata Theory and Formal Languages. LNCS, vol. 33, pp. 134–183. Springer, Berlin, Heidelberg (1975). https://doi.org/10.1007/3-540-07407-4_17
15. Dimitrova, R., Finkbeiner, B., Torfah, H.: Probabilistic hyperproperties of Markov decision processes. In: Hung, D.V., Sokolsky, O. (eds.) Automated Technology for Verification and Analysis. ATVA 2020. LNCS, vol. 12302, pp. 484–500. Springer, Cham (2020). https://doi.org/10.1007/978-3-030-59152-6_27

16. Dobe, O., Ábrahám, E., Bartocci, E., Bonakdarpour, B.: Model checking hyperproperties for markov decision processes. Inf. Comput. **289**, 104978 (2022)
17. Emerson, E.A., Halpern, J.Y.: "sometimes" and "not never" revisited: on branching versus linear time temporal logic. J. ACM **33**(1), 151–178 (1986)
18. Gerlach, L., Löding, C., Ábrahám, E.: A hyperlogic for strategies in stochastic games (extended version) (2025). https://arxiv.org/abs/2506.16775
19. Hansson, H., Jonsson, B.: A logic for reasoning about time and reliability. Form. Aspects Comput. **6**(5), 512–535 (1994)
20. Kwiatkowska, M., Norman, G., Parker, D., Santos, G.: Automated verification of concurrent stochastic games. In: McIver, A., Horvath, A. (eds.) Quantitative Evaluation of Systems. QEST 2018. LNCS, vol. 11024, pp. 223–239. Springer, Cham (2018). https://doi.org/10.1007/978-3-319-99154-2_14
21. Kwiatkowska, M., Norman, G., Parker, D., Santos, G.: Equilibria-based probabilistic model checking for concurrent stochastic games. In: ter Beek, M., McIver, A., Oliveira, J. (eds.) Formal Methods – The Next 30 Years. FM 2019. LNCS, vol. 11800, pp. 298–315. Springer, Cham (2019). https://doi.org/10.1007/978-3-030-30942-8_19
22. Wang, Y., Nalluri, S., Bonakdarpour, B., Pajic, M.: Statistical model checking for hyperproperties. In: 34th IEEE Computer Security Foundations Symposium, CSF 2021, pp. 1–16. IEEE (2021)

Time-Sensitive Importance Splitting

Gabriel Dengler[1](✉) , Carlos E. Budde[2] , Laura Carnevali[3] ,
and Arnd Hartmanns[4]

[1] Saarland University, Saarbrücken, Germany
dengler@depend.uni-saarland.de
[2] Technical University of Denmark, Lyngby, Denmark
[3] Department of Information Engineering,
University of Florence, Florence, Italy
[4] University of Twente, Enschede, The Netherlands

Abstract. State-of-the-art methods for rare event simulation of non-Markovian models face practical or theoretical limits if observing the event of interest requires prior knowledge or information on the timed behavior of the system. In this paper, we attack both limits by extending importance splitting with a time-sensitive importance function. To this end, we perform backwards reachability search from the target states, considering information about the lower and upper bounds of the active timers in order to steer the generation of paths towards the rare event. We have developed a prototype implementation of the approach for input/output stochastic automata within the MODEST TOOLSET. Preliminary experiments show the potential of the approach in estimating rare event probabilities for an example from reliability engineering.

1 Introduction

Quantitative evaluation methods provide model-driven guidance for the design, development, and operation of critical systems subject to RAMS requirements [24,39,44,52]. While the evaluation of Markovian models leverages efficient consolidated approaches [1,3,29,46], that of non-Markovian models faces notably harder challenges [26,28,54]. To overcome them, analytical methods provide numerical solutions under modeling restrictions, e.g., at most one non-exponential timer in each state [18,27,28,51], or perform semi-symbolic state space enumeration [33], suffering explosion issues with many concurrent non-exponential timers. Despite recent advances [2,6,7,16], simulation remains the

---

This work was partially funded by DFG grant 389792660 as part of TRR 248 CPEC, the European Union (EU) under the INTERREG North Sea project STORM_SAFE of the European Regional Development Fund, the EU's Horizon 2020 research and innovation programme under Marie Skłodowska-Curie grant agreement 101008233 (MISSION), the EU under the Italian National Recovery and Resilience Plan (NRRP) of NextGenerationEU, partnership on "Telecommunications of the Future" (PE00000001 program "RESTART"), and by NWO VIDI grant VI.Vidi.223.110 (TruSTy).

primary method to evaluate large non-Markovian models, as it can limit the computational effort beforehand.

In this context, notable attention has been paid to rare event simulation (RES), a method to accelerate crude Monte Carlo (MC) simulation towards the so-called *rare event*. One of the two popular RES methods is importance splitting (ISPLIT) [37], which layers (splits) the state space and computes the conditional probabilities of traversing layers until the rare event is reached. There is abundant research on the automatic computation of these levels, tackling complex real-world examples [9,11,47]. Especially compared to the other main RES method, importance sampling (IS), the precomputations do not require significant amount of human intuition to be made working for a specific example.

Despite this success, ISPLIT does not provide the overarching out-of-the-box solution for any non-Markovian model, as it heavily relies on a good importance function (IF) to ensure reaching one level from the next does not occur rarely and thus guarantees progress towards the rare event. Existing techniques, especially ISPLIT, fail for a potentially large class of models in which the rare event is caused by rare constellations of the sampled timer values, so that the discrete part of the state alone does not provide enough information to reliably measure the distance to the target [23]. While the method of [23], using first enumeration of stochastic state classes (SSCs) [33] and then MC simulation, can theoretically tackle these cases, it only works when the critical event happens near the root of the state-space. Furthermore, transient analysis with SSCs may suffer state space explosion and numerical issues for large and stiff models. Potential extensions to address these issues would require a heavy implementation effort, e.g., integrating a multilevel switch between SSCs and simulation whenever encountering a critical event, or exploiting regenerative analysis epochs like in [6].

Contributions. In this paper, we provide a conceptionally simple yet effective solution to these issues by introducing *time-sensitive importance splitting*. Compared to conventional ISPLIT implementations, it additionally considers lower and upper bounds of the active timers in each state. This allows us to check, e.g., if the target is still reachable, not only given the discrete part of the state but also under consideration of time. To this end, we employ an analysis of the timed, but not stochastic, behavior of the system by characterizing the reachability of states (encoding discrete values and timers) by state classes (SCs) [53]. As one of our main contributions, we show how classical SC analysis for forwards reachability from the initial state can be adapted to compute backwards reachability from the target states. To the best of our knowledge, our work is the first one to incorporate continuous timer information based on backwards reachability search to provide a finer-grained importance metric for ISPLIT. Our approach is attractive for two reasons:

– It is *simple*: During the simulation phase, it only requires to replace the time-agnostic IF with a time-sensitive IF. Consequently, we can exploit existing research on different ISPLIT methods like fixed effort (FE) [37] and RESTART [55] as well as threshold selection methods [10,17] with no or only minor modifications (see Sect. 5 for the latter).

- It is *effective*, because it provides a finer-grained estimation of how important a state is, which is especially apparent if the order of timers is relevant to reach a target state. This is an inevitable property in many dynamic fault tree (DFT) models, especially those with PAND gates, where ordered failures of the attached children must happen in order to trigger the PAND's failure.

Although our pre-calculations for the time-sensitive IF are more computationally expensive than just exploiting discrete information, especially due to the additional number of states to be considered, the computation times remain manageable for many examples. In fact, the considered timing constraints can be efficiently represented by difference bound matrices (DBMs) [4,53], requiring polynomial time in the number of timers for encoding and manipulation. As discussed in Sect. 5, we envision that, in future work, we can achieve both time-sensitive importance estimations and arbitrary scalability by combining our approach with compositional IF calculation methods from the literature [9,11].

Structure. We first review necessary background in Sect. 2, especially on non-deterministic timed analysis with SCs and RES. After that, we present the main contribution of our paper in Sect. 3, whose theoretical effectiveness is demonstrated with a simple toy example. The theoretical claims of our paper are supported by an experimental evaluation in Sect. 4 on a more complex model. We discuss the practicability of the proposed method in Sect. 5. Finally, we review related work in Sect. 6 and draw our conclusions in Sect. 7.

2 Background

2.1 Input/Output Stochastic Automata

An input/output stochastic automata (IOSA) with urgency [21] is a tuple $(\mathcal{L}, \mathcal{A}, \mathcal{T}, \rightarrow, T_0, l_0)$ consisting of a denumerable set of *locations* $\mathcal{L}$ with $l_0 \in \mathcal{L}$ the *initial location*; a denumerable set of *actions* $\mathcal{A}$ partitioned into *input* $\mathcal{A}^i$ and *output* $\mathcal{A}^o$ actions, where $\mathcal{A}^u \subseteq \mathcal{A}$ are called *urgent*; a finite set of *timers* $\mathcal{T}$ s.t. each $x \in \mathcal{T}$ has an associated continuous probability measure μ_x supported on $\mathbb{R}_{\geq 0}$, from which $T_0 \subseteq \mathcal{T}$ are called *initial*; and a transition function $\rightarrow \subseteq \mathcal{L} \times 2^{\mathcal{T}} \times \mathcal{A} \times 2^{\mathcal{T}} \times \mathcal{L}$ whose elements $\langle l, T, a, T', l' \rangle$ are written $l \xrightarrow{T,a,T'} l'$.

Elements from $\mathcal{L}$ are called «states» in [21]—we call them *locations* to emphasize their independence from timers. We further call elements from $\mathcal{T}$ *timers*, rather than «clocks», to align our semantics in Sect. 2.2 with the literature on stochastic time Petri nets (STPNs) and SCs. Urgent actions enable instantaneous communication in this timed setting: [21] gives sufficient conditions for the resulting (compositional) semantics to be weakly deterministic. In particular, Dirac probability density functions (PDFs) are not allowed, and each location $l \in \mathcal{L}$ has a unique set of enabled timers or «active clocks», which we denote with the set T_l. As a result, closed IOSA are fully stochastic, so e.g. reachability properties can be approximated by statistical estimators such as those given by MC and RES approaches [21].

We call *state* a tuple $\langle l, \tau \rangle \in \mathcal{L} \times (\mathbb{R}_{\geq 0} \,\dot\cup\, \{\bot\})^{\mathcal{T}} = \mathcal{S}$ that contains, besides the location, a valuation for all active timers, which is instantiated via the *timer value function* τ. As previously mentioned, for IOSA it is known that for each location l only one combination of active timers is possible [21,40].[1] Thus, for a location l, it holds for all associated states $s = \langle l, \tau \rangle$ and timers $t \in \mathcal{T}$ that $\tau(t) = \bot$ iff $t \notin T_l$, viz., when the timer is not active in the current state, it takes the special value $\bot$ that denotes an inactive timer in a location.

In the set $T \subseteq \mathcal{T}$ of a transition $l \xrightarrow{T,a,T'} l'$, the transition function semantically describes an external event (if a is an input action; then $T = \emptyset$) or the expiring of a timer (if a is an output action; then $|T| = 1$), which triggers action a and starts new timers T'. Input and output actions facilitate modeling concurrency: Synchronous compositions that are closed (no input actions left) result in a fully stochastic system, which is the focus of this contribution. For more details about the semantics, we refer the reader to [21]. Note that there are also other stochastic concurrency models that are heavily used for reliability assessment—one of the most popular being the STPN [45]—that are in essence models equally characterized through locations plus starting and elapsing timers.

2.2 Non-deterministic Analysis with State Classes

An SC collects states having the same location and different values of the active timers [4,53]. An SC $\Sigma = \langle l, D \rangle$ consists of a location $l \in \mathcal{L}$ and a joint domain D for the active timers in l, i.e., the timers in T_l. For an SC $\Sigma = \langle l, D \rangle$, let τ be the vector of random variables (RVs) containing the active timers in l.

Initial SC. The initial SC $\Sigma_0 = \langle l_0, D_0 \rangle$ collects the initial location l_0 and timer domain $D_0 = \times_{t_i \in T_{l_0}} [a_i, b_i] \subseteq (\mathbb{Q}_{\geq 0} \cup \{\infty\})^{|T_{l_0}|}$ with $a_i \in \mathbb{Q}_{\geq 0}$, $b_i \in \mathbb{Q}_{\geq 0} \cup \{\infty\}$ for all $t_i \in T_{l_0}$, i.e., D_0 is the Cartesian product of the initial domains[2] of the active timers, from which their values are independently sampled, these timers being in fact independent RVs. Thus, D_0 has a hyper-rectangular shape.

Successor SCs. The successor of an SC $\Sigma = \langle l, D \rangle$ via a transition $l \xrightarrow{T,a,T'} l'$ is an SC $\Sigma' = \langle l', D' \rangle$ if executing the transition from location l and timer vector τ supported over D yields location l' and timer vector τ' supported over D'. In particular, we derive domain D' from domain D through the following steps:

1. *Conditioning:* Let $T_l = \{t_1, t_2, \ldots, t_n\}$. In a fully deterministic resolved IOSA, the timer expiring first, say t_1, is contained in the singleton set T. We condition $\tau = \langle t_1, t_2, \ldots, t_n \rangle$ on t_1 expiring first, obtaining $\tau_\alpha := \langle t_1^\alpha, \ldots, t_n^\alpha \rangle = \tau \mid t_1 \leq t_i \, \forall i \in \{2, \ldots, n\}$[3], with support $D_\alpha = D \cap \{\tau(t_1) \leq \tau(t_i) \forall i \in \{2, \ldots, n\}\}$. Note that the conditioning step makes the elements of τ_α be dependent RVs, and their joint support D_α be a non-hyper-rectangular domain.

---

[1] Our approach also works in the general case, with an initial forwards analysis to find all sets of enabled timers in target locations, and then backwards analysis as in Sect. 3.
[2] We consider rational bounds for the initial value of each active timer to guarantee that the number of enumerated SCs is finite, as proved, e.g., by [33, Lemma 3.2].
[3] As in probability theory, we use the symbol | to denote the conditioning on an event.

2. *Time advancement:* We reduce the active timers by t_1^α and we drop t_1^α from τ_α, yielding $\tau_\beta := \langle t_2^\beta, \ldots, t_n^\beta \rangle = \langle t_2^\alpha - t_1^\alpha, \ldots, t_n^\alpha - t_1^\alpha \rangle$ with support $D_\beta = \{\langle \tau(t_2^\alpha), \ldots, \tau(t_n^\alpha) \rangle \text{ s.t. } \exists \tau(t_1^\alpha) \text{ s.t. } \langle \tau(t_1^\alpha), \tau(t_2^\alpha) + \tau(t_1^\alpha), \ldots, \tau(t_n^\alpha) + \tau(t_1^\alpha) \rangle \in D_\alpha \}$.
3. *Newly activating:* We add to τ_β the $m = |T'|$ timers newly activated in l' (i.e., active in l' but not in l), say $t_{n+1}, \ldots, t_{n+m}$, which are RVs independent of $t_1^\beta, \ldots, t_n^\beta$. Thus, we obtain $\tau' := \langle t_2', \ldots, t_n', t_{n+1}', \ldots, t_m' \rangle = \langle t_2^\beta, \ldots, t_n^\beta, t_{n+1}, \ldots, t_{n+m} \rangle$ with the new support $D' = D_\beta \times [a_{n+1}, b_{n+1}] \times \ldots \times [a_{n+m}, b_{n+m}]$.[4]

Enumeration of SCs from the initial SC Σ_0 yields a state class graph (SCG) representing the set of timed execution sequences of the IOSA. Note that—contrary to time Petri nets (TPNs) where the SC expansion method was originally developed for—an IOSA does not support the explicit deactivation of timers when traversing a transition. Each started timer has to elapse to be removed from the set of active timers. To theoretically incorporate the deactivation of timers in SC analysis, we drop between steps 2 and 3 from τ_β the p disabled timers, say $t_2, \ldots, t_{p+1}$, yielding the new timer vector $\tau_\gamma = \langle t_{p+2}^\beta, \ldots, t_n^\beta \rangle$ with support $D_\gamma = \{\langle \tau(t_{p+2}^\beta), \ldots, \tau(t_n^\beta) \rangle \text{ s.t. } \exists \tau(t_2^\beta), \ldots, \tau(t_{p+1}^\beta) \text{ s.t. } \langle \tau(t_2^\beta), \ldots, \tau(t_{p+1}^\beta), \tau(t_{p+2}^\beta), \ldots, \tau(t_n^\beta) \rangle \in D_\beta \}$. However, the repairable DFTs scenarios considered in this paper can be modeled in IOSA without deactivation of timers [40]. Therefore, we are not further considering the deactivating step in the following.

The domain D of each SC $\Sigma = \langle l, D \rangle$ turns out to be a DBM [4,53], i.e., the solution of a set of linear inequalities constraining the differences between pairs of timers (the hyper-rectangular domain D_0 of the initial SC Σ_0 is a special case of DBM). In addition, we use a fictitious timer t_* to denote the time at which Σ is entered, obtaining $D = \{\tau(t_i) - \tau(t_j) \leq b_{ij} \; \forall t_i, t_j \in T_l \cup \{t_*\} \text{ with } t_i \neq t_j\}$. We represent D by storing its coefficients b_{ij} in a $(|T_l|+1) \times (|T_l|+1)$ matrix, requiring polynomial complexity in the number $|T_l|$ of the active timers for manipulation and encoding. We obtain a unique representation (termed *normal*) of the DBM D by interpreting it as a transition matrix of a weighted graph and running the Floyd-Warshall (FW) algorithm to find the shortest paths between all pairs of nodes, which results in the resolution of all transitive implications [53].

2.3 Rare Event Simulation

Statistical model checking (SMC) can estimate quantitative properties in fully-stochastic formal models, without incurring state space explosion or relying on the memoryless property [20,56]. For instance, bounded reachability metrics boil down to estimating binomial proportions: Each simulation trace sampled, e.g. a sequence $x_i = \{\Sigma_j\}_{j=0}^{n_i}$ of SCs, either visits a goal location l_g or not. Then the sample mean for $y_i = 1$ <u>if</u> $\exists j. \Sigma_j = \langle l_g, D \rangle \in x_i$ <u>else</u> 0 is an unbiased estimator of the probability to reach l_g within the given bounds. For robustness, such

---

[4] In the steps of forwards state space analysis, the subscript i of each active timer of the domain being computed identifies a timer in the model with bounds a_i and b_i.

point estimates are given with a statistical correctness guarantee, typically a confidence interval (CI) of width $\pm\epsilon$ that contains the true value $(1-\delta)\cdot 100\,\%$ of the time [14]. The smaller the ϵ, the more *precise* (and "better") the estimation.

When the event of interest occurs with low probability, though, computing precise CIs needs infeasibly many samples [49]. RES can alleviate this, e.g., by partitioning the state space into layers surrounding the goal states $\mathcal{S}_g \subsetneq \mathcal{S}$, and computing the conditional probabilities of incrementally traversing layers from $\mathcal{S}_0$ to $\mathcal{S}_g$ [25,37]. This approach is called ISPLIT and its efficiency depends on the way $\mathcal{S}$ is layered. Such partition is given by an IF $f\colon \mathcal{S} \to \mathbb{N}_0$ that maps each state to an *importance* $f(s) \geq 0$ s.t. $f(s) > f(s')$ suggests a higher probability of observing the rare event when a simulation trace starts from s rather than s'.

For formal models and property queries, performant IF candidates can be automatically derived [9,12,47]. A general approach uses backwards reachability search to compute the number of transitions needed to reach a goal state. This yields a distance metric $d\colon \mathcal{S} \to \mathbb{N}_0$, and then $f(s) = f(\langle l, \tau \rangle) \doteq \max_{l' \in \mathcal{L}}\{d(l')\} - d(l)$. While this heuristics has been successfully applied to many case studies [10,11,13], it disregards stochastic information that is not explicit in $\mathcal{L}$. In particular, f ignores all timer values τ that in IOSA determine the successor states. Therefore, the algorithms from [9,12] are ineffective to study properties whose low probability depends on the unlikely occurrence of time events, e.g. an alarm that fails before a system error, but after the inspection that would have spotted it.

> Our contribution is a novel method to automatically derive a *time-sensitive* importance function that explicitly considers concrete valuations of (stochastic) timers.

3 Towards Time-Sensitive Importance Functions

We start by studying a small repairable DFT example in Fig. 1 inspired by [23]. It has two basic elements (BEs), UPS and AC, that are connected to a PAND gate. A top-level event (TLE)—i.e. general system failure—can only happen when UPS fails first and, before it gets repaired, AC fails. This example can be seen as a simple model of a highly reliable system powered by an unreliable grid, which has a UPS to remain operational during the recurrent blackouts. The UPS battery is replaced periodically: If during the replacement

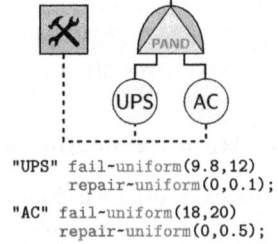

```
"UPS" fail-uniform(9.8,12)
      repair-uniform(0,0.1);
"AC"  fail-uniform(18,20)
      repair-uniform(0,0.5);
```

Fig. 1. Toy example

a blackout occurs, a system failure occurs. Both BEs are repaired by a single repair box that can repair one BE at a time, and repairs the one which fails first.

Code 1 shows the IOSA code for the UPS component. Its locations are given by the values of variables `inform` and `broken`. Actions `down` and `up` are output,

```
1   module UPS
2     fail, repair: clock;
3     inform: [0..2] init 0;
4     broken: [0..2] init 0;
5     // internal semantics for fail and repair, with (timed) output actions
6     [down!] broken=0 @ fail   -> (inform'=1) & (broken'=1);
7     [ up! ] broken=2 @ repair -> (inform'=2) & (broken'=0) & (fail'=uniform(9.8,12));
8     // RBOX triggers our repair, notifying with urgent input action
9     [ r?? ] broken=1 -> (broken'=2) & (repair'=uniform(0,0.1));
10    // notify state change to parent gates and RBOXes, with urgent output actions
11    [f_i!!] inform=1 -> (inform'=0);
12    [r_i!!] inform=2 -> (inform'=0);
13  endmodule
```

Code 1: IOSA for the UPS BE in Fig. 2 following the patterns from [40]

indicated by suffix !, and are broadcast upon the expiration of timers fail and repair, respectively. These timed events are denoted with @ followed by the timer that triggers the transition. Oppositely, urgent (untimed) actions are taken based on locations alone—IOSA operates in a maximal-progress regime. For example, the repair of UPS does not start immediately after a failure, but rather waits for the RBOX to trigger it via urgent input r on line 9. This represents a single RBOX that allows one repair at a time, thus implementing interdependent repairs among multiple components. Full details of these semantics are in [40].

To demonstrate how one could compute a time-sensitive IF in this example, we focus on the shortest path leading from the initial state to the system failure. Given the supports of the (failure and repair) timer distributions, the TLE happens when UPS has failed twice, with a failure time near the lower bound 9.8, while AC has failed once, with a failure time near the upper bound 20, and happening after the second failure of UPS. With this information one can already know in the initial state if the target is reachable on the shortest path. In contrast, using the location information alone to craft an IF cannot detect that UPS has to fail *twice* for the shortest path to the target. In the following, we explain how this information can be determined automatically in a generalized setting.

Timed Distance Metric. We define the timed distance metric $d : \mathcal{S} \to \mathbb{N}_0$ of a state $s \in \mathcal{S}$ as the distance of s to the target given the value of the active timers in s (rather than as a distance $d : \mathcal{L} \to \mathbb{N}_0$ based solely on the location of s). Thus, we obtain as IF $f : \mathcal{S} \to \mathbb{N}_0$ with $f(s) = f(\langle l, \tau \rangle) = \max_{s' \in \mathcal{S}} \{d(s')\} - d(s) = \max_{s' \in \mathcal{S}} \{d(s')\} - d(\langle l, \tau \rangle)$. To calculate our distance metric d, we use the theory of SCs and effectively run the analysis backwards from the target locations.

Target SCs. For each target location l', we build the largest possible domain D' for the active timers in l', considering only whether the target location l' is reachable from these evaluations of timers, not whether these evaluations of timers are reachable from an initial state. To this end, we limit each timer $t_i \in T_{l'}$ by its upper bound $b_i \in \mathbb{Q}_{\geq 0} \cup \{\infty\}$, obtaining the SC $\Sigma' = \langle l', D' \rangle$ with $D' = \times_{t_i \in T_{l'}} [0, b_i] \subseteq (\mathbb{Q}_{\geq 0} \cup \{\infty\})^{|T_{l'}|}$. Conversely, we do not limit t_i by its lower bound $a_i \in \mathbb{Q}_{\geq 0}$ given that, depending on the sequence of executed transitions, it may not be *newly* activated in l' (i.e., after the last incoming transition).

Predecessor SC. Given an SC $\Sigma' = \langle l', D' \rangle$ and an incoming transition $l \xrightarrow{T,a,T'} l'$, we compute the largest possible SC $\Sigma = \langle l, D \rangle$ such that for any state s in Σ, the action a leads to a state s' in Σ' (we could say that Σ is the *weakest precondition* of Σ' given a). At each step of this derivation, we normalize the DBM domain of the SC being computed using the FW algorithm [53] (as in forwards analysis), ensuring uniqueness of the result and correct resolution of all transitive restrictions. In the following derivation steps, let τ and τ' be the vectors of RVs containing the active timers in l and l', respectively, as in Sect. 2.2:

1. *Inverse of newly activating:* Let $\tau' := \langle t'_2, \ldots, t'_{n+m} \rangle$ and $t'_{n+1}, \ldots, t'_{n+m}$ be the m newly activated timers in l' (i.e., activated in l' but not in l). We limit each such timer by its lower bound (as no time elapses from newly activation to entrance in Σ'), obtaining $\tau_a := \langle t^a_2, \ldots, t^a_n, t^a_{n+1}, \ldots, t^a_{n+m} \rangle = \tau' \mid t'_{n+1}, \ldots, t'_{n+m} \in T'$, with support $D_a = D' \cap \{a_i \leq \tau(t'_i) \forall i \in \{n+1, \ldots, n+m\}\}$. After normalizing D_a, we remove $t^a_{n+1}, \ldots, t^a_{n+m}$ from τ_a, yielding $\tau_b := \langle t^b_2, \ldots, t^b_n \rangle = \langle t^a_2, \ldots, t^a_n \rangle$ with support $D_b = \{\langle \tau(t^a_2), \ldots, \tau(t^a_n) \rangle$ s.t. $\exists \tau(t^a_{n+1}), \ldots, \tau(t^a_{n+m})$ s.t. $\langle \tau(t^a_2), \ldots, \tau(t^a_n), \tau(t^a_{n+1}), \ldots, \tau(t^a_{n+m}) \rangle \in D_a\}$.
2. *Inverse of time advancement:* We increase the timers by the timer expiring first in Σ, say t^b_1, obtaining $\tau_c := \langle t^c_1, \ldots, t^c_n \rangle = \langle t^b_1, t^b_2 + t^b_1, \ldots, t^b_n + t^b_1 \rangle$ with support $D_c = \{\langle \tau(t^b_1), \tau(t^b_2), \ldots, \tau(t^b_n) \rangle$ s.t. $\tau(t^b_1) \in [0, \infty)$ s.t. $\langle \tau(t^b_2) - \tau(t^b_1), \ldots, \tau(t^b_n) - \tau(t^b_1) \rangle \in D_b\}$.
3. *Applying upper bounds of timers:* After normalizing D_c, we limit each active timer by its upper bound (to guarantee the timer does not exceed the bound due to the time retardation), obtaining $\tau := \langle t_1, \ldots, t_n \rangle = \tau_c \mid t^c_i \leq b_i \forall i \in \{1, \ldots, n\}$ with the final domain $D = D_c \cap \{\tau(t^c_i) \leq b_i \forall i \in \{1, \ldots, n\}\}$ (which is finally normalized, similarly to the previous passages).

Contrary to forwards analysis, there is no need for a conditioning step to guarantee that the timer t^b_1 expiring first actually elapses before the other timers. In fact, the non-negativity of the timer evaluations in domain D_b, i.e., $0 \leq \tau(t^b_i) - \tau(t^b_1)$ $\forall i \in \{2, \ldots, n\}$, already implies that $\tau(t^b_1) \leq \tau(t^b_i)$ $\forall i \in \{2, \ldots, n\}$.[5]

For each SC $\Sigma' = \langle l', D' \rangle$ and incoming transition $l \xrightarrow{T,a,T'} l'$, we compute the predecessor SC $\Sigma = \langle l, D \rangle$ and we derive its distance metric $\omega(\Sigma)$ as the minimum number of transitions needed to reach the target SCs from Σ. During simulation, we evaluate the distance metric of a state $s = \langle l, \tau \rangle$ as the minimum distance to the target among those of the SCs which s belongs to, i.e., $d(s) = \min_{\Sigma \mid s \in \Sigma} \omega(\Sigma)$. A state $s = \langle l, \tau \rangle$ belongs to an SC $\Sigma = \langle l', D \rangle$ (which we write as $s \in \Sigma$) if $l = l'$ and the timer evaluations defined by the timer evaluation function τ satisfy D (which can be easily checked by evaluating all inequalities encoded in D).

---

[5] Note that the conditioning step could also be omitted in forwards analysis when the lower bound of zero is applied to each timer after the time advancement step, as the conditioning step effectively ensures that every timer is non-negative after the time advancement step.

Optimizations. During calculation of SCs, we can eliminate redundant information, e.g., if for one location l we have two SCs $\Sigma_1 = \langle l, D_1 \rangle$ and $\Sigma_2 = \langle l, D_2 \rangle$ such that $D_1 \supseteq D_2$ and D_1 has a lower distance to the target than D_2, we can omit Σ_2. Furthermore, we can eliminate SCs that have a zero probability to be reached. This condition occurs if, for one non-urgent timer $t_i \in T_l$ (whose support is not a singleton by IOSA weakly determinism requirements, i.e., $a_i < b_i$), the timer evaluation $\tau(t_i)$ is bounded to a single value. Last but not least, as we use the distance metric $d(s)$ of a state s only as a heuristic, we can avoid enumerating the complete SC graph; we can, e.g., stop after reaching a given expansion depth.

Example Derivation. In the example of Fig. 1, we have timers t_{uf} for $\underline{U}PS$ $\underline{f}ail$, t_{ur} for $\underline{U}PS$ $\underline{r}epair$, t_{af} for $\underline{A}C$ $\underline{f}ail$, and t_{ar} for $\underline{A}C$ $\underline{r}epair$ (in the IOSA implementation [40], we also incorporate urgent actions to coordinate the communication between the two BEs, the repair box, and the PAND gate; for simplicity, we omit them here). We start in a target location and traverse the transitions of the IOSA backwards, which results in inversely following the elapsed timers:

- *Target SC:* We consider the first location where the target condition is met, i.e., both components have failed and UPS has failed first. As the repairing of UPS and AC is shared by a repair box, only UPS is actively repaired. Thus, the only active timer is t_{ur}, yielding an SC with domain $D_1 = \{0 \leq \tau(t_{ur}) \leq 0.1\}$.
- *Elapsing of t_{af}:* By following backwards the failure of AC, we have as active timers t_{ur} and t_{af}, obtaining an SC with timer domain:

$$D_2 = \{0 \leq \tau(t_{ur}) \leq 0.1 \land 0 \leq \tau(t_{af}) \leq 0.1 \land 0 \leq \tau(t_{ur}) - \tau(t_{af}) \leq 0.1\}$$

Note that the remaining failure time of t_{af} must be at maximum 0.1, as otherwise the gate UPS will already be repaired before AC can also fail.
- *Elapsing of t_{uf}:* Following the failure of UPS, we reach the initial location, where both the failure timers t_{uf} and t_{af} are newly activated:

$$D_3 = \{0 \leq \tau(t_{uf}) \leq 12 \land 0 \leq \tau(t_{af}) \leq 12.1 \land -0.1 \leq \tau(t_{uf}) - \tau(t_{af}) \leq 0\}$$

Note that, even though we have reached the initial location, we have not reached an initial state, as the remaining value of the timer t_{af} must be lower than 12.1, which is never the case when the timer is newly sampled. Thus, we need to execute further steps in our backwards enumeration of SCs.
- *Elapsing of t_{ur}:* We assume that UPS was repaired again, obtaining domain:

$$D_4 = \{0 \leq \tau(t_{ur}) \leq 0.1 \land 9.8 \leq \tau(t_{af}) \leq 12.2 \land \\ -12.1 \leq \tau(t_{ur}) - \tau(t_{af}) \leq -9.8\}$$

- *Elapsing of t_{uf}:* We reach the initial location again, with domain:

$$D_5 = \{0 \leq \tau(t_{uf}) \leq 10.2 \land 9.8 \leq \tau(t_{af}) \leq 20 \land \\ -12.2 \leq \tau(t_{uf}) - \tau(t_{af}) \leq -9.8\}$$

The condition $\tau(t_{uf}) - \tau(t_{af}) \leq -9.8 \iff \tau(t_{uf}) + 9.8 \leq \tau(t_{af})$ is of key importance. In fact, it expresses that, in the time difference between t_{uf} and t_{af}, an additional failure and repair of the UPS BE must take place, which takes at minimum 9.8 time units in total.

The resulting SC graph is obviously much larger, as multiple paths are possible to reach the target. However, we depicted this path to underline the effectiveness of how the ISPLIT technique can benefit substantially from this information.

4 Experimental Evaluation

We use our approach to estimate a (rare) total-failure probability that depends on an ordered sequence of failures and repairs of interdependent components.

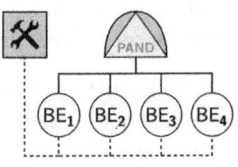

Fig. 2. DFT with sequential failures and repairs

```
1  toplevel "PAND1";
2  "PAND1" pand "BE1" "PAND2";
3  "PAND2" pand "BE2" "PAND3";
4  "PAND3" pand "BE3" "BE4";
5  "BE1" fail_uniform(1198,1218) repair_uniform(10,15);
6  "BE2" fail_uniform(530,595)   repair_uniform(10,45);
7  "BE3" fail_uniform(385,465)   repair_uniform(10,45);
8  "BE4" fail_uniform(1105,1205) repair_uniform(10,15);
9  "RBOX" rbox prio "BE1" "BE2" "BE3" "BE4";
```

Code 2: Kepler syntax for the DFT in Fig. 2 [13]

4.1 Model and Property

The system under study is depicted in Fig. 2: a synthetic repairable DFT whose TLE requires the sequential failure of components BE1 through BE4. Their repair and failure distributions make the system failure—i.e. the TLE—dependent on one failure and no repairs of BE1 and BE4, two failures and one repair of BE2, and three failures and two repairs of BE4.

Figure 3 depicts the possible order of these events based on the support of their corresponding uniform distributions. Since the top gate is a PAND gate, a TLE occurs when the time to the last failure of each of these components follows the sequence BE1 ≤ BE2 ≤ BE3 ≤ BE4.

The semantics of this non-Markovian system becomes fully stochastic when given in terms of an IOSA [40]. We study the transient property of observing a system failure before 1248 time units.

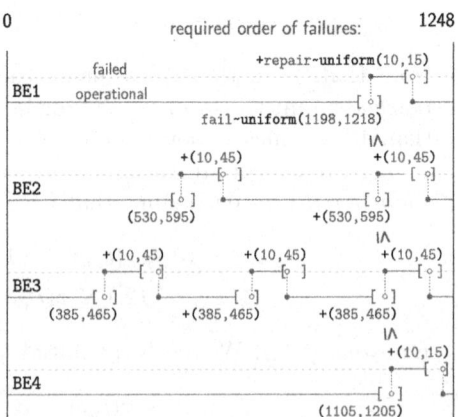

Fig. 3. Behavior of DFT in Code 2

$\varphi = \Pr(TLE \mid t_{age} \leq 1248)$, where t_{age} denotes the global system time. Given the absence of non-spurious nondeterminism, estimates $\hat{\varphi}$ for this property can be computed via SMC.

4.2 Experimental Support in the MODEST TOOLSET

We have extended the MODEST TOOLSET [30] to parse repairable DFT models written in Kepler, and compute time-sensitive IFs from models with stochastic timed semantics. With the exception of SPARE-gates and -BEs, the new support for repairable DFTs covers all gates from [13,40], including dynamic gates such as PAND (`pand` in Kepler) and functional dependence (`fdep`), as well as priority-ordered repair boxes (`rbox prio`). The constructed model analyzed by the MODEST TOOLSET is a parallel composition of multiple stochastic timed automata (STA) [5,19]. We mapped the semantics of IOSA into the corresponding subset of STA by following the patterns from [40]. As STA do not provide explicit support for timers, we implement a timer by using a pair of a clock c and a real-valued random sample x, reconstructing the timer value as the difference $x - c$. In Fig. 4, we depict the STA created for the UPS BE from Fig. 1. While `UPS__failure` and `UPS__up` correspond to output actions of the internal failure and repair timer, `RBOX__UPS__repair_start` is an input action issued by the repair box.

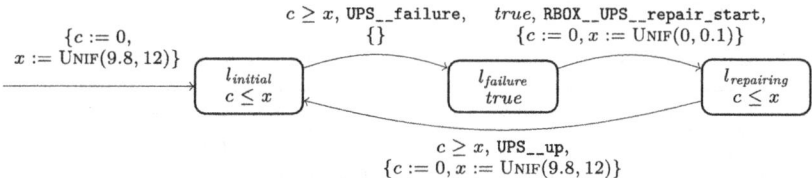

Fig. 4. Created STA for the UPS BE in Fig. 1/Code 1

As already mentioned in Sect. 3, we avoid enumerating the complete SC graph of the model, which would require significant runtime even for bounded (but high) expansion depths, and which is not required given that we use the IF solely as a heuristic to steer the simulation towards the rare event.

Furthermore, in the example in Fig. 2, there are only 7 failures and 3 repairs needed to trigger the system failure (see Fig. 3), making expansion depths below 10 not contributing more significant information. Still, to assess the impact of the expansion depth, we conduct an ablation study with varying expansion depth.

4.3 Experimentation and Results

We perform two types of experiments: bounding by execution runtime and by number of simulation runs, in both cases to estimate the (time-)bounded reachability property from Sect. 4.1. We estimate the ground truth $\varphi \approx 5.24\text{E-}7 \in$

[4.4E-7, 6.0E-7] by running 318410260 > 2^{28} independent MC runs (30 h of runtime). All experiments were performed on an AMD Ryzen 9 7950X3D CPU with 16 cores running Ubuntu 22.04. All RES simulations used the Fixed Effort ISPLIT method with an effort of 16 runs per layer [11,25].

Fixed Runtime. The first set of experiments fixes an execution wall-clock runtime of 30 min, and compares the normal approximation CI produced by different simulation approaches: CMC: crude MC; RES-notime: ISPLIT using the time-*agnostic* IF from [9,12]; and RES-time-*d*: ISPLIT using our new time-*sensitive* IF from Sect. 3 for various exploration depths d. Table 1a shows the results of these experiments, where the second and third columns show the computed estimate $\hat{\varphi}$ and the width of the corresponding CI (i.e. $\hat{\varphi} \pm \epsilon$), respectively; #runs is the number of simulation runs achieved in the given runtime; and φ? (resp. 0?) indicates whether $\varphi \in \hat{\varphi} \pm \epsilon$ (resp. $0 \in \hat{\varphi} \pm \epsilon$).

Generally speaking for statistical results, the smaller the ϵ the better the estimation. In Table 1a the best method in this respect was ISPLIT using the time-sensitive IF for an exploration depth of 10 (shadowed row). Also, all RES-time-*d* runs except $d = 22$ performed better than both CMC and RES-notime. Further quality criteria are whether the CI contains φ (good ✓), and whether it contains 0 (bad ✗, because $\varphi > 0$): All runs using a time-sensitive IF satisfied these criteria, again with the exception of RES-time-22.

Table 1. Results of experimentation on the DFT from Code 2

(a) Fixed execution runtime: 30 min

Method	$\hat{\varphi}$	ϵ	#runs	φ?	0?
CMC	1.8E-7	3.6E-7	5445630	✓	✗
RES-notime	7.1E-7	7.0E-7	320736	✓	
RES-time-6	5.2E-7	2.0E-7	332912	✓	
RES-time-10	4.3E-7	1.4E-7	334507	✓	
RES-time-14	5.3E-7	1.8E-7	316633	✓	
RES-time-18	5.3E-7	1.8E-7	227386	✓	
RES-time-22	1.1E-6	4.6E-7	151576		

(b) Fixed number of simulations: 50000

Method	$\hat{\varphi}$	ϵ	time	φ?	0?
CMC	0	0	16.4 s	✗	
RES-notime	0	0	277.4 s	✗	
RES-time-6	3.6E-7	4.6E-7	266.0 s	✓	✗
RES-time-10	3.3E-7	2.6E-7	270.7 s	✓	
RES-time-14	6.4E-7	5.3E-7	285.1 s	✓	
RES-time-18	8.7E-7	6.2E-7	403.3 s	✓	
RES-time-22	4.7E-7	4.3E-7	583.1 s	✓	

The under-performance of IF for exploration depths over 20 is caused by the low number of runs achieved, in turn caused by the simulation overhead derived from a splitting of the state space beyond the region of interest. As discussed in Sect. 4.2, the TLE occurs after 10 timer events, so exploration depths beyond 10 create layers in $\mathcal{S}$ that are irrelevant for φ. As a result, executing FE in those layers uses simulation budget that does not result in a reduced variance of the estimator. Such issues are typically alleviated by selecting *thresholds*: a subset of the IF values that the ISPLIT algorithms use for simulation. However, this does not help when the rare condition is to choose a few selected paths among many, all of the same length. This is the case for time-bounded properties, like in our example, which would instead benefit from adding the global time to the information used by the IF. We comment on these aspects in Sect. 5.

Fixed Number of Runs. The second set of experiments performs 50000 simulation runs per method, to see which IF deploys the most efficient RES under a very limited simulation budget, that also avoids the #runs imbalance observed in Table 1a. The simulation budget chosen is tight, to implement a stress test that evaluates how well the time-sensitive IF can observe rare events in a low number of trials. In this setting, the execution runtime becomes a relevant dependent variable: the faster the better. Results for experimentation with the same ISPLIT configurations as before are shown in Table 1b.

Here, CMC and RES-notime fail to reach the rare event, but the latter takes longer because the effort per simulation is higher in FE RES than in CMC. In contrast, all RES-time-d cases managed to produce useful results, in runtimes comparable (sometimes faster) than RES-notime. The best run was again obtained with RES-time-10. However, and in contrast to the previous results, here all RES-time-d runs produced CIs that contain the ground truth φ. We highlight that the simulation budget chosen is low enough to make even the time-sensitive IF produce bad runs on occasion. This is a real stress test to explore the boundaries of our implementation—e.g. for 20k runs the results were too unstable, while for 50k runs they appeared to be sufficiently stable.

5 Discussion

The experiments in Sect. 4 show how our time-sensitive IFs can help in practical scenarios to significantly increase the precision of estimates for time-bounded reachability properties under different simulation time budgets. However, our model was specifically designed for this aim, exacerbating different distances to the target (from the initial location) for different timer values. Moreover and albeit non-trivial, the model is quite small and rigid, e.g. displaying only uniform distributions. In this section, we discuss the most relevant limitations of our approach in its current implementation, devising posible ideas to tackle them.

State Space Explosion. In the given example, we limited the expansion depth to reduce the effects of the state space explosion. This issue is especially exacerbated when the bounds of the supports of the active timers contain many decimal places, as then they likely have a small common factor, leading to many resulting SCs. As discussed in Sect. 3, we can overcome this problem by resorting to partial SCG enumeration. However, in larger examples, the expansion depth might be too small for a meaningful IF. Ideas to mitigate this problem are:

- *Compositional IF calculation methods:* As the state space explosion problem also occurs in time-agnostic backwards search methods when considering large enough concurrent models, researchers have developed methods to extract reachability information from local components independently and

combine the local information of each component into one compositional IF [9], with applications to tandem queues, database systems, and oil pipelines. For recursive models like DFTs, the IF can also be computed in a recursive manner [11]. Similar techniques could also be applied to our setting. Contrary to the time-agnostic setting, the interesting information is usually detected when observing synchronizations of concurrent components (see the models in Sect. 3 and Sect. 4), making it natural to compute local importance functions based on the synchronizations of a few local components. It remains an open challenge to automatically detect which components should be synchronized.
- *Improved duplication detection methods:* As described in Sect. 3, for every SC $\langle l, D_1 \rangle$, we check if there is another SC $\langle l, D_2 \rangle$ with the same location l and lower distance to the target such that $D_1 \subseteq D_2$, and we remove $\langle l, D_1 \rangle$ if this condition occurs. Thus, we are only checking for the total inclusion of one DBM zone into another. However, it could also happen that there exists a set of SCs $\langle l, D_2 \rangle, \ldots, \langle l, D_k \rangle$ (with $k > 2$) with lower distance to the target than D_1 such that $D_1 \subseteq D_2 \cup \cdots \cup D_k$. Also in this case, domain D_1 becomes redundant and can be discarded, although it is not fully contained in any of the domains D_2 to D_k. Given that DBM zones are not closed under union operations, providing an efficient implementation is challenging.
- *Neglecting distribution bounds:* We can reduce the state space size by neglecting the bounds of timers and only considering the order in which the timers have to elapse to reach the target states. We can implement this approach by considering the bounds $[0, \infty)$ for each timer and running backwards analysis with SCs as described in Sect. 3. Although this solution is ineffective in many scenarios, including the example in Fig. 1, the approach can still be effective in timer value order-relevant decision-making with a significantly larger state space (e.g., for repairable DFT models featuring many PAND gates).

Timers with Infinite Support. In the current experiments, we only consider variables that follow the uniform distribution and thus have lower and upper bounds. Many distributions used in real-world scenarios have unbounded support, with the exponential distribution a prominent example. We could resort to the actual bound $[0, \infty)$ in the backwards SC calculations, though the efficiency of the time-sensitive IF lives from providing tight bounds for each timer.

To handle this problem, we observe that distributions with infinite support mostly sample on a small finite band. For demonstration, consider a random variable X that is exponentially distributed with rate 1. We can easily calculate that $\Pr(X \geq x) = 1 - P(x \leq X) = 1 - (1 - e^{-x}) = e^{-x}$, which means that the probability is exponentially decaying for larger selections of x. Given that already $Pr(X \geq 10) \approx 4.5 \cdot 10^{-5}$, we can say that almost certainly sampled values for X are in the domain $[0, 10]$. Moreover, the confidence of this statement can be strengthened by further increasing the value of x.

The above consideration does not account for cases where the rare event depends on heavy tails, or even e.g. sampling large values from exponential distributions. These cases cannot be reliably solved by ISPLIT in models where the simulation granularity stops at the level of sampling values from continuous

RVs. Instead, such problems can be tackled by modifying the distributions of the RV as in IS RES [38,48], or using analytical expansion with SSCs [23].

Time Until Simulation End. The current backwards reachability approach does not incorporate the elapsed global time and thus ignores the time until the end of the simulation in a time-bounded setting, like the one studied in our work. Especially when the remaining simulation time is low, it could be useful to know whether the target can still be reached given the time bound. For example, consider the example in Fig. 1 and assume that we want to reach the target from the initial location in 21 time units. We can achieve this objective only following the direct path to the target featuring two failures of UPS and then the failure of AC, as any other path would take longer and the target would not be reachable in time. To this end, we can incorporate a global timer t_{age} (with possible negative values) that is never removed from the computed SCs [33] and allows us to extract the minimum and maximum time needed until reaching the target. This solution has also been approached via *forcing* in IS, where the next firing values are sampled conditional to being smaller than the time bound [15,41].

Automatic Threshold Selection. An important precomputation step to yield performant ISPLIT implementations is threshold selection, viz. determining a subset of the importance values that will effectively be used to partition the state space. On the one hand, thresholds chosen too close together may incur a splitting overhead, and ultimately a runtime explosion. On the other hand, thresholds chosen too far away from each other can lead to an extinction phenomenon, e.g., no simulation path gets from one threshold to another [9,10].

Our experiments in Sect. 4 set a threshold for every IF value. As the computation effort using FE as ISPLIT method is limited by design, using all levels as thresholds is acceptable for a preliminary evaluation—in this sense, FE with static width 1 is also used as basis for the expected success threshold selection method [10]. Still, to increase the performance of ISPLIT techniques, the threshold-selection phase should be performed after the IF construction phase.

One challenge in this regard is that threshold selection methods start in the state with lowest importance. In most models, this condition simply applies to the initial states. However, in timed settings, different states belonging to the same location could have a different importance value (as demonstrated in Sect. 3). A possible way to tackle this problem is applying rejection sampling to determine states in the initial location with the lowest importance, and bootstrap the threshold selection algorithms from those states.

Confidence Intervals for Rare Events. One of the metrics used to compare the quality of the IFs in Sect. 4.3 is the width of the resulting CI. However, these were normal (central limit theorem) intervals, which are known to be unsound [14]. The problem with using sound intervals such as those based on the Dvoretzky–Kiefer–Wolfowitz (DKW) inequality is that they are largely uninformative of the quality of the rare event estimation, precisely by virtue of them providing safe bounds for any sample. In particular, DKW intervals were practically the same for cases like CMC and RES-time-10 (and -6) in Table 1b, even though RES-time-10 provides an estimate (and the normal CI contains φ), while CMC observed no rare event whatsoever. To the best of our knowledge, no sound CI provides relevant information in rare event regimes, hence we used the (unsound) standard intervals. We consider this an open problem worthy of investigation.

RES Algorithm. We performed our experiments using a single ISPLIT simulation algorithm, namely FE with 16 runs per layer. While a more comprehensive experimentation would be beneficial, the results would only be relevant in comparisons to CMC, which is anyway known to fail for extreme rare regimes. The main goal of our experimentation was to assess the quality of the (automatic) time-sensitive IF when compared against time-agnostic alternatives, for which the use of different ISPLIT simulation algorithms is only marginally relevant [10].

6 Related Work

Niehage et al. [43] recently defined a method to automatically obtain an IF for hybrid Petri nets with general transitions (HPnGs) based on the parametric location tree (PLT) [34,42], which is a symbolic representation of the embedded non-stochastic model's state space. Like SCs, each parametric location represents an uncountable set of states of the HPnG; they are connected via discrete events (e.g., firing of a transition). Similar to our work, the IF then considers the distance of each symbolic location to the target. However, where we *conceptually* start from the target states, [43] first performs a forwards expansion and then uses the information gained to search backwards in the computed PLT. This solution is less informative than our method, as it does not reflect that different parts of a node in the PLT could have a different distance to the target, e.g., the initial location, as we demonstrated in Sect. 3.

Our work is also related to the zone-based backwards reachability analysis for timed automata (TA) and probabilistic timed automata (PTA) [8]. Zones in TA and PTA are similar to SCs in the sense that they represent infinitely many configurations considering a location and can be efficiently stored with DBMs, but they work with models with increasing clocks instead of decreasing timers. Backwards reachability is, e.g., used in the model checking of timed computation tree logic (TCTL) formula against TA [32] or probabilistic TCTL (PTCTL) formula against PTA [31,36]. SMC with RES for stochastic TA-based models is also implemented in UPPAAL SMC [22], using IS [35] with a symbolic forwards exploration to identify states that cannot reach the goal.

We note that we effectively modeled the semantics of repairable DFTs with STA instead of IOSA, as that is what the MODEST TOOLSET's simulation engine works with (see Sect. 4.2). We thus defined timer values via pairs of clocks and stochastic variables (e.g., clock c with variable $x :=$ Uniform$(1, 2)$). One may be tempted to construct an IF based on the backwards search of an abstracted TA of the original STA, where the bounds of the support of a stochastic variable are used in location invariants and transition guards (e.g., turning $c \leq x$ and

$c \geq x$ into $c \leq 2$ and $c \geq 1$). Unfortunately, this would not incorporate the actual sampled values for stochastic variables like x, which is however crucial to estimating the distance to the target (as illustrated in Sect. 3).

7 Conclusion

In this paper, we introduced *time-sensitive importance splitting* as a novel extension of classical ISPLIT for the IF calculation, which incorporates the timer values in estimating the distance to the target states. We started by studying the necessity for considering timer information in ISPLIT techniques for specific scenarios where intricate timer constellations make the location information alone not sufficient to obtain an adequate distance measure to the target states. Based on the existing theory of SCs, we presented an automatic way to calculate this information in a backwards reachability manner. We implemented a prototype upon the MODEST TOOLSET. The empirical evaluation based on a repairable DFT example with PAND gates demonstrated very encouraging results.

In future work, we plan to realize some ideas to tackle the shortcomings of the current basic method as discussed in Sect. 5. Specifically, our first focus will be on providing an extended evaluation, considering larger and more realistic examples, potentially featuring also other gate types, e.g., SPARE gates. Apart from that, it will be a theoretically exciting avenue to study backwards reachability for other model classes, e.g. STA or hybrid systems.

Data Availability Statement. The artifact of this paper—a reproduction package for the experiments in Sect. 4—is available at DOI 10.5281/zenodo.15286766.

References

1. Ajmone Marsan, M., Conte, G., Balbo, G.: A class of generalized stochastic petri nets for the performance evaluation of multiprocessor systems. ACM Trans. Comput. Syst. **2**(2), 93–122 (1984). https://doi.org/10.1145/190.191
2. Amparore, E.G., Donatelli, S.: A component-based solution for reducible markov regenerative processes. Perform. Eval. **70**(6), 400–422 (2013)
3. Baier, C., Katoen, J.P.: Principles of Model Checking. MIT Press, Cambridge (2008)
4. Berthomieu, B., Diaz, M.: Modeling and verification of time dependent systems using time petri nets. IEEE Trans. Softw. Eng. **17**(3), 259 (1991)
5. Bertrand, N., et al.: Stochastic timed automata. Log. Methods Comput. Sci. **10**(4) (2014). https://doi.org/10.2168/LMCS-10(4:6)2014
6. Biagi, M., Carnevali, L., Paolieri, M., Papini, T., Vicario, E.: Exploiting non-deterministic analysis in the integration of transient solution techniques for Markov regenerative processes. In: Bertrand, N., Bortolussi, L. (eds.) Quantitative Evaluation of Systems - 14th International Conference, QEST 2017, Berlin, Germany, 5-7 September 2017, Proceedings. LNCS, vol. 10503, pp. 20–35. Springer, Cham (2017). https://doi.org/10.1007/978-3-319-66335-7_2

7. Biagi, M., Carnevali, L., Paolieri, M., Vicario, E.: Performability evaluation of the ERTMS/ETCS – Level 3. Transp. Res. Part C: Emerg. Technol. **82**, 314–336 (2017). https://doi.org/10.1016/j.trc.2017.07.002, https://www.sciencedirect.com/science/article/pii/S0968090X17301833
8. Bouyer, P.: An introduction to timed automata (2011). http://www.lsv.fr/~bouyer/files/mpri1112.pdf. Accessed 18 Apr 2025
9. Budde, C.E.: Automation of Importance Splitting Techniques for Rare Event Simulation. Ph.D. thesis (May 2017)
10. Budde, C.E., D'Argenio, P.R., Hartmanns, A.: Automated compositional importance splitting. Sci. Comput. Program. **174**, 90–108 (2019). https://doi.org/10.1016/J.SCICO.2019.01.006
11. Budde, C.E., D'Argenio, P.R., Hartmanns, A., Sedwards, S.: An efficient statistical model checker for nondeterminism and rare events. Int. J. Softw. Tools Technol. Transf. **22**(6), 759–780 (2020). https://doi.org/10.1007/S10009-020-00563-2
12. Budde, C.E., D'Argenio, P.R., Hermanns, H.: Rare event simulation with fully automated importance splitting. In: Beltrán, M., Knottenbelt, W., Bradley, J. (eds.), EPEW. LNCS, vol. 9272, pp. 275–290. Springer, Cham (2015). https://doi.org/10.1007/978-3-319-23267-6_18
13. Budde, C.E., D'Argenio, P.R., Monti, R.E., Stoelinga, M.: Analysis of non-markovian repairable fault trees through rare event simulation. Int. J. Softw. Tools Technol. Transf. **24**(5), 821–841 (2022). https://doi.org/10.1007/S10009-022-00675-X
14. Budde, C.E., Hartmanns, A., Meggendorfer, T., Weininger, M., Wienhöft, P.: Sound statistical model checking for probabilities and expected rewards. In: Gurfinkel, A., Heule, M. (eds) Tools and Algorithms for the Construction and Analysis of Systems. TACAS 2025. LNCS, vol. 15696, pp. 167–190. Springer, Cham (2025). https://doi.org/10.48550/arXiv.2411.00559, to appear, preprint available at
15. Budde, C.E., Ruijters, E., Stoelinga, M.: The dynamic fault tree rare event simulator. In: Gribaudo, M., Jansen, D.N., Remke, A. (eds.) Quantitative Evaluation of Systems. QEST 2020. LNCS, vol. 12289, pp. 233–238. Springer, Cham (2020). https://doi.org/10.1007/978-3-030-59854-9_17
16. Carnevali, L., German, R., Santoni, F., Vicario, E.: Compositional analysis of hierarchical uml statecharts. IEEE Trans. Softw. Eng. **48**(12), 4762–4788 (2021)
17. Cérou, F., Guyader, A.: Adaptive multilevel splitting for rare event analysis. Stoch. Anal. Appl. **25**(2), 417–443 (2007)
18. Choi, H., Kulkarni, V.G., Trivedi, K.S.: Markov regenerative stochastic petri nets. Perform. Eval. **20**(1–3), 337–357 (1994)
19. D'Argenio, P.R., Katoen, J.: A theory of stochastic systems part I: stochastic automata. Inf. Comput. **203**(1), 1–38 (2005). https://doi.org/10.1016/J.IC.2005.07.001, https://doi.org/10.1016/j.ic.2005.07.001
20. D'Argenio, P.R., Lee, M.D., Monti, R.E.: Input/Output stochastic automata - compositionality and determinism. In: Fränzle, M., Markey, N. (eds.) Formal Modeling and Analysis of Timed Systems. FORMATS 2016. LNCS, vol. 9884, pp. 53–68. Springer, Cham (2016). https://doi.org/10.1007/978-3-319-44878-7_4
21. D'Argenio, P.R., Monti, R.E.: Input/Output stochastic automata with urgency: confluence and weak determinism. In: Fischer, B., Uustalu, T. (eds.) Theoretical Aspects of Computing – ICTAC 2018. ICTAC 2018. LNCS, vol. 11187, pp. 132–152. Springer, Cham (2018). https://doi.org/10.1007/978-3-030-02508-3_8

22. David, A., Larsen, K.G., Legay, A., Mikucionis, M., Wang, Z.: Time for statistical model checking of real-time systems. In: Gopalakrishnan, G., Qadeer, S. (eds.) Computer Aided Verification. CAV 2011. LNCS, vol. 6806, pp. 349–355. Springer, Berlin, Heidelberg (2011). https://doi.org/10.1007/978-3-642-22110-1_27
23. Dengler, G., Carnevali, L., Budde, C.E., Vicario, E.: Transient evaluation of non-markovian models by stochastic state classes and simulation. In: Hillston, J., Soudjani, S., Waga, M. (eds.) Quantitative Evaluation of Systems and Formal Modeling and Analysis of Timed Systems. QEST+FORMATS 2024. LNCS, vol. 14996, pp. 213–232. Springer, Cham (2024). https://doi.org/10.1007/978-3-031-68416-6_13
24. Dongliang, Z., Kaiwen, Z., Chaofan, Z.: Reliability modeling and analysis of reactor protection system based on FPGA. Nucl. Power Eng. **42**(5), 173–177 (2021). https://doi.org/10.13832/j.jnpe.2021.05.0173
25. Garvels, M.J.J.: The splitting method in rare event simulation. Ph.D. thesis, University of Twente, Enschede, Netherlands (2000). http://eprints.eemcs.utwente.nl/14291/
26. German, R., Telek, M.: Formal relation of markov renewal theory and supplementary variables in the analysis of stochastic petri nets. In: Proceedings 8th International Workshop on Petri Nets and Performance Models (Cat. No.PR00331), pp. 64–73 (1999). https://doi.org/10.1109/PNPM.1999.796537
27. German, R.: Iterative analysis of markov regenerative models. Perform. Eval. **44**(1–4), 51–72 (2001)
28. German, R., Lindemann, C.: Analysis of stochastic petri nets by the method of supplementary variables. Perform. Eval. **20**(1-3), 317–335 (1994). https://doi.org/10.1016/0166-5316(94)90020-5
29. Grassmann, W.: Transient solutions in markovian queues: an algorithm for finding them and determining their waiting-time distributions. Eur. J. Oper. Res. **1**(6), 396–402 (1977). https://doi.org/10.1016/0377-2217(77)90049-2
30. Hartmanns, A., Hermanns, H.: The modest toolset: an integrated environment for quantitative modelling and verification. In: Ábrahám, E., Havelund, K. (eds.) Tools and Algorithms for the Construction and Analysis of Systems. TACAS 2014. LNCS, vol. 8413, pp. 593–598. Springer, Berlin, Heidelberg (2014). https://doi.org/10.1007/978-3-642-54862-8_51
31. Hartmanns, A., Kohlen, B.: Backwards reachability for probabilistic timed automata: a replication report. CoRR **abs/2208.11928** (2022). https://doi.org/10.48550/ARXIV.2208.11928, https://doi.org/10.48550/arXiv.2208.11928
32. Henzinger, T.A., Nicollin, X., Sifakis, J., Yovine, S.: Symbolic model checking for real-time systems. In: Proceedings of the Seventh Annual Symposium on Logic in Computer Science (LICS '92), Santa Cruz, California, USA, 22–25 June 1992, pp. 394–406. IEEE Computer Society (1992). https://doi.org/10.1109/LICS.1992.185551
33. Horváth, A., Paolieri, M., Ridi, L., Vicario, E.: Transient analysis of non-Markovian models using stochastic state classes. Perform. Eval. **69**(7-8), 315–335 (2012). https://doi.org/10.1016/J.PEVA.2011.11.002
34. Hüls, J., Pilch, C., Schinke, P., Niehaus, H., Delicaris, J., Remke, A.: State-space construction of hybrid petri nets with multiple stochastic firings. ACM Trans. Model. Comput. Simul. **31**(3), 13:1–13:37 (2021). https://doi.org/10.1145/3449353
35. Jégourel, C., Larsen, K.G., Legay, A., Mikucionis, M., Poulsen, D.B., Sedwards, S.: Importance sampling for stochastic timed automata. In: Fränzle, M., Kapur, D., Zhan, N. (eds.) Dependable Software Engineering: Theories, Tools, and Applications. SETTA 2016. LNCS, vol. 9984, pp. 163–178. Springer, Cham (2016). https://doi.org/10.1007/978-3-319-47677-3_11

36. Kwiatkowska, M., Norman, G., Sproston, J., Wang, F.: Symbolic model checking for probabilistic timed automata. Inf. Comput. **205**(7), 1027–1077 (2007). https://doi.org/10.1016/j.ic.2007.01.004, https://www.sciencedirect.com/science/article/pii/S0890540107000077
37. L'Ecuyer, P., Le Gland, F., Lezaud, P., Tuffin, B.: Splitting Techniques, chap. 3, pp. 39–61. In: Rubino and Tuffin [50] (2009). https://doi.org/10.1002/9780470745403.ch3
38. L'Ecuyer, P., Mandjes, M., Tuffin, B.: Importance Sampling in Rare Event Simulation, chap. 2, pp. 17–38. In: Rubino and Tuffin [50] (2009). https://doi.org/10.1002/9780470745403.ch2
39. Lee, J., Mitici, M.: Predictive aircraft maintenance: modeling and analysis using stochastic Petri nets. In: ESREL, pp. 146–153 (2021). https://doi.org/10.3850/978-981-18-2016-8_050-cd
40. Monti, R.E., Budde, C.E., D'Argenio, P.R.: A compositional semantics for Repairable Fault Trees with general distributions. In: LPAR 2020. EPiC Series in Computing, vol. 73, pp. 354–372. EasyChair (2020). https://doi.org/10.29007/P16V
41. Nicola, V., Shahabuddin, P., Nakayama, M.: Techniques for fast simulation of models of highly dependable systems. IEEE Trans. Reliab. **50**(3), 246–264 (2001). https://doi.org/10.1109/24.974122
42. Niehage, M., Remke, A.: Symbolic state-space exploration meets statistical model checking. Perform. Eval. **167**, 102449 (2025). https://doi.org/10.1016/J.PEVA.2024.102449
43. Niehage, M., da Silva, C., Remke, A., Hartmanns, A.: Rare event simulation for stochastic hybrid systems using symbolic importance functions. In: Dutle, A., Humphrey, L., Titolo, L. (eds.) NASA Formal Methods. NFM 2025. LNCS, vol. 15682, pp. 254–274. Springer, Cham (2025). https://doi.org/10.1007/978-3-031-93706-4_15
44. O'Connor, P.D.T., Kleyner, A.: Practical Reliability Engineering. John Wiley & Sons, Ltd., Hoboken (2011). https://doi.org/10.1002/9781119961260
45. Paolieri, M., Biagi, M., Carnevali, L., Vicario, E.: The oris tool: quantitative evaluation of non-markovian systems. IEEE Trans. Softw. Eng. **47**(6), 1211–1225 (2021). https://doi.org/10.1109/TSE.2019.2917202
46. Reibman, A., Trivedi, K.: Numerical transient analysis of markov models. Comput. Oper. Res. **15**(1), 19–36 (1988). https://doi.org/10.1016/0305-0548(88)90026-3
47. Reijsbergen, D., de Boer, P., Scheinhardt, W.R.W., Haverkort, B.R.: Automated rare event simulation for stochastic Petri nets. In: Joshi, K., Siegle, M., Stoelinga, M., D'Argenio, P.R. (eds.) Quantitative Evaluation of Systems. QEST 2013. LNCS, vol. 8054, pp. 372–388. Springer, Berlin, Heidelberg (2013). https://doi.org/10.1007/978-3-642-40196-1_31
48. Reijsbergen, D., de Boer, P., Scheinhardt, W.R.W., Juneja, S.: Path-ZVA: general, efficient, and automated importance sampling for highly reliable Markovian systems. ACM Trans. Model. Comput. Simul. **28**(3), 22:1–22:25 (2018). https://doi.org/10.1145/3161569
49. Rubino, G., Tuffin, B.: Introduction to Rare Event Simulation, chap. 1, pp. 1–13. In: Rubino and Tuffin [50] (2009). https://doi.org/10.1002/9780470745403.ch1
50. Rubino, G., Tuffin, B. (eds.): Rare Event Simulation Using Monte Carlo Methods. Wiley, Hoboken (2009). https://doi.org/10.1002/9780470745403
51. Telek, M., Horváth, A.: Transient analysis of age-mrspns by the method of supplementary variables. Perform. Eval. **45**(4), 205–221 (2001)

52. Trivedi, K.S., Bobbio, A.: Reliability and Availability Engineering: Modeling, Analysis, and Applications. Cambridge Univ. Press, Cambridge (2017)
53. Vicario, E.: Static analysis and dynamic steering of time-dependent systems. IEEE Trans. Softw. Eng. **27**(8), 728–748 (2001). https://doi.org/10.1109/32.940727
54. Vicario, E., Sassoli, L., Carnevali, L.: Using stochastic state classes in quantitative evaluation of dense-time reactive systems. IEEE Trans. Softw. Eng. **35**(5), 703–719 (2009)
55. Villén-Altamirano, M., Villén-Altamirano, J.: RESTART: a method for accelerating rare event simulations. Queueing, Performance and Control in ATM (ITC-13), pp. 71–76 (1991)
56. Younes, H.L.S., Simmons, R.G.: Probabilistic verification of discrete event systems using acceptance sampling. In: Brinksma, E., Larsen, K.G. (eds.) Computer Aided Verification. CAV 2002. LNCS, vol. 2404, pp. 223–235. Springer, Berlin, Heidelberg (2002). https://doi.org/10.1007/3-540-45657-0_17

Active Learning of Mealy Machines with Timers

Véronique Bruyère[1], Bharat Garhewal[2], Guillermo A. Pérez[3], Gaëtan Staquet[4(✉)], and Frits W. Vaandrager[2]

[1] University of Mons, Mons, Belgium
veronique.bruyere@umons.ac.be
[2] Radboud University, Nijmegen, The Netherlands
{b.garhewal,f.vaandrager}@cs.ru.nl
[3] University of Antwerp – Flanders Make, Antwerp, Belgium
guillermo.perez@uantwerpen.be
[4] Centre Inria de l'Université de Rennes, Rennes, France
gaetan.staquet@inria.fr

Abstract. We present the first algorithm for query learning Mealy machines with timers in a black-box context. Our algorithm is an extension of the $L^\#$ algorithm of Vaandrager et al. to a timed setting. We rely on symbolic queries which empower us to reason on untimed executions while learning. Similarly to the algorithm for learning timed automata of Waga, these symbolic queries can be realized using finitely many concrete queries. Experiments with a prototype implementation show that our algorithm is able to efficiently learn realistic benchmarks.

Keywords: Timed systems · model learning · active automata learning

1 Introduction

To understand and verify complex systems, we need accurate models that are understandable for humans or can be analyzed fully automatically. Such models are typically not available for legacy software and for AI systems constructed from training data. Model learning is a technology that potentially may fill this gap. In this work, we consider a specific kind of model learning: *active automata learning*, which is a black-box technique for constructing state machine models of software and hardware from information obtained through testing (i.e., providing inputs and observing the resulting outputs). It has been successfully used in many applications, e.g., for spotting bugs in implementations of major network protocols [13–17,28]. We refer to [24,31] for surveys and further references.

Gaëtan Staquet was a research fellow of the F.R.S.-FNRS. The research of B. Garhewal, F. Vaandrager, and G.A. Pérez was supported by NWO projects 612.001.852 "GIRLS" and OCENW.M.23.155 "EVI", and FWO project G0AH524N "SynthEx".

Timing plays a crucial role in many applications. However, extending model learning algorithms to a setting that incorporates quantitative timing information turns out to be challenging. Twenty years ago, the first papers on this subject were published [20,27], but we still do not have scalable algorithms for a general class of timed models. Consequently, in applications of model learning technology, timing issues still need to be artificially suppressed.

Several authors have proposed active learning algorithms for the popular framework of *timed automata* (TAs) [2], which extends DFAs with clock variables. Some of these proposals, for instance [21–23] have never been implemented. In recent years, however, several algorithms have been proposed and implemented that successfully learned realistic benchmark models. A first line of work restricts to subclasses of TAs such as deterministic one-clock TAs [4,35]. A second line of work explores synergies between active and passive learning algorithms. Aichernig et al. [1,30], for instance, employ a passive learning algorithm based on genetic programming to generate hypothesis models, which are subsequently refined using equivalence queries. A major result was obtained recently by Waga [34], who presents an algorithm for active learning of (general) deterministic TAs and shows the effectiveness of the algorithm on various benchmarks. Waga's algorithm is inspired by ideas of Maler & Pnueli [27]. Based on the notion of elementary languages of [27], Waga uses *symbolic queries* which are then performed using finitely many concrete queries. Notably, symbolic queries are also used for learning other families of automata, such as register automata [19].

A challenge for learning algorithms for TAs is the inference of the guards and resets that label transitions. Given those difficulties, Vaandrager et al. [32] propose to consider learning algorithms for models using *timers* instead of clocks, e.g., the class of models defined by Dill [11]. The value of a timer decreases when time advances, whereas the value of a clock increases. A timer can be set to integer values on transitions and may be stopped or times out (its value becomes 0) in later transitions. Each timeout triggers an observable output, allowing a learner to observe the occurrence of timeouts. In [32], the notion of Mealy machine with a *single timer* (MM1T) is introduced, such that the absence of guards and invariants simplifies learning. A learner still has to determine which transitions (re)start the timer, but this no longer creates a combinatorial blow-up. If a transition sets a timer, then slight changes in the timing of this transition will trigger the corresponding changes in the timing of the resulting timeout, allowing a learner to identify the exact cause of each timeout.

Even if many realistic systems can be modeled as MM1Ts (e.g., the benchmarks in [32] and the brick sorter and traffic controller examples in [10]), the restriction to a single timer is a serious limitation. Kogel et al. [26] propose Mealy machines with *local timers* (MMLTs), where multiple timers are subject to carefully chosen constraints to enable efficient learning. Although quite interesting, the constraints of MMLTs are too restrictive for many applications (e.g., the FDDI protocol [25,34]). Also, any MMLT can be converted to an equivalent MM1T. In [8], we explore a general extension of MM1Ts with *multiple timers* (MMTs), and show that for MMTs that are *race-avoiding*, the cause of a timeout event can be efficiently determined by "wiggling" the timing of input events. Finally, MMTs form a subclass of TAs, and thus, existing model checking algorithms and tools for TAs (such as UPPAAL[1]) can be used.

Our main contribution is a learning algorithm for the MMT model of [8], that is obtained by extending the $L^\#$ learning algorithm for Mealy machines of Vaandrager et al. [33] and by using the concept of symbolic queries as in [34]. Extending the $L^\#$ algorithm to multiple timers is not an easy task; it requires several new ideas. As this approach is different from Angluin's L^* framework [5], used in other timed model learning algorithms, we cannot immediately apply existing ideas. Moreover, to the best of our knowledge, there is no notion of minimal MMT, in contrast to the minimal representations of timed languages proposed in [6] and the minimal model for timed automata with integer resets of [12]. Experiments with a prototype implementation, written in Rust, show that our algorithm is able to efficiently learn realistic benchmarks. The missing proofs and further details can be consulted in the arXiv version [7].

2 Mealy Machines with Timers

A *Mealy machine* is a variant of the classical finite automaton that associates an output with each transition. It can be seen as a relation between input words and output words. Mealy machines with *timers* [32] can be used to enforce timing constraints over the behavior of the relation, e.g., if we send a message and do not receive the acknowledgment after d units of time, we resend the message.

We fix two non-empty finite sets I of *inputs* and O of *outputs*. A *Mealy machine with timers* uses a finite set X of *timers* that can trigger transitions when they time out. Intuitively, changes in the state of the machine will be driven both by reading inputs and by "timeouts". To formalize this, it is convenient to define the set $TO[X] = \{to[x] \mid x \in X\}$ of *timeouts* of X. We write $A(\mathcal{M})$ for the set $I \cup TO[X]$ of *actions* of $\mathcal{M}$: reading an input or processing a timeout. The set $U(\mathcal{M}) = (X \times \mathbb{N}^{>0}) \cup \{\bot\}$ contains the *updates* of $\mathcal{M}$, where (x, c) means that timer x is started with value c, and $\bot$ stands for no timer update. We impose constraints on the shape of MMTs, e.g., a timer x can time out only when it is active, to define the timed semantics in a straightforward manner.

Definition 1 (Mealy machine with timers). *A Mealy machine with timers (MMT, for short) is a tuple $\mathcal{M} = (X, Q, q_0, \chi, \delta)$ where:*

- *X is a finite set of timers,*
- *Q is a finite set of states, with $q_0 \in Q$ the initial state,*
- *$\chi : Q \to \mathcal{P}(X)$ assigns a set of active timers to each state, and*
- *$\delta : Q \times A(\mathcal{M}) \rightharpoonup Q \times O \times U(\mathcal{M})$ is a partial transition function.*

We write $q \xrightarrow{i/o}_u q'$ if $\delta(q, i) = (q', o, u)$. We require the following:

---

[1] https://uppaal.org/.

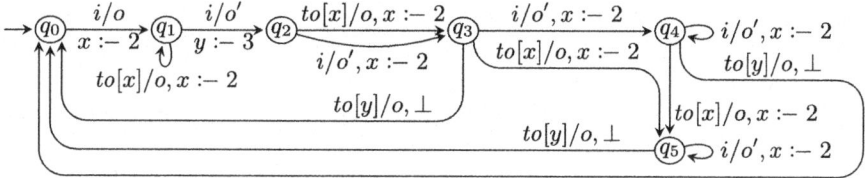

Fig. 1. An MMT with $\chi(q_0) = \emptyset$, $\chi(q_1) = \{x\}$, and $\chi(q) = \{x,y\}$ for all other q.

1. In the initial state, no timer is active, i.e., $\chi(q_0) = \emptyset$.
2. All active timers of the target state of a transition come from the source state, except at most one timer that may be started on the transition. That is, if $q \xrightarrow[\perp]{i/o} q'$, then $\chi(q') \subseteq \chi(q)$, and if $q \xrightarrow[(x,c)]{i/o} q'$, then $\chi(q') \setminus \{x\} \subseteq \chi(q)$.
3. If timer x times out, then x was active in the source state, i.e., if $q \xrightarrow[u]{to[x]/o} q'$ then $x \in \chi(q)$. Moreover, if $u \neq \perp$, then u must be (x,c) for some $c \in \mathbb{N}^{>0}$.

When needed, we add a superscript to indicate which MMT is considered, e.g., $Q^{\mathcal{M}}, q_0^{\mathcal{M}}$, etc. Missing symbols in $q \xrightarrow[u]{i/o} q'$ are quantified existentially. We say a transition $q \xrightarrow[u]{} q'$ *starts* (resp. *restarts*) a timer x if $u = (x,c)$ and x is inactive (resp. active) in q. We say that a transition[2] $q \xrightarrow{i} q'$ with $i \neq to[x]$ *stops* timer x if x is inactive in q'. The requirement that a $to[x]$-transition can only restart timer x is without loss of generality and a choice made to simplify the presentation (see [7, Appendix E.2]). An example of an MMT is given in Fig. 1.

A *run* π of $\mathcal{M}$ either consists of a single state p_0 or of a nonempty sequence of transitions $\pi = p_0 \xrightarrow[u_1]{i_1/o_1} p_1 \xrightarrow[u_2]{i_2/o_2} \cdots \xrightarrow[u_n]{i_n/o_n} p_n$. We denote by $\mathit{runs}(\mathcal{M})$ the set of runs of $\mathcal{M}$. We often write $q \xrightarrow{i} \in \mathit{runs}(\mathcal{M})$ to highlight that $\delta(q,i)$ is defined. We lift the notation to words $w = i_1 \cdots i_n$ as usual: $p_0 \xrightarrow{w} p_n \in \mathit{runs}(\mathcal{M})$ if there exists a run $p_0 \xrightarrow{i_1} \cdots \xrightarrow{i_n} p_n \in \mathit{runs}(\mathcal{M})$. Note that any run π is uniquely determined by state p_0 and word w, as $\mathcal{M}$ is deterministic. A run $p_0 \xrightarrow{w}$ is x-*spanning* (with $x \in X$) if it begins with a transition (re)starting x, ends with a $to[x]$-transition, and no intermediate transition restarts or stops x.

2.1 Timed Semantics

The semantics of an MMT $\mathcal{M}$ is defined via an infinite-state labeled transition system describing all possible configurations and transitions between them. A *valuation* is a partial function $\kappa \colon X \rightharpoonup \mathbb{R}^{\geq 0}$ that assigns nonnegative real numbers to timers. For $Y \subseteq X$, we write $\mathsf{Val}(Y)$ for the set of all valuations κ with $\mathsf{dom}(\kappa) = Y$. A *configuration* of $\mathcal{M}$ is a pair (q,κ) where $q \in Q$ and $\kappa \in \mathsf{Val}(\chi(q))$. The *initial configuration* is the pair (q_0, κ_0) where $\kappa_0 = \emptyset$ since

---

[2] Technically, it is the arrival at a state where the timer is marked as inactive that stops the timer. However, we find it convenient to reason about the transitions leading to such a state as being responsible for stopping the timer.

$\chi(q_0) = \emptyset$. If $\kappa \in \mathsf{Val}(Y)$ is a valuation in which all timers from Y have a value of at least $d \in \mathbb{R}^{\geq 0}$, then d units of time may *elapse*. We write $\kappa - d \in \mathsf{Val}(Y)$ for the resulting valuation that satisfies $(\kappa - d)(x) = \kappa(x) - d$, for all $x \in Y$. If the valuation κ contains a value $\kappa(x) = 0$ for some timer x, then x may *time out*. We define the transitions between configurations $(q, \kappa), (q', \kappa')$ as follows:

delay transition $(q, \kappa) \xrightarrow{d} (q, \kappa - d)$, with $\kappa(x) \geq d$ for every $x \in \chi(q)$,

discrete transition $(q, \kappa) \xrightarrow{i/o}_{u} (q', \kappa')$, with $q \xrightarrow{i/o}_{u} q'$ a transition, $\kappa'(y) = \kappa(y)$ for all $y \in \chi(q')$ except that $\kappa'(x) = c$ if $u = (x, c)$. Moreover, if $i = \mathsf{to}[x]$ then $\kappa(x) = 0$ and we call it a *timeout transition*; otherwise, we call it an *input transition*.

A *timed run* of $\mathcal{M}$ is a sequence of configuration transitions such that delay and discrete transitions alternate, beginning and ending with a delay transition. We say a configuration (q, κ) is *reachable* if there is a timed run ρ that starts with the initial configuration and ends with (q, κ). The *untimed projection* of ρ, noted $\mathsf{untime}(\rho)$, is the run obtained by omitting the valuations and delay transitions of ρ. A run π is said *feasible* if there is a timed run ρ such that $\mathsf{untime}(\rho) = \pi$.

A *timed word* over a set Σ is an alternating sequence of delays from $\mathbb{R}^{\geq 0}$ and symbols from Σ, such that it starts and ends with a delay. The length of a timed word w, noted $|w|$, is the number of symbols of Σ in w. Note that, when $\Sigma = A(\mathcal{M})$, a timed run reading a timed word w is uniquely determined by its first configuration and w. We thus write $(p, \kappa) \xrightarrow{w}$ for a timed run. A timed run ρ is called *x-spanning* (with $x \in X$) if $\mathsf{untime}(\rho)$ is x-spanning.

Example 1. Let $\mathcal{M}$ be the MMT of Fig. 1. The timed run reading the timed word $0.5 \cdot i \cdot 1 \cdot i \cdot 1 \cdot \mathsf{to}[x] \cdot 2 \cdot \mathsf{to}[y] \cdot 0$ and its untimed projection are:

$(q_0, \emptyset) \xrightarrow{0.5} (q_0, \emptyset) \xrightarrow{i/o}_{(x,2)} (q_1, x=2) \xrightarrow{1} (q_1, x=1) \xrightarrow{i/o'}_{(y,3)} (q_2, x=1, y=3) \xrightarrow{1} (q_2, x=0, y=2)$

$\xrightarrow{\mathsf{to}[x]/o}_{(x,2)} (q_3, x=y=2) \xrightarrow{2} (q_3, x=y=0) \xrightarrow{\mathsf{to}[y]/o}_{\bot} (q_0, \emptyset) \xrightarrow{0} (q_0, \emptyset),$

$\pi = q_0 \xrightarrow{i/o}_{(x,2)} q_1 \xrightarrow{i/o'}_{(y,3)} q_2 \xrightarrow{\mathsf{to}[x]/o}_{(x,2)} q_3 \xrightarrow{\mathsf{to}[y]/o}_{\bot} q_0.$

Hence, π is feasible, unlike the run $q_0 \xrightarrow{i \cdot i \cdot \mathsf{to}[y]}$, as the value of y in q_2 is always greater than the value of x, no matter the chosen delays. Observe that $q_1 \xrightarrow{i \cdot \mathsf{to}[x] \cdot \mathsf{to}[y]}$ is y-spanning, while $q_0 \xrightarrow{i \cdot i \cdot \mathsf{to}[x]}$ is x-spanning.

As $(q_2, x = 0, y = 2)$ is reachable, we say that x is *enabled* in q_2: it is possible to observe the timeout of x in q_2 along some timed run. However, y is not enabled in q_2, as it is impossible to reach a configuration $(q_2, x = \cdot, y = 0)$.

Formally, we write $\chi_0(q)$ for the set of all enabled timers of q, i.e.,

$$\chi_0(q) = \{x \in \chi(q) \mid \exists (q_0, \emptyset) \xrightarrow{w} (q, \kappa) : \kappa(x) = 0\}.$$

If q has at least one enabled timer then, just by waiting in q for long enough, we can force one timer to reach the value zero. A desirable property is for all such behaviors to have a corresponding timeout transition, which is the case for $\mathcal{M}$. We thus say that an MMT $\mathcal{N}$ is *complete* if for all $q \in Q$ and all $i \in I \cup TO[\chi_0(q)]$ we have $q \xrightarrow{i} \in runs(\mathcal{N})$.[3]

2.2 Symbolic Semantics

Our learning algorithm is based on an untimed, symbolic semantics that abstracts from timing delays and timer names. The idea is that, rather than the name of a timer in a $to[x]$ event, we record a pointer to the preceding transition that (re)started this timer, together with the value to which the timer was set in this preceding transition. Consider, for instance, the run

$$\pi = q_0 \xrightarrow[(x,2)]{i} q_1 \xrightarrow[(x,2)]{to[x]} q_1 \xrightarrow{to[x]} q_1$$

of the MMT of Fig. 1. Then the *symbolic input word* of π is $\mathtt{w} = i \cdot to[2,1] \cdot to[2,2]$, as the first action of π is the input i, the second action is a timeout that is caused by setting the timer to 2 in the first transition, and the third action is a timeout that is caused by setting the timer to 2 in the second transition. More generally, consider a run

$$\pi = p_0 \xrightarrow[u_1]{i_1/o_1} p_1 \xrightarrow[u_2]{i_2/o_2} \cdots \xrightarrow[u_n]{i_n/o_n} p_n$$

with $p_0 = q_0^{\mathcal{M}}$. For $k \in \{1,\ldots,n\}$, let π_k be the prefix of run π up to state p_k. Suppose that timer x is active in the last state of π_k, for some $k > 0$. Then we define $cs(\pi_k, x)$ as the pair (c,j), where j is the index of the last transition (re)starting x, and c is the value to which x is set in this transition:

$$cs(\pi_k, x) = \begin{cases} (c,k) & \text{if } u_k = (x,c), \\ cs(\pi_{k-1}, x) & \text{otherwise.} \end{cases}$$

Now we associate with π a *symbolic word* (sw, in short) $sw(\pi) = \mathtt{i}_1 \cdots \mathtt{i}_n$ over alphabet $\mathtt{A} = I \cup TO[\mathbb{N}^{>0} \times \mathbb{N}^{>0}]$ as follows. For every $k \in \{1,\ldots,n\}$, if $i_k \in I$ then $\mathtt{i}_k = i_k$, and if $i_k = to[x]$ then $\mathtt{i}_k = to[cs(\pi_{k-1}, x)]$. Similarly, we associate an *output word* $ow(\pi)$ with π, consisting of the sequence of output actions $o_1 \cdots o_n$ occurring in run π. We write $L_{sym}(\mathcal{M})$ to denote the *symbolic input language* of $\mathcal{M}$, defined as the set of symbolic input words associated with feasible runs of $\mathcal{M}$ starting from $q_0^{\mathcal{M}}$.

Notice that, since MMTs are deterministic, a feasible run π can be denoted via the word $w = i_1 \cdots i_n$, and vice-versa. Thus, for every word w labeling a feasible run π, we also associate a sw to w as $\overline{w} = sw(\pi)$. Moreover, since each transition of an MMT (re)starts at most one timer, at most one run can be associated with any symbolic input word. For instance, the symbolic input word $\mathtt{w} = i \cdot i \cdot to[2,1] \cdot to[3,2]$ induces the run $q_0 \xrightarrow[(x,2)]{i} q_1 \xrightarrow[(y,3)]{i} q_2 \xrightarrow{to[x]} q_3 \xrightarrow{to[y]} q_0$ in $\mathcal{M}$. If $\mathtt{w} \in L_{sym}(\mathcal{M})$, we write $out^{\mathcal{M}}(\mathtt{w})$ for the output word $ow(\pi)$ of the unique run π of $\mathcal{M}$ with $sw(\pi) = \mathtt{w}$.

We now define a notion of *symbolic equivalence* between two MMTs.

---

[3] Notice that we only require the timeout transitions to be defined for enabled timers.

Definition 2 (Symbolic equivalence). *Two complete MMTs $\mathcal{M}$ and $\mathcal{N}$ are symbolically equivalent, noted $\mathcal{M} \stackrel{sym}{\approx} \mathcal{N}$, if $L_{sym}(\mathcal{M}) = L_{sym}(\mathcal{N})$ and, for each $\mathtt{w} \in L_{sym}(\mathcal{M})$, $out^{\mathcal{M}}(\mathtt{w}) = out^{\mathcal{N}}(\mathtt{w})$.*

As outputs and updates at the start of spanning runs are the same, symbolic equivalence implies *timed equivalence* (defined as a bisimulation, see [7, Appendix A]), i.e., the MMTs produce the same output words on all timed input words.

2.3 Learning Framework

As usual for active learning algorithms, we rely on Angluin's framework [5]: A teacher knows an MMT $\mathcal{M}$, and a learner can query the teacher to obtain knowledge about $\mathcal{M}$. The set of MMTs $\mathcal{M}$ that we consider for our learning algorithm must be *s-learnable* (the s stands for symbolically), i.e., $\mathcal{M}$ is complete and every run of $\mathcal{M}$ is feasible. The MMT of Fig. 1 is s-learnable. From any complete MMT, one can construct an s-learnable MMT that is symbolically equivalent, by using *zones* (akin to the homonymous concept for timed automata, see [9]). The complete proof is given in [7, Appendix B].

Lemma 1. *For any complete MMT $\mathcal{M}$, there is an s-learnable MMT $\mathcal{N} \stackrel{sym}{\approx} \mathcal{M}$.*

We now define the *queries* the learner uses to gather knowledge about $\mathcal{M}$, the MMT of the teacher. For classical Mealy machines [29,33], there are two queries: *output queries* providing the sequence of outputs for a given input word, and *equivalence queries* asking whether a hypothesis $\mathcal{H}$ is correct. If it is not, a counterexample is returned, i.e., a word w inducing different outputs in $\mathcal{H}$ and in $\mathcal{M}$. In this work, we need to adapt those queries to encode the timed behavior induced by the timers of $\mathcal{M}$. As two MMTs do not use the same timers in general, we rely on sws, and adapt our queries accordingly. In order to deal with timed behavior, we also need a new type of query, called a *wait query*. Kogel et al. introduced a similar type of query, called timer query, in [26].

Definition 3 (Symbolic queries). *The learner uses three symbolic queries, parametrized by either a symbolic input word $\mathtt{w}$ or a complete MMT $\mathcal{H}$:*

OQ^s($\mathtt{w}$) *If $\mathtt{w} \in L_{sym}(\mathcal{M})$, then this query returns $out^{\mathcal{M}}(\mathtt{w})$.*
WQ^s($\mathtt{w}$) *If $\mathtt{w} \in L_{sym}(\mathcal{M})$, then this query returns the set of all pairs (c, j) with $\mathtt{w}\, to[c, j] \in L_{sym}(\mathcal{M})$.*
EQ^s($\mathcal{H}$) *This query returns yes if $\mathcal{H} \stackrel{sym}{\approx} \mathcal{M}$, and otherwise a symbolic input word $\mathtt{w}$ witnessing the non-equivalence: either $\mathtt{w} \in L_{sym}(\mathcal{H}) \setminus L_{sym}(\mathcal{M})$, or $\mathtt{w} \in L_{sym}(\mathcal{M}) \setminus L_{sym}(\mathcal{H})$, or $\mathtt{w} \in L_{sym}(\mathcal{H}) \cap L_{sym}(\mathcal{M})$ and $out^{\mathcal{H}}(\mathtt{w}) \neq out^{\mathcal{M}}(\mathtt{w})$.*

OQ^s and **EQ^s** are analogous to regular output and equivalence queries for Mealy machines, while **WQ^s** provides, for each timer x enabled at the end of the run induced by the symbolic word, the transition which last (re)started x and the constant c with which x was (re)started. (See Example 2 below.)

$$\rightarrow \underset{x_1 :- 2}{\textcircled{t_0}} \xrightarrow{i/o} \textcircled{t_1} \xrightarrow[i/o', \perp]{to[x_1]/o, x_1 :- 2} \textcircled{t_2} \xrightarrow[]{to[x_1]/o, \perp} \textcircled{t_4} \qquad \rightarrow \underset{y_1 :- 2}{\textcircled{t_0}} \xrightarrow{to[y_1]/o, y_1 :- 2} \textcircled{t_1} \circlearrowright i/o', \perp$$

Fig. 2. On the left, a tree $\mathcal{T}$ from which the hypothesis MMT on the right is constructed. Basis states of $\mathcal{T}$ have a gray background.

Similarly to the TA learning algorithm of Waga [34], these symbolic queries can be performed via concrete output and equivalence queries. This is possible under the assumption that $\mathcal{M}$ is *race-avoiding* [8], i.e., $\mathcal{M}$ allows runs to be observed deterministically, in the sense that any feasible run π is the untimed projection of a run ρ where all delays are non-zero and there are no two timers that time out at the same time along ρ.[4] Not all MMTs are race-avoiding and a decidable characterization of those MMTs that are is given in [8]. MMTs with one timer [32] and Mealy machines with local timers [26] (translated as MMTs) are all race-avoiding, this is also the case for all benchmarks in Sect. 4.

Lemma 2. *For race-avoiding MMTs, the three symbolic queries can be realized via a polynomial number of concrete output and equivalence queries.*

For intuition, for all symbolic queries, we can construct a system of linear temporal constraints whose solutions can be mapped to concrete input words. Technically, the system also depends on the tree we use to structure the result of queries we have made. See [7, Appendix C] for details.

3 Learning Algorithm

We now present our learning algorithm for MMTs, called $L_{\mathrm{MMT}}^{\#}$. Given the hidden s-learnable MMT $\mathcal{M}$ we want to learn, this algorithm gradually builds a tree-shaped MMT $\mathcal{T}$ that stores the observations obtained by membership and wait queries. It then constructs a hypothesis MMT $\mathcal{H}$ obtained from $\mathcal{T}$ whose certain states (the basis) are the states of $\mathcal{H}$ and certain transitions are redirected to the basis. An example of MMTs $\mathcal{T}$ and $\mathcal{H}$ is given in Fig. 2 for the MMT $\mathcal{M}$ of Fig. 1, such that the equivalence query fails. The main difficulties of the learning process are: *(i)* $\mathcal{T}$ uses its own timers without knowing those of $\mathcal{M}$ and for which it is necessary to discover when two distinct timers of $\mathcal{T}$ represent the same timer of $\mathcal{M}$ (see the concept of timer matching and functional simulation below); *(ii)* the basis of $\mathcal{T}$ used to define the states of $\mathcal{H}$ must be formed of states that correspond to distinct states of $\mathcal{M}$ despite partial knowledge of its timers (see the concept of apartness and Theorem 2 below). Before moving on to the details of the $L_{\mathrm{MMT}}^{\#}$ algorithm, let us state its complexity. Observe that the same bounds hold with concrete queries, by Lemma 2, and that the complexity becomes polynomial when $|X^{\mathcal{M}}|$ is fixed.

---

[4] We highlight that we do not require this to hold for any timed run ρ. Importantly, if $\mathcal{M}$ is race-avoiding, then we can always change the delays in the runs to avoid having concurrent timeouts.

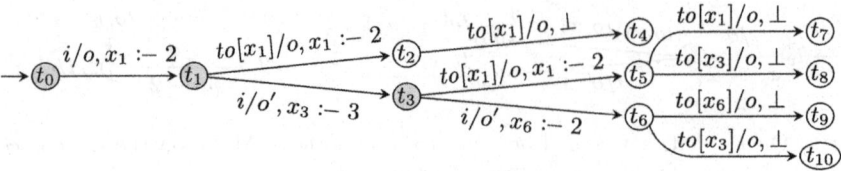

Fig. 3. Sample observation tree (we write x_i instead of x_{t_i} for all states t_i) with $\chi(t_1) = \chi(t_2) = \{x_1\}$, $\chi(t_3) = \chi(t_5) = \{x_1, x_3\}$, $\chi(t_6) = \{x_3, x_6\}$, and $\chi(t) = \emptyset$ for the other states t.

Theorem 1. *The $L^\#_{MMT}$ algorithm terminates and returns an MMT $\mathcal{N} \stackrel{sym}{\approx} \mathcal{M}$ s.t. $\mathcal{N}$ has a size polynomial in $|Q^\mathcal{M}|$ and factorial in $|X^\mathcal{M}|$. Its running time and number of symbolic queries of $L^\#_{MMT}$ are polynomial in $|Q^\mathcal{M}|, |I|$ and the length of the longest counterexample returned by the teacher, factorial in $|X^\mathcal{M}|$.*

3.1 Observation Tree

We describe the main data structure of $L^\#_{MMT}$: a modification of the *observation tree* used for $L^\#$ [33]. We impose that this tree $\mathcal{T}$ is a tree-shaped MMT whose every run is feasible. Each state q of $\mathcal{T}$ has its own timer x_q that can only be started by the incoming transition of q, and may be restarted only by a $to[x_q]$-transition. Due to the tree-shape nature of $\mathcal{T}$, we can impose strict constraints on the sets of active and enabled timers of its states, as described in the next definition. We only provide the main ideas, see [7, Appendix D] for details.

Definition 4 (Observation tree). *An MMT $\mathcal{M}$ is tree shaped if for every state $q \in Q^\mathcal{M}$ there is a unique run, denoted $run^\mathcal{M}(q)$, that starts in $q_0^\mathcal{M}$ and ends in q. An observation tree is a tree-shaped MMT $\mathcal{T} = (X, Q, q_0, \chi, \delta)$ such that $X = \{x_q \mid q \in Q \setminus \{q_0\}\}$, every run of $\mathcal{T}$ is feasible,*

- *for all $p \xrightarrow[(x,c)]{i} q$ with $i \in I$, we have $x = x_q$,*
- *for all $q \in Q, x \in X$, we have $x \in \chi(q)$ if and only if there is an x-spanning run traversing q, and*
- *for all $q \in Q, x \in X$, we have $x \in \chi_0(q)$ if and only if $q \xrightarrow{to[x]} \in runs(\mathcal{T})$.*

We explain how to use $\mathbf{OQ^s}$ and $\mathbf{WQ^s}$ to gradually grow $\mathcal{T}$ on an example.

Example 2. Let $\mathcal{M}$ be the MMT of Fig. 1 and $\mathcal{T}$ be the observation tree of Fig. 3, except that $t_3 \xrightarrow{i} \notin runs(\mathcal{T})$, i.e., the subtree rooted at t_6 is not present in the tree. We construct that subtree, via $\mathbf{OQ^s}$ and $\mathbf{WQ^s}$.

First, we create the $t_3 \xrightarrow{i}$ transition. Let $w = i \cdot i$ be the unique word such that $t_0 \xrightarrow{w} t_3$ and $\mathbf{w} = \overline{w} = i \cdot i$ be the corresponding sw. As $\mathcal{M}$ is s-learnable (and, thus, complete), it follows that $q_0^\mathcal{M} \xrightarrow{\mathbf{w} \cdot i} \in runs(\mathcal{M})$, i.e., $\mathbf{w} \cdot i \in L_{sym}(\mathcal{M})$. So, we can call $\mathbf{OQ^s}(\mathbf{w} \cdot i)$, which returns $o \cdot o' \cdot o'$. Hence, we create the transition

$t_3 \xrightarrow{i/o'}_{\perp} t_6$ in $\mathcal{T}$. We initially use a $\perp$ update (seen as a sort of wildcard) as we do not yet have any information about the potential update of the corresponding transition (here, $q_2 \xrightarrow{i}_{(x,2)} q_3$) in $\mathcal{M}$. Later on, we may discover that the transition must start a timer, in which case $\perp$ will be replaced by an actual update.

We then perform a symbolic wait query in t_6, i.e., call $\mathbf{WQ^s}(\mathbf{w} \cdot i)$, which returns the set $\{(2,3),(3,2)\}$ meaning that the second (resp. third) transition of the run $q_0^{\mathcal{T}} \xrightarrow{\mathbf{w} \cdot i}$ must (re)start a timer with constant 3 (resp. constant 2). So, the $\perp$ of the newly created transition is replaced by $(x_6, 2)$ (as the label of the transition is an input). It remains to create the $to[x_3]$- and $to[x_6]$-transitions from t_6 by performing two symbolic output queries. We call $\mathbf{OQ^s}(\mathbf{w} \cdot i \cdot to[2])$ and $\mathbf{OQ^s}(\mathbf{w} \cdot i \cdot to[3])$ and create the transitions. We obtain the tree of Fig. 3.

Functional Simulation. Each state of an observation tree can be *mapped to* a state of the hidden $\mathcal{M}$, by adapting the notion of *functional simulation* of $L^{\#}$ [33]. We have a function $f : Q^{\mathcal{T}} \to Q^{\mathcal{M}}$ that preserves the initial state and the behavior of transitions. In addition, since $\mathcal{T}$ and $\mathcal{M}$ may use *different* timers, we have a function $g : X^{\mathcal{T}} \to X^{\mathcal{M}}$ that maps timers of $\mathcal{T}$ to timers of $\mathcal{M}$. We lift g to actions such that $g(i) = i$ for every $i \in I$ and $g(to[x]) = to[g(x)]$ for every $x \in \mathsf{dom}(g)$. We require, for all $q, q' \in Q^{\mathcal{T}}, x, y \in X^{\mathcal{T}}, i \in I$ and $o \in O$:

1. f preserves the initial state, i.e., $f(q_0^{\mathcal{T}}) = q_0^{\mathcal{M}}$;
2. any timer that is active in q has a corresponding active timer in $f(q)$, i.e., if $x \in \chi^{\mathcal{T}}(q)$, then $g(x) \in \chi^{\mathcal{M}}(f(q))$;
3. two distinct timers that are active in q are mapped to two distinct timers in $\mathcal{M}$, i.e., if $x \neq y$ and both are active in q, then $g(x) \neq g(y)$;
4. for each $q \xrightarrow{i/o} q'$ in $\mathcal{T}$, there is a corresponding transition $f(q) \xrightarrow{g(i)/o} f(q')$ in $\mathcal{M}$;
5. using rule (4) we may construct, for each run π of $\mathcal{T}$, a corresponding run $\langle f, g \rangle(\pi)$ of $\mathcal{M}$. We require that the causality of timeouts is preserved, i.e., for each $q \xrightarrow{to[x]/o} q'$ in $\mathcal{T}$ with q the last state of π, $cs(\pi, x) = cs(\langle f, g \rangle(\pi), g(x))$.

One can show the following properties from the above constraints.

Lemma 3. *Let $\mathcal{M}$ be an s-learnable MMT, $\mathcal{T}$ be an observation tree, and f and g be the functions described above. Then, for any state $q \in Q^{\mathcal{T}}$, we have $|\chi^{\mathcal{T}}(q)| \leq |\chi^{\mathcal{M}}(f(q))|$ and for all $x \in \chi_0^{\mathcal{T}}(q)$, it holds that $g(x) \in \chi_0^{\mathcal{M}}(f(q))$.*

We say that a state q of $\mathcal{T}$ is *explored* once $\mathbf{WQ^s}(q)$ has been called. By definition of wait queries, it means that every enabled timer of $f(q)$ is identified in q, i.e., $|\chi_0(q)| = |\chi_0(f(q))|$. We thus obtain a *one-to-one correspondence* between $\chi_0^{\mathcal{T}}(q)$ and $\chi_0^{\mathcal{M}}(f(q))$. Define $\mathcal{E}^{\mathcal{T}}$ as the maximal set of explored states of $\mathcal{T}$. During the learning process, we will ensure that a wait query is performed only on states that are successors of an explored state to keep $\mathcal{E}^{\mathcal{T}}$ as a subtree of $\mathcal{T}$.

Example 3. In the tree of Fig. 3, the explored states are t_0, t_1, t_2, t_3, t_5, and t_6. In Example 2, t_0 to t_3, and t_5 were already explored and formed a subtree. The wait query over t_6 made it an explored state, i.e., $\mathcal{E}^{\mathcal{T}}$ is still a subtree.

Apartness. A key aspect of $L^\#$ for regular Mealy machines is the notion of *apartness* indicating which states of an observation tree have different behaviors [33]: states p, p' are *apart* if they have different output responses to the same input word. Here, we face a more complex situation as we need to handle the fact that different timers in $\mathcal{T}$ can represent the same timer in $\mathcal{M}$. Before defining apartness in the MMT setting, we need to introduce two concepts.

For two states p, p' of $\mathcal{T}$, we define a *matching* $m : \chi^\mathcal{T}(p) \rightharpoonup \chi^\mathcal{T}(p')$ as an injective partial function. This matching is meant to indicate that the timers x and $m(x)$ could represent the same timer in $\mathcal{M}$. By abuse of notation, it is denoted as $m : p \leftrightarrow p'$. A matching is *maximal* if m is total or surjective, i.e., m embeds $\chi^\mathcal{T}(p)$ in $\chi^\mathcal{T}(p')$ or m^{-1} embeds $\chi^\mathcal{T}(p')$ in $\chi^\mathcal{T}(p)$. Recall that if two distinct timers x and y are both active in the same state of $\mathcal{T}$, it must hold that $g(x) \neq g(y)$, i.e., they correspond to different timers in $\mathcal{M}$ (by functional simulation). Hence, we say that a matching m is *valid* if for all $x \in \text{dom}(m)$ and $q \in Q^\mathcal{T}$, we *do not* have $x, m(x) \in \chi^\mathcal{T}(q)$.

Second, given a run $\pi = p_0 \xrightarrow{i_1} p_1 \xrightarrow{i_2} \cdots \xrightarrow{i_n} p_n$ and a matching $m : p_0 \leftrightarrow p'_0$, we want to check whether it is possible to "read" the same word $i_1 \ldots i_n$ from p'_0 while extending m during this reading. If this is possible, this unique run, denoted $\text{read}^m_\pi(p'_0)$, is equal to $\pi' = p'_0 \xrightarrow{i'_1} p'_1 \xrightarrow{i'_2} \cdots \xrightarrow{i'_n} p'_n$ such that:

(i) If $i_j \in I$, then $i'_j = i_j$.
(ii) If $i_j = \text{to}[x]$ for some $x \in X^\mathcal{T}$, recall that $x \in \chi^\mathcal{T}(p_0)$ or $x = x_{p_k}$ for some k. Then, i'_j is either $\text{to}[m(x)]$, or $\text{to}[x_{p'_k}]$ with the same k.

When $\text{read}^m_\pi(p'_0)$ exists, we therefore have a matching between π and π', denoted $m^\pi_{\pi'} : \pi \leftrightarrow \pi'$, which extends m such that $m^\pi_{\pi'} = m \cup \{x_{p_k} \mapsto x_{p'_k} \mid k \leq n\}$.

While two states of a Mealy machine are apart when they have different output responses to the same input word, we have five different apartness cases in the MMT setting. An illustrative example is given after the definition.

Definition 5 (Apartness). *For two states p_0, p'_0 and a matching $m : p_0 \leftrightarrow p'_0$, we say that p_0, p'_0 are m-apart, denoted by $p_0 \#^m p'_0$, if there are $\pi = p_0 \xrightarrow{i_1} \cdots \xrightarrow{i_n/o} p_n$ and $\text{read}^m_\pi(p'_0) = \pi' = p'_0 \xrightarrow{i'_1} \cdots \xrightarrow{i'_n/o'} p'_n$ with $m^\pi_{\pi'} : \pi \leftrightarrow \pi'$, and*

Structural apartness *either $m^\pi_{\pi'}$ is invalid, or*
Behavioral apartness *one of the following cases holds:*

$$o \neq o' \quad \text{(outputs)}$$

$$u = (x, c) \wedge u' = (x', c') \wedge c \neq c' \quad \text{(constants)}$$

$$p_n, p'_n \in \mathcal{E}^\mathcal{T} \wedge |\chi^\mathcal{T}_0(p_n)| \neq |\chi^\mathcal{T}_0(p'_n)| \quad \text{(sizes)}$$

$$p_n, p'_n \in \mathcal{E}^\mathcal{T} \wedge \exists x \in \text{dom}(m^\pi_{\pi'}) : \big(x \in \chi^\mathcal{T}_0(p_n) \Leftrightarrow m^\pi_{\pi'}(x) \notin \chi^\mathcal{T}_0(p'_n)\big) \quad \text{(enabled)}$$

The word $w = i_1 \ldots i_n$ is called a witness of $p_0 \#^m p'_0$, denoted $w \vdash_0 \#^m p'_0$.

Example 4. Let $\mathcal{T}$ be the tree of Fig. 3 and $\pi = t_0 \xrightarrow[(x_1,2)]{i/o} t_1 \xrightarrow{to[x_1]/o} t_2 \in runs(\mathcal{T})$. Let us show that $read_\pi^\emptyset(t_3)$ exists ($\emptyset$ is the empty matching). As the first action in π is i, we take the transition $t_3 \xrightarrow[(x_6,2)]{i/o'} t_6$. As the second action in π is $to[x_1]$ with x_1 a fresh timer, we retrieve the corresponding fresh timer x_6 in the new run. Hence, $\pi' = read_\pi^\emptyset(t_3) = t_3 \xrightarrow[(x_6,2)]{i/o'} t_6 \xrightarrow{to[x_6]/o} t_9$ and $m_{\pi'}^\pi : \pi \leftrightarrow \pi'$ with $m_{\pi'}^\pi = \emptyset \cup \{x_1 \mapsto x_6\}$. Let us now show some cases of apartness. The states t_0, t_3 are $\emptyset$-apart as the first transition of π outputs o but the first transition of π' outputs $o' \neq o$, i.e., $i \vdash_0 \#^\emptyset t_3$ by (outputs). The states t_1, t_6 are also $\emptyset$-apart as $t_1, t_6 \in \mathcal{E}^\mathcal{T}$ and $|\chi_0(t_1)| = 1 \neq |\chi_0(t_6)| = 2$, i.e., $\varepsilon \vdash_1 \#^\emptyset t_6$ by (sizes). Now, with $m : x_1 \mapsto x_3$, let $\sigma = t_1 \xrightarrow[(x_3,3)]{i/o'} t_3 \xrightarrow{to[x_1]} t_5$ and $\sigma' = read_{\pi'}^m(t_3) = t_3 \xrightarrow[(x_6,2)]{i/o'} t_6 \xrightarrow{to[x_3]} t_{10}$. We have a structural apartness as $m_{\rho'}^\rho = m \cup \{x_3 \mapsto x_6\}$ is invalid ($\chi^\mathcal{T}(t_6) = \{x_3, x_6\}$). We also have $i \vdash_1 \#^m t_3$ by (constants). ⌡

Observe that any extension m' of m is such that $w \vdash \#^{m'} p'$, i.e., taking a larger matching does not break the apartness, as $read_{p \xrightarrow{w}}^{m'}(p') = read_{p \xrightarrow{w}}^{m'}(p') = p' \xrightarrow{w'}$. Moreover, we claim that the definition of apartness is reasonable: when $p \#^m p'$, then $f(p) \neq f(p')$ (the two states are really distinct) or $g(x) \neq g(m(x))$ for some x (m and g do not agree).

Theorem 2. *Let $\mathcal{T}$ be an observation tree for an s-learnable MMT with functional simulation $\langle f, g \rangle$, $p, p' \in Q^\mathcal{T}$, and $m, m' : p \leftrightarrow p'$ matchings. Then,*

- $w \vdash \#^m p' \wedge m \subseteq m' \Rightarrow w \vdash \#^{m'} p'$, *and*
- $p \#^m p' \Rightarrow f(p) \neq f(p') \vee \exists x \in dom(m) : g(x) \neq g(m(x))$.

The second implication of Theorem 2 is the one we leverage in our algorithm below. The reverse implication is false as the learned MMT can be smaller than that of the teacher (see the MMT in Fig. 4 learned from the MMT of Fig. 1).

3.2 Hypothesis Construction

In this section, we provide the construction of a hypothesis $\mathcal{H}$ from $\mathcal{T}$. In short, we extend the observation tree such that some conditions are satisfied and we define a subset of $Q^\mathcal{T}$, called the *basis*, that forms the set of states of $\mathcal{H}$. Similar to $L^\#$ [33], we then "fold" the tree; that is, some transitions $q \to r$ must be redirected to some state p of the basis. For MMTs, we also need to map *every* timer active in r to an active timer of p. Formally, we define:

- The *basis* $\mathcal{B}^\mathcal{T}$ is a subtree of $Q^\mathcal{T}$ such that $q_0^\mathcal{T} \in \mathcal{B}^\mathcal{T}$ and $p \#^m p'$ for any $p \neq p' \in \mathcal{B}^\mathcal{T}$ and maximal matching $m : p \leftrightarrow p'$. By Theorem 2, we thus know that $f(p) \neq f(p')$ or $g(x) \neq g(m(x))$ for some $x \in dom(m)$. As we have this for every maximal m, we *conjecture* that $f(p) \neq f(p')$. We may be wrong, i.e., $f(p) = f(p')$ but we need a matching that is currently unavailable, due to unknown active timers which will be discovered later.

- The *frontier* $\mathcal{F}^\mathcal{T} \subseteq Q^\mathcal{T}$ is the set of immediate non-basis successors of basis states. We say $p \in \mathcal{B}^\mathcal{T}$ and $r \in \mathcal{F}^\mathcal{T}$ are *compatible* under a maximal matching m if $\neg(p \#^m r)$. We write $compat^\mathcal{T}(r)$ for the set of all such pairs (p,m). Hence $compat^\mathcal{T}(r)$ indicates all the possible ways to redirect a transition $q \to r$ to some state $p \in \mathcal{B}^\mathcal{T}$ together with an adequate maximal matching m.

During the learning algorithm, the tree is extended such that a complete MMT $\mathcal{H}$ can be constructed from $\mathcal{T}$. We will ensure that:

(A) each basis and frontier state is explored, i.e. $\mathcal{B}^\mathcal{T} \cup \mathcal{F}^\mathcal{T} \subseteq \mathcal{E}^\mathcal{T}$, in order to discover timers as fast as possible,
(B) the basis is *complete*, i.e., $p \xrightarrow{i}$ is defined for every $i \in I \cup TO[\chi_0^\mathcal{T}(p)]$, and
(C) for every $r \in \mathcal{F}^\mathcal{T}$, $compat^\mathcal{T}(r) \neq \emptyset$ and $|\chi^\mathcal{T}(p)| = |\chi^\mathcal{T}(r)|$ for every $(p,m) \in compat^\mathcal{T}(r)$. The last constraint imposes the same number of active timers for both p and r since folding them using (p,m) will merge them.

These constraints are obtained via **OQ**^{s} and **WQ**^{s}, as we now illustrate.

Example 5. In order to simplify the explanations, we assume from now on that a call to **OQ**^{s}(w · i) with i $\in I$ automatically adds the corresponding transition to $\mathcal{T}$, and that a call to **WQ**^{s}(w) automatically calls **OQ**^{s}(w · $to[j]$), for every $to[j]$ deduced from the wait query, modifies updates accordingly, and adds the new explored states to $\mathcal{E}^\mathcal{T}$. Moreover, we let **OQ**$^s(q,i)$ denote **OQ**^{s}(w · i) and **WQ**$^s(q)$ denote **WQ**^{s}(w) with w such that $q_0^\mathcal{T} \xrightarrow{w} q \in runs(\mathcal{T})$.

Let $\mathcal{T}$ be the observation tree of Fig. 3 for the MMT $\mathcal{M}$ of Fig. 1. One can check that t_0, t_1, and t_3 are all pairwise apart under any maximal matching. So, $\mathcal{B}^\mathcal{T} = \{t_0, t_1, t_3\}$ (highlighted in gray in the figure) and $\mathcal{F}^\mathcal{T} = \{t_2, t_5, t_6\}$. Moreover, we have $compat^\mathcal{T}(t_2) = \{(t_1, x_1 \mapsto x_1), (t_3, x_1 \mapsto x_1)\}$, and $compat^\mathcal{T}(t_5) = compat^\mathcal{T}(t_6) = \emptyset$. We thus *promote* t_6 by moving it from the frontier to the basis, i.e., $\mathcal{B}^\mathcal{T} = \{t_0, t_1, t_3, t_6\}$. To satisfy *(B)*, we call **OQ**$^s(t_6, i)$ and add a new transition $t_6 \xrightarrow{i/o'}_\bot t_{11}$. We call **WQ**$^s(t_9)$, **WQ**$^s(t_{10})$, and **WQ**$^s(t_{11})$, which yield $\chi(t_9) = \chi_0(t_9) = \{x_3\}$, $\chi(t_{10}) = \emptyset$, and $\chi(t_{11}) = \chi_0(t_{11}) = \{x_3, x_{11}\}$. Hence, we get *(A)* with $\mathcal{F}^\mathcal{T} = \{t_2, t_5, t_9, t_{10}, t_{11}\}$.

Observe that *(C)* is not satisfied, due to $(t_3, m) \in compat^\mathcal{T}(t_2)$ with $m : x_1 \mapsto x_1$ but $|\chi(t_3)| = 2$ while $|\chi(t_2)| = 1$. To resolve this issue, we *replay*[5] the run $\pi = t_3 \xrightarrow{to[x_1]to[x_3]} t_8$ (witnessing that x_3 is active in t_3 and eventually times out) from the state t_2. That is, we extend the tree $\mathcal{T}$ in order to create $read_\pi^m(t_2)$ while extending m, if this is possible (see apartness paragraph). As $t_2 \xrightarrow{to[x_1]} t_4$ already exists, let us replay $t_5 \xrightarrow{to[x_3]}$ from t_4. We call **WQ**$^s(t_4)$ to learn $\chi_0(t_4) = \chi(t_4) = \{x_1\}$. We can stop the creation of $read_\pi^m(t_2)$. Indeed, as $\chi_0(t_5) = \{x_1, x_3\}$, $\chi_0(t_4) = \{x_1\}$, we conclude that $to[x_1] \vdash_3 \#^m t_2$ by (sizes). Hence, (t_3, m) is no longer a compatible pair and it is removed from $compat^\mathcal{T}(t_2)$.

---

[5] See [7, Appendix E.1] for details.

In general, replaying a run when *(C)* is not satisfied may have two results due to newly defined transitions: either finding a new active timer in a state, or a new apartness pair (as we illustrated). In any case, the set of compatible pairs of a state gets reduced. ⌟

We now explain, via an example, how to construct a hypothesis $\mathcal{H}$ such that $Q^{\mathcal{H}} = \mathcal{B}^{\mathcal{T}}$. The idea is to pick a $(p, m) \in compat^{\mathcal{T}}(r)$ for each frontier state r. Then, the unique transition $q \xrightarrow{i} r$ in $\mathcal{T}$ becomes $q \xrightarrow{i} p$ in $\mathcal{H}$, i.e., p and r are merged. We also globally rename the timers according to m.

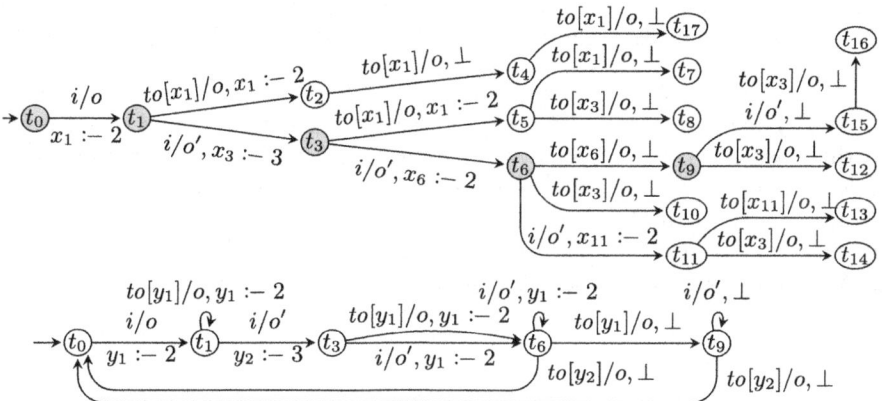

Fig. 4. On top, an observation tree from which the hypothesis MMT at the bottom is constructed, with $y_1 = [\![x_1]\!]_\equiv$ and $y_2 = [\![x_3]\!]_\equiv$. Basis states are highlighted with a gray background.

Example 6. Let $\mathcal{M}$ be the MMT of Fig. 1 and $\mathcal{T}$ be the observation tree of Fig. 4, with $\mathcal{B}^{\mathcal{T}} = \{t_0, t_1, t_3, t_6, t_9\}$ and $\mathcal{F}^{\mathcal{T}} = \{t_2, t_5, t_{10}, t_{11}, t_{12}, t_{15}\}$. Constraints *(A)* to *(C)* are satisfied and

$$compat^{\mathcal{T}}(t_2) = \{(t_1, x_1 \mapsto x_1)\}$$
$$compat^{\mathcal{T}}(t_5) = \{(t_6, x_6 \mapsto x_1, x_3 \mapsto x_3)\}$$
$$compat^{\mathcal{T}}(t_{10}) = compat^{\mathcal{T}}(t_{12}) = \{(t_0, \emptyset)\}$$
$$compat^{\mathcal{T}}(t_{11}) = \{(t_6, x_6 \mapsto x_{11}, x_3 \mapsto x_3)\}$$
$$compat^{\mathcal{T}}(t_{15}) = \{(t_9, x_3 \mapsto x_3)\}$$

We construct $\mathcal{H}$ with $Q^{\mathcal{H}} = \mathcal{B}^{\mathcal{T}}$. While defining the transitions $q \to q'$ is easy when $q, q' \in \mathcal{B}^{\mathcal{T}}$, we have to redirect the transition to some basis state when $q' \in \mathcal{F}^{\mathcal{T}}$. To do so, we first define a map $\boldsymbol{h} : \mathcal{F}^{\mathcal{T}} \to \mathcal{B}^{\mathcal{T}}$ (to redirect transitions into the basis), and an equivalence relation $\equiv$ over the set of active timers of the basis and the frontier (to declare which timers are equal). For each $r \in \mathcal{F}^{\mathcal{T}}$, we pick $(p, m) \in compat^{\mathcal{T}}(r)$, define $\boldsymbol{h}(r) = p$, and add $x \equiv m(x)$ for

every $x \in \text{dom}(m)$ (and compute the symmetric and transitive closure of $\equiv$).
Here,
$$\boldsymbol{h}(t_2) = t_1, \boldsymbol{h}(t_5) = \boldsymbol{h}(t_{11}) = t_6, \boldsymbol{h}(t_{10}) = \boldsymbol{h}(t_{12}) = t_0, \boldsymbol{h}(t_{15}) = t_9,$$
$$x_1 \equiv x_6 \equiv x_{11}, \text{and} x_3 \equiv x_3.$$

We check that the constructed relation $\equiv$ is *valid*, i.e., it does not lead to an undesirable situation $x \equiv y$ and $x, y \in \chi^\mathcal{T}(q)$ for some q (in which case, we restart again by picking some different pair $(p, m) \in \text{compat}^{\mathcal{T}}(r))$. In this example, $\equiv$ is valid, and we construct $\mathcal{H}$ by copying the transitions starting from a basis state (while folding the tree when required), except that a timer x is replaced by its equivalence class $[\![x]\!]_\equiv$. Figure 4 gives the resulting $\mathcal{H}$, which is symbolically equivalent to $\mathcal{M}$. ⌟

We highlight that it is *not* always possible to construct a valid $\equiv$. However, in this case, we can still construct a *generalized* MMT, in which the transitions can arbitrarily rename the active timers. The size of that machine is also $|\mathcal{B}^\mathcal{T}|$, and from it a classical MMT can be constructed of size $n! \cdot |\mathcal{B}^\mathcal{T}|$, with $n = \max_{p \in \mathcal{B}^\mathcal{T}} |\chi^\mathcal{T}(p)|$. See [7, Appendix E.2] for more information. We observed that a valid $\equiv$ can always be constructed for all our benchmarks.

3.3 Main Loop

We finally give the main loop of $L_{\text{MMT}}^{\#}$. We initialize $\mathcal{T}$ to only contain $q_0^\mathcal{T}$, $\mathcal{B}^\mathcal{T} = \mathcal{E}^\mathcal{T} = \{q_0^\mathcal{T}\}$, and $\mathcal{F}^\mathcal{T} = \emptyset$. The main loop is split into two parts:

Refinement loop The *refinement loop* extends the tree to obtain the conditions *(A)* to *(C)* (see Sect. 3.2 page 13), by performing the following operations, in this order, until no more changes are possible:

    Seismic If we discover a new active timer in a basis state, then it may be that $\neg(q \#^m q')$ for some $q, q' \in \mathcal{B}^\mathcal{T}$ and maximal m, due to the new timer. To avoid this, we reset the basis back to $\{q_0^\mathcal{T}\}$, as soon as a new timer is found, without removing states from $\mathcal{T}$.

    Promotion If $\text{compat}^\mathcal{T}(r)$ is empty for some frontier state r, then we know that $q \#^m r$ for every $q \in \mathcal{B}^\mathcal{T}$ and maximal matching $m : q \leftrightarrow r$. Hence, we promote r to the basis.

    Completion If an i-transition is missing from some basis state p, we complete the basis with that transition.

    Active timers We ensure $|\chi^\mathcal{T}(p)| = |\chi^\mathcal{T}(r)|$ for every $(p, \cdot) \in \text{compat}^\mathcal{T}(r)$.

Hypothesis and equivalence We call $\mathbf{EQ}^\text{s}(\mathcal{H})$ with $\mathcal{H}$ a hypothesis. If the answer is **yes**, we return $\mathcal{H}$. Otherwise, we process the counterexample, as we now explain in an example.

Example 7. Let $\mathcal{T}$ be the observation tree of Fig. 2 with $\mathcal{B}^\mathcal{T} = \{t_0, t_1\}$, $\mathcal{F}^\mathcal{T} = \{t_2, t_3\}$, and $\text{compat}^\mathcal{T}(t_2) = \text{compat}^\mathcal{T}(t_3) = \{(t_1, x_1 \mapsto x_1)\}$. Figure 2 also gives the MMT constructed from $\mathcal{T}$, which is not symbolically equivalent to the MMT of Fig. 1, with $\mathtt{w} = i \cdot i \cdot to[1] \cdot to[2]$ as a counterexample. We extend $\mathcal{T}$ such that $\mathtt{w}$ can be read in $\mathcal{T}$. That is, we call $\mathbf{WQ^s}(t_5)$ and discover that the transition from t_1 to t_3 must start the timer x_3. Hence, t_3 has no compatible state anymore (i.e., we found a new apartness pair) and gets promoted. After completing the tree and performing $\mathbf{WQ^s}$ on the frontier states, we get the tree from Fig. 3. ⌐

In our example, simply adding the symbolic word provided by the teacher was enough to discover a new apartness pair (meaning that the hypothesis can no longer be constructed). However, there may be cases where we need to *replay* (see Example 5) a part of the newly added run $t_0 \xrightarrow{\mathtt{w}}$: split the run into $t_0 \xrightarrow{u} t \xrightarrow{v}$ with $t \in \mathcal{F}^{\mathcal{T}}$ and replay $t \xrightarrow{v}$ from the state compatible with t that was selected to build the hypothesis. Repeat this principle until a compatible set is reduced.

Table 1. Experimental Results.

Model	$|Q|$	$|I|$	$|X|$	$|\mathbf{WQ^s}|$	$|\mathbf{OQ^s}|$	$|\mathbf{EQ^s}|$	Time[ms]
AKM	4	5	1	22	35	2	684
CAS	8	4	1	60	89	3	1344
Light	4	2	1	10	13	2	302
PC	8	9	1	75	183	4	2696
TCP	11	8	1	123	366	8	3182
Train	6	3	1	32	28	3	1559
MMT of Fig. 1	3	1	2	11	5	2	1039
FDDI 1-station	9	2	2	32	20	1	1105
Oven	12	5	1	907	317	3	9452
WSN	9	4	1	175	108	4	3291

4 Implementation and Experiments

We have implemented $L_{\mathrm{MMT}}^{\#}$ as an open-source tool.[6] Note that our prototype implementation only uses symbolic queries, with no explicit translation into concrete ones. As we do not yet have a timed conformance testing algorithm for checking equivalence between a hypothesis and the teacher's MMT, we utilize a BFS algorithm to check for symbolic equivalence between the two MMTs.[7]

We evaluated the performance of our tool on a selection of both real and synthetic benchmarks. We use the AKM, TCP and Train benchmarks from [32], and the CAS, Light and PC benchmarks from [1]. These have also been used for experimental evaluation by [26,32,34] and can be described as Mealy machines

---
[6] See: https://gitlab.science.ru.nl/bharat/mmt_lsharp and Zenodo [18].
[7] That is, seeking a difference in behavior in the product of the two MMTs.

with a single timer (MM1Ts). We introduce two additional benchmarks with two timers: a model of an FDDI station [25,34] (see [7, Appendix F] for details on our adaptation), and the MMT of Fig. 1. We did not include the FDDI 2 process benchmark from [34], as our implementation cannot (yet) handle the corresponding generalized MMTs. Finally, we learned instances of the Oven and WSN MMLTs benchmarks from [26]. We modified the timing parameters to generate smaller MM1Ts. For each experiment, we record the number of **OQ**^{s}, **WQ**^{s}, **EQ**^{s}, and the time taken to finish the experiment. Note, in practice, a **WQ**^{s}, in addition to returning the list of timeouts and their constants, also provides the outputs of the timeout transitions. This is straightforward, as a **WQ**^{s} must necessarily trigger the timeouts in order to observe them. Thus, we do not count the **OQ**^{s} associated with a **WQ**^{s}.[8]

The results are given in Table 1. Here, we highlight that the presented data is only meant to indicate that our algorithm is able to learn these models within a reasonable amount of time. The table also gives insight into which benchmarks require more or less (symbolic) queries to be learned with our algorithm.

Comparison with other learning algorithms for timed systems is complicated. First, we need to convert the numbers of symbolic $L_{MMT}^{\#}$ queries to concrete queries. As our prototype does not implement the concrete queries, we can only use the theoretical polynomial bounds given in Lemma 2. Note that for MM1Ts, each symbolic query can be implemented using a single concrete query (see Lemma 3 in [32]). For the FDDI protocol using two timers, we need at most 1282 concrete queries in total while around 118200 queries are used in [34]. Second, several algorithms presented in the literature learn TAs [1,4,34,35]. Typically, these models have different numbers of states and transitions than an MMT model: Mealy machines tend to be more compact than TAs, but the use of timers may lead to more states than a TA encoding. Therefore we cannot just compare numbers of queries. As a third complication, equivalence queries can be implemented in different ways, which may affect the total number of queries required for learning. Finally, concerning models close to our MMTs, MMLTs [26] can be converted to equivalent MM1Ts [32], but this may blow up the number of states. Since $L_{MMT}^{\#}$ learns the MM1Ts, it is less efficient than the MMLT learner of [26] which learns the more compact MMLT representations, or than the learner of [32] specially designed for MM1Ts. However, $L_{MMT}^{\#}$ can handle a larger class of models, as it can handle multiple timers.

5 Future Work

A major challenge in $L_{MMT}^{\#}$ is to infer the update on transitions. This would become easier if we know in advance that timers can only be started by specific inputs, akin to what is done in *event recording automata* (ERA) [3] for TAs. Although, as discussed in [32], the restrictions of ERAs make it hard to capture

---

[8] An upper bound on the number of **OQ**^{s} can be obtained by adding $|X||\mathbf{WQ}^s|$ to the values for $|\mathbf{OQ}^s|$.

the timing behavior of standard network protocols. It would be interesting to study the theoretical complexity of learning a system that can be modelled by ERAs as well as MMTs. A more interesting trail would be to allow transitions to start multiple timers, instead of a single one, which would permit more complex models (such as those resulting of parallel composition of simpler models) to be learned. It would also be interesting to explore the theory of generalized MMTs.

References

1. Aichernig, B.K., Pferscher, A., Tappler, M.: From passive to active: learning timed automata efficiently. In: Lee, R., Jha, S., Mavridou, A., Giannakopoulou, D. (eds.) NFM 2020. LNCS, vol. 12229, pp. 1–19. Springer, Cham (2020). https://doi.org/10.1007/978-3-030-55754-6_1
2. Alur, R., Dill, D.L.: A theory of timed automata. Theoret. Comput. Sci. **126**(2), 183–235 (1994). https://doi.org/10.1016/0304-3975(94)90010-8
3. Alur, R., Fix, L., Henzinger, T.A.: Event-clock automata: a determinizable class of timed automata. Theoret. Comput. Sci. **211**(1–2), 253–273 (1999). https://doi.org/10.1016/S0304-3975(97)00173-4
4. An, J., Chen, M., Zhan, B., Zhan, N., Zhang, M.: Learning one-clock timed automata. In: TACAS 2020. LNCS, vol. 12078, pp. 444–462. Springer, Cham (2020). https://doi.org/10.1007/978-3-030-45190-5_25
5. Angluin, D.: Learning regular sets from queries and counterexamples. Inf. Comput. **75**(2), 87–106 (1987). https://doi.org/10.1016/0890-5401(87)90052-6
6. Bojańczyk, M., Lasota, S.: A machine-independent characterization of timed languages. In: Czumaj, A., Mehlhorn, K., Pitts, A., Wattenhofer, R. (eds.) ICALP 2012. LNCS, vol. 7392, pp. 92–103. Springer, Heidelberg (2012). https://doi.org/10.1007/978-3-642-31585-5_12
7. Bruyère, V., Garhewal, B., Pérez, G.A., Staquet, G., Vaandrager, F.W.: Active Learning of Mealy Machines with Timers. CoRR abs/2403.02019 (2024). https://doi.org/10.48550/ARXIV.2403.02019
8. Bruyère, V., Pérez, G.A., Staquet, G., Vaandrager, F.W.: Automata with timers. In: Petrucci, L., Sproston, J. (eds.) Proceedings of the 21st International Conference Formal Modeling and Analysis of Timed Systems, FORMATS 2023. LNCS, vol. 14138, pp. 33–49. Springer, Cham (2023). https://doi.org/10.1007/978-3-031-42626-1_3
9. Clarke, E.M., Henzinger, T.A., Veith, H., Bloem, R. (eds.): Handbook of Model Checking. Springer, Cham (2018). https://doi.org/10.1007/978-3-319-10575-8
10. Dierl, S., et al.: Learning symbolic timed models from concrete timed data. In: Rozier, K.Y., Chaudhuri, S. (eds.) Proceedings of the 15th International Symposium NASA Formal Methods, NFM 2023. LNCS, vol. 13903, pp. 104–121. Springer, Cham (2023). https://doi.org/10.1007/978-3-031-33170-1_7
11. Dill, D.L.: Timing assumptions and verification of finite-state concurrent systems. In: Sifakis, J. (ed.) CAV 1989. LNCS, vol. 407, pp. 197–212. Springer, Heidelberg (1990). https://doi.org/10.1007/3-540-52148-8_17
12. Doveri, K., Ganty, P., Srivathsan, B.: A Myhill-Nerode style characterization for timed automata with integer resets. In: Barman, S., Lasota, S. (eds.) 44th IARCS Annual Conference on Foundations of Software Technology and Theoretical Computer Science, FSTTCS 2024, 16–18 December 2024, Gandhinagar, Gujarat, India. LIPIcs, vol. 323, pp. 21:1–21:18. Schloss Dagstuhl - Leibniz-Zentrum für Informatik (2024). https://doi.org/10.4230/LIPICS.FSTTCS.2024.21

13. Ferreira, T., Brewton, H., D'Antoni, L., Silva, A.: Prognosis: closed-box analysis of network protocol implementations. In: Kuipers, F.A., Caesar, M.C. (eds.) Proceedings of the ACM SIGCOMM 2021 Conference, pp. 762–774. ACM (2021). https://doi.org/10.1145/3452296.3472938
14. Fiterău-Broştean, P., Howar, F.: Learning-based testing the sliding window behavior of TCP implementations. In: Petrucci, L., Seceleanu, C., Cavalcanti, A. (eds.) FMICS/AVoCS -2017. LNCS, vol. 10471, pp. 185–200. Springer, Cham (2017). https://doi.org/10.1007/978-3-319-67113-0_12
15. Fiterău-Broştean, P., Janssen, R., Vaandrager, F.: Combining model learning and model checking to analyze TCP implementations. In: Chaudhuri, S., Farzan, A. (eds.) CAV 2016. LNCS, vol. 9780, pp. 454–471. Springer, Cham (2016). https://doi.org/10.1007/978-3-319-41540-6_25
16. Fiterau-Brostean, P., Jonsson, B., Merget, R., de Ruiter, J., Sagonas, K., Somorovsky, J.: Analysis of DTLS implementations using protocol state fuzzing. In: Capkun, S., Roesner, F. (eds.) Proceedings of the 29th USENIX Security Symposium, USENIX Security 2020, pp. 2523–2540. USENIX Association (2020). https://www.usenix.org/conference/usenixsecurity20/presentation/fiterau-brostean
17. Fiterau-Brostean, P., Lenaerts, T., Poll, E., de Ruiter, J., Vaandrager, F.W., Verleg, P.: Model learning and model checking of SSH implementations. In: Erdogmus, H., Havelund, K. (eds.) Proceedings of the 24th ACM SIGSOFT International SPIN Symposium on Model Checking of Software, pp. 142–151. ACM (2017). https://doi.org/10.1145/3092282.3092289
18. Garhewal, B.: L# MMT artifact (2024). https://doi.org/10.5281/ZENODO.10647627. https://zenodo.org/doi/10.5281/zenodo.10647627
19. Garhewal, B., Vaandrager, F., Howar, F., Schrijvers, T., Lenaerts, T., Smits, R.: Grey-box learning of register automata. In: Dongol, B., Troubitsyna, E. (eds.) IFM 2020. LNCS, vol. 12546, pp. 22–40. Springer, Cham (2020). https://doi.org/10.1007/978-3-030-63461-2_2
20. Grinchtein, O., Jonsson, B., Leucker, M.: Learning of event-recording automata. In: Lakhnech, Y., Yovine, S. (eds.) FORMATS/FTRTFT -2004. LNCS, vol. 3253, pp. 379–395. Springer, Heidelberg (2004). https://doi.org/10.1007/978-3-540-30206-3_26
21. Grinchtein, O., Jonsson, B., Leucker, M.: Learning of event-recording automata. Theoret. Comput. Sci. **411**(47), 4029–4054 (2010). https://doi.org/10.1016/J.TCS.2010.07.008
22. Grinchtein, O., Jonsson, B., Pettersson, P.: Inference of event-recording automata using timed decision trees. In: Baier, C., Hermanns, H. (eds.) CONCUR 2006. LNCS, vol. 4137, pp. 435–449. Springer, Heidelberg (2006). https://doi.org/10.1007/11817949_29
23. Henry, L., Jéron, T., Markey, N.: Active learning of timed automata with unobservable resets. In: Bertrand, N., Jansen, N. (eds.) FORMATS 2020. LNCS, vol. 12288, pp. 144–160. Springer, Cham (2020). https://doi.org/10.1007/978-3-030-57628-8_9
24. Howar, F., Steffen, B.: Active automata learning in practice. In: Bennaceur, A., Hähnle, R., Meinke, K. (eds.) Machine Learning for Dynamic Software Analysis: Potentials and Limits. LNCS, vol. 11026, pp. 123–148. Springer, Cham (2018). https://doi.org/10.1007/978-3-319-96562-8_5
25. Johnson, M.J.: Proof that timing requirements of the FDDI token ring protocol are satisfied. IEEE Trans. Comput. **35**(6), 620–625 (1987). https://doi.org/10.1109/TCOM.1987.1096832

26. Kogel, P., Klös, V., Glesner, S.: Learning mealy machines with local timers. In: Li, Y., Tahar, S. (eds.) Proceedings of the 24th International Conference on Formal Engineering Methods Formal Methods and Software Engineering, ICFEM 2023. LNCS, vol. 14308, pp. 47–64. Springer, Cham (2023). https://doi.org/10.1007/978-981-99-7584-6_4
27. Maler, O., Pnueli, A.: On recognizable timed languages. In: Walukiewicz, I. (ed.) FoSSaCS 2004. LNCS, vol. 2987, pp. 348–362. Springer, Heidelberg (2004). https://doi.org/10.1007/978-3-540-24727-2_25
28. de Ruiter, J., Poll, E.: Protocol state fuzzing of TLS implementations. In: Jung, J., Holz, T. (eds.) Proceedings of the 24th USENIX Security Symposium, USENIX Security 2015, pp. 193–206. USENIX Association (2015). https://www.usenix.org/conference/usenixsecurity15/technical-sessions/presentation/de-ruiter
29. Shahbaz, M., Groz, R.: Inferring mealy machines. In: Cavalcanti, A., Dams, D.R. (eds.) FM 2009. LNCS, vol. 5850, pp. 207–222. Springer, Heidelberg (2009). https://doi.org/10.1007/978-3-642-05089-3_14
30. Tappler, M., Aichernig, B.K., Larsen, K.G., Lorber, F.: Time to learn – learning timed automata from tests. In: André, É., Stoelinga, M. (eds.) FORMATS 2019. LNCS, vol. 11750, pp. 216–235. Springer, Cham (2019). https://doi.org/10.1007/978-3-030-29662-9_13
31. Vaandrager, F.W.: Model learning. Commun. ACM **60**(2), 86–95 (2017). https://doi.org/10.1145/2967606
32. Vaandrager, F., Bloem, R., Ebrahimi, M.: Learning mealy machines with one timer. In: Leporati, A., Martín-Vide, C., Shapira, D., Zandron, C. (eds.) LATA 2021. LNCS, vol. 12638, pp. 157–170. Springer, Cham (2021). https://doi.org/10.1007/978-3-030-68195-1_13
33. Vaandrager, F.W., Garhewal, B., Rot, J., Wißmann, T.: A new approach for active automata learning based on apartness. In: Fisman, D., Rosu, G. (eds.) Proceedings of the 28th International Conference Tools and Algorithms for the Construction and Analysis of Systems, TACAS 2022. LNCS, vol. 13243, pp. 223–243. Springer, Cham (2022). https://doi.org/10.1007/978-3-030-99524-9_12
34. Waga, M.: Active learning of deterministic timed automata with Myhill-Nerode style characterization. In: Enea, C., Lal, A. (eds.) Proceedings of the 35th International Conference Computer Aided Verification, CAV 2023. LNCS, vol. 13964, pp. 3–26. Springer, Cham (2023). https://doi.org/10.1007/978-3-031-37706-8_1
35. Xu, R., An, J., Zhan, B.: Active learning of one-clock timed automata using constraint solving. In: Bouajjani, A., Holík, L., Wu, Z. (eds.) Proceedings of the 20th International Symposium Automated Technology for Verification and Analysis, ATVA 2022. LNCS, vol. 13505, pp. 249–265. Springer, Cham (2022). https://doi.org/10.1007/978-3-031-19992-9_16

Formal Control for Uncertain Systems via Contract-Based Probabilistic Surrogates

Oliver Schön[1](✉)[iD], Sofie Haesaert[2][iD], and Sadegh Soudjani[3][iD]

[1] Newcastle University, Newcastle upon Tyne, UK
o.schoen2@ncl.ac.uk
[2] Eindhoven University of Technology, Eindhoven, The Netherlands
s.haesaert@tue.nl
[3] Max Planck Institute for Software Systems, Kaiserslautern, Germany
sadegh@mpi-sws.org

Abstract. The requirement for identifying accurate system representations has not only been a challenge to fulfill, but it has compromised the scalability of formal methods, as the resulting models are often too complex for effective decision making with formal correctness and performance guarantees. Focusing on probabilistic simulation relations and surrogate models of stochastic systems, we propose an approach that significantly enhances the scalability and practical applicability of such simulation relations by eliminating the need to compute error bounds directly. As a result, we provide an abstraction-based technique that scales effectively to higher dimensions while addressing complex nonlinear agent-environment interactions with infinite-horizon temporal logic guarantees amidst uncertainty. Our approach trades scalability for conservatism favorably, as demonstrated on a complex high-dimensional vehicle intersection case study.

Keywords: Policy Synthesis · Stochastic Systems · Situational Awareness · Simulation Relations · Temporal Logic · Model-Order Reduction

1 Introduction

To pave the way for the safe and responsible adoption of autonomous agents—especially in high-risk applications such as *cyber-physical systems* (CPSs)—the creation of formal techniques for decision making and control is proving to be a key puzzle piece [2,9,15,31,37]. In the pursuit of developing control software for agents that implements specifications with a certified probability, a main task is establishing a formal connection between the concrete environment $\mathbf{E}$ (see Fig. 1) in which these agents operate and (certification) results obtained via mathematical models (e.g., via surrogate models $\{\mathbf{S}_i\}_{i \in \mathcal{I}}$). Available approaches are challenged by the uncertain nature of actual CPS dynamics and model imperfections,

$$\forall i \in \mathcal{I}, \theta \in \Theta, \mathfrak{n} \in \mathfrak{N}: \left(\mathbb{P} \left[\begin{array}{c} \mathbf{C} \times \mathbf{A}(\theta) \\ \hline \mathbf{S}_i(\theta, \mathfrak{n}) \end{array} \right] \models \psi \right) \geq p_\mathrm{s} \overset{?}{\Longrightarrow} \left(\mathbb{P} \left[\begin{array}{c} \mathbf{C} \times \mathbf{A}(\theta^*) \\ \hline \mathbf{E}(\theta^*) \end{array} \right] \models \psi \right) \geq p_\mathrm{c}$$

Fig. 1. Design a robust controller $\mathbf{C}$ for an agent $\mathbf{A}$ based on finitely many surrogate models $\{\mathbf{S}_i\}_{i \in \mathcal{I}}$ and obtain formal guarantees for its performance (with respect to some specification ψ and satisfaction probability p_c) once deployed in the concrete environment $\mathbf{E}$. The model discrepancies arising from uncertainty and incomplete information are modeled via an uncertain parametrization $\theta \in \Theta$ and a nondeterministic adversary $\mathfrak{n} \in \mathfrak{N}$.

as indicated in Fig. 1, respectively, by an uncertain parametrization $\theta \in \Theta$ and an adversary[1] $\mathfrak{n} \in \mathfrak{N}$. The resulting *environmental ambiguity* renders a strict sampling-based cover of all possible scenarios $\{\mathbf{S}_i(\theta, \mathfrak{n}) | i \in \mathcal{I}, \theta \in \Theta, \mathfrak{n} \in \mathfrak{N}\}$ straight-out impossible.

Rooted in the control systems community, formal control synthesis approaches have been developed with the theoretical foundations of dynamical systems such as CPSs in mind [8,30,50]. For instance, *(stochastic) simulation relations* [6,21,22,43] are designed to bridge the gap between discrete-state models and their uncountable counterparts. This includes transferring guarantees that quantify the risk of undesired behavior even when the system behaves nondeterministically and available models are incomplete or imprecise [42].

A key challenge for formal methods such as abstraction-based approaches, however, is their limited scalability. CPSs are complex, high-dimensional systems and the semantics required to express their desired behavior can often only be handled over finite abstractions, which partition the underlying state space with a computational complexity exponential in the dimension of the system. To improve scalability, multiple studies have been conducted including the use of adaptive partitioning, *model-order reduction* (MOR), and compositional techniques [1,7,13,23,29,45,49]. These studies require appropriate assumptions, ranging from continuity to small-gain, input-to-state stability, bounded subsystem interaction, and decomposable specifications. By using the natural split between environment and agent in autonomous systems, we bypass these assumptions by leveraging the *situational awareness* (SA) block that is an integral part of the architecture of many modern autonomous systems [12]. This includes identifying which situation $\mathbf{E}_1, \mathbf{E}_2, \ldots$ the agent is in and provisioning a sufficiently rich situation-dependent model to aid in effective decision making. Consider the situations shown in Fig. 2. In essence, the SA block of an agent $\mathbf{A}$ processes (sensor) information, equipping it with a sufficient uncertainty/risk-aware understanding of its environment to make informed decisions. In this paper, we will mainly focus on the decision making/control synthesis part, assuming access to a situation supplied by the SA block. In the following,

---

[1] The adversary $\mathfrak{n}$—in the literature sometimes referred to as *nature*—compensates for model inadequacies by disturbing the system. We express its nondeterministic nature via the ambiguity set $\mathfrak{N}$.

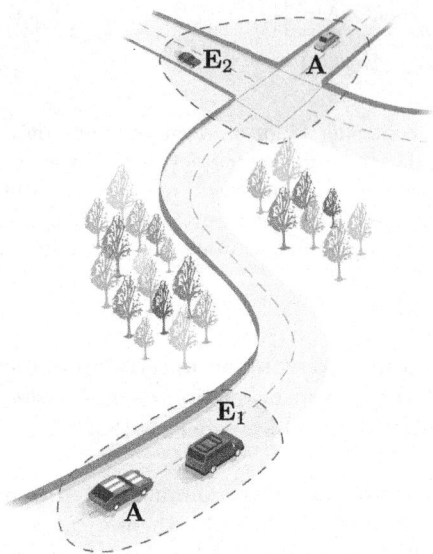

Fig. 2. Situational awareness: The environment can be broken down into situations $\mathbf{E}_1, \mathbf{E}_2, \ldots$ that individually describe its interaction with the agent $\mathbf{A}$ in a more concrete context. Note that, for the depicted interactions $\mathbf{A} \times \mathbf{E}_1$ and $\mathbf{A} \times \mathbf{E}_2$, models describing the longitudinal dynamics of the vehicles may be sufficient for effective decision making.

we assume that for each situation $i = 1, 2, \ldots$ the emanating interaction of the agent and the environment is entirely captured by the interconnected system $\mathbf{A} \times \mathbf{E}_i$. As an example, an autonomous vehicle (the agent $\mathbf{A}$) may encounter situations such as $\mathbf{E}_1$ and $\mathbf{E}_2$ illustrated in Fig. 2. Even though each system $\mathbf{E}_i$ may still be complex and high-dimensional, we will assume that the situations are defined sufficiently elemental such that for every situation $i = 1, 2, \ldots$ the encapsulated dynamics *relevant* to the interaction $\mathbf{A} \times \mathbf{E}_i$ and the specification ψ can be distilled into a much simpler form. For instance, following a car at a safe distance (see $\mathbf{A} \times \mathbf{E}_1$ in Fig. 2) concerns mainly the systems' longitudinal dynamics. Consequently, we abandon the family of potentially complex and high-dimensional environments $\mathbf{E}_1, \mathbf{E}_2, \ldots$ and replace it with a set of much simpler surrogate models $\mathbf{S}_1, \mathbf{S}_2, \ldots$ as identified by the agent's SA block [12]. For simplicity, we will call these surrogate models *simulators*. In the remainder of this work, we will focus our attention on situations in isolation by assuming that the system is in a single situation at a given time. Thus, we drop the subscript i and simply refer to $\mathbf{E}$ as the environment and $\mathbf{S}$ as the corresponding simulator. With this in mind, we make the following contributions.

Contributions. (1) We formalize a type of system interrelation called *behavioral inclusion* for stochastic systems *subject to model uncertainty*, that allows us to replace complex models with simpler surrogates by introducing an adversary. (2) We extend the definition of sub-simulation relations to systems subject to nondeterministic adversaries, by compensating the induced ambiguity in the stochastic coupling. (3) Based on the aforementioned two contributions, we establish an *assume-guarantee contract* (AGC) that allows us to reason about $\mathbf{A} \times \mathbf{E}$ via $\mathbf{A}_r \times \mathbf{S}$ as a proxy. As a result, we design robust controllers via the monolithic system $\mathbf{A}_r \times \mathbf{S}$ that can be readily deployed to $\mathbf{A} \times \mathbf{E}$. For simplicity, we limit ourselves to nonlinear systems with additive Gaussian noise. (4) We demonstrate the power of the proposed AGC framework on a high-dimensional traffic intersection case study. In particular, we use the new result to synthesize a controller for a 9-dimensional nonlinear stochastic system and a complex monolithic specification via a 4-dimensional linear proxy, resulting in a reduction in the number of required partitions by orders of magnitude. Notably, we emphasize that these contributions are for systems with inherent *model uncertainty*.

The remainder of this manuscript is organized as follows. After reviewing related work in Sect. 2, we present in Sect. 3 the general notation and the problem setup. In Sect. 4, we formalize the types of system interrelations used to provide the main assume-guarantee result in Sect. 5. We break down the steps necessary to establish the relations to design a controller along with robust guarantees that cover both environment and system in Sect. 6. The case study in Sect. 7 demonstrates the efficacy of the proposed AGC approach in obtaining results on a complex composed system. Section 8 concludes the paper. Due to space constraints, we defer the proofs to an extended version [40].

2 Related Work

Recent efforts in formal synthesis aim for scalable frameworks that address complex environmental interactions and uncertainties via strategies such as assume-guarantee reasoning and compositional MOR. While abstraction-free methods using barrier certificates are effective for low-level control, their extension to high-level decision making is challenged by temporal logic complexity and the trade-off between accuracy and efficiency—thereby prompting reliance on abstraction-based approaches [33,34]. In networked systems, scalability is often achieved by assuming weak subsystem interconnection between and local sub-specifications. For instance, [32] employ disturbance bisimulation, and [26] use reinforcement learning in an AGC setting; meanwhile, abstraction-free AGC has mainly targeted linear systems, e.g., [19] address linear time-variant systems via convex parametric contracts and [35] encode bounded stochastic STL contracts as mixed integer programs with chance constraints. Modeling real-world systems with probabilistic behavior introduces significant uncertainty, which many approaches tackle using variants of *robust Markov decision processes* (RMDPs)—including interval and parametric MDPs to robustify against epistemic uncertainty [5,24,39,42]—though high-dimensional extensions remain challenging. For instance, while simulation-relation-based synthesis via parametric MDPs with MOR error quantification has been explored [42,48], no known IMDP-based MOR approaches exist; similarly, distributionally robust RMDPs based on ambiguity sets are tractable only under fixed deterministic dynamics [20], with MOR for linear RMDPs explored in [38]. Similar to our work, integration of antagonistic non-determinism with stochastic simulation relations was pioneered in [51]. Abstraction-free methods that propagate reachable sets are typically limited to linear systems [3]. Recent work by [25] decouples deterministic and stochastic effects to compute reachable tubes via a deterministic proxy—an idea we adopt by replacing a rich stochastic system with a simpler nondeterministic proxy. Similarly, [14] relate system data to a nondeterministic transition system over probability measures using the scenario approach. [28] develop AGCs for verifying discrete-state systems exhibiting both probabilistic and nondeterministic behaviors. Complementary to our work—which assumes perfect perception—[4] establish contracts on environmental estimates from deterministic neural perception components, and [36] study a probabilistic variant.

3 Preliminaries and Problem Statement

Notation. The transpose of a matrix A is denoted by $A^\top$. We write $I_n \in \mathbb{R}^{n \times n}$ for the identity matrix and $0_{n \times m} \in \mathbb{R}^{n \times m}$ for the zero matrix. Let $(\mathbb{X}, \mathcal{B}(\mathbb{X}))$ be a *Borel measurable space*, with a *Polish* sample space $\mathbb{X}$ and a Borel σ-algebra $\mathcal{B}(\mathbb{X})$ [10]. A *measure* $\nu : \mathcal{B}(\mathbb{X}) \to \mathbb{R}_{\geq 0}$ is a *probability measure* if $\nu(\mathbb{X}) = 1$ and a *sub-probability measure* if $\nu(\mathbb{X}) \leq 1$. A *probability space* is given by $(\mathbb{X}, \mathcal{B}(\mathbb{X}), p)$, with realizations $x \sim p$. For two measurable spaces $(\mathbb{X}, \mathcal{B}(\mathbb{X}))$ and $(\mathbb{Y}, \mathcal{B}(\mathbb{Y}))$, a *(sub-)probability kernel* is a mapping $\mathbf{p} : \mathbb{X} \times \mathcal{B}(\mathbb{Y}) \to \mathbb{R}_{\geq 0}$ such that for each $x \in \mathbb{X}$, $\mathbf{p}(x, \cdot)$ is a measure on $(\mathbb{Y}, \mathcal{B}(\mathbb{Y}))$ and for each $B \in \mathcal{B}(\mathbb{Y})$ the function $x \mapsto \mathbf{p}(x, B)$ is measurable; we write $\mathbf{p}(\cdot | x)$ for the measure associated to x, which is a (sub-)probability measure when $\mathbf{p}(x, \cdot)$ is. The *Dirac delta* measure $\delta_a : \mathcal{B}(\mathbb{X}) \to \{0, 1\}$ at $a \in \mathbb{X}$ is defined by $\delta_a(A) = 1$ if $a \in A$ and $\delta_a(A) = 0$ otherwise. The Gaussian measure with mean $\mu \in \mathbb{R}^n$ and covariance matrix $\Sigma \in \mathbb{R}^{n \times n}$ is defined by $\mathcal{N}(dx | \mu, \Sigma) := dx / (\sqrt{(2\pi)^n |\Sigma|}) \exp(-\frac{1}{2}(x-\mu)^\top \Sigma^{-1}(x-\mu))$, with $|\Sigma|$ being the determinant of Σ. We denote the Gaussian measure truncated to a support set $A \in \mathcal{B}(\mathbb{X})$ as $\mathcal{N}_A(dx|\mu, \Sigma) := \mathbf{1}_A(x) \mathcal{N}(dx|\mu, \Sigma) / \int_A \mathcal{N}(d\xi|\mu, \Sigma)$, where the indicator function $\mathbf{1}_A(x)$ of a measurable set $A \in \mathcal{B}(\mathbb{X})$ evaluates to $\mathbf{1}_A(x) = 1$ if $x \in A$ and $\mathbf{1}_A(x) = 0$ otherwise. The *cumulative distribution function* (CDF) of a Gaussian distribution is cdf $(x) := \int_{-\infty}^{x} \frac{1}{\sqrt{2\pi}} \exp(-\xi^2/2) \, d\xi$. Given sets A, B, a relation $\mathcal{R} \subset A \times B$ relates $x \in A$ and $y \in B$ if $(x, y) \in \mathcal{R}$. A metric on $\mathbb{Y}$ is a function $\mathbf{d}_\mathbb{Y} : \mathbb{Y} \times \mathbb{Y} \to \mathbb{R}_{\geq 0}$ satisfying $\mathbf{d}_\mathbb{Y}(y_1, y_2) = 0$ iff $y_1 = y_2$, $\mathbf{d}_\mathbb{Y}(y_1, y_2) = \mathbf{d}_\mathbb{Y}(y_2, y_1)$, and $\mathbf{d}_\mathbb{Y}(y_1, y_3) \leq \mathbf{d}_\mathbb{Y}(y_1, y_2) + \mathbf{d}_\mathbb{Y}(y_2, y_3)$ for all $y_1, y_2, y_3 \in \mathbb{Y}$. An example is the norm function $\mathbf{d}_\mathbb{Y}(y_1, y_2) := \|y_1 - y_2\|$. Given two functions f, g with suitable (co-)domains, we indicate their composition as $f \circ g := f(g(\cdot))$.

3.1 Game Setup

We consider the interaction between an agent **A** and a complex environment **E**, where **E** is elemental. In essence, this means that the dynamics of **E** *relevant* to the considered situation (as defined below by virtue of the domain of **E**) can be distilled into a small subset of equations.

We cast the interaction as a $2\frac{1}{2}$-player game, where the two main players **A** and **E** are parameterized discrete-time nonlinear dynamical systems with uncertain parameters $\theta^{\mathbf{A}} \in \Theta^{\mathbf{A}}$ and $\theta^{\mathbf{E}} \in \Theta^{\mathbf{E}}$, respectively, subject to additive stochastic noise (the remaining $\frac{1}{2}$-player), i.e., systems of the form

$$\mathbf{A}(\theta^{\mathbf{A}}) : \begin{cases} x^{\mathbf{A}}_{t+1} = f^{\mathbf{A}}(x^{\mathbf{A}}_t, o^{\mathbf{A}}_t, u^{\mathbf{A}}_t; \theta^{\mathbf{A}}) + w^{\mathbf{A}}_t \\ y^{\mathbf{A}}_t = h^{\mathbf{A}}(x^{\mathbf{A}}_t) \end{cases}, \quad w^{\mathbf{A}}_t \sim \mathcal{N}(\cdot | 0, \Sigma^{\mathbf{A}}), \quad (1)$$

$$\mathbf{E}(\theta^{\mathbf{E}}) : \begin{cases} x^{\mathbf{E}}_{t+1} = f^{\mathbf{E}}(x^{\mathbf{E}}_t, o^{\mathbf{E}}_t; \theta^{\mathbf{E}}) + w^{\mathbf{E}}_t \\ y^{\mathbf{E}}_t = h^{\mathbf{E}}(x^{\mathbf{E}}_t) \end{cases}, \quad w^{\mathbf{E}}_t \sim \mathcal{N}_{\mathbb{W}^{\mathbf{E}}}(\cdot | 0, \Sigma^{\mathbf{E}}), \quad (2)$$

where the state, input, observation, noise, and output of system $\mathbf{M} \in \{\mathbf{A}, \mathbf{E}\}$ at the tth time step are denoted by $x^{\mathbf{M}}_t \in \mathbb{X}^{\mathbf{M}}$, $u^{\mathbf{A}}_t \in \mathbb{U}^{\mathbf{A}}$, $o^{\mathbf{M}}_t \in \mathbb{O}^{\mathbf{M}}$, $w^{\mathbf{M}}_t \in \mathbb{W}^{\mathbf{M}}$, and $y^{\mathbf{M}}_t \in \mathbb{Y}^{\mathbf{M}}$, respectively. The functions $f^{\mathbf{M}}$ and $h^{\mathbf{M}}$ specify, respectively, the parameterized state evolution of the system and the observation map. The process noise $w^{\mathbf{M}}_t \in \mathbb{W}^{\mathbf{M}}$ constitutes an independent, identically distributed (i.i.d.) noise sequence. Whilst the noise of the agent **A** can be unbounded (and even from an arbitrary, potentially $\theta^{\mathbf{A}}$-parametrized continuous distribution as shown in the paper [41]), we raise the following assumption on **E**.

Assumption 1 (Boundedness). *Let* $\mathbb{W}^{\mathbf{E}}$ *and* $\mathbb{X}^{\mathbf{E}}$ *be bounded.*

In this paper, we assume bounded noise to over-approximate the behavior of **E** in a simpler surrogate model. Furthermore, without loss of generality, we consider autonomous environments, i.e., $\mathbb{U}^{\mathbf{E}} := \emptyset$. Our task is to design a state-feedback controller **C** for the agent **A**, that is

$$\mathbf{C} : \mathbb{X}^{\mathbf{A}} \times \mathbb{X}^{\mathbf{E}} \to \mathbb{U}^{\mathbf{A}}.$$

The systems **A** and **E** are interconnected via the operator $\times : \mathbb{X}^{\mathbf{A}} \times \mathbb{X}^{\mathbf{E}} \to \mathbb{O}^{\mathbf{E}} \times \mathbb{O}^{\mathbf{A}}$, written as $\mathbf{A} \times \mathbf{E}$. Unless specified otherwise, we assume $\times$ is constructed with $o^{\mathbf{E}} = x^{\mathbf{A}}$ and $o^{\mathbf{A}} = x^{\mathbf{E}}$. We use the following running example.

*Running Example. Consider an interaction between two vehicles **A** and **E**. We assume that the dynamics of the agent are given by the stochastic linear system*

$$\mathbf{A}(\theta^{\mathbf{A}}) : \begin{cases} \begin{bmatrix} \xi^{\mathbf{A}}_{t+1} \\ v^{\mathbf{A}}_{t+1} \\ s^{\mathbf{A}}_{t+1} \end{bmatrix} = \begin{bmatrix} a_1 & 0 & 0 \\ a_2 & \theta^{\mathbf{A}} & 0 \\ a_3 & \tau & 1 \end{bmatrix} x^{\mathbf{A}}_t + \begin{bmatrix} b \\ \tau \\ 0 \end{bmatrix} u^{\mathbf{A}}_t + w^{\mathbf{A}}_t \\ y^{\mathbf{A}}_t = [v^{\mathbf{A}}_t, s^{\mathbf{A}}_t]^\top \end{cases}, \quad (3)$$

with state $x^{\mathbf{A}}_t := [\xi^{\mathbf{A}}_t, v^{\mathbf{A}}_t, s^{\mathbf{A}}_t]^\top \in \mathbb{X}^{\mathbf{A}}$, *control input* $u^{\mathbf{A}}_t \in \mathbb{U}^{\mathbf{A}}$, *process noise* $w^{\mathbf{A}}_t \sim p^{\mathbf{A}}_w(\cdot) := p^{\mathbf{A}}_{w,\xi} p^{\mathbf{A}}_{w,v} p^{\mathbf{A}}_{w,s}$, *time discretization* $\tau = 0.5$, *fixed parameters* $a_1, a_2, a_3, b \in \mathbb{R}$, *and an uncertain parameter* $\theta^{\mathbf{A}} \in \Theta^{\mathbf{A}} := [0.79, 0.81]$. *We defer further details to [40, Appendix C.1]. The ambiguity introduced by the stochastic noise* $w^{\mathbf{A}}_t$ *and the uncertain parameter* $\theta^{\mathbf{A}}$ *may arise from unmodeled dynamics, friction, aerodynamic drag, driver behavior, etc.*

We assume that the dynamics of the environment are described by an autonomous six-dimensional nonlinear stochastic system

$$\mathbf{E}(\theta^{\mathbf{E}}): \begin{cases} \begin{bmatrix} \beta^{\mathbf{E}}_{t+1} \\ d\Psi^{\mathbf{E}}_{t+1} \\ \Psi^{\mathbf{E}}_{t+1} \\ v^{\mathbf{E}}_{t+1} \\ s^{\mathbf{E}}_{x,t+1} \\ s^{\mathbf{E}}_{y,t+1} \end{bmatrix} = \begin{bmatrix} \beta^{\mathbf{E}}_t + \tau\dot{\beta}(x^{\mathbf{E}}_t) \\ d\Psi^{\mathbf{E}}_t + \tau\ddot{\Psi}(x^{\mathbf{E}}_t) \\ \Psi^{\mathbf{E}}_t + \tau d\Psi^{\mathbf{E}}_t \\ \theta^{\mathbf{E}} v^{\mathbf{E}}_t \\ s^{\mathbf{E}}_{x,t} + \tau v^{\mathbf{E}}_t \cos(\beta^{\mathbf{E}}_t + \Psi^{\mathbf{E}}_t) \\ s^{\mathbf{E}}_{y,t} + \tau v^{\mathbf{E}}_t \sin(\beta^{\mathbf{E}}_t + \Psi^{\mathbf{E}}_t) \end{bmatrix} + \begin{bmatrix} 0 & 0 & 0 \\ 0 & 0 & 0 \\ 0 & 0 & 0 \\ a_4 & 0 & 0 \\ a_5 & 0 & 0 \\ 0 & 0 & 0 \end{bmatrix} o^{\mathbf{E}}_t + w^{\mathbf{E}}_t \\ y^{\mathbf{E}}_t = [v^{\mathbf{E}}_t, s^{\mathbf{E}}_{x,t}]^\top \end{cases} \quad (4)$$

characterized by a state $x^{\mathbf{E}}_t := [\beta^{\mathbf{E}}_t, d\Psi^{\mathbf{E}}_t, \Psi^{\mathbf{E}}_t, v^{\mathbf{E}}_t, s^{\mathbf{E}}_{x,t}, s^{\mathbf{E}}_{y,t}]^\top \in \mathbb{X}^{\mathbf{E}}$, observation $o^{\mathbf{E}}_t \in \mathbb{X}^{\mathbf{A}}$, fixed parameters $a_4, a_5 \in \mathbb{R}$, uncertain parameter $\theta^{\mathbf{E}} \in \Theta^{\mathbf{E}} := [0.79, 0.81]$, and derivative terms $\dot{\beta}(x^{\mathbf{E}}_t)$ and $\ddot{\Psi}(x^{\mathbf{E}}_t)$ provided in [40, Appendix C.2]. As alluded to in Assumption 1, the environment's process noise $w^{\mathbf{E}}_t \sim p^{\mathbf{E}}_w(\cdot) := p^{\mathbf{E}}_{w,\beta} \cdots p^{\mathbf{E}}_{w,s_y}$ is constricted to a bounded support $\mathbb{W}^{\mathbf{E}} := \mathbb{W}^{\mathbf{E}}_\beta \times \cdots \times \mathbb{W}^{\mathbf{E}}_{s_y}$.

For brevity, the resulting composed system parametrized by $\theta := (\theta^{\mathbf{A}}, \theta^{\mathbf{E}})$ can be written as $(\mathbf{A} \times \mathbf{E})(\theta) := \mathbf{A}(\theta^{\mathbf{A}}) \times \mathbf{E}(\theta^{\mathbf{E}})$. Furthermore, in the remainder, we may drop the explicit parametrization for conciseness. With this, the game $(\mathbf{A} \times \mathbf{E})(\theta)$ unfolds as follows: In every iteration $t \in \mathbb{N}$ of the game, the agent $\mathbf{A}$ and environment $\mathbf{E}$ start from some initial states $x^{\mathbf{A}}_t, x^{\mathbf{E}}_t$. The controller $\mathbf{C}: \mathbb{X}^{\mathbf{A}} \times \mathbb{X}^{\mathbf{E}} \to \mathbb{U}^{\mathbf{A}}$ chooses a control input $u^{\mathbf{A}}_t \in \mathbb{U}^{\mathbf{A}}$. Based on $u^{\mathbf{A}}_t$ and the observation $o^{\mathbf{A}}_t$, the agent system $\mathbf{A}$ evolves to a successor state $x^{\mathbf{A}}_{t+1} = f^{\mathbf{A}}(x^{\mathbf{A}}_t, o^{\mathbf{A}}_t, u^{\mathbf{A}}_t) + w^{\mathbf{A}}_t$ by eliciting a noise realization $w^{\mathbf{A}}_t \sim \mathcal{N}(\cdot|0, \Sigma^{\mathbf{A}})$. Simultaneously, the environment $\mathbf{E}$ evolves to $x^{\mathbf{E}}_{t+1} = f^{\mathbf{E}}(x^{\mathbf{E}}_t, x^{\mathbf{A}}_t) + w^{\mathbf{E}}_t$ with $w^{\mathbf{E}}_t \sim \mathcal{N}_{\mathbb{W}^{\mathbf{E}}}(\cdot|0, \Sigma^{\mathbf{E}})$. In every iteration $t \in \mathbb{N}$, the systems emit outputs $y^{\mathbf{A}}_t = h^{\mathbf{A}}(x^{\mathbf{A}}_t)$ and $y^{\mathbf{E}}_t = h^{\mathbf{E}}(x^{\mathbf{E}}_t)$, which will be used to determine the behavior of the systems according to specifications defined in Subsect. 3.2. The setup can be generalized by defining a shared output mapping over the product space $\mathbb{X}^{\mathbf{A}} \times \mathbb{X}^{\mathbf{E}}$.

3.2 Temporal Logic Specifications

Consider a set of atomic propositions $AP := \{P_1, \ldots, P_L\}$ defining an *alphabet* $\Sigma := 2^{AP}$, where any *letter* $l \in \Sigma$ is composed of a set of atomic propositions. An infinite string of letters forms a *word* $\omega = l_0 l_1 l_2 \ldots \in \Sigma^\mathbb{N}$. Specifications imposed on the behavior of the system are defined as formulas composed of atomic propositions and operators. We consider the *co-safe* fragment of *linear-time temporal logic* properties (scLTL) [27], for which violation can be determined from a finite bad prefix of a word. The scLTL syntax is defined as follows:

$$\psi ::= \text{true} \mid P \mid \neg P \mid \psi_1 \vee \psi_2 \mid \psi_1 \wedge \psi_2 \mid \psi_1 \mathsf{U} \psi_2 \mid \bigcirc \psi,$$

where $P \in \mathsf{AP}$ is an atomic proposition. Using a labeling map $\mathcal{L}: \mathbb{Y} \to \Sigma$, we can define temporal logic specifications over the output of a system $\mathbf{M}$. Each output trace $\mathbf{y} = y_0, y_1, y_2, \ldots$ of $\mathbf{M}$ can be translated to a word via $\boldsymbol{\omega} = \mathcal{L}(\mathbf{y})$. A system $\mathbf{M}$ (e.g., $\mathbf{M} := \mathbf{A} \times \mathbf{E}$) satisfies the specification ψ with probability at least p if $\mathbb{P}(\mathbf{M} \vDash \psi) := \mathbb{E}[\mathcal{L}(\mathbf{y}) \vDash \psi] \geq p$, where $\mathbf{y}$ is any output trace of $\mathbf{M}$.

3.3 Simulated Environments

Instead of performing exhaustive computations on the potentially high-dimensional $\mathbf{E}$—in the spirit of SA—we replace $\mathbf{E}$ with a lower-dimensional and less complex dynamics model, which we aptly call the *simulator* $\mathbf{S}$. To compensate for the loss in dynamic resemblance, we invoke the concept of *red teaming*. More specifically, we equip the simulator $\mathbf{S}$ with an adversary $\mathfrak{n} : \mathbb{X}^\mathbf{S} \times \mathbb{O}^\mathbf{S} \times \mathcal{B}(\mathbb{D}^\mathbf{S}) \to [0,1]$ that perturbs $\mathbf{S}$ with adversarial disturbances $d_t^\mathbf{S} \sim \mathfrak{n}(\,\cdot\,|x_t^\mathbf{S}, o_t^\mathbf{S})$. The resulting system is of the form

$$\mathbf{S}(\theta^\mathbf{E}, \mathfrak{n}) : \begin{cases} x_{t+1}^\mathbf{S} = f^\mathbf{S}(x_t^\mathbf{S}, o_t^\mathbf{S}; \theta^\mathbf{E}) + d_t^\mathbf{S} + w_t^\mathbf{S} & w_t^\mathbf{S} \sim \mathcal{N}_{\mathbb{WS}}(\,\cdot\,|0, \Sigma^\mathbf{E}) \\ y_t^\mathbf{S} = h^\mathbf{S}(x_t^\mathbf{S}) & d_t^\mathbf{S} \sim \mathfrak{n}(\,\cdot\,|x_t^\mathbf{S}, o_t^\mathbf{S}) \end{cases}, \quad (5)$$

We will require that $x^\mathbf{S}$ retains all information relevant w.r.t. the output by assuming $h^\mathbf{E} = h^\mathbf{S} \circ P$ via some measurable map $P : \mathbb{X}^\mathbf{E} \to \mathbb{X}^\mathbf{S}$. Without loss of generality, we assume that the disturbance $d_t^\mathbf{S} \sim \mathfrak{n}(\,\cdot\,)$ is additive. Based on an ambiguity set $\mathfrak{N}$ of adversaries $\mathfrak{n} \in \mathfrak{N}$, (5) defines a family of simulator models $\mathbf{S}(\Theta, \mathfrak{N}) := \{\mathbf{S}(\theta^\mathbf{E}, \mathfrak{n}) \mid \theta^\mathbf{E} \in \Theta^\mathbf{E}, \mathfrak{n} \in \mathfrak{N}\}$. Analogous to $(\mathbf{A} \times \mathbf{E})(\theta)$, we define $(\mathbf{A} \times_a \mathbf{S})(\theta, \mathfrak{n}) := \mathbf{A}(\theta^\mathbf{A}) \times_a \mathbf{S}(\theta^\mathbf{E}, \mathfrak{n})$, where $\times_a : (x^\mathbf{A}, x^\mathbf{S}) \mapsto (F(x^\mathbf{A}), x^\mathbf{S}) = (o^\mathbf{S}, o^\mathbf{A})$ with some measurable map $F : \mathbb{X}^\mathbf{A} \to \mathbb{O}^\mathbf{S}$. The additional mapping F will grant us the flexibility to apply MOR to $\mathbf{A}$ (see Theorem 1).

Running Example. Assume that for the scenario at hand, only the longitudinal dynamics of $\mathbf{E}$ are relevant—e.g., as in the scenarios of Fig. 2. We thus substitute $\mathbf{E}$ with a simulator of the form[2].

$$\mathbf{S}(\theta^\mathbf{E}, \mathfrak{n}) : \begin{cases} \begin{bmatrix} v_{t+1}^\mathbf{S} \\ s_{t+1}^\mathbf{S} \end{bmatrix} = \begin{bmatrix} \theta^\mathbf{E} & 0 \\ 0 & 1 \end{bmatrix} x_t^\mathbf{S} + \begin{bmatrix} a_4 & 0 & 0 \\ a_5 & 0 & 0 \end{bmatrix} o_t^\mathbf{S} + \mathfrak{n}(\,\cdot\,) + w_t^\mathbf{S} \\ y_t^\mathbf{S} = x_t^\mathbf{S} \end{cases}, \quad (6)$$

with state $x_t^\mathbf{S} := [v_t^\mathbf{S}, s_t^\mathbf{S}]^\top \in \mathbb{X}^\mathbf{S}$, observation $o_t^\mathbf{S} \in \mathbb{X}^\mathbf{A}$, process noise $w_t^\mathbf{S} \sim p_{w,v}^\mathbf{E} p_{w,s_x}^\mathbf{E}$, and adversarial disturbance $d_t^\mathbf{S} = \mathfrak{n}(x_t^\mathbf{S}, o_t^\mathbf{S})$, elicited by a deterministic nonlinear adversary of the form

$$\mathfrak{n} \in \mathfrak{N} := \left\{ (x_t^\mathbf{S}, o_t^\mathbf{S}) \mapsto \begin{bmatrix} 0 \\ \tau v_t^\mathbf{S} \cos(\beta + \Psi) \end{bmatrix} \,\middle|\, \beta \in \mathbb{X}_\beta^\mathbf{E}, \Psi \in \mathbb{X}_\Psi^\mathbf{E} \right\}. \quad (7)$$

To further reduce complexity, we also choose a reduced-order model for $\mathbf{A}$ in (3):

$$\mathbf{A}_r(\theta^\mathbf{A}) : \begin{cases} \begin{bmatrix} v_{t+1}^{\mathbf{A}_r} \\ s_{t+1}^{\mathbf{A}_r} \end{bmatrix} = \begin{bmatrix} \theta^\mathbf{A} & 0 \\ \tau & 1 \end{bmatrix} x_t^{\mathbf{A}_r} + \begin{bmatrix} \tau \\ 0 \end{bmatrix} u_t^{\mathbf{A}_r} + w_t^{\mathbf{A}_r} \\ y_t^{\mathbf{A}_r} = x_t^{\mathbf{A}_r} \end{cases}, \quad (8)$$

---

[2] $\tau v_t^\mathbf{S}$ was absorbed into (7) as the disturbance would otherwise become multiplicative.

with state $x_t^{\mathbf{A}_r} := [v_t^{\mathbf{A}_r}, s_t^{\mathbf{A}_r}]^\top \in \mathbb{X}_v^{\mathbf{A}_r} \times \mathbb{X}_s^{\mathbf{A}_r}$, control input $u_t^{\mathbf{A}_r} \in \mathbb{U}^{\mathbf{A}}$, and process noise $w_t^{\mathbf{A}_r} \sim p_{w,v}^{\mathbf{A}} p_{w,s}^{\mathbf{A}}$.

To obtain end-to-end guarantees, establishing a formal connection between the simulator $\mathbf{S}$ and the environment $\mathbf{E}$ is crucial. Whilst suitable notions are available for generic MDPs, our setting is more complex. We build on the work [42], which formulates stochastic simulation relations for a parametrized MDP $\mathbf{M}(\theta)$ by establishing a partial coupling between a set $\{\mathbf{M}(\theta) \mid \theta \in \Theta\}$ and a nominal model $\mathbf{M}(\hat{\theta})$. Here, we intentionally extend $\mathbf{M}$ by an ambiguous adversary $\mathfrak{n} \in \mathfrak{N}$, i.e., $\mathbf{M}(\theta, \mathfrak{n})$, as it allows us to simplify $\mathbf{M}$ itself. As a result, the nominal $\mathbf{M}(\hat{\theta}, \hat{\mathfrak{n}})$ used for control synthesis can be much simpler.

3.4 Problem Formulation

Consider an agent $\mathbf{A}$ in an environment $\mathbf{E}$. Our goal is to design a controller $\mathbf{C}$ for $\mathbf{A}$ such that the composed system $(\mathbf{C} \times \mathbf{A}) \times \mathbf{E}$ satisfies a given specification ψ with a probability not lower than some fixed $p \in (0,1)$. To make the control design scalable, we leverage the notion of SA. In other words, although the concrete environment $\mathbf{E}$ may be a complex, high-dimensional system, we will assume that the dynamics *relevant* to the interaction with the agent $\mathbf{A}$ in a particular situation $i \in \mathcal{I} := \{1, 2, \ldots\}$ can be distilled in a much simpler form. For instance, we can coarsely partition the domain $\mathbb{X}^{\mathbf{E}}$ of $\mathbf{E}$ to define finitely many situation-dependent $\mathbf{E}_i$ as restrictions of $\mathbf{E}$ to $\mathbb{X}^{\mathbf{E}_i}$ such that $\cup_{i \in \mathcal{I}} \mathbb{X}^{\mathbf{E}_i} = \mathbb{X}^{\mathbf{E}}$. Then, for each partition $\mathbb{X}^{\mathbf{E}_i}$, we design a simulator model $\mathbf{S}_i$ alongside a set of adversaries $\mathfrak{N}$ such that $\mathbf{S}_i(\Theta, \mathfrak{N})$ encapsulates a valid overapproximation of all the dynamical features of $\mathbf{E}_i$ when operating in $\mathbb{X}^{\mathbf{E}_i}$. We may therefore also call $\mathbb{X}^{\mathbf{E}_i}$ the *region of validity* (RoV) for the simulator $\mathbf{S}_i$. In fact, it might suffice for $\mathbf{S}_i$ not only to be much less complex and lower dimensional than $\mathbf{E}_i$, but modeling all stochastic features of $\mathbf{E}_i$ precisely might not be necessary either. We will call such an environment $\mathbf{E}_i$ that associates a simpler simulator $\mathbf{S}_i$ on RoV $\mathbb{X}^{\mathbf{E}_i}$ *elemental*. In this work, we focus on a singular elemental environment $\mathbf{E}$ and establish a probabilistic relation that allows us to compute a robust controller for $\mathbf{A} \times \mathbf{E}$ via $\mathbf{A} \times_a \mathbf{S}$ as a proxy.

> **Problem 1.** Let a simulator $\mathbf{S}$ for an elemental environment $\mathbf{E}$ with RoV $\mathbb{X}^{\mathbf{E}}$ be given such that $\mathbf{S}$ exhibits a richer behavior than $\mathbf{E}$. For a global scLTL specification ψ and a corresponding threshold $p \in (0,1)$, design a controller $\mathbf{C}$ via $\mathbf{A} \times_a \mathbf{S}$ such that $\mathbb{P}((\mathbf{C} \times \mathbf{A}) \times \mathbf{E} \vDash \psi) \geq p$.

4 System Interrelations

In this work, we use two different types of relations between dynamical systems, *behavioral inclusions* (BIs) and *sub-simulation relations* (SSRs), that are introduced in Subsects. 4.1 and 4.2, respectively. In order to formalize these notions, we model the agent, environment, and simulator (1)-(5) as *general Markov decision processes* (gMDPs).

Definition 1 (General Markov Decision Process (gMDP)). *A gMDP is a tuple $(\mathbb{X}, \mathbb{X}_0, \mathbb{U}, \mathbb{O}, \mathbb{D}, \mathbf{t}, \mathbb{Y}, h)$, comprising $\mathbb{X}$, a measurable state space of states $x \in \mathbb{X}$; $\mathbb{X}_0 \subset \mathbb{X}$, a set of initial states $x_0 \in \mathbb{X}_0$; $\mathbb{U}$, a measurable input space of inputs $u \in \mathbb{U}$; $\mathbb{O}$, a measurable set of observations $o \in \mathbb{O}$; $\mathbb{D}$, a measurable set of disturbances $d \in \mathbb{D}$; $\mathbf{t} : \mathbb{X} \times \mathbb{U} \times \mathbb{O} \times \mathbb{D} \times \mathcal{B}(\mathbb{X}) \to [0,1]$, a probability kernel; $\mathbb{Y}$, a measurable output space of outputs $y \in \mathbb{Y}$, decorated with a metric $\mathbf{d}_\mathbb{Y}$; and $h : \mathbb{X} \to \mathbb{Y}$, a measurable output map.*

We will overload the notation and omit tuple elements whenever they are nil. For instance, in the remainder of the manuscript, we will treat the agent and environment as gMDPs of the form $\mathbf{A} = (\mathbb{X}^\mathbf{A}, \mathbb{X}_0^\mathbf{A}, \mathbb{U}^\mathbf{A}, \mathbb{O}^\mathbf{A}, \mathbf{t}^\mathbf{A}, \mathbb{Y}^\mathbf{A}, h^\mathbf{A})$ and $\mathbf{E} = (\mathbb{X}^\mathbf{E}, \mathbb{X}_0^\mathbf{E}, \mathbb{O}^\mathbf{E}, \mathbf{t}^\mathbf{E}, \mathbb{Y}^\mathbf{E}, h^\mathbf{E})$, respectively. As we focus on autonomous environments, we omit the input set $\mathbb{U}^\mathbf{E} := \emptyset$. Furthermore, we assume that both $\mathbf{A}$ and $\mathbf{E}$ do not receive any additional disturbances, i.e., $\mathbb{D}^\mathbf{A} = \mathbb{D}^\mathbf{E} = \emptyset$. The transition kernels $\mathbf{t}^\mathbf{A}, \mathbf{t}^\mathbf{E}$ are obtained from (1)-(2), respectively, via

$$\mathbf{t}^\mathbf{M}(dx_+^\mathbf{M}|z^\mathbf{M};\theta^\mathbf{M}) := \int_{\mathbb{W}^\mathbf{M}} \delta_{f^\mathbf{M}(z^\mathbf{M};\theta^\mathbf{M})+w^\mathbf{M}}(x_+^\mathbf{M})\, \mathcal{N}_{\mathbb{W}^\mathbf{M}}(dw^\mathbf{M}|0, \Sigma^\mathbf{M}),$$

where $z^\mathbf{M} := (x^\mathbf{M}, u^\mathbf{M}, o^\mathbf{M})$ and $\mathbf{M} \in \{\mathbf{A}, \mathbf{E}\}$. Note that the resulting (parametrized) transition kernels are conditioned on uncertain parameters $\theta^\mathbf{M} \in \Theta^\mathbf{M}$. The simulator (5) is captured as a gMDP $\mathbf{S} = (\mathbb{X}^\mathbf{S}, \mathbb{X}_0^\mathbf{S}, \mathbb{O}^\mathbf{S}, \mathbb{D}^\mathbf{S}, \mathbf{t}^\mathbf{S}, \mathbb{Y}^\mathbf{S}, h^\mathbf{S})$, receiving additional disturbances $d^\mathbf{S} \in \mathbb{D}^\mathbf{S}$ from an ambiguous adversary $\mathfrak{n} \in \mathfrak{N}$.

4.1 Behavioral Inclusion

We now formalize the relation that we will use to relate environments and simulators, utilizing the ambiguous adversary $\mathfrak{n} \in \mathfrak{N}$.

Definition 2 (Behavioral Inclusion (BI)). *Consider autonomous gMDPs $\mathbf{M} = (\mathbb{X}, \mathbb{X}_0, \mathbb{O}, \mathbf{t}, \mathbb{Y}, h)$, $\widehat{\mathbf{M}} = (\hat{\mathbb{X}}, \hat{\mathbb{X}}_0, \hat{\mathbb{O}}, \hat{\mathbb{D}}, \hat{\mathbf{t}}, \mathbb{Y}, \hat{h})$, an ambiguity set $\hat{\mathfrak{N}}$ of adversaries $\hat{\mathfrak{n}} : \hat{\mathbb{X}} \times \hat{\mathbb{O}} \times \mathcal{B}(\hat{\mathbb{D}}) \to [0,1]$, and measurable maps $P : \mathbb{X} \to \hat{\mathbb{X}}$ and $F : \mathbb{O} \to \hat{\mathbb{O}}$. For the relation $\mathcal{R} := \{(\hat{x}, x) \in \hat{\mathbb{X}} \times \mathbb{X} \mid \hat{x} = P(x)\}$, if*

(a) $\hat{\mathbb{X}}_0 = P(\mathbb{X}_0)$;
(b) $\forall (\hat{x}, x) \in \mathcal{R}, \forall o \in \mathbb{O}, \exists \hat{\mathfrak{n}} \in \hat{\mathfrak{N}}, \forall \hat{Q} \in \mathcal{B}(\hat{\mathbb{X}}) : \hat{\mathbf{t}}(\hat{Q}|\hat{x}, F(o), \hat{\mathfrak{n}}(\hat{d}|\hat{x}, F(o))) = \mathbf{t}(Q|x,o)$, with $Q := P^{-1}(\hat{Q})$; and
(c) $\forall (\hat{x}, x) \in \mathcal{R} : h(x) = \hat{h}(\hat{x})$;

then, we say that the behavior of $\mathbf{M}$ is included in $\widehat{\mathbf{M}}$, denoted as $\mathbf{M} \sqsubseteq \widehat{\mathbf{M}}(\hat{\mathfrak{N}})$.

Intuitively, the behavior of a system $\mathbf{M}$ is included in $\widehat{\mathbf{M}}$ if the following conditions are met: (a) All initial conditions of $\mathbf{M}$ are matched by $\widehat{\mathbf{M}}$. (b) For all state pairs in the relation and all observations $o \in \mathbb{O}$, there exists an adversary $\hat{\mathfrak{n}} \in \hat{\mathfrak{N}}$ resolving the nondeterminism of $\widehat{\mathbf{M}}$ such that the resulting probability measure $\hat{\mathbf{t}}(\,\cdot\,|\hat{x}, F(o), \hat{\mathfrak{n}}(\hat{d}|\hat{x}, F(o)))$ assigns a probability to an event $\hat{Q}$ that matches the probability of all equivalent events $Q := P^{-1}(\hat{Q})$ of $\mathbf{M}$, as quantified by $\mathbf{t}(\,\cdot\,|x,o)$. In other words, $Q := P^{-1}(\hat{Q})$ represents an *equivalence class* of rich-in-detail events that are mapped to the same abstract event $\hat{Q}$ via P. Finally, (c) whilst in the relation, both systems emit equivalent outputs.

At a first glance, the BI bears conceptual resemblance to the relations used in [46, Definition 4.1] and for perception in [4]. However, their formulations are substantially different as they are restricted to deterministic systems and do not specify how the relation can be established for systems of different dimensionality. More similarly, our condition (b) can be seen as a generalization of the bisimulation for nondeterministic labeled Markov processes in [11] to antisymmetric relations $\mathcal{R}$. In fact, the map P in the BI acts as a *zigzag morphism* [16,18], preserving transition probabilities across the two labeled Markov processes $\mathbf{M}, \widehat{\mathbf{M}}$. We refer to the early [17] and [22, Appendix C], which are situated in a similar relational context. In effect, the BI imposes the strict ordering of $\mathbf{M}$ and $\widehat{\mathbf{M}}$ according to their level of abstraction, as outlined in the intuition above.

Running Example. It is easy to verify that for the chosen RoV $\mathbb{X}^{\mathbf{E}}$—in particular, tight bounds on the slip angle $\beta^{\mathbf{E}} \in \mathbb{X}_\beta^{\mathbf{E}}$ and yaw angle $\Psi^{\mathbf{E}} \in \mathbb{X}_\Psi^{\mathbf{E}}$—the lateral dynamics of $\mathbf{E}$ become negligible and the simulator $\mathbf{S}(\theta^{\mathbf{E}}, \mathfrak{n})$ in (6) with the ambiguous adversary $\mathfrak{n} \in \mathfrak{N}$ in (7) subsumes all the behavior of $\mathbf{E}(\theta^{\mathbf{E}})$ in (4), i.e., we have $\mathbf{E}(\theta^{\mathbf{E}}) \sqsubseteq \mathbf{S}(\theta^{\mathbf{E}}, \mathfrak{N})$ via the map $P : [0_{2\times 3}, I_2, 0_{2\times 1}]x^{\mathbf{E}} \mapsto x^{\mathbf{S}}$. Note, that in contrast to the six-dimensional nonlinear $\mathbf{E}$ the simulator $\mathbf{S}$ is only two-dimensional, linear, and lifts the need for accurate knowledge of the underlying stochastic laws on $\beta^{\mathbf{E}}, \mathrm{d}\Psi^{\mathbf{E}}, \Psi^{\mathbf{E}},$ and $s_y^{\mathbf{E}}$. In practice, a basic model such as (6) is often supplied by the SA component of an autonomous agent [12]. Reachability techniques can be used to construct the corresponding adversary.

Remark 1 (Non-probabilistic simulators). The adversary in (7) can be augmented to absorb the noise $w_t^{\mathbf{S}} \in \mathbb{W}_{w,v}^{\mathbf{E}} \times \mathbb{W}_{w,s_x}^{\mathbf{E}}$ also, lifting the need for *any* information on the stochastic laws of $\mathbf{E}$ and yielding a fully non-probabilistic simulator $\mathbf{S}$. This would further trade scalability for conservatism.

4.2 Stochastic Simulation Relation

Before providing the definition of an SSR, we recall the instrumental definition of a sub-probability coupling between two probability measures [42, Definition 5].

Definition 3 (Sub-probability coupling). *Given $\hat{p} \in \mathcal{P}(\hat{\mathbb{X}})$ and $p \in \mathcal{P}(\mathbb{X})$, $\mathcal{R} \subset \mathcal{B}(\hat{\mathbb{X}} \times \mathbb{X})$, and a value $\delta \in [0,1]$, a sub-prob. measure v over $(\hat{\mathbb{X}} \times \mathbb{X}, \mathcal{B}(\hat{\mathbb{X}} \times \mathbb{X}))$ with $v(\hat{\mathbb{X}} \times \mathbb{X}) \geq 1 - \delta$ is a sub-probability coupling of $\hat{p}$ and p over $\mathcal{R}$ if*

(a) $v(\hat{\mathbb{X}} \times \mathbb{X}) = v(\mathcal{R})$, i.e., v's probability mass is located on $\mathcal{R}$;
(b) $\forall A \in \mathcal{B}(\mathbb{X}): v(\hat{\mathbb{X}} \times A) \leq p(A);$ and
(c) $\forall A \in \mathcal{B}(\hat{\mathbb{X}}): v(A \times \mathbb{X}) \leq \hat{p}(A).$

With this, we extend the original SSR Definition by [42] to accommodate a nondeterministic adversary.

Definition 4 ((ε, δ)-Sub-Simulation Relation (SSR)). *Consider two gMDPs $\widehat{\mathbf{M}} = (\widehat{\mathbb{X}}, \widehat{\mathbb{X}}_0, \widehat{\mathbb{U}}, \widehat{\mathbb{D}}, \hat{\mathbf{t}}, \hat{h}, \mathbb{Y})$ and $\mathbf{M} = (\mathbb{X}, \mathbb{X}_0, \mathbb{U}, \mathbb{D}, \mathbf{t}, h, \mathbb{Y})$, a set $\widehat{\mathfrak{N}}$ of adversaries $\hat{\mathfrak{n}}: \widehat{\mathbb{X}} \times \mathcal{B}(\widehat{\mathbb{D}}) \to [0, 1]$, an adversary $\mathfrak{n}: \mathbb{X} \times \mathcal{B}(\mathbb{D}) \to [0, 1]$, a measurable relation $\mathcal{R} \subset \mathcal{B}(\widehat{\mathbb{X}} \times \mathbb{X})$, and an interface function $\mathbf{i}_u: \widehat{\mathbb{X}} \times \mathbb{X} \times \widehat{\mathbb{U}} \times \mathcal{B}(\mathbb{U}) \to [0, 1]$. If there exists a sub-prob. kernel $\mathbf{v}: \widehat{\mathbb{X}} \times \mathbb{X} \times \widehat{\mathbb{U}} \times \widehat{\mathbb{D}} \times \mathbb{D} \times \mathcal{B}(\widehat{\mathbb{X}} \times \mathbb{X}) \to [0, 1]$ s.t.*

(a) $\forall x_0 \in \mathbb{X}_0, \exists \hat{x}_0 \in \widehat{\mathbb{X}}_0: (\hat{x}_0, x_0) \in \mathcal{R}$;

(b) $\forall (\hat{x}, x) \in \mathcal{R}, \forall \hat{u} \in \widehat{\mathbb{U}}$, there exists an $\hat{\mathfrak{n}} \in \widehat{\mathfrak{N}}$ such that $\mathbf{v}(d\hat{x}_+ \times dx_+ | \hat{x}, x, \hat{u}, \hat{\mathfrak{n}}(d|\hat{x}), \mathfrak{n}(d|x))$ is a sub-probability coupling of $\hat{\mathbf{t}}(d\hat{x}_+ | \hat{x}, \hat{u}, \hat{\mathfrak{n}}(d|\hat{x}))$ and $\mathbf{t}(dx_+ | x, \mathbf{i}_u(u|\hat{x}, x, \hat{u}), \mathfrak{n}(d|x))$ over $\mathcal{R}$ with respect to δ; and

(c) $\forall (\hat{x}, x) \in \mathcal{R}: \mathbf{d}_\mathbb{Y}(\hat{h}(\hat{x}), h(x)) \leq \varepsilon$;

then, $\mathbf{M}$ is in an SSR with $\widehat{\mathbf{M}}$, denoted as $\widehat{\mathbf{M}}(\widehat{\mathfrak{N}}) \preceq_\delta^\varepsilon \mathbf{M}(\mathfrak{n})$.

This extended SSR definition recovers the previous [42, Definition 6] for the undisturbed case, where $\widehat{\mathfrak{N}} = \{0\}$ and $\mathfrak{n} = 0$. In the following, with a slight abuse of notation, we may write $\widehat{\mathbf{M}}(\hat{\mathfrak{n}}) \preceq_\delta^\varepsilon \mathbf{M}(\mathfrak{n})$ to indicate that $\widehat{\mathfrak{N}} = \{\hat{\mathfrak{n}}\}$.

The intuition behind the three conditions of Definition 4 on a composed system $\widehat{\mathbf{M}} \times \mathbf{M}$ evolving on the product space $\widehat{\mathbb{X}} \times \mathbb{X}$ is as follows. Note that $\mathcal{R}$ defines a subset of $\widehat{\mathbb{X}} \times \mathbb{X}$. According to condition (a), both systems start in $\mathcal{R}$ upon initialization. In any subsequent time step, once in $\mathcal{R}$, condition (b) certifies that the systems remain in $\mathcal{R}$ for the next time step with a probability of at least $(1 - \delta)$ for all control inputs $\hat{u} \in \widehat{\mathbb{U}}$ and adversaries $\hat{\mathfrak{n}} \in \widehat{\mathfrak{N}}$. Finally, given the two systems are in $\mathcal{R}$, the corresponding outputs $\hat{y} := \hat{h}(\hat{x})$ and $y := h(x)$ are ε-close (condition (c)).

5 Assume-Guarantee Contracts

In this section, we associate the previously introduced systems and system interrelations within the framework of *assume-guarantee contracts* (AGCs). To this end, *assumptions* and *guarantees* resemble qualitative types of dynamical systems—conceptually similar to [44].

Definition 5 (Assumption). *Given gMDPs $\mathbf{M}$ and $\widehat{\mathbf{M}}$, $\widehat{\mathbf{M}}(\widehat{\mathfrak{N}})$ is an assumption of $\mathbf{M}$ (on the RoV $\mathbb{X}^\mathbf{M}$) if $\mathbf{M} \sqsubseteq \widehat{\mathbf{M}}(\widehat{\mathfrak{N}})$.*

In essence, an assumption of a system $\mathbf{M}$ is a model $\widehat{\mathbf{M}}$ that subsumes all its behavior (on a given RoV). Thus, a simulator $\mathbf{S}$ is as an assumption of a complex environment $\mathbf{E}$ if $\mathbf{E} \sqsubseteq \mathbf{S}(\widehat{\mathfrak{N}})$—*qualifying* $\mathbf{S}$ as a formally valid surrogate.

We proceed with the definition of (ε, δ)-*guarantees*, which, conversely, are specified for composed systems, and render the systems qualified *proxies*.

Definition 6 ((ε, δ)-Guarantee). *Let $\varepsilon \geq 0$ and $\delta \in [0, 1]$. Given gMDPs $\mathbf{M}_1$ and $\mathbf{M}_2$, a (ε, δ)-guarantee of $\mathbf{M}_1$ in $\mathbf{M}_2$ is a gMDP $\mathbf{G}$ such that $\mathbf{M}_1 \times \mathbf{M}_2 \preceq_\delta^\varepsilon \mathbf{G}$.*

We are ready to present the main result of this paper: We cast a formal AGC that *assumes* that the simulator $\mathbf{S}(\mathfrak{N})$ is a valid surrogate for $\mathbf{E}$ (on the RoV $\mathbb{X}^{\mathbf{E}}$) and renders a reduced-order system $\mathbf{A}_r \times \mathbf{S}$ a proxy of $\mathbf{A} \times \mathbf{E}$—the *guarantee*. For a given relation $\mathcal{R} \subset \mathcal{B}((\mathbb{X}^{\mathbf{A}_r} \times \mathbb{X}^{\mathbf{S}}) \times (\mathbb{X}^{\mathbf{A}_r} \times \mathbb{X}^{\mathbf{S}}))$, we define composite relations on $\mathcal{B}((\mathbb{X}^{\mathbf{A}_r} \times \mathbb{X}^{\mathbf{S}}) \times (\mathbb{X}^{\mathbf{A}} \times \mathbb{X}^{\mathbf{S}}))$ and $\mathcal{B}((\mathbb{X}^{\mathbf{A}_r} \times \mathbb{X}^{\mathbf{S}}) \times (\mathbb{X}^{\mathbf{A}} \times \mathbb{X}^{\mathbf{E}}))$, i.e.,

$$\mathcal{R}_a := \mathcal{R} \circ \times_a, \qquad \times_a : (x^{\mathbf{A}}, x^{\mathbf{S}}) \mapsto (F(x^{\mathbf{A}}), x^{\mathbf{S}}) \qquad \text{and}$$
$$\mathcal{R}_b := \mathcal{R} \circ \times_b, \qquad \times_b : (x^{\mathbf{A}}, x^{\mathbf{E}}) \mapsto (F(x^{\mathbf{A}}), P(x^{\mathbf{E}})),$$

respectively, with maps $F : \mathcal{B}(\mathbb{X}^{\mathbf{A}}) \to \mathbb{X}^{\mathbf{A}_r}$ and $P : \mathcal{B}(\mathbb{X}^{\mathbf{E}}) \to \mathbb{X}^{\mathbf{S}}$.

Theorem 1 (Proxy). *Let gMDPs $\mathbf{A}$, $\mathbf{A}_r$, and $\mathbf{E}$ be given. Let $\mathbf{S}(\mathfrak{N})$ be an assumption of $\mathbf{E}$, i.e., $\mathbf{E} \sqsubseteq \mathbf{S}(\mathfrak{N})$ via P. Then, for $\varepsilon \geq 0$ and $\delta \in [0,1]$:*

$$\forall \mathfrak{n} \in \mathfrak{N} : (\mathbf{A}_r \times \mathbf{S})(\mathfrak{n})) \preceq_\delta^\varepsilon (\mathbf{A} \times_a \mathbf{S})(\mathfrak{n}) \tag{9a}$$
$$\implies (\mathbf{A}_r \times \mathbf{S})(\mathfrak{N}) \preceq_\delta^\varepsilon \mathbf{A} \times \mathbf{E}, \tag{9b}$$

with relations $\mathcal{R}_a$ and $\mathcal{R}_b$, respectively.

Theorem 1 allows us to reason about $\mathbf{A} \times \mathbf{E}$ via $\mathbf{A}_r \times \mathbf{S}$ as a proxy without requiring us to compute the error bounds ε, δ considering the potentially high-dimensional domain of $\mathbf{A} \times \mathbf{E}$. Instead, we compute the error bounds for the relation between $\mathbf{A} \times_a \mathbf{S}$ and $\mathbf{A}_r \times \mathbf{S}$—both involving the simpler simulator $\mathbf{S}$.

Remark 2 (Reduced conservatism). Consider $\mathbf{A}_r \preceq_{\delta_1}^{\varepsilon_1} \mathbf{A}$ and $\mathbf{S} \preceq_{\delta_2}^{\varepsilon_2} \mathbf{E}$ for some $\varepsilon_1, \varepsilon_2 \geq 0$ and $\delta_1, \delta_2 \in [0,1]$, which also implies (9b) with $\varepsilon = \varepsilon_1 + \varepsilon_2$ and $\delta = 1 - (1 - \delta_1)(1 - \delta_2)$ [41, Theorem 1]. In comparison, our result is less restrictive, since they assume the worst-case behavior of the other system to establish local simulation relations.

Since for each $\mathfrak{n} \in \mathfrak{N}$ the simulator $\mathbf{S}(\mathfrak{n})$ in Theorem 1 is a (potentially deterministic) low-dimensional representation of the environment $\mathbf{E}$, establishing the antecedent (9a) is much easier than the consequent (9b). Theorem 1 then provides the guarantees that the closed-loop agent system will behave reliably in the actual environment without ever having to compute their interaction directly. In contrast to traditional compositional results relying on assumptions of dissipativity, low gain, or contractivity, we only require the assumption $\mathbf{E} \sqsubseteq \mathbf{S}(\mathfrak{N})$ to hold. Thus, our AGC can allow for computationally tractable control synthesis.

Remark 3 (Networks of subsystems). Most work on compositionality is focused on networks of $N \gg 2$ subsystems, where the goal is to compute local controllers for each subsystem [19,26,29,32,35,41]. We present our AGC result for the two subsystems $\mathbf{A}$ and $\mathbf{E}$ whilst remarking that it can be easily applied to $\mathbf{A}$ in a network of autonomous subsystems, by defining the rest of the network as its environment. This is the case, for example, when designing a controller for an agent interacting with other autonomous entities. For the more general case of multi-agent control, prior work computes local controllers on the decoupled subsystems of dimensionality d assuming weak interconnection, resulting in a complexity of $\mathcal{O}(Nn^{2d})$, where d is the maximum dimension of the subsystems. We can apply the same setup as before whilst adjusting local controllers iteratively, which scales with $\mathcal{O}(Nn^{2(d+d')})$, where d' is the maximum dimension of the reduced-order model of the environment for each active subsystem.

6 Bridging the Gap

In this section, we provide an abstraction-based solution to Problem 1. For this, we leverage the AGC result in Theorem 1 to establish a formal link between the original system $(\mathbf{A} \times \mathbf{E})(\theta)$ and a *finite* nominal model $(\mathbf{A}_r \times \mathbf{S})_f(\hat{\theta}, \hat{\mathfrak{n}})$. We synthesize a controller via $(\mathbf{A}_r \times \mathbf{S})_f(\hat{\theta}, \hat{\mathfrak{n}})$ that can be readily applied to $\mathbf{A} \times \mathbf{E}$, by robustifying it against parametric uncertainty, adversarial disturbance, and the discretization error.

6.1 Situational Awareness

Instead of performing computations on the actual system $(\mathbf{A} \times \mathbf{E})(\theta)$, we replace it with the less complex simulated system $(\mathbf{A}_r \times \mathbf{S})(\theta, \mathfrak{N})$. We assume that the simulator $\mathbf{S}$ is constructed such that $\mathbf{E} \sqsubseteq \mathbf{S}(\mathfrak{N})$. Theorem 1 establishes a formal link between the two systems of the form $(\mathbf{A}_r \times \mathbf{S})(\theta, \mathfrak{N}) \preceq^\varepsilon_\delta (\mathbf{A} \times \mathbf{E})(\theta)$ if the antecedent $(\mathbf{A}_r \times \mathbf{S})(\theta, \mathfrak{n}) \preceq^\varepsilon_\delta (\mathbf{A} \times_a \mathbf{S})(\theta, \mathfrak{n})$ holds for all $(\theta, \mathfrak{n}) \in \Theta \times \mathfrak{N}$. To establish the antecedent, we can use a *coupling compensator* for model-order reduction (MOR). We propose the following partial compensator.

Proposition 1 (Partial MOR coupling compensator). *Consider two gMDPs* $\mathbf{M}(\theta, \mathfrak{n}) = (\mathbb{X}, \mathbb{X}_0, \mathbb{U}, \mathbb{D}, \mathbf{t}, h, \mathbb{Y})$ *and* $\mathbf{M}_r(\theta, \mathfrak{n}_r) = (\mathbb{X}_r, \mathbb{X}_{r,0}, \mathbb{U}_r, \mathbb{D}_r, \mathbf{t}_r, h_r, \mathbb{Y})$ *with a mapping* $F : \mathbb{X} \to \mathbb{X}_r$ *such that* $x_r = F(x)$, *an interface function* $u_t \sim \mathbf{i}_u(\,\cdot\,|x_t, x_{r,t}, u_{r,t})$, *and a relation* $\mathcal{R}_r := \{(x_r, x) \in \mathbb{X}_r \times \mathbb{X} \mid \|F(x) - x_r\|_{D_r} \leq \varepsilon_r\}$, *where* $\|z\|_{D_r} := \sqrt{z^\top D_r z}$ *denotes the two norm with weight matrix* D_r. *If there exist* $\varepsilon_r \geq 0$ *and* $\delta_r : \mathbb{X} \times \mathbb{U}_r \to [0,1]$ *such that*

- $\varepsilon_r \geq \sup_{(x_r, x) \in \mathcal{R}_r} \|h(x) - h_r(x_r)\|$; *and*
- *there exists a sub-prob. kernel* $\boldsymbol{v} : \mathbb{X}_r \times \mathbb{X} \times \mathbb{U}_r \times \mathcal{B}(\mathbb{X}_r \times \mathbb{X}) \to [0,1]$, *such that* $\forall (x_r, x) \in \mathcal{R}_r$ *and* $\forall u_r \in \mathbb{U}_r$ *the resulting measure* $\boldsymbol{v}(dx_{r+} \times dx_+ | x_r, x, u_r)$ *is a sub-prob. coupling of* $\mathbf{t}_r(dx_{r+}|x_r, u_r, \mathfrak{n}_r(d_r|\,\cdot\,); \theta)$ *and* $\mathbf{t}(dx_+|x, \mathbf{i}_u(u|\,\cdot\,), \mathfrak{n}(d|\,\cdot\,); \theta)$ *over* $\mathcal{R}_r$ *with* $\boldsymbol{v}(\mathcal{R}_r) \geq 1 - \delta_r(x, u_r)$;

then, $\mathbf{M}_r(\theta, \mathfrak{n}_r) \preceq^{\varepsilon_r}_{\delta_r} \mathbf{M}(\theta, \mathfrak{n})$.

The above result provides conditions for the computation of the error parameters ε_r and δ_r of the simulation relation between $\mathbf{M}$ and $\mathbf{M}_r$. We give an examination of its distinction to prior works in [40, Appendix B]. Note that the resulting δ_r is dependent on the original state x. This dependence can be resolved by taking the worst case w.r.t. $x \in \mathbb{X}$ (if feasible) or—given the relevant variables from x are observed upon runtime—adding (a subset of) x as an input to $\mathbf{M}_r$.

Running Example. Based on Proposition 1, we establish $(\mathbf{A}_r \times \mathbf{S})(\theta, \mathfrak{n}) \preceq_{\delta_1}^{\varepsilon_1} (\mathbf{A} \times_a \mathbf{S})(\theta, \mathfrak{n})$, $(\theta, \mathfrak{n}) \in \Theta \times \mathfrak{N}$, for parameters $a_1 = 0.2$, $b = 0.001$, and $a_2 = a_3 = a_4 = a_5 = 0$, with $\varepsilon_1 = 0$ and $\delta_1 = 0\%$. See [40, Appendix C.3] for more details. We summon Theorem 1 to infer the relation $(\mathbf{A}_r \times \mathbf{S})(\theta, \mathfrak{N}) \preceq_{\delta_1}^{\varepsilon_1} (\mathbf{A} \times \mathbf{E})(\theta)$.

6.2 Ambiguity Compensation

The system $(\mathbf{A}_r \times \mathbf{S})(\theta, \mathfrak{n})$ incorporates two types of ambiguity, arising from the parametric uncertainty $\theta \in \Theta$ and the adversary $\mathfrak{n} \in \mathfrak{N}$. In fact, our aim is to relate a nominal system $(\mathbf{A}_r \times \mathbf{S})(\hat{\theta}, \hat{\mathfrak{n}}) := (\hat{\mathbb{X}}, \hat{\mathbb{X}}_0, \hat{U}, \hat{\mathbb{D}}, \hat{\mathfrak{n}}, \hat{\mathfrak{t}}, \hat{\mathbb{Y}}, \hat{h})$ with a fixed parameter estimate $\hat{\theta}$ and a (potentially simplified) nominal adversary $\hat{\mathfrak{n}} : \hat{\mathbb{X}} \times \mathcal{B}(\hat{\mathbb{D}}) \to [0, 1]$ to the ambiguity set of systems $(\mathbf{A}_r \times \mathbf{S})(\Theta, \mathfrak{N})$. To this end, we start by rewriting the dynamics of $(\mathbf{A}_r \times \mathbf{S})(\theta, \mathfrak{n})$ as follows:

$$x_{t+1} = \underbrace{\hat{f}(\hat{x}_t, \hat{u}_t; \hat{\theta}) + \hat{\mathfrak{n}}(\cdot | \hat{x}_t)}_{\text{nominal dynamics}} + \underbrace{\Delta(\cdot) + w_t}_{\text{disturbance}}, \qquad w_t \sim p_w(\cdot),$$

$$= \hat{f}(\hat{x}_t, \hat{u}_t; \hat{\theta}) + \hat{\mathfrak{n}}(\cdot | \hat{x}_t) + \hat{w}_t, \qquad \hat{w}_t := w_t + \Delta(\cdot),$$

where the noise distribution p_w is offset by the error dynamics

$$\Delta(\cdot) := (f(x, u; \theta) + \mathfrak{n}(\cdot | x)) - (\hat{f}(\hat{x}, \hat{u}; \hat{\theta}) + \hat{\mathfrak{n}}(\cdot | \hat{x})). \tag{10}$$

Based on the work by [42], we establish the following extended result to compensate both the parametric and adversarial ambiguity.

Theorem 2 (Ambiguity compensation). *For the models $(\mathbf{A}_r \times \mathbf{S})(\theta, \mathfrak{n})$ and $(\mathbf{A}_r \times \mathbf{S})(\hat{\theta}, \hat{\mathfrak{n}})$ with $(\theta, \mathfrak{n}) \in \Theta \times \mathfrak{N}$, we have that $(\mathbf{A}_r \times \mathbf{S})(\hat{\theta}, \hat{\mathfrak{n}}) \preceq_\delta^\varepsilon (\mathbf{A}_r \times \mathbf{S})(\theta, \mathfrak{n})$ with interface function $u_t = \hat{u}_t$, relation $\mathcal{R} := \{(\hat{x}, x) \in \hat{\mathbb{X}} \times \mathbb{X} \mid \hat{x} = x\}$, state mapping $\hat{x}_+ = x_+$, parameter $\varepsilon = 0$, and with the offset Δ defined in (10)*

$$\delta(\hat{x}, \hat{u}; \hat{\theta}, \hat{\mathfrak{n}}) = 1 - 2\,\mathrm{cdf}\left(-\frac{1}{2} \sup_{\theta \in \Theta, \mathfrak{n} \in \mathfrak{N}} \left\| \Delta(\hat{x}, \hat{u}; \hat{\theta}, \theta, \hat{\mathfrak{n}}, \mathfrak{n}) \right\| \right). \tag{11}$$

Remark 4 (Disturbance refinement). The nominal disturbance model $\hat{\mathfrak{n}}$ can reduce the conservativeness of the approach by acting as a virtual disturbance feedback. To make the approach even less conservative, the actual disturbance based on $x^{\mathbf{E}}$ can be inferred and can be fed back into the nominal model via an additional input, as stated below Proposition 1.

Running Example. We select a nominal model $(\mathbf{A}_r \times \mathbf{S})(\hat{\theta}, \hat{\mathfrak{n}})$ by fixing $\hat{\theta} := (\hat{\theta}^{\mathbf{A}}, \hat{\theta}^{\mathbf{E}}) = (0.8, 0.8)$ and a linearized nominal adversary $\hat{\mathfrak{n}} : (x_t^{\mathbf{S}}, o_t^{\mathbf{S}}) \mapsto [0, \tau v_t^{\mathbf{S}}]^\top$. Using Theorem 2, we obtain $(\mathbf{A}_r \times \mathbf{S})(\hat{\theta}, \hat{\mathfrak{n}}) \preceq_{\delta_2}^{\varepsilon_2} (\mathbf{A}_r \times \mathbf{S})(\theta, \mathfrak{n})$, with $\varepsilon_2 = 0$ and δ_2 as a function of the initial state of $(\mathbf{A}_r \times \mathbf{S})(\hat{\theta}, \hat{\mathfrak{n}})$ with a maximum of 3.64%. See [40, Appendix C.4] for further details.

6.3 Embedding with Situational Awareness

Given an SA block discriminating between situations $\{\mathbf{E}_i\}_{i \in \mathcal{I}}$ probabilistically, we may want to combine guarantees from multiple simulators $\{\mathbf{S}_i\}_{i \in \mathcal{I}}$ (see Fig. 1).

Theorem 3 (Situational awareness). *Let $P_1, P_2, \ldots$ be the probabilities of the agent $\mathbf{A}$ operating in the situations $\mathbf{E}_1, \mathbf{E}_2, \ldots$, and let $\{\mathbf{S}_i(\mathfrak{N}_i)\}_{i \in \mathcal{I}}$ be assumptions of $\{\mathbf{E}_i\}_{i \in \mathcal{I}}$. Let p_i and $\mathbf{C}_i$, respectively, be the robust satisfaction probability and robust controller of $\mathbf{A} \times \mathbf{E}_i$ obtained via Theorem 1. Then, we have for all $i \in \mathcal{I}$ that $\mathbb{P}((\mathbf{C}_i \times \mathbf{A}) \times \mathbf{E}_i \models \psi) \geq p^*$, where $p^* := \sum_{i \in \mathcal{I}} P_i p_i$.*

Fig. 3. Intersection alongside the regions associated with collision P_C and target P_T.

7 Case Study

Consider the scenario shown in Fig. 3, involving two vehicles—$\mathbf{A}$ and $\mathbf{E}$—at a traffic intersection. The corresponding dynamics are given in (3)-(4). Our goal is to design a controller for the agent vehicle $\mathbf{A}$, which wants to pass the intersection to continue its journey. It has to yield to the environment vehicle $\mathbf{E}$, which is traveling on the priority road. Based on the behavior of $\mathbf{E}$, the agent may have time to pass the intersection *before* $\mathbf{E}$, or it may go *after* $\mathbf{E}$ to avoid a collision. In scLTL, this specification is written as $\psi := (P_S \wedge \neg P_C) \cup P_T$, where P_S is a safety condition on the bounded domain $\mathbb{X}^{\mathbf{A}} \times \mathbb{X}^{\mathbf{E}}$. Note, that ψ can *not* be decomposed into local sub-specifications for $\mathbf{A}$ and $\mathbf{E}$, rendering most traditional compositional approaches inapplicable.

Abstraction-Based Control Design. Throughout the running example, we established a quantified relation between the nominal model $(\mathbf{A}_r \times \mathbf{S})(\hat{\theta}, \hat{\mathbf{n}})$ and the concrete system $(\mathbf{A} \times \mathbf{E})(\theta)$. The following steps are performed using the toolbox SySCoRe [47]. We discretize the spaces $\hat{\mathbb{X}}$ and $\hat{\mathbb{U}}$ of $(\mathbf{A}_r \times \mathbf{S})(\hat{\theta}, \hat{\mathbf{n}})$ (into 327×10^6 and 5 partitions, respectively) to obtain a finite gMDP $(\mathbf{A}_r \times \mathbf{S})_f$ and establish a simulation relation $(\mathbf{A}_r \times \mathbf{S})_f \preceq_{\delta_3}^{\varepsilon_3} (\mathbf{A}_r \times \mathbf{S})(\hat{\theta}, \hat{\mathbf{n}})$ with $\varepsilon_3 = 0.2$ and $\delta_3 = 14.96\%$. Note that the magnitude of the error bounds $(\varepsilon_3, \delta_3)$ is linked directly to the precision of the chosen discretization, which is an inherent aspect of abstraction-based methods. This underscores the importance of the new proxy framework, as it enables the computation of SSRs for high-dimensional systems. The values of $(\varepsilon_3, \delta_3)$ can be reduced by refining the abstraction or adopting an adaptive gridding strategy. As the development of such techniques is orthogonal to our work, we use an equidistant gridding to simplify the presentation. We perform controller synthesis on $(\mathbf{A}_r \times \mathbf{S})_f$ with $\varepsilon := \sum_{i=1}^{3} \varepsilon_i = 0.2$ and $\delta := \sum_{i=1}^{3} \delta_i$ (max. 18.60%) using the transitivity property of (ε, δ)-stochastic simulation relations [42, Theorem 3] to obtain a robust controller $\mathbf{C}$ w.r.t. the specification ψ. The corresponding robust satisfaction probability provided by SySCoRe is shown in Fig. 4 as a function of the initial state of $\mathbf{A} \times \mathbf{E}$.

Performance. To compare the performance of $\mathbf{C}$, we simulate the system ($\mathbf{C} \times \mathbf{A}) \times \mathbf{E}$ from two different initial states. Note that the specification includes an implicit time horizon, as the safety requirement P_S mandates the systems to remain within the bounded state domain till P_T is reached. Thus, a policy enforcing $\mathbf{A}$ to wait for $\mathbf{E}$ to pass may not always be optimal. We select the initial state $x_0 = (2.960, 1.387, 2.296, 0.014)$ and run the system 1000 times. During runtime, it is verified whether the state of the environment $\mathbf{E}$ remains within the RoV $\mathbb{X}^\mathbf{E}$, rendering the simulator $\mathbf{S}$ a valid representation according to $\mathbf{E} \sqsubseteq \mathbf{S}(\mathfrak{N})$. Whence $x_t^\mathbf{E} \notin \mathbb{X}^\mathbf{E}$ for some $t \in \mathbb{N}$, the AGC is violated. We then stop the corresponding run and discard it. As shown in Fig. 4(a), under the control policy obtained for the selected x_0, $\mathbf{A}$ crosses the intersection *before* $\mathbf{E}$. The associated robust satisfaction probability obtained via SySCoRe is $p = 58\%$, whereas the average experimental performance indicates an actual satisfaction probability closer to $p_{\text{real}} \approx 67\%$. We repeat the experiment for an initial state $x_0 = (1.122, 0.794, 3.520, 1.065)$, yielding a policy of waiting (see Fig. 4(b)). Due to the long time horizon, the robust satisfaction probability is low ($p = 19\%$),

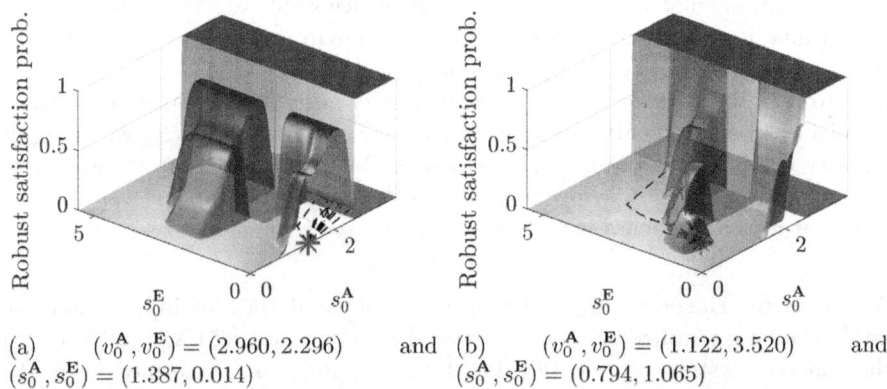

(a) $(v_0^\mathbf{A}, v_0^\mathbf{E}) = (2.960, 2.296)$ and (b) $(v_0^\mathbf{A}, v_0^\mathbf{E}) = (1.122, 3.520)$ and
$(s_0^\mathbf{A}, s_0^\mathbf{E}) = (1.387, 0.014)$ $(s_0^\mathbf{A}, s_0^\mathbf{E}) = (0.794, 1.065)$

Fig. 4. Robust satisfaction probability as a function of the initial position of $\mathbf{A} \times \mathbf{E}$, for initial velocities $(v_0^\mathbf{A}, v_0^\mathbf{E})$. Additionally, 10 random trajectories (in black) starting from $(s_0^\mathbf{A}, s_0^\mathbf{E})$ are shown, together with the collision P_C (in magenta) and target P_T (in blue) regions. (Color figure online)

whereas the average experimental performance indicates an actual satisfaction probability closer to $p_{\text{real}} \approx 41\%$. This illustrates the conservatism introduced in bridging the gap between the surrogate and the concrete system (Fig. 1).

8 Conclusion

We presented an assume-guarantee approach by establishing an adversarial probabilistic simulation relation between the surrogate and concrete closed-loop systems without directly computing concrete-system error parameters. By obviating domain discretization altogether, our approach enables scalable robust control synthesis for agents operating under environmental uncertainty. By extension, this work takes a decisive step toward scalable solutions for partially observed systems. We demonstrated its efficacy on a high-dimensional nonlinear case study featuring tight agent-environment interactions.

Acknowledgments. This work is supported by the following grants: EIC 101070802 and ERC 101089047.

References

1. Alur, R., Grosu, R., Lee, I., Sokolsky, O.: Compositional refinement for hierarchical hybrid systems. In: Hybrid Systems: Computation and Control, pp. 33–48. Springer, Heidelberg (2001)
2. Annaswamy, A.M., Johansson, K.H., Pappas, G.: Control for societal-scale challenges: Road map 2030. IEEE Control Syst. Mag. **44**(3), 30–32 (2024)
3. Arcari, E., Iannelli, A., Carron, A., Zeilinger, M.N.: Stochastic MPC with robustness to bounded parameteric uncertainty. IEEE Trans. Autom. Control **68**(12), 7601–7615 (2023)
4. Astorga, A., Hsieh, C., Madhusudan, P., Mitra, S.: Perception contracts for safety of ML-enabled systems. Proc. ACM Program. Lang. **7**(OOPSLA2) (2023)
5. Badings, T., Romao, L., Abate, A., Jansen, N.: Probabilities are not enough: formal controller synthesis for stochastic dynamical models with epistemic uncertainty. In: Proceedings of the AAAI Conference on Artificial Intelligence, vol. 37, no. 12, pp. 14701–14710 (2023)
6. Baier, C., Katoen, J.P.: Principles of Model Checking. MIT Press, Cambridge (2008)
7. Banse, A., Romao, L., Abate, A., Jungers, R.M.: Data-driven abstractions via adaptive refinements and a Kantorovich metric. In: 2023 62nd IEEE Conference on Decision and Control (CDC), pp. 6038–6043 (2023)
8. Belta, C.: Formal synthesis of control strategies for dynamical systems. In: 2016 IEEE 55th Conference on Decision and Control (CDC), pp. 3407–3431 (2016)
9. Bengio, Y., et al.: International scientific report on the safety of advanced AI. Technical report, Department for Science, Innovation and Technology (2024)
10. Bogachev, V.I.: Measure Theory. Springer, Heidelberg (2007)

11. Budde, C.E., D'Argenio, P.R., Sánchez Terraf, P., Wolovick, N.: A theory for the semantics of stochastic and non-deterministic continuous systems. In: Remke, A., Stoelinga, M. (eds.) Stochastic Model Checking. Rigorous Dependability Analysis Using Model Checking Techniques for Stochastic Systems. LNCS, vol. 8453, pp. 67–86. Springer, Heidelberg (2014). https://doi.org/10.1007/978-3-662-45489-3_3
12. Casablanca, E., Zhang, Z., Marchesini, G., Haesaert, S., Dimarogonas, D.V., Soudjani, S.: SymAware: a software development framework for trustworthy multi-agent systems with situational awareness. arXiv preprint arXiv:2409.14833 (2024)
13. Chou, Y., Chen, X., Sankaranarayanan, S.: A study of model-order reduction techniques for verification. In: Abate, A., Boldo, S. (eds.) NSV 2017. LNCS, vol. 10381, pp. 98–113. Springer, Cham (2017). https://doi.org/10.1007/978-3-319-63501-9_8
14. Coppola, R., Peruffo, A., Romao, L., Abate, A., Mazo, M.: Enhancing data-driven stochastic control via bundled interval MDP. IEEE Control Syst. Lett. (2024)
15. Dalrymple, D., et al.: Towards guaranteed safe AI: a framework for ensuring robust and reliable AI systems. arXiv preprint arXiv:2405.06624 (2024)
16. Desharnais, J., Edalat, A., Panangaden, P.: Bisimulation for labelled Markov processes. Inf. Comput. **179**(2), 163–193 (2002)
17. Desharnais, J., Gupta, V., Jagadeesan, R., Panangaden, P.: Approximating labelled Markov processes. Inf. Comput. **184**(1), 160–200 (2003)
18. Edalat, A.: Semi-pullbacks and bisimulation in categories of Markov processes. Math. Struct. Comput. Sci. **9**(5), 523–543 (1999)
19. Ghasemi, K., Sadraddini, S., Belta, C.: Compositional synthesis via a convex parameterization of assume-guarantee contracts. In: Proceedings of the 23rd International Conference on Hybrid Systems: Computation and Control, pp. 1–10 (2020)
20. Gracia, I., Boskos, D., Laurenti, L., Mazo Jr, M.: Distributionally robust strategy synthesis for switched stochastic systems. In: Proceedings of the 26th ACM International Conference on Hybrid Systems: Computation and Control, pp. 1–10 (2023)
21. Haesaert, S., Chen, F., Abate, A., Weiland, S.: Formal control synthesis via simulation relations and behavioral theory for discrete-time descriptor systems. IEEE Trans. Autom. Control **66**(3), 1024–1039 (2021)
22. Haesaert, S., Soudjani, S., Abate, A.: Verification of general Markov decision processes by approximate similarity relations and policy refinement. SIAM J. Control. Optim. **55**(4), 2333–2367 (2017)
23. Hahn, E.M., Hartmanns, A., Hermanns, H., Katoen, J.P.: A compositional modelling and analysis framework for stochastic hybrid systems. Formal Methods Syst. Des. **43**(2), 191–232 (2013)
24. Hahn, E.M., Hashemi, V., Hermanns, H., Lahijanian, M., Turrini, A.: Interval Markov decision processes with multiple objectives: from robust strategies to Pareto curves. ACM Trans. Model. Comput. Simul. (TOMACS) **29**(4), 1–31 (2019)
25. Jafarpour, S., Chen, Y.: Probabilistic reachability of stochastic control systems: a contraction-based approach. arXiv preprint arXiv:2407.12225 (2024)
26. Kazemi, M., Perez, M., Somenzi, F., Soudjani, S., Trivedi, A., Velasquez, A.: Assume-guarantee reinforcement learning. In: Proceedings of the AAAI Conference on Artificial Intelligence, vol. 38, pp. 21223–21231 (2024)
27. Kupferman, O., Vardi, M.Y.: Model checking of safety properties. Formal Methods Syst. Des. **19**(3), 291–314 (2001)
28. Kwiatkowska, M., Norman, G., Parker, D., Qu, H.: Assume-guarantee verification for probabilistic systems. In: Esparza, J., Majumdar, R. (eds.) TACAS 2010. LNCS, vol. 6015, pp. 23–37. Springer, Heidelberg (2010). https://doi.org/10.1007/978-3-642-12002-2_3

29. Lavaei, A., et al.: Compositional reinforcement learning for discrete-time stochastic control systems. IEEE Open J. Control Syst. **2**, 425–438 (2023)
30. Lavaei, A., Soudjani, S., Abate, A., Zamani, M.: Automated verification and synthesis of stochastic hybrid systems: a survey. Automatica **146**, 110617 (2022)
31. Luckcuck, M.: Using formal methods for autonomous systems: five recipes for formal verification. Proc. Inst. Mech. Engineers Part O: J. Risk Reliabil. **237**(2), 278–292 (2023)
32. Mallik, K., Schmuck, A.K., Soudjani, S., Majumdar, R.: Compositional synthesis of finite-state abstractions. IEEE Trans. Autom. Control **64**(6), 2629–2636 (2018)
33. Matni, N., Ames, A.D., Doyle, J.C.: A quantitative framework for layered multirate control: toward a theory of control architecture. IEEE Control Syst. Mag. **44**(3), 52–94 (2024)
34. Nayak, S.P., Egidio, L.N., Della Rossa, M., Schmuck, A.K., Jungers, R.M.: Context-triggered abstraction-based control design. IEEE Open J. Control Syst. **2**, 277–296 (2023)
35. Nuzzo, P., Li, J., Sangiovanni-Vincentelli, A.L., Xi, Y., Li, D.: Stochastic assume-guarantee contracts for cyber-physical system design. ACM Trans. Embed. Comput. Syst. (TECS) **18**(1), 1–26 (2019)
36. Păsăreanu, C.S., et al.: Closed-loop analysis of vision-based autonomous systems: a case study. In: International Conference on Computer Aided Verification, pp. 289–303. Springer, Heidelberg (2023). https://doi.org/10.1007/978-3-031-37706-8_15
37. Provan, G.: Formal methods for autonomous vehicles. IT Prof. **26**(1), 50–56 (2024)
38. Pulch, R.: 10 Model Order Reduction in Uncertainty Quantification, pp. 321–344. De Gruyter, Berlin (2021)
39. Puterman, M.L.: Markov Decision Processes: Discrete Stochastic Dynamic Programming. John Wiley & Sons, Hoboken (2014)
40. Schön, O., Haesaert, S., Soudjani, S.: Formal control for uncertain systems via contract-based probabilistic surrogates (extended version). arXiv preprint (2025)
41. Schön, O., van Huijgevoort, B., Haesaert, S., Soudjani, S.: Verifying the unknown: correct-by-design control synthesis for networks of stochastic uncertain systems. In: 2023 62nd IEEE Conference on Decision and Control (CDC), pp. 7035–7042. IEEE (2023)
42. Schön, O., van Huijgevoort, B., Haesaert, S., Soudjani, S.: Bayesian formal synthesis of unknown systems via robust simulation relations. IEEE Trans. Autom. Control (2024)
43. Segala, R., Lynch, N.: Probabilistic simulations for probabilistic processes. Nordic J. Comput. **2**(2), 250–273 (1995)
44. Shali, B.M., Heidema, H.M., van der Schaft, A.J., Besselink, B.: Series composition of simulation-based assume-guarantee contracts for linear dynamical systems. In: 2022 IEEE 61st Conference on Decision and Control (CDC), pp. 2204–2209 (2022)
45. Soudjani, S., Abate, A.: Adaptive and sequential gridding procedures for the abstraction and verification of stochastic processes. SIAM J. Appl. Dyn. Syst. **12**(2), 921–956 (2013)
46. Tabuada, P.: Verification and Control of Hybrid Systems: A Symbolic Approach. Springer, Boston (2009)
47. Van Huijgevoort, B., Schön, O., Soudjani, S., Haesaert, S.: SySCoRe: synthesis via stochastic coupling relations. In: Proceedings of the 26th ACM International Conference on Hybrid Systems: Computation and Control, pp. 1–11 (2023)
48. van Huijgevoort, B.C., Haesaert, S.: Similarity quantification for linear stochastic systems: a coupling compensator approach. Automatica **144**, 110476 (2022)

49. Wang, R., Sun, Z., Haesaert, S.: Unraveling tensor structures in correct-by-design controller synthesis. arXiv preprint arXiv:2503.24085 (2025)
50. Yin, X., Gao, B., Yu, X.: Formal synthesis of controllers for safety-critical autonomous systems: developments and challenges. Ann. Rev. Control. **57**, 100940 (2024)
51. Zhong, B., Lavaei, A., Zamani, M., Caccamo, M.: Automata-based controller synthesis for stochastic systems: a game framework via approximate probabilistic relations. Automatica **147**, 110696 (2023)

Statistical Model Checking Beyond Means: Quantiles, CVaR, and the DKW Inequality

Carlos E. Budde[1], Arnd Hartmanns[2(✉)], Tobias Meggendorfer[3], Maximilian Weininger[4], and Patrick Wienhöft[5,6]

[1] Technical University of Denmark, Lyngby, Denmark
[2] University of Twente, Enschede, The Netherlands
a.hartmanns@utwente.nl
[3] Lancaster University Leipzig, Leipzig, Germany
[4] Ruhr University Bochum, Bochum, Germany
[5] Dresden University of Technology, Dresden, Germany
[6] Centre for Tactile Internet with Human-in-the-Loop (CeTI), Dresden, Germany

Abstract. Statistical model checking (SMC) randomly samples probabilistic models to approximate quantities of interest with statistical error guarantees. It is traditionally used to estimate probabilities and expected rewards, i.e. means of different random variables on paths. In this paper, we develop methods using the Dvoretzky-Kiefer-Wolfowitz-Massart inequality (DKW) to extend SMC beyond means to compute quantities such as quantiles, conditional value-at-risk, and entropic risk. The DKW provides confidence bounds on the random variable's entire cumulative distribution function, a much more versatile guarantee compared to the statistical methods prevalent in SMC today. We have implemented support for computing new quantities via the DKW in the MODES simulator of the MODEST TOOLSET. We highlight the implementation and its versatility on benchmarks from the quantitative verification literature.

1 Introduction

Statistical model checking (SMC) [1,32,34,48] avoids the state space explosion problem of classic probabilistic model checking approaches (PMC) [4,5] that explore and numerically analyse a model's entire state space [28]: SMC instead *samples k* random paths from the model to *estimate* the value of the quantity of interest. As a simulation-based approach, it applies to any effectively executable model, including non-Markovian [20] and hybrid [23,40] ones. An SMC result

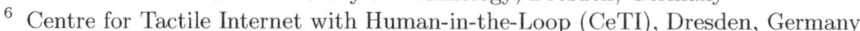

This work was supported by the DFG through the Cluster of Excellence EXC 2050/1 (CeTI, project ID 390696704, as part of Germany's Excellence Strategy) and the TRR 248 (see perspicuous-computing.science, project ID 389792660), by the ERC Starting Grant DEUCE (101077178), by the European Union's Horizon 2020 research and innovation programme under Marie Skłodowska-Curie grant agreement 101008233 (MISSION), by the Interreg North Sea project STORM_SAFE, and by NWO VIDI grant VI.Vidi.223.110 (TruSTy).

© The Author(s), under exclusive license to Springer Nature Switzerland AG 2026
P. Prabhakar and A. Vandin (Eds.): QEST+FORMATS 2025, LNCS 16143, pp. 83–94, 2026.
https://doi.org/10.1007/978-3-032-05792-1_5

comes with a *statistical* correctness guarantee, often expressed as a *confidence interval* $[l, u]$ that contains the true result $(1 - \delta) \cdot 100\%$ of the times [15].

The most fundamental quantities computed by PMC and SMC are reachability probabilities and expected rewards [15,28]. SMC estimates these quantities using statistical methods like the Clopper-Pearson confidence interval [17] for probabilities (i.e. binomial proportions) and Hoeffding's inequality [30] for means of bounded distributions, or compares them to each other [21] or to thresholds using Wald's sequential probability ratio test [46]. In the past decade, PMC has been extended to compute several other quantities of interest, such as quantiles/percentiles/value-at-risk [31, 42, 45], conditional value at risk [33], and entropic risk [6]. However, the application of SMC has so far been limited to probabilities and expected rewards, i.e. only the *means* of distributions associated to different random variables on sampled paths. To the best of our knowledge, no SMC approaches or tools support quantities other than means yet.

In this paper, we show how to extend SMC to estimate non-mean quantities using the Dvoretzky-Kiefer-Wolfowitz-Massart inequality (DKW) [22,38]. The DKW provides a sound *simultaneous confidence band* around the cumulative distribution function (cdf), i.e. upper and lower bound functions completely enveloping the (unknown) cdf $(1 - \delta) \cdot 100\%$ of the times (see Fig. 1). This is a stronger statement compared to the currently-used statistical methods for estimation mentioned above, as it applies to the entire cdf rather than a single point or pointwise. From the DKW, we can again derive a confidence interval for the mean [3,15], but equally (and simultaneously) obtain confidence intervals on other quantities as well. We show how to do so in particular for higher moments, quantiles, conditional value-at-risk, and entropic risk. We have implemented these DKW-based computations in the MODES statistical model checker [13], part of the MODEST TOOLSET [27]. MODES can now also export the empirical cdf and DKW confidence band for plotting and further analysis by the user. We highlight our implementation and its versatility using several models from the Quantitative Verification Benchmark Set (QVBS) [29] in Sect. 4.

2 Preliminaries

A *probability distribution* over a non-empty, countable set S is a function $\mu \colon S \to [0, 1]$ such that $\sum_{s \in S} \mu(s) = 1$. The set of all distributions over S is denoted by $\mathcal{D}(S)$. The *cumulative distribution function* (cdf) of a random variable X is given by $F_X(x) \stackrel{\text{def}}{=} \mathbb{P}(X \leq x)$. A random variable X *stochastically dominates* another random variable Y, written $Y \precsim_{SD} X$, if $F_Y(x) \geq F_X(x)$ for all x (i.e. for any x, obtaining a value less than or equal to x is more likely for Y than for X; intuitively, X yields larger values). If $Y \precsim_{SD} X$, then $\mathbb{E}(Y) \leq \mathbb{E}(X)$.

Definition 1. *A discrete-time Markov chain (DTMC) is a tuple $\langle S, R, T, s_I \rangle$ of a finite set of states S, a reward function $R \colon S \to \mathbb{R}_{\geq 0}$, an initial state $s_I \in S$, and a transition function $T \colon S \to \mathcal{D}(S)$ mapping each state to a probability distribution over successor states. A (finite) path π is (a prefix of) an infinite sequence $\pi = s_0 s_1 \ldots \in S^\omega$ such that $s_0 = s_I$ and $\forall i \colon T(s_i)(s_{i+1}) > 0$.*

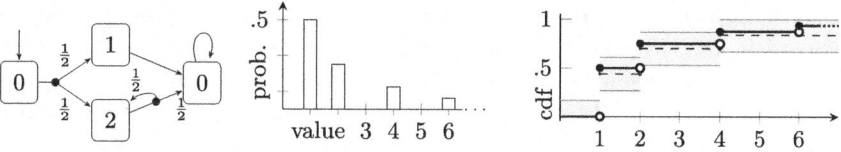

Fig. 1. Example of a DTMC (left) together with the probability distribution over possible reward outcomes (middle) and the corresponding cdf (right, solid line). The states in the DTMC are (only) labelled by their rewards. The right figure also includes an empirical cdf (dashed) and corresponding confidence band (gray) obtained from the DKW inequality (with $\delta = 0.1$ and $k = 50$).

See Fig. 1 (left) for an example of a DTMC. A DTMC induces a unique probability measure $\mathbb{P}$ over sets of paths that, intuitively, corresponds to multiplying the probabilities along the path (see e.g. [7, Chapter 10]).

Properties. Properties typically consist of two parts: First, a random variable X assigning a value to each path. For our results, the choice of X is largely irrelevant; we only require it to yield non-negative finite values. Concretely, we consider *total* and *reachability rewards*, i.e. $\mathsf{TR}(\pi) = \sum_{i=0}^{\infty} R(\pi_i)$ and the same sum cut off at the first goal state, respectively; see [15, Sec. 2] for details, and Fig. 1 (middle/right) for the distribution and cdf of TR on the example DTMC.

Second, a property comes with an *aggregation function* to "summarize" X into a single value, traditionally the expected value/mean $\mathbb{E}(X)$ (w.r.t. $\mathbb{P}$). Recently, alternative aggregations have gained popularity, for example

- higher *moments* (around 0), which are of the form $\mathbb{E}(X^n)$ for $n > 1$;
- the *t-quantile* (a.k.a. the *value-at-risk*) for $t \in (0,1)$, which is the smallest value v such that X is less than or equal to v with probability t [31,42]:

$$Q_t(X) \stackrel{\text{def}}{=} \inf \{\, v \mid \mathbb{P}(X \leq v) \geq t \,\};$$

- the *conditional value-at-risk* (a.k.a. expected shortfall, expected tail loss, average value-at-risk), which is the expectation over the t-quantile, i.e.

$$\mathrm{CVaR}_t(X) \stackrel{\text{def}}{=} \frac{1}{t}(P \cdot \mathbb{E}[X \mid X < v] + (t - P) \cdot v)$$

where $t \in (0,1)$, $v = Q_t(X)$ and $P = \mathbb{P}(X < v)$ [33,43]; and
- the *entropic risk*, which with $\gamma > 0$ is [6,24]

$$\mathrm{ERisk}_\gamma(X) \stackrel{\text{def}}{=} -\frac{1}{\gamma}\log(\mathbb{E}(e^{-\gamma X})).$$

We illustrate these for the DTMC of Fig. 1 in [16, App. A]. Additionally, as in [15], we distinguish whether X has a known upper bound (i.e. some U such that $\mathbb{P}(X \leq U) = 1$, the *bounded* case) or not (the *general* case).

Statistical model checking is, at its core, Monte Carlo simulation for formal models and properties: randomly generate a (predetermined) number k of paths, or *simulations*, from the model that give rise to samples $X_1, \ldots, X_k$ of the random variable X; and from that draw statistical conclusions on the property. While PMC approaches exist for all of the aforementioned properties, SMC so far exclusively focused on means as follows: compute the empirical mean

$$\hat{X} \stackrel{\text{def}}{=} \frac{1}{k} \sum_{i=1}^{k} X_i,$$

and perform a *statistical evaluation* to obtain a confidence interval $I = [l, u] \ni \hat{X}$ at a predetermined *confidence level* δ, so that with (a priori) probability $1 - \delta$ we have $\mathbb{E}(X) \in I$. That is, if we repeat the SMC procedure m times to obtain confidence intervals $I_1, \ldots, I_m$, we may find some of them (up to $\delta \cdot 100\%$ on average) incorrect, i.e. $\mathbb{E}(X) \notin I_i$ for some i. Occasionally obtaining an "incorrect" result is the nature of a statistical approach based on sampling. In this work, we develop statistical methods for other aggregations beyond the mean.

3 Statistical Guarantees Beyond Means

Before we introduce our approach, we formalise the exact kind of guarantees we aim to give. Observe that simply returning confidence intervals $[0, \infty]$ is always sound. However, we also want SMC procedures to yield "small" intervals. To formalize this requirement, we say a procedure yields *effective bounds* if (i) it produces correct intervals with high confidence, and (ii) for a large enough number k of samples, the intervals produced by the procedure are smaller than any ε and still correct with high confidence; see Definition 2 for the formal definition. We note that this is related to the notion of *consistent* estimators [2] from statistics, as the mid-point of effective intervals is a consistent estimator. However, we pose a stronger requirement since we require correct bounds to be produced.

Definition 2 (effective bounds). *Let X be a random variable and $\mathcal{F}$ an aggregator, mapping random variables to real numbers. An SMC procedure $\mathcal{A}$ yields effective bounds on $\mathcal{F}(X)$ if, for any confidence $\delta > 0$, the following two conditions hold: (i) For a collection of independent samples Ξ drawn from X, we have $\mathbb{P}(\mathcal{F}(X) \in \mathcal{A}(\Xi, \delta)) \geq 1 - \delta$. (ii) For any precision $\varepsilon > 0$, there exists a threshold k_0 such that for a collection of independent samples Ξ drawn from X with $|\Xi| \geq k_0$, we have $\mathbb{P}(\mathcal{F}(X) \in \mathcal{A}(\Xi, \delta) \wedge |\mathcal{A}(\Xi, \delta)| \leq \varepsilon) \geq 1 - \delta$.*

Remark 1. Some works consider the dual problem of gathering enough samples until a given precision is reached. They seek so-called *probably approximately correct* (PAC) guarantees: Given confidence level δ and precision ε, gather enough samples to return I with $|I| \leq 2\varepsilon$. We focus on deriving intervals given a fixed k, and in [16, App. B] describe how our methods extend to the dual problem.

As already observed in [15], obtaining two-sided bounds sometimes is infeasible (depending on the nature of the DTMC, random variable X, and aggregator $\mathcal{F}$). However, we may still be able to derive statistically sound, "converging" lower bounds. Thus, we extend the definition of "limit-PAC" from [15, Def. 3] and say an SMC procedure yields *effective lower bounds* if the value it produces is, with high confidence, (i) always a lower bound and (ii) close to the true value if given enough samples. Formally:

Definition 3 (effective lower bounds). *Let X be a random variable and $\mathcal{F}$ an aggregator, mapping random variables to real numbers. An SMC procedure $\mathcal{A}$ yields* effective lower bounds *on $\mathcal{F}(X)$ if, for any confidence $\delta > 0$, the following two conditions hold: (i) For a collection of independent samples Ξ drawn from X, we have $\mathbb{P}(\mathcal{A}(\Xi, \delta) \leq \mathcal{F}(X)) \geq 1 - \delta$. (ii) For any precision $\varepsilon > 0$, there exists a threshold k_0 such that for a collection of independent samples Ξ drawn from X with $|\Xi| \geq k_0$, we have $\mathbb{P}(\mathcal{F}(X) - \varepsilon \leq \mathcal{A}(\Xi, \delta) \leq \mathcal{F}(X)) \geq 1 - \delta$.*

3.1 DKW: The Dvoretzky-Kiefer-Wolfowitz-Massart Inequality

Our key to obtain effective bounds is the *Dvoretzky-Kiefer-Wolfowitz-Massart inequality* (DKW), which relates the cdf of the unknown distribution of X to the *empirical cdf* $\hat{F}(x) = \frac{1}{k} |\{ X_i \mid X_i \leq x \}|$ by

$$\mathbb{P}\left(\sup_{x \in \mathbb{R}} |\hat{F}(x) - F_X(x)| > \Delta\right) \leq \delta \quad \text{where } \Delta = \sqrt{\log(\delta/2)/(-2k)}.$$

Note that F_X is fixed but unknown, while $\hat{F}$ depends on the samples drawn from X. Intuitively, the DKW gives a *confidence band* in which the true cdf lies with high probability; see Fig. 1 (right) for an illustration. There, $\hat{F}$ is drawn dashed, and the gray area around $\hat{F}$ shows the confidence band (with width 2Δ). We refer to the bounds of this band as $\underline{F}(x) \stackrel{\text{def}}{=} \min\{\hat{F}(x) + \Delta, 1\}$ and $\overline{F}(x) \stackrel{\text{def}}{=} \max\{0, \hat{F}(x) - \Delta\}$, respectively. We denote the random variables that $\underline{F}$, $\hat{F}$, and $\overline{F}$ correspond to as $\underline{X}$, $\hat{X}$, and $\overline{X}$, respectively. Clearly, $\underline{X} \precsim_{SD} \hat{X} \precsim_{SD} \overline{X}$ and the DKW implies that $\underline{X} \precsim_{SD} X \precsim_{SD} \overline{X}$ with high confidence. In general, $\underline{X}$ and $\overline{X}$ yield 0 and ∞ with probability Δ, respectively. In the bounded case, we have $F_X(U) = 1$ and hence $\overline{X}$ would instead yield U with probability Δ. In [15], this is used to derive (lower and upper, in the bounded case, and lower, in the general case) bounds on expected rewards.

3.2 Obtaining Effective Bounds

As it turns out, computing the aggregations for $\underline{X}$ (and $\overline{X}$) gives effective (lower) bounds for all considered properties. We implicitly assume that the DKW condition holds and prove (below) that we then obtain correct (and converging) estimates. This means in general we get such estimates with high confidence. Moreover, as all results only depend on the DKW condition holding, we can give guarantees on *all* aggregations simultaneously, without splitting the confidence budget, which is particularly useful for e.g. multi-objective queries [31,33,42].

Moments. For higher-order moments, note that $Y \stackrel{\text{def}}{=} X^n$ is non-negative and has finite expectation if X satisfies these assumptions. Thus, the results of [15, Thm. 1] are directly applicable, which state that then the DKW yields effective lower bounds in the general case. In the bounded case, we naturally obtain effective bounds by direct application of the DKW (see Sect. 3.1).

Quantiles. By their definition, quantiles are monotone w.r.t. stochastic dominance, i.e. if $Y \precsim_{SD} X$, then $Q_t(Y) \leq Q_t(X)$. Thus, we also have $Q_t(\underline{X}) \leq Q_t(X) \leq Q_t(\overline{X})$, ensuring correctness of the computed values. While we can always obtain lower and upper bounds, even in the general case (by choosing k so that $t \in (\Delta, 1 - \Delta])$, only the lower bounds may be effective: Consider an X with distribution $\{1 \mapsto \frac{1}{2}, 2 \mapsto \frac{1}{2}\}$. We have $Q_{0.5}(X) = 1$, but any sound statistical upper bound on the cdf of X will have $\overline{F}(1) < F_X(1) = 0.5$, and thus always yield a 0.5-quantile of 2. This is a fundamental property of quantiles: they are not continuous w.r.t. small changes in the distribution. This already happens for the simple example in Fig. 1, as we illustrate in [16, App. A]. Thus, in general we cannot provide effective bounds. X is always discrete for DTMC as per Definition 1 and $X = \mathsf{TR}$; we can have non-discrete X if we allow e.g. continuously-distributed random rewards, use other models like continuous-time Markov chains (CTMCs), or other properties. Then, if X is continuous or, at least, if F_X is continuous at $Q_t(X)$, we get effective bounds.

Conditional Value-at-Risk. CVaR is a *distortion risk measure* (as is Q_t), which are monotone w.r.t. stochastic dominance [47]. Thus we again immediately get $\mathrm{CVaR}_t(\underline{X}) \leq \mathrm{CVaR}_t(X) \leq \mathrm{CVaR}_t(\overline{X})$. In contrast to general expectations, the bounded and general case do not differ: By assumption, we have $X < \infty$, hence there exists T such that $F_X(T) > 1 - \frac{t}{2}$. For a large enough k, we have $\Delta < \frac{t}{2}$, and $\overline{F}(T) \geq 1 - t$. Then, we know (with high confidence) that $X \leq T$ with probability t, i.e. $Q_t(\overline{X}) \leq T$ and therefore $\mathrm{CVaR}_t(\overline{X}) \leq T < \infty$. Thus, we can directly bound $|\mathrm{CVaR}_t(\underline{X}) - \mathrm{CVaR}_t(\overline{X})|$ by $T \cdot 2\Delta$, which goes to 0 for large enough k. Together, we obtain effective bounds in the general case.

Entropic Risk. First, observe that if $Y \precsim_{SD} X$, then $e^{-\gamma X} \precsim_{SD} e^{-\gamma Y}$ (the order reverses as $e^{-\gamma x}$ is decreasing). Consequently, $\mathbb{E}(e^{-\gamma X}) \leq \mathbb{E}(e^{-\gamma Y})$, and thus $\mathrm{ERisk}_\gamma(Y) \leq \mathrm{ERisk}_\gamma(X)$ (recall that $\mathrm{ERisk} = -1/\gamma \cdot \ldots$). Hence, we get $\mathrm{ERisk}_\gamma(\underline{X}) \leq \mathrm{ERisk}_\gamma(X) \leq \mathrm{ERisk}_\gamma(\overline{X})$. While there is no strict "cut-off" as for CVaR, we argue that we can still bound the overall difference between X and the bounds $\underline{X}$ and $\overline{X}$ in general. We have

$$\mathrm{ERisk}_\gamma(\overline{X}) - \mathrm{ERisk}_\gamma(X) = -1/\gamma \cdot \log(\mathbb{E}(e^{-\gamma \overline{X}})/\mathbb{E}(e^{-\gamma X})).$$

We now apply two useful general facts about cdfs, namely that (i) $\mathbb{E}(X) = \int_x (1 - F_X(x)) \cdot x \, dx$ and (ii) for a positive, continuous, strictly decreasing function f we have $F_{f(X)}(x) = 1 - F_X(f^{-1}(x))$. We get

$$\mathbb{E}(e^{-\gamma \overline{X}})/\mathbb{E}(e^{-\gamma X}) = \int \overline{F}(-\frac{1}{\gamma}\log(x))e^{-\gamma x}\,dx/\mathbb{E}(e^{-\gamma X}).$$

Table 1. Estimates and DKW confidence intervals for the examples' properties.

example	k	expected value $\hat{X}$	conf. int.	0.3-quantile est.	conf. int.	CVaR$_{0.3}$ est.	conf. int.
coupon	100	13.11	$[10.08, \infty)$	10	$[9, 11]$	8.90	$[4.56, 10.14]$
	1000	12.90	$[11.82, \infty)$	10	$[10, 11]$	8.71	$[7.28, 9.21]$
leader_sync	100	1.17	$[0.89, \infty)$	1	$[1, 1]$	1.00	$[0.55, 1.00]$
embedded	100	0.35	n/a	0.13	$[0.10, 0.20]$	0.35	n/a
	1000	0.33	n/a	0.17	$[0.15, 0.19]$	0.33	n/a

Recall that $F_X(x) - \Delta \leq \overline{F}(x)$. Hence,

$$\int \overline{F}(-\tfrac{1}{\gamma}\log(x))e^{-\gamma x}\, dx / \mathbb{E}(e^{-\gamma X}) \geq \int (F_X(-\tfrac{1}{\gamma}\log(x)) - \Delta)e^{-\gamma x}\, dx / \mathbb{E}(e^{-\gamma X})$$
$$= 1 - \Delta \int e^{-\gamma x}\, dx / \mathbb{E}(e^{-\gamma X}).$$

Consequently, for $\Delta \to 0$ this expression converges to 1, and thus $\mathrm{ERisk}_\gamma(\overline{X}) - \mathrm{ERisk}_\gamma(X) = -\tfrac{1}{\gamma}\log(\ldots) \to 0$. The proof for $\underline{X}$ is analogous.

4 Tool Implementation

The MODES SMC tool [13] was recently extended with sound statistical methods for estimating means, including the DKW [15]. Now, in version 3.1.287, we added syntax for quantile and CVaR properties to the parsers for its input languages, MODEST [9,26] and JANI [14], and extended its implementation of the DKW to estimate and provide bounds for such properties. Additionally, empirical cdfs can be exported to CSV and Excel files for plotting and further analysis.

To demonstrate the new tool features, we use three examples from the QVBS selected for diversity in cdfs: (1) the *coupon* model with parameters N = 15, DRAWS = 4, B = 5 (a DTMC of 17 billion states, to which SMC is agnostic) and the random variable underlying property *exp_draws*; (2) *leader_sync* with N = 5, K = 4 (DTMC, 4244 states) and *time*; and (3) *embedded* with MAX_COUNT = 8, T = 12 (a continuous-time Markov chain of 8548 states) and *danger_time*. The original properties query for expected reachability rewards; we add properties querying for the 0.3-quantile and CVaR$_{0.3}$ of the same reward specification, i.e. the same random variable on paths. We run MODES on each example with $k = 100$ simulations, and on *coupon* and *embedded* additionally with $k = 1000$. In addition to obtaining DKW-based confidence intervals, we use MODES' new --cdf parameter to export empirical CDFs with DKW confidence bands.

In Table 1, we show the results that MODES obtains for the properties. As reachability rewards fall into the general case, we can only obtain lower bounds for the expected values [15]. For quantiles and CVaR, as per Sect. 3.2, the DKW allows us to obtain (for CVaR effective) lower and upper bounds. On *embedded*, MODES cannot apply the DKW to expectation and CVaR because its syntactic

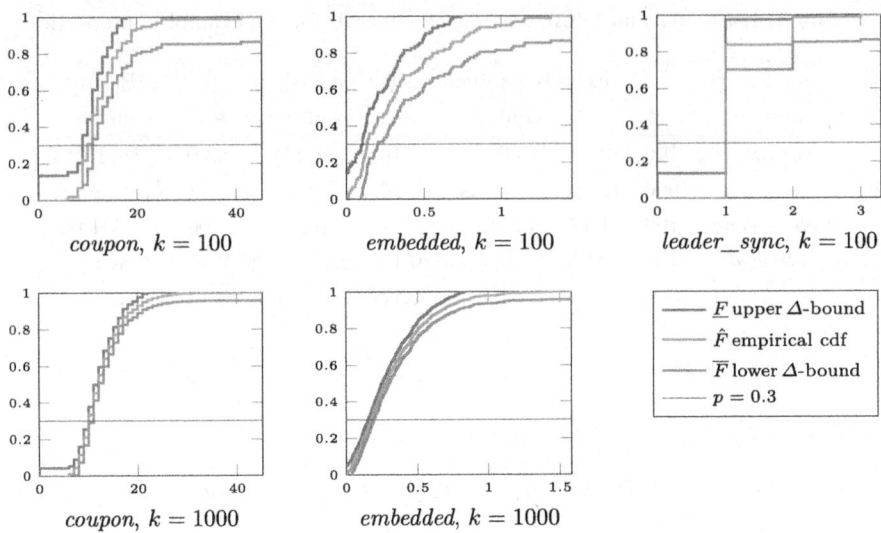

Fig. 2. DKW CDF confidence bands obtained on the example benchmarks.

procedure to find a lower bound for the rewards fails as they are encoded via an unbounded real-valued variable. For quantiles, the absence of bounds on the distribution is no hindrance. The DKW can produce rather asymmetric confidence intervals, which we see for *leader_sync*'s CVaR property.

We plot the empirical cdfs and associated DKW confidence bounds MODES delivered in Fig. 2. Graphically, the confidence interval for the quantile is the $p = 0.3$ line's segment between the bound curves, while the CVaR estimate and confidence interval stem from the curves cut off at that line. The DTMCs' reward distributions are necessarily discrete: *leader_sync* has only 3 possible outcomes of non-negligible probability (we thus omit $k = 1000$); increasing k for *coupon* does not smoothen the curve much, because most elements of the distribution's support were already sampled at $k = 100$—only the confidence band gets much thinner. For the CTMC *embedded*, the distribution is visibly continuous.

We do not report runtimes because the overhead of using the DKW—for collecting all samples to finally compute the intervals, instead of incremental averaging plus evaluation based on k and δ only as for traditional methods for the mean—was negligible in these experiments. Aside from DTMCs and CTMCs, MODES also supports more complex and expressive formalisms up to stochastic hybrid automata [25]; the new methods to check and bound quantiles and CVaRs work independent of the model type. They equally combine orthogonally with MODES' features for rare event simulation [12] as well as learning and scheduler sampling for nondeterministic models [18, 19, 39].

5 Conclusion

In this work, we have shown how the DKW inequality can be used to derive bounds on various aggregation functions beyond the classical expectation/mean, closing a significant gap of SMC compared to traditional verification. Moreover, as all our estimations are based on the DKW inequality, our methods can estimate all values simultaneously. Our experimental evaluation confirms the effectiveness of our methods, allowing for scalable estimation of such aggregation values for large systems. For future work, we believe that our approach should also be applicable to other risk measures such as variance [10,36], variance-penalized expected payoff [8,35,37,41], or cumulative prospect theory [11,44].

Data Availability. The MODES tool is available at modestchecker.net. An artifact for this paper—a reproduction package with the models and commands for the experiments in Sect. 4—is available at DOI 10.5281/zenodo.15286509.

References

1. Agha, G., Palmskog, K.: A survey of statistical model checking. ACM Trans. Model. Comput. Simul. **28**(1), 6:1–6:39 (2018). https://doi.org/10.1145/3158668
2. Amemiya, T.: Advanced Econometrics. Harvard University Press, Boston (1985)
3. Anderson, T.W.: Confidence limits for the value of an arbitrary bounded random variable with a continuous distribution function. Bull. Int. Stat. Inst. **43**, 249–251 (1969)
4. Baier, C.: Probabilistic model checking. In: Esparza, J., Grumberg, O., Sickert, S. (eds.) Dependable Software Systems Engineering, NATO Science for Peace and Security Series – D: Information and Communication Security, vol. 45, pp. 1–23. IOS Press (2016). https://doi.org/10.3233/978-1-61499-627-9-1
5. Baier, C., de Alfaro, L., Forejt, V., Kwiatkowska, M.: Model checking probabilistic systems. In: Clarke, E.M., Henzinger, T.A., Veith, H., Bloem, R. (eds.) Handbook of Model Checking, pp. 963–999. Springer, Heidelberg (2018). https://doi.org/10.1007/978-3-319-10575-8_28
6. Baier, C., Chatterjee, K., Meggendorfer, T., Piribauer, J.: Entropic risk for turn-based stochastic games. Inf. Comput. **301**, 105214 (2024). https://doi.org/10.1016/J.IC.2024.105214
7. Baier, C., Katoen, J.P.: Principles of Model Checking. MIT Press, Cambridge (2008). https://mitpress.mit.edu/books/principles-model-checking
8. Baier, C., Piribauer, J., Starke, M.: Risk-averse optimization of total rewards in Markovian models using deviation measures. In: CONCUR. LIPIcs, vol. 311, pp. 9:1–9:20. Schloss Dagstuhl - Leibniz-Zentrum für Informatik (2024)
9. Bohnenkamp, H.C., D'Argenio, P.R., Hermanns, H., Katoen, J.P.: Modest: a compositional modeling formalism for hard and softly timed systems. IEEE Trans. Softw. Eng. **32**(10), 812–830 (2006). https://doi.org/10.1109/TSE.2006.104
10. Brázdil, T., Chatterjee, K., Forejt, V., Kucera, A.: Trading performance for stability in Markov decision processes. J. Comput. Syst. Sci. **84**, 144–170 (2017)
11. Brihaye, T., Chatterjee, K., Mohr, S., Weininger, M.: Risk-aware Markov decision processes using cumulative prospect theory. In: LICS (2025). https://arxiv.org/abs/2505.09514

12. Budde, C.E., D'Argenio, P.R., Hartmanns, A.: Automated compositional importance splitting. Sci. Comput. Program. **174**, 90–108 (2019). https://doi.org/10.1016/J.SCICO.2019.01.006
13. Budde, C.E., D'Argenio, P.R., Hartmanns, A., Sedwards, S.: An efficient statistical model checker for nondeterminism and rare events. Int. J. Softw. Tools Technol. Transf. **22**(6), 759–780 (2020). https://doi.org/10.1007/S10009-020-00563-2
14. Budde, C.E., Dehnert, C., Hahn, E.M., Hartmanns, A., Junges, S., Turrini, A.: JANI: quantitative model and tool interaction. In: Legay, A., Margaria, T. (eds.) 23rd International Conference on Tools and Algorithms for the Construction and Analysis of Systems (TACAS). Lecture Notes in Computer Science, vol. 10206, pp. 151–168. Springer, Heidelberg (2017). https://doi.org/10.1007/978-3-662-54580-5_9
15. Budde, C.E., Hartmanns, A., Meggendorfer, T., Weininger, M., Wienhöft, P.: Sound statistical model checking for probabilities and expected rewards. In: TACAS (1). Lecture Notes in Computer Science, vol. 15696, pp. 167–190. Springer, Heidelberg (2025). https://doi.org/10.1007/978-3-031-90643-5_9
16. Budde, C.E., Hartmanns, A., Meggendorfer, T., Weininger, M., Wienhöft, P.: Statistical model checking beyond means: Quantiles, CVaR, and the DKW inequality (extended version). CoRR (2025). https://doi.org/10.48550/arXiv.2509.11859
17. Clopper, C., Pearson, E.: The use of confidence or fiducial limits illustrated in the case of the binomial. Biometrika **26**(4), 404–413 (1934). https://doi.org/10.1093/biomet/26.4.404
18. D'Argenio, P.R., Fraire, J.A., Hartmanns, A., Raverta, F.D.: Comparing statistical, analytical, and learning-based routing approaches for delay-tolerant networks. ACM Trans. Model. Comput. Simul. **35**(2) (2025). https://doi.org/10.1145/3665927
19. D'Argenio, P.R., Hartmanns, A., Sedwards, S.: Lightweight statistical model checking in nondeterministic continuous time. In: Margaria, T., Steffen, B. (eds.) 8th International Symposium on Leveraging Applications of Formal Methods, Verification and Validation (ISoLA). Lecture Notes in Computer Science, vol. 11245, pp. 336–353. Springer, Heidelberg (2018). https://doi.org/10.1007/978-3-030-03421-4_22
20. D'Argenio, P.R., Monti, R.E.: Input/output stochastic automata with urgency: confluence and weak determinism. In: Fischer, B., Uustalu, T. (eds.) 15th International Colloquium on Theoretical Aspects of Computing (ICTAC). Lecture Notes in Computer Science, vol. 11187, pp. 132–152. Springer, Heidelberg (2018). https://doi.org/10.1007/978-3-030-02508-3_8
21. David, A., Larsen, K.G., Legay, A., Mikucionis, M., Wang, Z.: Time for statistical model checking of real-time systems. In: Gopalakrishnan, G., Qadeer, S. (eds.) 23rd International Conference on Computer Aided Verification (CAV). Lecture Notes in Computer Science, vol. 6806, pp. 349–355. Springer, Heidelberg (2011). https://doi.org/10.1007/978-3-642-22110-1_27
22. Dvoretzky, A., Kiefer, J., Wolfowitz, J.: Asymptotic minimax character of the sample distribution function and of the classical multinomial estimator. Ann. Math. Stat. **27**(3), 642–669 (1956). https://doi.org/10.1214/aoms/1177728174
23. Ellen, C., Gerwinn, S., Fränzle, M.: Statistical model checking for stochastic hybrid systems involving nondeterminism over continuous domains. Int. J. Softw. Tools Technol. Transf. **17**(4), 485–504 (2015). https://doi.org/10.1007/S10009-014-0329-Y
24. Föllmer, H., Schied, A.: Convex measures of risk and trading constraints. Finan. Stochast. **6**, 429–447 (2002)

25. Fränzle, M., Hahn, E.M., Hermanns, H., Wolovick, N., Zhang, L.: Measurability and safety verification for stochastic hybrid systems. In: Caccamo, M., Frazzoli, E., Grosu, R. (eds.) 14th ACM International Conference on Hybrid Systems: Computation and Control (HSCC). pp. 43–52. ACM (2011). https://doi.org/10.1145/1967701.1967710
26. Hahn, E.M., Hartmanns, A., Hermanns, H., Katoen, J.P.: A compositional modelling and analysis framework for stochastic hybrid systems. Formal Methods Syst. Des. **43**(2), 191–232 (2013). https://doi.org/10.1007/S10703-012-0167-Z
27. Hartmanns, A., Hermanns, H.: The Modest Toolset: an integrated environment for quantitative modelling and verification. In: Abraham, E., Havelund, K. (eds.) 20th International Conference on Tools and Algorithms for the Construction and Analysis of Systems (TACAS). Lecture Notes in Computer Science, vol. 8413, pp. 593–598. Springer, Heidelberg (2014). https://doi.org/10.1007/978-3-642-54862-8_51
28. Hartmanns, A., Junges, S., Quatmann, T., Weininger, M.: A practitioner's guide to MDP model checking algorithms. In: Sankaranarayanan, S., Sharygina, N. (eds.) 29th International Conference on Tools and Algorithms for the Construction and Analysis of Systems (TACAS). Lecture Notes in Computer Science, vol. 13993, pp. 469–488. Springer, Heidelberg (2023). https://doi.org/10.1007/978-3-031-30823-9_24
29. Hartmanns, A., Klauck, M., Parker, D., Quatmann, T., Ruijters, E.: The quantitative verification benchmark set. In: Vojnar, T., Zhang, L. (eds.) 25th International Conference on Tools and Algorithms for the Construction and Analysis of Systems (TACAS). Lecture Notes in Computer Science, vol. 11427, pp. 344–350. Springer, Heidelberg (2019). https://doi.org/10.1007/978-3-030-17462-0_20
30. Hoeffding, W.: Probability inequalities for sums of bounded random variables. J. Am. Stat. Assoc. **58**(301), 13–30 (1963). https://doi.org/10.1080/01621459.1963.10500830
31. Krähmann, D., Schubert, J., Baier, C., Dubslaff, C.: Ratio and weight quantiles. In: Italiano, G.F., Pighizzini, G., Sannella, D. (eds.) 40th International Symposium on Mathematical Foundations of Computer Science 2015 (MFCS). Lecture Notes in Computer Science, vol. 9234, pp. 344–356. Springer, Heidelberg (2015). https://doi.org/10.1007/978-3-662-48057-1_27
32. Křetínský, J.: Survey of statistical verification of linear unbounded properties: model checking and distances. In: ISoLA (1). Lecture Notes in Computer Science, vol. 9952, pp. 27–45. Springer, Heidelberg (2016). https://doi.org/10.1007/978-3-319-47166-2_3
33. Kretínský, J., Meggendorfer, T.: Conditional value-at-risk for reachability and mean payoff in Markov decision processes. In: Dawar, A., Grädel, E. (eds.) 33rd Annual ACM/IEEE Symposium on Logic in Computer Science (LICS), pp. 609–618. ACM (2018).https://doi.org/10.1145/3209108.3209176
34. Legay, A., Lukina, A., Traonouez, L.M., Yang, J., Smolka, S.A., Grosu, R.: Statistical model checking. In: Steffen, B., Woeginger, G.J. (eds.) Computing and Software Science – State of the Art and Perspectives, Lecture Notes in Computer Science, vol. 10000, pp. 478–504. Springer, Heidelberg (2019). https://doi.org/10.1007/978-3-319-91908-9_23
35. Ma, S., Ma, X., Xia, L.: A unified algorithm framework for mean-variance optimization in discounted Markov decision processes. Eur. J. Oper. Res. **311**(3), 1057–1067 (2023)
36. Mannor, S., Tsitsiklis, J.N.: Mean-variance optimization in Markov decision processes. In: ICML, pp. 177–184. Omnipress (2011)

37. Markowitz, H.M.: Foundations of portfolio theory. J. Finan. **46**(2), 469–477 (1991)
38. Massart, P.: The tight constant in the Dvoretzky-Kiefer-Wolfowitz inequality. Ann. Probab. **18**(3), 1269–1283 (1990). https://doi.org/10.1214/aop/1176990746
39. Niehage, M., Hartmanns, A., Remke, A.: Learning optimal decisions for stochastic hybrid systems. In: Arun-Kumar, S., Méry, D., Saha, I., Zhang, L. (eds.) 19th ACM-IEEE International Conference on Formal Methods and Models for System Design (MEMOCODE), pp. 44–55. ACM (2021). https://doi.org/10.1145/3487212.3487339
40. Pilch, C., Remke, A.: Statistical model checking for hybrid Petri nets with multiple general transitions. In: 47th Annual IEEE/IFIP International Conference on Dependable Systems and Networks (DSN), pp. 475–486. IEEE Computer Society (2017). https://doi.org/10.1109/DSN.2017.41
41. Piribauer, J., Sankur, O., Baier, C.: The variance-penalized stochastic shortest path problem. In: ICALP. LIPIcs, vol. 229, pp. 129:1–129:19. Schloss Dagstuhl - Leibniz-Zentrum für Informatik (2022)
42. Randour, M., Raskin, J.F., Sankur, O.: Percentile queries in multi-dimensional Markov decision processes. Formal Methods Syst. Des. **50**(2–3), 207–248 (2017). https://doi.org/10.1007/S10703-016-0262-7
43. Rockafellar, R.T., Uryasev, S.: Conditional value-at-risk for general loss distributions. J. Bank. Finan. **26**(7), 1443–1471 (2002)
44. Tversky, A., Kahneman, D.: Advances in prospect theory: cumulative representation of uncertainty. J. Risk Uncertain. **5**(4), 297–323 (1992)
45. Ummels, M., Baier, C.: Computing quantiles in Markov reward models. In: Pfenning, F. (ed.) Foundations of Software Science and Computation Structures - 16th International Conference, FOSSACS 2013, Held as Part of the European Joint Conferences on Theory and Practice of Software, ETAPS 2013, Rome, Italy, 16–24 March 2013. Proceedings. Lecture Notes in Computer Science, vol. 7794, pp. 353–368. Springer, Heidelberg (2013). https://doi.org/10.1007/978-3-642-37075-5_23
46. Wald, A.: Sequential tests of statistical hypotheses. Ann. Math. Stat. **16**(2), 117–186 (1945). https://doi.org/10.1214/aoms/1177731118
47. Wirch, J.L., Hardy, M.R.: Distortion risk measures: coherence and stochastic dominance. In: International Congress on Insurance: Mathematics and Economics, pp. 15–17. Citeseer (2001)
48. Younes, H.L.S., Simmons, R.G.: Probabilistic verification of discrete event systems using acceptance sampling. In: Brinksma, E., Larsen, K.G. (eds.) 14th International Conference on Computer Aided Verification (CAV). Lecture Notes in Computer Science, vol. 2404, pp. 223–235. Springer, Heidelberg (2002). https://doi.org/10.1007/3-540-45657-0_17

Signal Sampling and Optimisation Under Symbolic Timed Automata Constraints

Benoît Barbot[1](✉), Nicolas Basset[2], Thao Dang[2],
Alexandre Donzé[2], Marco Esposito[1], and Dejan Nickovic[3]

[1] Univ Paris Est Creteil, LACL, 94010 Creteil, France
benoit.barbot@lacl.fr
[2] CNRS, VERIMAG, Univ Grenoble Alpes, Saint-Martin-d'Hères, France
[3] AIT Vienna, Vienna, Austria

Abstract. We consider the problem of generating input signals for validation of cyber-physical systems. The class of signals that we are interested in is subject to complex constraints in both time and value domains, which are captured by Symbolic Timed Automata (STA). We propose a method for uniform sampling of traces recognised by STA. The proposed procedure is general and works for arbitrary alphabets equipped with a probability space and a computable measure. We devise a concrete method for linear alphabets. We then show how this method can be used to enhance falsification based on both sampling and optimisation. The results are validated on the case study of an artificial pancreas.

1 Introduction

Cyber-Physical Systems (CPS) have profoundly impacted our society with applications like smart industry and autonomous vehicles. The complexity of CPS that seamlessly connect physical and digital worlds creates significant challenges to ensure their safety. Algorithmic verification does not scale to the size of real-life CPS and is therefore often replaced with more pragmatic testing activities. Falsification testing is a popular and widely used approach which attempts to disprove the correctness of the system by searching for an execution that violates its requirements. Falsification testing typically involves two phases: exploration and exploitation. During the exploration phase, the focus is on broadly searching the input space to uncover regions where issues are more likely to occur. Once promising areas are identified, the exploitation phase focuses the search in these regions, applying targeted methods such as global optimisation to generate tests that witness system failures, if any. Balancing these two phases is of utmost importance, as exploration allows diverse coverage, while exploitation guides the search towards failure witnesses or facilitates the discovery of edge cases that may

---

This work has been partly supported by ANR projects MAVeriQ (ANR-20-CE25-0012), the joint ANR-JST project CyPhAI and the Auvergne-Rhône-Alpes Region project DET-AI.

be otherwise overlooked. Nevertheless, good coverage in the exploitation phase is challenging to achieve when input traces are subject to complex constraints in both time and value domains.

Let us consider a falsification problem, which is to find an input signal to a cyber-physical system that makes it violate some requirement φ. The system is too complex and heterogeneous to admit a mathematical model and be analysed by model-based techniques. However, under a given input signal u, we can simulate or execute the system to obtain the output signal y_u[1]. Let $\theta_\varphi(u, y_u)$ be a function that evaluates the satisfaction of the input/output behaviours with respect to the requirement φ. The falsification problem can be formulated as:

$$\operatorname*{argmin}_{u \in U} \theta_\varphi(u, y_u) \qquad (1)$$

where U is the input signal space. Typically, an input u is considered a *falsifying* one if $\theta_\varphi(u, y_u) < 0$. In general, the input signal space is subject to constraints. In a closed-loop artificial pancreas [34] treated in this paper, the input to the system is a schedule of meals with corresponding ingested glucose amounts. The constraints on the meal timing and the quantity of glucose should reflect realistic eating behaviours of patients. These constraints are complex with nested temporal orders and real-valued constraints, and thus require an expressive specification formalism. We propose to use symbolic timed automata (STA) to describe constraints on the input signals. The challenge then is to efficiently perform uniform sampling (for exploration purposes) and optimisation (for exploitation purposes) on the resulting signal space constrained by the STA. This is the problem we address in the paper.

STA extends finite state automata (FSA) with (1) dense-time clock variables that measure time between transitions, and (2) alphabets given by Boolean algebras defined over infinite (or possibly large finite) domains. While every transition in an FSA has a single associated letter of the alphabet, transitions in STA are labelled with predicates that define sets of domain elements (alphabet letters) and with clock constraints. STA thus provide an appropriate and expressive language for modelling complex environments in which CPS operate.

We introduce in this paper a new method for *uniform sampling* of traces recognised by STAs. Our sampler computes traces of a given length accepted by the STA with the property that any two traces of the same length are sampled with the same probability. The procedure is general and works with arbitrary alphabets equipped with a probability space and a computable measure. Furthermore, the constraint transformation employed by this sampling method can be exploited to solve optimisation problems for falsification. With these properties, our sampler provides the perfect foundation for falsification testing of CPS, in both the exploration and exploitation phases.

---

[1] We write u as the subscript of y to emphasise that the output is induced by the input through the dynamics of the system.

Related Work. Falsification testing introduced in [28] consists in using metaheuristic optimisation algorithms to guide the generation of test inputs that steer the model executions towards the violation of the system's requirements. Formal specifications with quantitative semantics, such as Signal Temporal Logic (STL) [17,25], used to formalize the system requirements play a central role in falsification testing as they provide an indication how far an execution is from violating the requirements and is hence used as an objective function fed to the optimization engine to guide the search for a falsifying test. There is a vast literature on falsification testing (see [12]), with the majority of the work focusing on the exploration of different search strategies. This is in contrast to our work, in which we investigate how to do falsification testing under sophisticated environment models constraining the input space. This problem has been partially addressed in [28] by suggesting the use of hit-and-run sampling [32] to explore the set of constrained input parameters. Falsification testing under constrained input spaces given in the form of Shape Expressions [29] is studied in [13], but without giving guarantees on the uniformity of the sampled inputs. The method presented in [27] supports sampling and enumerating scenarios constrained via finite memory monitors from large but discrete spaces, but not falsification. In [11], timed automata are used as the environment model from which input traces are uniformly generated for the falsification of CPS. This work served as our main inspiration, and in this paper, we extend this idea to the more expressive symbolic timed automata model that allows natural encoding of traces that take into account both time and real-valued data. The theory of [11] is itself based on the uniform sampling for timed automata introduced in [6] that uses the notion of volume of timed languages from [3]. Statistical model-checking (SMC) is another relevant approach based on sampling system model executions and using statistical inference to estimate the probability that the system satisfies a formalized requirement (see [21] for an overview of the approach). In SMC, large part of research is devoted to the detection of rare events addressed by methods such as importance sampling [10] and importance splitting [18].

Paper Organisation. In Sect. 2 we discuss uniform sampling for timed automata with weighted alphabets (TAWA). This model is an intermediate step to address symbolic timed automata (STA). In Sect. 3, we focus on STA with linear alphabets and describe a method for uniform sampling of their traces. This method combines a weighted uniform sampling algorithm for TAWA and an algorithm for uniform sampling of points in a polyhedron. We show how the underlying transformation employed by these algorithms is used to perform optimisation on signal spaces constrained by STA. These algorithms were implemented, and we report in Sect. 4 the evaluation of sampling performance of our algorithms on an example, and the case study of an artificial pancreas [34]. The existing CPS falsification tools cannot handle STA constraints, and this case study aims at showing how our algorithms can be integrated with these tools to enhance their performance.

2 Uniform Sampling for TAWA

We first define timed languages, weighted timed languages, and what is weighted uniform sampling for a desired length of timed words. We then discuss timed automata (TA) and timed automata with weighted alphabet (WATA). The theory presented here extends that of [3,6] to the weighted case. This will be used as the foundation for sampling symbolic timed automata (STA).

2.1 Timed Languages and (Weighted) Uniform Sampling

A *timed word* of *length* n on a finite alphabet Σ is an alternating sequence $\nu = t_1 a_1 \cdots t_n a_n$ of *delays* $t_i \in \mathbb{R}_{\geq 0}$ and of *letters* a_i. We also consider *untimed words* of the form $w = a_1 \cdots a_n \in \Sigma^n$. We say that w is the label of ν. A *timed language* L is a set of timed words. Given $n \in \mathbb{N}$, we denote by L_n the restriction of L to timed words of length n.

Probability Density Function on a Timed Language. We are interested in sums of functions over timed languages. To do so, we have to do discrete summation over letters and continuous summation over time delays. Formally, given a timed languague L and $n \in \mathbb{N}$, the integral of a function $f : L_n \mapsto \mathbb{R}_{\geq 0}$ is

$$\int_{L_n} f(\nu) d\nu = \sum_{a_1 \in \Sigma} \cdots \sum_{a_n \in \Sigma} \int_{\mathbb{R}_{\geq 0}} \cdots \int_{\mathbb{R}_{\geq 0}} 1_{t_1 a_1 \cdots t_n a_n \in L_n} f(t_1 a_1 \cdots t_n a_n) dt_1 \cdots dt_n.$$

When $f = 1$ we get the *volume* of L_n (as defined in [3]): $\mathrm{Vol}(L_n) = \int_{L_n} 1 d\nu$. The volume can also be defined by summing the classical volume (aka. Lebesgue measure) of subsets of $\mathbb{R}^n$, one per untimed word. More precisely we define $V_w^L = \mathrm{Vol}(\{(t_1, \ldots, t_n) \mid t_1 w_1 \cdots t_n w_n \in L\})$ and we have $\mathrm{Vol}(L_n) = \sum_{w \in \Sigma^n} V_w^L$.

Example 1 (Running example). Consider the timed language on $\Sigma = \{a, b\}$ of timed words such that each block of consecutive a never last more than 1 time unit, and the same holds for b. For instance the untimed word $aabbbab$ leads to constraint on delays $t_1 + t_2 < 1$, $t_3 + t_4 + t_5 < 1$, $t_6 < 1$ and $t_7 < 1$. The polyhedra associated to aa and bb are $\{(t_1, t_2) \mid t_1, t_2 \geq 0 \wedge t_1 + t_2 < 1\}$ which is a triangle. The volume of this triangle is $\frac{1}{2}$. The polyhedra associated to ab and ba are the square $[0, 1]^2$ whose volume is 1. Overall, the volume of L_2 is $2 \times \frac{1}{2} + 2 \times 1 = 3$.

When $\int_{L_n} f(\nu) d\nu = 1$ we say that f is a *probability density function* (pdf). Given a function f such that $\int_{L_n} f(\nu) d\nu < \infty$, the normalised function $f / \int_{L_n} f(\nu) d\nu$ is a pdf. In particular, the uniform pdf $\nu \mapsto 1/\mathrm{Vol}(L_n)$ is obtained by normalisation of the function $\nu \mapsto 1$. In the sequel we generalize this by defining other weight functions on timed words.

Weighted Timed Languages and Weighted Uniform Sampling. A *weighted timed language* is a couple (L, μ) of a timed language L together with a *weight function* $\mu : \Sigma \to \mathbb{R}_{\geq 0}$. This weight function extends to words and timed words by taking the product of the weights of their letters: $\mu(a_1 \cdots a_n) = \mu(t_1 a_1 \cdots t_n a_n) = \mu(a_1) \ldots \mu(a_n)$.

The *weighted uniform sampling* consists in random sampling timed words according to the pdf

$$\nu \mapsto \frac{\mu(\nu)}{\int_{L_n} \mu(\nu) d\nu} \tag{2}$$

We call *weighted volume* the normalising constant $\int_{L_n} \mu(\nu) d\nu$ and denote it by $V^L_{\mu,n}$. It should be noted that the pdf is uniform among timed words having the same label, that is, every two timed words with the same label have the same value from the pdf. Moreover, given two timed words, the ratio of their probability is the ratio of their weight.

Proposition 1. *The weighted volume is the weighted sum of volumes V^L_w associated to untimed words:*

$$V^L_{\mu,n} = \int_{L_n} \mu(\nu) d\nu = \sum_{w \in \Sigma^n} \mu(w) V^L_w \tag{3}$$

Example 2. We add weights to the running example (Example 1) with $\mu(a) = \frac{1}{2}$ and $\mu(b) = \frac{5}{2}$. So $\mu(aa) = \frac{1}{4}$, $\mu(ab) = \mu(ba) = \frac{5}{4}$ and $\mu(bb) = \frac{25}{4}$. The weighted volume $V^L_{\mu,2}$ is $\frac{1}{4} \times \frac{1}{2} + 2 \times \frac{5}{4} \times 1 + \frac{25}{4} \times \frac{1}{2} = \frac{46}{8} = \frac{23}{4}$.

2.2 Timed Automata

We first define "classical" timed automata [1] before adding weights.

Definition 1 (Timed Automata). *A timed automaton is a tuple $A = (\Sigma, X, \mathcal{Q}, q_0, \mathcal{F}, \Delta)$ where Σ is a finite alphabet, X is a finite set of clocks, $\mathcal{Q}$ is a finite set of locations, $q_0 \in \mathcal{Q}$ is an initial location, $\mathcal{F} \subseteq \mathcal{Q}$ is a set of final locations, and $\Delta \subseteq \mathcal{Q} \times \Sigma \times \Phi \times 2^X \times \mathcal{Q}$ is the transition relation where Φ is the set of clock constraints (guards), and 2^X is the power set of X (clock resets).*

A *clock constraint* $\varphi \in \Phi$ is a finite conjunction of atomic clock constraints of the form $x \bowtie c$, where $x \in X$ is a clock, $\bowtie \in \{<, \leq, =, \geq, >\}$, and $c \in \mathbb{N}$. A *configuration* is a tuple $(q, \vec{x})$ where $q \in \mathcal{Q}$ is a location, and the vector $\vec{x} \in [0, \infty)^X$ is a *clock valuation*. The initial configuration is $(q_0, \vec{0})$.

For a *transition* $\delta = (q, a, \varphi, \mathfrak{r}, q_1) \in \Delta$, $q \in \mathcal{Q}$ is the *origin* location, $a \in \Sigma$ is the label, $\varphi \in \Phi$ is the clock guard, $\mathfrak{r}$ is the *reset function* determined by a subset $B \subseteq X$ of the clocks, and q_1 is the *target* location. Given a valuation $\vec{x}$, the valuation $\mathfrak{r}(\vec{x})$ is the valuation obtained from $\vec{x}$ by assigning 0 to the clock in B and keeping values of the other clocks unchanged. For a valuation $\vec{x}$ and a delay $t \geq 0$, we denote by $\vec{x} + t$ the valuation where t is added to all clock values of $\vec{x}$. Additional clock constraints can be added to each location; these are known as invariants that specify the conditions on the clock values for the automaton to stay in the location. In our setting, an automaton without invariant is obtained by transferring the invariant constraints of a location to the guards of all the outgoing transition from the location.

A *run* is an alternating sequence $(q_0, \vec{0}) \xrightarrow{t_1, a_1, \delta_1} (q_1, \vec{x}_1) \cdots \xrightarrow{t_n, a_n, \delta_n} (q_n, \vec{x}_n)$ where for all i, the valuation $\vec{x}_{i-1} + t_i$ satisfies the clock guard φ_i, and $\vec{x}_i = \mathsf{r}_i(\vec{x}_{i-1} + t_i)$. We say that the timed word $t_1 a_1 \cdots t_n a_n$ labels such a run. The *language A* is the set of timed words labelling *accepting runs*, that is runs which start in the initial configuration and end in a final location $q_n \in F$. We consider only *deterministic TA*, that is, TA for which every pair of transitions starting from the same location and having the same label must have disjoint guards.

Example 3. A TA for the timed language of the running example (Example 1) is depicted in Fig. 1. The clock y, initially equal to 0, measures the time since the last occurrence of a, and similarly for x and b.

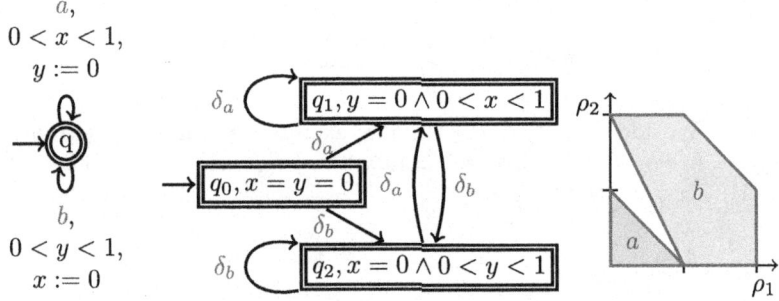

Fig. 1. Left: a TA with two clocks x and y, and two letters a and b. Center: a TA for the same language after splitting (Remark 1), the entry zone is depicted in each location, and transitions labelled by a and b are the same as for the previous TA but denoted δ_a and δ_b for brevity. These automata are also STA when we interpret letters as the following predicates $a := \rho_1 + \rho_2 < 1 \wedge 0 < \rho_1, \rho_2 < 2$, $b := 2\rho_1 + \rho_2 < 2 \wedge \rho_1 + \rho_2 < 3 \wedge 0 < \rho_1, \rho_2 < 2$ where ρ_1, ρ_2 are real-valued symbolic variables with values in $[0, 2]$. Right: the polyhedral sets of values for predicates a and b.

2.3 Timed Automata with Weighted Alphabets

A *timed automaton with weighted alphabet* (TAWA) is a couple $A_\mu = (A, \mu)$ of a timed automaton A together with a *weight function* $\mu : \Sigma \to \mathbb{R}_{\geq 0}$. Its weighted timed language is $(L(A), \mu)$.

2.3.1 Weighted Volume Computation for TAWA

The formula (3) characterizing the weighted volume involves a sum of exponentially many terms (one per untimed word). It is however possible to compute the weighted volume in polynomial time with respect to n by using the automaton structure rather than enumerating all words of length n. We adapt the operator approach in [3,6] to the weighted case.

We consider timed language parameterized by configurations of the TA. Given a configuration $(q, \vec{x})$, the language of the TA when the initial configuration is set to $(q, \vec{x})$ is denoted by $L_{(q,\vec{x})}$. In particular, $L(A) = L_{(q_0,\vec{0})}$. We denote by $v_{\mu,n}(q, \vec{x})$ the n-th weighted volume $V_{\mu,n}^{L(q,\vec{x})}$ and show in the following how to compute it. The functions $v_{\mu,k}$ ($k \leq n$) produced in the iterative computation of $v_{\mu,n}$ is used later for weighted uniform sampling of timed words.

In the base case for $n = 0$, we have that $v_{\mu,0} = 1_F$, the function such that $v_{\mu,0}(q, \vec{x}) = 1$ if the location q is final and 0 otherwise. Theorem 1 below states how the function $v_{\mu,n}$ for $n > 0$ can be computed by iterating an operator on functions defined over configurations. We first recall from [3,6], the definition of the operator ψ_δ for each transition $\delta = (q, a, \varphi, \mathfrak{r}, q')$. The operator takes as input a function f defined on an arrival configuration $(q', \vec{x}')$ after the transition and transforms it into a function defined on a departure configuration $(q, \vec{x})$ before the transition. It integrates, over all possible delays t such that $\vec{x} + t$ satisfies the guard φ, the value of f on the arrival configuration $(q', \mathfrak{r}(\vec{x}+t))$ whose clock valuation is obtained from $\vec{x}+t$ by applying the reset of the transition. Formally,

$$\psi_\delta(f)(q, \vec{x}) = \int_{t \mid \vec{x}+t \models \varphi} f(q', \mathfrak{r}(\vec{x}+t))dt \qquad (4)$$

Remark 1. To compute this integral, we use the splitting procedure described in [6]. This transformation of the TA preserves the language and is such that the integrals $\int_{t \mid \vec{x}+t \models \varphi}$ are of the form $\int_{a-x_i}^{b-x_j}$ where $a, b \in \mathbb{N}$ and x_i, x_j are the i-th and j-th coordinates of $\vec{x}$ or 0. Moreover, the set of possible valuations $\vec{x}$ when a location is entered is defined by a clock constraint, called *entry zone* (see Fig. 1-center, for an example).

We can now define our weighted operator that will be iterated to compute the $v_{\mu,n}$: $\Psi_\mu(f)(q, \vec{x}) = \sum_{\delta=(q,a,\varphi,\mathfrak{r},q')} \mu(a) \psi_\delta(f)(q, \vec{x})$. The operator in [3] can be obtained from ours by taking all weights equal to 1.

Theorem 1. *For all $n \in \mathbb{N}$, $v_{\mu,n} = \Psi_\mu(v_{\mu,n-1})$. For all $q \in \mathcal{Q}$, the functions $\vec{x} \mapsto v_{\mu,n}(q, \vec{x})$ are polynomials and can be computed in polynomial time.*

Example 4. We illustrate the volume computation for the STA in Fig. 1-center. Note that $v_0(q_i, \vec{x}) = 1$ (where the clock valuation $\vec{x} = (x, y)$) for every location q_i as they are all final. We focus first on q_1 where $y = 0$ in the entry zone:

$$v_1(q_1, (x, 0)) = \mu(a) \int_0^{1-x} v_0(q_1, (x+t, 0))dt + \mu(b) \int_0^1 v_0(q_2, (0, t))dt$$

$$= \frac{1}{2} \int_0^{1-x} 1 dt + \frac{5}{2} \int_0^1 1 dt = 3 - \frac{1}{2}x$$

In the first integral of the above equation, which corresponds to the transition with label a, the guard condition is $0 < x < 1$ and only the clock y is reset, thus t ranges from 0 to $1-x$. The coefficients of the integrals in the sum are the volumes of the predicates a and b. Similarly we have

$$v_1(q_2,(0,y)) = \mu(a)\int_0^1 v_0(q_1,(t,0))dt + \mu(b)\int_0^{1-y} v_0(q_2,(0,y+t))dt$$

$$= \frac{1}{2}\int_0^1 1 dt + \frac{5}{2}\int_0^{1-y} 1 dt = 3 - \frac{5}{2}y$$

Using the function v_1 computed above, we can proceed:

$$v_2(p_1,(x,0)) = \mu(a)\int_0^{1-x} v_1(p_1,(x+t,0))dt + \mu(b)\int_0^1 v_1(p_2,(0,t))dt$$

$$= \frac{1}{2}\int_0^{1-x} 3 - \frac{1}{2}(x+t)dt + \frac{5}{2}\int_0^1 3 - \frac{5}{2}tdt = \frac{23}{4} - 2x + \frac{1}{8}x^2$$

When $x = 0$, $v_2(p_1,(0,0)) = \frac{23}{4}$ and similarly we have $v_2(p_0,(0,0)) = \frac{23}{4}$ which is the weighted volume $V^L_{\mu,2}$ as in Example 2.

2.3.2 Weighted Uniform Sampling Algorithm for TAWA

Here we explain how to sample a timed word of length n. The method requires to pre-compute all the polynomials $\vec{x} \mapsto v_k(q,\vec{x})$ for $k \leq n$ and $q \in Q$. It follows a recursive scheme and assumes an input n and a configuration $(q,\vec{x})$ such that $v_{n,\mu}(q,\vec{x}) > 0$. It outputs a timed word ν according to the pdf $\nu \mapsto \dfrac{\mu(\nu)}{v_{n,\mu}(q,\vec{x})}$.

- Base case: if $n = 0$, output the empty word.
- When $n > 0$, starting from configuration $(q,\vec{x})$, a transition $\delta = (q,a,\varphi,\mathfrak{r},q')$ is randomly chosen according to the discrete probability distribution

$$\delta \mapsto \mu(a)\frac{\psi_\delta(v_{n-1})(q,\vec{x})}{v_n(q,\vec{x})} \tag{5}$$

Then the delay t is drawn according to the pdf

$$t \mapsto \frac{v_{n-1}(q',\mathfrak{r}(\vec{x}+t))}{\psi_\delta(v_{n-1})(q,\vec{x})} \tag{6}$$

The timed word starts with the delay t and the label a of δ and the rest of the timed word ν' (of length $n-1$) is sampled from the configuration $(q',\mathfrak{r}(\vec{x}+t))$ according to the pdf $\nu' \mapsto \dfrac{\mu(\nu')}{v_{n-1}(q',\mathfrak{r}(\vec{x}+t))}$ using a recursive call.

Theorem 2. *The algorithm presented above performs weighted uniform sampling of words of the input length n from the input configuration $(q, \vec{x})$ of the input TAWA (A, μ), that is, the sampling is according to the pdf $\nu \mapsto \dfrac{\mu(\nu)}{v_{n,\mu}(q,\vec{x})}$.*

Proof. We present a recursive proof. Assume that the result holds for words of length $n - 1$ for some $n > 0$. A timed word ν of length n drawn from $(q, \vec{x})$ is of the form $\nu = ta\nu'$ where a corresponds to the unique transition δ whose label is a and whose guard φ is such that $\vec{x} + t \models \varphi$ (the uniqueness of δ is due to determinism). The density of probability of $ta\nu'$ is hence

$$\left(\mu(a) \frac{\psi_\delta(v_{n-1,\mu})(q,\vec{x})}{v_{n,\mu}(q,\vec{x})}\right) \left(\frac{v_{n-1,\mu}(q', \mathfrak{r}(\vec{x}+t))}{\psi_\delta(v_{n-1,\mu})(q,\vec{x})}\right) \left(\frac{\mu(\nu')}{v_{n-1,\mu}(q', \mathfrak{r}(\vec{x}+t))}\right)$$

where the factors correspond to the random choice of δ (using (5)), of t (using (6)) and of ν' (using the recursive hypothesis and (2)). After simplification we obtain the desired density of probability $\dfrac{\mu(a)\mu(\nu')}{v_{n,\mu}(q,\vec{x})} = \dfrac{\mu(\nu)}{v_{n,\mu}(q,\vec{x})}$. □

3 Symbolic Timed Automata: Sampling and Optimisation

A symbolic timed automaton is a TA whose letters are associated to predicates on a large alphabet. In this work, we are interested in large alphabets defined by effective Boolean algebras with algorithmically computable Boolean operations and decidable equality and order relations. An *effective Boolean algebra* $\mathcal{A}$ is a tuple $(\mathcal{D}, \Psi, [\![\cdot]\!], \top, \bot, \vee, \wedge, \neg)$, where $\mathcal{D}$ is a set of *domain elements*, Ψ is a set of predicates closed under Boolean operations, $[\![\cdot]\!] : \Psi \to 2^\mathcal{D}$ is a *denotation function* such that for all $a, a' \in \Psi$, $[\![\bot]\!] = \emptyset$, $[\![\top]\!] = \mathcal{D}$, $[\![a \vee a']\!] = [\![a]\!] \cup [\![a']\!]$, $[\![a \wedge a']\!] = [\![a]\!] \cap [\![a']\!]$, and $[\![\neg a]\!] = \mathcal{D} \setminus [\![a]\!]$. Our framework requires that the problem of checking whether $[\![a]\!] \neq \emptyset$ for an arbitrary $a \in \Psi$ is decidable. In addition, we restrict our attention to effective Boolean algebras $\mathcal{A}$ with a set $\mathcal{D}$ of domain elements that is equipped with a measure $(\mathcal{D}, \Sigma_\mathcal{D}, \mu)$.

In the remainder of the paper we focus on *real linear alphabets* $\mathcal{A}_L$, which are effective Boolean algebras defined over bounded real domains and allow predicates of the form: $\psi := (\Sigma_{i=1}^m c_i \rho_i) \leq c \mid \neg a \mid a \vee a'$, where $\rho_1, \ldots, \rho_m$ are real valued variables, and $c, c_i \in \mathbb{R}$ are constants for all $1 \leq i \leq m$. We denote by $\vec{\rho} = (\rho_1, \ldots, \rho_m) \in \mathbb{R}^m$ a vector assigning values to the variables ρ_i. The measure μ on $\mathcal{D} \subseteq \mathbb{R}^m$ is the volume, *i.e.*, Lebesgue measure.

Given an STA A, its *symbolic timed language* is the language $L(A)$ where the labels are the predicates on the transitions. The *concrete timed language* of A is

$$\tilde{L}_n(A) = \{t_1\vec{\rho}_1 \cdots t_n\vec{\rho}_n \mid \exists a_1, \ldots, a_n : t_1 a_1 \cdots t_n a_n \in L(A) \wedge \forall i \leq n, \vec{\rho}_i \models a_i\}.$$

We consider only STA where in every location q, for every pair of transitions δ and δ' from q, the sets of couples $(\rho, \vec{x})$ respectively satisfying $a \times \varphi$ and $a' \times \varphi'$ are disjoint (where φ, φ' are the respective transition guards). This condition implies that the TA is deterministic (when predicates are seen as labels).

To each path of an STA corresponds a *delay-value polyhedron* each of which is in the subset of $\tilde{L}_n(A)$ of accepted timed word that follows the path. Each of them is the Cartesian product of a *value-polyhedron* which is the product of all predicates along the path and of a *delay-polyhedron* whose points are the sequences of delays, and which is defined by the timing constraints gathered along the path. The concrete timed language $\tilde{L}_n(A)$ is thus a union of polyhedra in $\mathbb{R}^{n(m+1)}$, whose volume is denoted by $\text{Vol}(\tilde{L}_n(A))$. This union of polyhedra is too complex as domain of optimisation variables and cannot be directly handled by the state-of-the-art optimisation algorithms. It is thus desirable to have a transformation from an efficiently manageable domain (such as a unit box) to this union of polyhedra. In Sect. 3.1, we present a method for uniform sampling *concrete timed words* from the concrete timed language of an STA which will also provide a key transformation from the unit box, useful for the optimisation (Sect. 3.2).

3.1 STA Uniform Sampling

The input of our sampling procedure is an STA defined by a timed automaton $A = (\mathcal{A}_L, X, \mathcal{Q}, q_0, \mathcal{F}, \Delta)$ where $\mathcal{A}_L$ is a real linear alphabet defined over a bounded set $\mathcal{D}_\mathbb{R}$ together with a word length $n \in \mathbb{N}$, the output is a *uniformly sampled concrete timed word of length n*. In other words every point of the union of polyhedra $\tilde{L}_n(A) \subseteq \mathbb{R}^{n(m+1)}$ has the same density of probability $\frac{1}{\text{Vol}(\tilde{L}_n(A))}$ of being sampled. The main steps of our sampling procedure are:

1. Computing the volumes of the predicates which are used to define the weight function μ for the corresponding TAWA (A, μ).
2. Sample a symbolic timed word $t_1 a_1 \cdots t_n a_n$ using the weighted uniform sampling for (A, μ), using the algorithm in Sect. 2.3.
3. Sample uniformly $\vec{\rho}_i$ from a_i for $i \in \{1, \ldots, n\}$.
4. Construct a concrete timed word $t_1 \vec{\rho}_1 \cdots t_n \vec{\rho}_n$.

Theorem 3. *The volume of the concrete timed language $\tilde{L}_n(A)$ is equal to the weighted volume of the weighted language of the TAWA (A, μ). The algorithm presented above performs uniform sampling from $\tilde{L}_n(A)$.*

Proof (Sketch). From (3) we have $V_{\mu,n}^{L(A)} = \sum_{w \in \Sigma^n} \mu(w) V_w^{L(A)}$. The union of delay-value polyhedra associated to paths labelled by w has a volume $\mu(w) V_w^{L(A)}$ because this is a Cartesian product of the value-polyhedron of volume $\mu(w)$ and of the union of delay-polyhedra of volume $V_w^{L(A)}$.

Now we take a concrete timed-word $t_1 \vec{\rho}_1 \cdots t_n \vec{\rho}_n \in \tilde{L}_n(A)$ and show that its density of probability is $\frac{1}{V_{\mu,n}^{L(A)}}$. The symbolic predicates a_i that contain the $\vec{\rho}_i$ form an untimed word w. First the symbolic timed word $t_1 a_1 \cdots t_n a_n$ is randomly chosen with probability $\frac{\mu(w)}{V_{\mu,n}^{L(A)}}$ (step 2) and then the concrete values $\vec{\rho}_i \in a_i$ are chosen uniformly with probability $\frac{1}{\mu(a_i)}$ (step 3). At the end the probability to generate the concrete timed word (step 2–4) is $\frac{\mu(w)}{V_{\mu,n}^{L(A)}} \frac{1}{\mu(a_1)} \cdots \frac{1}{\mu(a_n)} = \frac{1}{V_{\mu,n}^{L(A)}}$.

3.1.1 Sampling from a Polyhedron

Since the predicates of the alphabet define bounded polyhedra in $\mathbb{R}^m$, we discuss the problem of sampling uniformly from a bounded polyhedron. This problem has been addressed in various applications, such as randomized approximations of volumes of polytopes and convex bodies, and randomized algorithms for convex optimisation. Markov Chain Monte Carlo (MCMC) algorithms (such as Ball walk, hit-and-run, Dikin walk, Vaidya walk) were developed and some are specialized for convex polyhedra [14,15,19,20,22–24]. In this work, we propose a method based on simplicial partitions of bounded polyhedra. This is motivated by the fact that well-established methods are available for the major ingredients of this method. Another motivation for this solution is that simplicial partitions allow encoding STA temporal and symbolic constraints with simple box constraints, as will be shown in Sect. 3.2. The main steps of our method for uniform sampling in polyhedra are summarized as follows:

1. Partition the polyhedron P in $\mathbb{R}^m$ into simplices $\{S_1, \ldots, S_l\}$
2. Sampling one simplex S from the triangulation, based on their volumes, that is, the probability that S_i is chosen is $\text{Prob}(S_i) = \frac{\text{Vol}(S_i)}{\sum_{j=1}^{l} \text{Vol}(S_j)}$.
3. Draw a uniform sample from the simplex chosen in the second step.

We employ the library Qhull [5] for its support of exact polyhedral volume calculations and simplicial decomposition. The procedure for uniform sampling of points inside a simplex, required in the third step, is explained in the following.

3.1.1.1 Sampling from a Simplex

We consider a simplex S in $\mathbb{R}^m$ with $(m+1)$ vertices $\{z_1, \ldots, z_{m+1}\}$. A point inside S can be described by a linear combination of its $(m+1)$ barycentric coordinates $(\beta_1, \ldots, \beta_{m+1})$ as follows $\zeta = \sum_{i=1}^{m+1} \beta_i z_i$ where $\sum_{i=1}^{m+1} \beta_i = 1$ and $\beta_i \geq 0$ for all $i \in \{1, \ldots, m+1\}$.

We focus first on the problem of sampling a point in the unit simplex S_u in $\mathbb{R}^m$, defined by the following constraints: $\zeta_1 + \ldots + \zeta_m = 1$ and $\forall i \in \{1, \ldots, m\}, \zeta_i \in [0,1]$. Sampling from S_u in $\mathbb{R}^m$ is equivalent to sampling from a set S_p in $\mathbb{R}^{m-1}$ such that $\zeta_1 + \ldots + \zeta_{m-1} \leq 1$ then picking $\zeta_m = 1 - \sum_{1}^{m-1} \zeta_i$. To sample in S_p, we use a transformation that maps from the $(m-1)$-dimensional unit cube (within which uniform sampling is trivial) to S_p that preserves uniformity. This transformation is defined using the well-known method from [16,31], as follows: we first sample ζ uniformly at random in the unit cube $[0,1]^{m-1}$. Next, we sort the coordinates of ζ to obtain ζ' and add $\zeta'_0 = 0$ and $\zeta'_m = 1$ to the sorted sequence to obtain a new sequence: $0 = \zeta'_0 \leq \zeta'_1 \leq \zeta'_2 \leq \ldots \leq \zeta'_{m-1} \leq \zeta'_m = 1$. Now we define $\beta_i = \zeta'_i - \zeta'_{i-1}$ for all $i \in \{1, \ldots, m\}$. It is easy to see that $\sum_{i=1}^{m} \beta_i = 1$ and every β_i is positive; therefore $\beta \in S_u$. This transformation from the unit box to S_u preserves uniformity, that is the points $(\beta_1, \ldots, \beta_m)$ sampled in this manner is uniformly distributed over S_u [16].

We now go back to our problem of sampling uniformly a point ζ in the original simplex S in $\mathbb{R}^m$. To do so, we first sample uniformly a point $(\beta_1, \ldots, \beta_{m+1})$ in the unit simplex in $\mathbb{R}^{m+1}$ which then serves as the barycentric coordinates of the point ζ in S, that is $\zeta = \sum_{i=1}^{m+1} \beta_i z_i$.

3.1.1.2 Sampling from a Simplicial Partition

Note that sampling within a simplex reduces to sampling from the unit cube in $\mathbb{R}^m$. We can integrate this with the sampling of a simplex from the simplicial partition $\{S_1, \ldots, S_k\}$ by coding the sampling probabilities of the simplices in one of the dimensions of the sample from the unit cube in $\mathbb{R}^m$.

Let $(\lambda_1, \ldots, \lambda_k)$ be a vector of accumulated simplex volumes that is $\lambda_1 = \text{Vol}(S_1)$ and $\lambda_i = \sum_{j=1}^{i} \text{Vol}(S_i)$ for $i \in \{1, \ldots, k\}$. Note that λ_k equals the volume of the entire polyhedron.

Given a point $(\eta_1, \ldots, \eta_m)$ sampled uniformly from the unit box in $\mathbb{R}^m$. Let η_1 be uniformly sampled in $[0, \lambda_k]$ and then used to determine which simplex S_i to pick, by the following mapping: if $\eta_1 \in [0, \lambda_1)$, then $i = 1$, and if $\eta_1 \in [\lambda_j, \lambda_{j+1})$, then $i = j$. Then, η_1 is scaled back to the interval $[0,1]$ by: $\eta_1 = \lambda_i + \eta_1(\lambda_{i+1} - \lambda_i)$ (where i is the index of the chosen simplex). Then, together with the remaining coordinates $\eta_2, \ldots, \eta_m$ sampled in the unit box in $\mathbb{R}^{m-1}$, the resulting point $(\eta_1, \eta_2, \ldots, \eta_m)$ is used to map to a point inside the chosen simplex S_i, as described earlier in this section.

3.2 Optimisation over STA-Constrained Signal Space

So far we have considered uniform sampling on constrained signal spaces. Uniform sampling has the advantage of facilitating coverage as well as statistical correctness guarantees. It is however, as mentioned in the introduction, important to combine it with the optimisation approaches aiming at finding faulty behaviours by minimising some quantitative measures of system robustness or performance. Consider again the falsification problem (1), now formulated as:

$$\underset{\nu \in \tilde{L}_n(A)}{\arg\min}\, \theta_\varphi(u(\nu), y_u) \qquad (7)$$

where the function $u(\nu)$ converts a concrete timed word of length n (accepted by the STA A) to a continuous-time input signal. This conversion is needed because most existing CPS simulators operate in continuous time. Since the system dynamics is not known, one needs to use black-box optimisation, the computational complexity of which depends on the complexity of the involved constraints. It is clear that no existing black-box optimisation solvers can directly handle STA constraints without resorting to a rejection step. We show how to encode STA constraints into simple box constraints that most solvers can handle.

Indeed, we have seen in the previous section the box-to-polyhedron transformation that preserves the uniformity. A similar transformation can also be defined for the delay variables in timed words, as in [7], and is sketched as follows. We sample a delay t using the pdf in (6) using inverse sampling. It consists in sampling uniformly a random number r in $[0, 1]$ and then find (by a numerical scheme, such as the Newton method) the delay t such that $\int_0^t p(u)du = r$ (where p is the pdf in (6)). This means that our sampling algorithm maps a uniformly sampled vector $(r_1, \ldots, r_n)$ in the unit box in $\mathbb{R}^n$ to n delays of the timed word.

To choose a transition using the discrete probability distribution in (5), we can also use one uniform random number $r \in [0, 1]$ and make the discrete choice, similarly to the choice of a simplex in Sect. 3.1. Overall, we need $2n$ random numbers for choosing delays and transitions and further nm random number to choose the points in the polyhedra. This means we have a mapping from uniformly distributed points $\vec{r}$ in $[0,1]^{n(m+2)}$ to uniformly distributed concrete timed words ν of the concrete timed languages $\tilde{L}_n(A)$ of the STA A. We denote this transformation by $\nu(\vec{r})$, then the optimisation problem (7) becomes an optimisation problem with box constraints: $\operatorname{argmin}_{\vec{r} \in [0,1]^{n(m+2)}} \theta_\varphi(u(\nu(\vec{r})), y_u)$. Figure 2 shows the steps of our approach implemented in a prototype tool which makes use of an extension of Wordgen [9] for sampling Weighted TA and computing the above-mentioned transformation for delay variables.

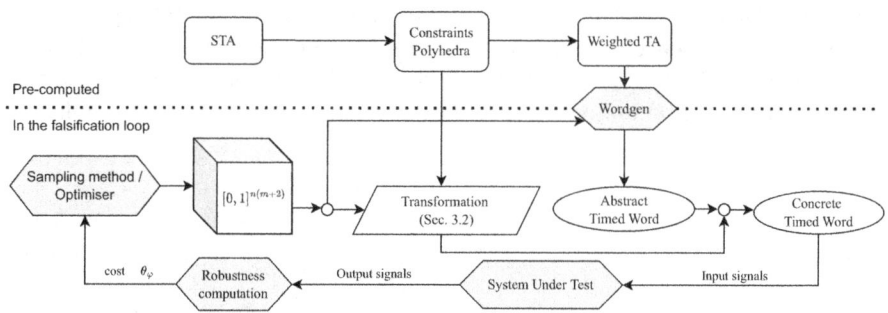

Fig. 2. Workflow for STA uniform sampling and optimization.

4 Experimental Evaluation

We demonstrate in a case study how STA and its associated algorithms can enhance falsification testing (Sect. 4.1). We first validate the implementation of our sampling algorithm on our running example, for which we can compute volumes and probabilities associated to each discrete path of the STA. We compare our uniform sampling with a simple, *isotropic sampling* method. The latter is used in several works, and was compared to uniform sampling for classical TA (see [7]). An artifact to reproduce all experiments and figures is available on Zenodo [8].

Example 5. To visualise the uniformity of sampled words, we determine the probability of sampling concrete timed words having the same untimed word. Concerning the automaton A of the running example of Fig. 1, its concrete timed language $\tilde{L}_5$ for words of length 5 is a subset of $\mathbb{R}^{15}$, where there are 5 dimensions for the delays and 5×2 dimensions for the symbolic values. $\tilde{L}_5$ is a union of 32 polyhedra, called delay-value polyhedra, one per untimed word $w \in \{a,b\}^5$. Every such delay-value polyhedron associated to an untimed word w is the Cartesian product[2] of the polyhedra of symbolic values (called value polyhedra) associated to the letters of w and of the polyhedron of delays (called delay

---

[2] Here and below, it is a Cartesian product up to a re-ordering of the variables.

polyhedron). The volume of the delay-value polyhedron corresponding to w is thus $\mu(w_1)\cdots\mu(w_n)V_w^L = \mu(w)V_w^L$ (we recall that V_w^L is the volume of the delay polyhedron P_w of w). Dividing this volume by the whole volume $\mathrm{Vol}(\tilde{L}_n(A))$ gives the (exact) probability that a uniformly sampled timed word falls in each delay-value polyhedron. For instance, the delay-value polyhedron associated to the untimed word $w = bbabb$ is $[\![a]\!] \times [\![b]\!]^4 \times P_w$ where the polyhedron P_w defined by $t_1 + t_2 \leq 1 \wedge t_3 \leq 1 \wedge t_4 + t_5 \leq 1$ has volume $\frac{1}{4}$. The overall volume is $\left(\frac{1}{2}\right) \times \left(\frac{5}{2}\right)^4 \times \left(\frac{1}{4}\right) = \frac{625}{128}$. We remark that a large $\mu(w)$ is obtained by having many b while a large volume of P_w is obtained by alternating a and b. The maximal probability is obtained with $bbabb$ that makes a trade-off between these two requirements, while purely alternating words (like $babab$) and the word $bbbbb$ (with only letters b) have lower probability. The exact probability distribution is shown in Fig. 3 with the empirical distribution obtained from sampled words. We compare with isotropic sampling, that at each step choose the next letter l proportionally to $\mu(l)$ and the next delay uniformly among those available.

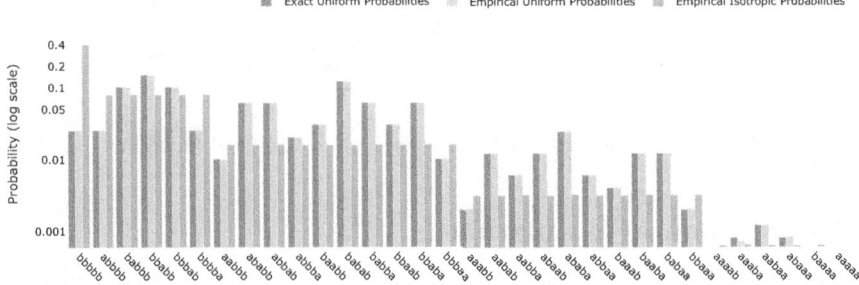

Fig. 3. Histograms for Example 5. Red bars: exact probabilities. Blue bars: empirical probabilities using our tool with 10^6 samples, generated in 2 m 9.6 s. Green bars: empirical probabilities using isotropic sampling with 10^6 samples, in 1 m 22 s. As expected blue and red bars are very close (max difference in the order of 10^{-4}) while green bars are significantly different since each green bar height is proportional to the weight $\mu(w)$ (*i.e.*, the value-polyhedron volume) while the true uniform distribution on $\tilde{L}_n$ gives bar height proportional to $\mu(w)V_w^L$ (*i.e.*, the delay-value polyhedron volume). (Color figure online)

4.1 Case Study: Closed-Loop Artificial Pancreas System

We show the effectiveness of our approach with Simglucose [34], an open-source Python implementation of the UVA/Padova Type 1 Diabetes Mellitus (T1DM) metabolic simulator [26], based on a widely accepted mathematical model of glucose-insulin dynamics. This CPS model is highly hybrid, with nonlinear algebraic differential equations and switching dynamics. Simglucose provides a programmable interface to the FDA-accepted UVA/Padova simulator, enabling *in silico* testing of glucose control strategies through customisable patient models, meal scenarios, and sensor/actuator configurations. The package includes 30 patient profiles with varying physiological parameters derived from clinical data.

The problem is to check if a closed-loop artificial pancreas system can successfully control the blood glucose of a patient who respects legal meal plans, to satisfy the following STL (Signal Temporal Logic) requirement [17,25] $\varphi_{BG} := \square_{[0,H]} \, BG > 70 \wedge BG < 350$ where BG is the output blood glucose signal and the time horizon H is 1440 min, *i.e.*, one day. The formula states that the blood glucose value must stay in $(70, 350)$ for all time instants in $[0, H]$. In all the experiments, we used the profile of patient adolescent#001. A legal meal plan, denoted by μ, is the input to our system under test, and the blood glucose BG is the output. Our falsification problem can be formulated as: $\mathrm{argmin}_{\mu} \theta_{\varphi_{BG}}(\mu, BG_{\mu})$, where $\theta_{\varphi_{BG}}$ is the robustness of φ_{BG}. The challenge in this optimisation problem is to capture accurately the complex temporal and quantitative constraints of legal meals. To this end, we use the following STA. The constraints over legal meal plans are modelled with an STA of Fig. 4, with 7 locations, 2 clocks, and one symbolic variable. The clocks, h and d, are used, respectively, to represent the hour of the day and the time delay between main meals (breakfast, lunch, and dinner). The symbolic variable m represents the amount of glucose ingested in each meal. Following [34], we consider meal plans with at most six meals per day: breakfast, morning snack, lunch, afternoon snack, dinner, and evening snack. We defined constraints that involve the timing and amount of glucose for each meal, as well as the conditions under which the snacks can be skipped. Legal meal plans, with respect to our constraints, have a balanced amount of glucose throughout the day and limit the fasting periods.

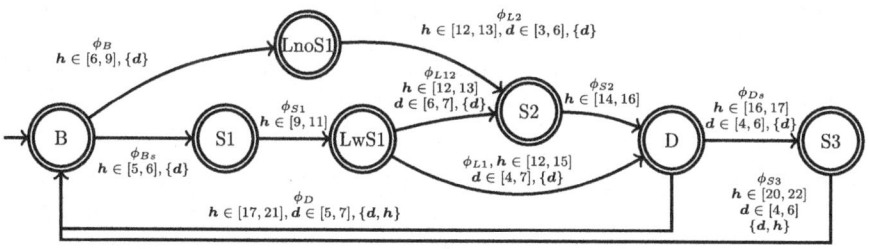

Fig. 4. STA for Simglucose. Two clocks, h for the time of day, d for the delay between main meals.

4.1.1 Falsification Experiments

To show how our STA model and sampling of the constrained input space can enhance the existing falsification methods, we considered the optimisation algorithms from three tools: Ψ-taliro [33], NOMAD [4], and Nevergrad [30]. Ψ-taliro is the Python variant of TaLiRo [2], a CPS falsification tool. NOMAD and Nevergrad are black-box optimisation tools, based, respectively, on the Mesh Adaptive Direct Search (MADS) algorithm and on an ensemble of gradient-free algorithms. We ran the following experiments.

- Experiment 1: We sampled meal plans with our STA-based method and isotropic sampling.
- Experiment 2: We ran Ψ-taliro without constraints.
- Experiment 3: Note that no existing optimiser can directly handle the STA constraints. Hence, we tried to capture the meal plan constraints using penalties in the objective function when running Ψ-taliro, and using black-box constraint functions when running NOMAD and Nevergrad.
- Experiment 4: We ran all the tools with our STA-based representation of the input signal space.

The first experiment was run 10 times with a budget of 1000 simulations. The second experiment was run 100 times. All other experiments were repeated for 10 runs with 10 different random seeds, each time with a budget of 500 simulations. For Ψ-taliro, we employed the Dual Annealing algorithm, while we used the default settings for NOMAD and Nevergrad.

Experiment 1: Falsification via Random Sampling. We used isotropic sampling and our STA-based uniform sampling to falsify the STL formula φ_{BG} with meal plans that satisfy the constraints by construction.

Experiment 2: Unconstrained falsification with Ψ-taliro. We used Ψ-taliro to falsify the STL formula φ_{BG} without any meal plan constraints; we simply set the domains of every variable so that every meaningful meal plan could be included. This yielded a 12-dimensional box-constrained problem as there are two variables for each of the 6 meals of the day, *i.e.*, one for the meal time and one for the ingested glucose amount.

Experiment 3: Constrained Falsification with Penalties and Black-Box Functions. The Dual Annealing algorithm implemented in Ψ-taliro does not explicitly support constraints, hence we reflect constraints as penalties on the objective function. In particular, for each constraints we implemented a Python function that computes its degree of violation for a given meal plan (for example, the excess glucose amount at lunch). Such functions return 0 if the constraint is satisfied and a positive number plus a large positive constant value otherwise. All these values are then added to the robustness value $\theta_{\varphi_{BG}}$ of the meal plan, to define the objective value. On the other hand, NOMAD and Nevergrad support the black-box constraints, so we passed the Python implementations of the meal plan constraints as black-box constraints, while the objective function is the robustness of the meal plan with respect to the property φ_{BG}. In both cases, the problem dimension is the same as the unconstrained relaxed version, *i.e.*, 12.

Experiment 4: STA-Based Falsification. We used our method to represent the input signal space via the STA modelling of Fig. 4. Thanks to the transformation in Sect. 3.2, the domain of the optimisation variables is a box in which every point represents an input signal that satisfies all the relevant constraints. We were thus able to directly use all three tools. As we considered a single day and at most 6 meals, the search space dimension is $6(2 + 1) = 18$.

4.1.2 Results and Discussion

In Experiment 1, although our uniform sampling method allowed finding less robust behaviours, compared to the isotropic sampling (minimal value 2.47 vs 3.62), neither of the methods found a falsifying meal plan within 10,000 simulations. Our method generated 10,000 meal plans in 1.27 s, while isotropic sampling took 0.045 s. Indeed, only by sampling it is hard to find a faulty behaviour since the requirement is not taken into account. It is thus of interest to combine it with optimisation, in particular when system simulation is costly as for this case study (~ 52 min in total).

Table 1. F and $F\checkmark$: n. of falsifying runs and n. of validated falsifying runs, $\overline{S}$ and $\tilde{S}$: mean n. and median n. of sim., I: infeasible iterations ratio.

Method	F	$F\checkmark$	$\overline{S}$	$\tilde{S}$	I
Ψ-taliro (unconstrained)	100	0	19.8 ± 22.9	11.0	99.5%
Ψ-taliro + penalties	1	1	493.2 ± 20.4	500.0	70.1%
NOMAD + bl-box constr	7	7	383.3 ± 123.1	436.5	67.9%
Nevergrad + bl-box constr	1	1	480.1 ± 59.7	500.0	93.5%
Ψ-taliro + STA	10	10	7.1 ± 4.5	8.0	0%
NOMAD + STA	10	10	31.4 ± 16.35	35.5	0%
Nevergrad + STA	10	10	24.6 ± 10.4	21.0	0%

Table 1 shows the results of Experiments 2–4, in which we used optimisation tools for falsification. Ψ-taliro was always able to falsify the property in the relaxed unconstrained problem formulation in just a few steps. However, none of the falsifying meal plans that it found satisfy the meal plan constraints. In terms of performance, we note a significant advantage of the STA-based falsification. Indeed, it improved the performance of all the optimisation tools we considered. Even if the STA approach yields a higher-dimensional (18 vs 12 dim) search space, i.e. the unit hypercube, it enables a more efficient optimisation process, and there is no risk of encountering infeasible meal plans during the search.

On the other hand, Ψ-taliro with penalties as well as NOMAD and Nevergrad with black-box constraints employed a large part of the simulation budget (respectively 70.1%, 67.9%, and 93.5%) simulating infeasible meal plans. Note that it is nonetheless desirable that infeasible meal plans are simulated and not simply discarded, as the optimisers can use their robustness to guide the search.

5 Conclusions and Future Work

We presented a method for generating uniformly signals from symbolic timed automata (STA). Uniformity favours fair exploration, and increases coverage of the sampled signal space. The method can also be used in combination with optimisation-based falsification, which avoids rejection sampling by encoding STA constraints into simple box constraints that can be directly handled by optimisation solvers. We demonstrated the advantages of our approach on an artificial pancreas case study. We have so far restricted our attention to deterministic STA with linear alphabets for which we can compute exact volumes. We will investigate STA with other kinds of alphabets (*e.g.* Boolean). We also plan to use our sampling methods for learning STA from data.

References

1. Alur, R., Dill, D.L.: A theory of timed automata. Theor. Comput. Sci. **126**(2), 183–235 (1994)
2. Annpureddy, Y., Liu, C., Fainekos, G., Sankaranarayanan, S.: S-TaLiRo: a tool for temporal logic falsification for hybrid systems. In: Abdulla, P.A., Leino, K.R.M. (eds.) TACAS 2011. LNCS, vol. 6605, pp. 254–257. Springer, Heidelberg (2011). https://doi.org/10.1007/978-3-642-19835-9_21
3. Asarin, E., Basset, N., Degorre, A.: Entropy of regular timed languages. Inf. Comput. **241**, 142–176 (2015)
4. Audet, C., Digabel, S.L., Montplaisir, V.R., Tribes, C.: Algorithm: nomad version 4: nonlinear optimization with the mads algorithm. ACM Trans. Math. Softw. **48**(3), 1–15 (2022)
5. Barber, C.B., Dobkin, D.P., Huhdanpaa, H.: The quickhull algorithm for convex hulls. ACM Trans. Math. Softw. **22**(4), 469–483 (1996)
6. Barbot, B., Basset, N., Beunardeau, M., Kwiatkowska, M.: Uniform Sampling for Timed Automata with Application to Language Inclusion Measurement. In: Agha, G., Van Houdt, B. (eds.) QEST 2016. LNCS, vol. 9826, pp. 175–190. Springer, Cham (2016). https://doi.org/10.1007/978-3-319-43425-4_13
7. Barbot, B., Basset, N., Dang, T.: Generation of signals under temporal constraints for CPS testing. In: Badger, J.M., Rozier, K.Y. (eds.) NFM 2019. LNCS, vol. 11460, pp. 54–70. Springer, Cham (2019). https://doi.org/10.1007/978-3-030-20652-9_4
8. Barbot, B., Basset, N., Dang, T., Donze, A., Esposito, M., Nickovic, D.: Signal sampling and optimisation under symbolic timed automata constraints (2025). (Artifact) https://doi.org/10.5281/zenodo.15285372
9. Barbot, B., Basset, N., Donzé, A.: Wordgen : a timed word generation tool. In: Proceedings of the 26th ACM International Conference on Hybrid Systems: Computation and Control, HSCC 2023, San Antonio, TX, USA, 9–12 May 2023, pp. 16:1–16:7. ACM (2023)
10. Barbot, B., Haddad, S., Picaronny, C.: Coupling and importance sampling for statistical model checking. In: Flanagan, C., König, B. (eds.) TACAS 2012. LNCS, vol. 7214, pp. 331–346. Springer, Heidelberg (2012). https://doi.org/10.1007/978-3-642-28756-5_23
11. Barbot, B., Basset, N., Dang, T., Donzé, A., Kapinski, J., Yamaguchi, T.: Falsification of cyber-physical systems with constrained signal spaces. In: NASA Formal Methods, Moffett Field, United States, pp. 420–439 (2020)

12. Bartocci, E., et al.: Specification-based monitoring of cyber-physical systems: a survey on theory, tools and applications. In: Bartocci, E., Falcone, Y. (eds.) Lectures on Runtime Verification. LNCS, vol. 10457, pp. 135–175. Springer, Cham (2018). https://doi.org/10.1007/978-3-319-75632-5_5
13. Basset, N., Dang, T., Gigler, F., Mateis, C., Nickovic, D.: Sampling of shape expressions with shapex. In: Arun-Kumar, S., Méry, D., Saha, I., Zhang, L. (eds.) MEMOCODE '21: 19th ACM-IEEE International Conference on Formal Methods and Models for System Design, Virtual Event, China, 20–22 November 2021, pp. 118–125. ACM (2021)
14. Bertsimas, D., Vempala, S.: Solving convex programs by random walks. J. ACM (JACM) **51**(4), 540–556 (2004)
15. Chen, Y., Dwivedi, R., Wainwright, M.J., Yu, B.: Vaidya walk: a sampling algorithm based on the volumetric barrier. In: 2017 55th Annual Allerton Conference on Communication, Control, and Computing (Allerton), pp. 1220–1227 (2017)
16. Devroye, L.: Non-Uniform Random Variate Generation. Springer-Verlag, New York (1986)
17. Donzé, A., Maler, O.: Robust satisfaction of temporal logic over real-valued signals. In: Chatterjee, K., Henzinger, T.A. (eds.) FORMATS 2010. LNCS, vol. 6246, pp. 92–106. Springer, Heidelberg (2010). https://doi.org/10.1007/978-3-642-15297-9_9
18. Jegourel, C., Legay, A., Sedwards, S.: Importance splitting for statistical model checking rare properties. In: Sharygina, N., Veith, H. (eds.) CAV 2013. LNCS, vol. 8044, pp. 576–591. Springer, Heidelberg (2013). https://doi.org/10.1007/978-3-642-39799-8_38
19. Kannan, R., Lovász, L., Simonovits, M.: Random walks and an o*(n5) volume algorithm for convex bodies. Rand. Struct. Algor. **11**(1), 1–50 (1997)
20. Kannan, R., Narayanan, H.: Random walks on polytopes and an affine interior point method for linear programming. Math. Oper. Res. **37**(1), 1–20 (2012)
21. Legay, A., Delahaye, B., Bensalem, S.: Statistical model checking: an overview. In: Barringer, H., et al. (eds.) RV 2010. LNCS, vol. 6418, pp. 122–135. Springer, Heidelberg (2010). https://doi.org/10.1007/978-3-642-16612-9_11
22. Lovász, L.: Hit-and-run mixes fast. Math. Prog. **86**, 443–461 (1998)
23. Lovász, L., Vempala, S.S.: Hit-and-run from a corner. SIAM J. Comput. **35**(4), 985–1005 (2006)
24. Lovász, L., Vempala, S.S.: Simulated annealing in convex bodies and an $o^*(n^4)$ volume algorithm. J. Comput. Syst. Sci. **72**(2), 392–417 (2006)
25. Maler, O., Nickovic, D.: Monitoring temporal properties of continuous signals. In: Lakhnech, Y., Yovine, S. (eds.) FORMATS/FTRTFT -2004. LNCS, vol. 3253, pp. 152–166. Springer, Heidelberg (2004). https://doi.org/10.1007/978-3-540-30206-3_12
26. Man, C.D., Micheletto, F., Lv, D., Breton, M., Kovatchev, B., Cobelli, C.: The UVA/PADOVA type 1 diabetes simulator. J. Diab. Sci. Technol. **8**(1), 26–34 (2014)
27. Mancini, T., Melatti, I., Tronci, E.: Any-horizon uniform random sampling and enumeration of constrained scenarios for simulation-based formal verification. IEEE Trans. Softw. Eng. **48**(10), 4002–4013 (2022)
28. Nghiem, T., Sankaranarayanan, S., Fainekos, G., Ivancic, F., Gupta, A., Pappas, G.J.: Monte-carlo techniques for falsification of temporal properties of non-linear hybrid systems. In: Johansson, K.H., Yi, W. (eds.) Proceedings of the 13th ACM International Conference on Hybrid Systems: Computation and Control, HSCC 2010, Stockholm, Sweden, 12–15 April 2010, pp. 211–220. ACM (2010)

29. Ničković, D., Qin, X., Ferrère, T., Mateis, C., Deshmukh, J.: Shape expressions for specifying and extracting signal features. In: Finkbeiner, B., Mariani, L. (eds.) RV 2019. LNCS, vol. 11757, pp. 292–309. Springer, Cham (2019). https://doi.org/10.1007/978-3-030-32079-9_17
30. Rapin, J., Teytaud, O.: Nevergrad - a gradient-free optimization platform (2018). https://GitHub.com/FacebookResearch/Nevergrad
31. Rubin, D.B.: The bayesian bootstrap. Annals Stat. 130–134 (1981)
32. Smith, R.L.: Efficient Monte Carlo procedures for generating points uniformly distributed over bounded regions. Oper. Res. **32**(6), 1296–1308 (1984)
33. Thibeault, Q., Anderson, J., Chandratre, A., Pedrielli, G., Fainekos, G.: PSY-TaLiRo: a python toolbox for search-based test generation for cyber-physical systems. In: Lluch Lafuente, A., Mavridou, A. (eds.) FMICS 2021. LNCS, vol. 12863, pp. 223–231. Springer, Cham (2021). https://doi.org/10.1007/978-3-030-85248-1_15
34. Xie, J.: Simglucose v0.2.1 (2018). https://github.com/jxx123/simglucose. Accessed 30 Jan 2025

Programming and Reasoning in Partially Observable Probabilistic Environments

Tobias Gürtler[1](✉) and Benjamin Lucien Kaminski[1,2]

[1] Saarland University Saarland Informatics Campus, Saarbrücken, Germany
{guertler,kaminski}@cs.uni-saarland.de
[2] University College London, London, UK

Abstract. Probabilistic partial observability is a phenomenon occuring when computer systems are deployed in environments that behave probabilistically and whose exact state cannot be fully observed. In this work, we lay the theoretical groundwork for a probabilistic belief programming language – pBLIMP – which maintains a probability distribution over the possible environment states, called a belief state. pBLIMP has language features to symbolically model the behavior of and interaction with the partially observable environment and to condition the belief state based on explicit observations. In particular, pBLIMP programs can perform state estimation and base their decisions (i.e. the control flow) on the likelihood that certain conditions hold in the current state. Furthermore, pBLIMP features unbounded loops, which sets it apart from many other probabilistic programming languages. For reasoning about pBLIMP programs and the situations they model, we present a weakest-precondition-style calculus (wp) that is capable of reasoning about *unbounded* loops. Soundness of our wp calculus is proven with respect to an operational semantics. We further demonstrate how our wp calculus reasons about (unbounded) loops with loop invariants.

Keywords: partial observability · belief programming · weakest preexpectation · probabilistic programming

1 Introduction

Probabilistic partial observability is a phenomenon which occurs when computer systems are deployed in an environment that behaves probabilistically and whose exact state cannot be fully observed [32]. Partial observability arises naturally when computer systems interact with the real world: For example, spacecrafts have to estimate their current position and speed based on inaccurate or noisy measurements [6], a problem which also occurs in autonomous robot navigation [31] or target tracking [5]. Beyond cyber-physical systems, partial observability also affects fields like medicine, where it is unobservable whether a patient is truly infected with a disease or not. Instead, one has to rely on tests which may randomly give wrong results.

All in all, we are dealing with scenarios with probabilistic partial observability: Not only is the true state of the system influenced by randomness, but moreover can the true state not be fully observed. A common approach to partial observability in probabilistic domains is to model such systems by a *partially observable Markov decision process* (POMDP) [8], but these approaches require the state space of the system to be modeled explicitly.

Probabilistic programming languages (PPLs) are well suited to model partial observability [13] compactly, as they can usually reason about conditional probabilities. PPLs such as Anglican [36], Dice [19], Psi [15], WebPPL [17] or Stan [7] – amongst many others – offer a symbolic way to model partially observable environments, and they offer a wide variety of inference algorithms to reason about the state of the unobservable environment. However, modern PPLs largely lack tools to soundly reason about the behavior of unbounded loops, and the exact analysis of probabilistic programs with unbounded loops remains an open problem [21]. To address this, we build upon the (non-probabilistic) work of Atkinson & Carbin [1] and present the probabilistic belief programming language pBLIMP which allows us to model controllers for partially observable environments. Additionally, we present a novel predicate transformer style wp-calculus, which can symbolically reason about pBLIMP programs featuring unbounded loops.

pBLIMP can be categorized as a model PPL with both nested inference and unbounded loops for discrete probabilistic environments. At runtime, the program dynamically keeps track of a probabilistic belief state, i.e. a *probability distribution* over true states. In order to model partially observable probabilistic environments, pBLIMP offers several dedicated language features.

In particular, pBLIMP features sampling from discrete probability distributions to model the behavior of the unobservable environment. pBLIMP also includes a statement to make sound inferences about the state of the environment, and to use the result of those inferences to guide the control flow of the belief program. Finally, under partial observability, it is still possible to obtain partial information about the true state of the environment, and so pBLIMP includes the ability to explicitly model partial observations of the unobservable environment. Making such observations has two effects: For one, the execution randomly branches into a successor belief state for each possible observation, and secondly each successor belief state is conditioned with the corresponding observation. This branching behavior differentiates pBLIMP from other PPLs, as probabilistic programs usually reason about a single, fixed observation. In contrast, pBLIMP allows developers to verify the behavior of a program across all possible sequences of observations which the program may encounter.

To reason about pBLIMP programs, we develop our main contribution: A partial-observability-tailored weakest preexpectation (wp) calculus à la McIver and Morgan [24], allowing us to reason about the expected value of some quantity at termination. This weakest preexpectation calculus offers a novel way to symbolically reason about probabilistic programs with nested inference, and furthermore enables reasoning using *loop invariants*. To our knowledge our weakest preexpectation calculus is the first to handle nested inference.

Contributions. Our work lays the theoretical groundwork to lift belief programming [1] to probabilistic settings, and enables further research on the symbolic analysis of partial observability. We summarize our contributions as follows:

- *Probabilistic Belief Programming:* We present pBLIMP, a model probabilistic programming language featuring nested inference and unbounded loops.
- *Weakest Preexpectation Calculus:* We present a predicate-transformer-style calculus (wp) to reason about the correctness of pBLIMP programs.
- *Effective wp Calculations:* We introduce an expressive grammar for the properties that the above-mentioned wp calculus uses. Under certain reasonable assumptions, this grammar allows for an effective calculation of wp.

2 An Overview of pBLIMP

Consider the (ongoing) treatment of a patient suffering from the fictional disease *exemplitis*. Due to prior data, we know that the patient is sick with a probability of 90%, and we further know that the medicine we prescribe has a 25% chance of curing the disease. Our goal is to prescribe medicine until the patient is cured.

Unfortunately, the true status of having exemplitis is unobservable. Instead, we have to rely on a test which returns correct results with 95% probability. This imperfect test naturally introduces probabilistic partial observability to our problem, as the test only reveals partial information about the disease. Our goal can hence only be to prescribe medicine until we are "reasonable certain" that the patient is cured, i.e. we continue treatment until the chance for the patient to be sick is below some parameter q. We assume that this is the only criterion to discharge the patient, and thus they could remain in medical care arbitrarily long. Figure 1 models the scenario above as a pBLIMP program.

```
1  d = sample(0.9|1> + 0.1|0>);
2  inCare = true;
3  while(inCare){
4    infer(p(d = 1) > q) {
5      d = sample(0.75|d> + 0.25|0>);
6      t = sample(0.95|d> + 0.05|1-d>);
7      observe t;
8    } else {
9      inCare = false;
10   }
11 }
```

Fig. 1. Probabilistic belief program with a loop modeling the treatment of a patient until they are likely cured from a disease.

Sampling. The true state of having exemplitis is modeled by the variable d. Line 1 uses a sample statement to model that the patient initially actually has exemplitis with a probability of 90%. The value of d is not directly accessible to the program. Instead, the pBLIMP program updates the current belief state to a probability distribution where the variable assignment $\{d \mapsto 1\}$ has probability $\frac{9}{10}$ and $\{d \mapsto 0\}$ has probability $\frac{1}{10}$.

Notably, the program is still *deterministically* in a *single* belief state. We then set a boolean flag inCare to represent that the patient is in medical care, and enter an unbounded loop which models their treatment.

Inference. In each loop iteration we decide whether the patient should continue treatment or be discharged. This is modeled with an infer statement, which can be seen as a generalized if statement: Infer statements use the current belief state to determine the probability that a condition holds, and branch based on that value. In Fig. 1, the infer statement in Line 4 infers the likelihood that d = 1 holds, i.e. the probability that the patient is still sick. If the inferred probability is higher than the parameter q, then the patient continues treatment in Lines 5–7, otherwise we terminate the loop by setting inCare to false in Line 9.

Observation. Next, Lines 5–7 model the treatment of the patient. In Line 5, the patient is given medicine which has a 25% chance of curing exemplitis and setting d to 0. Afterwards, Line 6 and 7 model an imperfect test: Line 6 describes the test's behavior, namely it has a 95% chance to return the true status of having exemplitis, and a 5% chance of a wrong result. By nature of the sample statement, the true value of neither d nor t is observable. In Line 7, we now *explicitly observe* the test result t, which has two effects: First, the execution *probabilistically branches* into two possible belief states, one per possible outcome of the test. Second, in both belief states the value of t is now *determined* and in each belief state the underlying probability that the patient is sick is automatically *conditioned* on the observed test outcome.

This branching behavior is non-standard. In most PPLs, observe statements would only condition the program behavior with a single test result, i.e. the developer has to choose if there will be a positive or a negative test. In contrast, pBLIMP branches and considers all possible test results and their likelihood to occur. Hence, we can evaluate the effectiveness of our treatment strategy across all possible test results, and even all possible sequences of test results.

Verification. We can now ask an important question about the program in Fig. 1: Across all possible sequences of test results, how likely are we to discharge a sick patient? Put differently, what is the probability that the program terminates with d = 1? Many tools have been developed to answer such queries, but our example includes several features which complicate inference:

For one, the infer statement forces us to perform *nested inference*, meaning we need to perform inference about programs which may themselves perform inferences. In practice, nested queries naturally arise from programs which perform probabilistic inference as a sub-routine, for example to track moving objects [10,28], or to perform state estimation [35]. Reasoning about programs with such nested queries in turn requires nested inference. While tools for nested inference have been developed [14,17,36], it remains non-standard; e.g. Dice [19] or SOGA [28] do not support nested queries. Additionally, our example included the *parameterized* threshold q, requiring to perform *symbolic* inference and the result of our analysis needs to be parameterized in q. This leads us to PPLs like PSI [15], which implements exact symbolic inference methods and can answer our query. However, PSI does not support unbounded loops. In fact, many, if not most PPLs do not support unbounded loops and instead opt to approximating

$$
\begin{array}{rcl}
E_o & ::= & x_o \in \mathsf{Vars}_o \mid E_o \oplus E_o \mid c \in \mathbb{N} \qquad\qquad\qquad \oplus \in \{+,-,\cdot,/\} \\
E_u & ::= & x_u \in \mathsf{Vars}_u \mid E_u \oplus E_u \mid E_o \\
P_o & ::= & E_o \sim E_o \mid P_o \wedge P_o \mid P_o \vee P_o \mid !P_o \mid b \in \{1,0\} \quad \sim \in \{<,\leq,=,\neq,\geq,>\} \\
P_u & ::= & E_u \sim E_u \mid P_u \wedge P_u \mid P_u \vee P_u \mid !P_u \mid P_o \\[4pt]
C & ::= & \texttt{skip} \mid C\,\mathring{;}\,C \mid x_o := E_o \mid \texttt{if}\,(P_o)\,\{C\}\,\texttt{else}\,\{C\} \mid \texttt{while}\,(P_o)\,\{C\} \\
 & & \mid\ y_o := \texttt{observe}\,(x_u) \mid \texttt{infer}\,(p(P_u) \in i)\,\{C\}\,\texttt{else}\,\{C\} \\
 & & \mid\ x_u := \texttt{sample}\,(f)\ ,\quad \text{where } f\colon \varSigma \to \mathcal{D}(\mathbb{N})
\end{array}
$$

Fig. 2. Formal definition of the model language pBLIMP.

them via bounded loops. This, however, yields no guarantees on the quality of the approximation [37].

To address this gap, we present a weakest preexpectation calculus wp à la McIver and Morgan [24], which can perform exact symbolic inference on programs with loops. The wp-calculus enables reasoning about the expected value of some quantity at termination. Furthermore, programs with unbounded loops can be analyzed using *loop invariants*, which provide sound upper bounds on the behavior of pBLIMP programs. For the program from Fig. 1, we can formally verify that the probability to discharge a sick patient is always less than q.

3 pBLIMP – Syntax and Semantics

3.1 Syntax of pBLIMP

Figure 2 describes the syntax of pBLIMP. Notably, we consistently delineate between *observable* and *unobservable* constructs, as seen e.g. with variables: Variables are partitioned into observable variables $x_o \in \mathsf{Vars}_o$ and unobservable variables $x_u \in \mathsf{Vars}_u$. Observable variables are like variables in most programming languages: their values are freely accessible. Values of unobservable variables are not freely accessible, and not all program statements may use them.

Expressions and Propositions. Similarly to variables, we use E_o to refer to observable expressions and E_u to refer to unobservable expressions. Observable expressions are either a numeric constant $c \in \mathbb{N}$, an *observable* variable x_o, or are recursively constructed using the standard binary operators $+,-,\cdot,/$ where subtraction truncates at 0. Notably, no *un*observable variable may occur in an observable expression. Unobservable expressions are constructed in the same way, but they *may* contain *un*observable variables x_u.

Propositions are Boolean expressions comparing (un)observable expressions. Analogous to expressions, we distinguish observable propositions P_o and unobservable propositions P_u, see Fig. 2.

pBLIMP Statements. The syntax of pBLIMP statements is defined in Fig. 2, and we use C as a metavariable to range over pBLIMP statements. pBLIMP is based on the imperative model programming language IMP [39], and hence contains all standard imperative programming concepts such as skipping, assignments, loops, etc. Notably, all statements which pBLIMP inherits from IMP are restricted to observable expressions or propositions. This syntactically guarantees that all standard statements behave in a completely standard way (apart from being restricted to positive numbers). In addition to IMP, pBLIMP offers three additional, non-standard, statements over which we go in the following.

Sampling. A *sampling* statement $x_u \coloneqq \texttt{sample}\,(f)$ assigns a random value to an *un*observable variable x_u using a function f of type $\Sigma \to \mathcal{D}(\mathbb{N})$ which maps a variable assignment $\sigma \in \Sigma$ to a discrete *probability distribution* over possible values in $\mathbb{N}$. Here, $f(\sigma)(c)$ describes the probability to sample the value $c \in \mathbb{N}$, assuming that the current state aligns with the variable assignment σ. We use the notation $f(\sigma, c) = f(\sigma)(c)$ for readability.

Observing. $y_o \coloneqq \texttt{observe}\,(x_u)$ allows to *explicitly observe (or measure) the value of an unobservable variable* x_u, thus deliberately lifting the veil of partial observability from the value of x_u. For technical reasons, x_u remains an *un*observable variable even after the observe statement, as otherwise x_u would be typed ambiguously after executing `if (P) {observe x} else {skip}`. To anyway have access to the observed value of x_u, that value is assigned to an *observable* variable y_o. We use $\texttt{observe}\,(x_u)$ as syntactic sugar for $y_o \coloneqq \texttt{observe}\,(x_u)$, where y_o is never used in the program (see e.g. Line 7 in Fig. 1).

Inferring. The infer statement $\texttt{infer}\,(\,p(P_u) \in i\,)\,\{\,C_1\,\}\,\texttt{else}\,\{\,C_2\,\}$ can be understood as a generalized if-statement. Rather than being restricted to observable propositions as branching conditions, infer statements can branch on an unobservable proposition P_u. The infer statement guides the control flow by determining the probability that P_u holds, and executes either C_1 or C_2, depending on whether or not that probability falls into the interval i or not.

Remark 1 (Restrictions on Unobserved Variables). We would like to justify the restrictions on unobservable variables with an example: Anna and Bob are playing a game. Anna tosses a coin so that Bob cannot see the result, and Bob has to guess the outcome. If Bob guesses correctly, he wins the game, otherwise Anna wins. If Anna is not lying and the coin is fair, then every possible guessing strategy of Bob should have a win-rate of 50%. Assume now that we allow unobservable propositions as conditions of if-statements. Then Bob could model his strategy and the outcome of the game with the following program:

```
1 coin = sample(0.5|Heads> + 0.5|Tails>);
2 if (coin = Heads) {guess = Heads;} else {guess = Tails;}
3 actual = observe coin;
```

Here, `0.5|Heads> + 0.5|Tails>` is a distribution which assigns the value `Heads` to `coin` with a probability of $\frac{1}{2}$, and similarly for `Tails`. Bob makes his guess in Line 2 using an if-statement, which uses coin in its condition to ignore the partial observability which was introduced by the sample statement. Afterwards, we observe the actual outcome of the coin, and we find that the program will always terminate in a state where actual and guess are equal, thus giving Bob an impossible win-rate of 100%.   ◁

3.2 Operational Semantics of pBLIMP

Before defining operational semantics of pBLIMP, we first need to define the notions of variable assignments and belief states.

Variable Assignments. A variable assignment $\sigma \in \Sigma$ assigns a value $c \in \mathbb{N}$ to every (observable and unobservable) variable x, written as $\sigma(x) = c$. We denote by $\sigma(E)$ the evaluation of an expression E under σ (to a number in $\mathbb{N}$), and by $\sigma(P)$ the evaluation of a proposition P under σ (to a Boolean in $\{0, 1\}$).

Belief States. A belief state $\beta \in \mathcal{D}(\Sigma)$ is a probability distribution over variable assignments $\sigma \in \Sigma$ and we denote by $\beta(\sigma)$ the probability of σ under β. Furthermore, we will use bra-ket notation $\beta = p_1|\sigma_1\rangle + \ldots + p_n|\sigma_n\rangle$ to denote a belief state with $\beta(\sigma_i) = p_i$. Belief states capture the inherent uncertainty in pBLIMP programs which have no direct access to the values of the unobservable variables, i.e. $\beta(\sigma)$ represents the likelihood that σ is the true variable assignment.

We say that σ is *possible* within β if $\beta(\sigma) > 0$. We call β *consistent* if all possible variable assignments within β agree on the values of the *observable* variables. For the remainder of this paper, we will only consider consistent belief states. Notably, all operations on belief states we present will preserve consistency. We can *evaluate a proposition P under a belief state β* as follows:

$$Pr_\beta(P) := \sum_{\sigma \in \Sigma} \beta(\sigma) \cdot \sigma(P) \qquad (\star)$$

$Pr_\beta(P)$ yields a value in $[0, 1]$ and can be understood as the likelihood that P holds in the true state if β is our current belief state. If β is consistent and P is an observable proposition, then either $Pr_\beta(P) = 0$ or $Pr_\beta(P) = 1$ holds.

Configurations. A program configuration is a quadruple (C, β, σ, p), where:

- C is either a pBLIMP statement that remains to be executed or the special symbol $\top$ that indicates termination.
- β is the current belief state.
- σ is the current (true) variable assignment, which captures the true state of both the observable and unobservable variables.
- p stores the probability of the execution path up to this configuration.

Here, C and β capture the state of the program, as well as its current knowledge about the true state of the (unobservable) environment. However, the program directly interacts with the environment during observations, and hence σ is included in the program configuration to model the unobservable environment. We now provide a small-step operational semantics that steps through the program by applying the transition rules in Fig. 3. Let us go over these rules.

$$\frac{\beta\left[x_o \mapsto E_o\right](\sigma') := \sum_{\rho[x_o \mapsto E_o] \,=\, \sigma'} \beta(\rho)}{(x_o := E_o, \beta, \sigma, p) \,\triangleright\, (\top, \beta\left[x_o \mapsto E_o\right], \sigma\left[x_o \mapsto E_o\right], p)} \qquad \overline{(\texttt{skip}, \beta, \sigma, p) \,\triangleright\, (\top, \beta, \sigma, p)}$$

$$\frac{f(\sigma, c) \,=\, q \qquad q \,>\, 0}{(x_u := \texttt{sample}\,(f), \beta, \sigma, p) \,\triangleright\, (\top, \beta\left[x_u \mapsto f\right], \sigma\left[x_u \mapsto c\right], p \cdot q)}$$

$$\frac{\sigma(x_u) \,=\, c}{(y_o := \texttt{observe}\,(x_u), \beta, \sigma, p) \,\triangleright\, (\top, \beta|_{x_u = c}\left[y_o \mapsto c\right], \sigma\left[y_o \mapsto c\right], p)}$$

$$\frac{Pr_\beta(P_o) \,=\, 1}{(\texttt{if}\,(P_o)\,\{C_1\}\,\texttt{else}\,\{C_2\}, \beta, \sigma, p) \,\triangleright\, (C_1, \beta, \sigma, p)}$$

$$\frac{Pr_\beta(P_o) \,=\, 0}{(\texttt{if}\,(P_o)\,\{C_1\}\,\texttt{else}\,\{C_2\}, \beta, \sigma, p) \,\triangleright\, (C_2, \beta, \sigma, p)}$$

$$\overline{(\texttt{while}\,(P_o)\{C\}, \beta, \sigma, p) \,\triangleright\, (\texttt{if}\,(P_o)\,\{C\,\texttt{;}\,\texttt{while}\,(P_o)\{C\}\}\,\texttt{else}\,\{\texttt{skip}\}, \beta, \sigma, p)}$$

$$\frac{Pr_\beta(P_u) \,\in\, i}{(\texttt{infer}\,(P_u \in i)\,\{C_1\}\,\texttt{else}\,\{C_2\}, \beta, \sigma, p) \,\triangleright\, (C_1, \beta, \sigma, p)} \qquad \frac{(C_1, \beta, \sigma, 1) \,\triangleright\, (C_1', \beta', \sigma', q) \qquad C_1' \,\neq\, \top}{(C_1\,\texttt{;}\,C_2, \beta, \sigma, p) \,\triangleright\, (C_1'\,\texttt{;}\,C_2, \beta', \sigma', p \cdot q)}$$

$$\frac{Pr_\beta(P_u) \,\notin\, i}{(\texttt{infer}\,(P_u \in i)\,\{C_1\}\,\texttt{else}\,\{C_2\}, \beta, \sigma, p) \,\triangleright\, (C_2, \beta, \sigma, p)} \qquad \frac{(C_1, \beta, \sigma, 1) \,\triangleright\, (\top, \beta', \sigma', q)}{(C_1\,\texttt{;}\,C_2, \beta, \sigma, p) \,\triangleright\, (C_2, \beta', \sigma', p \cdot q)}$$

Fig. 3. Operational transition rules for configurations of the form (C, β, σ, p). $\beta\left[x_u \mapsto f\right]$ and $\beta|_{x_u = c}$ are defined in † and ‡ respectively.

Assignments. $x := E_o$ deterministically modifies the value of the observable variable x_o to the observable expression E_o. We thus need to update the value of x_o in the true variable assignment σ and in all possible variable assignments within β. The updated variable assignment is given by:

$$\sigma\left[x \mapsto E\right](y) := \begin{cases} \sigma(y) & , x \neq y \\ \sigma(E) & , x = y \end{cases}$$

$\beta\left[x_o \mapsto E_o\right]$ is obtained from β by applying $\left[x_o \mapsto E_o\right]$ to all possible variable assignments within β and possibly summing the probability masses should two or more formerly different variable assignments coincide after applying $\left[x_o \mapsto E_o\right]$.

Sampling. The sample statement models the behavior of the environment by assigning a new value to a variable x_u via sampling from the distribution $f(\sigma)$. At this point, the configuration has multiple possible successors, each with a different true variable assignment. To account for the program's uncertainty about the true variable assignment, the belief state splits each possible variable assignment ρ within β in accordance with the distribution $f(\rho)$ to create a new belief state. Formally, the resulting belief state is defined as

$$\beta\left[x \mapsto f\right] := \lambda \sigma \boldsymbol{.} \sum_{\rho \in \Sigma, \rho[x \mapsto \sigma(x)] = \sigma} \beta(\rho) \cdot f(\rho, \sigma(x)) \boldsymbol{.} \tag{†}$$

Here, the likelihood of a variable assignment σ *after* sampling, is described by considering all variable assignments ρ *before* sampling, so that ρ aligns with σ

if x is set to $\sigma(x)$. The updated probability for σ can then be calculated as the sum over all such variable assignments ρ, where we consider the probability to "start" in ρ, i.e. $\beta(\rho)$, and to then sample the value $\sigma(x)$, which is $f(\rho, \sigma(x))$. The premise $q > 0$ prevents division by 0 in the rule for the observe statement.

Observing. $y_o := $ observe (x_u) has two effects: Determinizing the value of x_u within the belief state β, and assigning the value of x_u to the variable y_o. The latter effect is analogous to the assignment, and we will thus not elaborate further on the effect it has on the belief state and the true variable assignment.

The true variable assignment σ models the state of the environment which is hidden behind partial observability, and thus σ fixes the true value of x_u. The observe statement passes the true value of x_u to the program, and uses it to condition the belief state: As we now know the true value of x_u from σ, some of the variable assignment within β are potentially no longer possible, simply because they map x_u to the wrong value. We incorporate this information into β by filtering out variable assignments which do not match our observation, and then re-normalizing the distribution. Formally, this is described as:

$$\beta|_{x=c} := \lambda\sigma \cdot \begin{cases} \frac{\beta(\sigma) \cdot (\sigma(x=c))}{Pr_\beta(x=c)}, & \text{if } Pr_\beta(x=c) > 0 \\ 0, & \text{otherwise.} \end{cases} \quad (\ddagger)$$

For example, observing x on belief state $\frac{1}{3}|0\,0\rangle + \frac{1}{3}|1\,0\rangle + \frac{1}{6}|0\,1\rangle + \frac{1}{6}|1\,1\rangle$ will branch with probability $\frac{1}{3} + \frac{1}{6} = \frac{1}{2}$ into $\frac{2}{3}|0\,0\rangle + \frac{1}{3}|0\,1\rangle$ and with probability $\frac{1}{3} + \frac{1}{6} = \frac{1}{2}$ into $\frac{2}{3}|1\,0\rangle + \frac{1}{3}|1\,1\rangle$. This branching is not directly represented in the operational semantics above, but it is the result of the sample statements which set up the believe state *prior* to the observation and branched out into multiple configurations with the same belief state. However, for the program, which only has access to the belief state, it appears as though the execution branches at the observe statement.

Conditional Branching and Inferring. Both the if- and the infer-statement pick a branch depending on their condition. For the if-statement, the condition is an observable proposition P_o. The configuration transitions to C_1 if $Pr_\beta(P_o) = 1$, otherwise the configuration transitions to C_2. We always have either $Pr_\beta(P_o) = 1$ or $Pr_\beta(P_o) = 0$ for all observable propositions, as we only consider consistent belief states. For the infer-statement, the condition is an unobservable proposition P_u which is evaluated under the belief state β (cf. $(\star)$ on p. 7). The configuration transitions to C_1 if the value of $Pr_\beta(P_u)$ falls into the interval i, otherwise the configuration transitions to C_2.

While Loop, Composition and Skip. The remaining statements follow their typical semantics. The while loop is unrolled into if-statements until the loop guard is no longer satisfied. The composed statement $C_1 \mathbin{;} C_2$ first executes C_1 step-wise until termination, i.e. until $\top; C_2$ would be reached, and then behaves like C_2. The skip statement terminates with no effect.

3.3 Execution and Termination

Starting in a configuration (C, β, σ, p), we can exhaustively apply the transition rules to create a computation tree which includes all possible execution paths. In this tree, the leaves are terminal configurations of the form $(\top, \beta', \sigma', p')$. If the root configuration of the tree satisfies $p = 1$, then p' encodes the probability to reach that particular leaf. However, note that the tree may contain multiple copies of this leaf, and thus p' by itself is *not* the probability to reach the configuration $(\top, \beta', \sigma', p')$. We can take the sum over all leaves of the tree to get a sub-distribution over the terminal configurations. We consider sub-distributions, i.e. probability distributions which possibly add up to less than 1, as there may be diverging paths in the computation tree. We will write $[C]^{\beta,\sigma}$ for the sub-distribution created by the tree starting from the configuration $(C, \beta, \sigma, 1)$.

$[C]^{\beta,\sigma}$ can be used to determine the probability to reach a terminal configuration containing the belief state β', written as $[C]^{\beta,\sigma}(\beta')$, by considering all terminal configurations in which β' occurs, and summing up the probabilities to reach those configurations. Analogously, we can also determine the probability to terminate in configuration containing the belief state β' and the variable assignment σ', written as $[C]^{\beta,\sigma}(\beta', \sigma')$. We can consider execution from an initial belief state as the expected outcome over all possible initial configurations with different true variable assignments σ. Thus, we define the probability to reach a terminal state containing β' starting from a belief state β as:

$$[C]^{\beta}(\beta') := \sum_{\sigma \in \Sigma} \beta(\sigma) \cdot [C]^{\beta,\sigma}(\beta')$$

Belief State Accuracy. We can now ask how the belief state β and the true variable assignment σ relate. In particular, if we terminate with a belief state β, we expect the probability to also terminate with the true system state σ to be $\beta(\sigma)$. This ensures that we can make statements about the true variable assignment based only on the belief state, which is necessary as the programmer only has access to the latter. We can formalize this property as follows:

Theorem 1 (Belief Accuracy Preservation). *Let C be a program, β_0 be the initial belief state, β be a belief state, and σ be a variable assignment. Then we have:*

$$[C]^{\beta_0}(\beta, \sigma) = \beta(\sigma) \cdot [C]^{\beta_0}(\beta)$$

Proof. By structural induction on C. We refer to the extended version [18]. □

This theorem guarantees that the belief state accurately represents the unobservable environment. Due to this, it is possible to drop the true variable assignment from the semantics altogether, and to define a semantics where configurations only maintain the belief state. Notably, program executions then branch at the observe statement, whereas the sample statement updates the belief state in a deterministic manner. For details on this we refer to the extended version [18].

4 Weakest Preexpectations for pBLIMP

This section introduces a weakest preexpectation calculus for pBLIMP programs. The calculus enables verification of quantitative properties on pBLIMP programs, thus yielding hard guarantees on the execution of such programs.

4.1 Weakest Preexpectation Calculus

Usually, predicates are functions mapping to 1 or 0, depending on whether the predicate holds or not. To account for the probabilistic uncertainty within pBLIMP, we generalize to quantitative predicates which map belief states to any non-negative value (or infinity), i.e. we define a predicate F to be a function of type $F\colon \mathcal{D}(\Sigma) \to \mathbb{R}_{\geq 0}^{\infty}$. Thanks to Theorem 1, these predicates still allow us to reason about the true variable assignment, e.g. the predicate $F(\beta) := Pr_\beta(x > 3)$ describes the probability that $x > 3$ holds in the true variable assignment.

Including infinity in the co-domain of predicates is a technical requirement for obtaining a *complete lattice* of predicates. We adhere to the convention $0 \cdot \infty = 0$ and moreover $\frac{a}{0} = 0$ for all a including ∞. We denote by $Pr(P)$ the predicate $\lambda\beta.\ Pr_\beta(P)$, and we define predicate manipulations such as $F[x \mapsto E] := \lambda\beta.\ F(\beta[x \mapsto E])$. Similarly, we define $F[x \mapsto f]$ (cf. (†) on p. 8) and $F|_{x=c}$ (cf. (‡) on p. 9). To reason about predicates in the context of belief programs, we present a predicate transformer function wp, which is designed in the style of McIver & Morgan [24]. For our purpose, wp is of type

$$\mathsf{wp}\colon \quad \mathsf{pBLIMP} \quad \to \quad \bigl(\mathcal{D}(\Sigma) \to \mathbb{R}_{\geq 0}^{\infty}\bigr) \quad \to \quad \bigl(\mathcal{D}(\Sigma) \to \mathbb{R}_{\geq 0}^{\infty}\bigr)\ .$$

The rules which define wp can be found in Fig. 4. To our knowledge, Olmedo et al. [27] and Symczakk & Katoen [33] (and in a broader sense also Nori et al. [26]) are the only works which present weakest preexpectation calculi that include conditioning statements. Still, none of those works support infer-statements.

The function wp determines the weakest preexpectation of a given belief program C with respect to a predicate F. Intuitively, this can be understood as "predicting" the value of F in the terminal state which is reached after executing the program C. But, belief programs are probabilistic, meaning there may not be just a single terminal state to reach. Hence, wp instead determines the *expected value* of F after executing C, hence also the term weakest pre*expectation*. Formally, $\mathsf{wp}\,[\![C]\!]\,(F)\,(\beta)$ can be understood as the expected value of the predicate F on all terminal states after executing the program C starting in the belief state β. We now consider the rules from Fig. 4 with this intuition in mind.

Assignment and Sampling. As described above, wp determines the expected value of the predicate F after executing the program C. For the assignment, there is only a single possible terminal belief state, namely $\beta[x \mapsto E]$. We can thus determine the expected value of F at termination by evaluating F in the belief state $\beta[x \mapsto E]$, which amounts to the predicate $F[x \mapsto E]$. Similarly, the sample statement also only has a single reachable terminal belief state, namely $\beta[x \mapsto f]$, and thus its weakest preexpectation is $F[x \mapsto f]$.

C	wp $[\![C]\!]\,(F)$
skip	F
$x := E$	$F\,[x \mapsto E]$
if $(P_o)\,\{C_1\}$ else $\{C_2\}$	$Pr(P_o) \cdot$ wp $[\![C_1]\!]\,(F) + (1 - Pr(P_o)) \cdot$ wp $[\![C_2]\!]\,(F)$
while $(P_o)\{C\}$	lfp $X.\ (1 - Pr(P_o)) \cdot F\ +\ Pr(P_o) \cdot$ wp $[\![C]\!]\,(X)$
$C_1 \mathbin{;} C_2$	wp $[\![C_1]\!]\,($wp $[\![C_2]\!]\,(F))$
$x := \mathtt{sample}\,(f)$	$F\,[x \mapsto f]$
$y := \mathtt{observe}\,(x)$	$\displaystyle\sum_{c \in \mathbb{N}} Pr(x = c) \cdot F|_{x=c}\,[y \mapsto c]$
$\mathtt{infer}\,(p(P) \in i)\,\{C_1\}$ else $\{C_2\}$	$[Pr(P) \in i] \cdot$ wp $[\![C_1]\!]\,(F)\ + [Pr(P) \notin i] \cdot$ wp $[\![C_2]\!]\,(F)$

Fig. 4. Rules for the weakest preexpectation calculus wp for pBLIMP.

Observing. Consider the execution of an observe statement from the initial belief state β. Notably, if we observe x to have value c, which occurs with probability $Pr_\beta\,(x = c)$ by Theorem 1, then we always terminate with the belief state $\beta|_{x=c}\,[y \mapsto c]$. In line with the intuition for wp, the weakest preexpectation thus determines the expected value of F over all such observations: When we observe x to be c, which happens with probability $Pr(x = c)$, then we terminate with the belief state $\beta|_{x=c}\,[y \mapsto c]$, which will yield the "reward" $F|_{x=c}\,[y \mapsto c]$.

Conditionals and Inference. Both the if- and the infer-statement execute either C_1 or C_2, depending on whether or not their condition holds. If the condition holds, then C_1 is executed and we can predict the expected value of F with wp $[\![C_1]\!]\,(F)$, otherwise C_2 is executed and we can use wp $[\![C_2]\!]\,(F)$. What remains is to distinguish the two cases: For the if-statement, the condition P_o is an observable proposition and $Pr_\beta\,(P_o)$ evaluates to either 0 or 1. We can thus multiply with $Pr(P_o)$ to "select" the correct preexpectation. For the infer-statement, the same principle applies, but we use the Iverson bracket $[Pr(P) \in i]$ to distinguish the two cases, where we define $[\varphi]$ as:

$$[\varphi]\,(\beta) := \lambda\,\beta.\ \begin{cases} 1, & \text{if } \beta \text{ satisfies } \varphi \\ 0, & \text{otherwise} \end{cases}$$

While Loop. For the while-loop, wp is defined as the least fixed point over the loop's characteristic function $\Phi(X) = (1 - Pr(P_o)) \cdot F + Pr(P_o) \cdot$ wp $[\![C]\!]\,(X)$. Φ is closely related to the weakest preexpectation of if-statements, as while loops are unrolled into if-statements during execution. Calculating the least fixed point is undecidable [29], which means that determining wp $[\![C]\!]\,(F)$ is also undecidable in general. Still, the fixed point can be lower-bounded by repeatedly applying Φ to the predicate $\mathbf{0} := \lambda\,\beta.\ 0$. This can be seen as approximating the weakest preexpectation of a general while loop with the weakest preexpectation of bounded while loops with an ever increasing number of iterations. Formally, we even have:

$$\text{wp} [\![\texttt{while} \, (\, P_o \,) \, \{ \, C \, \}]\!] \, (F) = \text{lfp} \, \Phi = \lim_{n \to \infty} \Phi^n(\mathbf{0})$$

Composition and Skip. Both statements follow the classical rules for wp. The composition $C_1 \, \fatsemi \, C_2$ applies wp step-wise by first determining the weakest preexpectation for C_2 with the predicate F, and then uses the result as the predicate for the weakest preexpectation of C_1. The skip statement has no effect.

4.2 Soundness

As previously mentioned, our weakest preexpectation calculus shall capture the expected value of F after executing the program C. This idea is formally captured in the following soundness theorem:

Theorem 2 (Soundness). *For all programs C, initial belief states β_0, and predicates F, we have*

$$\text{wp} [\![C]\!] \, (F) \, (\beta_0) \quad = \quad \sum_{\beta \in \mathcal{D}(\Sigma)} [C]^{\beta_0}(\beta) \cdot F(\beta) \; .$$

Proof. By structural induction over C. See the extended version [18] for details. □

Note that the soundness theorem refers to the sub-distribution $[C]^{\beta_0}$. Thus, wp only considers terminal configurations, and we cannot gain significant insights into phenomenons which are not captured by terminal states. However, divergence still affects the value of $\text{wp} [\![C]\!] \, (F) \, (\beta_0)$. The missing probability mass in $[C]^{\beta_0}$ amounts to the probability that C diverges, and is automatically weighted at 0 in the soundness theorem, regardless of F.

Additionally, we will briefly discuss two useful properties which allows us to calculate wp more easily. We refer to the extended version [18] for a more formal introduction of these properties. The first property is *linearity*, which is characterized by the following equation for predicates F, G and constant $\alpha \in \mathbb{R}_{\geq 0}$:

$$\text{wp} [\![C]\!] \, (\alpha \cdot F + G) \quad = \quad \alpha \cdot \text{wp} [\![C]\!] \, (F) \; + \; \text{wp} [\![C]\!] \, (G)$$

In particular, linearity enables *decomposing* predicates, making wp computations amenable to parallelization. The second property is *independence*, which allows one to "skip" calculating $\text{wp} [\![C]\!] \, (F)$ for a predicate if the program C does not affect the value of F. For example, one can show that if the loop-free program C does not modify any variables in the predicate P, then we have:

$$\text{wp} [\![C]\!] \, (Pr(P)) \quad = \quad Pr(P)$$

$$E ::= x \in \mathsf{Vars}_o \cup \mathsf{Vars}_u \mid c \in \mathbb{N} \mid [P_u] \mid E \oplus E \qquad \oplus \in \{+, -, \cdot, /\}$$
$$\Phi ::= Ex(E) \sim q \cdot Ex(E) \qquad \sim \in \{<, \leq, =, \neq, \geq, >\}$$
$$G ::= Ex(E) \mid G + G \mid \alpha \cdot G \mid [\Phi] \cdot G$$

$$Ex(E) := \lambda \beta. \sum_{\sigma \in \Sigma} \beta(\sigma) \cdot \sigma(E)$$

Fig. 5. An expressive grammar G for predicates.

4.3 Structuring Predicates

There is another hurdle to an efficient computation of wp: If we apply the rules from Fig. 4 naively, then the predicates we push backwards through the program would accumulate modifiers such as $[t \mapsto f]$ and we would end up with terms like $Pr(d = 1)|_{t=0} [t \mapsto f] [d \mapsto g]$, which quickly becomes infeasible for larger programs. To address this issue, we present the set of predicates G as defined by the grammar in Fig. 5, which can simplify terms like $F [x \mapsto f]$ efficiently. Here, $E \colon \Sigma \to \mathbb{N}$ is a G-expression which additionally may include propositions P, which will be evaluated to either 0 or 1 in accordance with $\sigma(P)$. The predicate $Ex(E)$ is defined as the expected value of E within β. Note that $Pr(P) = Ex([P])$, which will be useful when we consider wp. Furthermore, G is expressive with respect to wp for loop-free programs, as formalized below:

Theorem 3 (Expressiveness of G). *For any syntactically expressible loop-free program C and predicate $F \in G$ we have* $\mathsf{wp}[\![C]\!](F) \in G$

Proof. By structural induction on C, see the extended version [18]. □

There are two restrictions on in Theorem 3, namely the programs need to be *loop-free* and *syntactically expressible*. A grammar for programs with loops is beyond the scope of this paper, but loop-free programs still allow us to reason about bounded loops or loop invariants. Being syntactically expressible does not significantly restrict programs, but it still guarantees that programs are sufficiently structured to easily resolve predicate modifiers. We consider C to be *syntactically expressible*, if it adheres to the following restrictions:

- The conditions of infer statements may only feature such intervals i that can be expressed using a comparison operator.
- The functions for sample statements must be expressible as finite sums $f(\sigma) = \sum p_i | \sigma(E_i) \rangle$ for some fixed G-expressions E_i and probabilities p_i.

Notably, Theorem 3 guarantees that any predicate modifier which is introduced during the computation of wp will be removed. To do so, every modifier applied to a predicate in G will first be passed down recursively until we reach the base case $Ex(E)$. Then, any given modifier can be resolved syntactically as follows:

Lemma 1 (Resolution). *Let E, E' be G-expressions, x be a variable, $c \in \mathbb{N}$ be a value and $f = \sum p_i | E_i \rangle$ be the function of a sample statement. Then we have:*

$$Ex(E)[x \mapsto E'] = Ex(E[x/E']) \quad \text{(Assign)}$$

$$Ex(E)|_{x=c} = \frac{Ex(E \cdot [x=c])}{Ex([x=c])} \quad \text{(Observe)}$$

$$Ex(E)[x \mapsto f] = Ex(E \leftarrow f) \quad \text{(Sample)}$$

where $E[x/E']$ is the G-expression E with all occurrences of x replaced by E', and $E \leftarrow f := \sum_{i \in \mathbb{N}} p_i \cdot E[x/E_i]$

Proof. Equation (Assign) follows from definition of $Ex(E)$ and $\sigma[x \mapsto E'](E) = \sigma(E[x/E'])$. For Eq. (Observe) and Eq. (Sample) we refer to the extended version [18]. Here, $E \leftarrow f$ builds upon Jacobs [20] to describe the outcome of a sampling as the expectation over all possible samples.

4.4 Reasoning About Loops with Loop Invariants for wp

We now showcase how wp can soundly reason about unbounded loops, a task which remains challenging for most PPLs. The weakest preexpectation of a loop with respect to postexpectation f is defined as the least fixed point of the loops characteristic function Φ_f. This fixed point can be under-approximated by repeatedly applying Φ_f to the predicate **0**, where each application yields a tighter *lower* bound on the least fixed point, until we eventually converge (after possibly infinitely many iterations). For *upper* bounds, we can use a loop invariant: Verifying that I is an invariant for the loop while (P_o) { C } is computationally cheaper than iteration, as we only apply Φ_f once and check following inequality:

$$I \geq \Phi_f(I) = (1 - Pr(P_o)) \cdot f + Pr(P_o) \cdot \text{wp} [\![C]\!] (I)$$

```
1  d = sample(0.9|1> + 0.1|0>);
2  inCare = true;
3  while(inCare){
4    infer(p(d = 1) > q) {
5      d = sample(0.75|d> + 0.25|0>);
6      t = sample(0.95|d> + 0.05|1-d>);
7      observe t;
8    } else {
9      inCare = false;
10   }
11 }
```

Fig. 6. Probabilistic belief program C_w

Now, we reconsider the introductory example, restated in Fig. 6 for convenience. The program models a strategy for the ongoing treatment of a (possibly) sick patient, and the goal is to only release the patient once they are cured. Hence, the probability to release a sick patient is highly relevant to us, and we can determine it through the weakest preexpectation of the predicate $Pr(d = 1)$. However, determining the value of $\mathsf{wp} \, [\![C_w]\!] \, (Pr(d = 1))$ is difficult due to the loop, so instead we can show that the following predicate I is an invariant of the loop:

$$I := (1 - Pr(inCare)) \cdot Pr(d = 1) \; + \; Pr(inCare) \cdot q$$

This invariant encodes the two possible outcomes when entering the loop: Either *inCare* is true and the patient takes treatment until the likelihood of the disease is at most q, or the patient never enters treatment and the likelihood of the disease is unchanged. We show I to be an invariant in the extended version [18].

5 Related Work

A very common approach to modeling partial observability is to use a POMDP [12,23,32]. However, POMDPs are based on an explicit representation of the state space, which is in stark contrast to the symbolic approach of pBLIMP and other PPLs. In particular, the symbolic approach of PPLs allows one to compactly model and reason about state spaces without having to consider each state individually. Furthermore, Evans et al. [13] demonstrated how probabilistic programs comparable to pBLIMP can model POMDPs. Using a similar approach, pBLIMP can be used as a language to encode certain POMDPs. We will not compare POMDPs and PPLs as a means to model partial observability in detail, but we refer to Evans et al. [13] for an in-depth description. Instead, we would like to focus more thoroughly on the relation between pBLIMP and other PPLs.

wp-calculi in the style of [24] have long been used as a tool for formal verification of program correctness, and our work builds upon this idea. In the context of PPLs, our wp can be considered an inference method for probabilistic programs with nested inference, i.e. inference about programs which themselves perform inferences. However, not all PPLs feature nested inference, and some languages [19,28] trade off the lower expressiveness for faster inference. Still, PPLs with nested inference can often model the same systems as pBLIMP, even though their observe statement behaves differently. Additionally, these languages commonly support continuous distributions, and they may feature richer interaction with the inferred distribution than pBLIMP's infer statement [3,16,38].

There is a variety of PPLs which use sampling algorithms to perform *approximate* nested inference [3,17,34,36]. Sampling is often faster than exact inference, but it is difficult to get sound bounds on the quality of the approximation [9,11].

In contrast to sampling, the wp calculus performs exact symbolic inference, which can also be performed with Bhat et al. [4] or tools like Hakaru [25], and some PPLs like WebPPL [17] support inference for discrete distributions by enumerating all paths. But, to our knowledge, the only exact symbolic inference tool with nested inference is PSI [15]. PSI is able to analyze programs containing both discrete and continuous distributions, while providing simple descriptions of the inferred distribution. PSI (and most PPLs) does not support unbounded loops, thus programs like Fig. 1 need to be approximated with bounded loops.

This yields either an under-approximation or no guarantees at all, and bounded loops provide no guarantees regarding the accuracy of the final result [37].

Therefore, loop invariants as seen with Fig. 6 are a key distinction between wp and existing inference methods for PPLs. Loop invariants are well-established in the field of program analysis [22], but to our knowledge they have not been applied to probabilistic programs with both conditioning and nested inference. Loop invariants yield a sound over-approximation of the result, but they usually need to be provided by the developer. Still, loop invariants for probabilistic programs without conditioning have successfully been automatically synthesized [2].

6 Conclusion and Future Work

Our work thoroughly laid the theoretical groundwork for probabilistic belief programming, a methodology to write programs for partially observable environments. For this, we introduced the imperative probabilistic belief programming language pBLIMP and provided operational semantics for it. We introduced a weakest preexpectation calculus – sound with respect to the operational semantics – which can be used to reason about quantitative properties of pBLIMP programs. Furthermore, we demonstrated how weakest preexpectations can be calculated in an effective manner, including a treatment of loops. Nonetheless, there are still challenges for a practical implementation of our work. In particular, the automatic derivation of loop invariants [2,30] is a crucial feature for the analysis of loops. In future work we would like to address this gap and implement the efficient and fully automatic analysis of pBLIMP programs.

References

1. Atkinson, E., Carbin, M.: Programming and reasoning with partial observability. Proc. ACM Program. Lang. **4**(OOPSLA), 1–28 (2020)
2. Batz, K., Chen, M., Junges, S., Kaminski, B.L., Katoen, J.P., Matheja, C.: Probabilistic program verification via inductive synthesis of inductive invariants. In: International Conference on Tools and Algorithms for the Construction and Analysis of Systems, pp. 410–429. Springer, Cham (2023)
3. Baudart, G., Mandel, L., Atkinson, E., Sherman, B., Pouzet, M., Carbin, M.: Reactive probabilistic programming. In: Proceedings of the 41st ACM SIGPLAN Conference on Programming Language Design and Implementation, pp. 898–912 (2020)
4. Bhat, S., Borgström, J., Gordon, A.D., Russo, C.: Deriving probability density functions from probabilistic functional programs. In: Tools and Algorithms for the Construction and Analysis of Systems: 19th International Conference, TACAS 2013, Held as Part of the European Joint Conferences on Theory and Practice of Software, ETAPS 2013, Rome, Italy, 16–24 March 2013. Proceedings 19, pp. 508–522. Springer, Cham (2013)
5. Bilik, I., Tabrikian, J.: Maneuvering target tracking in the presence of glint using the nonlinear gaussian mixture Kalman filter. IEEE Trans. Aerosp. Electron. Syst. **46**(1), 246–262 (2010)

6. Burkhart, P.D., Bishop, R.H.: Adaptive orbit determination for interplanetary spacecraft. J. Guid. Control. Dyn. **19**(3), 693–701 (1996)
7. Carpenter, B., et al.: Stan: a probabilistic programming language. J. Stat. Softw. **76** (2017)
8. Cassandra, A.R.: A survey of pomdp applications. In: Working Notes of AAAI 1998 Fall Symposium on Planning with Partially Observable Markov Decision Processes, vol. 1724 (1998)
9. Chatterjee, S., Diaconis, P.: The sample size required in importance sampling. Ann. Appl. Probab. **28**(2), 1099–1135 (2018)
10. Cheng, E.Y., Atkinson, E., Baudart, G., Mandel, L., Carbin, M.: Inference plans for hybrid particle filtering. Proc. ACM Program. Lang. **9**(POPL), 271–299 (2025)
11. Cowles, M.K., Carlin, B.P.: Markov chain Monte Carlo convergence diagnostics: a comparative review. J. Am. Stat. Assoc. **91**(434), 883–904 (1996)
12. Durbin, R., Eddy, S.R., Krogh, A., Mitchison, G.: Biological Sequence Analysis: Probabilistic Models of Proteins and Nucleic Acids. Cambridge University Press (1998)
13. Evans, O., Stuhlmüller, A., Salvatier, J., Filan, D.: Modeling Agents with Probabilistic Programs (2017). http://agentmodels.org. Accessed 11 Mar 2024
14. Gehr, T., Misailovic, S., Vechev, M.: PSI: exact symbolic inference for probabilistic programs. In: Chaudhuri, S., Farzan, A. (eds.) CAV 2016. LNCS, vol. 9779, pp. 62–83. Springer, Cham (2016). https://doi.org/10.1007/978-3-319-41528-4_4
15. Gehr, T., Steffen, S., Vechev, M.: λPSI: exact Inference for Higher-Order Probabilistic Programs. In: Proceedings of the 41st ACM SIGPLAN Conference on Programming Language Design and Implementation. ACM (2020). https://doi.org/10.1145/3385412.3386006
16. Goodman, N., Mansinghka, V., Roy, D.M., Bonawitz, K., Tenenbaum, J.B.: Church: a language for generative models. arXiv preprint arXiv:1206.3255 (2012)
17. Goodman, N.D., Stuhlmüller, A.: The Design and Implementation of Probabilistic Programming Languages (2014). http://dippl.org. Accessed 11 Mar 2024
18. Gürtler, T., Kaminski, B.L.: Programming and reasoning in partially observable probabilistic environments (2025). https://arxiv.org/abs/2506.13491
19. Holtzen, S., Van den Broeck, G., Millstein, T.: Scaling exact inference for discrete probabilistic programs. Proc. ACM Program. Lang. **4**(OOPSLA), 1–31 (2020)
20. Jacobs, B.: The mathematics of changing one's mind, via Jeffrey's or via pearl's update rule. J. Artif. Intell. Res. **65**, 783–806 (2019)
21. Junges, S., Katoen, J.P., Sanner, S., Van den Broeck, G., Salmani, B.: Scalable analysis of probabilistic models and programs (dagstuhl seminar 23241) (2024)
22. Kaminski, B.L.: Advanced weakest precondition calculi for probabilistic programs. Ph.D. thesis, RWTH Aachen University (2019)
23. Kress-Gazit, H., Fainekos, G.E., Pappas, G.J.: Temporal-logic-based reactive mission and motion planning. IEEE Trans. Rob. **25**(6), 1370–1381 (2009)
24. McIver, A., Morgan, C.: Abstraction, Refinement and Proof for Probabilistic Systems. Monographs in Computer Science. Springer (2005)
25. Narayanan, P., Carette, J., Romano, W., Shan, C., Zinkov, R.: Probabilistic inference by program transformation in hakaru (system description). In: Kiselyov, O., King, A. (eds.) FLOPS 2016. LNCS, vol. 9613, pp. 62–79. Springer, Cham (2016). https://doi.org/10.1007/978-3-319-29604-3_5
26. Nori, A., Hur, C.K., Rajamani, S., Samuel, S.: R2: An efficient MCMC sampler for probabilistic programs. In: Proceedings of the AAAI Conference on Artificial Intelligence, vol. 28 (2014)

27. Olmedo, F., Gretz, F., Jansen, N., Kaminski, B.L., Katoen, J.P., McIver, A.: Conditioning in probabilistic programming. ACM Trans. Program. Lang. Syst. (TOPLAS) **40**(1), 1–50 (2018)
28. Randone, F., Bortolussi, L., Incerto, E., Tribastone, M.: Inference of probabilistic programs with moment-matching gaussian mixtures. Proc. ACM Program. Lang. **8**(POPL), 1882–1912 (2024)
29. Rice, H.G.: Classes of recursively enumerable sets and their decision problems. Trans. Am. Math. Soc. **74**(2), 358–366 (1953)
30. Schröer, P., Batz, K., Kaminski, B.L., Katoen, J., Matheja, C.: A deductive verification infrastructure for probabilistic programs. Proc. ACM Program. Lang. **7**(OOPSLA2), 2052–2082 (2023)
31. Simmons, R., Koenig, S.: Probabilistic robot navigation in partially observable environments (1995)
32. Smallwood, R.D., Sondik, E.J.: The optimal control of partially observable Markov processes over a finite horizon. Oper. Res. **21**(5), 1071–1088 (1973)
33. Szymczak, M., Katoen, J.-P.: Weakest preexpectation semantics for Bayesian inference. In: Bowen, J.P., Liu, Z., Zhang, Z. (eds.) SETSS 2019. LNCS, vol. 12154, pp. 44–121. Springer, Cham (2020). https://doi.org/10.1007/978-3-030-55089-9_3
34. Tavares, Z., Zhang, X., Minaysan, E., Burroni, J., Ranganath, R., Lezama, A.S.: The random conditional distribution for higher-order probabilistic inference. arXiv preprint arXiv:1903.10556 (2019)
35. Thrun, S.: Probabilistic robotics. Commun. ACM **45**(3), 52–57 (2002)
36. Tolpin, D., van de Meent, J.W., Yang, H., Wood, F.: Design and implementation of probabilistic programming language Anglican. In: Proceedings of the 28th Symposium on the Implementation and Application of Functional programming Languages, pp. 1–12 (2016)
37. Torres-Ruiz, M., Piedeleu, R., Silva, A., Zanasi, F.: On iteration in discrete probabilistic programming. In: 9th International Conference on Formal Structures for Computation and Deduction (FSCD 2024). Schloss Dagstuhl–Leibniz-Zentrum für Informatik (2024)
38. Tran, D., Hoffman, M.D., Saurous, R.A., Brevdo, E., Murphy, K., Blei, D.M.: Deep probabilistic programming. arXiv preprint arXiv:1701.03757 (2017)
39. Winskel, G.: The Formal Semantics of Programming Languages: An Introduction. MIT Press (1993)

PyDSMC: Statistical Model Checking for Neural Agents Using the Gymnasium Interface

Timo P. Gros[1,2](✉), Arnd Hartmanns[3], Ivo Hoese[2], Joshua Meyer[1,2],
Nicola J. Müller[1,2], and Verena Wolf[1,2]

[1] German Research Center for Artificial Intelligence (DFKI),
Saarbrücken, Germany
{timo.gros,joshua.meyer,nicola.mueller,
verena.wolf}@dfki.de
[2] Saarland University, Saarland Informatics Campus,
Saarbrücken, Germany
ivho00001@stud.uni-saarland.de
[3] University of Twente, Enschede, The Netherlands
a.hartmanns@utwente.nl

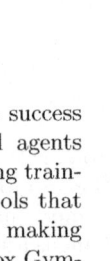

Abstract. Artificial intelligence (AI) has achieved remarkable success in sequential decision-making. However, evaluating its neural agents remains challenging, as current methods often rely on interpreting training curves only, overlooking key statistical factors. Existing tools that allow a formal evaluation also require white-box formal models, making them impractical for most AI benchmarks based on the black-box Gymnasium interface. We introduce **PyDSMC**, a lightweight and easy-to-use Python tool for statistical model checking of neural agents on arbitrary Gymnasium environments. **PyDSMC** automates the selection of statistical methods to compute confidence intervals, supporting both convergence-based and resource-limited evaluation settings. We empirically demonstrate the importance of rigorous agent evaluation and showcase **PyDSMC**'s capabilities to more reliably judge and report an AI agent's performance.

1 Introduction

Artificial intelligence (AI) exerts a significant impact on both everyday life and contemporary research. Deep learning, in particular, is increasingly utilized for *sequential decision-making* (SDM) where it achieved considerable success in areas such as the training of large language models [1,17], improving the prediction of protein foldings [61], or mastering complex computer games such as StarCraft [68] or Dota 2 [12]. We call such AI-based decision-makers *neural agents*.

Despite these great achievements, a critical gap remains in accurately measuring the performance of neural agents. Even landmark papers introducing state-of-the-art algorithms like DQN [52], PPO [60], SAC [41], RND [22], and DreamerV2 [42] typically assess agent performance only through training curves, i.e.,

© The Author(s), under exclusive license to Springer Nature Switzerland AG 2026
P. Prabhakar and A. Vandin (Eds.): QEST+FORMATS 2025, LNCS 16143, pp. 134–156, 2026.
https://doi.org/10.1007/978-3-032-05792-1_8

the accumulated sum of rewards/scores over training time. These assessments typically fail to account for important influences, such as the system's variance, the limited number of samples used for each score in the training curve, or the effects of exploration strategies [35,36]. This calls for a more formal approach.

In recent years, the gap between the AI and verification communities was bridged by tools that adapt and extend verification methods to neural agents. Notably, COOL-MC [40] and MoGym [34] provide ways to apply probabilistic [6,7] and statistical model checking (SMC) [3,49,69], respectively, to evaluate the behavior of neural agents. These tools leverage the capabilities of established model checkers—Storm [28], and MODES [19] of the MODEST TOOLSET [43], respectively—to enhance the rigor of evaluations.

However, they crucially require a formal model of the environment, encoded in modeling languages such as JANI [20] or PRISM [48], while the benchmarks most commonly used for SDM in the AI context follow the Farama Foundation's Gymnasium [65] interface (the successor to OpenAI Gym [15]). COOL-MC and MoGym implement this interface for formal models, giving Gymnasium-based tools access to formal models, but not the reverse. Thus, formal evaluation methods remain inaccessible to the majority of AI community benchmarks—e.g., MuJoCo [63], PROCGEN [26], or Atari 2600 [10]—that are based on arbitrary (black-box) simulations.

With this paper, we aim to bridge this AI-verification gap from the other side: We introduce PyDSMC (publicly available at github.com/neuro-mechanistic-modeling/PyDSMC), a lightweight, easy-to-install, and easy-to-use tool that facilitates SMC of neural agents (in prior work called *Deep Statistical Model Checking* (DSMC) [35]) across any environment conforming to the Gymnasium interface, irrespective of the underlying implementation. PyDSMC provides predefined trajectory-based properties to evaluate, such as accumulated rewards, the number of steps until termination, or the goal-reaching probability. In addition, a simple interface allows users to define custom properties. PyDSMC computes confidence intervals to either (1) achieve a predefined error margin and level of confidence, or (2) report the error margin given a confidence level once a specified resource limit (runtime or number of samples) is reached. For every property, PyDSMC automatically selects the appropriate statistical method. Thus, it provides accurate and adaptable statistical verification capabilities.

PyDSMC is designed for ease of use and compatibility with a broad range of environments using the Gymnasium interface. Notably, using SMC as the underlying analysis technique allows PyDSMC to remain agnostic w.r.t. the environment's underlying implementation. We believe that these features will facilitate greater adoption of SMC within the AI community and improve the state of the art in how the performance of neural agents is measured and reported.

Related Work. COOL-MC and MoGym make AI algorithms available for use with formal models—whereas PyDSMC makes formal evaluation available for use with AI agents. A similar goal is achieved by MultiVeStA for economic agent-based models [66,67]. The area of neural network verification (see, e.g., [4,9,27,54,59]) is a wide field that includes constraint- and abstraction-based symbolic and explicit verification methods and tools, typically aimed at ensuring the correctness or safety [5,39,47] of a neural agent w.r.t. a specification. In contrast, PyDSMC's purpose is specifically to provide a toolbox for the AI practitioner to easily evaluate the performance of their agent in a sound, formally justified way.

Outline. We elaborate the motivation to use SMC for neural agents in Sect. 2. Then Sects. 3 and 4 introduce Gymnasium and the statistical methods used in our work, respectively. In Sect. 5 we present the PyDSMC Python package and demonstrate some exemplary results in Sect. 6.

2 The Importance of Verifying Neural Agents

This section highlights the critical role of verification in assessing the performance of neural agents effectively.

2.1 Current State of the Art of Reporting Agents' Performances

Several papers evaluate the performance of their neural agent using the training curve. The training curve plots the number of training steps on the x-axis against the estimated expected return, i.e., the cumulative sum of (discounted) rewards, on the y-axis. Broadly, there are three common ways how the learning curve's data is used for agent evaluation:

Single Random Seed. It is common for papers—including those introducing influential algorithms such as DQN [52], DreamerV2 [42], PPO [60], or Rainbow [44]— to present the training curve of a single training run/random seed. Typically, this curve is smoothed using a sliding mean over the most recent steps. Regularly,

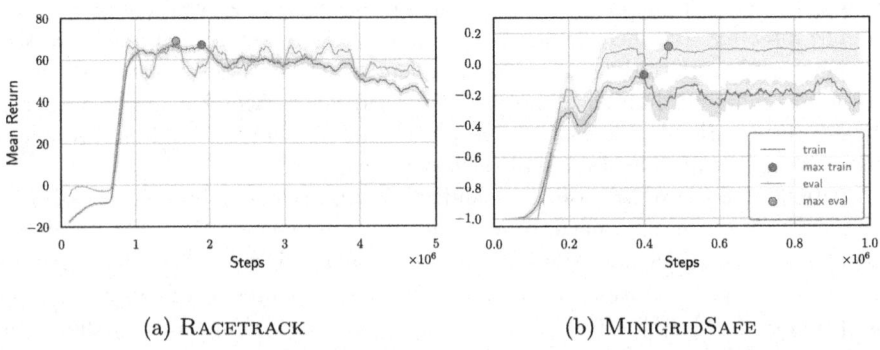

(a) RACETRACK   (b) MINIGRIDSAFE

Fig. 1. Mean return on two common benchmarks during training for a single seed. The blue curve depicts the values observed during training, whereas the orange curve was additionally computed after every training step through PyDSMC by using additional evaluation runs. The shaded area represents the 95% confidence interval over the sliding window (training) or over the evaluation samples (evaluation), respectively. (Color figure online)

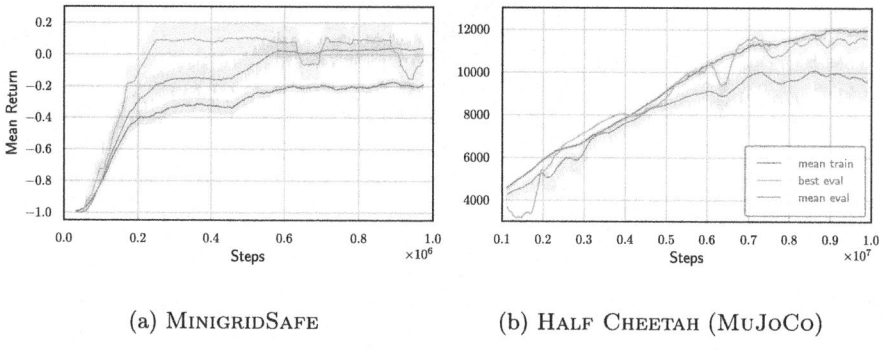

(a) MINIGRIDSAFE   (b) HALF CHEETAH (MUJOCO)

Fig. 2. Mean return on two common benchmarks during training over 10 seeds. The blue curve depicts the values observed during training, whereas the red and orange curves were additionally computed after every training step using PyDSMC by performing additional evaluation runs. The orange curve was computed by using the best random seed, while the red curve averages over all used random seeds. The shaded area represents the 95% confidence interval over the 10 different random seeds. (Color figure online)

this is done without providing any confidence interval or any additional information beyond the training curve itself. Consider the blue curve of Fig. 1, which exemplarily provides the training curves for two different benchmarks (left RACETRACK [8,31,38], right MINIGRIDSAFE [57]) trained with PPO. The shaded blue area here additionally provides the 95% confidence interval computed with the sequential Student's-t method (details in Sect. 4) using the samples from the sliding mean.

Mean of Multiple Random Seeds with Additional Confidence Information. Another common approach is to report the mean over several training runs, for example, done by Burda et al. [22], Duan et al. [29], or Agarwal et al. [2]. In addition to the mean, these papers typically report either the standard deviation [2], or the 95% confidence interval across these multiple runs [29].

Consider Fig. 2, which provides an example for such a training curve in blue for two different benchmarks (MINIGRIDSAFE trained with PPO and Half Cheetah (MuJoCo) [63] trained with SAC [41]). The shaded blue area provides the 95% confidence interval over the 10 different random seeds used.

Mean of Multiple Random Seeds with Additional Min/Max Information. Alternatively, some papers report the minimum and maximum values observed across multiple runs, e.g., SAC [41]. Consider Fig. 3, providing such an example for RACETRACK and MINIGRIDSAFE. While the blue curve still represents the mean, the shaded blue area now represents the range of observed values across the different training runs.

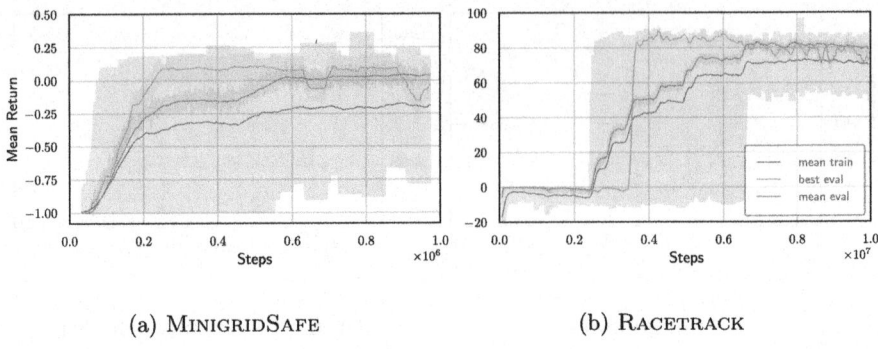

(a) MINIGRIDSAFE   (b) RACETRACK

Fig. 3. Mean return on two common benchmarks during training over 10 seeds. The blue curve depicts the values observed during training, whereas the red and orange curves were additionally computed after every training step through PyDSMC by using additional evaluation runs. The orange curve was computed by using the best random seed, while the red curve averages over all used random seeds. The shaded area represents the 95% confidence interval (red and orange) or the range of minimal and maximal observed values (blue), respectively, over the 10 different random seeds. (Color figure online)

2.2 Failure to Assess the Agents' Real Performances

Although all three approaches described to measure a neural agent's performance are common, we will now show that they fail to capture their real performance. To measure their actual performance, we use PyDSMC throughout the training to perform additional evaluation runs, i.e., runs without additional exploration influences that are exclusively used to measure the current performance and not for training. In Fig. 1, the orange curve depicts the results of these additional evaluation runs, with the shaded area indicating the 95% confidence interval. In Fig. 2 and Fig. 3, we provide two additional curves computed with PyDSMC: the orange curve depicts the performance of the best-performing seed, while the red curve shows the average performance across all seeds. For both curves, the shaded area indicates the 95% confidence interval computed over the used evaluation samples. We measure both the best and the average performance to enable a more comprehensive comparison: while most papers report the average over several training runs, in practice, one naturally selects the best agent for (real-world) deployment *after* training. Thus, we compare against both the actual best and the actual average performance.

Two major discrepancies become apparent when comparing the performance captured by PyDSMC (orange, red) with the training curves (blue):

Discrepancy in Performance Measurement. The substantial gap between the training and the evaluation curves underscores the inadequacy of training curves as reliable indicators of the expected return. Mostly, the training curves do not even lie within the confidence interval of the PyDSMC's calculated value, and vice versa. This is evident from the difference between the orange and blue curves for a single seed in Fig. 1, as well as when comparing in the multiple seed setting: both the orange curve (best seed) and the red curve (mean over seeds) differ significantly from the training curve (blue), mostly falling outside each other's confidence interval.

Comparing the confidence interval of the best (orange) and average (red) performance to the min/max shaded area (blue) in Fig. 3 area reveals that the min/max information does not provide any insights about agents' performances.

Therefore, the blue training curve must be considered insufficient for accurately measuring an agent's performance.

Variability in Optimal Performance Timing. There is considerable variation in the timing of peak returns between training and evaluation. While this occurs in both single-seed and multiple-seed settings, it can most clearly be observed when considering a single seed (Fig. 1).

For both RACETRACK and MINIGRIDSAFE, we compare (i) the time x_t where the training curve (blue) reaches its maximum—corresponding to the point where the agent would typically be selected for deployment after training—with (ii) the time x_e at which the evaluation curve (orange) reaches its maximum, indicating the agent's true peak performance. These two points in time, x_t and x_e, differ significantly. Moreover, the evaluation curve (orange) reveals that the performance at agent selection time x_t is substantially lower than at its actual peak x_e. This implies that the selected agent is far from optimal at the point it would be chosen based on the training curve alone.

2.3 Consequence for Training Neural Agents

These experiments[1] reveal a significant discrepancy in performance measurement. The training curve is affected by additional factors such as exploration, limited samples (especially in the multiple-seed setting), and policy changes during sampling due to ongoing learning steps. Therefore, it does not reflect the true performance, which can only be determined through rigorous verification.[2]

Additionally, the observed variability in optimal performance timing indicates a substantial risk of selecting suboptimal agents when relying solely on training performance. Therefore, selecting the best-performing agent and accurately assessing its performance requires repeated verification throughout and after the training process.

---

[1] It is important to note that the experiments presented in this section—involving evaluation after every learning step—are computationally expensive and might (depending on the domain) therefore be impractical. Nevertheless, they clearly demonstrate the importance of incorporating verification *at regular intervals* during training.

[2] It is worth noting that, in rare cases that we were especially looking for, the training curve was expressive enough for performance measures. We observed either only negligible differences between the training and evaluation curves, i.e., the training curve resembled the actual performance, or that the extraction point x_t yielded performance comparable to that at x_e. However, the fact that such discrepancies can occur—even if not always—demonstrates the necessity of rigorous verification, as otherwise the actual performance will remain unknown.

These observations underscore the necessity of employing robust verification tools like PyDSMC to ensure that the reported performance of neural agents accurately reflects their true capabilities.

3 The Gymnasium Interface

The Farama Foundation's Gymnasium [64], the successor to OpenAI Gym [16], is an open-source library designed to standardize the interface between neural agents and their environments. To this end, it provides a feature-rich and extensible API to abstract the agent-environment interactions, enabling easy interoperability between libraries and tools. Its widespread adoption across various libraries, tools, and environments serves as evidence of its success.

Available Tools. Stable Baselines3 [58], a popular reinforcement learning framework, allows users to train agents based on a broad variety of algorithms, such as PPO [60] and DQN [52], given that they follow the Gymnasium interface. Other training libraries supporting the Gymnasium interface include CleanRL [46], which provides clean and minimalistic implementations of learning algorithms, and Dopamine [23], a research framework developed by Google that focuses on reproducibility and simplicity.

Having access to numerous training algorithms naturally requires compatible environments. While Gymnasium already bundles example environments like MOUNTAINCAR, and MUJOCO [63], there is no shortage of custom environments implementing the Gymnasium interface either. Examples include the Atari Learning Environment [11], running original Atari 2600 games on an emulator, and MINIGRID [24], providing easy access to implement arbitrary grid-based environments.

Interface. An environment striving to be Gymnasium API-compatible has to implement two methods: `step` and `reset`. Both return a tuple consisting of an *observation*, a *reward*, a *termination* flag, a *truncated* flag, and an *info* dictionary, defining additional (possibly environment-specific) information.

On top of that, keeping efficiency in mind, Gymnasium provides easy vectorization functionality by grouping environments together, enabling batch inference. By default, each contained environment can either be managed by its proper subprocess, or by the main thread itself. The better choice depends on many factors, including the environment's complexity.

4 Statistical Methods

The core of SMC is Monte Carlo simulation: Generate k *simulation runs* (i.e. random executions of the neural agent acting within its Gymnasium environment), which give rise to samples $X_1, \ldots, X_k$ of the random variable X of the property of interest, and return the sample mean $\hat{X} = \frac{1}{k}\sum_{i=1}^{k} X_i$ as an estimate for the value of the property. For example, if the property concerns the probability p to reach a goal, then X is 1 for every run that reaches a goal and 0 for all others (i.e. we estimate a binomial proportion), so $p = \mathbb{P}(X = 1) = \mathbb{E}(X)$; if the property queries for the expected accumulated discounted reward r, then X for a run is the discounted sum of the rewards along that run, and $r = \mathbb{E}(X)$. In the latter case, if we know that rewards are non-negative and the maximum reward of any step is $\leq r_{max}$, we can derive that the support of the distribution of X lies within $[0, \frac{r_{max}}{1-\gamma}]$ for discount factor $\gamma < 1$, and we say that X is *bounded*. Without knowledge of r_{max}, or for $\gamma = 1$, we have to assume unbounded X.

SMC additionally provides a formal statistical guarantee on the correctness of its results. PyDSMC uses *confidence intervals* (CIs) as an easy-to-understand way to express its guarantee: In addition to $\hat{X}$, it returns an interval $I = [\ell, u]$ such that, in $(1-\kappa) \cdot 100 \%$ of the times such an interval is returned by PyDSMC, the (unknown) true value $x = \mathbb{E}(X)$ lies within I. Additionally, we ensure that $u - \ell \leq 2\varepsilon$ (absolute width) or $u - \ell \leq \varepsilon \cdot (\ell + u)$ (relative width). The significance level κ must always be specified. The absolute or relative error bound ε *can* be specified by the user. If it is, then PyDSMC generates runs until it can deliver an ε-interval with confidence $1-\kappa$ (the *sequential* setting). If ε is not given, k must be specified—either directly, or indirectly via a bound on the runtime after which to stop generating runs. Then, once k runs have been collected, PyDSMC returns an interval with confidence $1-\kappa$ (the *fixed-runs* setting), and ε is implicitly given by the interval's half-width (and thus the distinction between absolute and relative ε does not apply). PyDSMC uses the sequential setting with relative $\varepsilon = 0.05$ and $\kappa = 0.05$ (i.e. a $\pm 5\%$ error with 95% confidence) by default.

The statistics literature provides many different statistical methods (SMs) to obtain CIs. PyDSMC implements the most widely-used "standard" SMs as well as a set of state-of-the-art "sound" methods recommended in recent surveys [21,51]. An important aspect of an SM is its *coverage probability* p_{cov}: the fraction of intervals that in the limit, if we perform SMC again and again to generate a sequence of independent CIs, contain the true value. A sound SM guarantees $p_{cov} \geq 1 - \kappa$, no matter what parameters we use and what distribution (from those supported by the SM) we sample from.

Standard Methods. The most widely-used SMs, in fields ranging from psychology over medical sciences to economics as well as in many SMC tools [21, Table 1], rely on the central limit theorem (CLT) and assume that they are used with a "large enough" number of samples. These CLT-based **standard SMs are not sound** [21,51]: They only attain $p_{cov} \approx 1 - \kappa$ *on average* over the supported distributions—e.g. on average when ranging over $p \in (0, 1)$ for binomial proportion intervals. We nevertheless include them in PyDSMC as they are the de-facto standard in statistical evaluation of results, flawed as they may be, and require few runs. Notably, for unbounded distributions, they are the only methods available [21, Section 4]. In PyDSMC, we offer the following standard methods:

- **Normal intervals** approximate the distribution of error by a normal distribution: $I = [\hat{X} - \frac{zs}{\sqrt{k}}, \hat{X} + \frac{zs}{\sqrt{k}}]$ where z is a $1 - \frac{1}{2}\kappa$ quantile from the standard

normal distribution and s is the sample standard deviation. As they require k, normal intervals apply to the fixed-runs setting.
- **Student's-t intervals** use a quantile from the Student's-t distribution with $k - 1$ degrees of freedom instead, which works a little better for small k.
- **Chow-Robbins' method** keeps generating runs until the normal interval for the current set of runs has half-width at most ε (absolute) or $\varepsilon \cdot \hat{X}$. Chow and Robbins showed that this method attains coverage $1 - \kappa$ *in the limit* as $\varepsilon \to 0$ [25]. For any concrete $\varepsilon > 0$, p_{cov} may be much lower than $1 - \kappa$.
- **Sequential Student's-t intervals** work the same way as the Chow-Robbins method but use Student's-t instead of normal intervals.

Sound Methods. If we know we estimate a binomial proportion, or the distribution's support is bounded to $[a, b]$, then **sound SMs** are available for all settings. In general, these require more runs than standard methods—but are arguably the methods of choice to evaluate any (safety-)critical application of AI. Finding efficient sound SMs is an area of active research [55,56]; in PyDSMC, we provide the methods recommended by [21] for the fixed-runs setting and for the sequential setting with absolute error, and EBStop [53] for relative error:

- **Clopper-Pearson** intervals [18] apply to binomial proportions only, where the method guarantees coverage $\geq 1 - \kappa$. It requires around 2× as many runs as normal intervals in our experiments using PyDSMC's defaults. For the sequential setting, we precompute k via exponential and binary search assuming the worst-case of $p = 0.5$ [51] before any runs are performed.
- **Hoeffding's inequality** [45] gives the relation $k \geq (b-a)^2 \cdot (\ln 2/\delta)/2\varepsilon^2$. We can thus precompute k to solve the sequential setting with absolute error, or solve for ε instead for the fixed-runs setting. Due to the quadratic influence of the range of the distribution, k may become very large.
- **DKW** uses the Dvoretzky-Kiefer-Wolfowitz(-Massart) inequality [30,50] to obtain CIs on the mean in the fixed-runs setting for bounded distributions as described in [21, Section 4.1]. This method delivers smaller CIs than Hoeffding's inequality that are usually asymmetric: the worst case of DKW coincides with Hoeffding's inequality; in the best case, intervals are half as wide.
- **EBStop** [53] is a truly sequential SM (i.e., it does not precompute k but determines whether to stop after every run) for the relative-error case based on Bernstein's inequality [13,14] used with an estimation of X's variance.

Choice of SM. Figure 4 provides an overview of the SMs available in PyDSMC and the decision tree that it employs to select the method to use. The choice depends on the setting (fixed-runs or sequential), the kind of distribution underlying the property (binomial, bounded, or unbounded), how the interval width ε is specified (absolute or relative), and whether the user requests a sound method to be used. By default, PyDSMC uses the standard methods.

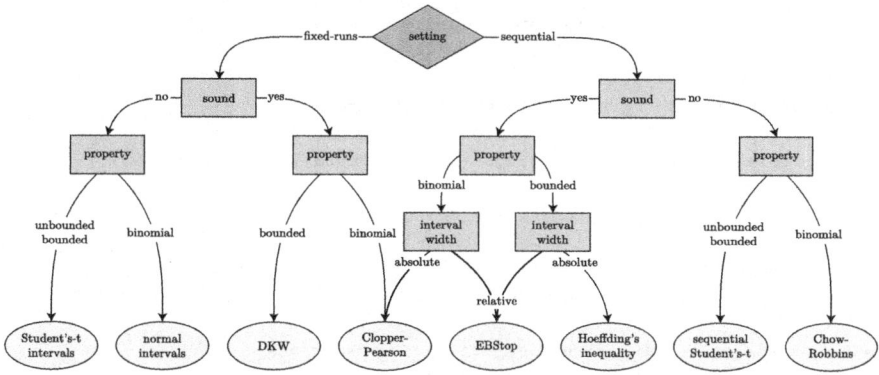

Fig. 4. Automated statistical method selection.

5 The PyDSMC Python Package

PyDSMC is implemented as a Python package that offers numerous SMC techniques and, thus, can easily be integrated into existing training pipelines. To access these functions, users only need to interact with two classes: First, the Property class, which allows specifying arbitrary trajectory-based properties, and second, the Evaluator class, which handles the sampling and checking of the properties.

Properties. PyDSMC offers various predefined properties, e.g., the return, the episode length, or the goal-reaching probability. These properties may be parameterized, for instance, by the discount factor used in the return calculation. Further, users can easily define custom properties by providing functions that check the property. The properties are thereby assumed to be trajectory-based, i.e., they can use the Gymnasium-provided information for all steps of the trajectory. If the user specifies ε to be None, PyDSMC uses the fixed-runs setting, and the sequential setting otherwise. Since the statistical methods depend on property-specific attributes (e.g., bounded, binomial), these have to be set during property initialization. For each property, the user can additionally set: (i) a name, (ii) a flag whether sound methods should be used, and (iii) a flag to toggle between absolute and relative error ε.

Evaluator. Having defined the properties, they have to be registered in an Evaluator instance, which manages the environment and the logging directory. The evaluation can then be started by calling eval on the Evaluator and providing the agent to verify. Additionally, eval takes the following arguments: (i) an optional resource limit (time, number of samples, or both), (ii) the number of samples taken between convergence checks, (iii) a flag whether to stop on convergence of all properties or to run until the specified resources are exhausted, (iv) a logging interval, (v) a flag whether to store every sample of every property, and (vi) the number of threads used for parallelization.

```python
from pydsmc import Evaluator, property as prop

# Create a predefined property
return_property = prop.create_predefined_property(
    property_id='return', epsilon=0.025, kappa=0.05,
    relative_error=True, bounds=(-1, 1), sound=True, gamma=0.99)
# Create a custom property
collision_property = prop.create_custom_property(
    name='obstacle_collision_prob',
    check_fn=lambda self, t: float(t[-1][2] == -1), epsilon=0.05,
    kappa=0.05, relative_error=False, bounds=(0, 1), binomial=True)

# Create the evaluator and register the properties
evaluator = Evaluator(env=env, log_dir="./example_logs")
evaluator.register_properties([return_property, collision_property])

# Evaluate the agent with respect to the registered properties
results = evaluator.eval(agent=agent, save_every_n_episodes=1000,
    time_limit=150, stop_on_convergence=True, num_threads=2)
```

Fig. 5. Example usage of PyDSMC.

Storing Results. Within the logging directory, PyDSMC creates a subdirectory for each property, where all files containing its evaluation results and its parameters are stored. PyDSMC additionally saves files storing the evaluation parameters and the utilized resources corresponding to the number of episodes and runtime.

Parallelization. To decrease runtime, PyDSMC supports vectorized environments and multithreading, where each thread manages a separate vector environment. We observed that vectorization significantly accelerates PyDSMC's execution.

Example. Consider the example in Fig. 5 which shows how PyDSMC can be used to evaluate an agent on a given MINIGRID environment(see Fig. 6a). We first create a predefined return property with parameters epsilon=0.025, kappa=0.05, and the property-specific discount factor gamma=0.99, where ε describes the maximum tolerated relative error. Since the rewards in the sample environment lie within the interval $[-1, 1]$, we can use a bounded SM by setting bounds=(-1, 1).

Second, we define a custom, environment-specific property that corresponds to the agent's probability of colliding with an obstacle, which, in this domain, can be identified by a negative reward in the trajectory's last step. The remaining arguments specify that we want to evaluate the absolute error of this binomial property. As the sound flag is not specified, PyDSMC defaults to an unsound SM.

Afterward, the evaluator is initialized and the properties are registered. We limit the evaluation to 2.5 h, but stop early if all properties have converged while using two threads.

6 Analyzing Neural Agents Using PyDSMC

We demonstrate PyDSMC by evaluating neural agents trained using state-of-the-art deep reinforcement learning algorithms on eight Gymnasium benchmarks, including four from MuJuCo [63].

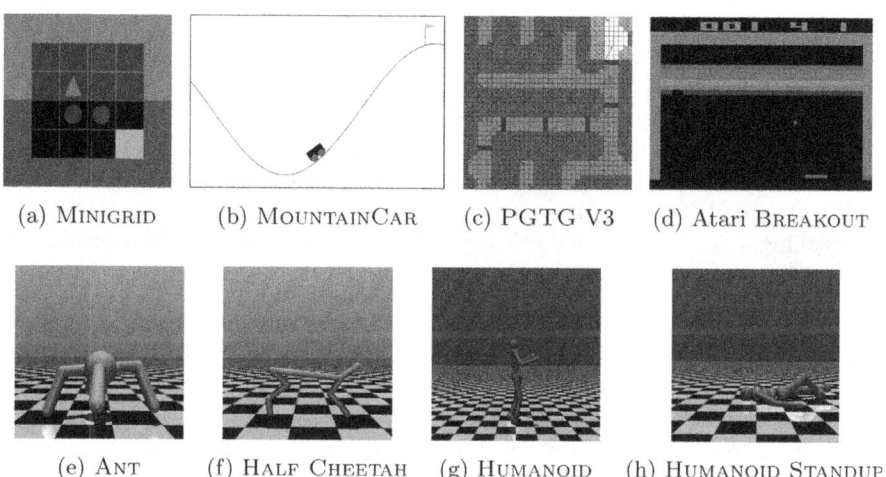

Fig. 6. The eight Gymnasium benchmark environments used in this section. The second row depicts the four MuJoCo Benchmarks, all in version V5.

6.1 Benchmarks

We present exemplary results on eight benchmarks commonly used in the AI community: (a) MINIGRID [24], (b) MOUNTAINCAR [65], (c) ProcGrid Traffic Gym [33], (d) BREAKOUT [10], and four MUJOCO benchmarks (e) ANT, (f) HALF CHEETAH, (g) HUMANOID, and (h) HUMANOID STANDUP. Figure 6 shows the eight benchmarks.

MINIGRID. A MINIGRID environment corresponds to a 2D navigation task, where the agent has to traverse a grid to reach a goal cell. The available actions are moving forward by one cell, rotating by 90° to the left or right, picking up objects, and interacting with objects (e.g., opening doors). Based on its current direction, the agent can only observe a limited section of the grid.

We focus on the MINIGRID Dynamic Obstacles environment, where, starting from a random cell, the agent has to reach the goal cell while avoiding three randomly moving obstacles. A reward of $1 - 0.9 \cdot \frac{steps}{144}$ is given when reaching the goal, -1 when colliding with an obstacle, and 0 otherwise.

MOUNTAINCAR. MOUNTAINCAR is a classic control benchmark, where a car is randomly placed in a valley. The goal is to reach the top of the right hill by accelerating either to the left or the right. The actions represent the car's directional force, which is within the range $[-1; 1]$. The agent observes its current location and velocity. At each time step, a negative reward of $-0.1 \cdot action^2$ is given, with an additional positive reward of 100 if the agent reaches the goal. The (undiscounted) return bounds are $[-999; 100]$, and the episode length bounds are $[1; 999]$.

ProcGrid Traffic Gym. ProcGrid Traffic Gym (PGTG) is an extension of the popular RACETRACK benchmark. The task is a rough simplification of autonomous driving, where the agent needs to drive from a starting line to a goal line across a randomly generated racetrack. Additionally, environments can be customized with different features such as ice, sand, or traffic. The observation is limited to the agent's surroundings, where green lines function as guidance toward the goal line. A reward of $\frac{100}{\text{number of subgoals}}$ is given when reaching a subgoal for the first time, a reward of 150 is given when reaching the goal line, a reward of -100 is given for crashing, and a reward of 0 otherwise. The (undiscounted) return bounds are $[-100; 150]$, and the episode length bounds are $[1; 100]$.

BREAKOUT. BREAKOUT is a classic Atari 2600 game in which the agent moves a paddle horizontally to bounce a ball such that it destroys the blocks at the top. Whenever the ball touches the bottom, the agent loses a life and a new ball spawns. The game is over when the agent has either lost all of its 5 lives or has destroyed all blocks. A reward is given when a block is destroyed, with the values ranging from 1 for blue blocks and 7 for red blocks. The (undiscounted) return bounds are $[0; 864]$, but the episode length is unbounded as no time step limit is given.

ANT. In this MuJoCo environment, a four-legged 3D robot is tasked to move forward from a random initial state. The actions correspond to applying torque to the ant's joints, and an episode ends when the height of the ant's torso is outside a predefined range. ANT features a dense reward function, where the agent is rewarded for moving forward, and keeping its torso at a certain height, and is penalized for applying too much torque to the joints or when the external contact forces on the ant's body parts are too high. We do not assume any bounds on the return for all MuJoCo benchmarks.

HALF CHEETAH. This environment features a 2D robot resembling a cheetah, that ought to run as fast as possible from a random initial state. The action space represents the torque applied to the cheetah's joints. All episodes end after a fixed number of time steps. The agent's rewards are based on how fast it moves forward, and it is penalized for applying too much torque.

HUMANOID. A 3D robot resembling a human needs to quickly walk forward without falling from a random initial state. The actions are the torques applied to the robot's legs, arms, and torso, and an episode ends when the torso's height is outside a predefined range. The agent is rewarded for moving forward and keeping its torso at a certain height, and it is penalized for applying too much torque or when the external contact forces are too high.

HUMANOID STANDUP. This environment is similar to HUMANOID, with the difference that the robot does not have to move forward but starts lying on the ground and the task is to stand up. Unlike in humanoid, the episodes are not terminated early.

6.2 Properties

We evaluate the agent by using some standard properties. To further highlight PyDSMC's flexibility, we also define a custom property for each environment.

Standard Properties. We evaluate the return with a discount factor of $\gamma = 0.99$, the undiscounted return (i.e., $\gamma = 1.0$), the average episode length, and the goal-reaching probability. For all properties, we use a significance level $\kappa = 0.05$, and an error bound $\varepsilon = 0.025$, which is relative for all properties but the goal-reaching probability, where we use the absolute setting.

Custom Properties. We define a custom property for each benchmark. In MINI-GRID, we define the custom property *Collision Prob.* as the probability of the agent moving onto a cell that is already occupied. In MOUNTAINCAR we consider the average acceleration *Avg. Acceleration*. For PGTG, we examine the probability of crashing into a wall *(Crash Prob)*. In BREAKOUT, we analyze the number of steps until the agent loses its first life *First Life Lost*. In ANT, we consider *Sum. Control Cost* as the sum of all obtained penalties for applying too much torque. For HALF CHEETAH, we analyze the probability of ending an episode with a negative return (*Neg. Return Prob*). In HUMANOID and HUMANOID STANDUP, we customize *Sum. Contact Cost* as the sum of penalties that was given because of too much contact force.

6.3 Exemplary Evaluation

Setup. All experiments were performed on a single machine with an AMD Ryzen Threadripper PRO 5965WX 24 Core CPU, an NVIDIA RTX A6000 GPU, and 512 GB of memory.[3] We provide details about the used hyperparameters in Appendix A.

Results. Table 1 provides all results obtained with PyDSMC. For all benchmarks, we provide the automatically selected SM, the approximated mean with its standard deviation, and the confidence interval. The **Conv.** column reports the number of samples that were needed to achieve the target confidence interval. Since we used the fixed-runs setting with 10,000 episodes for BREAKOUT, the **Conv.** column is empty. We marked those properties where we enforced a sound SM.

---

[3] Note that memory was never an issue and the machine used does not need this large amount of memory.

Table 1. Evaluation results. Properties marked with * are custom properties, whereas † denotes the usage of a sound statistical method. For BREAKOUT, the **Conv.** column is empty since the fixed-runs setting was used. The reported time corresponds to the total runtime of all listed properties analyzed simultaneously, rounded to the nearest full minute.

Property	Stat. Method	Mean	St.D.	C.I.	Conv.
MINIGRID (⏱ ≈ 45m)					
Return† (γ=0.99)	EBStop	0.14	0.21	[0.14;0.15]	249000
Return† (γ=1.0)	EBStop	0.25	0.3	[0.25;0.26]	153000
Episode Length	Student's-t	112.2	41.73	[112.04;112.37]	1000
Goal-reaching Prob.	Normal Interval	0.49	0.5	[0.49;0.49]	2000
Collision Prob.*	Normal Interval	0.0	0.0	[0.0;0.0]	1000
MOUNTAINCAR (⏱ ≈ 3m)					
Return† (γ=0.99)	EBStop	41.43	2.05	[40.39;42.46]	80000
Return† (γ=1.0)	EBStop	95.12	0.49	[92.78;97.53]	31000
Episode Length	Student's-t	81.21	4.48	[81.18;81.24]	1000
Goal-reaching Prob.	Normal Interval	1.0	0.0	[1.0;1.0]	1000
Avg. Acceleration*	Student's-t	0.08	0.08	[0.08;0.08]	1000
PGTG (⏱ ≈ 2m)					
Return† (γ=0.99)	EBStop	170.49	77.3	[166.22;174.74]	23000
Return† (γ=1.0)	EBStop	208.79	97.48	[203.56;214.0]	23000
Episode Length	Student's-t	31.69	19.77	[31.43;31.95]	3000
Goal-reaching Prob.	Normal Interval	0.84	0.36	[0.84;0.85]	1000
Crash Prob.*	Normal Interval	0.08	0.28	[0.08;0.09]	1000
BREAKOUT (⏱ ≈ 10h 38m)					
Return† (γ=0.99)	DKW	1.55	0.43	[1.52;13.28]	—
Return† (γ=1.0)	DKW	25.81	15.41	[24.84;37.54]	—
Episode Length	Student's-t	26484.0	15207.3	[26186;26782]	—
First Life Lost*	Student's-t	259.26	353.02	[252.34;266.18]	—
ANT (⏱ ≈ 33m)					
Return (γ=0.99)	Student's-t	137.75	31.60	[137.27;138.22]	1000
Return (γ=1.0)	Student's-t	1017.36	497.49	[1009.88;1024.84]	2000
Episode Length†	EBStop	779.53	325.81	[760.08;799.03]	17000
Sum. Control Cost*	Student's-t	-502.42	208.53	[-505.6;-499.3]	2000
HALF CHEETAH (⏱ ≈ 2m)					
Return (γ=0.99)	Student's-t	247.67	101.01	[243.24;252.10]	2000
Return (γ=1.0)	Student's-t	3247.40	1146.31	[3197.13;3297.67]	2000
Neg. Return Prob.*	Normal Interval	0.009	0.09444	[0.0049;0.0131]	1000
HUMANOID (⏱ ≈ 12m)					
Return (γ=0.99)	Student's-t	485.97	9.64	[485.72;486.22]	1000
Return (γ=1.0)	Student's-t	4782.39	886.35	[4759.96;4804.82]	1000
Episode Length†	EBStop	937.04	172.30	[913.59;960.41]	6000
Sum. Contact Cost*	Student's-t	-119.48	22.34	[-120.0;-118.9]	1000
HUMANOID STANDUP (⏱ ≈ 2m)					
Return (γ=0.99)	Student's-t	13940.7	285.2	[13928.2;13953.2]	1000
Return (γ=1.0)	Student's-t	146413.7	6684.6	[146120;146707]	1000
Sum. Contact Cost*	Student's-t	-51.97	3.96	[-52.14;-51.79]	1000

Further Insights. Additionally analyzing other properties and not only the standard objective, i.e., the accumulated discounted return, can provide deeper insights about the agent.

For example, consider the MINIGRID results. Despite the low return ($\gamma = 0.99$) of 0.14, the goal-reaching probability is 49%, indicating that the agent reaches the goal almost every other try. By also taking into account the episode length of 112 and the obstacle collision probability of 0, we can conclude that the poor performance is due to the agent frequently standing still until the step limit is reached.

As another example, consider MOUNTAINCAR. We observe a goal-reaching probability of 100%, indicating that the agent has learned to always reach the top of the hill. The undiscounted return of 95.12 suggests that it does so with little acceleration, which is further confirmed by the small average acceleration of 0.08.

In ANT, the high standard deviation of the episode lengths suggests that sometimes the agent quickly fails to keep its torso at the required height. This can be explained by the large value of the summed control penalties, showing that the agent tends to apply large torques, which can lead to situations where it is impossible to prevent early termination.

Lastly, consider the results of HALF CHEETAH. The discounted return already indicates that the agent is performing well. While the additional information of the undiscounted reward already strengthens this finding, the custom property *Neg. Return Prob.* additionally shows that the agent rarely uses too much torque, which indicates that the agent has learned to precisely control the joints.

7 Conclusion and Future Work

In this paper, we presented PyDSMC, a Python tool for applying statistical model checking to arbitrary neural agents in any Gymnasium environment, independent of the underlying implementation. We highlighted the importance of statistical model checking for neural agents, as standard evaluation methods like training curves fail to capture key influences, potentially leading to suboptimal agent selection and misleading performance assessments. We demonstrated PyDSMC's usage and illustrated how it can provide critical insights into agents' behaviors.

For the future, we have planned several extensions of PyDSMC. These include expanding the set of predefined properties and extending compatibility to *Petting-Zoo* [62], the standard interface for multi-agent reinforcement learning. Additionally, we aim to adapt PyDSMC for symbolic AI approaches such as planning. The algorithms *DSMC Evaluation Stages* [32,37] and *RARE* [39] have already integrated DSMC results into the training procedure to improve the performance of neural agents. In the future, we plan to integrate PyDSMC into these algorithms.

With an increasing range of applications, we also plan to integrate additional statistical methods to enhance both evaluation accuracy and efficiency.

Funding Information. This work was partially supported by the German Federal Ministry of Education and Research (BMBF) as part of project MAC-MERLin (grant agreement no. 01IW24007), by the German Research Foundation (DFG) under grant no. 389792660, as part of TRR 248, see https://perspicuous-computing.science and as GRK 2853/1 "Neuroexplicit Models of Language, Vision, and Action" (project no. 471607914), by the European Regional Development Fund (ERDF) and Saarland within the scope of (To)CERTAIN, by the European Union's Horizon 2020 research and innovation programme under Marie Skłodowska-Curie grant agreement no. 101008233 (MISSION), by the Interreg North Sea project STORM_SAFE, and by NWO VIDI grant VI.Vidi.223.110 (TruSTy).

Data Availibility Statement. The models, scripts, and tools to reproduce our experimental evaluation are archived and publicly available at DOI 10.5281/zenodo.15267298.

A Training Hyperparameters

All agents were trained using the algorithms provided by Stable Baselines 3. In Table 2, we list the training hyperparameters for each environment, where unspecified hyperparameters are set to their default values.

Table 2. Training hyperparameters for all environments.

Parameter	Value
MINIGRID	
Algorithm	DQN
Learning Rate	0.0001
Total Time Steps	50,000
Initial Time Steps	1,000
Update Frequency	10
Wrappers	FlatObsWrapper
MOUNTAINCAR	
Algorithm	SAC
Learning Rate	0.0003
Total Time Steps	100,000
Update Frequency	32
Entropy Coefficient	0.1
Gamma	0.9999
Tau	0.01
Gradient Steps	32
Hidden Sizes	64, 64

(*continued*)

Table 2. (*continued*)

Parameter	Value
PGTG	
Algorithm	DQN
Wrappers	TimeLimit (100 steps), FlattenObservation
Breakout	
Algorithm	PPO
Learning Rate	0.00025
Total Time Steps	10,000,000
Update Frequency	128
Entropy Coefficient	0.01
Value Function Coefficient	0.5
Wrappers	AtariWrapper
Ant	
Algorithm	PPO
Total Time Steps	1,000,000
Wrappers	NormalizeObservation, TimeFeatureWrapper
Link to the Evaluated Agent	huggingface.co/sb3/ppo-ant-v3
Half Cheetah	
Algorithm	SAC
Total Time Steps	1,000,000
Initial Time Steps	10,000
Wrappers	NormalizeObservation
Humanoid	
Algorithm	SAC
Total Time Steps	2,000,000
Initial Time Steps	10,000
Parallel Environments	16
Humanoid Standup	
Algorithm	SAC
Total Time Steps	2,000,000
Initial Time Steps	10,000
Parallel Environments	16

References

1. Achiam, J., et al.: GPT-4 Technical Report. arXiv preprint arXiv:2303.08774 (2024). https://doi.org/10.48550/arXiv.2303.08774
2. Agarwal, R., Schuurmans, D., Norouzi, M.: An optimistic perspective on offline reinforcement learning. In: Proceedings of the 37th International Conference on

Machine Learning, ICML 2020, 13–18 July 2020, Virtual Event. Proceedings of Machine Learning Research, vol. 119, pp. 104–114. PMLR (2020). http://proceedings.mlr.press/v119/agarwal20c.html
3. Agha, G., Palmskog, K.: A survey of statistical model checking. ACM Trans. Model. Comput. Simul. **28**(1), 6:1–6:39 (2018). https://doi.org/10.1145/3158668
4. Albarghouthi, A.: Introduction to neural network verification. Found. Trends Program. Lang. **7**(1–2), 1–157 (2021). https://doi.org/10.1561/2500000051
5. Alshiekh, M., Bloem, R., Ehlers, R., Könighofer, B., Niekum, S., Topcu, U.: Safe reinforcement learning via shielding. In: Proceedings of the AAAI Conference on Artificial Intelligence, vol. 32 (2018)
6. Baier, C.: Probabilistic model checking. In: Esparza, J., Grumberg, O., Sickert, S. (eds.) Dependable Software Systems Engineering, NATO Science for Peace and Security Series – D: Information and Communication Security, vol. 45, pp. 1–23. IOS Press (2016). https://doi.org/10.3233/978-1-61499-627-9-1
7. Baier, C., de Alfaro, L., Forejt, V., Kwiatkowska, M.: Model checking probabilistic systems. In: Handbook of Model Checking, pp. 963–999. Springer, Cham (2018). https://doi.org/10.1007/978-3-319-10575-8_28
8. Baier, C., et al.: Lab conditions for research on explainable automated decisions. In: Heintz, F., Milano, M., O'Sullivan, B. (eds.) TAILOR 2020. LNCS (LNAI), vol. 12641, pp. 83–90. Springer, Cham (2021). https://doi.org/10.1007/978-3-030-73959-1_8
9. Banerjee, D., Xu, C., Singh, G.: Input-relational verification of deep neural networks. Proc. ACM Program. Lang. **8**(PLDI), 1–27 (2024). https://doi.org/10.1145/3656377
10. Bellemare, M.G., Naddaf, Y., Veness, J., Bowling, M.: The arcade learning environment: an evaluation platform for general agents. J. Artif. Intell. Res. **47**, 253–279 (2013)
11. Bellemare, M.G., Naddaf, Y., Veness, J., Bowling, M.: The arcade learning environment: an evaluation platform for general agents. J. Artif. Intell. Res. **47**, 253–279 (2013). https://doi.org/10.1613/jair.3912
12. Berner, C., et al.: Dota 2 with large scale deep reinforcement learning. arXiv preprint arXiv:1912.06680 (2019). https://doi.org/10.48550/arXiv.1912.06680
13. Bernstein, S.: On a modification of Chebyshev's inequality and of the error formula of Laplace. Ann. Sci. Inst. Sav. Ukraine, Sect. Math **1**(4), 38–49 (1924)
14. Bernstein, S.: Theory of Probability, 2 edn. (1934)
15. Brockman, G., et al.: Openai gym (2016)
16. Brockman, G., et al.: OpenAI Gym. arXiv preprint arXiv:1606.01540 (2016). https://doi.org/10.48550/arXiv.1606.01540
17. Brown, T., et al.: Language models are few-shot learners. In: Advances in Neural Information Processing Systems, vol. 33, pp. 1877–1901. Curran Associates, Inc. (2020). https://proceedings.neurips.cc/paper/2020/hash/1457c0d6bfcb4967418bfb8ac142f64a-Abstract.html
18. Bu, H., Sun, M.: Clopper-pearson algorithms for efficient statistical model checking estimation. IEEE Trans. Softw. Eng. **01**, 1–20 (2024). https://doi.org/10.1109/TSE.2024.3392720
19. Budde, C.E., D'Argenio, P.R., Hartmanns, A., Sedwards, S.: An efficient statistical model checker for nondeterminism and rare events. Int. J. Softw. Tools Technol. Transf. **22**(6), 759–780 (2020). https://doi.org/10.1007/S10009-020-00563-2
20. Budde, C.E., Dehnert, C., Hahn, E.M., Hartmanns, A., Junges, S., Turrini, A.: JANI: quantitative model and tool interaction. In: Legay, A., Margaria, T. (eds.)

TACAS 2017. LNCS, vol. 10206, pp. 151–168. Springer, Heidelberg (2017). https://doi.org/10.1007/978-3-662-54580-5_9
21. Budde, C.E., Hartmanns, A., Meggendorfer, T., Weininger, M., Wienhöft, P.: Sound statistical model checking for probabilities and expected rewards. In: 31st International Conference on Tools and Algorithms for Construction and Analysis of Systems (TACAS). Lecture Notes in Computer Science. Springer, Heidelberg (2025). https://doi.org/10.1007/978-3-031-90643-5_9
22. Burda, Y., Edwards, H., Storkey, A., Klimov, O.: Exploration by random network distillation. arXiv preprint arXiv:1810.12894 (2018)
23. Castro, P.S., Moitra, S., Gelada, C., Kumar, S., Bellemare, M.G.: Dopamine: a research framework for deep reinforcement Learning. arXiv preprint arXiv:1812.06110 (2018). https://doi.org/10.48550/arXiv.1812.06110
24. Chevalier-Boisvert, M., et al.: Minigrid & miniworld: modular & customizable reinforcement learning environments for goal-oriented tasks. CoRR arxiv:2306.13831 (2023)
25. Chow, Y.S., Robbins, H.: On the asymptotic theory of fixed-width sequential confidence intervals for the mean. Ann. Math. Stat. **36**(2), 457–462 (1965). https://doi.org/10.1214/aoms/1177700156
26. Cobbe, K., Hesse, C., Hilton, J., Schulman, J.: Leveraging procedural generation to benchmark reinforcement learning. In: International Conference on Machine Learning, pp. 2048–2056. PMLR (2020)
27. Corsi, D., Marchesini, E., Farinelli, A.: Formal verification of neural networks for safety-critical tasks in deep reinforcement learning. In: de Campos, C.P., Maathuis, M.H., Quaeghebeur, E. (eds.) 37th Conference on Uncertainty in Artificial Intelligence (UAI). Proceedings of Machine Learning Research, vol. 161, pp. 333–343. AUAI Press (2021). https://proceedings.mlr.press/v161/corsi21a.html
28. Dehnert, C., Junges, S., Katoen, J.-P., Volk, M.: A storm is coming: a modern probabilistic model checker. In: Majumdar, R., Kunčak, V. (eds.) CAV 2017. LNCS, vol. 10427, pp. 592–600. Springer, Cham (2017). https://doi.org/10.1007/978-3-319-63390-9_31
29. Duan, J., Guan, Y., Li, S.E., Ren, Y., Sun, Q., Cheng, B.: Distributional soft actor-critic: off-policy reinforcement learning for addressing value estimation errors. IEEE Trans. Neural Networks Learn. Syst. **33**(11), 6584–6598 (2022). https://doi.org/10.1109/TNNLS.2021.3082568
30. Dvoretzky, A., Kiefer, J., Wolfowitz, J.: Asymptotic minimax character of the sample distribution function and of the classical multinomial estimator. Ann. Math. Stat. **27**(3), 642–669 (1956). https://doi.org/10.1214/aoms/1177728174
31. Gros, T.P.: Tracking the race: analyzing racetrack agents trained with imitation learning and deep reinforcement learning. Master's thesis **5** (2021)
32. Gros, T.P., et al.: Dsmc evaluation stages: fostering robust and safe behavior in deep reinforcement learning-extended version. ACM Trans. Model. Comput. Simul. **33**(4), 1–28 (2023)
33. Gros, T.P., Groß, D., Kamp, J., Gumhold, S., Hoffman, J.: Visual analysis of action policy behavior: a case study in grid-world driving. In: World Conference on Explainable Artificial Intelligence. Springer, Heidelberg (2025)
34. Gros, T.P., Hermanns, H., Hoffmann, J., Klauck, M., Köhl, M.A., Wolf, V.: Mogym: using formal models for training and verifying decision-making agents. In: International Conference on Computer Aided Verification, pp. 430–443. Springer, Heidelberg (2022). https://doi.org/10.1007/978-3-031-13188-2_21

35. Gros, T.P., Hermanns, H., Hoffmann, J., Klauck, M., Steinmetz, M.: Deep statistical model checking. In: Gotsman, A., Sokolova, A. (eds.) FORTE 2020. LNCS, vol. 12136, pp. 96–114. Springer, Cham (2020). https://doi.org/10.1007/978-3-030-50086-3_6
36. Gros, T.P., Hermanns, H., Hoffmann, J., Klauck, M., Steinmetz, M.: Analyzing neural network behavior through deep statistical model checking. Int. J. Softw. Tools Technol. Transfer **25**(3), 407–426 (2023)
37. Gros, T.P., Höller, D., Hoffmann, J., Klauck, M., Meerkamp, H., Wolf, V.: Dsmc evaluation stages: fostering robust and safe behavior in deep reinforcement learning. In: Abate, A., Marin, A. (eds.) QEST 2021. LNCS, vol. 12846, pp. 197–216. Springer, Cham (2021). https://doi.org/10.1007/978-3-030-85172-9_11
38. Gros, T.P., Höller, D., Hoffmann, J., Wolf, V.: Tracking the race between deep reinforcement learning and imitation learning. In: Gribaudo, M., Jansen, D.N., Remke, A. (eds.) QEST 2020. LNCS, vol. 12289, pp. 11–17. Springer, Cham (2020). https://doi.org/10.1007/978-3-030-59854-9_2
39. Gros, T.P., Müller, N., Höller, D., Hoffmann, J., Wolf, V.: Safe reinforcement learning through regret and state restorations in evaluation stages. Currently in publication (2024)
40. Gross, D., Jansen, N., Junges, S., Pérez, G.A.: Cool-mc: a comprehensive tool for reinforcement learning and model checking. In: International Symposium on Dependable Software Engineering: Theories, Tools, and Applications, pp. 41–49. Springer, Heidelberg (2022). https://doi.org/10.1007/978-3-031-21213-0_3
41. Haarnoja, T., Zhou, A., Abbeel, P., Levine, S.: Soft actor-critic: off-policy maximum entropy deep reinforcement learning with a stochastic actor. In: International Conference on Machine Learning, pp. 1861–1870. PMLR (2018)
42. Hafner, D., Lillicrap, T., Norouzi, M., Ba, J.: Mastering atari with discrete world models. arXiv preprint arXiv:2010.02193 (2020)
43. Hartmanns, A., Hermanns, H.: The Modest Toolset: an integrated environment for quantitative modelling and verification. In: Ábrahám, E., Havelund, K. (eds.) TACAS 2014. LNCS, vol. 8413, pp. 593–598. Springer, Heidelberg (2014). https://doi.org/10.1007/978-3-642-54862-8_51
44. Hessel, M., et al.: Rainbow: combining improvements in deep reinforcement learning. In: Proceedings of the AAAI Conference on Artificial Intelligence, vol. 32 (2018)
45. Hoeffding, W.: Probability inequalities for sums of bounded random variables. J. Am. Stat. Assoc. **58**(301), 13–30 (1963). https://doi.org/10.1080/01621459.1963.10500830
46. Huang, S., et al.: CleanRL: high-quality single-file implementations of deep reinforcement learning algorithms. J. Mach. Learn. Res. **23**(274), 1–18 (2022). http://jmlr.org/papers/v23/21-1342.html
47. Jansen, N., Könighofer, B., Junges, S., Serban, A., Bloem, R.: Safe reinforcement learning using probabilistic shields. In: 31st International Conference on Concurrency Theory (CONCUR 2020). Schloss-Dagstuhl-Leibniz Zentrum für Informatik (2020)
48. Kwiatkowska, M., Norman, G., Parker, D.: Prism 4.0: verification of probabilistic real-time systems. In: Gopalakrishnan, G., Qadeer, S. (eds.) CAV 2011. LNCS, vol. 6806, pp. 585–591. Springer, Heidelberg (2011). https://doi.org/10.1007/978-3-642-22110-1_47
49. Legay, A., Lukina, A., Traonouez, L.M., Yang, J., Smolka, S.A., Grosu, R.: Statistical model checking. In: Steffen, B., Woeginger, G.J. (eds.) Computing and Software Science – State of the Art and Perspectives, Lecture Notes in Computer Science,

vol. 10000, pp. 478–504. Springer, Heidelberg (2019). https://doi.org/10.1007/978-3-319-91908-9_23
50. Massart, P.: The tight constant in the dvoretzky-kiefer-wolfowitz inequality. Ann. Probab. **18**(3), 1269–1283 (1990). https://doi.org/10.1214/aop/1176990746
51. Meggendorfer, T., Weininger, M., Wienhöft, P.: What are the odds? Improving the foundations of statistical model checking. CoRR arxiv:2404.05424 (2024). https://doi.org/10.48550/ARXIV.2404.05424
52. Mnih, V., et al.: Human-level control through deep reinforcement learning. Nature **518**(7540), 529–533 (2015)
53. Mnih, V., Szepesvári, C., Audibert, J.Y.: Empirical Bernstein stopping. In: Cohen, W.W., McCallum, A., Roweis, S.T. (eds.) 25th International Conference on Machine Learning (ICML). ACM International Conference Proceeding Series, vol. 307, pp. 672–679. ACM (2008). https://doi.org/10.1145/1390156.1390241
54. Narodytska, N.: Formal verification of deep neural networks. In: Bjørner, N.S., Gurfinkel, A. (eds.) 18th Conference on Formal Methods in Computer Aided Design (FMCAD). IEEE (2018). https://doi.org/10.23919/FMCAD.2018.8603017
55. Parmentier, M., Legay, A.: Adaptive stopping algorithms based on concentration inequalities. In: Steffen, B. (ed.) 2nd International Conference on Bridging the Gap Between AI and Reality (AISoLA). Lecture Notes in Computer Science, vol. 15217, pp. 336–353. Springer, Heidelberg (2024). https://doi.org/10.1007/978-3-031-75434-0_23
56. Phan, M., Thomas, P.S., Learned-Miller, E.G.: Towards practical mean bounds for small samples. In: Meila, M., Zhang, T. (eds.) 38th International Conference on Machine Learning (ICML). Proceedings of Machine Learning Research, vol. 139, pp. 8567–8576. PMLR (2021). http://proceedings.mlr.press/v139/phan21a.html
57. Pranger, S.: Minigridsafe: an extension of the minigrid library for safe reinforcement learning (2025). https://github.com/PrangerStefan/MinigridSafex
58. Raffin, A., Hill, A., Gleave, A., Kanervisto, A., Ernestus, M., Dormann, N.: Stable-Baselines3: reliable reinforcement learning implementations. J. Mach. Learn. Res. **22**(268), 1–8 (2021). http://jmlr.org/papers/v22/20-1364.html
59. Schlüter, M., Steffen, B.: Affinitree: a compositional framework for formal analysis and explanation of deep neural networks. In: Huisman, M., Howar, F. (eds.) 18th International Conference on Tests and Proofs (TAP). Lecture Notes in Computer Science, vol. 15153, pp. 148–167. Springer, Heidelberg (2024). https://doi.org/10.1007/978-3-031-72044-4_8
60. Schulman, J., Wolski, F., Dhariwal, P., Radford, A., Klimov, O.: Proximal policy optimization algorithms. arXiv preprint arXiv:1707.06347 (2017)
61. Senior, A.W., et al.: Improved protein structure prediction using potentials from deep learning. Nature **577**(7792), 706–710 (2020)
62. Terry, J., et al.: Pettingzoo: gym for multi-agent reinforcement learning. Adv. Neural. Inf. Process. Syst. **34**, 15032–15043 (2021)
63. Todorov, E., Erez, T., Tassa, Y.: MuJoCo: a physics engine for model-based control. In: 2012 IEEE/RSJ International Conference on Intelligent Robots and Systems, pp. 5026–5033 (2012). https://doi.org/10.1109/IROS.2012.6386109
64. Towers, M., et al.: Gymnasium: a standard interface for reinforcement learning environments. arXiv preprint arXiv:2407.17032 (2024). https://doi.org/10.48550/arXiv.2407.17032
65. Towers, M., et al.: Gymnasium: a standard interface for reinforcement learning environments. arXiv preprint arXiv:2407.17032 (2024)

66. Vandin, A.: Statistical model checking of python agent-based models: an integration of multivesta and mesa. In: Steffen, B. (ed.) 2nd International Conference on Bridging the Gap Between AI and Reality (AISoLA). Lecture Notes in Computer Science, vol. 15217, pp. 398–419. Springer, Heidelberg (2024). https://doi.org/10.1007/978-3-031-75434-0_26
67. Vandin, A., Giachini, D., Lamperti, F., Chiaromonte, F.: Multivesta: statistical analysis of economic agent-based models by statistical model checking. In: Bowles, J., Broccia, G., Pellungrini, R. (eds.) 10th International DataMod Symposium – From Data to Models and Back. Lecture Notes in Computer Science, vol. 13268, pp. 3–6. Springer, Heidelberg (2021). https://doi.org/10.1007/978-3-031-16011-0_1
68. Vinyals, O., et al.: Grandmaster level in starcraft ii using multi-agent reinforcement learning. Nature **575**(7782), 350–354 (2019)
69. Younes, H.L.S., Simmons, R.G.: Probabilistic verification of discrete event systems using acceptance sampling. In: Brinksma, E., Larsen, K.G. (eds.) 14th International Conference on Computer Aided Verification (CAV). Lecture Notes in Computer Science, vol. 2404, pp. 223–235. Springer, Heidelberg (2002). https://doi.org/10.1007/3-540-45657-0_17

Minimal Per-Flow Backlog Bounds at an Aggregate FIFO Server Under Piecewise-Linear Arrival Curves

Lukas Wildberger[✉], Anja Hamscher, and Jens B. Schmitt

DISCO Lab, RPTU Kaiserslautern-Landau, 67663 Kaiserslautern, Germany
lukas.wildberger@cs.rptu.de, {hamscher,jschmitt}@cs.uni-kl.de

Abstract. Network Calculus (NC) is a versatile methodology based on min-plus algebra to derive worst-case *per-flow* performance bounds in networked systems with many concurrent flows. In particular, NC can analyze many scheduling disciplines; yet, somewhat surprisingly, an aggregate FIFO server is a notoriously hard case due to its min-plus *non-linearity*. A resort is to represent the FIFO residual service by a family of functions with a free parameter instead of just a single curve. For simple token-bucket arrival curves, literature provides optimal choices for that free parameter to minimize delay and backlog bounds. In this paper, we tackle the challenge of more general arrival curves than just token buckets. In particular, we derive residual service curves resulting in minimal backlog bounds for general piecewise-linear arrival curves. To that end, we first show that a backlog bound can always be calculated at a breakpoint of either the arrival curve of the flow of interest or its residual service curve. Further, we define a set of curves that characterize the backlog for a fixed breakpoint, depending on the free parameter of the residual service curve. We show that the backlog-minimizing residual service curve family parameter corresponds to the largest intersection of those curves with the arrival curve. In more complex scenarios finding this largest intersection can become inefficient as the search space grows in the number of flows. Therefore, we present an efficient heuristic that finds, in many cases, the optimal parameter or at least a close conservative approximation. This heuristic is evaluated in terms of accuracy and execution time. Finally, we utilize these backlog-minimizing residual service curves to enhance the DiscoDNC tool and observe considerable reductions in the corresponding backlog bounds.

Keywords: Network Calculus · FIFO Scheduling · Backlog Bound

1 Introduction

First-In First-Out (FIFO) is a popular scheduling policy for networked systems due to its simplicity and low cost of implementation. In various network analysis methods, it is an interesting policy to analyze. One such analysis method is

Network Calculus (NC) [5], which is a versatile methodology for deriving performance bounds in networked systems [1,6,7]. In particular, we are interested in the derivation of backlog bounds, which are important for sizing queues or buffers appropriately. It is straightforward to obtain a tight backlog bound at a FIFO node when considering all incoming traffic as a aggregate flow [7], such as in switches with a shared queue. A challenge arises when we are instead interested in a *per-flow* backlog bound. This means that we want to analyze how much data can queue up for an individual flow, rather than analyzing the aggregate traffic of the flows. This setting is considered for separate FIFO queues like in input-buffered switches with virtual output queues (VOQ) [10], or big data processing such as in Hadoop, where a job queue is fed data from a distributed file system through switches [13].

With NC, per-flow backlog bounds can be calculated using a so-called *residual service curve*, which represents the residual service that is exclusively available to a specific flow of interest (foi). However, due to the min-plus non-linearity of FIFO-scheduled systems [9], computing a residual service curve is hard. Defining a *family of functions* with a free parameter θ, rather than a single FIFO residual service curve, is a way of dealing with this. The family of FIFO residual service curves [6] is given by

$$\beta_\theta^1(t) = [\beta(t) - \alpha_2(t - \theta)]^+ \wedge \delta_\theta(t), \text{with } \theta \geq 0,$$

where β models the service available to the traffic aggregate and α_2 is an upper bound on the traffic of all the cross flows that are multiplexed into the aggregate together with the foi. Each value of θ results in a different residual service curve β_θ^1. Figure 1 shows such different β_θ^1, illustrating its dependence on the chosen θ. This has an effect on the resulting backlog bound that is obtained (drawn in Fig. 1 as vertical lines).

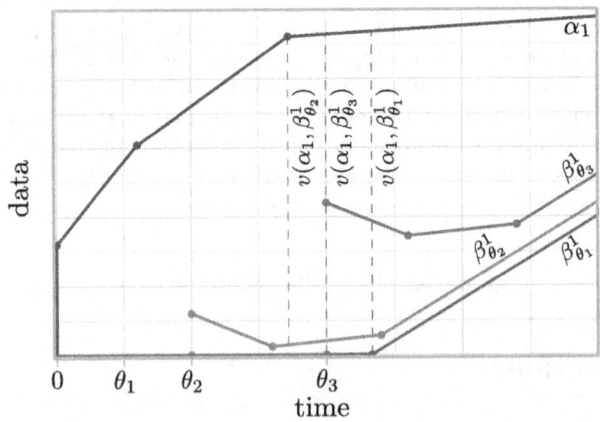

Fig. 1. FIFO residual service curves for different values of θ, and resulting backlog bounds.

Our goal is to obtain the smallest possible backlog bound. Due to β_θ^1 and the backlog bound changing depending on the chosen θ, there exists an optimal value of θ that provides the smallest backlog bound over all possible residual curves. In fact, a closed form for this θ exists when only considering token-bucket-constrained arrivals [1,7]. Yet, for more general functions such as piecewise linear (PWL) curves, to the best of our knowledge there exist no per-flow backlog bound results in literature regarding the calculation of θ values. Somewhat the only exception is [4], where an output bound for PWL arrival curves has been derived, yet without using NC. Nevertheless, it is interesting as the burst term of an output bound is also a bound on the backlog. The analysis is however restricted to the case of a constant rate server, whereas we deal with PWL service curves. Hence, in this paper we are considering PWL curves and are interested in closed form for θ that minimizes the backlog bound for such curves. To that end, we make the following contributions in this paper:

- In Sect. 4, we derive an exact method to find the backlog-minimizing θ value for PWL-constrained arrival and service curves.
- An efficient heuristic that determines the backlog-minimizing parameter θ in most cases is presented in Sect. 5. We evaluate its accuracy and execution time in comparison to the exact method.
- Finally, we show in Sect. 6 that the parameter θ, derived using our new methods, poses a significant accuracy improvement for backlog bounds computed by the DiscoDNC tool over its current default setting.

2 Network Calculus Background

Let $\mathbb{R}^+$ be the set of non-negative real numbers. $\mathcal{F} := \{f : \mathbb{R}^+ \to \mathbb{R} \cup \{+\infty\}\}$ is the set of (min, plus) functions. Based on $\mathcal{F}$, we let $\mathcal{F}^\uparrow$ be the set of non-decreasing functions $f \in \mathcal{F}$, and $\mathcal{F}_0^\uparrow$ be the set of functions in $\mathcal{F}^\uparrow$ with $f(0) = 0$.

Definition 1 (Basic Operators [3]). *Let $f, g \in \mathcal{F}$. The min-plus convolution of f and g is defined as $f \otimes g(t) := \inf_{0 \leq s \leq t}\{f(t-s) + g(s)\}$. The (max-plus) deconvolution is defined as $f \overline{\oslash} g(t) := \inf_{s \geq 0}\{f(t+s) - g(s)\}$. We define the $\wedge$ operator as $f \wedge g(t) := \min\{f(t), g(t)\}$.*

The *impulse function* $\delta_T(t)$ is defined as $\delta_T(t) = \infty$, if $t > T$ and 0 otherwise. The *indicator function* $\mathbb{1}_A$ is defined as $\mathbb{1}_A = 1$, if A is true and 0 otherwise. For a given $\beta \in \mathcal{F}$, the *lower non-decreasing closure* is defined as the largest non-decreasing function with $\beta_\downarrow \leq \beta$, given by $\beta_\downarrow := \beta \overline{\oslash} 0$ [3, p. 107].

Definition 2 (Pseudo-Inverse [3]). *Let $f \in \mathcal{F}$ be a non-negative and non-decreasing function. The pseudo-inverse f^{-1} is for all $x \in \mathbb{R}^+$ given by*

$$f^{-1}(x) = \inf\{t \mid f(t) \geq x\}.$$

Next, we define various notions that are used to model a network and derive its performance bounds. Let $A, D \in \mathcal{F}_0^\uparrow$ be the *cumulative arrival* and *departure process* of a flow in the network, assuming causality $A \geq D$. Furthermore, we assume the system to be lossless. We define the most important performance measures for such a system:

Definition 3 (Virtual Delay at Time t). *The* virtual delay *of data arriving at system $\mathcal{S}$ at time t is the time until this data would be served, assuming FIFO-per-flow order of service,*

$$d_{A,D}(t) = \inf\{d \geq 0 : A(t) \leq D(t+d)\} = D^{-1}(A(t)) - t \quad (1)$$

Definition 4 (Backlog at Time t). *The* backlog *of system $\mathcal{S}$ at time t is the vertical distance between arrival process A and departure process D at time t,*

$$q_{A,D}(t) := A(t) - D(t). \quad (2)$$

Arrival and service curves are central for the performance analysis using NC.

Definition 5 (Arrival Curve). *Let $\alpha \in \mathcal{F}_0^\uparrow$. We say that α is an* arrival curve *for arrival process A if it holds for all $0 \leq s \leq t$ that*

$$A(t) - A(s) \leq \alpha(t-s) \iff A = A \otimes \alpha.$$

An example is a *token-bucket* arrival curve $\gamma_{r,b}(t) = b + rt$ if $t > 0$, $\gamma_{r,b}(0) = 0$.

Definition 6 (Service Curve). *Let a flow with arrival process A and departure process D traverse a system $\mathcal{S}$. The system offers a* min-plus service curve β *to the flow if $\beta \in \mathcal{F}$ and it holds for all $t \geq 0$ that*

$$D(t) \geq A \otimes \beta \ (t) = \inf_{0 \leq s \leq t} \{A(t-s) + \beta(s)\}.$$

An example is a *rate-latency* curve $\beta_{R,T}(t) := R \cdot [t - T]^+$, $[x]^+ := \max\{x, 0\}$. We define two characteristic distances between functions.

Definition 7. *Let $f, g \in \mathcal{F}$. The* horizontal deviation *between f and g is*

$$h(f,g) := \sup_{t \geq 0}\{\inf\{d \geq 0 \mid f(t) \leq g(t+d)\}\}$$

and the vertical deviation *between f and g is*

$$v(f,g) := \sup_{t \geq 0}\{f(t) - g(t)\}.$$

Using these concepts, one can derive a backlog bound [3, p. 115], [7, p. 118].

Theorem 8 (Backlog Bound). *Assume a single flow with arrival process A, with arrival curve $\alpha \in \mathcal{F}_0^\uparrow$, and departure process D traverses a system $\mathcal{S}$. Let the system $\mathcal{S}$ offer a service curve $\beta \in \mathcal{F}_0^\uparrow$. The backlog $q(t)$ satisfies for all t*

$$q_{A,D}(t) \leq v(\alpha, \beta).$$

A FIFO residual service curve can be calculated as follows.

Theorem 9 (Residual Service Curve for FIFO [6]). *Let $t \geq 0$. Consider a system $\mathcal{S}$ that multiplexes two flows f_1 and f_2 using FIFO scheduling. The arrivals of f_2, A_2, are constrained by α_2. Further, assume that $\mathcal{S}$ guarantees a service curve β to the aggregate of the flows. Then, for any $\theta \geq 0$, the residual service of f_1 is*

$$\beta_\theta^1(t) = [\beta(t) - \alpha_2(t - \theta)]^+ \wedge \delta_\theta(t) \tag{3}$$

Definition 10 (PWL Concave Normal Form [3]). *Let $r_i, b_i \in \mathbb{R}^+$ and set $\gamma_i = \gamma_{r_i, b_i}$. The piecewise linear concave function $f = \min\{\gamma_i\}$ is said to be in normal form, if γ_i are sorted by a decreasing rate and no γ_i can be removed without modifying the minimum:*

$$i < j \Rightarrow r_i > r_j, \tag{4}$$
$$\forall i, \exists t > 0, \forall j \neq i, \gamma_i(t) < \gamma_j(t). \tag{5}$$

If $f = \min\{\gamma_i\}, i \in \{1, \ldots, n\}$ is in normal form, then there is a sequence of a_i of respective intersections of the linear functions γ_i and γ_{i+1}. These intersections, denoted by a_i, are also called breakpoints of f.

Definition 11 (PWL Convex Normal Form [3]). *Let $R_i, T_i \in \mathbb{R}^+$ and set $\beta_i = \beta_{R_i, T_i}$. The PWL convex function $f = \max\{\beta_i\}$ is said to be in normal form, if β_i are sorted by an increasing rate and no β_i can be removed without modifying the maximum:*

$$i < j \Rightarrow R_i < R_j \;\wedge\; \forall i, \exists t > 0, \forall j \neq i, \beta_i(t) > \beta_j(t).$$

If $f = \max\{\beta_i\}, i \in \{1, \ldots, n\}$ is in normal form, then there is a sequence of s_i of respective intersections of the linear functions β_i and β_{i+1}. These intersections, denoted by s_i, are also called breakpoints of f.

Definition 12. *Let the linear segment of a given PWL (concave or convex) function f at time t be called f^t, with $f^t = \beta_{R,T}$ or $f^t = \gamma_{r,b}$. The rate, r or R, and the y-axis intercept b or x-axis intercept T of this linear segment is then also referred to as r^t, R^t, b^t, T^t.*

Definition 13. *Let α be a PWL concave curve in normal form and β a PWL convex curve in normal form. Let A and B be the sets of breakpoints of α and β, respectively. Let I_A be the set of all breakpoints mapped to α, defined as $I_A := A \cup \{\alpha^{-1}(\beta(s)) \mid s \in B\}$. Then the first point in time for which the corresponding rate of α is less than or equal to the corresponding rate of β, mapped to α, is called $a^*_{\alpha,\beta}$. We define $a^*_{\alpha,\beta}$ as follows:*

$$a^*_{\alpha,\beta} := \min\{i \in I_A : r^i \leq R^{\beta^{-1}(\alpha(i))}\}.$$

3 System Model

In this section, we introduce the system setting considered in this paper. We also discuss the challenge associated with this setting and how to deal with it.

Figure 2 illustrates the system model under investigation. Here, each (distributed) flow (or job) $f_1, \ldots, f_n$ is associated with its own flow queue, each with its own size. These flows send their requests into a single (task) queue, which ensures the FIFO order between the individual flows when processed by system $\mathcal{S}$, for which we know a service curve β. The central question of our work now is: how to appropriately size the individual flow queues, or, in other words, how to calculate per-flow backlog bounds.

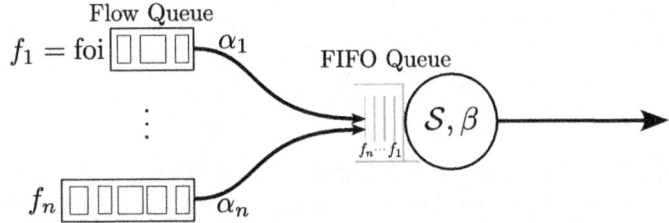

Fig. 2. Distributed per-flow queues served by single FIFO system.

We point out that per-flow queues in total generally require more space than an aggregate shared queue, i.e., we incur a *segregation penalty*. The extent of that penalty needs to weighed against advantages from the distributed setting as given in Fig. 2. We come back to that issue in Sect. 6.

Throughout the following sections, we assume that the flows $f_1 = \text{foi}, \ldots, f_n$ are constrained by PWL concave arrival curves $\alpha_1, \ldots, \alpha_n \in \mathcal{F}$. We aggregate all cross flows arrival curves $\alpha_2, \ldots, \alpha_n$, and simply call it α_2. The system $\mathcal{S}$ that multiplexes the flows according to FIFO, offers a PWL convex service curve $\beta \in \mathcal{F}$ to the flow aggregate.

When applying NC to compute per-flow backlog bounds under PWL arrival and service curves, an issue arises: the residual service curve β_θ^1 may contain a finite number of segments with negative slopes, which violates the non-decreasing property. While some literature formally requires that $\beta_\theta^1 \in \mathcal{F}_0^\uparrow$ [7], this constraint is not always reflected in the conditions of the corresponding theorems. Anyway, the following lemma states that this violation does not impact the calculation of the vertical deviation (which we need to compute backlog bounds), since we can make use of the lower non-decreasing closure.

Lemma 14. *Let $\alpha \in \mathcal{F}$ and $\beta \in \mathcal{F}$ be given. Then it holds that*

$$v(\alpha, \beta) = v(\alpha, \beta_\downarrow). \tag{6}$$

The proof for this and all subsequent lemmas can be found in the arXiv version of this paper [14].

4 Derivation of the θ Parameter for Minimal Per-Flow Backlog Bounds

In this section, we derive the value θ_{opt} that minimizes per-flow backlog bounds. In order to derive such a value of θ, we begin by formulating the problem mathematically as

$$\theta_{\text{opt}} = \arg\min_{\theta \geq 0}\{v(\alpha_1, \beta_\theta^1)\}. \tag{7}$$

In the following, we will transform this problem step by step until we are able to determine a solution. At first, in order to constrain the choice of θ, we make use of the following lemma:

Lemma 15. *Let α_1 and α_2 be PWL concave arrival curves and let β be a PWL convex service curve under FIFO multiplexing. Let $h(\alpha_2, \beta)$ be the horizontal deviation of α_2 and β. Then, it holds for $0 \leq \theta \leq h(\alpha_2, \beta)$ that*

$$v(\alpha_1, \beta_\theta^1) \geq v(\alpha_1, \beta_{h(\alpha_2,\beta)}^1).$$

Note that for token-bucket arrival curves and rate-latency service curves, the result of Lemma 15 has already been stated in [1,3,7,8].

Based on Lemma 15, we conclude that the search space for θ can be restricted to $\theta \geq h(\alpha_2, \beta)$. This allows us to transform the problem in Eq. (7) to

$$\theta_{\text{opt}} = \arg\min_{\theta \geq h(\alpha_2,\beta)}\{v(\alpha_1, \beta_\theta^1)\}.$$

To proceed, we define the family of functions $v_t(\theta)$ as the vertical distances between α_1 and β_θ^1 at an arbitrary but fixed time t:

$$v_t(\theta) := \begin{cases} \alpha_1(t) - \beta_\theta^1(t), & \text{if } h(\alpha_2,\beta) \leq \theta < t, \\ \alpha_1(\theta), & \text{if } \theta \geq t > h(\alpha_2,\beta), \end{cases}$$

$$= \begin{cases} \alpha_1(t) - \beta(t) + \alpha_2(t-\theta), & \text{if } h(\alpha_2,\beta) \leq \theta < t, \\ \alpha_1(\theta), & \text{if } \theta \geq t > h(\alpha_2,\beta). \end{cases}$$

Here, we have used that $\beta_\theta^1(t) = \beta(t) + \alpha_2(t-\theta)$, for $h(\alpha_2, \beta) \leq \theta < t$. Particularly, by definition we have $\beta_\theta^1 = [\beta(t) - \alpha_2(t-\theta)]^+ \wedge \delta_\theta(t)$. Since for $\theta < t$ it holds that $\delta_\theta(t) = 0$, and because $h(\alpha_2, \beta) \leq \theta$ ensures that $\beta(t) \geq \alpha_2(t-\theta)$ for all t (according to Definition 7), both the positive part and $\delta_\theta(t)$ can be omitted.

With this, the original problem in Eq. (7) can now be rewritten as:

$$\begin{aligned} \theta_{\text{opt}} &= \arg\min_{\theta \geq h(\alpha_2,\beta)}\{v(\alpha_1, \beta_\theta^1)\} \\ &= \arg\min_{\theta \geq h(\alpha_2,\beta)}\{\sup_{t \geq 0}\{\alpha_1(t) - \beta_\theta^1(t)\}\} \\ &= \arg\min_{\theta \geq h(\alpha_2,\beta)}\{\sup_{t \geq 0}\{v_t(\theta)\}\} \end{aligned} \tag{8}$$

Here, the supremum is over all $t \geq 0$. The following lemma will be instrumental to restrict the relevant values of t to a finite set.

Lemma 16. *Let α be a PWL concave function and β be a PWL convex function. Let A and B be the set of breakpoints of α and β, respectively. Then, the vertical deviation of α and β can always be calculated at some time $t \in A \cup B$.*

Remark 17. *The result of Lemma 16 also applies to the horizontal deviation.*

In the following, we denote the set of breakpoints of α_1, α_2, and β as A_1, A_2, and B, respectively. Lemma 16 ensures that only A_1 and the set of breakpoints of β_θ^1, induced by $A_2 \cup B$, need to be considered. A closer examination of these breakpoints reveals structural differences. In particular, the breakpoints of β_θ^1 need further attention, as they depend on θ. In contrast, the breakpoints of α_1, A_1, remain invariant with respect to θ. We call the set of breakpoints A_1 *absolute time set* of α_1. The absolute time set of the breakpoints of β_θ^1, including the shift by θ, is denoted by $\mathcal{T}_{\beta_\theta^1}^{\text{abs}}$. The set of breakpoints of β_θ^1, without being shifted by θ, is called *relative time set* and is given by

$$\mathcal{T}_{\beta_\theta^1}^{\text{rel}} = A_2 \cup B.$$

The relative time set $\mathcal{T}_{\beta_\theta^1}^{\text{rel}}$ allows us the usage of the breakpoints without having a dependency on θ. In the following, we derive from this relative time set an absolute time set of β_θ^1, which is also no longer dependent on θ. This is achieved by determining a single value of θ for each $t \in \mathcal{T}_{\beta_\theta^1}^{\text{rel}}$. Formally, for each $t \in \mathcal{T}_{\beta_\theta^1}^{\text{rel}}$, the corresponding value θ_t is defined as the solution to the following equation:

$$\alpha_1(\theta_t) = \alpha_1(t) - \beta_{\theta_t}^1(t). \tag{9}$$

As established in Lemma 16, only the breakpoints of the curves are relevant for computing the backlog bound. So we consider each relative point in time $t \in \mathcal{T}_{\beta_\theta^1}^{\text{rel}}$, assuming that it would be the one for which the vertical deviation is taken on. Then we obtain θ_t as the solution of Eq. (9), that minimizes this backlog bound under this assumption. In order to justify Eq. (9), let us consider an arbitrary, but fixed time $t \in \mathcal{T}_{\beta_\theta^1}^{\text{rel}}$. At this time t, the vertical distance between the two functions is given by $\alpha_1(t) - \beta_\theta^1(t)$, for $\theta < t$. This distance decreases as θ increases, so higher values of θ appear to be better for minimizing the backlog bound. However, this perspective neglects an important factor—namely, the behavior of the vertical distance at time θ. Specifically, as θ increases, the vertical distance at time θ, given by $\alpha_1(\theta)$, also increases. Consequently, selecting a larger value of θ to reduce the distance at time t simultaneously results in a larger distance at time θ. The optimal balance of this trade-off is given by θ_t as the solution of Eq. (9). For $\theta_t < t$, the respective θ_t can be derived as follows:

$$\begin{aligned}
&& \alpha_1(\theta_t) &= \alpha_1(t + \theta_t) - \beta_{\theta_t}^1(t + \theta_t) \\
\Leftrightarrow && \alpha_1(\theta_t) &= \alpha_1(t + \theta_t) - \beta(t + \theta_t) + \alpha_2(t) \\
\Leftrightarrow && \beta(t + \theta_t) - \alpha_1(t + \theta_t) + \alpha_1(\theta_t) &= \alpha_2(t) \\
\Leftrightarrow && \beta(t + \theta_t) - \alpha_1(t + \theta_t) + \alpha_1 \otimes \delta_t(t + \theta_t) &= \alpha_2(t) \\
\Leftrightarrow && (\beta - \alpha_1 + (\alpha_1 \otimes \delta_t))(t + \theta_t) &= \alpha_2(t), \tag{10}
\end{aligned}$$

$$\Rightarrow \theta_t = d_{\beta-\alpha_1+(\alpha_1\otimes\delta_t),\alpha_2}(t) \tag{11}$$
$$= ([\beta - \alpha_1 + (\alpha_1 \otimes \delta_t)]^+)^{-1}(\alpha_2(t)) - t \tag{12}$$

where we used for Eq. (11) that the horizontal distance is given by the shift θ_t in Eq. (10) (see also Eq. (1)). We apply the positive part in Eq. (12), since the pseudo-inverse requires wide-sense increasing functions and the horizontal distance for a positive function α_2 only requires non-negative functions.

Utilizing θ_t, the absolute time set is now explicitly given as

$$\mathcal{T}^{\text{abs}}_{\beta^1_\theta} = \mathcal{T}^{\text{rel}}_{\beta^1_\theta} + \theta_t = \{a_i^2 + \theta_{a_i^2} \mid a_i^2 \in A_2\} \cup \{s_i + \theta_{s_i} \mid s_i \in B\}.$$

Combining this with A_1, we obtain the complete set of absolute breakpoint times:

$$\mathcal{T} = A_1 \cup \mathcal{T}^{\text{abs}}_{\beta^1_\theta}.$$

This reduces the number of $v_t(\theta)$ curves to be evaluated from an uncountable set (for $t \geq 0$) to a finite set (for $t \in \mathcal{T}$), which can be computed given the parameters of the arrival and service curves. Since t is now an element of the finite set $\mathcal{T}$, we can replace the supremum by a maximum in Eq. (8), leading to the following transformed problem:

$$\theta_{\text{opt}} = \arg\min_{\theta \geq h(\alpha_2,\beta)} \{\max_{t \in \mathcal{T}}\{v_t(\theta)\}\}.$$

We can further restrict the domain of θ. According to Lemma 16, the backlog bound can always be calculated at a breakpoint of either α_1 or β^1_θ. Moreover, $\forall t \in \mathcal{T}$ and $\theta > t_{\max}$, with $t_{\max} = \max \mathcal{T}$, it holds that $v_t(\theta) = \alpha_1(\theta)$, which increases as θ increases. Hence, the backlog-minimizing value of θ cannot lie beyond $t_{\max}$. Additionally, since $\max_{t \in \mathcal{T}}\{v_t(\theta)\} = \max\{\max_{t \in \mathcal{T}}\{v_t(\theta)\}, \alpha_1(\theta)\}$ trivially holds, as $\alpha_1(\theta) = v_0(\theta) \in \{v_t(\theta) \mid t \in \mathcal{T}\}$, the problem can be transformed into

$$\theta_{\text{opt}} = \arg\min_{\theta \in [h(\alpha_2,\beta), t_{\max}]} \{\max\{\max_{t \in \mathcal{T}}\{v_t(\theta)\}, \alpha_1(\theta)\}\}. \tag{13}$$

Figure 3 gives a graphical representation of Eq. (13). Here, two v_t curves, v_{t_1} and v_{t_2} are plotted against $\alpha_1(\theta)$. Each v_t represents the vertical distance between α_1 and β^1_θ at a specific breakpoint $t \in \mathcal{T}$. According to Lemma 16, only these breakpoints need to be considered to compute backlog bounds. As θ varies, the vertical distances $v_t(\theta)$ decrease at some of these breakpoints and increase at others. In order to minimize the overall backlog bound, we aim to find the value of θ that balances this trade-off such that the largest vertical distance, at all breakpoints, is minimized. Formally, this means that this (optimal) value of θ (θ_{opt}), marked by a red dot, is obtained at the intersection of $\max_{t \in \mathcal{T}}\{v_t(\theta)\}$ and $\alpha_1(\theta)$.

It is useful to gain further insight into the problem, and examine the behavior of the two functions, $\max_{t \in \mathcal{T}}\{v_t(\theta)\}$ and $\alpha_1(\theta)$, at the boundary values of the interval for θ. The following lemma characterizes the relation between these functions at the interval's endpoints.

Lemma 18. *Let* $\max_{t \in \mathcal{T}}\{v_t(\theta)\}$ *and* $\alpha_1(\theta)$ *be defined as above. For the endpoints of the given interval* $[h(\alpha_2, \beta), t_{\max}]$ *it holds that*

$$\alpha_1(h(\alpha_2, \beta)) \leq \max_{t \in \mathcal{T}}\{v_t(h(\alpha_2, \beta))\} \text{ and } \alpha_1(t_{\max}) = \max_{t \in \mathcal{T}}\{v_t(t_{\max})\}.$$

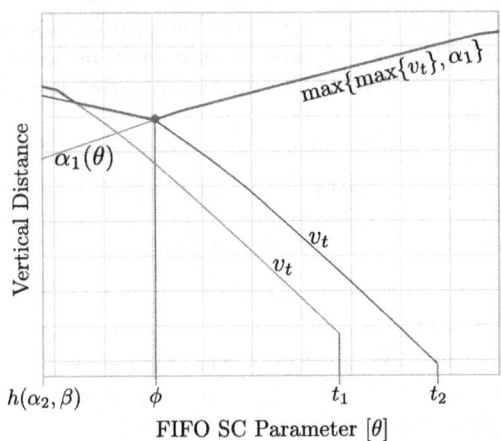

Fig. 3. Vertical distances $v_t(\theta)$ and $\alpha_1(\theta)$, for $\theta \geq h(\alpha_2, \beta)$.

To address the problem given by Eq. (13), we introduce the concept of a *first intersection* between $\max_{t \in \mathcal{T}}\{v_t(\theta)\}$ and $\alpha_1(\theta)$. This is essential for identifying the exact point in time where the function $\max_{t \in \mathcal{T}}\{v_t(\theta)\}$ drops onto the function $\alpha_1(\theta)$. We say that $\max_{t \in \mathcal{T}}\{v_t(\theta)\}$ and $\alpha_1(\theta)$ *first intersect* at

$$\phi = \min\{x \mid \alpha_1(x) = \max_{t \in \mathcal{T}}\{v_t(x)\}\}.$$

It is important to note that a first intersection between $\max_{t \in \mathcal{T}}\{v_t(\theta)\}$ and $\alpha_1(\theta)$ always exists. According to Lemma 18, these two functions are guaranteed to intersect at least at $\theta = t_{\max}$. Furthermore, by definition, every $v_t(\theta)$ curve, with $t \in \mathcal{T}$, is strictly decreasing for $\theta < t$ and strictly increasing for $\theta \geq t$. If $\max_{t \in \mathcal{T}}\{v_t(\theta)\}$ is taken at the respective strictly decreasing parts of the $v_t(\theta)$ curves, then the maximum function $\max_{t \in \mathcal{T}}\{v_t(\theta)\}$ itself is also strictly decreasing. The $\max_{t \in \mathcal{T}}\{v_t(\theta)\}$ is taken on the strictly decreasing parts until the maximum function $\max_{t \in \mathcal{T}}\{v_t(\theta)\}$ eventually equals $\alpha_1(\theta)$, which is strictly increasing. This is exactly the first intersection ϕ of $\max_{t \in \mathcal{T}}\{v_t(\theta)\}$ and $\alpha_1(\theta)$, after which $\max_{t \in \mathcal{T}}\{v_t(\theta)\} = \alpha_1(\theta)$ holds. So we conclude that $\max_{t \in \mathcal{T}}\{v_t(\theta)\}$ is strictly decreasing for $\theta < \phi$ and strictly increasing for $\theta \geq \phi$. Note that if $\max_{t \in \mathcal{T}}\{v_t(\theta)\}$ is directly assumed to be on the strictly increasing part of the $v_t(\theta)$ curves, then the first intersection is exactly at $h(\alpha_2, \beta)$.

The following theorem provides the solution to the problem stated in Eq. (13), given by the first intersection ϕ.

Theorem 19. Let $\max_{t\in\mathcal{T}}\{v_t(\theta)\}$ and $\alpha_1(\theta)$ be defined as above. Further, let the first intersection of $\max_{t\in\mathcal{T}}\{v_t(\theta)\}$ and $\alpha_1(\theta)$ be at ϕ. Then, it holds

$$\theta_{\text{opt}} = \underset{\theta\in[h(\alpha_2,\beta),t_{\max}]}{\arg\min}\{\max\{\max_{t\in\mathcal{T}}\{v_t(\theta)\},\alpha_1(\theta)\}\} = \phi.$$

Proof. According to Lemma 18 it holds that $\alpha_1(h(\alpha_2,\beta)) \leq \max_{t\in\mathcal{T}}\{v_t(h(\alpha_2,\beta))\}$ and $\alpha_1(t_{\max}) = \max_{t\in\mathcal{T}}\{v_t(t_{\max})\}$ so the following also holds:

$$\forall\, \theta < \phi : \alpha_1(\theta) \leq \max_{t\in\mathcal{T}}\{v_t(\theta)\} \,\wedge\, \forall\, \theta \geq \phi : \alpha_1(\theta) = \max_{t\in\mathcal{T}}\{v_t(\theta)\}.$$

So the maximum function can be described as follows:

$$\max\{\max_{t\in\mathcal{T}}\{v_t(\theta)\},\alpha_1(\theta)\} = \begin{cases} \max_{t\in\mathcal{T}}\{v_t(\theta)\}, & \text{if } \theta < \phi, \\ \alpha_1(\theta), & \text{if } \theta \geq \phi. \end{cases}$$

Suppose $\arg\min_{\theta\in[h(\alpha_2,\beta),t_{\max}]}\{\max\{\max_{t\in\mathcal{T}}\{v_t(\theta)\},\alpha_1(\theta)\}\}$ is assumed at ϕ', with $\phi' \in [h(\alpha_2,\beta),t_{\max}]$ and $\phi \neq \phi'$. So

$$\underset{\theta\in[h(\alpha_2,\beta),t_{\max}]}{\arg\min}\{\max\{\max_{t\in\mathcal{T}}\{v_t(\theta)\},\alpha_1(\theta)\}\} = \phi'$$

should hold. Consider two cases: $\phi' < \phi$ and $\phi' > \phi$.
Case I ($\phi' < \phi$): Since $\phi' < \phi$ and $\max\{\max_{t\in\mathcal{T}}\{v_t(\theta)\},\alpha_1(\theta)\} = \max_{t\in\mathcal{T}}\{v_t(\theta)\}$ is strictly decreasing for $\theta < \phi$, it holds that:

$$\max\{\max_{t\in\mathcal{T}}\{v_t(\phi')\},\alpha_1(\phi')\} > \max\{\max_{t\in\mathcal{T}}\{v_t(\phi)\},\alpha_1(\phi)\}. \quad \lightning$$

Case II ($\phi' > \phi$): Since $\phi' > \phi$ and $\max\{\max_{t\in\mathcal{T}}\{v_t(\theta)\},\alpha_1(\theta)\} = \alpha_1(\theta)$ is strictly increasing for $\theta \geq \phi$, it holds that:

$$\max\{\max_{t\in\mathcal{T}}\{v_t(\phi')\},\alpha_1(\phi')\} > \max\{\max_{t\in\mathcal{T}}\{v_t(\phi)\},\alpha_1(\phi)\}. \quad \lightning$$

So $\theta_{\text{opt}} = \arg\min_{\theta\in[h(\alpha_2,\beta),t_{\max}]}\{\max\{\max_{t\in\mathcal{T}}\{v_t(\theta)\},\alpha_1(\theta)\}\} = \phi$ holds. □

Remark 20. Note that the first intersection of $\max_{t\in\mathcal{T}}\{v_t(\theta)\}$ and $\alpha_1(\theta)$ is also the last of all intersections of $v_t(\theta)$ curves, $t \in \mathcal{T}$, with $\alpha_1(\theta)$. See again Fig. 3.

5 Efficient Calculation of Near-Optimal Backlog Bounds

In the following, we expand on the calculation of the FIFO SC parameter θ_{opt} and its corresponding minimal backlog bound $q_{\min}$. We strive for an efficient calculation of backlog bounds that are still near to the minimal ones from the previous section using Theorem 19. In particular, according to Remark 20, we calculate the v_t curves of all $t \in \mathcal{T}$, calculate their respective intersection points with α_1, then determine ϕ as the maximum of all intersections to obtain θ_{opt} and $q_{\min} = v(\alpha_1,\beta^1_{\theta_{\text{opt}}})$, where $q_{\min}$ is measured at θ_{opt}. This constitutes an exact method, but requires the calculation of the intersection of *all* possible breakpoints $t \in \mathcal{T}$ of α_1 and α_2. For an efficient calculation of more complex scenarios (in particular with more flows and more segments per flow), the question arises whether we can exclude certain $t \in \mathcal{T}$ in a first step to reduce the number of v_t curves and intersections we have to calculate. We first describe the central idea: Instead of considering α_1 as a whole, we take a *decomposition* approach. From Definition 10, we know that α_1 is the minimum of n token-bucket arrival curves γ_i. Hence, we can split α_1 into its token-bucket segments $S_i, 1 \le i \le n$. For a token-bucket foi, we can identify a particular breakpoint $\tau \in \mathcal{T}_{\beta_\theta^1}^{\text{rel}}$ where the backlog is taken on and find a closed form for θ_{opt} that minimizes its backlog $q_{\min}$.

Theorem 21. *Let the foi α_1 be a token-bucket arrival curve γ_{r_1,b_1} and let the cross-flow α_2 be a PWL concave arrival curve in normal form. Further, let β be a PWL convex service curve. The FIFO SC parameter that minimizes the backlog bound is given by:*

$$\theta_{\text{opt}} = \underset{\theta \in [h(\alpha_2,\beta),t_{\max}]}{\arg\min} \left\{ \max_{t \in \mathcal{T}} \{v_t(\theta)\} \right\}$$

$$= h(\alpha_2 + \gamma_{r_1,0}, \beta) = \frac{b_2^\tau + (r_2^\tau + r_1) \cdot \tau}{R^{\beta^{-1}(\alpha_2(\tau))}} + T^{\beta^{-1}(\alpha_2(\tau))} - \tau,$$

with $\tau = a^*_{\alpha_2+\gamma_{r_1,0},\beta}$.

Proof. According to Remark 20, the result of Theorem 19 also states that the optimal θ is given by the last of all intersections of $v_t(\theta)$ curves and $\alpha_1(\theta)$. In order to obtain the last intersection, let us consider the intersections of $\alpha_1(\theta)$ and $v_t(\theta)$, with $t \in \mathcal{T}_{\beta_\theta^1}^{\text{rel}}$. The optimal value θ_t for each $t \in \mathcal{T}_{\beta_\theta^1}^{\text{rel}}$ can be calculated as follows:

$$\alpha_1(\theta_t) = \alpha_1(t+\theta_t) - \beta(t+\theta_t) + \alpha_2(t)$$
$$\Leftrightarrow \quad r_1\theta_t + b_1 = r_1 t + r_1\theta_t + b_1 - \beta(t+\theta_t) + \alpha_2(t)$$
$$\Leftrightarrow \quad \beta(t+\theta_t) = r_1 t + \alpha_2(t)$$
$$\Leftrightarrow \quad \theta_t = \beta^{-1}(r_1 t + \alpha_2(t)) - t$$
$$\Leftrightarrow \quad \theta_t = d_{\alpha_2+\gamma_{r_1,0}}(t).$$

Now we need to find the time at which the backlog is maximal. As the backlog bound coincides with $\alpha_1(\theta_{\text{opt}})$ and α_1 is increasing, we look for the maximal θ_t:

$$\theta_{\text{opt}} = \sup_{t \in \mathcal{T}_{\beta_\theta^1}^{\text{rel}}} \{d_{\alpha_2+\gamma_{r_1,0},\beta}(t)\} \stackrel{(\text{Remark } 17)}{=} \sup_{t \ge 0}\{d_{\alpha_2+\gamma_{r_1,0},\beta}(t)\} = h(\alpha_2 + \gamma_{r_1,0}, \beta).$$

Using the piecewise linear nature of the curves, it is clear that this horizontal deviation is taken on at $\tau = a^*_{\alpha_2+\gamma_{r_1,0},\beta}$, and thus

$$\theta_{\text{opt}} = h(\alpha_2 + \gamma_{r_1,0}, \beta) = \frac{b_2^\tau + (r_2^\tau + r_1) \cdot \tau}{R^{\beta^{-1}(\alpha_2(\tau))}} + T^{\beta^{-1}(\alpha_2(\tau))} - \tau.$$

□

Using Theorem 21, we can find the optimal value θ^i_{opt} that minimizes $q^i_{\min}$ for each γ_i corresponding to segment S_i of α_1. Considering Eq. (5), we recognize that $q^i_{\min}$ may not be assumed at a valid point in time when considering α_1 as a whole. Indeed, each segment S_i is only defined over its respective interval $I_i := [a_i, a_{i+1})$. If now $\theta^i_{\text{opt}} \notin I_i$, we adjust its value as follows (with an arbitrarily small $\epsilon > 0$):

$$\tilde{\theta}^i_{\text{opt}} := \begin{cases} a_i, & \theta^i_{\text{opt}} < a_i, \\ a_{i+1} - \epsilon, & \theta^i_{\text{opt}} > a_{i+1}, \\ \theta^i_{\text{opt}}, & \text{otherwise.} \end{cases} \qquad (14)$$

This adjustment follows from the properties of the involved curves. We know that γ_i and α_2 are concave curves, their sum is hence also concave. In addition, the service curve β is convex. If we now assume θ^i_{opt} outside of the interval I_i, we can use these properties to find the largest backlog bound that is still assumed over I_i as follows: In case 1 of Eq. (14), if $\theta^i_{\text{opt}} < a_i$, we set $\tilde{\theta}^i_{\text{opt}} = a_i$. Since α_1 and α_2 are concave, their aggregate rate decreases in time, while due to the convexity of β, its rate increases in time. As such, measuring the backlog at any point $x \in I_i$, $x > a_i$, results in a smaller backlog than if measured at a_i. For case 2 it is similar, but we need to take the half-open nature of I_i into account. Using an arbitrarily small $\epsilon > 0$, and assuming the backlog at $a_{i+1} - \epsilon$ ensures that we can approximate the largest possible backlog that is still in I_i as close as we desire. This leaves us with two vectors of n values for θ^i_{opt} and $\tilde{\theta}^i_{\text{opt}}$. For purposes of the heuristic, we let $\Theta := [\theta^1_{\text{opt}}, \ldots, \theta^n_{\text{opt}}]$ and $\tilde{\Theta} := [\tilde{\theta}^1_{\text{opt}}, \ldots, \tilde{\theta}^n_{\text{opt}}]$. We proceed with two observations about θ^i_{opt} and $\tilde{\theta}^i_{\text{opt}}$:

Lemma 22. *It holds that $\theta^1_{\text{opt}} \geq \ldots \geq \theta^n_{\text{opt}}$.*

Lemma 23. *There is at most one θ^i_{opt} with $\theta^i_{\text{opt}} = \tilde{\theta}^i_{\text{opt}}$.*

Using these insights, we proceed with our decomposition heuristic:

1. Calculate $\Theta_{\text{diff}} := \Theta - \tilde{\Theta} = [\theta^1_{\text{opt}} - \tilde{\theta}^1_{\text{opt}}, \ldots, \theta^n_{\text{opt}} - \tilde{\theta}^n_{\text{opt}}]$.
2. Check whether there exists an entry with value $i = 0$ in Θ_{diff}. If yes, *return* $v(\alpha_1, \beta^1_{\theta^i_{\text{opt}}})$ *and break*.
3. If no such entry exists in Θ_{diff}, iterate over $t \in A_1$ and find the first intersection ϕ. Return $v(\alpha_1, \beta^1_\phi)$ and break.

Let us discuss the rationale and potential shortcomings of the heuristic: If a θ^i_{opt} is unchanged after applying Eq. (14), it may seem that we should have obtained the same value of θ_{opt} when using Theorem 19 on α_1 without decomposing the curve into its n segments. Consequently, setting $\theta_{\text{opt}} = \theta^i_{\text{opt}}$ seems to be the correct choice in this case. From Lem. 23, we know that the heuristic can find at most one such θ^i_{opt}, hence the result is unique. However, in fact, this does not yield the same result as using Theorem 19 in all cases. Whenever $\alpha_1(\tau)$ is assumed at a different segment than $\alpha_1(\tau + \theta_{\text{opt}})$, the heuristic calculates a larger θ_{opt}, and consequently larger backlog bound, as it only considers one segment of α_1. Furthermore, the heuristic can also falsely assume that there exists no entry $i = 0$ in Θ_{diff}. In step 3, we know that all $\tilde{\theta}^i_{\text{opt}}$ in Θ_{diff} are $t \in A_1$, as we have adjusted them to their respective interval limits using Eq. (14)[1]. For each $\tilde{\theta}^i_{\text{opt}}$, we calculate its $v_{\tilde{\theta}^i_{\text{opt}}}$ curve and obtain the intersection with α_1. From Remark 20, it follows that the largest of these intersections is equal to ϕ, hence the θ_{opt} that minimizes the backlog bound. We remark that the heuristic always returns a backlog bound, as it always finds a value for θ – it may, however, be conservative.

5.1 Evaluation of the Decomposition Heuristic

We continue with an evaluation of the decomposition heuristic with respect to its accuracy and efficiency. To this end, we calculate and compare the backlog bounds and execution times of the exact method using Theorem 19 and the decomposition heuristic. We consider the following scenario: Assume we have a variable number of 2 to 10 cross flows at the node, multiplexed in a FIFO aggregate with the foi. Let α_1 be a PWL concave arrival curve for the foi with either two or four segments (as typically found in literature, e.g. [15]). Each cross flow arrival curve α_2^i is modeled as a T-Spec curve, defined as $\alpha_2^i(t) := \min\{b_2^1 + r_2^1 t, b_2^2 + r_2^2 t\}$. We pick the packet size b_i^1 for each flow i randomly from 0.001 Mbit, the sustained rate r_i^2 from 1 Mbit/s–10 Mbit/s, and and the first breakpoint a_1^i from 50 ms – 500 ms. For α_1, we pick a random breakpoint spacing from 100 ms–500 ms that is added to a_1^1 to obtain further breakpoints. We set the peak rate for each arrival curve as $r_i^1 = r_i^2 \cdot 8$, similar to [7]. For α_1 with four segments, we set the rate of the fourth segment as sustained rate, and assign rate $6 \cdot r_i^1$ to segment two, and rate $3 \cdot r_i^1$ to segment three. The burst value of each subsequent token bucket is set as

$$b^i = b^{i-1} - (r^i - r^{i-1}) \cdot a_i. \tag{15}$$

The server offers a rate-latency service curve $\beta_{R,T}$, where the rate is set such that there is an 80% utilization of the server when considering the sum of sustained rates of all flows. The latency is set as $T = 1/R$ s.

The experiments were run on hardware using an Intel i7-1165G7 CPU with 4 cores at 2.8 GHz base frequency, 32 GB DDR4-3200 MHz RAM, a 512 GB NVMe SSD and the Ubuntu 22.04 LTS operating system. The experiments are implemented using the Nancy library [16] for C#, and are run in JetBrains Rider 2024.3 with the .NET SDK 9.0.102 using default compiler settings.

We perform 500 iterations, where in each one of them we iterate again over the number of cross flows from 2 to 10. For each run, we measure the execution time using *System.Diagnostics.Process.UserProcessorTime*. This method returns the CPU time of the process that is running the experiments, without including any other processing time. We only measure code fragments that are used to calculate one of the two methods, ignoring any other code parts. We take the execution time of both methods and calculate the percentage difference of both

---

[1] They could be off by ϵ, but we can let $\epsilon \to 0$.

measured times. Of course, we also compare the backlog bounds obtained by using the exact method and the heuristic.

Table 1. Comparison of exact method and heuristic for α_1 with two segments.

Cross [#]	2	3	4	5	6	7	8	9	10
μ_{ex}	15.1	15.7	16.1	16.4	16.6	16.7	16.9	17.0	17.1
CI(95%)	0.8	0.9	0.9	0.9	0.9	0.9	0.9	0.9	0.9
μ_{heu}	15.1	15.7	16.1	16.4	16.6	16.7	16.9	17.0	17.1
CI(95%)	0.8	0.9	0.9	0.9	0.9	0.9	0.9	0.9	0.9
Acc. [%]	98.4	99.6	99.6	100	100	100	100	100	100
Inc. [%]	1.6	6.7	3.3	0	0	0	0	0	0
CI(95%)	0.0	0.3	0.1	0	0	0	0	0	0
$t_{\text{ex}}^{\text{exec}}$ [ms]	976	1362	1742	2114	2520	2918	3335	3758	4202
$t_{\text{heu}}^{\text{exec}}$ [ms]	14.3	12.6	12.8	13.0	13.3	13.6	14.1	14.2	14.6
Speedup	68	108	136	163	189	215	237	265	288

The results of the experiments for two segments are shown in Table 1. The results for four segments an be found in the arXiv version of this paper [14]. The tables are organized as follows. We report the mean and confidence interval (CI) for the backlog bound (μ_{ex} and μ_{heu}) as well as execution times ($t_{\text{ex}}^{\text{exec}}$ and $t_{\text{heu}}^{\text{exec}}$) for each number of cross flows. The accuracy reports how often the resulting backlog bound using the heuristic is equal to the backlog bound of the exact method. We also calculate the percentage increase and its CI from the exact method to the heuristic in cases where the results are not identical. For the execution times, the speedup from the exact method to the heuristic is calculated as $t_{\text{ex}}^{\text{exec}}/t_{\text{heu}}^{\text{exec}}$. Furthermore, for 3 and 7 cross flows, we illustrate the relation of the obtained backlog bounds and execution times for both methods and segment counts in Figs 4 and 5. For both figures, the scenario numbers are ordered by the value of the respective heuristic results (execution times and backlog bounds), from smallest to largest. For the backlog bounds, whenever the exact method calculates a more accurate backlog bound, this is represented as a spike down in the plot.

We observe that, for two segments, the heuristic provides very accurate results compared to the exact method. For 2 to 4 cross flows, a small percentage of runs do not produce the same backlog bound. For these, we observe a percentage increase of up to 6.7% of the backlog bound for 3 cross flows. (Yet, the overall mean of the iterations is not affected by this.) The accuracy for four α_1 segments is worse, but the percentage increase stays low across all numbers of cross flows. For the execution times, we observe that it grows in the number of cross flows. Additionally, in Table 1 we observe that the ratio of the execution times (speedup) between the two methods grows significantly. For four segments, we observe the same behavior, with a slower speedup across cross flow numbers.

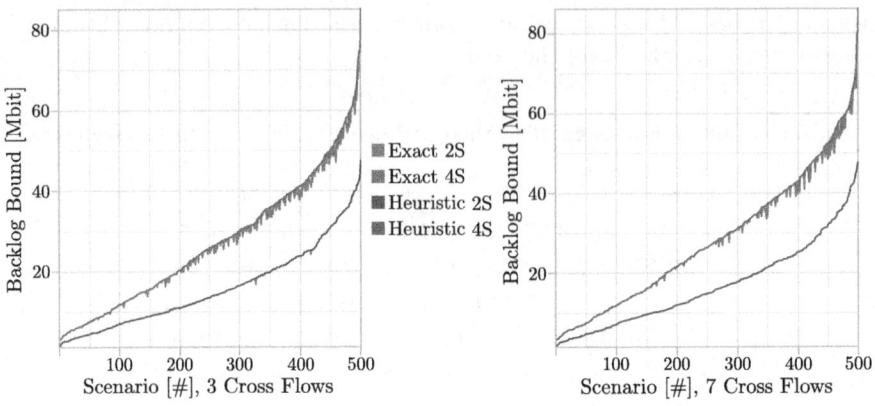

Fig. 4. Backlog bounds for varying numbers of cross flows and foi segments.

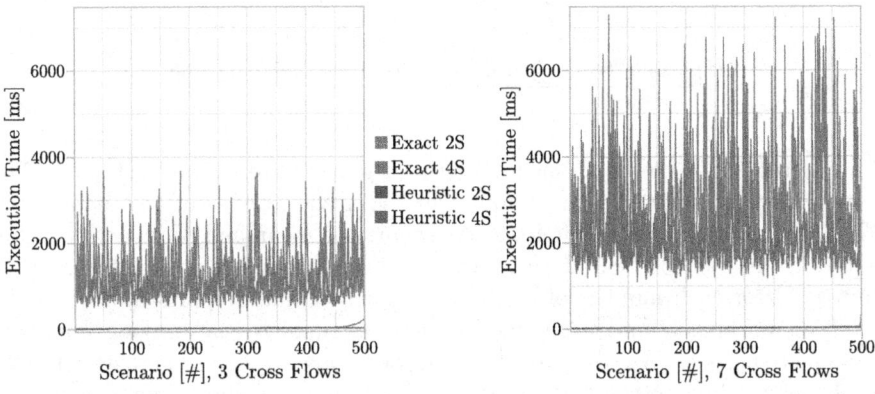

Fig. 5. Execution time for varying numbers of cross flows and foi segments.

In Fig. 5, we observe that the execution time of the exact method is faster for α_1 with four segments. We suspect that, due to the fact that α_1 is larger for four segments than it is for two, that we find ϕ earlier on average. This is supported by the fact that the calculation of the whole set of v_t curves results in similar execution times for both numbers of segments.

6 Usage of the Results in the DiscoDNC Tool

In the previous section, we have conducted our experiments using Nancy. Nancy is a general toolbox to support network calculus analyses but does not provide complete analysis methods by itself. Consequently, it does not mandate any specific value for θ, but instead we used it to compute such values. In contrast to Nancy, another open-source tool called DiscoDNC [2,11,12] provides complete network calculus analyses, in particular also for FIFO with a default setting of

θ as follows: $\theta_{\text{disco}} = \beta^{-1}\left(\sum_{i=2}^{n} b_1^i\right)$. In fact, this θ value coincides with the one that was used in Sect. 4 as a lower bound to the optimal theta value θ_{opt} *when all arrival curves are token buckets and the service curve is a rate-latency service curve*. For more complex curves, it holds that $\theta_{\text{disco}} < h(\alpha_2, \beta)$, however. Setting $\theta \leq h(\alpha_2, \beta)$ results in a β_θ^1 that is continuous in all points. It was consequently a reasonable first choice for an NC tool.

Yet, now we are able to adjust the θ value employed by DiscoDNC using our exact method and the decomposition heuristic. In the following, we are interested in the increase in quality of the calculated backlog bounds we can achieve when adjusting θ using the exact method as well as the heuristic.

We use the same experiment setup that we have used in Sect. 5.1. We compare the obtained backlog bound when using θ_{disco} against the backlog bound obtained using the exact method and heuristic. The results are illustrated in Fig. 6a. Here, we have calculated the mean backlog bounds and confidence intervals (CI) for the three methods for different numbers of cross flows. We observe that for all numbers of cross flows, the exact method and heuristic produce very similar backlog bounds (invisible difference in the figure). For $n = 10$, the backlog bound of both methods has the same mean and CI, with CI(95%) = 17.76±0.308. The backlog bound for the default value θ_{disco} has CI(95%) = 37.34±0.506. For two cross flows, the exact method and heuristic have CI(95%) = 15.8 ± 0.272, and the default value has CI(95%) = 22.3 ± 0.320.

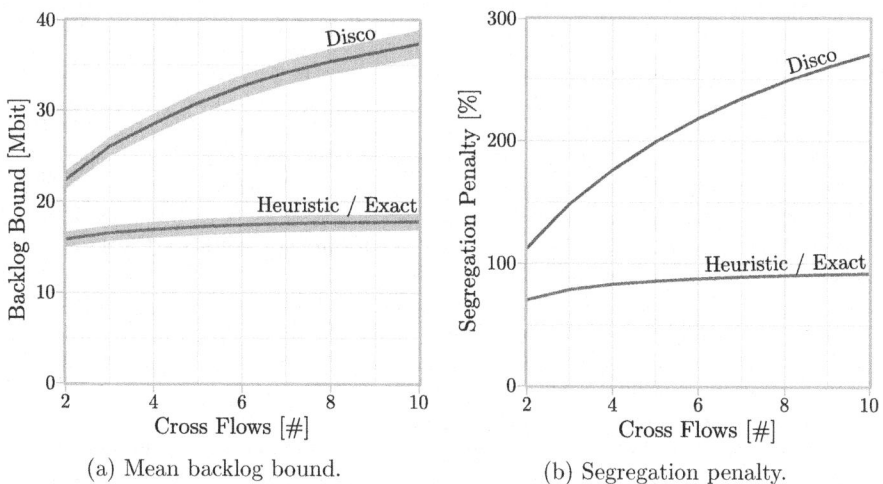

(a) Mean backlog bound.   (b) Segregation penalty.

Fig. 6. Comparisons of exact method and heuristic against the DiscoDNC tool.

Additionally, we study the segregation penalty (see Sect. 3) for this experimental setup. To this end, we calculate the segregation penalty as $(\sum q_{\min}^i - q_{\text{agg}})/q_{\text{agg}} \cdot 100$, where q_{agg} is the aggregate backlog bound. The results are given in Fig. 6b. We observe that the segregation penalty using the default θ value of

DiscoDNC grows significantly in the number of cross flows, in contrast to the new methods. This may potentially result in sub-optimal system design decisions.

To conclude, the backlog bound using θ_{disco} is significantly less accurate than the other two methods for any number of cross flows, with the difference becoming larger and larger the more cross flows traverse the server.

7 Conclusion

In this paper, we presented an exact method for deriving FIFO residual service curves that minimize backlog bounds for PWL concave arrival curves. While this method provides precise results, its computational inefficiency becomes apparent in more complex scenarios, such as settings with a large number of crossflows. To address this limitation, we introduced an efficient heuristic that provides an upper bound on the backlog-minimizing parameter of the FIFO residual service curve. Despite its approximate nature, we could show the heuristic yields backlog bounds that are close to those of the exact method, particularly in scenarios with a higher number of crossflows. Importantly, it achieves this with significantly reduced execution time. In a final experiment with the DiscoDNC tool, both the exact and heuristic approaches produced significantly more precise backlog bounds than the existing default setting of the tool.

References

1. Blanc, A.P.: Quality of service guarantees for FIFO queues with constrained inputs. University of California, San Diego (2006)
2. Bondorf, S., Schmitt, J.B.: The DiscoDNC v2 – a comprehensive tool for deterministic network calculus. In: Proceedings of the International Conference on Performance Evaluation Methodologies and Tools, ValueTools 2014, pp. 44–49 (2014). https://dl.acm.org/citation.cfm?id=2747659
3. Bouillard, A., Boyer, M., Le Corronc, E.: Deterministic Network Calculus: From Theory to Practical Implementation. Wiley (2018)
4. Cholvi, V., Echagüe, J., Boudec, J.Y.: Worst case burstiness increase due to FIFO multiplexing. Perform. Eval. **49**(1–4), 491–506 (2002)
5. Cruz, R.L.: A calculus for network delay. I. Network elements in isolation. IEEE Trans. Inf. Theory **37**(1), 114–131 (1991)
6. Cruz, R.L.: SCED+: efficient management of quality of service guarantees. In: Proceedings. IEEE INFOCOM 1998, the Conference on Computer Communications. Seventeenth Annual Joint Conference of the IEEE Computer and Communications Societies. Gateway to the 21st Century (Cat. No. 98. vol. 2, pp. 625–634. IEEE (1998)
7. Le Boudec, J.Y., Thiran, P.: Network Calculus: A Theory of Deterministic Queuing Systems for the Internet. Springer (2001). https://leboudec.github.io/netcal/
8. Lenzini, L., Mingozzi, E., Stea, G.: Delay bounds for FIFO aggregates: a case study. Comput. Commun. **28**(3), 287–299 (2005)
9. Liebeherr, J., Fidler, M., Valaee, S.: A system-theoretic approach to bandwidth estimation. IEEE/ACM Trans. Networking **18**(4), 1040–1053 (2009)

10. McKeown, N., Mekkittikul, A., Anantharam, V., Walrand, J.: Achieving 100% throughput in an input-queued switch. IEEE Trans. Commun. **47**(8), 1260–1267 (2002)
11. Scheffler, A., Bondorf, S.: Network calculus for bounding delays in feedforward networks of FIFO queueing systems. In: Abate, A., Marin, A. (eds.) QEST 2021. LNCS, vol. 12846, pp. 149–167. Springer, Cham (2021). https://doi.org/10.1007/978-3-030-85172-9_8
12. Schmitt, J.B., Zdarsky, F.A.: The disco network calculator: a toolbox for worst case analysis. In: Proceedings of the 1st International Conference on Performance Evaluation Methodologies and Tools, pp. 8–es (2006)
13. Shvachko, K., Kuang, H., Radia, S., Chansler, R.: The hadoop distributed file system. In: 2010 IEEE 26th Symposium on Mass Storage Systems and Technologies (MSST), pp. 1–10. IEEE (2010)
14. Wildberger, L., Hamscher, A., Schmitt, J.B.: Minimal per-flow backlog bounds at an aggregate FIFO server under piecewise-linear arrival curves (2025). https://arxiv.org/abs/2506.16914
15. Wrege, D.E., Knightly, E.W., Zhang, H., Liebeherr, J.: Deterministic delay bounds for VBR video in packet-switching networks: fundamental limits and practical trade-offs. IEEE/ACM Trans. Networking **4**(3), 352–362 (1996)
16. Zippo, R., Stea, G.: Nancy: an efficient parallel network calculus library. SoftwareX **19**, 101178 (2022)

Learning Mealy Machines with Sparse Observation Tables

Wolffhardt Schwabe[✉][iD], Paul Kogel[iD], and Sabine Glesner[iD]

TU Berlin, Berlin, Germany
{schwabe,p.kogel,sabine.glesner}@tu-berlin.de

Abstract. Active automata learning is a promising approach for the automated construction of system models in black-box settings. However, existing algorithms have issues scaling to large systems. Addressing this issue, we present L^s, a new learning algorithm for Mealy machines. It decouples state characterization from state identification to identify transition targets more efficiently. Its backbone is a new data structure, the *sparse observation table*. We define *constructiveness* as a sufficient criterion to infer unique hypotheses from this table and develop the *granularity* metric. This metric naturally translates into a heuristic for establishing constructiveness efficiently. We prove the correctness of the algorithm and experimentally evaluate it with models of real applications. While asymptotic worst-case complexity is not improved over existing algorithms, our experiments indicate performance gains in practice.

Keywords: Automata Learning · Model Learning · Mealy Machines

1 Introduction

The application of formal methods usually requires a model of the target system. But accurate models are often missing and difficult to obtain, especially when source code is unavailable. *Active Automata Learning* (AAL) approaches this issue by constructing accurate models of stateful systems in black-box settings. Through systematic interaction and observation of the input-output behavior, it infers the underlying structure. In this work, we focus on Mealy machines. They pose a particularly interesting target class, because they maintain a compact formalism, while covering many application domains from embedded control logic to network protocols. Many case studies successfully applied Mealy learning in practice, see for example [1–6]. However, existing algorithms have problems scaling to large systems.

Many fundamental concepts of AAL were introduced with the learning algorithm L^* by Dana Angluin [7]. Learning a Mealy machine requires inferring its states and identifying the transition targets. L^* captures the structure of the target system in an observation table. Rivest and Schapire then improved scalability by eliminating redundant rows from the table [8]. The table is split into rows

---

© The Author(s), under exclusive license to Springer Nature Switzerland AG 2026
P. Prabhakar and A. Vandin (Eds.): QEST+FORMATS 2025, LNCS 16143, pp. 176–194, 2026.
https://doi.org/10.1007/978-3-032-05792-1_10

that characterize states and rows that identify transition targets based on those characterizations. States are represented as prefixes and characterized by their outputs for different suffixes. Because every row uses the same set of suffixes, redundancy may occur *within* different rows. To reduce this type of redundancy, algorithms like TTT select suffixes by traversing a classification tree [9]. However, redundancy in state characterizations could provide valuable information for further increasing identification efficiency.

In this work, we present a new data structure that is partially redundant by design, with the aim to combine the advantages of tables and trees. This is driven by the insight that state characterization and state identification do not need to follow the same routine. Since complexity is dominated by identification, it is reasonable to accept reduced efficiency for characterization, if it helps with identification. Our structure encourages redundant information in state characterizations, but leaves flexibility in the identification of transition targets. From this structure, we then derive the new learning algorithm L^s. We prove correctness and analyze complexity. We also developed an implementation in *LearnLib* [10] and conduct an experimental evaluation, using models of real applications from the *Automata Wiki* benchmark collection [11].

Outline. The rest of this work is structured as follows. Sect. 2 covers related approaches. In Sect. 3, we explain the algorithmic context for table-based learning of Mealy machines. We then present our algorithm and prove its correctness in Sect. 4. Afterwards, we describe an experimental evaluation in Sect. 5 and discuss implementation details in Sect. 6. We conclude in Sect. 7 with an idea for future work.

2 Related Work

Since the integration of classification trees into AAL, many other optimizations have followed. Tackling redundancy at an even lower level than previous tree-based algorithms, Markus Frohme developed ADT [12]. It uses *Adaptive Distinguishing Sequences* (ADS) to identify states symbol-wise instead of suffix-wise. By contrast, we stick to suffix level and try to exploit the information available at this level more extensively.

Later, Vaandrager *et al.* showed that state identification does not require any fixed classification scheme [13]. While identifying a transition target, their algorithm $L^\#$ always tracks which states remain possible candidates. It then picks distinguishing suffixes on the fly. In this regard, $L^\#$ and L^s are similar. But $L^\#$ performs state characterization in a more modular way that makes it harder to optimize suffix selection.

Howar and Steffen then introduced L^λ [14]. It constructs state characterizations in a way that always guarantees minimal length for every suffix involved. While this is a significant result, it does not improve identification mechanics.

We return perspective to prior work. Before TTT, Isberner *et al.* presented an alternative approach for table-based learning with *State-Local Alphabet Abstraction (SLA)* [15]. This algorithm combines groups of transitions with identical source, output and target into abstract transitions. For each of these, it uses a single row to identify the corresponding target. This way, *SLA* reduces the number of rows in the table. However, it does not reduce the complexity at the level of individual rows. Furthermore, the abstraction comes with significant overhead. Thus, *SLA* is only useful if there is a sufficient degree of redundancy in the state graph. Inspired by this approach, our algorithm aims to improve table efficiency in a more general way by focusing on the mechanics at row level. Before we explain our approach, we next cover the relevant background.

3 Background

In this section, we formally define Mealy machines and associated concepts. Then, we describe the algorithmic framework for AAL. Lastly, we give a short introduction to table-based learning.

3.1 Mealy Machines

Definition 1. *A* Mealy machine *is a tuple* $M = (S, s_0, I, O, \delta, \lambda)$, *where S is the finite state set, $s_0 \in S$ the start state, I and O the finite input respectively output alphabet, $\delta : S \times I \to S$ the transition function and $\lambda : S \times I \to O$ the output function.*

We write ε for the empty word. Concatenation of two words u and v is denoted by $u \cdot v$ or uv. This operation can also be applied to sets of words to perform pairwise concatenation. For a word w, we denote its length by $|w|$. For words $iv \in I^*$ where $i \in I$, the extended output function $\lambda^* : S \times I^* \to O^*$ is defined for every $s \in S$ as follows:

$$\lambda^*(s, iv) = \lambda(s, i) \cdot \lambda^*(\delta(s, i), v)$$
$$\lambda^*(s, \varepsilon) = \varepsilon$$

The *behavior* of M for some $w \in I^*$ is then given by $M(w) = \lambda^*(s_0, w)$. When not further specified, the term refers to all input-output traces producible by M. For any $uv \in I^*$, we write $M(u, v)$ for the suffix of $M(uv)$ with length $|v|$. The extended output function is also important for reasoning about states.

Definition 2. *Let $s_1, s_2 \in S$. They are* **distinct** *iff $\lambda^*(s_1, v) \neq \lambda^*(s_2, v)$ for some $v \in I^*$. Any such v is called a* **discriminator**.

If every pair of non-identical states is distinct, then M is *minimal*. This means that no equivalent machine with fewer states exists.

Definition 3. *Let M_1, M_2 be Mealy machines. They are* **equivalent**, *written $M_1 \approx M_2$, iff $I_{M_1} = I_{M_2}$ and $M_1(w) = M_2(w)$ for all $w \in I^*$, where $I = I_{M_1}$.*

It follows that if $M_1 \not\approx M_2$ and $I_{M_1} = I_{M_2}$, there must be a word $e \in I^+$ for which $M_1(e) \neq M_2(e)$. Such e is a *counterexample* (to the equivalence of M_1 and M_2). It is apparent without proof that the structure of a minimal machine is uniquely determined by its behavior (modulo state labeling). This fact is essential for AAL, as explained in the next subsection.

3.2 MAT Framework

The goal of AAL is to infer an accurate state machine model of some black-boxed system, called the *System Under Learning* (SUL). The model class plus input alphabet are given, and the SUL can be executed with arbitrary input words. But nothing is known about its internal structure. Angluin defined a conceptual framework that models the inference process as an interaction between a *learner* and a *Minimally Adequate Teacher* (MAT) [7]. The key idea is to divide the process into two separate parts: hypothesis construction and hypothesis testing. These are implemented by the learner and the teacher, respectively. To simplify its task, the learner assumes that the teacher has perfect knowledge of the SUL. The learner never interacts with the SUL directly but only through the teacher. Let $\mathcal{M}$ be the Mealy machine implemented by the SUL. Without loss of generality, assume $\mathcal{M}$ is minimal. The teacher answers two types of queries:

Output Query (OQ): Given an input $w \in I_\mathcal{M}^*$, returns $\mathcal{M}(w)$.
For any decomposition $w = u \cdot v$, we write $OQ(u, v)$ for $\mathcal{M}(u, v)$.
Equivalence Query (EQ): For a Mealy machine $\mathcal{H}$ with $I_\mathcal{H} = I_\mathcal{M}$,
replies whether $\mathcal{H} \approx \mathcal{M}$. If $\mathcal{H} \not\approx \mathcal{M}$, provides a counterexample.

Initially, the learner has no information about $\mathcal{M}$, except its input alphabet. Through OQs, it systematically explores the SUL's behavior to construct some *hypothesis* $\mathcal{H}$, which is then subjected to an EQ. If the query reveals equivalence, learning terminates. Otherwise, the learner receives a counterexample. From this new observation, it then *refines* $\mathcal{H}$. Refinements and EQs alternate until no more counterexample is found.

To guarantee termination, refinements must follow some convergence criterion. The most common criterion is state count, which we follow in this work. Each time before the learner poses an EQ, it ensures that $\mathcal{H}$ is *unique* with respect to all its observations under the constraint of minimality. This way, the learner is forced to increase the number of states with each refinement. Since Mealy machines always have finite state count, this strict monotony guarantees termination of the learning process. Assuming that EQs are always answered correctly, $\mathcal{H}$ will converge to match the SUL's behavior.

The teacher's task is to live up to the learner's belief of its omniscience. An OQ can be answered by resetting the SUL, executing it with the given input and forwarding its output to the learner. EQs are harder, because the teacher can only compare $\mathcal{H}$ and $\mathcal{M}$ for specific inputs. Consequently, an EQ must be reduced to a set of OQs. The number of OQs required for verifying equivalence grows exponentially with SUL states. But in practice, equivalence checks can often be sufficiently approximated with more efficient testing methods [16].

Complexity analysis in AAL assumes that runtime is dominated by SUL execution. This idea is captured by two different metrics. *Query complexity* regards the number of OQs and accounts for the overhead of system resets. *Symbol complexity* regards the number of input symbols accumulated over all OQs. It accounts for the delay caused by input processing.

3.3 Observation Tables

Reasoning about observations requires a data structure. The structure used in our algorithm is based on the *Observation Table* (OT) of Rivest and Schapire [8]. We now briefly describe the OT in simplified notation before presenting our optimized version in the next section.

Learning $\mathcal{M}$ requires finding all its transitions and identifying their outputs and target states. Once a new transition is discovered, its output can easily be obtained with an OQ. But identifying its target is less simple. An OT guides this process and enables the construction of a state graph. Let T be an OT. It has a set of *prefixes* $U \subset I_\mathcal{M}^*$ and *suffixes* $V \subset I_\mathcal{M}^+$. Each $u \in U$ refers to some row and each $v \in V$ to some column. Together, they address the table cell defined by $T(u,v) = \mathcal{M}(u,v)$. Each prefix describes a transition path from the start state and represents the target state reached by it. A special symbol $\tau \notin I_\mathcal{M}$ denotes the output symbol of the final transition along that path. We define $T(u, \tau) = \mathcal{M}(p, i)$ if $u = p \cdot i$ with $i \in I_\mathcal{M}$. Since the empty prefix leading to the start state does not produce any output, $T(\varepsilon, \tau)$ is undefined. The suffixes serve as discriminators to distinguish the states that different prefixes represent.

To reflect the structure of a state graph, the prefix set U is split into two disjoint sets U_c and U_f. We call U_c the set of *core* prefixes, which define a spanning tree of the hypothesis states. U_c is *prefix-closed*; meaning for any $u_c \in U_c$, all of its prefixes are also contained. U_f is the set of *fringe* prefixes. It is defined as $U_f = (U_c \cdot I_\mathcal{M} - U_c)$, extending the spanning tree to cover all remaining transitions of the state graph. All core prefixes have distinct rows. That is, for all $u_1, u_2 \in U_c$ with $u_1 \neq u_2$, there is some $v \in V$ with $T(u_1, v) \neq T(u_2, v)$. By contrast, every fringe row must *match* some core row. That is, for all $u_f \in U_f$, there must be some $u_c \in U_c$ with $T(u_f, v) = T(u_c, v)$ for all $v \in V$. Intuitively, u_f then represents the same target state as u_c. If the table satisfies these conditions, it is called *closed*. Closedness implies that every transition in the state graph represented by T has a well-defined target. This in turn allows the learner to derive a unique hypothesis. Figure 1a shows an example model and Fig. 1b shows a closed OT that reflects its state graph. The table has $U_c = \{\varepsilon, \text{b}, \text{bb}\}$, $U_f = \{\text{a}, \text{ba}, \text{bba}, \text{bbb}\}$ and $V = \{\text{ba}, \text{bba}\}$. Core rows in the top represent bold transitions and fringe rows below represent dashed transitions. In the next section, we optimize the table structure to increase learning efficiency.

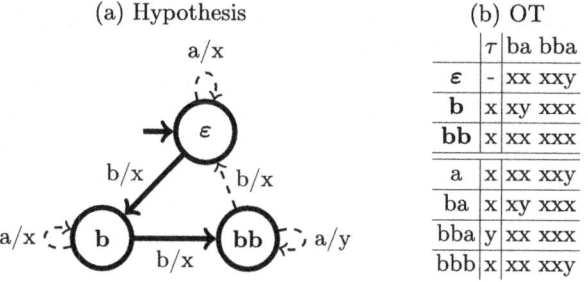

Fig. 1. Hypothesis derived from example OT. Bold edges mark the spanning tree.

4 Learning Algorithm

In this section, we introduce our new learning algorithm L^s. We start with a look at the scaling behavior of *Observation Tables* (OTs) to derive a more flexible type of table as our central structure. Next, we explain the general construction process for this table and then develop an optimization to reduce learning complexity. Afterwards, we complete the algorithm with an adapted refinement routine, before closing the section with an analysis of correctness and complexity.

4.1 Sparse Observation Tables

Deriving a unique hypothesis from an OT requires a unique mapping from the fringe rows U_f to the core rows U_c. Looking at the table structure, we make an important observation. Assume the *System Under Learning* (SUL) implements a Mealy machine $\mathcal{M}$ with $|S_\mathcal{M}| = n$ and $|I_\mathcal{M}| = k$. Because $U_f = (U_c \cdot I_\mathcal{M} - U_c)$, it follows that U_f has roughly k times as many elements as U_c. Since every cell is associated with a specific *Output Query* (OQ), about $\frac{k-1}{k}$ of OQs posed during table construction are caused by fringe rows. However, returning to the example from Fig. 1b, we can see that not all fringe cells contribute equally to the row mapping. Figure 1c shows a reduced table, where some cells remain undefined. The table still yields a unique row mapping, producing the same hypothesis (Fig. 1a). Omitting redundant fringe cells would allow for a more efficient hypothesis encoding. With this insight, we present our new data structure.

Definition 4. *For a target machine $\mathcal{M}$, a **Sparse Observation Table (SOT)** is a tuple $T = (V, U_c, U_f, \varphi)$, where $V \subset I_\mathcal{M}^+$ is the suffix set, $U_c \subset I_\mathcal{M}^*$ the core prefix set, $U_f = (U_c \cdot I_\mathcal{M} - U_c)$ the fringe prefix set and $\varphi : U_f \to 2^V$ the **fringe function**. For any $u_c \in U_c$ and $v \in V$, we set $T(u_c, v) = \mathcal{M}(u_c, v)$. For any $u_f \in U_f$, the value $\varphi(u_f)$ specifies a subset $V_f \subseteq V$. For any $v_f \in V_f$, we set $T(u_f, v_f) = \mathcal{M}(u_f, v_f)$. For every $u \in (U_c \cup U_f)$ with $u = p \cdot i$ where $i \in I_\mathcal{M}$, we set $T(u, \tau) = \mathcal{M}(p, i)$. All other cells in the table are undefined.*

SOTs treat core and fringe rows differently. Like previously, every core row uses the full suffix set, even if a subset would suffice to distinguish it from all other core rows. We later leverage this redundancy to the learner's advantage. In contrast to before, fringe rows may now have gaps. Consequently, checking closedness is not generally possible. But the learner still needs to construct unique hypotheses. We thus need another way to constrain the table so that all transitions have well-defined targets. For this purpose, we first introduce a few helpful concepts.

Definition 5. *Let $u_f \in U_f$ and $u_c \in U_c$ be two row prefixes. They are **incompatible** iff there is some suffix $v \in \varphi(u_f)$ with $T(u_f, v) \neq T(u_c, v)$. Otherwise, they are **compatible**. Iff there is exactly one compatible core prefix, we call u_f **classified**.*

Incompatibility implies that the corresponding states are distinct. By contrast, compatibility simply means that the table does not hold any evidence of those states being distinct. If u_f is classified and u_c the compatible core prefix, we can assume from the observations that u_f and u_c represent the same state. With these concepts, we derive a simple uniqueness criterion for the table.

Definition 6. *Iff every $u_f \in U_f$ is classified, T is **constructive** (to some unique hypothesis).*

Constructiveness guarantees a unique mapping from fringe to core rows. It can be interpreted as a relaxed notion of closedness, with the advantage of potentially requiring less information to reach a unique hypothesis. When building the hypothesis $\mathcal{H}$, we encode its states by the core prefixes, that is $S_\mathcal{H} = U_c$. For every transition $u_c \xrightarrow{i}$ from state $u_c \in U_c$ with input $i \in I_\mathcal{M}$, set the target as follows. If $u_c i \in U_c$, set it to $u_c i$. Otherwise, we know that $u_c i \in U_f$. In this case, set the target to the unique core prefix with which $u_c i$ is compatible. In either case, the transition output is defined by $T(u_c i, \tau)$.

During table construction, whenever we query the first suffix in an empty row with prefix $u_c i \in U_f$, we derive the transition output $T(u_c i, \tau)$ from the corresponding OQ. Only for the first hypothesis, additional OQs are required to obtain the transition outputs. Initially, we set $U_c = \{\varepsilon\}$, $U_f = I_\mathcal{M}$ and $V = \varnothing$. Thus, the first hypothesis consists of a single state with self-loops. Through refinements, prefixes and suffixes will gradually be extended, as we describe later. Before that, we next explain how to establish constructiveness. For simplicity, we omit transition outputs in the pseudocode.

4.2 Table Construction

Establishing constructiveness requires classification of the fringe prefixes. When a fringe row with prefix $u_f \in U_f$ is created, we set its suffix set $\varphi(u_f) = \varnothing$. Consequently, u_f is compatible with all core prefixes. When we fill a table cell $T(u_f, v)$ for some $v \in V$, any core prefix $u_c \in U_c$ with $T(u_c, v) \neq T(u_f, v)$ becomes incompatible. We then need to distinguish three cases:[1]

---

[1] An equivalent case distinction appears in the state identification of $L^\#$ [13]. However, we handle cases (ii) and (iii) differently to exploit the advantages of our SOT.

(i) Exactly one core prefix remains compatible.
(ii) More than one core prefix remains compatible.
(iii) No core prefix remains compatible.

In case (i), we can move on to the next row. In case (ii), we need to query another suffix to further constrain u_f and then repeat the analysis. In case (iii), u_f represents a new state we need to incorporate into the core rows. Algorithm 1 shows our classification procedure. It returns *true* iff u_f necessitates a core extension. The function pickSuffix is defined later.

Algorithm 1. Fringe Prefix Classification

  procedure classify(u_f)   ▷ $u_f \in U_f$
      $U_r \leftarrow \{u_c \in U_c \mid \forall\, v \in \varphi(u_f) : T(u_c, v) = T(u_f, v)\}$ ▷ remaining core rows
      while $|U_r| > 1$ ▷ case (ii)
            $v \leftarrow$ pickSuffix(U_r)
            $\varphi(u_f) \leftarrow \varphi(u_f) \cup \{v\}$
            $T(u_f, v) \leftarrow \mathrm{OQ}(u_f, v)$
            $U_r \leftarrow \{u_c \in U_r \mid T(u_c, v) = T(u_f, v)\}$

      if $|U_r| < 1$
            return *true*   ▷ case (iii)
      else
            return *false*   ▷ case (i)

When we discover a new state, we first complete its row. Then, we move its prefix from U_f to U_c and add its outgoing transitions to U_f. The new fringe prefixes are initially unclassified. Additionally, previously existing fringe prefixes are not guaranteed to be classified anymore, because the set of possible transition targets has grown. So anytime we extend the core, we must reexamine all fringe rows. Algorithm 2 shows the procedure for turning the table constructive. In principle, different fringe prefixes could be classified in parallel. For simplicity, we here stick with sequential iteration. The iteration order can impact performance. We prioritize newer prefixes, because transitions from new states are more likely to lead to other new states. This way, we expect faster core extension, which lets us classify the remaining prefixes more efficiently.

In the worst case, we end up filling every table cell and constructing a table without sparsity. However, the resulting table is guaranteed to be constructive. This is because we only extend the core when we find a new state distinguished by a suffix in V. Thus, the core rows are bounded by the number of states in $\mathcal{M}$. Once the core is complete, any fringe row without gaps necessarily matches a single core row. Now that we have a general method for establishing constructiveness, the next step is to find an effective suffix selection strategy.

Algorithm 2. Establishing Constructiveness

procedure constructTable
  for each $u_f \in U_f$
    if classify(u_f)   ▷ prefix leads to new state
      for each $v \in (V - \varphi(u_f))$ ▷ fill missing cells
        $T(u_f, v) \leftarrow \text{OQ}(u_f, v)$
      $U_f \leftarrow U_f - \{u_f\}$ ▷ move fringe row to core
      $U_c \leftarrow U_c \cup \{u_f\}$
      for each $i \in I_M$ ▷ add new fringe rows
        $U_f \leftarrow U_f \cup \{u_f i\}$
        $\varphi(u_f i) \leftarrow \varnothing$
    constructTable()   ▷ restart loop by tail recursion
    return

4.3 Suffix Selection

To minimize learntime, we need to maximize the number of gaps in the SOT. Accordingly, we want to keep the suffix set of each fringe row as small as possible while classifying its prefix. But in general, we cannot know the output of a suffix for a specific prefix before posing the corresponding query. This makes it impossible to compute a smallest suffix set in advance. Instead, starting from the empty set, it must be extended one by one, always considering the information gained by previous OQs. Here, the structure of the SOT comes into play.

In order to turn the table constructive, it is insignificant which suffix set each fringe row uses, as long as its prefix is classified. But picking suffixes effectively requires a suitable criterion to guide selection. We define the *utility* of a suffix as the number of core prefixes remaining compatible after filling its cell. Since utility depends on the output, it cannot be known prior to query. However, we can use a proxy measure to guide selection. For this, the redundancy within core rows proves useful. Their gaplessness allows us to compute a utility-related metric for each suffix, for any set U_r of remaining core prefixes.

Definition 7. *Let $U_r \subseteq U_c$ and $v \in V$. We define $D = \bigcup_{u_c \in U_r} \{T(u_c, v)\}$ as the set of distinct outputs produced by v across prefixes in U_r. The **granularity** of v with respect to U_r is* $\text{gran}(v, U_r) = \max_{o \in D} |\{u_c \in U_r : T(u_c, v) = o\}|$.

Assume we are classifying some fringe prefix $u_f \in U_f$ and U_r contains all core prefixes with which u_f currently remains compatible. For any suffix $v \in V$, $\text{gran}(v, U_r)$ tells us how many prefixes would remain compatible *at most* after querying $T(u_f, v)$. Specifically, the granularity corresponds to the size of a largest subset $Z \subseteq U_r$ in which all prefixes produce the same output for v. That is, $T(u_1, v) = T(u_2, v)$ for all $u_1, u_2 \in Z$. Put simply, the worst-case utility of a suffix scales inversely with its granularity. Consequently, granularity lends itself naturally as a criterion for suffix selection. For an example, we return to the table in Fig. 1c. If $U_r = \{\varepsilon, \text{b}, \text{bb}\}$, we have $\text{gran}(\text{ba}, U_r) = \text{gran}(\text{bba}, U_r) = 2$. But if $U_r = \{\text{b}, \text{bb}\}$, then $\text{gran}(\text{ba}, U_r) = 1 \neq \text{gran}(\text{bba}, U_r) = 2$.

To pick a suffix now, we compute the granularity for each $v \in V$. Then, we discard all suffixes with non-minimal granularity. Since all core rows are distinct, there must always be some $v \in V$ with $\texttt{gran}(v, U_r) < |U_r|$. This guarantees a strictly monotonic decrease in the number of remaining prefixes with each suffix. However, there may exist multiple suffixes with minimal granularity. In this case, we pick the one that was most recently added to V. The motivation for this rule is as follows. We need to classify u_f with respect to the current core. But to avoid querying further suffixes, we also want to increase the chance that u_f stays classified when future refinements extend the core. By always picking the youngest suffix, we spread the risk of choosing discriminators that badly generalize to future states. Algorithm 3 shows the suffix selection function. Now we have all the algorithmic components for efficient hypothesis construction.

Algorithm 3. Suffix Selection

   function `pickSuffix`(U_r) ▷ $U_r \subseteq U_c$
       $V_g \leftarrow \varnothing$ ▷ lowest-granularity suffixes
       $minRank \leftarrow \infty$
       for each $v \in V$
           $rank \leftarrow \texttt{gran}(v, U_r)$
           if $rank < minRank$
               $V_g \leftarrow \{v\}$ ▷ discard previous candidates
           else if $rank = minRank$
               $V_g \leftarrow V_g \cup \{v\}$ ▷ add to previous candidates
           $minRank \leftarrow \min(minRank, rank)$

       return $v \in V_g$ ▷ take youngest suffix

4.4 Refinement

Once we have a hypothesis, we submit it to an *Equivalence Query* (EQ). If the query returns a counterexample, we must perform a refinement. To simplify analysis, we truncate the counterexample after the first symbol revealing inequivalence. Then, we extend the table. As our table is a reduced form of the one by Rivest and Schapire [8], we transfer their refinement strategy and adapt it to exploit the structural flexibility of the SOT.

After truncation, assume we have the counterexample $e \in I_\mathcal{M}^+$. Since the hypothesis was unique, the existence of e proves its state set incomplete. Consequently, the spanning tree defined by U_c is incomplete and must be extended. As the spanning tree must be prefix-closed, any new core prefix can only originate from U_f. Thus, at least one fringe prefix is wrongly classified. Therefore, the hypothesis transition corresponding to that prefix is wrong. Since e disproved the hypothesis, it must have triggered such a transition. Rivest and Schapire gave a procedure for analyzing counterexamples and identifying wrongly mapped transitions. This procedure is transferable to the SOT, because it only requires knowledge of the correspondence between hypothesis states and core prefixes. This requirement is trivially satisfied, as our hypothesis states are directly encoded by the core prefixes ($S_\mathcal{H} = U_c$). The procedure has also been transferred to many other learning algorithms. For details, readers are referred to the generalized description and proof by Isberner and Steffen [17].

The analysis produces a decomposition $e = u_e \cdot i \cdot v_e$ with $i \in I_\mathcal{M}$. For any $w \in I_\mathcal{M}^*$, we write $\mathcal{H}[w]$ for the core prefix of the state in which $\mathcal{H}$ ends after processing w. The analysis guarantees that $\mathcal{H}[u_e] \cdot i$ is a fringe prefix with $\mathcal{M}(\mathcal{H}[u_e] \cdot i, v_e) \neq \mathcal{M}(\mathcal{H}[u_e i], v_e)$. Consequently, v_e is a discriminator for $\mathcal{H}[u_e] \cdot i$ and $\mathcal{H}[u_e i]$. This implies that the transition $\mathcal{H}[u_e] \xrightarrow{i}$ points to some new state not yet recognized. The state set is then extended by moving $\mathcal{H}[u_e] \cdot i$ into the core. But maintaining unique state prefixes requires that the distinctness of $\mathcal{H}[u_e] \cdot i$ and $\mathcal{H}[u_e i]$ is reflected in their rows. By above guarantee over the decomposition, this can be achieved by setting $V = V \cup \{v_e\}$.

Before adopting this procedure for the SOT, we should briefly consider the difference in table structure. In the OT of Rivest and Schapire, fringe rows have no gaps. Thus, each refinement must extend the suffix set to preserve distinctness of the core rows. But in the SOT, before moving a row to the core, we first query its undefined cells. If the resulting outputs turn the row distinct, no new suffix is required. Otherwise, we follow the original procedure and set $V = V \cup \{v_e\}$. By only adding suffixes when necessary, we avoid unnecessarily inflating the core. Reconsider the example from Fig. 1c. Assume the corresponding hypothesis in Fig. 1a is wrong and we receive a counterexample which reveals that the transition $b \xrightarrow{a}$ is not a self-loop. This means the prefix ba should be incompatible with b. In this case, we first query the missing cell in row ba before we decide whether to add a new suffix.

After turning the identified fringe row distinct, we use the procedure presented in Algorithm 2 to reestablish constructiveness, which moves the row to core and adds fringe rows accordingly. The full refinement procedure is shown in Algorithm 4. After each refinement, we build a new hypothesis. Then, we reexamine the counterexample to check if it reveals another flaw. If not, we pose a new EQ.

4.5 Correctness and Complexity

In the MAT framework, correctness of the learned model is guaranteed if the learner terminates with an EQ that returns no counterexample. In the following, we first prove an upper bound on the number of EQs posed by L^s. Subsequently, we derive an upper bound on its OQs and finally show that refinements always terminate. We use $n = |S_\mathcal{M}|$ and $k = |I_\mathcal{M}|$.

Algorithm 4. Table Refinement

procedure refine(e) ▷ $e \in I_\mathcal{M}^+$ is counterexample
    $u_e \cdot i \cdot v_e \leftarrow$ rivestSchapire(e) ▷ decompose counterexample
    $u_{\text{old}} \leftarrow \mathcal{H}[u_e i]$
    $u_{\text{new}} \leftarrow \mathcal{H}[u_e] \cdot i$

    for each $v \in (V - \varphi(u_{\text{new}}))$ ▷ fill missing cells
        $T(u_{\text{new}}, v) \leftarrow$ OQ(u_{new}, v)

    if $[\forall v \in V : T(u_{\text{new}}, v) = T(u_{\text{old}}, v)]$ ▷ u_{new} is not yet distinct
        $V \leftarrow V \cup \{v_e\}$ ▷ add suffix
        $T(u_{\text{new}}, v_e) \leftarrow$ OQ(u_{new}, v_e)
        for each $u_c \in U_c$
            $T(u_c, v_e) \leftarrow$ OQ(u_c, v_e)

    $\varphi(u_{\text{new}}) \leftarrow V$
    constructTable()

Theorem 1. *L^s will pose at most n EQs.*

Proof. L^s only poses an EQ when it established constructiveness. Thus, each hypothesis transition is backed by observations proving its target inequivalent to all other identified states. Consequently, if the hypothesis is wrong, a counterexample must traverse some SUL state that is not reflected in the hypothesis. The Rivest-Schapire decomposition then identifies a faulty hypothesis transition whose target gets split off to form a new state. With each refinement, the hypothesis grows in states. But L^s only adds states whose distinctness is backed by observation. Once it reaches n states, the hypothesis describes the SUL correctly and no more counterexamples exist.

Theorem 2. *Let m be the length of the longest counterexample received. L^s will pose at most $\mathcal{O}(kn^2 + n \log m)$ OQs, each with less than $(n + m)$ symbols.*

Proof. L^s poses OQs during table construction and counterexample analysis. To determine a limit for the former, we must consider the table's dimensions. Refinements never delete rows, as new core rows always emerge from existing fringe rows. The maximum number of core rows equals the final hypothesis size, that is $|U_c| = n$. Since $U_f = (U_c \cdot I_\mathcal{M} - U_c)$ and $\varepsilon \in U_c$, we then know that $|U_f| = n(k-1) + 1$. That implies a total row count of $(nk + 1)$. For column count, we need to consider the suffix set. Initially, $|V| = 0$. With each refinement, L^s adds at most one suffix. This implies $|V| < n$. Consequently, the OQs for table construction are limited by a cell bound of $\mathcal{O}(kn^2)$. To that, we must add the OQs from counterexample analysis. The number of counterexamples received before the final construction step never exceeds $(n - 1)$. This bounds analysis-related OQs by $\mathcal{O}(n \log m)$, as the Rivest-Schapire procedure uses a form of binary search to decompose counterexamples [17].

To verify the symbol bound, recall that any counterexample e is decomposed into $e = u_e \cdot i \cdot v_e$ with $i \in I_\mathcal{M}$ and $\mathcal{M}(\mathcal{H}[u_e] \cdot i, v_e) \neq \mathcal{M}(\mathcal{H}[u_e i], v_e)$. Since $S_\mathcal{H} = U_c$, we know that $\mathcal{H}[w] \in U_c$ for any $w \in I_\mathcal{M}^*$. As U_c is prefix-closed, $\varepsilon \in U_c$ and $|U_c| \leq n$, it follows that $|\mathcal{H}[w]| < n$. Because $|e| \leq m$, we also have $|v_e| < m$. Thus, any decomposition tested during counterexample analysis uses less than $(n + m)$ symbols. Since suffixes in V are always taken from counterexamples, this bound carries over to OQs from table construction: For any table cell (u, v) with $u \in (U_c \cup U_f)$ and $v \in V$, we know that $|u| \leq n$ and $|v| < m$.

Theorem 3. *L^s always terminates with the correct model.*

Proof. Having proved hypothesis convergence, we only need to verify that each construction step terminates. Since all loops in L^s iterate over finite structures, the only cause for non-termination could be the recursive call in Algorithm 2. But the recursion only triggers upon identifying a new state, which limits its depth in n.

The asymptotic worst-case complexity of L^s is on par with the Rivest-Schapire learner. We evaluate practical performance in the next section.

5 Evaluation

For a practical comparison with state-of-the-art learners, we implemented our algorithm in the open-source framework *LearnLib* [10], using the latest stable release 0.18.0. The source code is available in our artifact. LearnLib contains ready-to-use implementations of many learners, including Rivest-Schapire (*RS*) for comparison with our baseline. There are two versions of L^λ, using either an OT or building on the tree-based *TTT* learner. There is also an extended $L^\#$ variant with *Adaptive Distinguishing Sequences* (ADS) [13]. Of the algorithms mentioned in this work, only *State-Local Alphabet Abstraction* (*SLA*) was not available in the LearnLib release. Thus, we created our own implementation as described in the original publication [15], with one change. Originally, *SLA* produces abstract hypotheses that define only a subset of transitions. The remaining transitions are defined implicitly by a remapping of input symbols. To convert the hypotheses to fully defined Mealy machines, we precompute the abstraction during hypothesis construction.

All learners share the same setup. To eliminate repeated OQs, the SUL is wrapped with a query cache. We also use this cache to look for counterexamples in previous observations before executing EQs. Furthermore, we always truncate counterexamples before passing them to learners. We learned six models from different case studies, all of which are part of the *Automata Wiki* benchmark collection [11]. These models include two smartcards and four network protocols:

- Biometric Passport [1]
- Volksbank Bank Card [2]
- GNU TLS Client [3]

- Mosquitto MQTT Client [4]
- Ubuntu TCP Server [5]
- BitVise SSH Client [6]

All these models were also used in the comprehensive benchmark study of Aichernig et al. [16]. We obtained the model files from their repository.[2] The study compared many different combinations of learning and testing algorithms. For testing, *RandomWp* from LearnLib gave the best overall performance (with parameters $minSize = 0$ and $rndLen = 4$). Thus, we adopted this testing configuration for our experiments. Each model was learned to completion. For a clearer comparison of performance, we excluded the last EQ of each run from the statistics, because it never produces a counterexample and the learners have no influence on its complexity. As *RandomWp* is non-deterministic, we ran each combination of model and learner 100 times with distinct seeds. We measured query and symbol complexity, both for learning only, as well as the overall complexity including the OQs caused by hypothesis testing. Since in practice, automata learning is only applied once to a target system, we took the median from all runs for each quantity.

Tables 1 and 2 show the relative performance of L^s to other learners. Rows indicate models annotated with state count (n) and alphabet size (k), columns indicate learners. Each cell shows the median speedup of L^s in percent. For example in Table 1, for the passport model, the median query complexity of L^s was 35.4% lower than that of *RS*. Green cells indicate improvements, while uncolored cells indicate that L^s performed worse. Table 3 shows the median value of each performance metric for L^s. Table 4 shows the median number of EQs posed by each learner. All statistics exclude queries answered by cache.

During learning, L^s outperforms most other algorithms, except for *ADT*, $L^{\#}_{ADS}$ and L^{λ}_{TTT}. For the last, the advantage is mostly restricted to symbol complexity. This is because L^λ uses a special procedure to minimize the length of discriminators. The advantage of the other two is due to their use of ADS.

When considering overall performance, results seem more mixed at first glance. There is no model for which L^s exhibits the best performance among all learners. However, when reading the table column-wise, we see that L^s outperforms almost every other learner for the majority of models, the only exception being $L^{\#}_{ADS}$. But for the two largest models, L^s pulls ahead. The difference between learning-only and overall complexity is caused by testing. For example, *ADT* loses advantage by requiring more EQs. For the largest model (SSH), the best performance is achieved by *SLA*. This implies a large number of redundant transitions in the underlying state graph. To improve performance in such cases, L^s could be integrated with *SLA* [15], or more lightweight optimizations aimed at redundant transitions, like a symbol filter for self-looping transitions [18].

---

[2] gitlab.com/felixwallner/learn-combinations/-/tree/master/input/mealy (commit ad489b0f).

Table 1. Learning-only relative complexity savings of L^s in percent.

	Model	RS	SLA	TTT	ADT	L_{OT}^λ	L_{TTT}^λ	$L^\#$	$L_{ADS}^\#$
Queries	Passport ($n=4, k=19$)	+35.4	+27.4	+33.5	+25.7	+39.3	+14.2	+24.8	−6.6
	Bankcard ($n=7, k=14$)	+38.8	+38.1	+31.4	+22.6	+38.9	+0.3	+34.3	+3.1
	TLS ($n=15, k=12$)	+81.9	+43.5	+16.8	−2.9	+84.5	−3.6	+6.6	−28.5
	MQTT ($n=18, k=9$)	+65.1	+55.6	+23.2	+3.9	+63.7	+3.5	+22.1	−18
	TCP ($n=57, k=12$)	+75.9	+64.2	+29.4	−9.3	+79.4	+17.9	+36	−1.8
	SSH ($n=66, k=13$)	+82.3	+69.3	+27.2	−22.5	+84.9	+14.1	+26.2	−17.7
Symbols	Passport ($n=4, k=19$)	+31.2	+35.5	+17.2	+17.7	+21.6	+0.2	+2	−14.8
	Bankcard ($n=7, k=14$)	+33.9	+41.6	+18.3	+24	+28.5	−7.2	+22.4	−7.6
	TLS ($n=15, k=12$)	+83.5	+45.5	+3	−12.2	+86.4	−2.7	−11.2	−42.2
	MQTT ($n=18, k=9$)	+62.5	+57.3	−6.7	−11.4	+58.4	−22.6	+2.3	−24.9
	TCP ($n=57, k=12$)	+74.6	+62.6	+3.2	−22.2	+77.6	−6.5	+22.4	−13.8
	SSH ($n=66, k=13$)	+81.1	+67.2	+12.3	−37.6	+84.9	−0.6	+17.3	−32.3

Table 2. Overall relative complexity savings of L^s in percent.

	Model	RS	SLA	TTT	ADT	L_{OT}^λ	L_{TTT}^λ	$L^\#$	$L_{ADS}^\#$
Queries	Passport ($n=4, k=19$)	+22	+37.6	+22.6	+15.1	+26.5	+14.3	+16.5	−1.1
	Bankcard ($n=7, k=14$)	+27	+51.1	+22	+11.8	+29.4	−1.7	+25.3	−7.3
	TLS ($n=15, k=12$)	−6.2	+3.3	−1.3	−1.4	+0.3	−1.6	−1.4	−4
	MQTT ($n=18, k=9$)	+58.7	+59.6	+30.6	+12.7	+56.4	+19.6	+25.2	−4.9
	TCP ($n=57, k=12$)	+37.7	+42.1	+29.4	−2.9	+28.3	+36.3	+23.3	+23.4
	SSH ($n=66, k=13$)	+13.7	−18.6	+15.4	+14.7	+23.8	+3.4	+13.2	+11.8
Symbols	Passport ($n=4, k=19$)	+15.3	+43.1	+7.2	+5.3	+10.2	+0.3	−0.7	−9.5
	Bankcard ($n=7, k=14$)	+19.2	+56.7	+4.4	+11.4	+13.2	−16.9	+9.3	−25.1
	TLS ($n=15, k=12$)	−9.1	+2	−2	−1.7	−0.8	−2.1	−2	−5.1
	MQTT ($n=18, k=9$)	+57.2	+62.6	+20	+5.5	+51.2	+10.8	+16.9	−4.3
	TCP ($n=57, k=12$)	+35.6	+41.9	+27	−3.9	+23.5	+35.6	+19.9	+23.2
	SSH ($n=66, k=13$)	+12.2	−20	+15.9	+14.6	+23.6	+4.8	+13.2	+12.4

Table 3. Median performance values of L^s (LO = Learning-Only).

Model	LO OQs	LO Sym	Overall OQs	Overall Sym
Passport ($n=4, k=19$)	153	650	267	1500
Bankcard ($n=7, k=14$)	184	932	236	1362
TLS ($n=15, k=12$)	270	1439	41130	427146
MQTT ($n=18, k=9$)	432	3257	581	4587
TCP ($n=57, k=12$)	2618	33659	15276	199990
SSH ($n=66, k=13$)	2789	35239	63450	847470

Table 4. Median number of EQs for each learner.

Model	L^s	RS	SLA	TTT	ADT	L^λ_{OT}	L^λ_{TTT}	$L^\#$	$L^\#_{ADS}$
Passport ($n=4$, $k=19$)	2	2	8	3	2	3	3	3	2
Bankcard ($n=7$, $k=14$)	3	3	12	3	3	3	3	3	3
TLS ($n=15$, $k=12$)	10	9	14	12	11	11	12	11	12
MQTT ($n=18$, $k=9$)	8	7	14	10	10	6	11	10	8
TCP ($n=57$, $k=12$)	31	15	110	37	34	16.5	36	33	35
SSH ($n=66$, $k=13$)	35	17	93	42	37	21	42	36	36

6 Notes on Implementation

It is a central assumption in AAL that OQs take longer to execute than internal computations. However, there is no general threshold for the computational overhead that is acceptable within learning algorithms before violating this assumption. Thus, it is generally desirable to keep the overhead as small as possible.

L^s can be implemented as described in the pseudocode. This is the version that we used for our experiments and there were no scaling issues in our benchmark. However, the authors of $L^\#$ mentioned that they observed scaling issues with their learner for very large systems [13]. Their implementation failed to learn the model of a printer engine status manager with 3410 states [19]. We did not include this model in our benchmark, because its complexity is out of reach for the testing algorithms currently available in LearnLib. But to assess our learner's scalability, we performed test runs with a white-box oracle that derives counterexamples directly from the target model. And indeed, we found that the time between OQs increased significantly. Consequently, we spent great effort to identify computational shortcuts and created a second, thoroughly optimized implementation. The underlying algorithm is identical. Both implementations pose exactly the same OQs and construct equivalent hypotheses at each refinement. We used an *AMD Ryzen 7 PRO 7840U* for the test runs. With our optimized implementation, L^s is able to learn the printer model in under a minute. It took about 30µs per query, with a total of 1547570 OQs, 38941324 symbols and 1886 EQs. Among the learners from our benchmark, only TTT had a shorter runtime. It took about 15µs per query, with a total of 2210321 OQs, 47004933 symbols and 2176 EQs.

In the following, we sketch our optimizations. Our artifact contains both the simple and the optimized implementation of L^s, and a test of their equivalence.

6.1 Table Compression

By definition, in a constructive SOT, observations in a fringe row never conflict with its associated core row. Thus, all distinct suffix-output pairs occur within the core rows. To avoid storing identical pairs redundantly, we simply construct an index with a numerical identifier for each unique combination of suffix and output. The rows then store identifiers instead of suffix-output pairs. Besides saving memory, this also enables faster compatibility checking of rows, as the identifiers are cheaper to compare than output sequences. To avoid repetitive checks, fringe rows track established incompatibilities.

Additionally, we noticed there are often many fringe rows that contain the same set of identifiers. This happens when their prefixes lead to the same state and they are assigned the same suffixes during classification. For further compression, we group rows with identical identifiers and store them together.

6.2 Granularity Caching

Let $u_{f_1}, u_{f_2} \in U_f$ be prefixes of two empty fringe rows with $u_{f_1} \neq u_{f_2}$. Furthermore, assume that u_{f_1} and u_{f_2} both lead to the same SUL state as some core prefix $u_c \in U_c$. Since our classification procedure is deterministic, it repeats the exact same steps for both rows (except replacing their prefixes in OQs), as long as the core rows remain unchanged in between. That is, it performs the very same computations during suffix selection again. To avoid wasting execution time, we cache calls to the suffix selection function in a tree. For each call, we store the function input and the selected suffix. Before we execute a call to `pickSuffix` (Algorithm 3), we always check if the tree already contains a call with identical input. If so, we skip execution and reuse the result. To maintain consistency, we discard previous cache entries whenever a new suffix is added to the table.

7 Conclusion

We have presented L^s, an active automata learning algorithm that sacrifices performance in state characterization for faster state identification. To assess effectiveness, we conducted an experimental evaluation with models from the *Automata Wiki* benchmark collection [11]. In a pairwise comparison, L^s outperforms most state-of-the-art algorithms for the majority of models.

A central idea in L^s is to choose output queries with metric-based suffix selection. For future work, it would be interesting to explore if this idea can also improve testing efficiency. Testing reduces to querying a set of test words and checking if any output diverges from what the hypothesis predicts. Test word construction is often split into a prefix and a suffix component. The prefix describes a path through the state graph. The suffix checks if the state reached by the prefix exhibits unexpected behavior. Picking better suffixes might help reveal new states with fewer test words.

Acknowledgements. Much gratitude goes to Markus Frohme for maintaining *LearnLib*, without which this work would not exist. Julian Klein provided valuable feedback on the draft. The first author thanks Rudi Schneider for exciting discussions about data structures.

Code Availability. All source code of our experiments is archived and publicly available at DOI 10.5281/zenodo.15284082.

References

1. Aarts, F., Schmaltz, J., Vaandrager, F.: Inference and abstraction of the biometric passport. In: Margaria, T., Steffen, B. (eds.) ISoLA 2010. LNCS, vol. 6415, pp. 673–686. Springer, Heidelberg (2010). https://doi.org/10.1007/978-3-642-16558-0_54
2. Aarts, F., De Ruiter, J., Poll, E.: Formal models of bank cards for free. In: 2013 IEEE Sixth International Conference on Software Testing, Verification and Validation Workshops, pp. 461–468. IEEE (2013)
3. De Ruiter, J., Poll, E.: Protocol state fuzzing of TLS implementations. In: 24th USENIX Security Symposium (USENIX Security 2015), pp. 193–206 (2015)
4. Tappler, M., Aichernig, B. K., Bloem, R.: Model-based testing IoT communication via active automata learning. In: 2017 IEEE International Conference on Software Testing, Verification and Validation (ICST), pp. 276–287. IEEE (2017)
5. Fiterău-Broştean, P., Janssen, R., Vaandrager, F.: Combining model learning and model checking to analyze TCP implementations. In: Chaudhuri, S., Farzan, A. (eds.) CAV 2016. LNCS, vol. 9780, pp. 454–471. Springer, Cham (2016). https://doi.org/10.1007/978-3-319-41540-6_25
6. Fiterău-Broştean, P., Lenaerts, T., Poll, E., de Ruiter, J., Vaandrager, F., Verleg, P.: Model learning and model checking of SSH implementations. In: Proceedings of the 24th ACM SIGSOFT International SPIN Symposium on Model Checking of Software, pp. 142–151 (2017)
7. Angluin, D.: Learning regular sets from queries and counterexamples. Inf. Comput. **75**(2), 87–106 (1987)
8. Rivest, R., Schapire, R.: Inference of finite automata using homing sequences. Inf. Comput. **103**, 299–347 (1993)
9. Isberner, M., Howar, F., Steffen, B.: The TTT algorithm: a redundancy-free approach to active automata learning. In: Bonakdarpour, B., Smolka, S.A. (eds.) RV 2014. LNCS, vol. 8734, pp. 307–322. Springer, Cham (2014). https://doi.org/10.1007/978-3-319-11164-3_26
10. Isberner, M., Howar, F., Steffen, B.: The open-source LearnLib: a framework for active automata learning. In: Kroening, D., Păsăreanu, C.S. (eds.) CAV 2015. LNCS, vol. 9206, pp. 487–495. Springer, Cham (2015). https://doi.org/10.1007/978-3-319-21690-4_32
11. Margaria, T., Graf, S., Larsen, K.G.: Models, mindsets, meta: the what, the how, and the why not? In: Margaria, T., Graf, S., Larsen, K.G. (eds.) Models, Mindsets, Meta: The What, the How, and the Why Not? LNCS, vol. 11200, pp. 3–13. Springer, Cham (2019). https://doi.org/10.1007/978-3-030-22348-9_1
12. Frohme, M.: Active Automata Learning with Adaptive Distinguishing Sequences. arXiv preprint arXiv:1902.01139 (2019)
13. Vaandrager, F., Garhewal, B., Rot, J., Wißmann, T.: A new approach for active automata learning based on apartness. In: TACAS 2022. LNCS, vol. 13243, pp. 223–243. Springer, Cham (2022). https://doi.org/10.1007/978-3-030-99524-9_12
14. Howar, F., Steffen, B.: Active automata learning as black-box search and lazy partition refinement. In: A Journey from Process Algebra via Timed Automata to Model Learning: Essays Dedicated to Frits Vaandrager on the Occasion of His 60th Birthday, pp. 321–338. Springer, Cham (2022)
15. Isberner, M., Howar, F., Steffen, B.: Inferring automata with state-local alphabet abstractions. In: Brat, G., Rungta, N., Venet, A. (eds.) NFM 2013. LNCS, vol. 7871, pp. 124–138. Springer, Heidelberg (2013). https://doi.org/10.1007/978-3-642-38088-4_9

16. Aichernig, B.K., Tappler, M., Wallner, F.: Benchmarking combinations of learning and testing algorithms for automata learning. Formal Aspects Comput. **36**(1), 1–37 (2024)
17. Isberner, M., Steffen, B.: An abstract framework for counterexample analysis in active automata learning. In: International Conference on Grammatical Inference, pp. 79–93. PMLR (2014)
18. Kogel, P., Klös, V., Glesner, S.: TTT/ik: learning accurate mealy automata efficiently with an imprecise symbol filter. In: International Conference on Formal Engineering Methods, pp. 227–243. Springer, Cham (2022)
19. Smeenk, W., Moerman, J., Vaandrager, F., Jansen, D.N.: Applying automata learning to embedded control software. In: Butler, M., Conchon, S., Zaïdi, F. (eds.) ICFEM 2015. LNCS, vol. 9407, pp. 67–83. Springer, Cham (2015). https://doi.org/10.1007/978-3-319-25423-4_5

What Are the Odds? Improving Statistical Model Checking of Markov Decision Processes

Tobias Meggendorfer[1], Maximilian Weininger[2,3], and Patrick Wienhöft[4,5](✉)

[1] Lancaster University Leipzig, Leipzig, Germany
tobias@meggendorfer.de
[2] Ruhr University Bochum, Bochum, Germany
maximilian.weininger@rub.de
[3] Institute of Science and Technology Austria, Klosterneuburg, Austria
[4] Dresden University of Technology, Dresden, Germany
[5] Centre for Tactile Internet with Human-in-the-Loop (CeTI), Dresden, Germany
patrick.wienhoeft@tu-dresden.de

Abstract. Markov decision processes (MDPs) are a fundamental model of decision making which exhibit non-deterministic choice as well as probabilistic uncertainty. Traditionally, verification assumes exact knowledge of the probabilities that govern the behaviour of an MDP. However, this assumption often is unrealistic, e.g. when modelling cyber-physical systems or biological processes. There, we can employ statistical model checking (SMC) to obtain an estimate of the MDP's value (e.g. the maximal probability of reaching a goal state) that is close to the true value with high confidence (probably approximately correct). Model-based SMC algorithms sample the MDP and build a model of it by estimating all transition probabilities, essentially for every transition answering the question: "What are the odds?" However, so far the statistical methods employed by state-of-the-art SMC verification algorithms are quite naive or even compromise the correctness guarantees.

Our first contribution is to survey, categorize, and analyse statistical methods, identifying those few that are most efficient and that provide suitable guarantees for the verification setting. Secondly, we propose improvements that exploit structural knowledge of the MDP. Both contributions generalize to many types of problem statements as they are largely independent of the setting. Moreover, our experimental evaluation shows that they lead to significant gains, reducing the number of samples that an SMC algorithm has to collect by up to two orders of magnitude.

This work was supported by the European Union's Horizon 2020 research and innovation programme under the Marie Sklodowska-Curie grant agreement No 10103441, the ERC Starting Grant DEUCE (101077178) and the DFG through the Cluster of Excellence EXC 2050/1 (CeTI, project ID 390696704, as part of Germany's Excellence Strategy) and the DFG grant 389792660 as part of TRR 248 (see https://perspicuous-computing.science).

Keywords: Probabilistic verification · Statistical model checking · Markov decision processes · Confidence intervals

1 Introduction

Markov decision processes (MDPs) [74] are *the* classic modelling formalism for dynamic systems with probabilistic and nondeterministic behaviour. In essence, MDPs comprise several states and each state has an associated set of available actions to choose from. In order to capture the *aleatoric* uncertainty (the randomness of the process, e.g. a coin toss), each action corresponds to a distribution on the successor states rather than a single successor. The system evolves from a state by choosing an action, moving to a successor sampled from the corresponding distribution, and repeating this process ad infinitum. In such a model, we are (for example) interested in *infinite-horizon reachability*, where we want to know the *optimal* probability to eventually reach a given set of target states. Solving an MDP means to correctly computing this *value*, where the optimum is taken over the set of all *strategies*, i.e. ways to choose actions in every state.

Restricted Knowledge. Traditionally, verification procedures assume full knowledge of the MDP. However, this is often unrealistic in practice, as for example probabilities governing biological processes or cyber-physical systems are usually not known precisely, see also [27, Chp. 28.7.3] or [12], and we are now also dealing with *epistemic* uncertainty [12] (the lack of knowledge, e.g. unknown bias of a coin). Clearly, in this case we cannot give exact results, as we do not know the exact system we are dealing with. Nevertheless, the desire to obtain guarantees on the correctness of the result remains unchanged.

PAC Guarantees. With unknown transition probabilities, the goal of verification usually is to guarantee that the result is *probably approximately correct* (PAC): Given a confidence budget δ and a precision ε, we want a result that is *probably* (with probability greater than $1 - \delta$) *approximately correct* (at least ε-precise). For example, consider a coin with unknown bias. If we sample it 1000 times and see heads 800 times, its bias is likely around 0.8. More formally, for a required confidence of 95% (i.e. $\delta = 0.05$), using a statistical method such as Hoeffding's inequality [46] (see Sect. 3.2) yields that the coin's bias is within 0.8 ± 0.05.

Statistical Model Checking: Markov Chains. Markov chains effectively are MDPs where every state has exactly one available action. As such, the outcome is purely stochastic (without any non-determinism); such systems effectively describe a random variable. Here, obtaining PAC guarantees is straightforward: Intuitively, we gather simulations of the Markov chain and remember whether they reach a target state or not. Then, we can estimate the probability of reaching the target analogous to the above coin example. We refer to extensive surveys of statistical model checking for Markov chains with properties expressed in linear temporal or computational tree logic [3,52,60], or reward properties [25]. Notably, in the context of Markov chains, statistical approaches may even be preferred over traditional methods to combat state space explosion, as sampling may still be feasible even when the model is too large to be analysed precisely. However, for MDPs, state-of-the-art statistical approaches are far less efficient than traditional methods; and the main interest is to obtain PAC guarantees in the presence of epistemic uncertainty, where traditional methods are simply not applicable.

Statistical Model Checking: MDPs. Solving MDPs with unknown transition probabilities is fundamentally more difficult than solving purely stochastic Markov chains. This is because the satisfaction of the property depends on the *strategy*; already for reachability, a naive approach would have to check exponentially many of these. One way of tackling this problem [20,22,38,89] are *model-free* approaches (see [22, Rem. 5.1] for details). These algorithms require an astronomical number of samples to provide non-trivial bounds [20,22,89] (see [8, Sec. 4]) or knowledge of the mixing time of the MDP [38], which is as hard to compute as the value. Thus, state-of-the-art statistical verification of MDPs is *model-based*. Essentially, it proceeds by first learning the unknown transition probabilities in the MDP, thereby constructing an MDP with full knowledge that is probably correct, and then solving this using traditional methods. This way, PAC-guarantees can be obtained more efficiently in many settings, including MDPs that are communicating [10], continuous-time [2], or continuous-space [11], in stochastic games [8], and for properties that are ω-regular [73] or consist of multiple objectives [87].

Sample Efficiency. To gather samples, we need to simulate or interact with the system, which may require significant effort (e.g. for cyber-physical systems). Naturally, we thus are interested in obtaining PAC bounds efficiently, i.e. with as few samples as possible. For Markov chains, we refer again to [3,25,52,60], and in particular highlight the efforts for dealing with rare events [24,83]. For MDPs, many works focus on simpler cases of finite-horizon or discounted properties, bounding the sample complexity in terms of the horizon or discount factor, e.g. [1,44,50,57,78,84]. Notably, early works contained subtle mistakes compromising guarantees, see [52, Sec. 3.3]. Our focus in this work is on *infinite-horizon* objectives, focusing on reachability for simplicity, and detailing in [67, App. D.3] how our contributions extend to other objectives.

Our contribution is practically improving the sample efficiency of model-based statistical verification for MDPs with unknown transition probabilities. Colloquially speaking, when learning the transition probabilities, we provide a *more precise* answer to the question: "What are the odds?". The methods we propose make the most of the data, i.e. given the same simulations of the system, they provide better probability estimates than the state-of-the-art and as such also transparently improve the sample efficiency of these approaches. Concretely, our contribution is threefold:

Firstly, we discuss estimation of categorical distributions, the central component of model-based approaches (Sect. 3). To this end, we survey relevant literature and contrast numerous statistical methods. We identify most inequalities as being unsuitable due to insufficient guarantees, as well as several SMC approaches that claim correctness despite using them. Moreover, we prove a statement of independent interest about the maximum size of the confidence interval produced by Hoeffding's inequality and the Clopper-Pearson interval (Prop. 1), solving an open question from [23]. We empirically show that the former produces intervals significantly greater than the latter (Sect. 3.3). Thus, for the purpose of estimating transition probabilities, the Clopper-Pearson interval is always preferable.

Secondly, we provide techniques for exploiting the knowledge we have about the MDP and the property of interest (Sect. 4). They allow us to invest less or even no part of our confidence budget δ for certain state-action pairs. Many of them are based on already known observations that however – to our knowledge – have not yet been applied in this context. Moreover, we propose the novel technique of collapsing so-called **Fragments**.

Thirdly, we empirically evaluate the impact of our improvements (Sect. 5). We conclude that our methods always have a positive impact: They have practically no computational overhead and always reduce the number of samples necessary to achieve a given precision ε, in many cases by two orders of magnitude.

Relevance. Regarding *efficiency*, a collection of statistical methods and structural improvements is overdue: Most papers in the verification community overlook their potential, even those with the declared goal of improving the scalability and practical applicability of statistical approaches [2,8,11]. In particular, the state-of-the-art method for estimating categorical distributions is Hoeffding's inequality [2,8,13,73,87]. As mentioned, we empirically show that this is significantly worse than the (well-known *in statistics*) Clopper-Pearson interval (Sect. 3.3). Further, we prove that the method for estimating distributions developed in [11] is in fact a weaker version of the Clopper-Pearson interval ([67, App. A.3]). Moreover, even arguably trivial ways of exploiting the MDP structure (Sect. 4) such as **Small Support** and **Independence** have not been employed, let alone the more advanced techniques of **Equivalence Structures** and **Fragments**.

With respect to *soundness*, we show that several papers use methods that *compromise the PAC-guarantees* [15,24,81] (see [67, App. A]). Our survey clearly separates methods for estimating distributions that are unsuitable for our setting, and we provide proofs for the soundness of all structural improvements.

Finally, while our focus is on tackling MDPs, many of our findings are also applicable in purely probabilistic cases, such as Markov chains. In particular, building on the preprint of our work [67], the recent work [25, Tab. 1] shows that almost all tools that apply statistical methods in the context of Markov chains employ methods that are inefficient or unsound.

2   Preliminaries

A *probability distribution* over a countable set X is a mapping $d : X \to [0, 1]$, such that $\sum_{x \in X} d(x) = 1$. The set of all probability distributions on X is $\mathcal{D}(X)$.

A *Markov decision process (MDP)*, e.g. [74], is a tuple $\mathcal{M} = (S, A, \mathsf{P})$, where S is a finite set of states; A is a finite set of actions, overloaded to yield for each state $s \in S$ a non-empty set of *available actions* $A(s) \subseteq A$; and $\mathsf{P} : S \times A \rightharpoonup \mathcal{D}(S)$ is the (partial) transition function, that yields for each state $s \in S$ and $a \in A(s)$ the associated distribution over successor states $\mathsf{P}(s, a)$. For ease of notation, we write $\mathsf{P}(s, a, s')$ instead of $\mathsf{P}(s, a)(s')$.

The *semantics* of MDPs is defined in the usual way by means of paths, strategies and the probability measure in the induced Markov chain. We briefly recall this here and refer to [14, Chp. 10] for an extensive introduction. Let $\mathcal{M}$ be an MDP. An *infinite path* is a sequence of state-action pairs $\rho = s_1 a_1 s_2 a_2 \cdots \in (S \times A)^\omega$ with $\mathsf{P}(s_i, a_i, s_{i+1}) > 0$. We denote by $\rho(i)$ the i-th state s_i in a given path and by $\mathsf{Paths}_\mathcal{M}$ the set of all infinite paths. A (memoryless deterministic, MD) *strategy* is a mapping $\pi : S \to A$, choosing one enabled action in each state, i.e. $\pi(s) \in A(s)$. We write $\Pi_\mathcal{M}^{\mathsf{MD}}$ to refer to all MD strategies. Complementing an MDP with such a strategy and an initial state $\hat{s} \in S$ yields a Markov chain that induces a unique probability measure $\mathsf{Pr}_{\mathcal{M},\hat{s}}^\pi$ over infinite paths [14, Chp. 10.1].

An *objective* formalises the goal of the MDP. For simplicity, we focus on *reachability*, and in [67, App. D.3] explain how our methods extend to other objectives. Define $\lozenge T := \{ \rho \in \mathsf{Paths}_\mathcal{M} \mid \exists i.\ \rho(i) \in T \}$ as the set of paths that eventually reach T. The *value* of a state in an MDP is the maximum probability to achieve the objective, i.e. reach the goal states, under any strategy. Formally, the value of state s is defined as $\mathsf{V}_\mathcal{M}(s) := \max_{\pi \in \Pi_\mathcal{M}^{\mathsf{MD}}} \mathsf{Pr}_{\mathcal{M},s}^\pi [\lozenge T]$. Note that MD strategies are sufficient to maximise the reachability probability [14, Lem. 10.102].

2.1 Statistical Guarantees and Statistical Model Checking

In this work, we deal with MDPs where the transition function P is unknown. There are multiple approaches to tackle this problem, and we focus on *model-based statistical model checking* (SMC) (see Secs. 1 and 6 for discussion of other SMC approaches). Algorithms for model-based SMC, e.g. [2,8,11,13,81,87], comprise three conceptual steps: First, they obtain a finite number of samples from the MDP (see Sect. 6 for a discussion of sampling strategies). Then, from these samples they construct confidence intervals on each *transition* probability. Finally, they solve the induced interval MDP [40], yielding bounds on the true value.

Our work focuses on the second part of model-based SMC: Given a finite number of samples, what is the best way to estimate transition probabilities? Colloquially, we "make the most of the data" by obtaining "as small as possible" intervals.[1] By getting more precise estimates from the samples, we reduce the width of the confidence intervals on transition probabilities in the second phase of the SMC algorithms, and thus also improve their overall performance. We additionally assume knowledge of the topology, i.e. that we know the support $\{ s' \mid \mathsf{P}(s, a, s') > 0 \}$ of every distribution $\mathsf{P}(s, a)$, also called *grey-box setting* [2,8], a standard assumption for SMC. This is relevant for our structural

---

[1] We intentionally do not ask for the *minimum* interval size, because it is a random variable (as it depends on the sampling outcome) and minimising its expected value would require assuming a prior distribution over P, which we cannot justify.

improvements in Sect. 4. Rem. 2 discusses how our methods extend to the completely opaque *black-box setting* [8,30]. Together, we formalize our problem as follows.

> **PROBLEM STATEMENT: GREY-BOX SMC OF MDPS**
> **Input:** A confidence budget δ, an MDP $\mathcal{M}$ with unknown P, but known support of each distribution $P(s,a)$, and a sequence of random samples $(s,a,s') \in S \times A \times S$ from $\mathcal{M}$.
> **Output:** An interval $[\underline{v}, \overline{v}]$ such that $\underline{v} \leq \mathsf{V}_{\mathcal{M}}(\hat{s}) \leq \overline{v}$ with probability at least $1 - \delta$ and $\overline{v} - \underline{v}$ is as small as possible.

State-of-the-art methods [2,8,11,13,81,87] usually distribute the confidence budget δ uniformly over all transitions and apply *Hoeffding's inequality* (see Sect. 3.2) to get a confidence interval for each of them. Every transition then is correct with probability at least $1 - \frac{\delta}{|P|}$, where $|P|$ is the number of transitions. By union bound, the probability of all transition being correct is greater than $1 - \delta$. In [67, App. A], we comment on the few cases in verification literature where methods other than Hoeffding's inequality are employed for model-based SMC. These either compromise the PAC-guarantee [15,81] or we prove their inferiority [11].

3 Statistical Methods for Estimating Probabilities

As mentioned, estimating transition probabilities lies at the heart of model-based SMC. This section presents methods to estimate the distribution $P(s,a)$ for a single state-action pair (s,a). Section 3.1 discusses estimating the whole distribution ("a die") at once. Section 3.2 details how to estimate a single transition probability $P(s,a,s')$; essentially viewing the samples drawn from $P(s,a)$ as a coin toss, either reaching s' or not. We survey the literature, including methods used in verification, the concentration inequalities from [19] applicable in our setting, and methods from statistics literature [4,21,35,71,86]. Finally, Sect. 3.3 complements our survey with an efficiency analysis of the most relevant methods.

3.1 Estimating a Die

The most common method to estimate the distribution $P(s,a)$ as a whole is based on [88], and used by, e.g., MBIE [79], and UCRL2 [10]. It computes a maximum likelihood estimate of the probability distribution, i.e. the empirical average of each outcome, and then constructs the confidence region as an L^1-ball around this estimate. In other words, the confidence region contains all probability distributions whose L^1-distance from the empirical average is less than a certain value which depends on the number of samples and the number of successors [88].

The advantage of the L^1-ball is that it accounts for the dependence between the transition probabilities, i.e. that $\sum_{i=1}^{k} p_i = 1$, making it seem like a canonical choice. However, the method scales poorly with the number of successors of a state-action pair [91]. One can transform the model such that every state-action pair has two successors, minimising the number of samples required to prove a property [90, App. A]. However, then the method of [88] coincides with applying Hoeffding's inequality to each transition probability (see [67, App. B.1]) which we show to be a sub-optimal way to estimate these Sect. 3.3. Thus, we focus on the approach that estimates every transition probability individually.

3.2 Estimating a Coin

This section discusses the most basic SMC problem: estimating a single transition probability $\mathsf{P}(s, a, s')$ with confidence budget δ. The relevant information from the samples are (i) how often was (s, a) sampled and (ii) how often was the successor s' chosen, effectively a binomial distribution (sampling s' is a success). Throughout this section, we refer to a binomial distribution with success probability p (also called "binomial proportion") and a test sequence on it with n trials and k successes. We write $\hat{p} = \frac{k}{n}$ for the maximum likelihood estimate of p.

> **PROBABILITY ESTIMATION PROBLEM**
> **Input:** A confidence budget δ and n random samples drawn from a binomial distribution with *unknown* success probability p.
> **Output:** A confidence interval $[\underline{p}, \overline{p}]$ for which $\Pr[\underline{p} \leq p \leq \overline{p}] \geq 1 - \delta$.

Remark 1. Many works focus on confidence intervals centred around $\hat{p}$ (either additive $[\hat{p} - \varepsilon, \hat{p} + \varepsilon]$ or relative $[\hat{p} \cdot (1 - \varepsilon), \hat{p} \cdot (1 + \varepsilon)]$ for some ε). However, this excludes potentially tighter *asymmetrical* confidence intervals.

For soundness, we require that the *coverage probability* $\Pr[\underline{p} \leq p \leq \overline{p}]$ is consistently at least $1 - \delta$ *for all p*. (Recall that p is fixed but unknown and the random variable is $\hat{p}$.) Notably, this is different from requiring *average coverage*, i.e. if p were chosen uniformly at random (corresponding to a Bayesian approach assuming a uniform prior). The explicit goal of SMC is to achieve correctness guarantees for *all* models, without any prior assumptions. Also, note that the problem statement deliberately only requires correctness, allowing trivial solutions such as $[0, 1]$; we discuss minimizing the interval size in Sect. 3.3. We now present two methods solving the Probability Estimation Problem.

Hoeffding's Inequality. Hoeffding's seminal paper [46] provides a confidence interval for the sum of random variables. Several works have applied this to estimate the mean of binomial random variable X, i.e. $X \in \{0, 1\}$. Applying Hoeffding's inequality yields the $(1 - \delta)$ confidence interval $[\underline{p}_{ho}, \overline{p}_{ho}]$ where $\underline{p}_{ho} = \max\{0, \hat{p} - c_{ho}\}$, $\overline{p}_{ho} = \min\{1, \hat{p} + c_{ho}\}$, and $c_{ho} = \sqrt{\ln(2/\delta)/2n}$ [9].

This result is often referred to as Hoeffding's inequality, although Okamoto considered the special case of binomial random variables before [72]. Further,

Hoeffding's is a specialization of Chernoff's and Markov's inequality [19, Chp. 2]. Thus, variants of this bound can appear under some combinations of these names, e.g. Okamoto-Chernoff inequality in [15]. We call it Hoeffding's inequality, as that is common in verification literature. While the number of samples required by Hoeffding's inequality for a fixed ε is asymptotically optimal at $\mathcal{O}(1/\varepsilon^2)$ (see, e.g. [7]), we show in Sect. 3.3 that in practice we can do significantly better.

Clopper-Pearson Interval. A widely used confidence interval method was introduced by Clopper and Pearson [28], sometimes called the "exact method". The approach inverts the task and, given n and k, asks for the minimum (and maximum) p such that observing at least (at most) k out of n successes has a probability of $1 - \frac{\delta}{2}$. The $(1-\delta)$ confidence interval using the Clopper-Pearson interval can be represented in a closed form using the inverse regularised beta function $I_x^{-1}(a, b)$ [82] as $[\underline{p}_{cp}, \overline{p}_{cp}] = [I_{\delta/2}^{-1}(k+1, n-k), I_{1-\delta/2}^{-1}(k, n-k+1)]$ for $0 < k < n$. Additionally, for $k = 0$ (or $k = n$) the lower (or upper) bound needs to be taken as 0 (or 1). Note that, unlike Hoeffding's inequality, it is generally not centered around $\hat{p}$ but has its center shifted towards $\frac{1}{2}$.

Recently, two methods related to Clopper-Pearson appeared: a newly developed approach [11] that we show to be a strictly weaker version of the Clopper-Pearson interval, and a sequential variant [23]; we provide details in [67, App. A].

Theorem 1 (From [9, App. D] and [28]). *Hoeffding's inequality and the Clopper-Pearson interval both solve the Probability Estimation Problem.*

Further Confidence Methods. We now briefly discuss several other confidence interval methods and reasons why we do not recommend to use them in the context of SMC with PAC-guarantees for Markov systems.

Many statistical methods either fail to guarantee any coverage probability at all, or only provide an *average* coverage probability of $1 - \delta$ when p is uniform over $[0, 1]$. These include the (adjusted) central limit theorem [71], Wald interval [21,86], the Wilson score interval [21,86,92], the Agresti-Coull interval [4], the Arcsine interval [21], and the Logit interval [21]. For the Wilson score, Newcombe introduced a continuity corrected version with better coverage properties [70]. However, even with the continuity correction, the coverage is insufficient. We discuss this in more detail in [67, App. A].

Hypothesis tests, such as Student's t-test [80] or the sequential probability ratio test (SPRT) [85] are designed to distinguish between two hypotheses. As such, in the context of SMC they are widely used for answering threshold queries (see e.g. [59,75]), but they are not suitable for constructing confidence intervals.

Monte Carlo simulations, e.g. [31,47,48], provide an efficient ε-approximation for the mean of Bernoulli random variables. However, they require δ and ε to be given and then define a stopping condition which needs unlimited sampling access for every state-action pair. This does not fit the SMC algorithm structure: As the impact of a transition probability on the the value is unclear, we cannot fix the precision for individual transitions a-priori.

Jeffrey's interval [21] gives guarantees in a Bayesian sense by providing a *credible interval*, i.e. for given a prior distribution on p an interval such that the probability of p being within that interval is at least $1 - \delta$. We avoid credible intervals as we do not have priors in our setting. However, such Bayesian methods may be considered if prior distributions are known.

Several methods provide proper coverage probability, but are either not applicable to our setting or provably suboptimal: The Massart bound improves Hoeffding's inequality for $p \neq \frac{1}{2}$ [49], but we do not know whether this assumption is satisfied for a given transition. Similarly, Bennett's inequality [16] improves Hoeffding's inequality by including information about the variance of the random variable which is unknown in our setting. Using the trivial upper bound of $\frac{1}{4}$, Benett's inequality is always worse than Hoeffding's inequality (see [67, App. B.2]). More advanced methods estimating the variance from data asymptotically outperform Hoeffding's inequality [65], but only for p away from $\frac{1}{2}$ and very large n. Bernstein's inequality [17] is a conservative relaxation of Bennett's inequality, providing strictly wider confidence intervals. The Dvoretzky-Kiefer-Wolfowitz-Massart inequality (DKW) [37,64] yields a *confidence band* around the empirical distribution from which bounds on the mean can be derived. However, since all sample values are extreme values in the binary case (i.e. 0 or 1), DKW coincides with Hoeffding's inequality when estimating probabilities [25].

Lastly, we summarize the discussion of [29,77] about further methods with proper coverage probability: The Blyth-Still(-Casella) interval [18,26], inverted exact likelihood ratio test interval [5] and Duffy-Santner interval [36] all deliver intervals similar to the Clopper-Pearson interval; they are only slightly smaller for p close to 0 or 1 in exchange for slightly larger intervals if $p \approx \frac{1}{2}$. Since they are computationally expensive and, unlike the Clopper-Pearson interval, not readily available in a lot of statistical libraries, we do not consider them in this paper.

3.3 Sample Complexity

In this section, we compare the two methods from Sect. 3.2 with respect to their (worst-case) sample complexity: We provide an a-priori bound on the number of samples required to obtain a given precision ε (i.e. a confidence interval of width at most ε) as well as compare the required number of samples empirically.

> **PROBABILITY SAMPLE COMPLEXITY PROBLEM**
> **Input:** Confidence budget δ, precision ε, and *unknown* binomial parameter p.
> **Output:** The minimum number of samples n to guarantee
> $$\Pr[\underline{p} \leq p \leq \overline{p} \mid \hat{p} = \tfrac{k}{n}] \geq 1 - \delta \quad \text{where} \quad \overline{p} - \underline{p} \leq \varepsilon$$

At first, this problem seems difficult to solve, as the interval width not only depends on the sample size n and confidence budget δ, but also on the empirical success rate $\hat{p}$. However, we show in [67, App. B.3] that both Hoeffding's

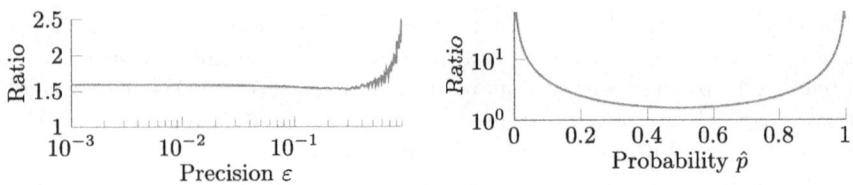

Fig. 1. Left: Ratio of worst-case sample complexity ($\hat{p} = 0.5$) between Hoeffding bound and Clopper-Pearson interval for confidence $\delta = 0.01$ and varying precision ε. Right: Ratio of sample complexity between Hoeffding bound and Clopper-Pearson interval for varying $\hat{p}$, precision $\varepsilon = 0.01$, and confidence $\delta = 0.01$. Note the logarithmic scale for the X-axis on the left and Y-axis on the right.

inequality and the Clopper-Pearson interval maximise the interval size when $\hat{p} = \frac{1}{2}$.

Proposition 1. *Given a confidence budget $\delta \in (0,1]$, the confidence interval $[\underline{p}, \overline{p}]$ obtained by Hoeffding's inequality and the Clopper-Pearson interval from a sequence $\mathcal{S}$ of n samples of a binomial distribution maximises $\overline{p} - \underline{p}$ when $\mathcal{S}$ contains equally many positive and negative samples (i.e. $\hat{p} = \frac{1}{2}$).*

It is generally well known that the width of most confidence intervals, including the Clopper-Pearson interval, tends to be largest when the sample variance is maximised, i.e. when $\hat{p} = \frac{1}{2}$. However, to the best of our knowledge, we provide the first proof that this statement holds for *all* sample sizes n and confidence budgets δ. We highlight that this result is conjectured but not proven in [23, Hypoth. 1] and required for completing their soundness proof [23, Thm. 1]. Further, Prop. 1 is the basis for providing sequential SMC algorithms in [25, Sec. 3.2].

Empirical Analysis. We complement this result with an experimental analysis. We solve the Probability Sample Complexity Problem by considering the worst-case $\hat{p} = \frac{1}{2}$ and computing the number of required samples by binary search on n for varying precisions ε. Figure 1 (left) shows the *ratio* of required samples between using Hoeffding's inequality and Clopper-Pearson. The ratio rather consistently is at ≈ 1.5, indicating that the Clopper-Pearson interval only requires around two thirds of the samples that Hoeffding's inequality needs to guarantee the same precision for $\hat{p} = \frac{1}{2}$.

While the worst-case sample complexity is the only sound a-priori bound, for practical purposes the sample complexity when $\hat{p} \neq \frac{1}{2}$ is also relevant. Thus, in Fig. 1 (right) we instead fix a precision of $\varepsilon = 0.01$ and confidence budget $\delta = 0.1$ and now vary $\hat{p}$. We again consider the ratio between the samples required to obtain a confidence interval of width $\leq \varepsilon$ using Hoeffding or Clopper-Pearson. When $\hat{p}$ deviates from $\frac{1}{2}$, the advantage of Clopper-Pearson interval becomes even larger, often requiring an order of magnitude less samples, especially for $\hat{p}$ close to 0 or 1. This is not surprising, since the width using Hoeffding's inequality is largely independent of $\hat{p}$ (except when $\hat{p} \pm \varepsilon \notin [0,1]$)) whereas $\hat{p}$ directly

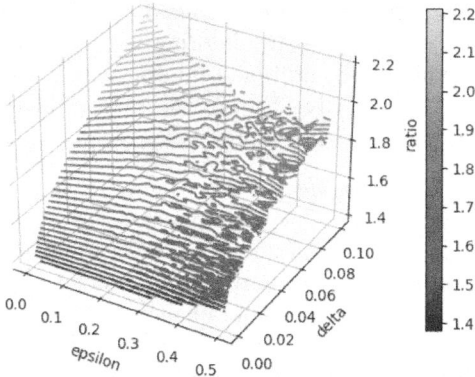

Fig. 2. Ratio of required samples for Hoeffding bound and Clopper-Pearson interval for varying precision requirement ε and confidence budgets δ in the worst-case ($\hat{p} = \frac{1}{2}$).

impacts the width of the Clopper-Pearson interval. This also is relevant for our application, as transition probabilities in MDPs often are away from $\frac{1}{2}$.

To address how different confidence budgets δ affect the interval width we plot the ratios between the sample complexity of the two methods while varying both ε and δ in Fig. 2 First of all, it shows once again that the Hoeffding bound always requires more samples than Clopper-Pearson interval in the worst-case, as the ratio between the sample complexity of the Hoeffding bound and the Clopper-Pearson interval is larger than 1 for all values of ε and δ. Moreover, the ratio is not monotonically related to ε, but it slightly shifts in favour of the Clopper-Pearson interval for smaller ε and larger δ. Finally, we observe a general positive correlation of the improvement factor with δ. This means the potential benefit of using the Clopper-Pearson interval over Hoeffding's inequality is larger with a higher δ per transition.

> **KEY TAKEAWAY**
> Algorithms estimating binomial probabilities (as in standard model-based SMC approaches) should refrain from using the commonly employed Hoeffding's inequality, as it is consistently outperformed by the Clopper-Pearson interval, most notably for probabilities close to 0 or 1.

4 Structural Improvements

After investigating the basic problem of estimating single transitions, we now turn to utilizing the structure of the MDP and property of interest. State-of-the-art SMC algorithms naively distribute the confidence budget among all transitions and estimate the probability for each of them. However, especially in the grey-box setting, we actually have a lot of structural information that we can utilise to improve estimates or even conclude that estimation is not needed at all. In Sect. 4.1, we explain how to use information about the model structure, and in Sect. 4.2, we exploit the additional information about the property. The overall goal of both is to identify parts of the MDP which require less, or no confidence, leaving more confidence budget for relevant transition, synergizing.

Remark 2 (Applicability in the black-box setting). In the black-box setting (as defined in [8,22,30]), we do not know the support of any distribution, but instead only have a lower bound on the minimum occurring transition probability $p_{\min}$. However, even given only that information, we can infer the topology using the methods from [8,30]: Assume we have obtained confidence intervals on the transition probabilities of all successor states in a distribution $\mathsf{P}(s,a)$ observed in our samples. Intuitively, when the sum of all lower bounds of these confidence intervals is greater than $1 - p_{\min}$, the probability of having overlooked a successor is less than our confidence budget. Thus, from that point onward, we know the support of $\mathsf{P}(s,a)$ with sufficient confidence and can apply structural improvements. Using the Clopper-Pearson interval (Sect. 3.2) also assists here, since smaller confidence intervals lead to knowing the topology sooner.

4.1 Using Information About the Model

Small Support. If a distribution only has a single successor, we can trivially conclude that the transition probability equals 1. If a distribution has two successors, it suffices to estimate only one of the two probabilities: Upon estimating $p_1 \in \hat{p} \pm c$, we obtain $p_2 \in (1-\hat{p}) \pm c$. Notably, the confidence in the estimation of p_1 transfers to p_2: The estimation of p_2 is correct exactly when p_1 is correct. Thus, we only need to learn one transition probability instead of two, saving budget. While this observation seems rather trivial, it is not exploited by state-of-the-art SMC approaches [2,8,11,73,87].

Remark 3. This reasoning does not easily extend to larger distributions: Given k successors, one could estimate $k-1$ probabilities and infer the k-th from the rest. However, the resulting error bound for the k-th estimate is worse: In a three-successor distribution where $p_1 \in \hat{p_1} \pm c$ and $p_2 \in \hat{p_2} \pm c$ we only obtain $p_3 \in (1-\hat{p_1}-\hat{p_2}) \pm 2c$. This also relates to Sect. 3.1, where we argue that focusing on individual probabilities rather than entire distributions seems to be beneficial.

Independence. By their nature, the transition distributions in Markov systems are independent. Thus, the events of correctly estimating the probabilities for different state-action pairs are independent. Consequently, instead of dividing the confidence budget additively, we can divide it multiplicatively over the distributions (only utilizing the union bound for the transitions comprising each single state-action pair). We prove in [67, App. D.1]:

Proposition 2. *Let $\mathcal{D}$ be the set of all distributions to be learnt, δ the confidence budget and δ_d the confidence budget of a single distribution d. If $\prod_{d \in \mathcal{D}}(1-\delta_d) \geq \delta$, then the probability of correctly estimating all distributions is larger than δ.*

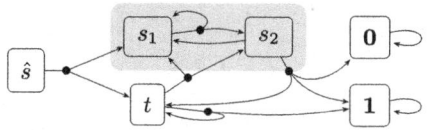

Fig. 3. A small MDP to illustrate several potential savings that can be obtained through `Equivalence Structures`. The boxed states form an end component, **1** denotes the designated target state. We omit transition probabilities, as these are also not visible to our algorithm. We also omit action labels for readability.

4.2 Using Information About the Property

Now, we discuss optimizations specific to exploiting the structure of the given systems *relative to the given property*.

Equivalence Structures. Classical verification uses several graph structural analyses which so far have been neglected for SMC. A very general analysis is the *never-worse relation* [76, Def. 5]. Intuitively, a state s_1 is never worse than s_2 if we have $V_\mathcal{M}(s_1) \geq V_\mathcal{M}(s_2)$ independently of the exact transition probabilities. If additionally s_2 is also never worse than s_1, these states form an equivalence class and can be merged [76, Thm. 1], reducing the number of transitions. Since deciding this relation is coNP-complete even on Markov chains [76, Thm. 4], we utilize special cases which we can identify efficiently in our setting [76, Sec. 3.2]: attractors, maximal end components, and dominated states.

Example 1. We illustrate `Equivalence Structures` and its synergy with other optimization using the example in Fig. 3. Firstly, state t is never worse than the goal state **1**: using the lower action, **1** is reached almost surely, independent of the exact probabilities. Consequently, we do not need to estimate *any* action of t. Computing such value-1 (and value-0) states is a well-known graph analysis, see e.g. [14, Chp. 10.6.1]. Secondly, as t and **1** have the same value, for the lower action of s_2 we only need to estimate the probability of moving to $\{t, \mathbf{1}\}$, not each individually. Combined with `Small Support`, we can from this infer the probability of moving to **0**. Thirdly, s_1 and s_2 can mutually reach each other with probability 1 and hence achieve the same value (they form an end component [34, Chp. 3.3]). Thus, we do not need to learn their "internal" transitions (in particular, the upper action of s_1). Overall, instead of dividing the confidence budget over all twelve transitions (as the state-of-the-art does), we can focus on only two transitions (one of $\hat{s}$ and the one from s_2 to $\{t, \mathbf{1}\}$). The higher confidence budget per transition mean the Clopper-Pearson interval is more effective (see Sect. 3.3).

Fragments. We introduce a new optimization which identifies sets of states where the "internal" behaviour is not relevant, and thus can be abstracted.

Example 2. Consider the MDP in Fig. 4 (left). The marked area of the state space has a lot of internal structure; however, we only care about how we could

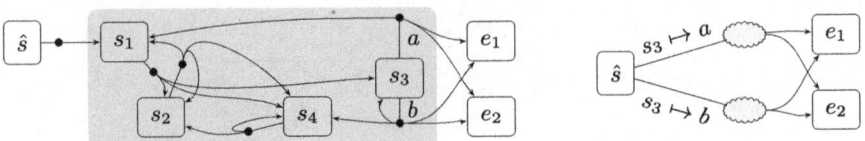

Fig. 4. Left: An MDP with a fragment R highlighted. Right: Fragment-quotient.

leave this area. In this example, there are only two possibilities, namely through the actions a or b in state s_3. Intuitively, we only are interested in the "big step" behaviour of the system as depicted in Fig. 4 (right). So, once we identify this fragment, we only need to estimate two probabilities, namely the probability to reach e_1 under a and b, respectively, instead of > 10 distinct values.

Generally, for a set of states $R \subseteq S$ (a fragment), we replace all internal transitions as follows: Let $\Pi_\mathcal{M}^{\mathsf{MD}}(R)$ be all MD-strategies restricted to R, i.e. all possible choices of actions for states in R. We replace every transition entering R, by a "macro action" for every strategy $\pi \in \Pi_\mathcal{M}^{\mathsf{MD}}(R)$ that immediately leads to the states outside the fragment. (Practically, this means sampling a path under the internal behaviour π until one of the exits is reached.) Thus, we omit all internal transitions, instead having a transition for the "big-step" behaviour. Formally:

Definition 1. *Fix an MDP $\mathcal{M}$, a goal set T, and a set of states $R \subseteq S \setminus T$. The fragment-quotient $\mathcal{M}_R$ is obtained as follows. Let $In \subseteq R$ all entries to R, i.e. $\{\hat{s}\} \cup \bigcup_{s \in S \setminus R, a \in A(s)} \mathrm{supp}(\mathsf{P}(s,a)) \cap R$ and set $S_R = (S \setminus R) \cup In$. For all states in $S \setminus R$, keep their actions and transitions unchanged. For all $s \in In$, set $A(s) = \Pi_\mathcal{M}^{\mathsf{MD}}(R)$ and for every $\pi \in \Pi_\mathcal{M}^{\mathsf{MD}}(R)$ and $s' \in S \setminus R$, define $\mathsf{P}_R(s, \pi, s')$ as the probability of reaching s' from s when following π until R is left, i.e. $\mathrm{Pr}_{\mathcal{M},s}^\pi[\{\varrho \mid \varrho_0 = s \wedge \exists i > 0. \ \varrho_i = s' \wedge \forall 0 \leq j < i. \ \varrho_j \in R\}]$.*

For correctness, we require that for any strategy the probability to eventually leave R is 1, i.e. it is not an end component. (This case can be detected and treated by using `Equivalence Structures`.) We prove in [67, App. D.2]:

Proposition 3. *Fix an MDP $\mathcal{M}$ with goal set T and set of states $R \subseteq S \setminus T$ which does not contain an end component, and let $\mathcal{M}_R$ be the fragment-quotient. For all states $s \in S_R$, we have $\mathsf{V}_\mathcal{M}(s) = \mathsf{V}_{\mathcal{M}_R}(s)$.*

It remains to decide which set R is a good candidate for a fragment. The number of transitions for R in the fragment quotient is the number of internal strategies multiplied with the number of entries and exits. Naively, we can solve a global optimization problem to find the fragments which minimize the overall number of transitions. However, since we want the improvements to be easy to implement and fast, we instead utilize two other candidates for fragments: Chains (states with a single predecessor), and strongly connected components (SCC, a well-known graph theoretic notion), both of which can be determined in linear time. In [67, App. D.2], we provide more details, and in [67, App.D 3] we outline how to extend the construction to objectives other than reachability.

5 Evaluation

In this section, we experimentally evaluate the effectiveness of our methods. Our implementation is available at https://doi.org/10.5281/zenodo.15231337.

5.1 Setup

We implemented our statistical methods as a prototype in Python and use the Partial Exploration Tool (PET) [66] for graph analysis and preprocessing. As models, we consider reachability instances from the PRISM benchmark suite [55], removing "trivial" ones, i.e. where the result is equal to 0 or 1. These can be immediately solved by graph analysis (Sect. 4.1), yielding "infinite" improvement.

We fix a confidence of $\delta = 0.1$ and choose the desired precision ε depending on the model such that the baseline also terminated reasonably quickly, in order to keep the evaluation time manageable (see [67, App. D] for further details). We solve the models using SMC as described in Sect. 2.1. We employ a uniform sampling strategy, i.e. starting from the initial state we choose a random action, sample a successor, and repeat, until we reach a goal or sink state. Then we estimate transition probabilities and solve the resulting interval MDP with interval iteration [41,68]. For the second step, we use different approaches: (i) the baseline, using the state-of-the-art model-based SMC technique of splitting the confidence budget uniformly over all transitions and applying Hoeffding's inequality to every transition probability, as in e.g. [2,8,11]; (ii) our approach, utilizing all improvements from Secs. 3 and 4 (Table 1); and (iii) our approach, but leaving out single improvements for an ablation study (Tab. 2). For Fragments we only use Chain Fragments since we did not find any SCCs that would reduce the number of transitions. We run every approach 10 times for each model and record the average number of sampled paths for each approach (see [67, App. E] for full results). Note that we intentionally quantify the sample complexity in terms of paths rather than transitions since the improvements of Sect. 4.2 modify the graph structure. Hence the number of samples per transition is not directly comparable.

5.2 Results and Discussion

Runtime. Our analysis focuses on sample count rather than runtime, as in most realistic scenarios the cost of gathering samples dominates other computations. Still, we mention that even in our prototype the overhead introduced by our improvements (e.g. removing chains and collapsing equivalent states) is negligible: for our largest model, it took roughly 12 s. As we report in [67, App. E] for all models in our evaluation, our improvements reduced the runtime of the SMC algorithm, with the speed-up ranging from approximately 1.2 up to 80. Primarily, this is due to the structural improvements of Sect. 4.2 reducing the size of the model, thereby speeding up solving the inferred models. Moreover, the baseline requires more samples, and thus also more time for gathering these. These savings massively outweigh the computational overhead of structural improvements.

Table 1. Summary of our improvements' effects on the complete SMC-algorithm. For each instance, we report the number of transitions in the model; the number of sampled runs needed for the precision ε with the baseline approach and with our improvements; and the improvement factor, i.e. ratio of the previous two values.

model	objective	ε	transitions	baseline	ours	ratio
consensus	disagree	0.3	484	38 090	14 766	2.6
csma	all_before	0.1	1276	64 310	783	82.1
firewire_dl	deadline	0.05	17 417	1 685 965	46 286	36.4
pacman	crash	0.1	84	5 048	181	28.0
wlan	collisions_max	0.7	5444	354 949	9 118	38.9
wlan_dl	deadline	0.9	326 883	21 083 954	201 536	104.6
zeroconf	correct	0.05	953	18 202	164	111.1
zeroconf_dl	deadline	0.05	4777	48 248	64	749.2

Table 2. Ablation study for our improvements with minimum, average (geometric mean), and maximum sample reduction factor when removing each improvement. The final line shows overall improvements over the baseline approach.

Ablated	min	avg	max	Ablated	min	avg	max
Clopper-Pearson	1.09	2.90	14.71	Structural impr.	1.06	13.79	252.42
Small Support	1.17	1.48	1.79	↪ Equiv. Str.	1.00	8.20	143.39
Independence	1.01	1.23	1.82	↪ Chain Frag.	1	1.71	7.89
Baseline	2.58	54.07	749.19				

All Improvements. Table 1 shows that our improvements always reduce the number of required samples. The number of samples as compared to the baseline approach is less than 50% on all models, less than 10% on all but one, and even less than 1% on three (two orders of magnitude improvement).

Ablation Study. Investigating individual effects of improvements, we report the ratio of paths when leaving a single improvement out over all improvements for all models in [67, App. E]. Table 2 aggregates over models, reporting minimum, geometric average and maximum of the improvement factors. Note that, intuitively, one might expect that the improvement factors simply multiply, however they are not independent and may synergise super-linearly, e.g. as we have discussed in Ex. 1. Thus all improvement factors in the ablation study are generally larger than if one were to apply each improvement on its own. As predicted in Sect. 3.3, Clopper-Pearson outperforms Hoeffding, with the effect being most pronounced for models where small transition probabilities are present, in line with Fig. 1 (right). Independence and Small Support consistently reduce sample count with no computational overhead. The structural improvements (Equivalence Structures and Fragments) of course depend on the structure of the particular system at hand. They are significant in most cases and they positively impact each other, e.g. with `Equivalence Structures` giving rise to additional chains. Additionally, all our suggested improvements lead to a smaller (or equivalent) sample complexity, i.e. *never make it worse*.

6 Related Work and Extensions

Related results of statistics are discussed in-depth in Sect. 3. The recent [25] builds on a preprint of our work to categorize existing tools based on their usage of statistical methods, utilizing our insights from Sect. 3. However, their focus lies on purely stochastic Markov chains and they mainly use statistical methods to tackle extremely large models, while we explicitly focus on non-determinism. For a history of statistical approaches dealing with stochastic systems, we refer to the surveys [3,52,60].

We detail one direction of related work, namely algorithms with weaker guarantees than PAC. Approaches for MDP such as [42,43,45,58,63] heuristically find promising strategies but in general cannot guarantee the chosen strategy is close to the optimum, thus only providing a lower bound on the value. The model-free approach of lightweight scheduler sampling [32,33,61,62] only guarantees that the estimate is close to the optimum if good strategies are frequent and also only provides lower bounds. The reinforcement learning algorithms in [39,91] guarantee an improvement over the policy that is used for obtaining the samples. These and our paper work in a setting where we are allowed to obtain samples from any part the unknown system, even unsafe states. In contrast, in a "safe online" setting, a good policy has to be computed *while executing the system*, and unsafe states have to be avoided. There, we refer to works on PAC online learning [53,54], shielding [6], and regret minimization [84].

Finally, we discuss how our improvements are (nearly) universally applicable, including methods where SMC is invoked as an inner step such as parameter synthesis [51,69], and *largely independent of the setting*. Concretely, there exist several variants of our problem, depending on the assumptions about the unknown MDP, where our improvements also are applicable:

Topology Knowledge. Either we know at least the topology of the underlying graph of the MDP (called grey-box) or even that has to be inferred from the simulations (called black-box) [8, Def. 2]. We focus on the grey-box, but detail in Remark 2 how our methods are also applicable in the black-box setting.

Sampling Access. Assumptions about the sampling access to the MDP include, from least to most restrictive: sampling any state-action pair [38,93], running simulations through the system that can only be restarted in the initial state [2,8,20,81], or batch learning, where we only get a fixed data set of past interactions [39,56,78,91]. Our methods are agnostic of the sampling method and just "make the most of the data" they get.

Objectives. MDPs can be complemented with various objectives, e.g., reachability, mean payoff, (discounted) total reward, or linear temporal logic. Most of our improvements are independent of the objective, and we sketch in [67, App. D.3] how to generalize the few that do utilise information about the objective.

7 Conclusion

We presented several improvements for the foundations of statistical model checking. Overall, we suggest to use the Clopper-Pearson interval for estimating single transitions, and to exploit knowledge about the structure of the MDP and the property of interest where possible. Our presented methods all are fast and often significantly improve the precision of the resulting intervals. Thus, every implementation of a model-based SMC algorithm should use them.

In settings where samples are very expensive, we can employ more time-consuming improvements, for example computing the full never-worse relation [76] or employing a global search to find the optimal fragments (see Sect. 4.2). As future work, we want to show effectiveness of our methods on further models, including reinforcement learning and model learning benchmarks. Additionally, we aim to develop heuristics for focusing the confidence budget on "important" states. Concretely, we plan to use stochastic gradient descent to find those states where giving them more of the confidence budget increases the precision for the overall value the most.

References

1. Agarwal, A., Kakade, S.M., Yang, L.F.: Model-based reinforcement learning with a generative model is minimax optimal. In: COLT. Proceedings of Machine Learning Research, vol. 125, pp. 67–83. PMLR (2020). http://proceedings.mlr.press/v125/agarwal20b.html
2. Agarwal, C., Guha, S., Křetínskỳ, J., Muruganandham, P.: PAC statistical model checking of mean payoff in discrete-and continuous-time MDP. In: Shoham, S., Vizel, Y. (eds.) CAV 2022. LNCS, vol. 13372, pp. 3–25. Springer, Cham (2022). https://doi.org/10.1007/978-3-031-13188-2_1
3. Agha, G., Palmskog, K.: A survey of statistical model checking. ACM Trans. Model. Comput. Simul. **28**(1), 6:1–6:39 (2018). https://doi.org/10.1145/3158668
4. Agresti, A., Coull, B.A.: Approximate is better than "exact" for interval estimation of binomial proportions. Am. Stat. **52**(2), 119–126 (1998)
5. Aitkin, M., Anderson, D., Francis, B., Hinde, J.: Statistical modelling in GLIM, pp. 112–118 (1988)
6. Alshiekh, M., Bloem, R., Ehlers, R., Könighofer, B., Niekum, S., Topcu, U.: Safe reinforcement learning via shielding. In: AAAI, pp. 2669–2678. AAAI Press (2018)
7. Anthony, M., Bartlett, P.L.: General Lower Bounds on Sample Complexity, pp. 59–73. Cambridge University Press (1999)
8. Ashok, P., Křetínský, J., Weininger, M.: PAC statistical model checking for Markov decision processes and stochastic games. In: Dillig, I., Tasiran, S. (eds.) CAV 2019. LNCS, vol. 11561, pp. 497–519. Springer, Cham (2019). https://doi.org/10.1007/978-3-030-25540-4_29
9. Ashok, P., Kretínský, J., Weininger, M.: PAC statistical model checking for Markov decision processes and stochastic games. CoRR abs/1905.04403 (2019). http://arxiv.org/abs/1905.04403
10. Auer, P., Jaksch, T., Ortner, R.: Near-optimal regret bounds for reinforcement learning. In: Advances in Neural Information Processing Systems, vol. 21 (2008)

11. Badings, T.S., et al.: Robust control for dynamical systems with non-gaussian noise via formal abstractions. J. Artif. Intell. Res. **76**, 341–391 (2023)
12. Badings, T.S., Simão, T.D., Suilen, M., Jansen, N.: Decision-making under uncertainty: beyond probabilities. Int. J. Softw. Tools Technol. Transf. **25**(3), 375–391 (2023)
13. Baier, C., Dubslaff, C., Wienhöft, P., Kiebel, S.J.: Strategy synthesis in Markov decision processes under limited sampling access. In: Rozier, K.Y., Chaudhuri, S. (eds.) NFM 2023. LNCS, vol. 13903, pp. 86–103. Springer, Cham (2023). https://doi.org/10.1007/978-3-031-33170-1_6
14. Baier, C., Katoen, J.: Principles of Model Checking. MIT Press (2008). https://mitpress.mit.edu/books/principles-model-checking
15. Bazille, H., Genest, B., Jegourel, C., Sun, J.: Global PAC bounds for learning discrete time Markov chains. In: Lahiri, S.K., Wang, C. (eds.) CAV 2020. LNCS, vol. 12225, pp. 304–326. Springer, Cham (2020). https://doi.org/10.1007/978-3-030-53291-8_17
16. Bennett, G.: Probability inequalities for the sum of independent random variables. J. Am. Stat. Assoc. **57**(297), 33–45 (1962)
17. Bernstein, S.: On a modification of Chebyshev's inequality and of the error formula of Laplace. Ann. Sci. Inst. Sav. Ukraine Sect. Math **1**(4), 38–49 (1924)
18. Blyth, C.R., Still, H.A.: Binomial confidence intervals. J. Am. Stat. Assoc. **78**(381), 108–116 (1983)
19. Boucheron, S., Lugosi, G., Massart, P.: Concentration Inequalities: A Nonasymptotic Theory of Independence. Oxford University Press (2013) https://doi.org/10.1093/acprof:oso/9780199535255.001.0001
20. Brázdil, T., et al.: Verification of Markov decision processes using learning algorithms. In: Cassez, F., Raskin, J.-F. (eds.) ATVA 2014. LNCS, vol. 8837, pp. 98–114. Springer, Cham (2014). https://doi.org/10.1007/978-3-319-11936-6_8
21. Brown, L.D., Cai, T.T., DasGupta, A.: Interval estimation for a binomial proportion. Stat. Sci. **16**(2), 101–133 (2001). https://doi.org/10.1214/ss/1009213286
22. Brázdil, T., et al.: Learning algorithms for verification of Markov decision processes. TheoretiCS **4**, 10 (2025). https://doi.org/10.46298/theoretics.25.10
23. Bu, H., Sun, M.: Clopper-Pearson algorithms for efficient statistical model checking estimation. IEEE Trans. Software Eng. **50**(7), 1726–1746 (2024). https://doi.org/10.1109/TSE.2024.3392720
24. Budde, C.E., D'Argenio, P.R., Hartmanns, A.: Better automated importance splitting for transient rare events. In: Larsen, K.G., Sokolsky, O., Wang, J. (eds.) SETTA 2017. LNCS, vol. 10606, pp. 42–58. Springer, Cham (2017). https://doi.org/10.1007/978-3-319-69483-2_3
25. Carlos, E.B., Arnd, H., Tobias, M., Maximilian, W., Patrick, W.: Sound statistical model checking for probabilities and expected rewards. In: Arie, G., Marijn., H. (ed.) Tools and Algorithms for the Construction and Analysis of Systems-31st International Conference, Held as Part of the International Joint Conferences on Theory and Practice of Software, Hamilton, ON, Canada, Proceedings, Part I, Lecture Notes in Computer Science, vol. 15696, pp.167—190. Springer, (2025). https://doi.org/10.1007/978-3-031-90643-5_9
26. Casella, G., McCulloch, C.E.: Confidence intervals for discrete distributions (1984)
27. Clarke, E.M., Henzinger, T.A., Veith, H., Bloem, R. (eds.): Handbook of Model Checking. Springer, Cham (2018). https://doi.org/10.1007/978-3-319-10575-8
28. Clopper, C.J., Pearson, E.S.: The use of confidence or fiducial limits illustrated in the case of the binomial. Biometrika **26**(4), 404–413 (1934)

29. Corcoran, C., Mehta, C.: Comment on "interval estimation for a binomial proportion". Stat. Sci. **16**(2), 120–122 (2001). https://doi.org/10.1214/ss/1009213286
30. Daca, P., Henzinger, T.A., Kretínský, J., Petrov, T.: Faster statistical model checking for unbounded temporal properties. ACM Trans. Comput. Log. **18**(2), 12:1–12:25 (2017)
31. Dagum, P., Karp, R., Luby, M., Ross, S.: An optimal algorithm for Monte Carlo estimation. In: Proceedings of IEEE 36th Annual Foundations of Computer Science, pp. 142–149 (1995). https://doi.org/10.1109/SFCS.1995.492471
32. D'Argenio, P.R., Hartmanns, A., Sedwards, S.: Lightweight statistical model checking in nondeterministic continuous time. In: Margaria, T., Steffen, B. (eds.) ISoLA 2018. LNCS, vol. 11245, pp. 336–353. Springer, Cham (2018). https://doi.org/10.1007/978-3-030-03421-4_22
33. D'Argenio, P.R., Legay, A., Sedwards, S., Traonouez, L.: Smart sampling for lightweight verification of Markov decision processes. Int. J. Softw. Tools Technol. Transf. **17**(4), 469–484 (2015). https://doi.org/10.1007/S10009-015-0383-0
34. De Alfaro, L.: Formal verification of probabilistic systems. Ph.D. thesis, Stanford University (1997). https://exhibits.stanford.edu/cs/catalog/xz681dy7227
35. Dean, N., Pagano, M.: Evaluating confidence interval methods for binomial proportions in clustered surveys. J. Surv. Stat. Methodol. **3**(4), 484–503 (2015)
36. Duffy, D.E., Santner, T.J.: Confidence intervals for a binomial parameter based on multistage tests. Biometrics 81–93 (1987)
37. Dvoretzky, A., Kiefer, J., Wolfowitz, J.: Asymptotic minimax character of the sample distribution function and of the classical multinomial estimator. Ann. Math. Stat. **27**(3), 642–669 (1956). https://doi.org/10.1214/aoms/1177728174
38. Fu, J., Topcu, U.: Probably approximately correct MDP learning and control with temporal logic constraints. In: Robotics: Science and Systems (2014). https://doi.org/10.15607/RSS.2014.X.039
39. Ghavamzadeh, M., Petrik, M., Chow, Y.: Safe policy improvement by minimizing robust baseline regret. In: Lee, D.D., Sugiyama, M., von Luxburg, U., Guyon, I., Garnett, R. (eds.) Advances in Neural Information Processing Systems 29: Annual Conference on Neural Information Processing Systems 2016, December 5-10, 2016, Barcelona, Spain, pp. 2298–2306 (2016). https://proceedings.neurips.cc/paper/2016/hash/9a3d458322d70046f63dfd8b0153ece4-Abstract.html
40. Givan, R., Leach, S., Dean, T.: Bounded-parameter Markov decision processes. Artif. Intell. **122**(1–2), 71–109 (2000)
41. Haddad, S., Monmege, B.: Interval iteration algorithm for MDPs and IMDPs. Theor. Comput. Sci. **735**, 111–131 (2018). https://doi.org/10.1016/j.tcs.2016.12.003
42. Hahn, E.M., Perez, M., Schewe, S., Somenzi, F., Trivedi, A., Wojtczak, D.: Omega-regular objectives in model-free reinforcement learning. In: Vojnar, T., Zhang, L. (eds.) TACAS 2019. LNCS, vol. 11427, pp. 395–412. Springer, Cham (2019). https://doi.org/10.1007/978-3-030-17462-0_27
43. Hahn, E.M., Perez, M., Schewe, S., Somenzi, F., Trivedi, A., Wojtczak, D.: Mungojerrie: linear-time objectives in model-free reinforcement learning. In: Sankaranarayanan, S., Sharygina, N. (eds.) TACAS 2023. LNCS, vol. 13993, pp. 527–545. Springer, Cham (2023). https://doi.org/10.1007/978-3-031-30823-9_27
44. HasanzadeZonuzy, A., Bura, A., Kalathil, D.M., Shakkottai, S.: Learning with safety constraints: sample complexity of reinforcement learning for constrained MDPs. In: AAAI, pp. 7667–7674. AAAI Press (2021). https://doi.org/10.1609/AAAI.V35I9.16937

45. Henriques, D., Martins, J.G., Zuliani, P., Platzer, A., Clarke, E.M.: Statistical model checking for Markov decision processes. In: Ninth International Conference on Quantitative Evaluation of Systems, QEST 2012, London, United Kingdom, 17–20 September 2012, pp. 84–93. IEEE Computer Society (2012). https://doi.org/10.1109/QEST.2012.19
46. Hoeffding, W.: Probability inequalities for sums of bounded random variables. J. Am. Stat. Assoc. **58**(301), 13–30 (1963)
47. Huber, M.: An unbiased estimate for the mean of a 0, 1 random variable with relative error distribution independent of the mean. CoRR abs/1309.5413 (2013). https://arxiv.org/abs/1309.5413
48. Huber, M.: Tight relative estimation in the mean of Bernoulli random variables. CoRR abs/2210.12861 (2022). https://doi.org/10.48550/ARXIV.2210.12861
49. Jégourel, C., Sun, J., Dong, J.S.: Sequential schemes for frequentist estimation of properties in statistical model checking. ACM Trans. Model. Comput. Simul. **29**(4), 25:1–25:22 (2019). https://doi.org/10.1145/3310226
50. Jin, C., Krishnamurthy, A., Simchowitz, M., Yu, T.: Reward-free exploration for reinforcement learning. In: ICML. Proceedings of Machine Learning Research, vol. 119, pp. 4870–4879. PMLR (2020). http://proceedings.mlr.press/v119/jin20d.html
51. Klein, J., Phung, H., Hajnal, M., Šafránek, D., Petrov, T.: Combining formal methods and Bayesian approach for inferring discrete-state stochastic models from steady-state data. PLOS ONE **18**(11), 1–26 (2023). https://doi.org/10.1371/journal.pone.0291151
52. Křetínský, J.: Survey of statistical verification of linear unbounded properties: model checking and distances. In: Margaria, T., Steffen, B. (eds.) ISoLA 2016. LNCS, vol. 9952, pp. 27–45. Springer, Cham (2016). https://doi.org/10.1007/978-3-319-47166-2_3
53. Křetínský, J., Michel, F., Michel, L., Pérez, G.A.: Finite-memory near-optimal learning for Markov decision processes with long-run average reward. In: UAI. Proceedings of Machine Learning Research, vol. 124, pp. 1149–1158. AUAI Press (2020)
54. Křetínský, J., Pérez, G.A., Raskin, J.F.: Learning-based mean-payoff optimization in an unknown MDP under omega-regular constraints. In: CONCUR, pp. 8:1–8:18. Dagstuhl (2018)
55. Kwiatkowska, M.Z., Norman, G., Parker, D.: The PRISM benchmark suite. In: QEST, pp. 203–204. IEEE Computer Society (2012). https://doi.org/10.1109/QEST.2012.14
56. Lange, S., Gabel, T., Riedmiller, M.A.: Batch reinforcement learning. In: Wiering, M., van Otterlo, M. (eds.) Reinforcement Learning. Adaptation, Learning, and Optimization, vol. 12, pp. 45–73. Springer, Heidelberg (2012). https://doi.org/10.1007/978-3-642-27645-3_2
57. Lassaigne, R., Peyronnet, S.: Approximate planning and verification for large Markov decision processes. In: SAC, pp. 1314–1319. ACM (2012). https://doi.org/10.1145/2245276.2231984
58. Lassaigne, R., Peyronnet, S.: Approximate planning and verification for large Markov decision processes. Int. J. Softw. Tools Technol. Transf. **17**(4), 457–467 (2015). https://doi.org/10.1007/S10009-014-0344-Z
59. Legay, A., Delahaye, B., Bensalem, S.: Statistical model checking: an overview. In: Barringer, H., et al. (eds.) RV 2010. LNCS, vol. 6418, pp. 122–135. Springer, Heidelberg (2010). https://doi.org/10.1007/978-3-642-16612-9_11

60. Legay, A., Lukina, A., Traonouez, L.M., Yang, J., Smolka, S.A., Grosu, R.: Statistical model checking. In: Steffen, B., Woeginger, G. (eds.) Computing and Software Science. LNCS, vol. 10000, pp. 478–504. Springer, Cham (2019). https://doi.org/10.1007/978-3-319-91908-9_23
61. Legay, A., Sedwards, S., Traonouez, L.-M.: Scalable verification of Markov decision processes. In: Canal, C., Idani, A. (eds.) SEFM 2014. LNCS, vol. 8938, pp. 350–362. Springer, Cham (2015). https://doi.org/10.1007/978-3-319-15201-1_23
62. Legay, A., Sedwards, S., Traonouez, L.-M.: Plasma lab: a modular statistical model checking platform. In: Margaria, T., Steffen, B. (eds.) ISoLA 2016. LNCS, vol. 9952, pp. 77–93. Springer, Cham (2016). https://doi.org/10.1007/978-3-319-47166-2_6
63. Lukina, A., et al.: ARES: adaptive receding-horizon synthesis of optimal plans. In: Legay, A., Margaria, T. (eds.) TACAS 2017, Part II. LNCS, vol. 10206, pp. 286–302. Springer, Heidelberg (2017). https://doi.org/10.1007/978-3-662-54580-5_17
64. Massart, P.: The tight constant in the Dvoretzky-Kiefer-Wolfowitz inequality. Ann. Probab. **18**(3), 1269–1283 (1990). https://doi.org/10.1214/aop/1176990746
65. Maurer, A., Pontil, M.: Empirical Bernstein bounds and sample-variance penalization. In: COLT (2009). http://www.cs.mcgill.ca/%7Ecolt2009/papers/012.pdf#page=1
66. Meggendorfer, T., Weininger, M.: Playing games with your PET: extending the partial exploration tool to stochastic games. In: Gurfinkel, A., Ganesh, V. (eds.) CAV 2024. LNCS, vol. 14683, pp. 359–372. Springer, Cham (2024). https://doi.org/10.1007/978-3-031-65633-0_16
67. Meggendorfer, T., Weininger, M., Wienhöft, P.: What are the odds? Improving the foundations of statistical model checking. CoRR abs/2404.05424 (2024). https://doi.org/10.48550/ARXIV.2404.05424
68. Meggendorfer, T., Weininger, M., Wienhöft, P.: Solving robust Markov decision processes: generic, reliable, efficient. In: Proceedings of the AAAI Conference on Artificial Intelligence, vol. 39, pp. 26631–26641 (2025). https://ojs.aaai.org/index.php/AAAI/article/download/34865/37020
69. Molyneux, G.W., Abate, A.: ABC(SMC)^2: simultaneous inference and model checking of chemical reaction networks. In: Abate, A., Petrov, T., Wolf, V. (eds.) CMSB 2020. LNCS, vol. 12314, pp. 255–279. Springer, Cham (2020). https://doi.org/10.1007/978-3-030-60327-4_14
70. Newcombe, R.G.: Two-sided confidence intervals for the single proportion: comparison of seven methods. Stat. Med. **17**(8), 857–872 (1998). https://doi.org/10.1002/(SICI)1097-0258(19980430)17:8<857::AID-SIM777>3.0.CO;2-E
71. Office of Emergency and Remedial Response U.S. Environmental Protection Agency: Calculating upper confidence limits for exposure point concentrations at hazardous waste sites (2002)
72. Okamoto, M.: Some inequalities relating to the partial sum of binomial probabilities. Ann. Inst. Stat. Math. **10**(1), 29–35 (1959). https://doi.org/10.1007/bf02883985
73. Perez, M., Somenzi, F., Trivedi, A.: A PAC learning algorithm for LTL and omega-regular objectives in MDPs. In: AAAI, pp. 21510–21517. AAAI Press (2024). https://doi.org/10.1609/AAAI.V38I19.30148
74. Puterman, M.L.: Markov Decision Processes: Discrete Stochastic Dynamic Programming. Wiley (1994)
75. Reijsbergen, D., de Boer, P., Scheinhardt, W.R.W., Haverkort, B.R.: On hypothesis testing for statistical model checking. Int. J. Softw. Tools Technol. Transf. **17**(4), 377–395 (2015). https://doi.org/10.1007/S10009-014-0350-1

76. Le Roux, S., Pérez, G.A.: The complexity of graph-based reductions for reachability in Markov decision processes. In: Baier, C., Dal Lago, U. (eds.) FoSSaCS 2018. LNCS, vol. 10803, pp. 367–383. Springer, Cham (2018). https://doi.org/10.1007/978-3-319-89366-2_20
77. Santner, T.J.: Comment on "interval estimation for a binomial proportion". Stat. Sci. **16**(2), 126–128 (2001). https://doi.org/10.1214/ss/1009213286
78. Shi, L., Li, G., Wei, Y., Chen, Y., Chi, Y.: Pessimistic Q-learning for offline reinforcement learning: towards optimal sample complexity. In: ICML. Proceedings of Machine Learning Research, vol. 162, pp. 19967–20025. PMLR (2022). https://proceedings.mlr.press/v162/shi22c.html
79. Strehl, A.L., Littman, M.L.: An empirical evaluation of interval estimation for Markov decision processes. In: 16th IEEE International Conference on Tools with Artificial Intelligence, pp. 128–135. IEEE (2004)
80. Student: the probable error of a mean. Biometrika 1–25 (1908)
81. Suilen, M., Simão, T.D., Parker, D., Jansen, N.: Robust anytime learning of Markov decision processes. In: NeurIPS (2022)
82. Temme, N.M.: Asymptotic inversion of the incomplete beta function. J. Comput. Appl. Math. **41**(1–2), 145–157 (1992)
83. Villén-Altamirano, M., Villén-Altamirano, J.: The rare event simulation method RESTART: efficiency analysis and guidelines for its application. In: Kouvatsos, D.D. (ed.) Network Performance Engineering. LNCS, vol. 5233, pp. 509–547. Springer, Heidelberg (2011). https://doi.org/10.1007/978-3-642-02742-0_22
84. Wagenmaker, A.J., Simchowitz, M., Jamieson, K.: Beyond no regret: instance-dependent PAC reinforcement learning. In: COLT. Proceedings of Machine Learning Research, vol. 178, pp. 358–418. PMLR (2022). https://proceedings.mlr.press/v178/wagenmaker22a.html
85. Wald, A.: Sequential tests of statistical hypotheses. Ann. Math. Stat. **16**(2), 117–186 (1945). https://doi.org/10.1214/aoms/1177731118
86. Wallis, S.: Binomial confidence intervals and contingency tests: mathematical fundamentals and the evaluation of alternative methods. J. Quant. Linguistics **20**(3), 178–208 (2013). https://doi.org/10.1080/09296174.2013.799918
87. Weininger, M., Grover, K., Misra, S., Kretínský, J.: Guaranteed trade-offs in dynamic information flow tracking games. In: CDC, pp. 3786–3793. IEEE (2021). https://doi.org/10.1109/CDC45484.2021.9683447
88. Weissman, T., Ordentlich, E., Seroussi, G., Verdu, S., Weinberger, M.J.: Inequalities for the L1 deviation of the empirical distribution. Technical report, Hewlett-Packard Labs (2003)
89. Wen, M., Topcu, U.: Probably approximately correct learning in adversarial environments with temporal logic specifications. IEEE Trans. Autom. Control **67**(10), 5055–5070 (2022). https://doi.org/10.1109/TAC.2021.3115080
90. Wienhöft, P., Suilen, M., Simão, T.D., Dubslaff, C., Baier, C., Jansen, N.: More for less: safe policy improvement with stronger performance guarantees. arXiv preprint arXiv:2305.07958 (2023)
91. Wienhöft, P., Suilen, M., Simão, T.D., Dubslaff, C., Baier, C., Jansen, N.: More for less: safe policy improvement with stronger performance guarantees. In: Proceedings of the International Joint Conference on Artificial Intelligence (IJCAI) (2023). https://doi.org/10.24963/ijcai.2023/490
92. Wilson, E.B.: Probable inference, the law of succession, and statistical inference. J. Am. Stat. Assoc. **22**(158), 209–212 (1927)

93. Younes, H.L.S., Simmons, R.G.: Probabilistic verification of discrete event systems using acceptance sampling. In: Brinksma, E., Larsen, K.G. (eds.) CAV 2002. LNCS, vol. 2404, pp. 223–235. Springer, Heidelberg (2002). https://doi.org/10.1007/3-540-45657-0_17

A Product-Form Model for Systems with Aging Objects and Similarities

Andrea Marin[1], Diletta Olliaro[1(✉)], Sabina Rossi[1], and Daniel Menasché[2]

[1] Department of Environmental Sciences, Informatics and Statistics, Università Ca' Foscari Venezia, Venice, Italy
`{marin,diletta.olliaro,sabina.rossi}@unive.it`
[2] Institute of Computing, Federal University of Rio de Janeiro, Rio de Janeiro, Brazil
`sadoc@ic.ufrj.br`

Abstract. In this paper, we present a stochastic model for a system consisting of aging objects. The state of each object is an integer between 0 and T, where T indicates that the object has expired, while state $t < T$ denotes its age (i.e., the object is alive). The object's state increments from t to $t + 1$ at each clock tick until it expires. If an object receives a request, it jumps to state 0 if it is alive, or to a random state drawn from its limiting distribution if it has expired. When an expired object is requested, the request is instantaneously routed to another similar object. We show that, under standard exponential assumptions, the model admits a product-form steady-state distribution, and that relevant performance indices can be efficiently computed. Application domains include the analysis of similarity-based time-to-live (TTL) caches and the routing of distributed computations.

Keywords: Product-form analysis · TTL cache networks · aging objects

1 Introduction

Caching mechanisms are foundational components of modern computing systems. From processor memory hierarchies and content delivery networks [27] to serverless platforms [11] and inference-serving pipelines [28], caching reduces latency and alleviates load on backend infrastructure. Two common principles shape these systems: (1) objects typically *age* over time, losing freshness or relevance unless explicitly refreshed; and (2) objects are often *related*—semantically, topologically, or functionally—so that a request that misses one object may be rerouted to a related object.

Time-to-live (TTL) caches are an example of systems that combine aging and similarity-based redirection, where each object remains available for a limited duration unless refreshed. Upon a miss, systems frequently employ routing or fallback mechanisms that exploit object similarity or alternate cache locations. Understanding the interplay between object aging, redirection policies, and request patterns is essential to optimizing hit rates, occupancy, and forwarding overhead [1,23,26].

Modern infrastructures amplify the importance of such mechanisms. In large language model (LLM) serving, key-value caching stores past attention vectors for reuse; CacheGen [18] shows how such caches can be compressed and streamed efficiently, modeling query traces as Markov chains. In serverless Function-as-a-Service (FaaS) environments, functions are cached in memory when "warm" and evicted when "idle", mimicking TTL behavior [11]. In inference-serving systems like Amazon SageMaker [28], models are loaded and evicted on demand, and missed requests trigger loading or redirection. Similar fallback patterns appear in multi-level CDNs and similarity-aware cache networks [23,27].

In this paper, we introduce a stochastic model that captures these behaviors in a unified framework. Our model generalizes TTL caching and includes redirection via a probabilistic routing matrix. We show that the system admits a closed-form *product-form* solution for the steady-state distribution under exponential assumptions. Based on this foundation, we design efficient algorithms for both simulation and performance evaluation.

In summary, our main contributions are as follows:

- We propose a stochastic model for systems where objects age and can forward requests to related objects when expired. This captures key dynamics of TTL caches, similarity-aware storage, and redirect-on-miss infrastructure (Sects. 2 and 3).
- We prove that the model admits a product-form stationary distribution and derive explicit expressions for hit probabilities and object occupancy (Sect. 3).
- We develop three algorithms: the first two algorithms account for a single object, illustrating implementations of the considered caching mechanism under general and exponentially distributed time between clock ticks, respectively (Algorithms 1 and 2), while the third algorithm is a fixed-point solver to determine occupancies and hit probabilities for multi-object systems (Algorithm 3), enabling scalable numerical analysis.
- We show how the model applies to hierarchical caching, and demonstrate how it can be used to analyze cache hit rates, occupancy levels, and forwarding load (Sect. 4).

The rest of this paper is organized as follows. Section 2 introduces the model for a single aging object and derives its steady-state distribution. Section 3 presents the product-form analysis for networks of interacting objects and provides an algorithm for solving the associated rate equations. Section 4 illustrates the application of our model in a content delivery context. Section 5 reviews related work, and Sect. 6 concludes.

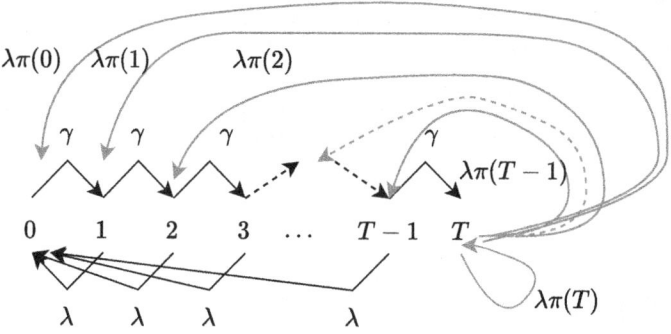

Fig. 1. Representation of a single aging object

2 Single Aging Object

In this section, we describe the behavior of a single aging object, introduce the Markovian assumptions, and derive the expression of its stationary distribution.

2.1 Object Description

Let us consider a single aging object with $T+1$ states numbered from 0 to T. States i with $i \in \{0,\ldots,T-1\}$ represent an available object with age i, while state T denotes an unavailable object. The passage from age i to $i+1$ occurs after an independent exponentially distributed random time with rate $\gamma \geq 0$, where the case $\gamma = 0$ indicates that the object is always alive. The requests for the object follow an independent Poisson process with intensity λ.

When the object is in state $0,\ldots,T-1$ and receives a request, we have a *hit* and its timer is set to 0. If the object is in state T and receives a request, then we have a *miss* and the next state is chosen probabilistically according to the stationary distribution of the object, i.e., the next state will be state j with probability $\pi(j)$, where $\pi(j)$ is the stationary probability of state j. In Fig. 1, we show a graphical representation of the model.

Notice that thanks to the exponential assumptions, the model is a continuous time Markov chain (CTMC) with irreducible and finite state space. As a consequence, there exists a unique limiting distribution and the probabilities $\pi(i)$, $i = 1,\ldots,T$ are well defined.

One of the singular characteristics of this model is that the colored transitions shown in Fig. 1 depend on the stationary state distribution. In order to clarify how this can be implemented in practice and why it is reasonable, we show a pseudo code for the simulation of a single aging object. Notice that the purpose of this simulation code is to show a possible practical implementation of this aging policy rather than computing the steady-state probabilities since, as we will see in Sect. 2.2, these are known.

The simulation code in Algorithm 1 maintains a vector called *times* that accumulates the residence times of the object in each of its $T+1$ states (lines 8

Algorithm 1. Simulation of a single aging object

```
 1: procedure SIMULATION(λ, γ, T, events)
 2:     times ← (0, . . . , 0)                                    ▷ Vector of T + 1 0s
 3:     simtime ← 0
 4:     fel ← (SAMPLE_ARRIVAL(λ), SAMPLE_AGING(γ))               ▷ Future event list
 5:     state ← 0
 6:     for i ← 1 . . . events do
 7:         if fel[0] < fel[1] then                                ▷ Request event
 8:             times[state] ← times[state] + fel[0] − simtime    ▷ Update time in state
 9:             simtime ← fel[0]
10:             if state < T then
11:                 state ← 0                                      ▷ Refresh the timer
12:             else:
13:                 state ←SAMPLE_EMPIRICAL_DISCRETE(times, simtime)
14:             fel[0] ← simtime+ SAMPLE_ARRIVAL(λ)
15:         else                                                    ▷ Aging event
16:             times[state] ← times[state] + fel[1] − simtime    ▷ Update time in state
17:             state ← MIN(state + 1, T)
18:             simtime ← fel[1]
19:             fel[1] ← simtime+ SAMPLE_AGING(γ)
       return times/simtime
```

and 16). When a request occurs and the object is expired, the next state is chosen with probability proportional to the state residence times measured up to that instant (line 13). This empirical sampling approach implicitly favors states in which the object tends to persist longer, thereby discouraging premature rejuvenation to state 0 for objects that are rarely requested, e.g., one-hit wonders [19], that are likely to expire again before the next request. This behavior aligns with the principles behind adaptive cache policies like q-LRU [7,24], where items are inserted with a fixed probability q, effectively prioritizing frequently requested content and reducing pollution from transient items. After Sect. 2.2, we will discuss a new version of this algorithm that takes advantage of the analytical expression of the stationary distribution, avoiding the need to track residence times explicitly and enabling more efficient simulation or implementation using only hit/miss counters and a geometric sampling mechanism.

2.2 Stationary Solution

In this subsection, we compute the stationary distribution of this model.

Lemma 1 (Steady-state distribution of single aging objects). *If $\gamma/\lambda < T$, the stationary probabilities have truncated geometric distribution, i.e.:*

$$\pi(i) = \begin{cases} (1-z)z^i & \text{if } 0 \leq i < T \\ z^T & \text{if } i = T \end{cases},$$

for some $z \in [0,1]$ to be determined. If $\gamma/\lambda \geq T$ the chain has transient states, i.e.:
$$\pi(i) = \begin{cases} 0 & \text{if } 0 \leq i < T \\ 1 & \text{if } i = T \end{cases}.$$

Proof. The case $\gamma = 0$ is trivial, so we can assume $\gamma > 0$. Let us write the balance equations for states $0 < i < T$:
$$\pi(i)(\lambda + \gamma) = \pi(i-1)\gamma + \pi(T)\lambda\pi(i).$$

By replacing the expression of π and dividing by $\pi(i)$, we obtain:
$$\lambda + \gamma = \frac{1}{z}\gamma + z^T \lambda.$$

Notice that this equation is independent of i and can be rewritten as:
$$\lambda z^{T+1} - (\lambda + \gamma)z + \gamma = 0. \tag{1}$$

Now we use the following theorem from [3][Thm. 2.1]:

Theorem 1 (Zeros of trinomials [3]). *Let $a > b > 0$ be real numbers and $n > m > 0$ be integers. Then, the number of zeros of:*
$$P(z) = bz^n - az^m + a - b,$$
strictly inside the unit circle is:
$$\begin{cases} m - \gcd(m,n) & \text{if } a/b \geq n/m, \\ m & \text{if } a/b < n/m. \end{cases}$$

Notice that the trinomial $P(z)$ corresponds to our trinomial for $b = \lambda$, $a = \lambda + \gamma$, $n = T+1$, $m = 1$ and we clearly have $a > b > 0$ and $n > m$. Suppose that $a/b \geq T+1$, i.e., $\gamma/\lambda \geq T$. Then the number of roots in the unit circle is $1 - \gcd(T+1, 1) = 0$. If $\gamma/\lambda < T$, then we have $m = 1$ root in the unit circle.

In the latter case, we can prove that this root is real. In fact, if we consider the derivative of the polynomial at the left-hand side of Eq. (1), we have:
$$\frac{d}{dz}\left(\lambda z^{T+1} - (\lambda + \gamma)z + \gamma\right) = (T+1)\lambda z^T - (\lambda + \gamma),$$

and notice that under the hypothesis $\gamma/\lambda < T$, it admits a minimum between 0 and 1. Then, observe that the polynomial in 0 assumes value γ and in 1 is 0. By continuity, it must have a real root between 0 and 1.

We need to check the balance equation of state T that is different from those of states $1, \ldots, T-1$:
$$\pi(T)\lambda = \pi(T-1)\gamma + \pi(T)^2 \lambda.$$

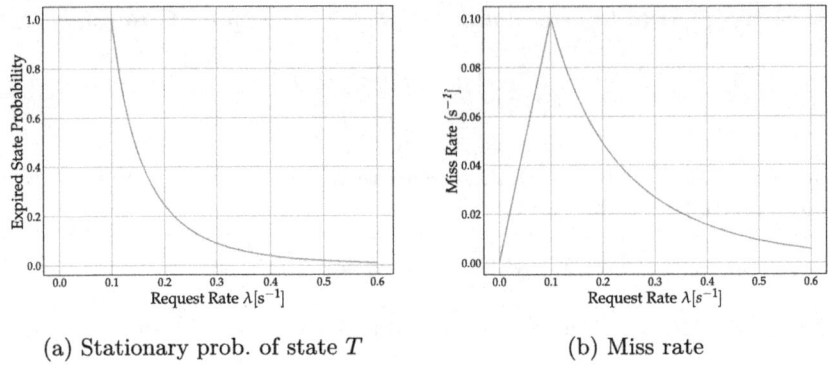

(a) Stationary prob. of state T (b) Miss rate

Fig. 2. Aging object with $T = 10$ and $\gamma = 1\,\mathrm{s}^{-1}$: performance measures as a function of λ. If $\lambda < \gamma/T$, the object is never inserted into the cache in steady state. Otherwise, the miss rate decreases as λ increases.

Dividing both hand sides by z^{T-1}, we have:
$$z\lambda = (1-z)\gamma + z^{T+1}\lambda,$$
which can be rearranged to obtain Eq. (1), as required. □

To illustrate the behavior of a single aging object, Fig. 2 shows the miss probability, $\pi(T)$, and the miss rate, $\pi(T)\lambda$, as a function of λ. The condition $\gamma/\lambda < T$ has a natural interpretation: rewriting it as $\lambda^{-1} < T\gamma^{-1}$ reveals that the average time between requests is shorter than the object's expected lifetime. When this condition is not satisfied, the object is never inserted into the cache in steady state (plateau in Fig. 2(a) and linear increase in Fig. 2(b)). Otherwise, the miss probability and the miss rate decrease as λ increases.

The following corollary is important for the product-form result that will be presented in the next section.

Corollary 1. *The reversed rate of all transitions originating from state T is independent of the destination state and equal to $\pi(T)\lambda$.*

Proof. According to [16], the reversed rate $\overline{q}_{iT}$ of the transition from state T to state i, i.e., the rate from i to T in the reversed process, is given by:
$$\overline{q}_{iT} = \frac{\pi(T)}{\pi(i)} q_{Ti} = \frac{\pi(T)}{\pi(i)} \cdot \lambda \cdot \pi(i) = \pi(T)\lambda. \tag{2}$$

Here, q_{Ti} is the original transition rate from T to i. This concludes the proof. □

Notice that the transitions outgoing from state T are colored in Fig. 1. The rate $\pi(T)\lambda$ is the rate at which we observe a *miss event*, i.e., the expired object is requested. However, the theorem states something stronger than the derivation of the miss rate, i.e., it states that each state of the model can be reached by a cache miss event and that the transition rate from each of these states to T in the reversed process equals the miss rate.

Thanks to Lemma 1, we know that the steady-state distribution of a single aging object depends only on a parameter z which is the minimal real root in $[0, 1]$ of a certain polynomial. We can take advantage of this observation to simplify the aging and rejuvenation mechanism of an object as illustrated in Algorithm 2. We stress on the fact that the role of the simulation is that of suggesting a possible implementation of the aging policy. The code takes advantage of the Poisson Arrivals See Times Average (PASTA) property (see, e.g., [12, Ch. 13]), i.e., the state observed by the request arriving according to a Poisson process has the same statistics of the state seen by a random observer. In this way, the stationary distribution of the object can be inferred only by counting how many hits and misses have occurred until a certain time epoch. The fraction of misses is z^T by Lemma 1 and hence the probabilities of all other states can be retrieved by observing that we have to sample from a truncated geometric distribution (line 15 of Algorithm 2). Notice that the aging events no longer require to keep track of the amount of time spent in each state.

Algorithm 2. Simulation of a single aging object (with exponential assumption)

1: **procedure** SIMULATION(λ, γ, T, $events$)
2:     $hits \leftarrow 0$ ▷ Number of hits
3:     $misses \leftarrow 0$ ▷ Number of misses
4:     $simtime \leftarrow 0$
5:     $fel \leftarrow$ (SAMPLE_EXP(λ), SAMPLE_EXP(γ))   ▷ Future event list
6:     $state \leftarrow 0$
7:     **for** $i \leftarrow 1 \ldots events$ **do**:
8:         **if** $fel[0] < fel[1]$ **then**   ▷ Request event
9:             $simtime \leftarrow fel[0]$
10:            **if** $state < T$ **then**:
11:                $hits \leftarrow hits + 1$
12:                $state \leftarrow 0$ ▷ Refresh the timer
13:            **else**:
14:                $misses \leftarrow misses + 1$
15:                $state \leftarrow \min(\text{SAMPLEGEOMETRIC}((\frac{misses}{misses+hits})^{1/T}), T)$
16:            $fel[0] \leftarrow simtime+$ SAMPLE_EXP(λ)
17:        **else**   ▷ Aging event
18:            $state \leftarrow \text{MIN}(state + 1, T)$
19:            $simtime \leftarrow fel[1]$
20:            $fel[1] \leftarrow simtime+$ SAMPLE_EXP(γ)
        return $times/simtime$

3 Product-Form Analysis

Let us consider a model consisting of N objects whose aging rate is $\gamma_i \geq 0$, $i \in \{1, \ldots, N\}$, and each of which can be in one of the states between 0 and T. If $\gamma_i = 0$, the object is always alive. Let n_i be the state of object i, and λ_i the intensity of the independent Poisson process modeling the requests to object i. When a request is received by object i and its state is $n_i \in \{0, \ldots, T-1\}$, then its state is reset to 0. If the request occurs when $n_i = T$, then the request is instantaneously forwarded to object j with probability $P(i,j)$. Object j receives this request and treats it like an exogenous one; that is, it may in turn trigger another request to a different object. To capture this behavior, we introduce a stochastic matrix $\mathbf{P} = P(i,j)$ $1 \leq i,j \leq N$, referred to as the *routing matrix*. Henceforth, we assume that $\mathbf{P}$ is irreducible and $P(i,i) = 0$ for all $i \in \{0, \ldots, N\}$. The following analysis remains valid even if each object has a different number of states. However, for simplicity of notation, we assume a common T for all objects and use γ_i to differentiate their lifetimes.

We say that an instance of the model is *non-trivial* if at least one object has a strictly positive steady-state probability of being alive. In what follows, we consider such a non-trivial instance. Given the assumption of irreducible routing, any chain of forwarded requests terminates with probability 1 and has finite expected length. To see this, suppose object j has a non-zero steady-state probability of being alive. Then, a request arriving at an expired object will eventually be forwarded to another object, possibly reaching object j. Although object j may be expired at that moment, it is reset according to its stationary distribution, which includes a non-zero probability of being in an alive state. The request may continue through other objects, but once it returns to object j, there is always a non-zero probability that it will be served without further forwarding.

3.1 Preliminaries on Product-Form Analysis

A Markovian model is said to be in product-form if its steady-state distribution can be expressed as normalized product of functions that depend only on a partial information of the state. For example, if we consider a system with N aging objects, we desire to find N functions $g_1, \ldots, g_N$ where g_i depend solely on the state of job i. While there are several ways to find the expressions of g_i (when they exist), we describe the approach introduced in [13,15] because thanks to its modularity, it greatly simplify the proofs.

Consider each aging object in isolation. In principle, we are unable to compute its marginal stationary distribution because it receives some inputs from the other objects of the network that we are unable to characterize statistically. However, under some conditions (see the proof of Theorem 2), we solve a set of possibly non-linear equations, namely the rate equations, which return the rate at which these inputs occur. At this point, we solve the isolated systems *as if* the stochastic processes governing the input events were independent Poisson processes whose rates are given by the solution of the rate equations. The marginal distributions obtained in this way are functions g_i.

Notice that the product-form solution does not require the objects to be statistically independent because the product-form holds only in steady-state (i.e., after a long time since the initial epoch) and we also have to account for inter-dependence captured by the rate equation solutions. The advantage of product-form solutions is clear: instead of solving a Markov chain with very large state space (in our example N^T) we solve N small chains. In our system, the solution of the joint model would require a number of operations $\mathcal{O}(N^{3T})$ while the product-form reduces this to $\mathcal{O}(NT^3)$, assuming a Gaussian elimination approach for the solution of the linear system of equations.

3.2 Product-Form Theorem for Networks of Aging Objects

The main result that we prove in this subsection is the product-form of a network of aging objects that interact as described above.

Theorem 2. *The model with N objects has product-form steady-state distribution*

$$\pi(n_1, \ldots, n_N) = \prod_{i=1}^{N} g_i(n_i). \tag{3}$$

where:

-
$$g_i(n_i) = \begin{cases} (1 - z_i) z_i^{n_i} & \text{if } 0 \leq n_i < T \\ z_i^T & \text{if } n_i = T \end{cases},$$

- z_i *is the minimum real root such that $0 \leq z_i \leq 1$ of the polynomial:*

$$P_i(z) = x_i z^{T+1} - (x_i + \gamma_i) z + \gamma_i, \tag{4}$$

- x_i *is the solution of the following system of rate equations:*

$$x_i = \lambda_i + \sum_{\substack{j=1 \\ j \neq i}}^{N} z_j^T x_j P(j, i). \tag{5}$$

The model is non-trivial if and only if there exists at least one z_i such that $0 \leq z_i < 1$.

Proof. In order to prove the product-form result, we resort to the Reversed Compound Agent Theorem (RCAT) in the version given in [15] where multiway synchronizations among objects are considered. The proofs based on RCAT are modular, i.e., we can study each object in isolation as shown in Fig. 3. Notice that the process underlying a single object is *not* reversible (as, e.g., in Jakson's product-form), and this makes the result more intriguing.

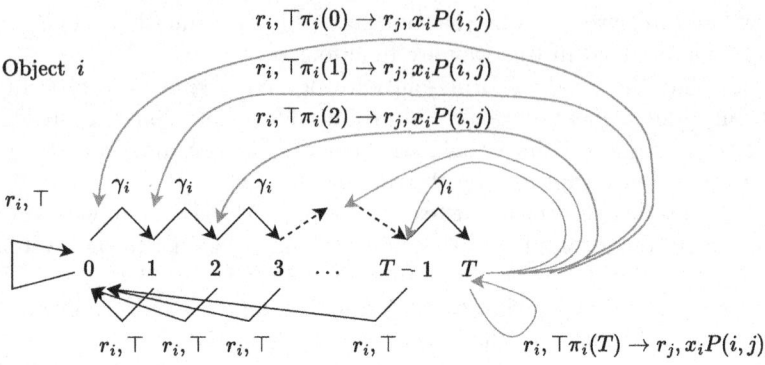

Fig. 3. Synchronizing single aging object. Arrows labeled γ_i represent aging; $r_i, \top$ denote local request handling; $r_j, x_i P(i,j)$ indicate forwarded requests from object i to object j at rate $x_i P(i,j)$.

We have to verify three conditions. (i) Whenever an object receives a request from the outside or from another object, it performs a transition, possibly to the same state it was just before the request. This is modeled in Fig. 3 by the transitions labeled $r_i, \top$ for all states $1, \ldots, T-1$. As for state T, notice that the synchronization is probabilistic, in the sense that the target state is chosen according to the distribution π_i and $\sum_{n=1}^{T} \pi_i(n) = 1$. (ii) The second condition requires that each state must be reachable by one transition that forwards a request to another object. These transitions are colored in purple in Fig. 3 and the condition is easily verified. (iii) The third condition requires all transitions directed to an object j to have the same reversed rate. By Corollary 1, we know that the sum of all reversed rates of the purple transitions going from state T to state n is independent of n and equal to $\pi_i(T) x_i$, where x_i is the rate at which the object receives requests. As shown in [13], the reversed rate of the purple transitions from object i to object j is proportional to their forward rate, i.e., they are also independent of n and equal to $\pi_i(T) x_i P(i,j)$.

Since the three conditions of RCAT are satisfied, the unnormalized stationary probability has product-form and we can use Lemma 1 to obtain the marginal distribution of a single object. Rate Eq. (5) give the parameterization of a single object when isolated from the rest of the network. It remains to prove that the sum of all $\pi(n_1, \ldots, n_N)$ on the state space gives 1. It is sufficient to observe that the joint process is given by the Cartesian product of the state spaces of the single components. This is true because the transitions with rate γ_i are independent and hence if state 0 of object i has positive stationary probability than all other states are reachable independently of other objects. The case $\gamma_i = 0$ is trivial since the only positive recurrent state is 0. If state 0 has zero stationary probability, i.e., it is transient, then the only positive recurrent state of object i is T, therefore it is trivial to observe that the joint state space is the Cartesian product of the recurrent state spaces of the single objects. □

Corollary 2. *A sufficient condition for the model to be non-trivial is that $\gamma/\lambda_i < T$ for at least one $i \in \{1, \ldots, N\}$*

Proof. It is sufficient to notice that $x_i \geq \lambda_i$ for all $i = 1, \ldots, N$. □

3.3 Solution of the System of Rate Equations

In this subsection, we present the solution to the system of rate Eq. (5). As this system is non-linear, we adopt an iterative numerical method to compute its solution. The problem can be formulated as a fixed-point iteration, and it is shown in [14, Thm. 2] that a unique fixed point exists for this system.

The proposed solution is presented in Algorithm 3. Note that the fixed point uniqueness does not guarantee that the iterative scheme shown in Algorithm 3 converges for an arbitrary initial point. Algorithm 3 uses a simple bisection method to find the real root of $f_i(z)$ in the interval $(0,1)$, when such a root exists (line 15). As shown in the proof of Lemma 1, the polynomial $f_i(z)$ defined in line 14 is monotonically decreasing over the interval used as initial bounds for the bisection method. The remainder of the algorithm consists of a simple iterative method over the system of non-linear rate equations.

Algorithm 3. Fixed Point Solver for Multi-Object System with Aging

---
1: **procedure** FIXEDPOINTSOLVER(N, T, $\boldsymbol{\gamma}$, $\boldsymbol{\lambda}$, $\mathbf{P}$, θ)
2:    $\mathbf{x}^{new} \leftarrow \boldsymbol{\lambda}$
3:    $\mathbf{z}^{new} \leftarrow$ array of zeros of size N
4:    $\varepsilon \leftarrow \infty$ ▷ Current precision
5:
6:    **while** $\varepsilon > \theta$ **do**
7:      $\mathbf{x}^{prev} \leftarrow \mathbf{x}^{new}$
8:      $\mathbf{z}^{prev} \leftarrow \mathbf{z}^{new}$
9:      **for** $i \leftarrow 1 \ldots N$ **do** ▷ Step 1: Compute z_i^{new} for each i
10:        **if** $\gamma_i = 0$ **then** ▷ The object does not age
11:          $z_i^{new} \leftarrow 0$
12:        **else**
13:          **if** $x_i > 0$ & $\gamma_i/x_i < T$ **then**
14:            $f_i(z) = x_i^{prev} z^{T+1} - (x_i^{prev} - \gamma_i)z + \gamma_i$
15:            $z_i^{new} \leftarrow$ BISECTIONMETHOD$\left(f_i(z),\ 0,\ \left(\frac{x_i^{prev}+\gamma_i}{x_i^{prev}(T+1)}\right)^{1/T}\right)$
16:          **else** ▷ The object is required rarely
17:            $z_i^{new} \leftarrow 1$
18:      **for** $i \leftarrow 1 \ldots N$ **do** ▷ Step 2: Update x_i^{new} for each i
19:        $x_i^{new} \leftarrow \lambda_i + \sum_{\substack{j=1 \\ j \neq i}}^{N} (z_j^{new})^T \cdot x_j^{prev} \cdot \mathbf{P}(j,i)$
20:      $\varepsilon \leftarrow \max |\mathbf{z}^{new} - \mathbf{z}^{prev}|$
      return $\mathbf{x}, \mathbf{z}$
---

4 Evaluation: TTL Cache Network

In this section, we consider the instantiation of the proposed model in the realm of a CDN with three layers: edge, midgress, and origin server.

We consider a collection of O objects stored in a TTL cache network. Each object o may be requested at any cache and this is either present (hit event) or evicted (miss event). In the case of miss event, the request is forwarded to another cache or to a central repository and the cache decides if it will store the copy that it has just received or not according to the policy described in Sect. 2.

Table 1. Notation

Symbol	Description
$\lambda_{c,o}$	Exogenous request rate at cache c for object o
$x_{c,o}$	Total arrival rate (exogenous + forwarded) at cache c for object o
$z_{c,o}$	Aging parameter (geometric base) for object o at cache c
O_c	Expected number of objects present at cache c (occupancy)
H_c	Hit probability at cache c
D_c	Hop distance: expected number of hops to serve requests arriving at cache c

Every object is always available at the central repository. The repository never forwards requests for object retrieval, noting that this setup does not violate the assumption of ergodic routing in our model. The TTL cache network is shown in Fig. 4. The green boxes in the figure assign the numbers to the cache objects of our model, i.e., $N = 6$.

Notation is summarized in Table 1, noting that subscripts refer to the cache c and the object o. Cache 1 is the repository and never receives requests from the outside, i.e., $\lambda_{1,o} = 0$. Moreover, $\gamma_{1,o} = 0$ since the object must be always available. At the repository, object o always remains in state 0; therefore, the repository never forwards requests to other objects. As a consequence, we can model its routing arbitrarily, e.g., uniformly among the other objects. We assume that the routing matrix $\mathbf{P}$ is object independent.

Caches 2 and 3 are Level 1 (L1) caches; they receive both exogenous requests, i.e., $\lambda_{2,o}, \lambda_{3,o} > 0$, and forwarded requests from Level 2 (L2) caches. In particular, cache 5 forwards miss-induced requests to either cache 2 or cache 3 with equal probability. Suppose an exogenous request arrives at cache 5 for object o. If the object is in state $\{0, \ldots, T-1\}$, the timer is reset to 0 and the request results in a cache hit. If the object is in state T, the request is forwarded to cache 2 or 3 with equal probability 0.5. Assuming forwarding to cache 2, if the object is not cached at cache 2, the request is then forwarded to the central repository, where it is always served (i.e., guaranteed hit). The object is subsequently sent back to cache 2, which may store it with an age sampled according to its stationary distribution (reflecting the request frequency), or decide not to store it at all. The same behavior applies to cache 5.

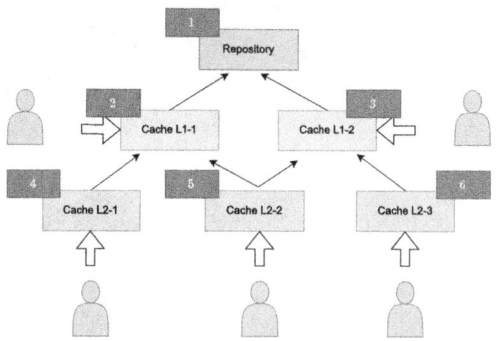

Fig. 4. TTL cache hierarchy with three levels.

In general, requests for object o arrive at cache c according to an independent Poisson process with rate $\lambda_{c,o}$. This corresponds to the well-known *Independent Reference Model*.

Let $x_{c,o}$ and $z_{c,o}$ denote the solution to the rate equation and aging parameter, respectively, for object o at cache c, as defined in Sect. 3. Then, we can compute the following performance metrics.

Occupancy. The expected occupancy of cache c is $O_c = \sum_{o=1}^{O} \left(1 - z_{c,o}^T\right)$.

Hit probability. The overall hit probability (considering both exogenous and forwarded requests) is given by: $H_c = \sum_{o=1}^{O} \left(1 - z_{c,o}^T\right) \cdot x_{c,o} / \sum_{p=1}^{O} x_{c,p}$.

Hop Distance. Let D_c denote the expected number of hops to serve a request arriving at cache c. By the PASTA property, D_c is equal to the steady-state average number of hops to serve such requests. For caches directly connected to the origin server, a hit results in 0 hops, while a miss incurs 1 hop: $D_2 = 1 - H_2$, $D_3 = 1 - H_3$. For cache 4:

$$D_4 = \sum_{o=1}^{O} \left(z_{4,o}^T(1 - z_{2,o}^T) + 2z_{4,o}^T z_{2,o}^T\right) \cdot \frac{x_{4,o}}{\sum_{p=1}^{O} x_{4,p}}. \tag{6}$$

For cache 5 (equally splitting traffic to caches 2 and 3):

$$D_5 = \sum_{c \in \{2,3\}} \frac{1}{2} \sum_{o=1}^{O} \left(z_{5,o}^T(1 - z_{c,o}^T) + 2z_{5,o}^T z_{c,o}^T\right) \cdot \frac{x_{5,o}}{\sum_{p=1}^{O} x_{5,p}}. \tag{7}$$

The value D_6 is symmetric to D_4.

Equations (6) and (7) leverage the statistical independence of request paths, in accordance with the product-form stationary distribution. Indeed, the product-form stationary distribution allows us to multiply miss probabilities along paths. For example, in D_5, if a request at cache 5 misses, it is routed to

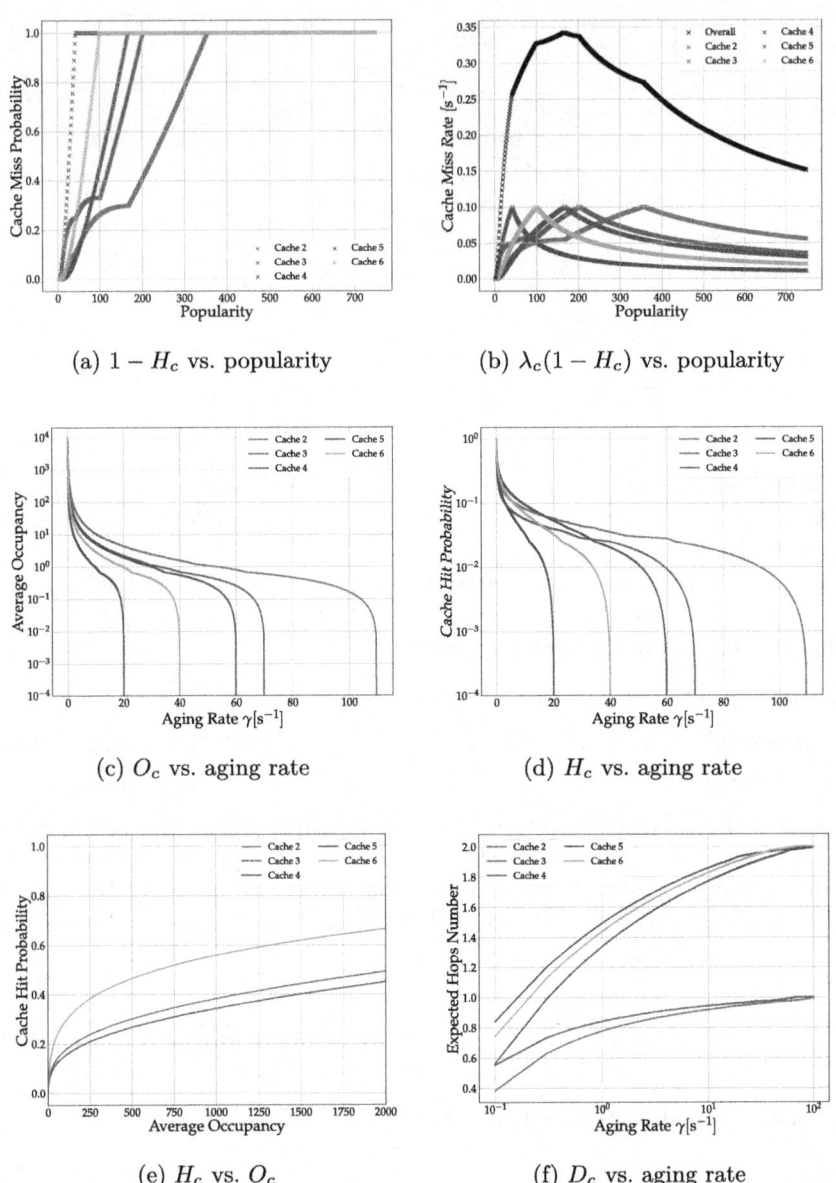

Fig. 5. TTL cache network: performance evaluation

either cache 2 or 3 (each with probability 0.5). Then, the request may cause a hit, which occurs with probability $z_{5,o}^T(1-z_{c,o}^T)$, $c \in \{2,3\}$, or miss again and be forwarded to the central repository, which occurs with probability $z_{5,o}^T z_{c,o}^T$, $c \in \{2,3\}$.

Numerical Evaluation. We consider a system with $10,000$ objects where the popularity of objects follows a Zipf distribution with exponent 0.8, i.e., $\lambda_{c,o} = \lambda_{c,1}/o^{0.8}$ and $\lambda_{2,1} = \lambda_{6,1} = 4s^{-1}$, $\lambda_{3,1} = \lambda_{5,1} = 2s^{-1}$, $\lambda_{4,1} = 6s^{-1}$. In addition, $\lambda_c = \sum_{o=1}^{O} \lambda_{c,o}$. Time is measured in seconds. We assume the same aging rate $\gamma = 1s^{-1}$ for all objects in caches $2, \ldots, 6$ unless otherwise specified.

Figures 5(a) and 5(b) depict the miss probability and miss rate per object across different caches, as a function of the popularity rank (with higher rank indicating lower popularity). Miss probabilities are notably higher at the lower levels of the hierarchy (i.e., L2 caches) compared to L1 caches, especially for contents beyond rank 100. Indeed, L2 caches receive only exogenous requests, and their limited traffic makes it less likely for less popular objects to be refreshed before they expire. In contrast, L1 caches receive exogenous traffic and also benefit from forwarded requests from L2, amplifying the request rate and improving the chance that objects are retained in cache. As a result, L1 caches exhibit significantly lower miss probabilities for a broader range of content.

The total miss rate is skewed toward the head of the popularity distribution. The rank of the content with maximum miss rate increases from L2 to L1 caches—it corresponds to objects that are requested often enough to generate misses, but not frequently enough to remain cached. In L2, this peak occurs at mid-ranked objects due to limited traffic. In L1, additional forwarded requests amplify access to less popular content, shifting the peak to higher ranks. This reflects the trade-off between request frequency and cache retention: very popular items are usually hits, while rare ones contribute little to the miss rate.

Figures 5(c) and 5(d) explore how cache performance is affected by the aging rate γ. As γ increases, the average occupancy of each cache decreases sharply (Fig. 5(c)). Simultaneously, hit probabilities (Fig. 5(d)) decline, especially for deeper caches, as fast aging reduces the likelihood that an object remains in cache between requests. Whereas for low γ all caches show similar occupancy and hit performance, this symmetry breaks under high expiration rates, highlighting the importance of per-cache tuning specially when caching resources are scarce and average occupancy is constrained by costs.

Figure 5(e) shows the relationship between hit probability and average occupancy. The concave shape of the curves confirms the classical "diminishing returns" property of caching: initial increases in occupancy lead to sharp gains in hit probability, but the marginal benefit decreases as more objects are stored.

Finally, Fig. 5(f) presents the expected number of hops to satisfy a request as a function of γ. For L1 and L2 caches, the expected hop count increases monotonically toward 1 and 2, respectively. In the latter case, the two-hop regime becomes dominant for $\gamma = 100$ while low expiration rates result in a significant fraction of requests being resolved locally or at the first forwarding hop.

5 Related Work

In this work, we evaluate the interplay between content routing and content occupancies in networked caches. Such interplay is crucial in the analysis of cache networks [27] and similarity-aware caching [23]. Classical models of cache networks have provided exact [1] and approximate [23,26] characterizations of content dynamics under various policies and topologies. For example, Berger et al. [1] present an exact analysis of TTL cache networks, focusing on how content lifetime and network topology affect performance metrics such as hit probability and latency. Rosensweig et al. [26] propose approximate models for general cache networks, where the impact of routing is captured through heuristics inspired by mean-field approximations.

In the context of similarity caching, Mazziane et al. [23] introduce a TTL-based model that incorporates similarities among items to infer content hit rates when the request for a content can be routed to another similar content in case of a miss. Similar to our approach, both [26] and [23] consider fixed-point formulations in which content occupancies and routed flows are inferred in an interdependent manner—with routing across caches in [26] and across contents in [23]. None of these works, however, consider product-form solutions.

Gast and Van Houdt [9] investigate the transient and steady-state behavior of list-based policies, including TTL variants, but without an explicit notion of similarity-driven routing or routing across network nodes. In this work, we extend these models by showing a simple closed-form solution for networks of TTL caches, where object aging and request redirection coexist under a tractable analytical framework.

Product-form models are widely used for performance engineering of various types of networks as shown, e.g., in [2,17,20,22,25]. Regarding the contribution on product-form networks, we observe that, for infinite capacity queues, the mechanism of resets (jumps to states chosen according to the steady-state distribution) and catastrophes (jumps to state 0) have been studied in the context of G-networks [6,10]. With respect to these results, our model is significantly different thanks to the instantaneous propagation of requests among objects and to their finite state spaces.

6 Conclusion

In this paper, we have presented a product-form model for systems with aging objects. The model belongs to the family of models that include the celebrated G-networks that require the solution of a system of non-linear traffic equations (see, e.g., [5,21]). Therefore, a fixed point iteration scheme in the style of the one presented in [14] must be employed. According to our numerical experiments, we observed that the fixed point computation has always converged in a few iterations but the proof of convergence is left for future work.

It is quite interesting to observe that, in our model, the rate at which a cache i receives the requests for a certain object from cache j is given by the product of three terms: the request rate of the object at cache j, the probability of a cache miss at j, and the routing probability $P_{j,i}$. This is quite intuitive, and previous works use this approach to approximate the analysis of networks of TTL caches by a factorization of the model [4,8,26]. However, our product-form analysis shows one case in which this is not an approximation but the correct result, i.e., when we handle the rejuvenation of an evicted object according to the policy described in Sect. 2.

Beside the analysis of convergence of the fixed point algorithm for the solution of the system of rate equations, we leave as future work the application of the model to the analysis of a real-world scenario and the comparison of our result with those obtained with other TTL policies.

Acknowledgement. This work was partially funded by CAPES, FAPERJ E-26/204.268/2024 and CNPq, Brazil, under grant 403601/2023.

References

1. Berger, D.S., Gland, P., Singla, S., Ciucu, F.: Exact analysis of TTL cache networks. Perform. Eval. **79**, 2–23 (2014)
2. Busic, A., Doncel, J., Fourneau, J.: Dynamic load balancing in energy packet networks. Perform. Eval. **165**, 102414 (2024)
3. Dilcher, K., Nulton, J.D., Stolarsky, K.B.: The zeros of a certain family of trinomials. Glasg. Math. J. **34**, 55–74 (1992)
4. Domingues, G., e Silva, E.D.S., Leão, R.M., Menasché, D.S., Towsley, D.: Enabling opportunistic search and placement in cache networks. Comput. Netw. **119**, 17–34 (2017)
5. Fourneau, J., Gelenbe, E., Surós, R.: G-networks with multiple classes of negative and positive customers. Theor. Comput. Sci. **155**, 141–156 (1996)
6. Fourneau, J., Kloul, L., Quessette, F.: Multiple class G-networks with jumps back to zero. In: 3rd International Workshop on Modeling, Analysis, and Simulation On Computer and Telecommunication (MASCOTS), pp. 28–32 (1995)
7. Garetto, M., Leonardi, E., Martina, V.: A unified approach to the performance analysis of caching systems. ACM Trans. Model. Perform. Eval. Comput. Syst. (TOMPECS) **1**(3), 1–28 (2016)
8. Garetto, M., Leonardi, E., Neglia, G.: Content placement in networks of similarity caches. Comput. Netw. **201**, 108570 (2021)
9. Gast, N., Houdt, B.V.: Transient and steady-state regime of a family of list-based cache replacement algorithms. Queueing Syst. **83**, 293–328 (2016)
10. Gelenbe, E., Fourneau, J.: G-networks with resets. Perform. Eval. **49**, 179–191 (2002)
11. Gias, A.U., Casale, G.: Cocoa: cold start aware capacity planning for function-as-a-service platforms. In: 28th International symposium on Modeling, Analysis, and Simulation of Computer and Telecommunication Systems (MASCOTS), pp. 1–8 (2020)
12. Harchol-Balter, M.: Performance Modeling and Design of Computer Systems: Queueing Theory in Action. Cambridge Press, Cambridge (2013)
13. Harrison, P.G.: Turning back time in Markovian process algebra. Theoret. Comput. Sci. **290**, 1947–1986 (2003)
14. Harrison, P.G., Lee, T.T.: Separable equilibrium state probabilities via time reversal in Markovian process algebra. Theoret. Comput. Sci. **346**, 161–182 (2005)
15. Harrison, P.G., Marin, A.: Product-forms in multi-way synchronizations. Comput. J. **57**(11), 1693–1710 (2014)

16. Kelly, F.: Reversibility and Stochastic Networks. Cambridge Press, Cambridge (1979)
17. Li, Y., Ioannidis, S.: Cache networks of counting queues. IEEE/ACM Trans. Networking **29**(6), 2751–2764 (2021)
18. Liu, Y., et al.: CacheGen: KV cache compression and streaming for fast large language model serving. In: Proceedings of the ACM SIGCOMM 2024 Conference, pp. 38–56 (2024)
19. Maggs, B.M., Sitaraman, R.K.: Algorithmic nuggets in content delivery. ACM SIGCOMM Comput. Commun. Rev. **45**(3), 52–66 (2015)
20. Mahdian, M., Moharrer, A., Ioannidis, S., Yeh, E.: Kelly cache networks. IEEE/ACM Trans. Networking **28**(3), 1130–1143 (2020)
21. Marin, A.: Product-form in G-networks. Probab. Eng. Inf. Sci. **30**, 345–360 (2016)
22. Marin, A., Meo, M., Sereno, M., Ajmone Marsan, M.: Queuing network models of multiservice RANs. ACM Trans. Model. Perform. Eval. Comput. Syst. **9**, 7:1–7:26 (2024)
23. Mazziane, Y.B., Alouf, S., Neglia, G., Menasche, D.S.: TTL model for an LRU-based similarity caching policy. Comput. Netw. **241**, 110206 (2024)
24. Neglia, G., Carra, D., Feng, M., Janardhan, V., Michiardi, P., Tsigkari, D.: Access-time-aware cache algorithms. ACM Trans. Model. Perform. Eval. Comput. Syst. (TOMPECS) **2**(4), 1–29 (2017)
25. Olliaro, D., Balbo, G., Marin, A., Sereno, M.: Computational algorithms and arrival theorem for non-conventional product-form solutions. Perform. Eval. **168**, 102469 (2025)
26. Rosensweig, E.J., Kurose, J., Towsley, D.: Approximate models for general cache networks. In: 2010 Proceedings IEEE INFOCOM, pp. 1–9 (2010)
27. Sundarrajan, A., Kasbekar, M., Sitaraman, R.K., Shukla, S.: Midgress-aware traffic provisioning for content delivery. In: 2020 USENIX Annual Technical Conference (USENIX ATC 2020), pp. 543–557 (2020)
28. Wu, J., Elango, V., Zamarin, S., Trikande, S.: Achieve High Performance at Scale for Model Serving Using Amazon SageMaker Multi-Model Endpoints with GPU (2023). AWS Machine Learning Blog. https://aws.amazon.com/blogs/machine-learning/achieve-high-performance-at-scale-for-model-serving-using-amazon-sagemaker-multi-model-endpoints-with-gpu/

Tightening the Frontier of Decidability for Decisiveness

Gaspard Fougea[1], Serge Haddad[1], Lina Ye[2(✉)], Shreyas Jain[3], and Alain Finkel[1]

[1] Université Paris-Saclay, CNRS, ENS Paris-Saclay, LMF, Gif-sur-Yvette, France
{gaspard.fougea,serge.haddad,alain.finkel}@lmf.cnrs.fr
[2] Université Paris-Saclay, CentraleSupelec, CNRS, ENS Paris-Saclay, LMF, Gif-sur-Yvette, France
lina.ye@centralesupelec.fr
[3] Indian Institute of Science Education and Research (IISER), Mohali, India
jshreyas@vt.edu

Abstract. While computing reachability probabilities in an infinite Mar-kov chain is a challenging problem, computing an approximation up to an arbitrary precision is possible when the chain is *decisive*. Some high-level probabilistic formalisms lead to decisive Markov chain but for most of them, the generated chain can be decisive or not depending on the specification. This raises the problem of deciding decisiveness given a high-level probabilistic formalism. In a previous work, some of the authors of this paper have studied the decisiveness status of several formalisms. Here we improve their work in two ways. It was shown that the decisiveness problem was decidable for homogeneous probabilistic one-counter machine (an extension of the quasi-birth-death processes) when some underlying finite Markov chain is irreducible. We show the problem remains decidable without requiring irreducibility. On the other side, it was shown that the decisiveness problem was undecidable for probabilistic Petri nets with polynomial weights for transitions. We show that the problem remains undecidable even when the weights of the transitions are specified by affine functions of a single place marking.

Keywords: infinite Markov chains · reachability probability · decisiveness

1 Introduction

Reachability in Markov Chains. With the development of quantitative verification and more specifically probabilistic verification, Markov chains have become one of the main low-level formalism for the modeling and analysis of probabilistic systems [14]. While computing reachability probabilities in finite

---

This work of Serge Haddad and Lina Ye has been supported by ANR project MAVeriQ (ANR-20-CE25-0012).

© The Author(s), under exclusive license to Springer Nature Switzerland AG 2026
P. Prabhakar and A. Vandin (Eds.): QEST+FORMATS 2025, LNCS 16143, pp. 237–255, 2026.
https://doi.org/10.1007/978-3-032-05792-1_13

Markov chains can be efficiently performed using linear algebra techniques such a computation remains a challenging problem in infinite Markov chains. More precisely the problem of *Computing the Reachability Probability up to an arbitrary precision* (CRP) is a central problem in quantitative verification and it has been studied by many authors [1,5,11,13].

CRP Problem. The CRP problem can be addressed in two opposite ways. First one studies the particular features of a given high-level formalism in order to design an ad-hoc CRP-algorithm. For instance in [5], the authors propose several algorithms for probabilistic pushdown automata and some of their restrictions. Second one exhibits a property of Markov chains that yields a generic algorithm for solving the CRP problem and then one looks for high-level formalisms fulfilling this property. For instance a Markov chain is *decisive* w.r.t. an initial state and a target set if almost surely a random path either reaches the target or the target becomes unreachable. Probabilistic lossy channel systems are decisive and probabilistic Petri nets (pPN) are decisive when the target set is upward-closed [1]. Another generic property that complements in some way decisiveness and also yields an algorithm for the CRP problem is the *divergence* of Markov chains [8].

Static Versus Dynamic Weights. In most of the works, the probabilistic models associate a constant (also called *static*) weight for transitions and get transition probabilities by normalizing these weights among the enabled transitions in the current state (except for some semantics of pLCS like in [9] where transition probabilities depend on the state due to the possibility of message losses). This forbids to model phenomena like congestion in networks (resp. performance collapsing in distributed systems) when the number of messages (resp. processes) exceeds some threshold leading to an increasing probability of message arrivals (resp. process creations) before message departures (resp. process terminations). In order to handle them, one needs to consider *dynamic* weights i.e., weights depending on the current state. More precisely, very often these weights are specified by polynomials over some state indices. For instance,

- in queuing networks, the policy of a server may be the infinite server policy leading to a linear weight;
- in biological and epidemiological models, the rate of some "synchronization" between two instances of some species is quadratic w.r.t. the size of the species.

Decisiveness in Presence of Dynamic Weights. With dynamic weights (even polynomial) whatever the high-level probabilistic formalism, decisiveness cannot ensured for all models of this formalism and thus raises the issue of the decidability status of the decisiveness problem depending on the formalism. In [6] some of the authors study (un)decidability of several formalisms and furthermore provide syntactical restrictions on some others that ensure decisiveness. In particular, they introduce probabilistic homogeneous counter machines (pHM) with polynomial weights which can be viewed both as restricted one-counter machines or extended quasi-birth-death processes [12]. They show that decisiveness in pHM is decidable in polynomial time. On the other side they show that in probabilistic Petri nets decisiveness become undecidable.

Our Contributions. However the two previous results have some limitations that we remove here:

- The decidability of decisiveness in pHM is based on the existence of a finite Markov chain which is supposed to be irreducible. We show that this requirement can be omitted and that the CRP problem is still being solvable in polynomial time;
- In probabilistic Petri nets the reduction that establishes the undecidability result uses polynomials of degree 4. We show that the CRP problem remains undecidable even when the weights are affine functions.

Organization. Section 2 recalls basic notions and decisive Markov chains. In Sect. 3, we study the decidability status of decisiveness for general pHMs. In Sect. 4, the indecidability status of decisiveness for pPNs is generalized from polynomial to affine weights. Finally in Sect. 5, we conclude and give some perspectives to this work.

2 Decisiveness

As usual, $\mathbb{N}$ and $\mathbb{N}^*$ denote respectively the set of non negative integers and the set of positive integers. The notations $\mathbb{Q}$, $\mathbb{Q}_{\geq 0}$ and $\mathbb{Q}_{>0}$ denote the set of rationals, non-negative rationals and positive rationals. Let $F \subseteq E$; when there is no ambiguity about E, $\overline{F}$ will denote $E \setminus F$.

2.1 Markov Chains: Definitions and Properties

A set S is *countable* if there exists an injective function from S to $\mathbb{N}$: hence it could be finite or countably infinite. Let S be a countable set of elements called states. Then $Dist(S) = \{\Delta : S \to \mathbb{Q}_{\geq 0} \mid \sum_{s \in S} \Delta(s) = 1\}$ is the set of *rational distributions* over S. Let $\Delta \in Dist(S)$, then $Supp(\Delta) = \Delta^{-1}(\mathbb{Q}_{>0})$.

Definition 1 (Effective Markov chain). *A Markov chain $\mathcal{M} = (S, p)$ is a tuple where:*

- *S is a countable set of states;*
- *p is the transition function from S to $Dist(S)$.*

When for all $s \in S$, $Supp(p(s))$ is finite with both $Supp(p(s))$ and the function $s \mapsto p(s)$ being computable, one says that $\mathcal{M}$ is effective.

When S is infinite, we say that $\mathcal{M}$ is *infinite* and we sometimes identify S with $\mathbb{N}$. We also denote $p(s)(s')$ by $p(s, s')$ and $p(s, s') > 0$ by $s \xrightarrow{p(s,s')} s'$. A Markov chain is also viewed as a transition system $(S, \to)$ whose transition relation $\to$ is defined by $s \to s'$ if $p(s, s') > 0$.

Example 1. Let $\mathcal{M}_1$ be the Markov chain of Fig. 1. In any state $i > 0$, the probability for going to the "right", $p(i, i+1) = \frac{f(i)}{f(i)+g(i)}$ and for going to the "left", $p(i, i-1) = \frac{g(i)}{f(i)+g(i)}$. In state 0, one goes to 1 with probability 1. $\mathcal{M}_1$ is effective if the functions f and g are computable.

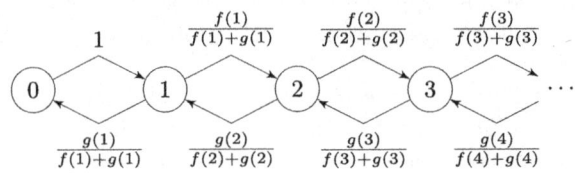

Fig. 1. A Markov chain $\mathcal{M}_1$ with for all $n \in \mathbb{N}$, $0 < f(n)$ and $0 < g(n)$.

We denote $\rightarrow^*$, the reflexive and transitive closure of $\rightarrow$ and we say that s' is *reachable from* s if $s \rightarrow^* s'$. We say that a subset $A \subseteq S$ is *reachable* from s if some $s' \in A$ is reachable from s and we denote $s \rightarrow^* A$.

Given an initial state s_0, a *run* of a Markov chain $\mathcal{M}$ is an *infinite random sequence of states* (i.e., a path) $\sigma = s_0 s_1 \ldots$ such that for all $i \geq 0$, $s_i \rightarrow s_{i+1}$. As usual, the corresponding σ-algebra is generated by the finite prefixes of infinite paths and the probability of a measurable subset Υ of infinite paths, given an initial state s_0, is denoted $\mathbf{Pr}_{\mathcal{M},s_0}(\Upsilon)$. In particular denoting $s_0 \ldots s_n S^\omega$ the set of infinite paths with $s_0 \ldots s_n$ as prefix, $\mathbf{Pr}_{\mathcal{M},s_0}(s_0 \ldots s_n S^\omega) = \prod_{0 \leq i < n} p(s_i, s_{i+1})$.

From now on, $\mathbf{G}$ (resp. $\mathbf{F}$, $\mathbf{X}$, $\mathbf{E}$) denotes the always (resp. eventual, next, existential) operator of temporal logics. Let $A \subseteq S$. We say that σ reaches A if $\exists i \in \mathbb{N}$ $s_i \in A$ which corresponds to $\sigma \models \mathbf{F}A$. Similarly $\sigma \models \mathbf{X}\mathbf{F}A$ if $\exists i \in \mathbb{N}^*$ $s_i \in A$. The probability that starting from s_0, the path σ reaches A is thus denoted by $\mathbf{Pr}_{\mathcal{M},s_0}(\mathbf{F}A)$.

The next definition states some properties of a Markov chain.

Definition 2 (Irreducibility, recurrence, transience). *Let $\mathcal{M} = (S, p)$ be a Markov chain and $s \in S$. Then:*

- *$\mathcal{M}$ is* irreducible *if for all $s, s' \in S$, $s \rightarrow^* s'$;*
- *s is* recurrent *if $\mathbf{Pr}_{\mathcal{M},s}(\mathbf{X}\mathbf{F}\{s\}) = 1$ otherwise s is* transient.

The next proposition states that in an irreducible Markov chain, all states are in the same category.

Proposition 1 ([10]). *Let $\mathcal{M} = (S, p)$ be an irreducible Markov chain and $s, s' \in S$. Then s is recurrent if and only if s' is recurrent.*

Thus an irreducible Markov chain will be said transient or recurrent depending on the category of its states (all states are in the same category). In the remainder of this section, we will relate this category with techniques for computing reachability probabilities.

Example 2. $\mathcal{M}_1$ of Fig. 1 is irreducible. Let us define $p_n = \frac{f(n)}{f(n)+g(n)}$. Then (see [7] for more details), $\mathcal{M}_1$ is recurrent if and only if $\sum_{n\in\mathbb{N}} \prod_{1\le m<n} \rho_m = \infty$ with $\rho_m = \frac{1-p_m}{p_m}$, and when transient, the probability that starting from i the random path reaches 0 is equal to $\frac{\sum_{i\le n} \prod_{1\le m<n} \rho_m}{\sum_{n\in\mathbb{N}} \prod_{1\le m<n} \rho_m}$.

2.2 Decisive Markov Chains

One of the goals of the quantitative analysis of infinite Markov chains is to approximately compute reachability probabilities. Let us formalize it. Given a finite representation of a subset $A \subseteq S$, one says that this representation is *effective* if one can decide the membership problem for A. With a slight abuse of language, we identify A with any effective representation of A.

The Computing of Reachability Probability (CRP) problem

> - Input: an effective Markov chain $\mathcal{M}$, an (initial) state s_0, an effective subset of states A, and a rational $\theta > 0$.
> - Output: an interval $[low, up]$ such that $up - low \le \theta$ and $\mathbf{Pr}_{\mathcal{M}, s_0}(\mathbf{F}A) \in [low, up]$.

In finite Markov chains, there is a well-known algorithm for exactly computing the reachability probabilities in polynomial time [2]. In infinite Markov chains, *decisiveness* is one way to address the CRP problem [1]. Given an initial state and a target set, a Markov chain is decisive if almost surely a random path reaches either the target or a state from which the target becomes unreachable.

Definition 3. *A Markov chain $\mathcal{M}$ is decisive w.r.t. $s_0 \in S$ and $A \subseteq S$ if:*

$$\mathbf{Pr}_{\mathcal{M}, s_0}(\mathbf{G}(\overline{A} \cap \mathbf{EF}A)) = 0$$

Notation. When $\mathcal{M}$ and A are fixed, one says that $s_0 \in S$ is decisive if $\mathcal{M}$ is decisive w.r.t. s_0 and A.

Under the hypotheses of decisiveness w.r.t. s_0 and A and decidability of the reachability problem w.r.t. A, an algorithm to solve the CRP problem has been proposed, whose correctness is formally proved in [1]. This algorithm explores the computation tree of the Markov chain, maintaining an interval including the exact reachability probability. It increases the lower bound when reaching the target set and decreases the upper bound when reaching the state from which the target set is not reachable. Finally it stops when the interval is smaller than θ. In [6], a strong relation was established between decisiveness and recurrence.

3 Homogeneous One-Counter Machines

In this section, we introduce and study a probabilistic version of one-counter machine with dynamic weights (which can also viewed as an extension of quasi-birth death processes). Let $\mathcal{X} = \{X_1, \ldots, X_k\}$ be a set of variables. A polynomial

P over $\mathcal{X}$ is defined by: $P = \sum_{\alpha_1,\ldots,\alpha_k \in \mathbb{N}} a_{\alpha_1,\ldots,\alpha_k} X_1^{\alpha_1} \cdots X_k^{\alpha_k}$ such that the set $\{\alpha_1,\ldots,\alpha_k \mid a_{\alpha_1,\ldots,\alpha_k} \neq 0\}$ is finite. In the sequel, we only consider integer coefficients and say that P is *positive* if $a_{0,\ldots,0} > 0$ and for all $\alpha_1,\ldots,\alpha_k \neq 0,\ldots,0$, $a_{\alpha_1,\ldots,\alpha_k} \geq 0$. When $\mathcal{X} = \{X\}$ one says that P is a polynomial in X.

We introduce here a probabilistic version of a one-counter machine. The transitions of this model are equipped with positive polynomials, where the variable is intended to be the value of the counter. Its semantics is defined as a countable Markov chain, where the states are the configurations of the machine and the probabilities of transitions are obtained by evaluating the polynomials w.r.t the current value of the counter and normalizing this value in order to get a probability distribution.

Definition 4 (p1CM). *A probabilistic one-counter machine (p1CM) is a tuple $\mathcal{C} = (Q, \Delta, W)$ where:*

- *Q is a finite set of control states;*
- *Δ is a finite subset of $Q \times \{>, =\} \times \{-1, 0, 1\} \times Q$ such that for all $(q, =, \delta, q') \in \Delta$, $\delta \geq 0$ and for all $q \in Q$, there exist some $(q, =, \delta', q'), (q, >, \delta'', q'') \in \Delta$;*
- *For all $t = (q, >, \delta, q') \in \Delta$, $W(t)$ is a positive polynomial in X and for all $t = (q, =, \delta, q') \in \Delta$, $W(t)$ is a positive integer.*

Observation. Quasi-birth death processes are simply p1CM with constant instead of polynomials.

Let $\mathcal{C}$ be a 1pCM. For all $q \in Q$, let $S_{q,\bowtie,\delta} = \sum_{t=(q,\bowtie,\delta,q') \in \Delta} W(t)$, $S_{q,\bowtie,q'} = \sum_{t=(q,\bowtie,\delta,q') \in \Delta} W(t)$, $S_{q,\bowtie} = \sum_{q' \in Q} S_{q,\bowtie,q'}$, and $\mathbf{M}_{\mathcal{C},\bowtie}$ be the $Q \times Q$ matrix defined by:

$$\mathbf{M}_{\mathcal{C},\bowtie}[q,q'] = \frac{S_{q,\bowtie,q'}}{S_{q,\bowtie}}.$$

Thus $\mathbf{M}_{\mathcal{C},>}[q,q']$ is a rational function (with variable X) from $\mathbb{N}$ to $\mathbb{Q}_{\geq 0}$ and $\mathbf{M}_{\mathcal{C},=}$ is a transition matrix.

Definition 5 (pHM). *A probabilistic homogeneous machine (pHM) is a p1CM $\mathcal{C} = (Q, \Delta, W)$ such that for all $q, q' \in Q$, $\mathbf{M}_{\mathcal{C},>}[q,q']$ is constant.*

Thus in a pHM, $\mathbf{M}_{\mathcal{C},>}$ is a transition matrix specifying the change of control states regardless of the counter value, when positive.

Notation. A transition $t = (q, \bowtie, \delta, q')$ is said a $\bowtie$-transition and also denoted $q \xrightarrow{W(t),\bowtie,\delta} q'$.

Example 3. The figure below depicts all $>$-transitions outgoing from q in some pHM. Here $\mathbf{M}_{\mathcal{C},>}[q,q'] = \mathbf{M}_{\mathcal{C},>}[q,q''] = \frac{X^2+X+1}{2(X^2+X+1)} = \frac{1}{2}$ fulfilling the homogeneous requirement.

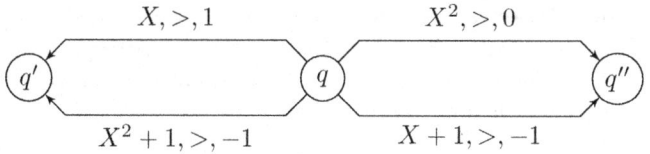

A *configuration* of $\mathcal{C}$ is an element of $Q \times \mathbb{N}$. Let $s = (q, \nu)$ be a configuration and $t = (q^-, \bowtie, \delta, q^+)$ be a transition in Δ. Then t is *enabled* in s if $\nu \bowtie 0$ and $q = q^-$; its *firing* leads to the configuration $s' = (q^+, \nu + \delta)$. One denotes the configuration change by: $s \xrightarrow{t} s'$. One denotes $En(s)$, the set of transitions enabled in s. From the definition of p1CM, $En(s)$ is always non empty. If $\nu = 0$ then let $Weight(s) = \sum_{t \in En(s)} W(t)$ else let $Weight(s) = \sum_{t \in En(s)} W(t)(\nu)$. Let $\sigma = t_1 \ldots t_n$ be a sequence of transitions. We define the enabling and the firing of σ by induction. The empty sequence is always enabled in s and its firing leads to s. When $n > 0$, σ is enabled if $s \xrightarrow{t_1} s_1$ and $t_2 \ldots t_n$ is enabled in s_1. The firing of σ leads to the configuration reached by $t_2 \ldots t_n$ from s_1. A configuration s is *reachable* from some s_0 if there is a firing sequence σ that reaches s from s_0. When Q is a singleton, one omits the control state in the definition of transitions and configurations.

We now provide the semantic of a p1CM as a countable Markov chain.

Definition 6. *Let $\mathcal{C} = (Q, \Delta, W)$ be a p1CM. Then the Markov chain $\mathcal{M}_\mathcal{C} = (S, p)$ is defined by:*

- *$S = Q \times \mathbb{N}$;*
- *For all $s = (q, \nu) \in S$, if $\nu = 0$ then $p(s, s') = Weight(s)^{-1} \sum_{s \xrightarrow{t} s'} W(t)$; Otherwise for all $s' \in S$: $p(s, s') = Weight(s)^{-1} \sum_{s \xrightarrow{t} s'} W(t)(\nu)$.*

Example 4. Let us describe a possible transition related to the figure in Example 3 given a configuration (q, ν) with $\nu > 0$, the probability to go to $(q', \nu + 1)$ is equal to $\frac{\nu}{2(\nu^2 + \nu + 1)}$.

We recall two useful results for establishing our main theorem.

Theorem 1 (Thm. 14 [6]). *The decisiveness problem w.r.t. s_0 and finite A for p1CM $\mathcal{C}$ with a single state is decidable in linear time.*

As can be deduced from the next definition, computing the d_q's allows to decide decisiveness w.r.t. any initial configuration and target $Q \times \{0\}$.

Definition 7. *Let $\mathcal{C}$ be a pHM. Then for all $q \in Q$, let $d_q \in \mathbb{N} \cup \{\infty\}$ be defined by:*

$$d_q = sup\{k \mid \mathcal{C} \text{ is decisive w.r.t } (q, k) \text{ and } Q \times \{0\}\}.$$

Notations and Observation. Since we are interested in decisiveness w.r.t. $Q \times \{0\}$, w.l.o.g. we assume that for all $q, q' \in Q$, $p((q,0),(q',0)) = \mathbf{M}_{\mathcal{C},>}[q,q']$ meaning that the counter always remains null and that the change between states of Q does not depend on the counter value.

Thus for any path starting from (q,k) with $k > 0$ reaching $(q',0)$, there is a path with same probability starting from $(q, k-1)$ reaching $(q',0)$. As a consequence, if (q,k) is decisive then for all $k' \leq k$, (q,k') is decisive. So $k \leq d_q$ if and only if (q,k) is decisive meaning that computing $(d_q)_{q \in Q}$ solves the decisiveness problem.

In order to alleviate the notations, $\mathbf{M}_{\mathcal{C},>}$ will be simply denoted $\mathbf{M}$ and it defines an underlying graph $G_\mathbf{M}$ where $q \to q'$ if $\mathbf{M}[q,q'] > 0$. Similarly $S_{q,>}$ (resp. $S_{q,>,q'}$, $S_{q,>,\delta}$) will be more simply denoted S_q (resp. $S_{q,q'}$, $S_{q,\delta}$).

The next definition does not depend on the polynomial weights.

Definition 8. *Let $\mathcal{C}$ be a p1CM and $q \in Q$. Then $r_q \in \mathbb{N} \cup \{\infty\}$ is defined by $r_q = \sup\{k \in \mathbb{N} \mid Q \times \{0\} \text{ is reachable from } (q,k)\}$.*

Proposition 2 (Prop. 17 [6]). *Let $\mathcal{C}$ be a p1CM. Then (1) for all q, $r_q \in \{0, \ldots, |Q|-1\} \cup \{\infty\}$, (2) one can compute in polynomial time r_q, and (3) $Q \times \{0\}$ is reachable from (q,k) iff $k \leq r_q$.*

We introduce two subsets: $Q_f = \{q \in Q \mid r_q < \infty\}$ and $Q_\infty = Q \setminus Q_f$.
First we present some results about paths between states in $G_\mathbf{M}$.

Lemma 1. *Let $q \to^* q'$ with $q \in Q_f$. Then $q' \in Q_f$.*

Proof. Using induction, it is enough to prove that if $q \xrightarrow{>,\delta} q'$ with $q \in Q_f$ then $q' \in Q_f$. Let $k > \max(r_q, -\delta)$. Then in $\mathcal{C}$, $(q,k) \to (q', k+\delta)$. Since $k > r_q$ this implies that $Q \times \{0\}$ is unreachable from $(q', k+\delta)$. Thus $r_{q'} < k + \delta$ and so $q' \in Q_f$. □

Due to this result, any strongly connected component of $G_\mathbf{M}$ is either included in Q_f or Q_∞. Let Q^B be the subset of bottom strongly connected components (BSCC) of $G_\mathbf{M}$ and for $X \subseteq Q$, let $X^B = X \cap Q^B$.

Lemma 2. *Let $q \to^* q'$ with $d_q = \infty$. Then $d_{q'} = \infty$.*

Proof. Using induction, it is enough to prove that if $q \xrightarrow{>,\delta} q'$ with $d_q = \infty$ then $d_{q'} = \infty$. Let $k > 1$, then $(q, k-\delta) \to (q', k)$ which implies that (q',k) is decisive. Since k is arbitrary one gets $d_{q'} = \infty$. □

Due to this result, in any strongly connected component of $G_\mathbf{M}$ either for all q, $d_q < \infty$ or for all q, $d_q = \infty$.

In the two next lemmas we establish particular cases of the main theorem.

Lemma 3. *Assume that $Q_f = Q$ and let $(q_0, k_0) \in Q \times \mathbb{N}$. Then $\mathcal{C}$ is decisive w.r.t. (q_0, k_0) and $Q \times \{0\}$ and so $d_{q_0} = \infty$.*

Proof. Since $r_{q_0} < \infty$, for all $k_0 > r_{q_0}$, (q_0, k_0) cannot reach $Q \times \{0\}$ and thus (q_0, k_0) is decisive. Now consider any configuration (q, k) with $k \leq r_q$. By definition there is a positive probability say $p_{(q,k)}$ to reach $Q \times \{0\}$ from (q, k) by considering a shortest path. Let $p_{\min} = \min(p_{(q,k)} \mid q \in Q \wedge k \leq r_q)$. Then for all (q, k) with $q \in Q, k \leq r_q$, there is a probability at least $p_{\min}$ to reach $Q \times \{0\}$ and thus to reach either $Q \times \{0\}$ or $\{(q, k) \mid q \in Q \wedge k > r_q\}$ by a path of length at most $\ell = \sum_{q \in Q}(r_q + 1)$. This implies that after $n\ell$ transitions the probability to reach either $Q \times \{0\}$ or $\{(q, k) \mid q \in Q \wedge k > r_q\}$ is at least $1 - (1 - p_{\min})^n$. Thus (q_0, k_0) is decisive. Summarizing for all (q_0, k_0),(q_0, k_0) is decisive. □

Lemma 4. *Assume that* $\mathbf{M}$ *is irreducible and* $Q_\infty = Q$. *Then* $(d_q)_{q \in Q}$ *can be computed in polynomial time.*

Proof. Since for all $(q, k) \in Q \times \mathbb{N}$, $Q \times \{0\}$ is reachable from (q, k), the decisiveness problem boils down to the almost sure reachability of $Q \times \{0\}$.
Since $\mathbf{M}$ is irreducible, there is a unique invariant distribution π_∞ (i.e., $\pi_\infty \mathbf{M} = \pi_\infty$) fulfilling for all $q \in Q$, $\pi_\infty(q) > 0$.
Let $(Q_n, N_n)_{n \in \mathbb{N}}$ be the stochastic process defined by $\mathcal{M}_\mathcal{C}$ with $N_0 = k$ for some k and for all $q \in Q$, $\mathbf{Pr}(Q_0 = q) = \pi_\infty(q)$. Due to the invariance of π_∞ and the choice of transitions for $Q \times \{0\}$, one gets by induction that for all $n \in \mathbb{N}$:

- $\mathbf{Pr}(Q_n = q) = \pi_\infty(q)$;
- for all $k > 0$, $\mathbf{Pr}(N_{n+1} = k + \delta \mid N_n = k) = \sum_{q \in Q} \pi_\infty(q) \frac{S_{q,\delta}(k)}{S_q(k)}$
$= \frac{\sum_{q \in Q} \pi_\infty(q) S_{q,\delta}(k) \prod_{q' \neq q} S_{q'}(k)}{\prod_{q' \in Q} S_{q'}(k)}$;
- $\mathbf{Pr}(N_{n+1} = 0 \mid N_n = 0) = 1$.

For $\delta \in \{-1, 0, 1\}$, let us define the polynomial P_δ by:

$$\sum_{q \in Q} \pi_\infty(q) \left(\prod_{q' \neq q} S_{q'}\right) \sum_{t=(q,>,\delta,q') \in \Delta_1} W(t)$$

Due to the previous observations, the stochastic process $(N_n)_{n \in \mathbb{N}}$ is the Markov chain defined below where the weights outgoing from a state have to be normalized:

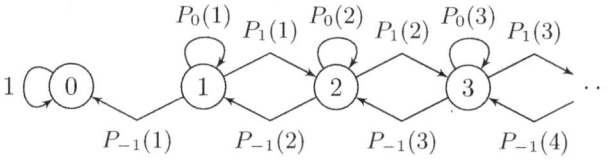

Using our hypothesis about reachability, P_{-1} is a positive polynomial (while P_1 could be null) and thus the decisiveness of this Markov chain w.r.t. state 0 is equivalent to the decisiveness of the Markov chain below:

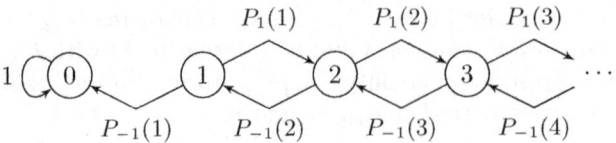

Due to Theorem 1, this problem is decidable (in linear time) and either (1) for all $k \in \mathbb{N}$ this Markov chain is decisive w.r.t k and 0 or (2) for all $k > 0$ this Markov chain is not decisive w.r.t k and 0. Let us analyze the two cases w.r.t. the Markov chain of the pHM.

Case (1) In the stochastic process $(Q_n, N_n)_{n \in \mathbb{N}}$, the initial (stationary) distribution has a positive probability for (q, k) for all $q \in Q$. This implies that for all q, (q, k) is decisive. Since k was arbitrary, this means that for all $(q, k) \in Q \times \mathbb{N}$, (q, k) is decisive.

Case (2) Choosing $k = 1$ and applying the same reasoning as for the previous case, there is some $(q, 1)$ which is not decisive (and so for all (q, k') with $k' > 0$). Let $q' \in Q$, since **M** is irreducible, consider a shortest sequence of $>$-transitions in Δ leading from q' to q whose length is at most $|Q| - 1$. Thus, for all (q', k') with $k' \geq |Q|$ there is a positive probability to reach some (q, k) with $k > 0$ and so (q', k) can be decisive only if $k < |Q|$. Therefore, only configurations included in $Q \times [0, |Q|[$ can be decisive. Define the set of decisive configurations to be $T = \{(q, k) \in Q \times \mathbb{N} \mid (q, k) \text{ is decisive}\}$. Then, Algorithm 1 computes d_q for each $q \in Q$. Let us describe it and prove its correctness:

- **Description.** The algorithm starts off with the set of all configurations which are possibly decisive by initialising $(x_q)_{q \in Q}$ to be $|Q| - 1$ since all configurations (q, k) with $k \geq |Q|$ are undecisive. It decreases x_q whenever there is a transition $(q, x_q) \to (q', x_q + \delta)$ with $x_q + \delta > x_{q'}$ since $(q', x_q + \delta)$ is undecisive and stops when there is no more such transitions.
- **Termination.** Since $|S| < \infty$, and each iteration (except the last one) decreases the size of S, the algorithm terminates.
- **Loop Invariant.** The invariant is : $T \subseteq S$. Initially $S = Q \times [0, |Q|[$ and as discussed above $T \subseteq S$. The 'if' statement decreases x_q if there exists a transition $q \xrightarrow{\delta} q'$ such that $x_q > x_{q'} - \delta$, which means that there is a positive probability to reach $(q', x_q + \delta)$ with $x_q + \delta > x_{q'}$ (i.e. a configuration out of T due to the invariant) from (q, x_q). So $T \subseteq S$ remains true after any iteration.
- **Correctness.** When it terminates, every transition outgoing from $s \in S$ leads to a state $s' \in S$. So $\mathcal{M}_C$ restricted to S is a finite Markov chain. Since $r_q = \infty$ for all $q \in Q$, from any $s \in S$ one almost surely reaches $Q \times \{0\}$ and thus s is decisive.

□

Tightening the Frontier of Decidability for Decisiveness

Algorithm 1: Computing d_q when the random walk is undecisive

for $q \in Q$ do $x_q \leftarrow |Q| - 1$
$S \leftarrow \{(q, k) \mid q \in Q, k \leq x_q\}$
do
$\quad$ $oldS \leftarrow S$; for $q \in Q$ do $oldx_q \leftarrow x_q$
$\quad$ for $(q, >, \delta, q') \in \Delta$ do
$\quad\quad$ | if $x_q > 0$ and $x_q + \delta > oldx_{q'}$ then $x_q \leftarrow x_q - 1$
$\quad$ **end**
$\quad$ $S \leftarrow \{(q, k) \mid q \in Q, k \leq x_q\}$
while $S \neq oldS$
return $(x_q)_{q \in Q}$

Illustration. Let us consider the $p1CM$ depicted below. Here all polynomials labelling the transitions are constant and there is at most one transition from one state to another. So the induced Markov chain has the same structure as the $p1CM$ and thus it is irreducible. The invariant distribution for (q_1, q_2, q_3) is equal to $(\frac{2}{3}, \frac{1}{6}, \frac{1}{6})$. Let us compute (w.r.t. the invariant distribution) the weight associated with the incrementation of the counter $\frac{2}{3} \times \frac{3}{4} = \frac{1}{2}$, and the weight associated with the decrementation of the counter $\frac{2}{3} \times \frac{1}{4} + \frac{1}{6} = \frac{1}{3}$. Since the former weight is greater than the latter one, the corresponding random walk is transient implying that all d_q's are finite. Let us examine Algorithm 1. Since $|Q| = 3$, $(x_{q_1}, x_{q_2}, x_{q_3})$ is initialized to $(2, 2, 2)$. After the first iteration due to the loop transition around q_1, x_{q_1} is decremented yielding $(1, 2, 2)$. During the second iteration both x_{q_1} and x_{q_3} are decremented yielding $(0, 2, 1)$. Afterwards the successive values of the d_q's are $(0, 2, 0)$ and then $(0, 1, 0)$ which is returned.

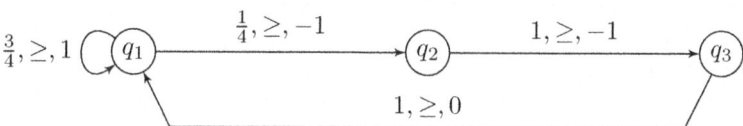

Due to Lemmas 3 and 4, as soon as $\mathcal{C}_Q$ is irreducible one can compute $(d_q)_{q \in Q}$ in polynomial time. The next theorem removes this hypothesis.

Theorem 2. *Let $\mathcal{C}$ be a pHM. Then the decisiveness problem w.r.t $s_0 = (q_0, k_0) \in Q \times \mathbb{N}$ and $Q \times \{0\}$ is decidable in polynomial time.*

Proof. Let us denote $\mathcal{M}_Q$, the Markov chain over Q defined by $\mathbf{M}$. The proof is based on the following partition of Q:

1. The set of terminal components of $\mathcal{M}_Q$, i.e., the recurrent states of $\mathcal{M}_Q$;
2. The set of transient states of $\mathcal{M}_Q$.

Computing this partition can be performed in linear time by Tarjan's algorithm for strongly connected components in $G_{\mathbf{M}}$.

Case q_0 belongs to a BSCC B of $\mathcal{M}_Q$. As said before such a BSCC can be viewed as an irreducible Markov chain. Thus applying Lemma 3 and 4, one can compute $(d_q)_{q \in B}$ in polynomial time.

Let $Y_f \subseteq Q^B$ (resp. $Y_\infty \subseteq Q^B$) be the set of states q of the BSCCs such that $d_q < \infty$ (resp. $d_q = \infty$).

Case q_0 does not belong to a BSCC of $\mathcal{M}_Q$ and Y_f is unreachable from q_0. Then almost surely q_0 reaches Y_∞ which implies $d_{q_0} = \infty$. The set of such states can be computed in polynomial time by a backward saturation algorithm.

Case q_0 does not belong to a BSCC of $\mathcal{M}_Q$ and Y_f is reachable from q_0. Here again the set of such states can be computed in polynomial time by a backward saturation algorithm. Let $q \in Y_f$ be reachable from q_0 thus by a path of length at most $|Q| - 1$. This implies that from $(q_0, d_q + |Q|)$ one can reach some (q, k) with $k > d_q$ and thus $(q_0, d_q + |Q|)$ is not decisive. Since $d_q < |Q|$, one gets that $d_{q_0} \leq 2(|Q| - 1)$.

Let Z be the set of such initial q_0. In order to compute the exact value of d_q for all $q \in Z$, we first define the Markov chain $\mathcal{M}'$ as follows: the set of states of $\mathcal{M}'$ is $(Y_f \cup Z) \times \mathbb{N} \cup q_\perp$ where $q_\perp$ is an absorbing state and for all $q \in Y_f \cup Z$ and transition $(q, k) \to (q', k')$ in $\mathcal{M}_C$, if $q' \in Y_f \cup Z$ there is the same transition in $\mathcal{M}'$ otherwise there is a transition $(q, k) \to q_\perp$ in $\mathcal{M}'$ with same probability (cumulating the probabilities if a transition occurs several times). Observe that for all $(q, k) \in (Y_f \cup Z) \times \mathbb{N}$, $\mathcal{M}'$ is decisive w.r.t. (q, k) and $(Y_f \cup Z) \times \{0\}$ iff $\mathcal{M}_C$ is decisive w.r.t. (q, k) and $Q \times \{0\}$. Indeed a transition $(q, k) \to (q', k')$ is 'replaced' by a transition $(q, k) \to q_\perp$ if $d_{q'} = \infty$ which means that from (q', k'), $\mathcal{M}_C$ almost surely reaches $Q \times \{0\}$ or $Q \times \{0\}$ becomes unreachable. Let T be the set of decisive configurations of $\mathcal{M}'$ w.r.t. $(Y_f \cup Z) \times \{0\}$. Then, Algorithm 2 computes d_q for all $q \in Z$. Let us describe it and prove its correctness:

- **Description.** The algorithm begins by computing $(d_q)_{q \in Y_f}$ using Algorithm 1 and then starts off with the set of all configurations which are possibly decisive by initialising $(x_q)_{q \in Z}$ to be $2(|Q| - 1)$ since all configurations (q, k) with $k > 2(|Q| - 1)$ are undecisive. It decreases x_q whenever there is a transition $(q, x_q) \to (q', x_q + \delta)$ with $x_q + \delta > x_{q'}$ since $(q', x_q + \delta)$ is undecisive and stops when there is no more such transitions.
- **Termination.** Since $|S| < \infty$, and each iteration (except the last one) decreases the size of S, the algorithm terminates.
- **Loop Invariant.** The invariant is : $T \subseteq S$. Initially as discussed above $T \subseteq S$. The 'if' statement decreases x_q if there exists a transition $q \xrightarrow{\delta} q'$ such that $x_q > x_{q'} - \delta$, which means that there is a positive probability to reach $(q', x_q + \delta)$ with $x_q + \delta > x_{q'}$ (i.e. a configuration out of T due to the invariant) from (q, x_q). So $T \subseteq S$ remains true after any iteration.
- **Correctness.** When it terminates, every transition from $s \in S$ leads to $s' \in S$. So $\mathcal{M}'$ restricted to S is a finite Markov chain. From any $s = (q_0, k_0)$ with $k_0 \leq d_{q_0}$, one almost surely reaches the set of decisive configurations $\{(q, k) \mid q \in Y_f \wedge k \leq d_q\} \cup \{q_\perp\}$ which is the set of recurrent states of $\mathcal{M}'$ restricted to S. So $\mathcal{M}'$ is decisive w.r.t. s and $(Y_f \cup Z) \times \{0\}$.

□

Algorithm 2: Computing d_q for $q \in Z$

Compute $(d_q)_q \in Y_f$ using Algorithm 1
$(x_q)_{q \in Y_f} \leftarrow (d_q)_{q \in Y_f}$
for $q \in Z$ **do** $x_q \leftarrow 2(|Q|-1)$
$S \leftarrow \{(q,k) \mid q \in Y_f \cup Z, k \leq x_q\} \cup \{q_\perp\}$
do
  $oldS \leftarrow S$; **for** $q \in Y_f \cup Z$ **do** $oldx_q \leftarrow x_q$
  for $(q, >, \delta, q') \in \Delta$ with $q \in Z$ and $q' \in Y_f \cup Z$ **do**
    | **if** $x_q > 0$ and $x_q + \delta > oldx_{q'}$ **then** $x_q \leftarrow x_q - 1$
  end
  $S \leftarrow \{(q,k) \mid q \in Y_f \cup Z, k \leq x_q\} \cup \{q_\perp\}$
while $S \neq oldS$
return $(x_q)_{q \in Z}$

For both algorithms, the number of iterations of do-while loop belongs to $O(|Q|^2)$ and every iteration is performed in polynomial time yielding polynomial time algorithms.

4 Undecidability of Decisiveness for Petri Nets

Petri nets are counter machines without zero test entailing decidability of some standard properties like reachability and coverability. As in the previous section, we enlarge Petri nets with dynamic weights and define a Markov chain semantics. The transitions of the net are labeled with positive polynomials, where the variables represent the value of the places. Its semantics is a countable Markov chain, where the states are the markings and the probabilities of transitions are obtained by evaluating the polynomials w.r.t the current value of the places and normalizing this value in order to get a probability distribution.

Definition 9 (Polynomial probabilistic Petri nets). *A polynomial probabilistic Petri net (pPN) is a tuple* $\mathcal{N} = (P, T, \mathbf{Pre}, \mathbf{Post}, W, \mathbf{m}_0)$ *where:*

- *P is a finite set of places;*
- *T is a finite set of transitions with $P \cap T = \emptyset$;*
- **Pre** *(resp.* **Post***) from $P \times T$ to $\mathbb{N}$ is the backward (resp. forward) incidence matrix;*
- *W is the weight function from T to the set of positive polynomials over the set of variables P;*
- *$\mathbf{m}_0 \in \mathbb{N}^P$ is the initial marking.*

Notations. Let $\mathcal{N}$ be a pPN, a *marking* $\mathbf{m}$ of $\mathcal{N}$ is an item of $\mathbb{N}^P$. Let $t \in T$, we denote $\mathbf{Pre}(t)$ (resp. $\mathbf{Post}(t)$) the vector $(\mathbf{Pre}(p,t))_{p \in P}$ (resp. $(\mathbf{Post}(p,t))_{p \in P}$) of $\mathbb{N}^P$. Let $\mathbf{m}$ be a marking of $\mathcal{N}$. Then t is *enabled* in $\mathbf{m}$ if $\mathbf{m} \geq \mathbf{Pre}(t)$. Its firing leads to the marking $\mathbf{m}' = \mathbf{m} - \mathbf{Pre}(t) + \mathbf{Post}(t)$. One denotes the change of marking by $\mathbf{m} \xrightarrow{t} \mathbf{m'}$. One denotes $En(\mathbf{m})$ the set of enabled transitions in $\mathbf{m}$ and $Weight(\mathbf{m}) = \sum_{t \in En(\mathbf{m})} W(t)(\mathbf{m})$. Enabling and firing of a sequence of transitions and reachability of a marking are defined as in the previous section.

We now provide the semantics of a pPN of a countable Markov chain.

Definition 10. Let $\mathcal{N}$ be a pPN. Then the Markov chain $\mathcal{M}_\mathcal{N} = (S, p)$ is defined by:

- $S = \mathbb{N}^P$
- For all $\mathbf{m} \in S$, if $En(\mathbf{m}) = \varnothing$, then $p(\mathbf{m}, \mathbf{m}) = 1$. Otherwise, for all $\mathbf{m'} \in S$,

$$p(\mathbf{m}, \mathbf{m'}) = Weight(\mathbf{m})^{-1} \sum_{\mathbf{m} \xrightarrow{t} \mathbf{m'}} W(t)(\mathbf{m})$$

The undecidability of the decisiveness problem of pPNs w.r.t a finite or upward closed set has been established in [6] even in the case of polynomials over a single variable. The authors proved the undecidability of decisiveness for pPNs by a reduction from the halting problem of counter machines (CMs). This reduction requires that the associated pPN has some polynomials of degree 4. Here, we refine the reduction in order to get affine weights (i.e., $aX + b$). First we give a sketch of the proof of the original result on pPNs before adapting the proof to establish the new result.

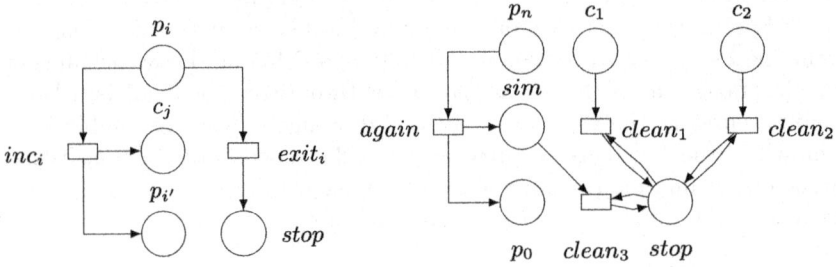

Fig. 2. (a) $i : c_j \leftarrow c_j + 1;\, \mathbf{goto}\ i'$ (b) halt instruction and cleaning stage

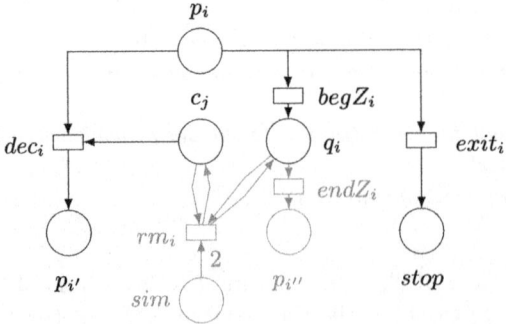

Fig. 3. $i :$ **if** $c_j > 0$ **then** $c_j \leftarrow c_j - 1;\, \mathbf{goto}\ i'$ **else goto** i''

Theorem 3. ([6]). *The decisiveness problem of pPNs w.r.t. a finite or upward closed set is undecidable even in the case of polynomials over a single variable.*

Proof (sketch). The proof is based on a reduction of the halting problem for normalized two counter programs (a variant of CM) to the decisiveness problem of pPNs. A normalized counter program resets the counters at the beginning of the computation and on termination. The halting problem for normalized two counter programs asks, given such a program $\mathcal{P}$, whether $\mathcal{P}$ eventually halts.

Given $\mathcal{P}$, the reduction builds a pPN $\mathcal{N}$ that performs a repeated *weak* simulation of $\mathcal{P}$, with a possibility to exit from the simulation at (almost) every transition firing. The instructions of $\mathcal{P}$ are numbered from 0 to n. The set of places of $\mathcal{N}$ is $P = \{p_i \mid 0 \leq i \leq n\} \cup \{q_i \mid i \text{ is a test instruction}\} \cup \{c_j \mid 1 \leq j \leq 2\} \cup \{sim, stop\}$. The initial marking is $\mathbf{m}_0 = p_0$ (i.e. the marking with one token in p_0 and no token elsewhere). Given a configuration of $\mathcal{P}$, the corresponding marking $\mathbf{m}$ is defined by for all $i \leq n$ with $\mathbf{m}(p_i) = 1$ if i is the next instruction to be executed, for all $j \leq 2$, $\mathbf{m}(c_j)$ is the current value of counter c_j and all places except possibly sim (whose meaning is explained below) are unmarked.

Figure 2(a) shows the simulation of instruction i in case of an incrementation of c_j leading to instruction i'. The transition $exit_i$ stops the simulation by unmarking p_i and marking place $stop$.

Figure 2(b) shows the simulation of the halting instruction (i.e., the instruction n). As one wants to repeat the simulation transition *again* that unmarks p_n and marks p_0 (the first instruction of $\mathcal{P}$). In addition, it increases the marking of sim which is a lower bound of the number of performed simulations (see below). This figure also shows the cleaning stage when place $stop$ is marked in order to reach the target marking related to the decisiveness with only a token in place $stop$.

Figure 3 shows the simulation of instruction i in case of a test with a possible decrementation of c_j leading to instruction i' when the counter of c_j is positive and to instruction i'' when c_j is null. As for incrementation simulation there is possible exit of the simulation (via transition $exit_i$). As the zero test cannot be specified in Petri nets, transition $begZ_i$ can be fired regardless of the counter value. However the possible firing of rm_i 'punishes' the erroneous simulation (i.e., firing $begZ_i$ when $c_j > 0$) by decrementing the counter sim which explains why sim is a lower bound of the number of performed simulations.

We specify weight functions as follows[1]: for all incrementation $i, W(inc_i,)(\mathbf{m}) = (\mathbf{m}(sim) + 2)^2 - 1$. For all decrementation i, $W(dec_i, \mathbf{m}) = (\mathbf{m}(sim) + 2)^4 - (\mathbf{m}(sim) + 2)^2$, $W(begZ_i, \mathbf{m}) = (\mathbf{m}(sim) + 2)^2 - 1$, and $W(rm_i, \mathbf{m}) = 2$. The remaining weights are equal to the constant function 1.

We claim that $\mathcal{N}$ is decisive w.r.t $\mathbf{m}_0$ and $\{stop\}$ iff $\mathcal{P}$ does not halt.

- Assume that $\mathcal{P}$ halts. Consider the (single) infinite run corresponding to the repeated exact simulation of $\mathcal{P}$. It can been shown that the probability of this

---

[1] The weights specified here are slightly different from the ones in [6]. However the positive probability of this infinite path is based on the fact that $\sum_{n \in \mathbb{N}} \frac{1}{P(n)} < \infty$ whenever P is a positive polynomial of degree at least 2 and thus the exact choice of the weights is irrelevant.

run is positive. In more details, during the $n+1^{th}$ simulation the marking of sim is equal to n. Thus the probability of an exact simulation of an instruction is greater or equal to $e^{-\frac{1}{P(n)}}$ for some polynomial P of degree two. Let ℓ be the length of the execution of $\mathcal{P}$. Then the probability of this simulation is greater or equal to $e^{-\frac{\ell}{P(n)}}$ implying that the probability of the infinite run is greater or equal to $e^{-\ell \sum_{n \geq 0} \frac{1}{P(n)}} > 0$. Since $\{stop\}$ is always reachable but is never marked by this infinite run, this proves the undecisiveness of $\mathcal{N}$.

- Assume that $\mathcal{P}$ does not halt. Consider the set of infinite runs that never reach marking $stop$. Either such a run performs a finite number of (erroneous) simulations (say k) without ever ending the last one or it performs an infinite number of simulations. Let E_k be the set of runs that does not achieve the kth simulation. During this simulation the marking of sim is less than k implying a positive lower bound for the probability of exiting the kth simulation after every instruction simulation. So the set E_k has null probability and so $\bigcup_k E_k$ has also null probability which corresponds to the first case. It can be shown that a run performing an infinite number of simulations almost surely performs such simulation with the marking of sim equal to 1. Thus by a similar argument, this set of runs has also null probability. So $\mathcal{N}$ is decisive w.r.t. marking $stop$.

The result of the theorem remains valid by substituting the singleton $\{stop\}$ with the set of markings greater than or equal to $stop$. □

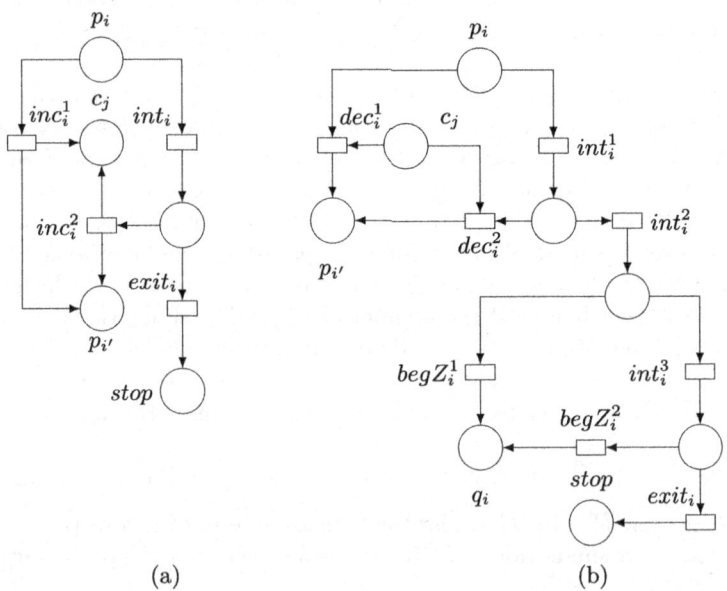

Fig. 4. (a) New incrementation simulation. (b) New decrementation simulation

We are now ready to extend this result.

Theorem 4. *The decisiveness problem of affine pPNs w.r.t. a finite or upward closed set is undecidable.*

Proof. We build an affine pPN that is an exact probabilistic simulation for every instruction simulation of the polynomial pPN described in the sketch of the previous proof. This means that after such a simulation the probabilities of the next markings are unchanged.

Since the non constant weights only occur in the simulation of an incrementation and decrementation, we focus on these instructions.

- **Case when instruction i is an incrementation of c_j leading to instruction i'.** The specification of the new subnet is depicted in Fig. 4(a). Let $\mathbf{m}$ be the current marking then the weights associated with the transitions of this subnet are defined as follows: $W(inc_i^1)(\mathbf{m}) = W(inc_i^2)(\mathbf{m}) = (\mathbf{m}(sim) + 1)$, and the other weights are equal to 1. Let $\mathbf{m}' = \mathbf{m} - p_i + p_{i'} + c_j$ and $\mathbf{m}'' = \mathbf{m} - p_i + stop$, the possible markings after the simulation. In the subnet, the probability to reach $\mathbf{m}''$ from $\mathbf{m}$ is equal to:

$$\frac{1}{\mathbf{m}(sim) + 2} \times \frac{1}{\mathbf{m}(sim) + 2} = \frac{1}{(\mathbf{m}(sim) + 2)^2}$$

  which is equal to the probability to reach $\mathbf{m}''$ from $\mathbf{m}$ in the subnet of Fig. 2(a) implying the equality of the probabilities to reach $\mathbf{m}'$ from $\mathbf{m}$ in the two subnets.

- **Case when instruction i is a test with a possible decrementation of c_j leading to instruction i' when the counter of c_j is positive and to instruction i'' when c_j is null.** The specification of the new subnet is depicted in Fig. 4(b). We have not reproduced the gray part (i.e., the possible punishing part) of Fig. 3 since it is identical. Let $\mathbf{m}$ be the current marking then the weights associated with the transitions of this subnet are defined as follows: for $k \in \{1, 2\}$, $W(dec_i^k)(\mathbf{m}) = W(begZ_i^k)(\mathbf{m}) = \mathbf{m}(sim) + 1$, the weight of rm_i is 2, and the other weights are equal to 1.
  - **Subcase $\mathbf{m}(c_j) = 0$.** Let $\mathbf{m}' = \mathbf{m} - p_i + q_i$ and $\mathbf{m}'' = \mathbf{m} - p_i + stop$, the possible markings after the simulation. In the subnet, the probability to reach $\mathbf{m}''$ from $\mathbf{m}$ is equal to:

$$\frac{1}{\mathbf{m}(sim) + 2} \times \frac{1}{\mathbf{m}(sim) + 2} = \frac{1}{(\mathbf{m}(sim) + 2)^2}$$

    which is equal to the probability to reach $\mathbf{m}''$ from $\mathbf{m}$ in the subnet of Fig. 3 implying the equality of the probabilities to reach $\mathbf{m}'$ from $\mathbf{m}$ in the two subnets.
  - **Subcase $\mathbf{m}(c_j) > 0$.** In addition to $\mathbf{m}'$ and $\mathbf{m}''$, marking $\mathbf{m}^* = \mathbf{m} - p_i - c_j + p_{i'}$ is the third possible marking after the simulation. In the subnet, the probability to reach $\mathbf{m}''$ from $\mathbf{m}$ is equal to:

$$\frac{1}{\mathbf{m}(sim) + 2} \times \frac{1}{\mathbf{m}(sim) + 2} \times \frac{1}{\mathbf{m}(sim) + 2} \times \frac{1}{\mathbf{m}(sim) + 2} = \frac{1}{(\mathbf{m}(sim) + 2)^4}$$

which is equal to the probability to reach $\mathbf{m}''$ from $\mathbf{m}$ in the subnet of Fig. 3 In the subnet, the probability to reach $\mathbf{m}^*$ from $\mathbf{m}$ is equal to:

$$\frac{\mathbf{m}(sim)+1}{\mathbf{m}(sim)+2} + \frac{1}{\mathbf{m}(sim)+2} \times \frac{\mathbf{m}(sim)+1}{\mathbf{m}(sim)+2} = 1 - \frac{1}{(\mathbf{m}(sim)+2)^2}$$

which is equal to the probability to reach $\mathbf{m}^*$ from $\mathbf{m}$ in the subnet of Fig. 3 implying the equality of the probabilities to reach $\mathbf{m}'$ from $\mathbf{m}$ in the two subnets.

This concludes the proof. □

5 Conclusion

In this paper, we have tightened the frontier between formalisms with decidable/undecidable decisiveness problem. There are two directions that we want to undertake. From a theoretical point of view, we will study the decisiveness problem for probabilistic pushdown automata and probabilistic Petri nets with constant weights. From a practical point of view, for formalisms where decisiveness is undecidable, we will look for sufficient conditions for decisiveness that can be efficiently checked. Finally in the tool COSMOS [3], there is already an implementation of the algorithm for decisive Markov chains in some high-level formalisms [4]. So we plan to integrate our algorithms for pHMs in COSMOS.

References

1. Abdulla, P.A., Henda, N.B., Mayr, R.: Decisive Markov chains. Log. Methods Comput. Sci. **3**(4) (2007). https://doi.org/10.2168/LMCS-3(4:7)2007
2. Baier, C., Katoen, J.: Principles of model checking. MIT Press (2008)
3. Ballarini, P., Barbot, B., Duflot, M., Haddad, S., Pekergin, N.: HASL: a new approach for performance evaluation and model checking from concepts to experimentation. Perform. Eval. **90**, 53–77 (2015). https://doi.org/10.1016/J.PEVA.2015.04.003
4. Benoît Barbot, Patricia Bouyer, S.H.: Beyond decisiveness of infinite Markov chains. In: FSTTCS 2024, pp. 8:1–8:22 (2024). https://doi.org/10.4230/LIPICS.FSTTCS.2024.8
5. Brázdil, T., Esparza, J., Kiefer, S., Kucera, A.: Analyzing probabilistic pushdown automata. Formal Methods Syst. Des. **43**(2), 124–163 (2013). https://doi.org/10.1007/S10703-012-0166-0
6. Finkel, A., Haddad, S., Ye, L.: About decisiveness of dynamic probabilistic models. In: Pérez, G.A., Raskin, J. (eds.) 34th International Conference on Concurrency Theory, CONCUR 2023, 18–23 September 2023, Antwerp, Belgium. LIPIcs, vol. 279, pp. 14:1–14:17. Schloss Dagstuhl - Leibniz-Zentrum für Informatik (2023). https://doi.org/10.4230/LIPICS.CONCUR.2023.14
7. Finkel, A., Haddad, S., Ye, L.: About decisiveness of dynamic probabilistic models. CoRR (2023). https://doi.org/10.48550/ARXIV.2305.19564

8. Finkel, A., Haddad, S., Ye, L.: Introducing divergence for infinite probabilistic models. In: Bournez, O., Formenti, E., Potapov, I. (eds.) Reachability Problems - 17th International Conference, RP 2023, Nice, France, 11–13 October 2023, Proceedings. Lecture Notes in Computer Science, vol. 14235, pp. 127–140. Springer (2023). https://doi.org/10.1007/978-3-031-45286-4_10
9. Iyer, S.P., Narasimha, M.: Probabilistic lossy channel systems. In: Theory and Practice of Software Development (TAPSOFT), 7th International Joint Conference CAAP/FASE. LNCS, vol. 1214, pp. 667–681. Springer (1997). https://doi.org/10.1007/BFB0030633
10. Kemeny, J., Snell, J., Knapp, A.: Denumerable Markov Chains, 2nd edn. Springer (1976)
11. Kucera, A., Esparza, J., Mayr, R.: Model checking probabilistic pushdown automata. Log. Methods Comput. Sci. **2**(1) (2006). https://doi.org/10.2168/LMCS-2(1:2)2006
12. Latouche, G., Pearce, C.E.M., Taylor, P.G.: Invariant measures for quasi-birth-and-death processes. Commun. Stat. Stochast. Models **14**, 443–460 (1998). https://doi.org/10.1080/15326349808807481
13. Rabinovich, A.: Quantitative analysis of probabilistic lossy channel systems. In: Baeten, J.C.M., Lenstra, J.K., Parrow, J., Woeginger, G.J. (eds.) ICALP 2003. LNCS, vol. 2719, pp. 1008–1021. Springer, Heidelberg (2003). https://doi.org/10.1007/3-540-45061-0_78
14. Rutten, J.J.M.M., Kwiatkowska, M.Z., Norman, G., Parker, D., Panangaden, P.: Mathematical techniques for analyzing concurrent and probabilistic systems, CRM Monograph Series, vol. 23. American Mathematical Society (2004)

Fuzzy Fault Trees: the Fast and the Formal

Thi Kim Nhung Dang[1], Benedikt Peterseim[2]($\boxtimes$), Milan Lopuhaä-Zwakenberg[2], and Mariëlle Stoelinga[2,3]

[1] Enschede, The Netherlands
[2] University of Twente, Enschede, The Netherlands
{benedikt.peterseim,m.a.lopuhaa,m.i.a.stoelinga}@utwente.nl
[3] Radboud University, Nijmegen, The Netherlands
m.stoelinga@cs.ru.nl

Abstract. We provide a rigorous framework for handling uncertainty in quantitative fault tree analysis based on fuzzy theory. We show that any algorithm for fault tree unreliability analysis can be adapted to this framework in a fully general and computationally efficient manner. This result crucially leverages both the α-cut representation of fuzzy numbers and the coherence property of fault trees. We evaluate our algorithms on an established benchmark of synthetic fault trees, demonstrating their practical effectiveness.

Keywords: Fault trees · reliability analysis · fuzzy numbers · uncertainty · directed acyclic graphs

1 Introduction

Fuzzy fault trees are a tool to assess the dependability of safety-critical systems, while simultaneously quantifying the uncertainty that enters these models through their parameters. To achieve this, they aim to combine classical *fault tree analysis* with concepts from *fuzzy theory*. Our goal is to make the idea of fuzzy fault tree analysis rigorous, and to provide computationally fast methods for carrying it out.

Fault Trees. Fault tree analysis (FTA) is a popular method in reliability engineering [30,33]. It is widely used in industry to assess and improve the dependability of, amongst others, nuclear power plants, self-driving cars, and aeroplanes. FTA is recommended by several ISO standards and certification bodies, such as the Federal Aviation Administration (FAA). A key aspect of FTA is quantitative assessment, calculating essential *dependability* or *risk metrics*, such as *unreliability*, *availability*, and *mean time to failure*. This paper studies the so-called mission-time reliability model [34]. Here each basic event b is assigned a probability p_b, representing its probability to fail within mission time. From these, one can compute the failure probability of the top event, i.e. the probability of the system to fail within its mission time, called the *system unreliability*.

T. K. N. Dang—Independent Scholar.

Fault Trees Under Uncertainty. Reliability analysis presupposes the availability of precisely known failure probabilities p_b. However, in practice this assumption may be unrealistic due to conflicting expert opinions, or the lack of reliable data. In such situations, *uncertainty quantification* enables a precise assessment of the confidence in the estimated unreliability. While there are other approaches to uncertainty quantification in fault trees, as discussed in Sect. 8, we will focus on devising precise and efficient foundations for an approach rooted in *fuzzy theory*.

Fuzzy Theory. Fuzzy theory has successfully been applied in numerous domains, including control systems [1], medical imaging, economic risk assessment, decision trees [4], and machine learning [6]. Its application to fault trees yields *fuzzy fault trees*. These handle parameter uncertainty by taking the failure probabilities of basic events to be *fuzzy numbers*. Whereas most works on fuzzy fault trees are case studies with an emphasis *how* fuzzy probabilities of basic events are obtained in practice (see, for example, [37]), we assume that these basic fuzzy probabilities are given. Instead, our focus will be to rigorously define the *fuzzy unreliability* in a principled way, and how to compute it efficiently.

Challenges. Despite its successful application in many case studies (see Sect. 8), fuzzy FTA still lacks a rigorous mathematical foundation. Unclear or ad-hoc definitions of system unreliability in fuzzy fault trees impede the interpretability of the resulting risk metric and are hence not an acceptable means for decision-making in safety-critical situations. In addition, to the best of our knowledge, no generally applicable, precise and efficient algorithm for quantitative fuzzy FTA has so far been presented. This paper aims to close both of these gaps.

One major obstacle in fuzzy fault tree analysis is that performing exact arithmetic operations on fuzzy numbers is generally computationally expensive. Most notably, common classes, or "shapes", of fuzzy numbers such as *triangular* and *trapezoidal* fuzzy numbers are not closed under basic ("fuzzified") arithmetic operations [4,18,35]. For example, if we multiply two triangular fuzzy numbers (via the canonical *Zadeh extension*), then the result is no longer triangular [18].

Contributions. To overcome these challenges, this paper contributes:

1. A well-motivated, principled and mathematically rigorous definition of the *fuzzy unreliability* metric;
2. A fast bottom-up algorithm based on α-cuts for computing fuzzy unreliability in *tree-structured* fault trees in a simple and intuitive way;
3. A correctness result showing that, even in the general case of *DAG-structured* FTs, *any* unreliability algorithm can be extended to the fuzzy case, assuming a mild regularity condition on fuzzy numbers that holds for all classes of fuzzy numbers used in practice; see Theorem 2;
4. An empirical evaluation of our algorithms using the model checker Storm [12].

Our main result, Theorem 2, enables a simple implementation of unreliability analysis in fuzzy fault trees using existing tools, making our algorithms readily applicable in practice. The reason this works is a delicate interplay between two crucial assumptions underlying both fault tree analysis and fuzzy theory. The main property of fault trees used is that they are *coherent*: the failure of any basic event never *decreases* the failure probability of the top event. On the other hand, the main common assumption on fuzzy numbers we use is that their α-cuts (i.e. α-upper level sets) are intervals. The insight that fuzzy probabilities are faithfully and efficiently represented and computed by their α-cuts is the final ingredient which makes our methods work.

2 Fault Trees

Fault trees (FTs) are hierarchical diagrams whose top event represents system failure, and whose leaves, called *basic events* (BEs), represent atomic failures. Intermediate gates are AND- or OR-gates, and propagate failures according to the status of their inputs; see Fig. 1.

Definition 1. *A fault tree (FT) is a tuple $T = (V, E, t)$, where (V, E) is a rooted directed acyclic graph (DAG), and t is a map $t: V \to \{\text{BE}, \text{OR}, \text{AND}\}$ such that for all $v \in V$, $t(v) = \text{BE}$ if and only if v is a leaf. The set of basic events is written $BE_T = \{v \in V | t(v) = \text{BE}\}$. The root of T is denoted R_T. For a node $v \in V$, we write $ch(v)$ for the set of all children of v.*

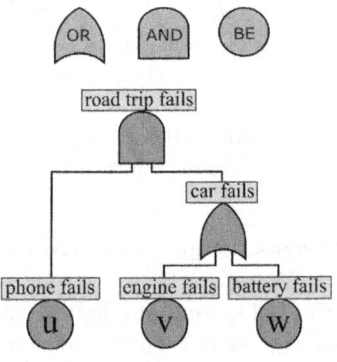

Fig. 1. Fault tree for a road trip. The road trip fails if both the phone fails and the car fails; the latter happens when either the engine or the battery fails. Its structure function equals $S_T(\vec{b}) = b_u \wedge (b_v \vee b_w)$.

A fault tree $T = (V, E, t)$ need not be a tree in the graph-theoretic sense. If the underlying DAG (V, E) forms a tree (each node has a unique parent), it is *tree-structured*; otherwise, it is *DAG-structured*.

A binary vector $\vec{b} \in \mathbb{B}^{BE_T}$ is called a *status vector*, where $\mathbb{B} = \{0, 1\}$ is the set of Booleans, where 1 indicates failure, and 0 means operational. A *probabilistic status vector* is a probability vector $\vec{p} \in [0, 1]^{BE_T}$. Whether or not the overall system fails given a status vector is determined by the *structure function*:

Definition 2. *Let T be a FT. The structure function $S_T: V \times \mathbb{B}^{BE_T} \to \mathbb{B}$ of T is defined, for a node $v \in V$ and a status vector $\vec{b} = (b_w)_{w \in BE_T} \in \mathbb{B}^{BE_T}$, by*

$$S_T(v, \vec{b}) = \begin{cases} \bigvee_{w \in ch(v)} S_T(w, \vec{b}) & \text{if } t(v) = \text{OR}, \\ \bigwedge_{w \in ch(v)} S_T(w, \vec{b}) & \text{if } t(v) = \text{AND}, \\ b_v & \text{if } t(v) = \text{BE}. \end{cases}$$

We write $S_T(\vec{b}) := S_T(R_T, \vec{b})$ for the structure function of T at its roots. A status vector $\vec{b}$ that reaches the root R_T i.e., $S_T(\vec{b}) = 1$ is called a cut set. The set of all cut sets of T is denoted $\mathcal{C}_T$.

2.1 Fault Tree Reliability Analysis

The *system unreliability* is the probability that a system fails within its mission time. In fault tree analysis, the unreliability is obtained as the probability that the top event occurs, if each basic event v is assigned a failure probability p_v, i.e. the probability that this BE fails within its mission time. The latter are given as a *probabilistic status vector* $\vec{p} \in [0,1]^{\mathrm{BE}_T}$.

Definition 3. *Let $T = (V, E, t)$ be a FT with probabilistic status vector $\vec{p} = (p_v)_{v \in V}$. The unreliability of T with respect to $\vec{p}$ is defined as*

$$U_T(\vec{p}) := \mathbb{P}\left[S_T\left(\vec{B}\right) = 1\right],$$

where $\vec{B} = (B_v)_{v \in \mathrm{BE}_T}$ is a random vector whose components B_v are all independent and Bernoulli-distributed with probability p_v.

This definition assumes all basic event failures to be independent—a standard assumption, since all dependencies are captured by the FT gates. Using the definition of $\mathcal{C}_T$, we see that $U_T(\vec{p})$ is equivalently given by,

$$U_T(\vec{p}) = \sum_{\vec{b} \in \mathcal{C}_T} \prod_{v \in \mathrm{BE}_T} p_v^{b_v} \cdot (1 - p_v)^{(1 - b_v)}. \tag{1}$$

3 Fuzzy Numbers

Fuzzy theory was proposed in [40] to reason about vagueness in a precise way. Figure 2 illustrates the fuzzy number x = "approximately 0.3". Here, x is not a single value, but rather a function, called the *fuzzy membership degree*. That is, x[x] indicates how much x resembles 0.3: At 0.3, the membership degree is 1, indicating that "0.3 is unequivocally 0.3".

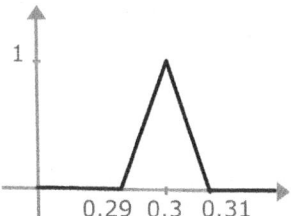

Fig. 2. "Approximately 0.3"

As we move farther from 0.3, the likeness to "approximately 0.3" diminishes, and at 0.31 or 0.29, it is clear these numbers are *not* "approximately 0.3". Alternatively, membership functions can represent trust, where x[x] denotes our trust in x equalling x. If x[x] = 0, then trust is zero; if x[x] = 1, trust is maximal. This notion can be expressed for members of any set, leading to the definition of *fuzzy elements*.

Definition 4. *A fuzzy element of a set X is a function $X \to [0,1]$. The set of fuzzy elements of X is denoted $\mathbf{F}(X)$. A fuzzy number is a fuzzy element of $\mathbb{R}$.*

Classes of Fuzzy Numbers. Several common types of fuzzy numbers exist. For real numbers $a \leq b \leq c \leq d$, the *trapezoidal fuzzy number* $\mathsf{trap}_{a,b,c,d} \in \mathbf{F}(\mathbb{R})$ is defined as (see Fig. 3):

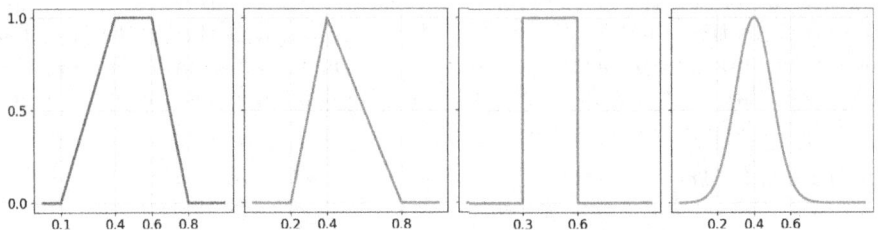

Fig. 3. Membership functions, from left to right, of a *trapezoidal* $\mathsf{trap}_{0.1,0.4,0.6,0.8}$, *triangular* $\triangle_{0.2,0.4,0.8}$, *interval* $\mathbb{1}_{[0.3,0.6]}$, and *Gaussian* $\mathsf{gauss}_{0.4,0.1}$ fuzzy number.

$$\mathsf{trap}_{a,b,c,d}[x] := \begin{cases} \frac{x-a}{b-a}, & \text{if } a < x < b, \\ 1, & \text{if } b \leq x \leq c, \\ \frac{d-x}{d-c}, & \text{if } c < x < d, \\ 0, & \text{otherwise.} \end{cases}$$

Trapezoidal fuzzy numbers generalize *triangular fuzzy numbers*, which are defined as $\triangle_{a,b,d} := \mathsf{trap}_{a,b,b,d}$, and interval fuzzy numbers ($\mathbb{1}_{[a,b]} := \mathsf{trap}_{a,a,b,b}$). The latter allows for the treatment of *imprecise probability* in fault trees [13, 14] as a special case of fuzzy fault tree analysis. *Gaussian fuzzy numbers* are characterized by their mean m and standard deviation d:

$$\mathsf{gauss}_{m,d}[x] := \exp\left(\frac{-(x-m)^2}{2d^2}\right).$$

3.1 Zadeh's Extension Principle

Zadeh's extension principle [15, 40] lifts any function $f : X \to Y$ to a function of fuzzy elements $\widetilde{f} : \mathbf{F}(X) \to \mathbf{F}(Y)$. To understand how this works, assume first that f is injective and that $y \in Y$ with $f(x) = y$. Then we set $\widetilde{f}(\mathsf{x})[y] = \mathsf{x}[x]$, as our trust for f to be equal to y at x should be the same as our trust for x to be equal to x. Also, for $y \in Y$ with $f^{-1}(y) = \varnothing$, we set $\widetilde{f}(\mathsf{x})[y] = 0$, since f never takes on the value y. Now, if f is not injective, then there are multiple values x with $f(x) = y$. Each of these values $x \in f^{-1}(y)$ has a fuzzy membership degree $\mathsf{x}[x]$. Zadeh's extension principle takes the highest possible membership degree among these. The underlying idea is as follows: Consider a fuzzy number x with

$\mathsf{x}[-2] = 0.3$ and $\mathsf{x}[2] = 0.9$ consider the function $f(x) = x^2$. How much trust do we have that "$x^2 = 4$", i.e. what is the value of $\tilde{f}(\mathsf{x})[4]$? We can defend a trust degree of 0.9, by assuming that the value $x^2 = 4$ was obtained by taking as input $x = 2$, which has trust degree $\mathsf{x}[2] = 0.9$.

Finally, for functions of multiple arguments, membership degrees of independent arguments are combined by taking their minimum. This is justified as follows: to trust in the value of a multivariate function f with a degree of α, our trust at *all* of its arguments must be at least α.

Definition 5 (Zadeh's Extension Principle). *Let $f : X_1 \times \cdots \times X_n \to Y$ be a function. The Zadeh extension of f is defined as the function,*

$$\tilde{f} \colon \mathbf{F}(X_1) \times \cdots \times \mathbf{F}(X_n) \to \mathbf{F}(Y),$$

$$\tilde{f}(\vec{\mathsf{x}})[y] := \sup \left\{ \min_{i=1,\ldots n} \mathsf{x}_i[x_i] \;\middle|\; \vec{x} \in f^{-1}(y) \right\},$$

for all $y \in Y$, $\vec{\mathsf{x}} \in \mathbf{F}(X_1) \times \cdots \times \mathbf{F}(X_n)$.

On certain families of fuzzy elements, addition and subtraction operations can be performed in a straightforward manner. For example, for two trapezoidal fuzzy numbers we have

$$\mathsf{trap}_{a_1,a_2,a_3,a_4} \tilde{+} \mathsf{trap}_{b_1,b_2,b_3,b_4} = \mathsf{trap}_{a_1+b_1,a_2+b_2,a_3+b_3,a_4+b_4},$$
$$\mathsf{trap}_{a_1,a_2,a_3,a_4} \tilde{-} \mathsf{trap}_{b_1,b_2,b_3,b_4} = \mathsf{trap}_{a_1-b_4,a_2-b_3,a_3-b_2,a_4-b_1}.$$

In general, however, there are no such simple formulas for the Zadeh extension of arithmetic operations. In particular, arithmetic operations do not preserve the shape of the fuzzy numbers: For example, the product of two trapezoidal fuzzy numbers is *not* a trapezoidal fuzzy number, in general.

4 Fuzzy Fault Trees

Fuzzy failure probabilities arise when the failure probabilities of basic events as fuzzy numbers. Then, the fuzzy unreliability is obtained by the Zadeh extension of the system unreliability function.

Example 1. Consider again the FT T from Fig. 1. By Eq. (1) its unreliability function $U_T \colon [0,1]^3 \to [0,1]$ is given by

$$\begin{aligned} U_T(p_a, p_b, p_c) &= p_a p_b p_c + p_a(1 - p_b)p_c + p_a p_b(1 - p_c) \\ &= p_a p_c + p_a p_b - p_a p_b p_c. \end{aligned}$$

Now, suppose T is equipped with a *fuzzy* failure probabilistic status vector $\vec{p} = (\mathsf{p}_a, \mathsf{p}_b, \mathsf{p}_c)$. The fuzzy unreliability $\widetilde{U}_T(\mathsf{p}_a, \mathsf{p}_b, \mathsf{p}_c)$ is then defined as the Zadeh extension of U_T.

$$\widetilde{U}_T(\vec{p})[y] = \sup_{p_a, p_b, p_c} \min\{\mathsf{p}_a[p_a], \mathsf{p}_b[p_b], \mathsf{p}_c[p_c]\},$$

where the supremum is taken over the set of all $p_a, p_b, p_c \in [0, 1]$ such that $p_a p_c + p_a p_b - p_a p_b p_c = y$, for all $y \in [0, 1]$.

Suppose that p_b and p_c are crisp numbers with $\mathsf{p}_b = 0.1$ and $\mathsf{p}_c = 0.4$. Assume that p_a takes probability 0.5 or 0.8 with fuzzy membership values 0.7 and 1, respectively.

$$\widetilde{U}_T(\vec{p}) = \begin{cases} 1, & \text{if } y = 0.368, \\ 0.7, & \text{if } y = 0.23, \\ 0, & \text{otherwise.} \end{cases}$$

Generalising this example, we obtain our main definition.

Definition 6 (Fuzzy unreliability, fuzzy fault trees). *Let T be a FT.*

1. *A fuzzy probabilistic status vector is an element $\vec{p}$ of $\mathbf{F}([0,1])^{\mathrm{BE}_T}$.*
2. *The fuzzy unreliability of T given $\vec{p}$ is defined as $\widetilde{U}_T(\vec{p})$, where*

$$\widetilde{U}_T : \mathbf{F}([0,1])^{\mathrm{BE}_T} \to \mathbf{F}([0,1])$$

is the Zadeh extension of the function U_T from Definition 3.

More concretely, $\widetilde{U}_T(\vec{p})$ is the fuzzy element of $[0,1]$ defined by,

$$\widetilde{U}_T(\vec{p})[y] = \sup \left\{ \min_{v \in \mathrm{BE}_T} \mathsf{p}_v[p_v] \,\middle|\, \vec{p} \in [0,1]^{\mathrm{BE}_T}, U_T(\vec{p}) = y \right\}, \tag{2}$$

for all $y \in [0,1]$. A fuzzy fault tree is a fault tree equipped with a fuzzy probabilistic status vector.

We will see a concrete, illustrative computation of the fuzzy unreliability in Example 3. More realistic examples will be provided in the experiments in Sect. 7. Figure 4 illustrates the fuzzy probability of top-level event failure for a particular fuzzy fault tree $(T, \vec{p})$, which we have randomly selected from the benchmark used in Sect. 7.2. From the figure, we see that the unreliability shows a considerable skew and

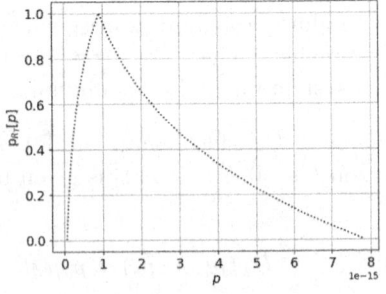

Fig. 4. Fuzzy unreliability $\widetilde{U}_T(\vec{p})$ for specific T and $\vec{p}$ from Sect. 7.2.

non-linearity, despite all basic events being equipped with symmetric triangular fuzzy probabilities. We hence obtain a much more informative and complete picture of the uncertainty in the probability of top-level system failure, going beyond both point-estimates and probability intervals.

5 Computing Fuzzy Unreliability I: Tree-Structured Case

When T is tree-structured, the fuzzy unreliability $\tilde{U}_T(\vec{p})$ can be found using a bottom-up algorithm that proceeds exactly as for ordinary fault trees, replacing arithmetic operations by their Zadeh extensions. We first review the ordinary, "crisp" case. The general DAG-structured case will be treated in Sect. 6.

Crisp Case. Already in the crisp case, computing the unreliability for a fault tree is generally difficult (in fact, NP-hard [21]). Naively applying Eq. (1) requires a summation over the entire set $\mathcal{C}_T$ of cut sets, and is therefore computationally infeasible for large FTs. When T is tree-structured, the unreliability can instead be computed in a bottom-up fashion, assigning a probability p_v to each node along the way. For a node v with children $v_1, \ldots, v_n$, we write

$$p_v := \begin{cases} 1 - \prod_{i=1}^n (1 - p_{v_i}) & \text{if } t(v) = \text{OR}, \\ \prod_{i=1}^n p_{v_i} & \text{if } t(v) = \text{AND}. \end{cases} \quad (3)$$

Then, if T is tree-structured, $U(T) = p_{R_T}$.

Example 2. Consider the FT from Fig. 1; see also below. Let $p_u = 0.8, p_v = 0.1$, and $p_w = 0.4$. The top-level failure probability is

$$p_{R_T} = p_u \cdot \left(1 - (1 - p_v) \cdot (1 - p_w)\right) = 0.368.$$

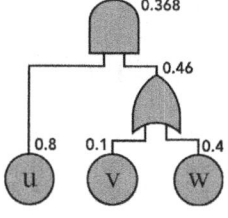

The bottom-up algorithm is fast, but does not extend to general, DAG-structured FTs. The reason is that (3) only computes the failure probability of v correctly when the v_i all represent independent events, which will not generally be true if they share children. For DAG-structured FTs the state-of-the-art approach is to translate the FT to a binary decision diagram [29], on which a bottom-up algorithm is run. The worst-case time complexity of this approach is exponential, but is very fast in practice [2].

Fuzzy Case. To compute the *fuzzy* unreliability, we now replace arithmetic operations by their Zadeh extensions, letting

$$\mathsf{p}_v = \begin{cases} 1 \, \tilde{-} \, \tilde{\prod}_{w \in ch(v)} (1 \, \tilde{-} \, \mathsf{p}_w) & \text{if } t(v) = \text{OR}, \\ \tilde{\prod}_{w \in ch(v)} \mathsf{p}_w & \text{if } t(v) = \text{AND}. \end{cases} \quad (4)$$

Here, we write "1" for the ("crisp") fuzzy number whose membership function is 1 at the number 1 and vanishes everywhere else. The following result then states that the fuzzy unreliability can be computed recursively using Eq. (4). It is proved analogously to a similar result for attack trees in [8].

Theorem 1. *Let T be a tree-structured FT, and let $\vec{p} \in \mathbb{F}([0,1])^{BE_T}$ be a vector of fuzzy probabilities. Then $\mathsf{p}_{R_T} = \tilde{U}_T(\vec{p})$.*

In the subsequent illustrative example, we employ the following compact notation for fuzzy probabilities that take only finitely many values,

$$\{x_1 \mapsto a_1, \ldots, x_n \mapsto a_n\}[x] := \begin{cases} a_i & \text{if } x = x_i \text{ for some } i \in \{1, \ldots, n\}, \\ 0 & \text{otherwise}, \end{cases}$$

for any $x_1, \ldots, x_n, a_1, \ldots, a_n, x \in [0,1]$ with $x_1, \ldots, x_n$ distinct.

Example 3. We apply the algorithm to Example 1. Given

$$\mathsf{p}_a := \{0.5 \mapsto 0.7, 0.8 \mapsto 1\},\ \mathsf{p}_b := \{0.1 \mapsto 1\},\ \mathsf{p}_c := \{0.4 \mapsto 1\},$$

we calculate the unreliability as follows:

$$\begin{aligned}\mathsf{p}_{\mathrm{OR}(b,c)} &= 1 \tilde{-} (1 \tilde{-} \mathsf{p}_b) \tilde{\cdot} (1 \tilde{-} \mathsf{p}_c) \\ &= 1 \tilde{-} \{0.9 \mapsto 1\} \tilde{\cdot} \{0.6 \mapsto 1\} \\ &= \{0.46 \mapsto 1\},\end{aligned}$$

and similarly for $\mathsf{p}_{R_T} = \mathsf{p}_a \tilde{\cdot} \mathsf{p}_{\mathrm{OR}(b,c)}$. This way, we obtain

$$\tilde{U}_T(\vec{p}) = \mathsf{p}_{R_T} = \{0.23 \mapsto 0.7, 0.368 \mapsto 1\}.$$

6 Computing Fuzzy Unreliability II: General Case

When a fault tree is not tree-structured, but a general DAG-structured FT, then the bottom-up approach no longer computes $\tilde{U}_T(\vec{p})$ correctly. This is because the probabilities of the children of a node may no longer be independent. In fact, this problem already arises for crisp probability values [30], so it is no surprise that the same holds in the fuzzy setting.

For general DAG-structured fault trees, we instead leverage the α-cut representation of fuzzy numbers.

6.1 The α-Cut Representation of Fuzzy Numbers

Using fuzzy membership functions that take only finitely many non-zero values to represent fuzzy probabilities, as in Example 3, is conceptually simple. However, this representation is inefficient for computing the Zadeh extension. This is because, in general, the number of non-zero values of the fuzzy unreliability's membership function will grow exponentially with the number of basic events.

If, instead, we choose to work with fuzzy numbers of a particular shape, such as trapezoidal fuzzy numbers, we face the issue that arithmetic operations do not preserve the shape of the fuzzy numbers.

One way to mitigate these problems is to perform arithmetic operations on the given fuzzy numbers' α-*cuts* [4]. The α-cut of a fuzzy element x is the set of all elements of X whose membership degree is at least α.

Definition 7. *Let* $\mathsf{x} \in \mathbf{F}(X)$ *and* $\alpha \in [0,1]$. *The* α-*cut* $\mathsf{x}^{(\alpha)}$ *of* x *is*

$$\mathsf{x}^{(\alpha)} := \{x \in X \mid \mathsf{x}[x] \geq \alpha\}. \tag{5}$$

Example 4. The α-cuts of a trapezoidal fuzzy number are given by

$$\mathsf{trap}^{(\alpha)}_{a,b,c,d} = [(b-a) \cdot \alpha + a, \ d - (d-c) \cdot \alpha], \tag{6}$$

for all real numbers $a \leq b \leq c \leq d$ and all $\alpha \in (0,1]$ (see Fig. 5).

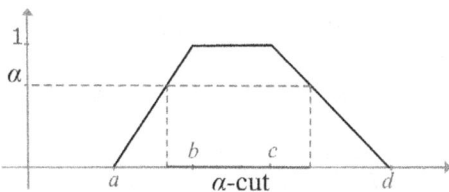

Fig. 5. α−cut of trapezoidal fuzzy number $\mathsf{trap}_{a,b,c,d}$.

Regular Fuzzy Numbers. In the literature, the term "fuzzy number" almost universally refers to fuzzy elements of $\mathbb{R}$ that also satisfy certain regularity conditions, where the precise conditions vary across the literature [10].

We consider *regular* fuzzy numbers, whose α-cuts are all intervals. Regularity ensures that Zadeh extensions can be computed at the endpoints of these intervals, under mild assumptions; see Lemma 1.

Definition 8. *A regular fuzzy number is a fuzzy element* $\mathsf{x} \in \mathbf{F}(\mathbb{R})$ *such that:*

1. *For all* $\alpha \in (0,1]$, *the* α-*cut* $\mathsf{x}^{(\alpha)}$ *is a compact interval, i.e. there exist* $l, r \in \mathbb{R}$ *satisfying* $\mathsf{x}^{(\alpha)} = [l, r]$.
2. *The fuzzy membership function* $\mathsf{x} : \mathbb{R} \to [0,1]$ *is compactly supported, i.e. the set of all* $x \in \mathbb{R}$ *with* $\mathsf{x}[x] \neq 0$ *is bounded.*

We write $\mathbb{F}$ for the set of all regular fuzzy numbers. A fuzzy number x is nonnegative if $\mathsf{x}[x] = 0$ for all $x < 0$.

Our definition is more general than e.g. the one given in [11] and covers most classes of fuzzy numbers used in practice. In particular, common parametrized families of fuzzy numbers such as triangular, trapezoidal, and interval fuzzy numbers are fuzzy numbers in the above sense. Note that Gaussian fuzzy numbers are *not* regular, as $\mathsf{gaus}_{m,d}[x] > 0$ for all $x \in \mathbb{R}$. Note, however, that since they represent fuzzy *probabilities*, all fuzzy numbers appearing in fuzzy fault trees will have fuzzy membership functions vanishing outside of $[0, 1]$, therefore automatically satisfying the second condition from Definition 8.

The regularity of a fuzzy number allows us to describe it in two alternative ways: either by its fuzzy membership function $\mathsf{x}\colon \mathbb{R} \to [0,1]$, or by the two functions $(0,1] \to \mathbb{R}$ that assign to α the endpoints of the α-cut of x. The advantage of the second viewpoint is that Zadeh extensions become significantly easier to compute: it turns out that for nonnegative fuzzy numbers, Zadeh extensions of arithmetic operations can simply be computed at the endpoints of the α-cuts.

More generally, we have the following result, which we will later also use to compute the fuzzy unreliability of a fault tree.

Lemma 1. *Let $I \subseteq \mathbb{R}$ be a closed interval, let $\vec{\mathsf{x}} = (\mathsf{x}_1, \ldots, \mathsf{x}_n) \in \mathbb{F}^n$ be a tuple of regular fuzzy numbers each of whose fuzzy membership function is supported in I (i.e. $\mathsf{x}[x] = 0$ for all $x \notin I$), and let $f : I^n \to \mathbb{R}$ be a continuous function. Then:*

1. *For all $\alpha \in [0,1]$, the α-cut of the Zadeh extension of f is,*

$$\widetilde{f}(\vec{\mathsf{x}})^{(\alpha)} = f\left[\mathsf{x}_1^{(\alpha)} \times \cdots \times \mathsf{x}_n^{(\alpha)}\right],$$

   i.e. the image of the product of the α-cuts of each $\mathsf{x}_1, \ldots, \mathsf{x}_n$ under f.
2. *Assume, moreover, that f is either monotonically non-decreasing or non-increasing, and write each α-cut as an interval,*

$$\mathsf{x}_i^{(\alpha)} = \left[\mathsf{x}_i^{(\alpha,l)}, \mathsf{x}_i^{(\alpha,r)}\right],$$

   for all $i \in \{1, \ldots, n\}$, $\alpha \in (0,1]$. Then, for all $\alpha \in (0,1]$,

$$\widetilde{f}(\vec{\mathsf{x}})^{(\alpha)} = \left[f\left(\vec{\mathsf{x}}^{(\alpha,l)}\right), f\left(\vec{\mathsf{x}}^{(\alpha,r)}\right)\right],$$

   where $\vec{\mathsf{x}}^{(\alpha,l)} = (\mathsf{x}_1^{(\alpha,l)}, \ldots, \mathsf{x}_n^{(\alpha,l)})$ and, similarly, $\vec{\mathsf{x}}^{(\alpha,r)} = (\mathsf{x}_1^{(\alpha,r)}, \ldots, \mathsf{x}_n^{(\alpha,r)})$. In other words, the Zadeh extension of f can be computed at the endpoints of each α-cut.

The proof of Lemma 1 is given in the appendix of the extended version [9] of this paper. Its first part is essentially well known [24, Proposition 5.1], while the second part follows easily from the assumption of monotonicity.

Applying Lemma 1 to the addition and multiplication functions $+, \cdot : [0,\infty)^2 \to [0,\infty)$ as well as to $x \mapsto 1-x$, we see that the Zadeh extension of all arithmetic operations that appear in elementary probability calculations can be computed on the level of α-cuts.

Example 5. Consider two triangular fuzzy numbers $\triangle_{1,2,3}, \triangle_{3,4,6} \in \mathbb{F}$. We wish to calculate the (Zadeh-extended) product of these two fuzzy numbers using their α-cuts. By Eq. (6), for any $\alpha \in (0,1]$,

$$\triangle_{1,2,3}^{(\alpha)} = [1+\alpha, 3-\alpha], \quad \triangle_{3,4,6}^{(\alpha)} = [3+\alpha, 6-2\alpha].$$

Therefore, by Lemma 1,

$$(\triangle_{1,2,3} \tilde{\cdot} \triangle_{3,4,6})^{(\alpha)} = [\alpha^2 + 4\alpha + 3, 2\alpha^2 - 12\alpha + 18]$$

Clearly, these α-cuts are not the ones of a triangular fuzzy number. To determine the fuzzy membership function of the above (Zadeh-extended) product of fuzzy numbers, we instead need to solve two nonlinear equations: $\alpha^2 + 4\alpha + 3 = x$, and $2\alpha^2 - 12\alpha + 18 = x$. The solutions in $[0,1]$ then yield the fuzzy membership function

$$(\triangle_{1,2,3} \tilde{\cdot} \triangle_{3,4,6})[x] = \begin{cases} -2 + \sqrt{1+x}, & \text{if } 3 \leq x \leq 8, \\ \frac{6-\sqrt{2x}}{2}, & \text{if } 8 < x \leq 18, \\ 0, & \text{otherwise.} \end{cases} \quad (7)$$

This provides an example of how working with α-cuts tends to be simpler than working fuzzy membership functions directly.

6.2 Computations via α-Cuts

To make our algorithms for computing the fuzzy unreliability applicable in practice, we first choose a finite number of α-cuts n_{cuts}. We then represent a (regular) fuzzy number x by n_{cuts} pairs $(\mathsf{x}^{(\alpha,l)}, \mathsf{x}^{(\alpha,r)})$ for each $\alpha \in \left\{\frac{1}{n_{\text{cuts}}}, \frac{2}{n_{\text{cuts}}}, \cdots, 1\right\}$. The interpretation is that $\mathsf{x}^{(\alpha)} = [\mathsf{x}^{(\alpha,l)}, \mathsf{x}^{(\alpha,r)}]$. In this way, we can represent common families of fuzzy numbers such as triangular or trapezoidal fuzzy numbers at a good degree of accuracy in a finite array of a fixed size. Moreover, the fuzzy operations appearing in the bottom-up algorithm, see (4), can now be performed level-wise using Lemma 1. Hence, at each node, we need to make $\mathcal{O}(n_{\text{cuts}})$ crisp arithmetic operations, for a total time complexity of $\mathcal{O}(n_{\text{cuts}} \cdot |V|)$, where $|V|$ is the number of nodes of the given fault tree.

We will exploit α-cuts in the same way in the general DAG-structured case.

6.3 Computing Fuzzy Unreliability via α-Cuts

In fact, the following result allows us to compute the α-cuts of the fuzzy unreliability using *any* algorithm for the unreliability of ordinary DAG-structured fault trees.

Theorem 2. *Let T be a fault tree and let $\vec{\mathsf{p}} = (\mathsf{p}_v) \in \mathbf{F}([0,1])^{BE_T}$ be a fuzzy probabilistic status vector. Moreover, assume that for all $v \in BE_T$, p_v is a regular fuzzy number in the sense of Definition 8 and write*

$$\mathsf{p}_v^{(\alpha)} = [\mathsf{p}_v^{(\alpha,l)}, \mathsf{p}_v^{(\alpha,r)}],$$

as well as, $\vec{\mathsf{p}}^{(\alpha,l)} := (\mathsf{p}_v^{(\alpha,l)})_{v \in BE_T}$ and $\vec{\mathsf{p}}^{(\alpha,r)} := (\mathsf{p}_v^{(\alpha,r)})_{v \in BE_T}$, for all $\alpha \in (0,1]$. Then

$$\widetilde{U}_T(\vec{\mathsf{p}})^{(\alpha)} = \left[U\left(\vec{\mathsf{p}}^{(\alpha,l)}\right), U\left(\vec{\mathsf{p}}^{(\alpha,r)}\right) \right],$$

for all $\alpha \in (0,1]$.

In other words, the α-cuts of the fuzzy unreliability can be computed at the endpoints of the intervals that constitute α-cuts of the given fuzzy probabilistic status vector. The key reason for this result is the monotonicity of the unreliability function; a full proof of Theorem 2 is given in the appendix of the extended version [9].

Algorithms for Fuzzy Unreliability in the General Case. As a direct consequence of Theorem 2, we can adapt any unreliability algorithm (see, for example, [29]) to also compute the fuzzy unreliability $\widetilde{U}_T(\vec{\mathsf{p}})$ in its α-cut representation, by applying it to the endpoints of the α-cuts of the given fuzzy probabilities. If we represent fuzzy numbers using N α-cuts, we will hence need to run our chosen unreliability algorithm $2N$ times to compute the fuzzy unreliability, introducing a merely constant overhead. Therefore, we may conclude that any unreliability algorithm is efficient for fuzzy fault trees whenever it is efficient for ordinary ones. Using the state-of-the-art (modularised) BDD-based unreliability algorithm [2], we confirm this conclusion with the experiments presented in the subsequent section.

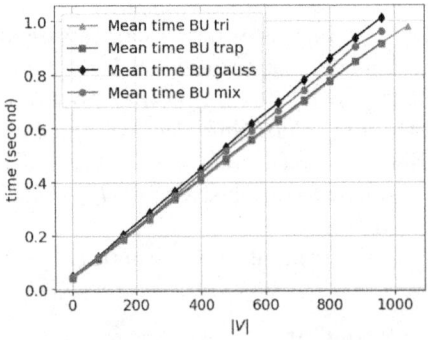

Fig. 6. Mean time (in seconds) of performing bottom-up algorithm. Groups of generated trees are determined by the number of nodes $|V|$.

Table 1. FTs from the benchmarks used as subtrees for generation.

Source	$|V|$
[33] Fig. 12-7	10
[33] Fig. 10-14	11
[23] Alt A1	17
[23] Alt A2	22
[23] Alt A6	31
[38] Fig. 3	35
[23] Alt A12	39
[28] Fig. 2	42
[23] Alt A11	45
[32] Fig. 5	50

7 Experiments

We evaluate the performance of our two methods for fuzzy unreliability: the linear-time bottom-up algorithm for the special case of tree-structured FTs, and the general approach for DAG-structured FTs based on Theorem 2.

7.1 Performance Evaluation for Tree-Shaped FTs

By the linear-time complexity of the bottom-up method, computing the unreliability for tree-structured FTs should be feasible even for very large sizes. To confirm this hypothesis, we therefore need to consider particularly large fault trees. However, obtaining large FTs for real-world systems is infeasible due to confidentiality reasons, and the fact that we require tree-structured and non-dynamic FTs for this experiment. For this reason, we generate a custom benchmark of large tree-structured FTs, using the set of ten FTs from the literature shown in Table 1. This method is inspired by the one used in [20].

Generation of Large Tree-Structured FTs. We generate large FTs using two ways of combining two FTs T_1, T_2 as shown in Fig. 7, starting from the ten FTs shown in Table 1:

1. *Horizontal combination.* We introduce a new root node with a random gate and add two edges: one edge from the new root to R_{T_1} and another one from the new root to R_{T_2}. The resulting system is a larger system consisting of two subsystems side-by-side.
2. *Vertical combination.* We randomly pick a basic event v from T_1 and replace v with T_2. The resulting system is a larger system in which the subsystem T_2 becomes part of T_1.

To obtain a benchmark of *fuzzy* FTs, we fuzzify the probabilities of all basic events using triangular, trapezoidal, truncated Gaussian fuzzy numbers, as well as a mixture of these, all centred at the original probabilities.

The experiment was then conducted as follows. For each number $k \in \{1, \ldots, N\}$, we randomly create large FTs using the two aforementioned combinations such that their number of nodes $|V|$ is at least k. These large FTs are then divided into groups by their size. Herein, we take $N = 1000$, $P = 80$ so the trees belong to group $\lceil |V_k|/P \rceil$. We run the bottom-up algorithm for every FT in each group and then derive the mean time for every group of large FTs. The mean time for running the bottom-up method for different BE fuzzy types is displayed in Fig. 6. As expected from Sect. 6.2, computation time is linear in FT size. Furthermore, computation is very fast,

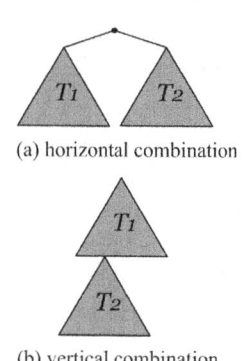

(a) horizontal combination

(b) vertical combination

Fig. 7. Combining FTs.

with even very large FTs taking only approximately 1s to calculate. Therefore, the proposed bottom-up algorithm is not only accurate, but also very efficient.

7.2 Performance Evaluation for General DAG-Structured FTs

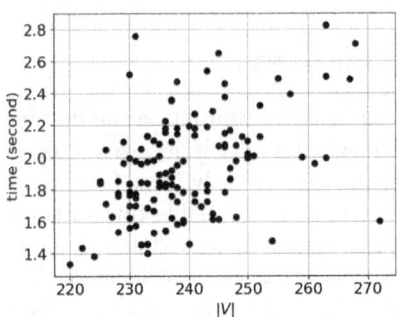

Fig. 8. Performance of our method for general DAG-structured FTs.

In this experiment, we employ Storm [12], a state-of-the-art model checker that calculates FT unreliability via a BDD-based approach with modularization [2]. Theorem 2 allows us to conveniently make use of this off-the-shelf tool to also compute the *fuzzy* unreliability; see Sect. 6.3. Since we do not need particularly large FTs in this case, we evaluate our method on an established benchmark of 125 randomly generated fault trees from [3]. The FTs in the benchmark have 239 nodes on average. The DAG-structured FTs in this set have *crisp* probabilistic status vectors. In order to obtain a benchmark of *fuzzy* fault trees, we equip each basic event b with the triangular fuzzy number $\triangle_{0.8p_b, p_b, 1.2p_b}$, given the crisp probability p_b. The number of α-cuts is set to 10. Figure 8 shows the performance of our approach, as measured by the runtime of each instance. We observe that the elapsed time of computation for each tree is generally fast (less than 3 s).

8 Related Work

In the 1980s, fuzzy fault tree analysis was pioneered by Tanaka et al. [35] and has since been extensively studied. Reviews are reported in [16, 22, 30].

Efficient Fuzzy Arithmetic. Various studies offer solutions to the problem of efficiently computing the fuzzy arithmetic operations. When dealing with the multiplication of triangular or trapezoidal fuzzy numbers, it is common to use approximations that yield a fuzzy number of the same type as the original operands [18, 26, 35]. However, these approaches yield approximate solutions that generally overestimate the fuzzy probability of the top event. While over-approximations are conservative, there are no formal guarantees on their accuracy, rendering this method less informative in general. Alternatively, [36] provides an efficient way to calculate fuzzy arithmetic operations using the α-cut representation of fuzzy numbers.

Uncertainty Quantification in FTs. Alternative frameworks for uncertainty quantification in fault trees include *sensitivity analysis* [30, 31], *imprecise probability* (or "probability intervals") [13, 14], as well as Bayesian approaches [27], in which uncertainty in parameters is treated by viewing these parameters as random variables themselves. The approach using imprecise probability is subsumed by fuzzy fault tree analysis using *interval fuzzy numbers*; see Sect. 3. Sensitivity analysis considers how sensitively a given risk metric responds to varying model parameters. In contrast to fuzzy fault tree analysis, it is not concerned with a precise quantification of uncertainty in these parameters themselves. Finally, the Bayesian method of [27] relies on Monte-Carlo simulation, which is an important further difference to fuzzy fault tree analysis, in addition to its differing conceptual foundation.

Defuzzification Approaches. A widely studied approach [17,19] to fuzzy fault tree analysis is to first obtain fuzzy probabilities for all basic events from expert opinions, then "defuzzify" these to a crisp values, and finally compute the probability of the top-level event as usual. Here, fuzzy theory is only used as an intermediate step for converting and aggregating expert knowledge to precise numerical probability values, and does not provide any uncertainty quantification at the system level.

Linguistic Variables. The concept of *linguistic variables* is used in fuzzy fault tree analysis to obtain fuzzy probabilities of basic events from expert opinions [5,19,25,39]. A linguistic variable is a variable whose values range over a finite set of natural language descriptions. Each such description is then assigned a fuzzy number, modelling its inherent ambiguities. In the case of fuzzy fault trees, the linguistic variables of interest concern the probabilities of basic events, ranging from "very low" over "medium" to "very high".

Case Studies. An overview of case studies using uncertainty quantification in fault trees, and fuzzy fault trees in particular, is given in [37].

9 Conclusion and Future Work

Based on our rigorous definition of fuzzy fault trees (Definition 6), Theorem 2 enables extending any fault tree unreliability algorithm to this setting—in a simple, correct and efficient manner. In addition, the efficacy of this method is confirmed by the experiments presented in Sect. 7.

There remain several open problems for future work. Most importantly, our rigorous formulation of fuzzy fault tree analysis may serve as a basis for a thorough and critical comparison to other frameworks for uncertainty quantification—which framework is most appropriate in which context? Moreover, finding efficient algorithms for purely probabilistic ("Bayesian") uncertainty quantification in fault trees remains an open direction. Finally, an interesting avenue for future research is to extend our results for fuzzy fault tree analysis to related risk models, such as dynamic fault trees (DFTs), as well as general DAG-structured attack trees.

Acknowledgements. This work was partially funded by the NWO grants NWA.1160.18.238 (PrimaVera), and KICH1.ST02.21.003 (ZORRO), the European Union's Horizon 2020 research and innovation programme under the Marie Skłodowska-Curie grant agreement No 101008233, the ERC Proof-of-Concept grant 101187945 (*RUBICON*), and the ERC Consolidator Grant 864075 (*CAESAR*).

Data Availibility Statement. Scripts to reproduce our experimental evaluation are archived and publicly available at [7].

References

1. Barros, L.C., Bassanezi, R.C., Lodwick, W.A.: The extension principle of Zadeh and fuzzy numbers. In: Barros, L.C., Bassanezi, R.C., Lodwick, W.A. (eds.) A First Course in Fuzzy Logic, Fuzzy Dynamical Systems, and Biomathematics. SFSC, vol. 347, pp. 23–41. Springer, Heidelberg (2017). https://doi.org/10.1007/978-3-662-53324-6_2
2. Basgöze, D., Volk, M., Katoen, J.P., Khan, S., Stoelinga, M.: BDDs strike back: efficient analysis of static and dynamic fault trees. In: Deshmukh, J.V., Havelund, K., Perez, I. (eds.) NFM 2022. LNCS, vol. 13260, pp. 713–732. Springer, Heidelberg (2022). https://doi.org/10.1007/978-3-031-06773-0_38
3. Basgöze, D., Volk, M., Katoen, J.P., Khan, S., Stoelinga, M.: Artifact for "BDDs Strike Back - Efficient Analysis of Static and Dynamic Fault Trees". Zenodo (2022)
4. Basiura, B., et al.: Fuzzy numbers, pp. 1–26. Springer, Cham (2015)
5. Bowles, J.B., Pelaez, C.E.: Application of fuzzy logic to reliability engineering. Proc. IEEE **83**(3), 435–449 (1995)
6. Couso, I., Borgelt, C., Hullermeier, E., Kruse, R.: Fuzzy sets in data analysis: from statistical foundations to machine learning. IEEE Comput. Intell. Mag. **14**(1), 31–44 (2019)
7. Dang, T.K.N.: Artefact for "fuzzy fault trees: the fast and the formal" (2025). https://doi.org/10.5281/zenodo.15337097
8. Dang, T.K.N., Lopuhaä-Zwakenberg, M., Stoelinga, M.: Fuzzy quantitative attack tree analysis. In: Beyer, D., Cavalcanti, A. (eds.) FASE 2024. LNCS, vol. 14573, pp. 210–231. Springer, Cham (2024). https://doi.org/10.1007/978-3-031-57259-3_10
9. Dang, T.K.N., Peterseim, B., Lopuhaä-Zwakenberg, M., Stoelinga, M.: Extended version (with appedndix) of "Fuzzy fault trees: the fast and the formal" (2025). https://doi.org/10.5281/zenodo.15720558
10. Dijkman, J., Van Haeringen, H., De Lange, S.: Fuzzy numbers. J. Math. Anal. Appl. **92**(2), 301–341 (1983)
11. Dubois, D., Prade, H.: Operations on fuzzy numbers. Int. J. Syst. Sci. **9**(6), 613–626 (1978)
12. Hensel, C., Sebastian, J., Katoen, J.P., Quatmann, T., Volk, M.: The probabilistic model checker storm. Int. J. Softw. Tools Technol. Transfer **24**, 589–610 (2022)
13. Jacob, C., Dubois, D., Cardoso, J.: Uncertainty handling in quantitative BDD-based fault-tree analysis by interval computation. In: Benferhat, S., Grant, J. (eds.) SUM 2011. LNCS (LNAI), vol. 6929, pp. 205–218. Springer, Heidelberg (2011). https://doi.org/10.1007/978-3-642-23963-2_17
14. Jacob, C., Dubois, D., Cardoso, J.: From imprecise probability laws to fault tree analysis. In: Hüllermeier, E., Link, S., Fober, T., Seeger, B. (eds.) SUM 2012. LNCS (LNAI), vol. 7520, pp. 525–538. Springer, Heidelberg (2012). https://doi.org/10.1007/978-3-642-33362-0_40

15. Jezewski, M., Czabanski, R., Leski, J.: Introduction to fuzzy sets. In: Prokopowicz, P., Czerniak, J., Mikołajewski, D., Apiecionek, Ł., Ślęzak, D. (eds.) Theory and Applications of Ordered Fuzzy Numbers. Studies in Fuzziness and Soft Computing, vol. 356, pp. 3–22. Springer, Cham (2017). https://doi.org/10.1007/978-3-319-59614-3_1
16. Kabir, S.: An overview of fault tree analysis and its application in model based dependability analysis. Expert Syst. Appl. **77**, 114–135 (2017)
17. Kumar, M., Singh, K.: Fuzzy fault tree analysis of chlorine gas release hazard in chlor-alkali industry using alpha-cut interval-based similarity aggregation method. Appl. Soft Comput. **125**, 109199 (2022)
18. Liang, G.S., Wang, M.J.J.: Fuzzy fault-tree analysis using failure possibility. Microelectron. Reliab. **33**(4), 583–597 (1993)
19. Lin, C.T., Wang, M.J.J.: Hybrid fault tree analysis using fuzzy sets. Reliabil. Eng. Syst. Saf. **58**(3), 205–213 (1997)
20. Lopuhaä-Zwakenberg, M., Stoelinga, M.: Attack time analysis in dynamic attack trees via integer linear programming. In: Ferreira, C., Willemse, T.A.C. (eds.) SEFM 2023. LNCS, vol. 14323, pp. 165–183. Springer, Cham (2023). https://doi.org/10.1007/978-3-031-47115-5_10
21. Lopuhaä-Zwakenberg, M.: Fault tree reliability analysis via squarefree polynomials (2023)
22. Mahmood, Y.A., Ahmadi, A., Verma, A.K., Srividya, A., Kumar, U.: Fuzzy fault tree analysis: a review of concept and application. Int. J. Syst. Assur. Eng. Manage. **4**, 19–32 (2013)
23. Mentes, A., Helvacioglu, I.H.: An application of fuzzy fault tree analysis for spread mooring systems. Ocean Eng. **38**(2), 285–294 (2011)
24. Nguyen, H.T.: A note on the extension principle for fuzzy sets. J. Math. Anal. Appl. **64**(2), 369–380 (1978)
25. Pan, N.F., Wang, H.: Assessing failure of bridge construction using fuzzy fault tree analysis. In: Fourth International Conference on Fuzzy Systems and Knowledge Discovery (FSKD 2007), vol. 1, pp. 96–100. IEEE (2007)
26. Peng, Z., Xiaodong, M., Zongrun, Y., Zhaoxiang, Y.: An approach of fault diagnosis for system based on fuzzy fault tree. In: Proceedings of the 2008 International Conference on MultiMedia and Information Technology, MMIT 2008, pp. 697–700. IEEE Computer Society, USA (2009)
27. Prabhu, S., Ehrett, C., Javanbarg, M., Brown, D.A., Lehmann, M., Atamturktur, S.: Uncertainty quantification in fault tree analysis: estimating business interruption due to seismic hazard. Nat. Hazard. Rev. **21**(2), 04020015 (2020)
28. R. Sadiq, E.S.M., Kleiner, Y.: Predicting risk of water quality failures in distribution networks under uncertainties using fault-tree analysis. Urban Water J. **5**(4), 287–304 (2008)
29. Rauzy, A.: New algorithms for fault trees analysis. Reliabi. Eng. Syst. Saf. **40**(3), 203–211 (1993)
30. Ruijters, E., Stoelinga, M.: Fault tree analysis: a survey of the state-of-the-art in modeling, analysis and tools. Comput. Sci. Rev. **15–16**, 29–62 (2015)
31. Rushdi, A.M.: Uncertainty analysis of fault-tree outputs. IEEE Trans. Reliab. **34**(5), 458–462 (1985)
32. Shu, M.H., Cheng, C.H., Chang, J.R.: Using intuitionistic fuzzy sets for fault-tree analysis on printed circuit board assembly. Microelectron. Reliab. **46**(12), 2139–2148 (2006)
33. Stamatelatos, M., Vesely, W., Dugan, J., Fragola, J., Minarick, J., Railsback, J.: Fault tree handbook with aerospace applications. NASA, Washington, DC (2002)

34. Stoelinga, M., Ruijters, E., Krčál, P.: Concise Guide to Fault Tree Analysis: Models, Methods and Algorithms. Springer, Cham (2025)
35. Tanaka, H., Fan, L.T., Lai, F.S., Toguchi, K.: Fault-tree analysis by fuzzy probability. IEEE Trans. Reliabil. **R-32**(5), 453–457 (1983)
36. Dong, W.M., Shah, H.C., Wongt, F.S.: Fuzzy computations in risk and decision analysis. Civ. Eng. Syst. **2**(4), 201–208 (1985)
37. Yazdi, M., Kabir, S., Walker, M.: Uncertainty handling in fault tree based risk assessment: state of the art and future perspectives. Process Saf. Environ. Prot. **131**, 89–104 (2019)
38. Yazdi, M., Nikfar, F., Nasrabadi, M.: Failure probability analysis by employing fuzzy fault tree analysis. Int. J. Syst. Assur. Eng. Manag. **8**, 1177–1193 (2017)
39. Yuhua, D., Datao, Y.: Estimation of failure probability of oil and gas transmission pipelines by fuzzy fault tree analysis. J. Loss Prev. Process Ind. **18**(2), 83–88 (2005)
40. Zadeh, L.: Fuzzy sets. Inf. Control **8**(3), 338–353 (1965)

Positive Almost-Sure Termination of Polynomial Random Walks

Lorenz Winkler(✉) and Laura Kovács

TU Wien, Vienna, Austria
lorenz.winkler@tuwien.ac.at

Abstract. The number of steps until termination of a probabilistic program is a random variable. Probabilistic program termination therefore requires qualitative analysis via almost-sure termination (AST), while also providing quantitative answers via positive almost-sure termination (PAST) on the expected number of steps until termination. While every program which is PAST is AST, the converse is not true. The symmetric random walk with constant step size is a prominent example of a program that is AST but not PAST.

In this paper we show that a more general class of polynomial random walks is PAST. Our random walks implement a step size that is polynomially increasing in the number of loop iterations and have a constant probability p of choosing either branch. We show that such programs are PAST when the degree of the polynomial is higher than both the degree of the drift and a threshold $d_{\min}(p)$. Our approach does not use proof rules, nor auxiliary arithmetic expressions, such as martingales or invariants. Rather, we establish an inductive bound for the cumulative distribution function of the loop guard, based on which PAST is proven. We implemented the approximation of this threshold, by combining genetic programming, algebraic reasoning and linear programming.

1 Introduction

Probabilistic programs extend programs written in classical programming languages by statements that draw samples from stochastic distributions, such as Normal and Bernoulli. The output as well as the number of steps until termination of a probabilistic program are random variables [14], which makes the analysis even harder than in the nonprobabilistic setting [12]. *In this paper we focus on probabilistic termination and introduce a class of loops for which we provide a new sufficient condition for positive almost-sure termination.*

Probabilistic loop termination [4,8] requires qualitative arguments via almost-sure termination (AST), and quantitative answers via positive almost-sure termination (PAST) on the expected number of steps until termination. While every program which is PAST is AST, the converse is not true. The symmetric random walk with constant step size is a prominent example of a program that is AST but not PAST. Existing works for proving PAST and/or AST rely on proof rules that need auxiliary arithmetic expressions, such as invariants and martingales, over program variables [17]. In particular, ranking supermartingales (RSMs) or lexicographical RSMs are commonly used ingredients in (P)AST analysis [1]. For probabilistic programs without nondeterminism, RSMs are a sound and complete method for proving termination [3,5].

However, finding an RSM is challenging, as shown in Fig. 1. Here, $\leftarrow$ denotes variable assignments and $\oplus_p$ captures probabilistic choice: the expression on the left hand side of $\oplus_p$ is chosen with probability p, and the one on the right hand side of $\oplus_p$ is chosen with probability $1-p$. While the program has a finite stopping time, to the best of our knowledge, no RSM of this program has been found until now. As such, existing methods would fail proving PAST for Fig. 1.

$n \leftarrow 0$
$s \leftarrow 0$
while $s \geq 0$ **do**
$\quad n \leftarrow n+1$
$\quad s \leftarrow s - n + 3 \oplus_{\frac{1}{2}} s + n + 5$
end while

Fig. 1. Program with non-trivial RSM

Our Approach. We overcome challenges of RSM inference in PAST analysis, by identifying a class of loops, called *polynomial random walks* (Sect. 3), for which PAST can be proven without martingale/invariant synthesis. While polynomial random walks are less expressive than arbitrary polynomial loops, their PAST analysis can be automated using genetic programming, algebraic reasoning and linear programming (Sect. 4).

Our approach relies on bounding the tails of random loop variables in order to guarantee that a random variable is close to its mean. Tail bounds [11] are important in providing guarantees on the probability of extreme outcomes. Our work analyzes tail probabilities $\mathbb{P}(X \geq t)$ of the random variable X summing up the steps, to establish that the probability that X exceeds some value t decreases fast with increasing t, and the deviation from 0 is in some sense "controlled". Key to our approach is that variables of polynomial random walks converge to random variables with almost Normal distributions (Lemma 5), whose tail bounds can be approximated (Lemma 6). We therefore transform polynomial random walks into programs with larger expected stopping time, where several steps are accumulated before the loop guard is checked. In other words, we connect polynomial random walk analysis to stochastic processes over random variables with almost Normal distributions whose variance is exponentially growing (Sect. 2). By summing up random variables of such processes, we prove that a sub-Gaussian tail bound is preserved. The cumulative distribution function of this summation can tightly be approximated using an *inductive bound* over random variables.

By proving that an inductive bound always exists, we find that constant-probability polynomial random walks are PAST when the step size grows fast enough (Theorem 2). Our PAST result holds whenever (i) the degree d of the polynomials is larger than the degree of the expected value of the increments, and (ii) d is larger than a threshold $d_{\min}(p)$ parametrized by the probabilistic choice probability p. Our work establishes that Fig. 1 satisfies this threshold (Table 1), implying thus PAST of Fig. 1 without the need of an RSM.

We implemented our PAST analysis over polynomial random walks in extension of the algebraic program analysis tool Polar [22]. To this end, we use linear programming, by relying on OR-Tools [24] and the Gurobi-solver [10], to derive inductive bounds for fixed parameters of the program transformation over polynomial random walks. We further combined Polar with a genetic algorithm, to find the best values for those parameters. Our experimental results in Sect. 4, give practical evidence on the tightness of our inductive bounds on polynomial random walks. Existence of these inductive bounds implies thus PAST of the programs that are analyzed.

Our Contributions. We translate the problem of verifying PAST into the problem of tightly approximating tail bounds of random loop variables. We bring the following contributions[1]:

- We introduce the class of polynomial random walks for which we provide sufficient conditions to determine PAST. These conditions do not require user-provided invariants and/or martingales. We determine PAST above a threshold $d_{\min}(p)$, which in turn only depends on the polynomial degree d of the loop updates and the probabilistic choice p in the loop (Sect. 3).
- We show that such a threshold $d_{\min}(p)$ always exists by transforming polynomial random walks into stochastic processes over almost Normal variables (Sect. 2). We prove that such processes admit an inductive bound over their cumulative distribution function, allowing us to tightly approximate our threshold $d_{\min}(p)$.
- We implemented our approach to approximate $d_{\min}(p)$, and hence conclude PAST, in extension of the Polar framework (Sect. 4). Our experiments showcase the tightness of our approximation, implying thus PAST.

2 Almost Normal Variables and Conditioning

This section introduces a special class of stochastic processes (Sect. 2.1) and establishes bound properties for such processes. The stochastic processes we consider are given in Fig. 2, for which we show that they induce random variables with sub-Gaussian tail bounds (Sect. 2.2) whose inductive bounds can tightly be approximated (Sect. 2.3). In Sect. 3 we then prove that polynomial random walks can be transformed into the stochastic process of Fig. 2, allowing to conclude PAST of polynomial random walks (Theorem 2).

$n \leftarrow 0$
$z \sim \mathcal{N}_{c_0}^{\delta_1, C_1}(0, \sigma_0)$
$s \leftarrow s_0 + z$
while $s \geq F_{S_n}^{-1}(\epsilon)$ **do**
　　$n \leftarrow n + 1$
　　$z \sim \mathcal{N}_{c_0}^{\delta_1, C_1}(0, \sigma_n)$
　　$s \leftarrow s + z$
end while

Fig. 2. Probabilistic programs summing up almost normally distributed variables

---
[1] with detailed proofs of our results given in [27].

In the sequel, we respectively denote by $\mathbb{N}, \mathbb{R}, \mathbb{R}^+$ the set of natural, real, and positive real (including zero) numbers. We reserve $n \in \mathbb{N}$ for the loop iteration (counter). We assume familiarity with probabilistic programs and their semantics, and refer to [7,11] for details. The probability measure is denoted by $\mathbb{P}$, while we use $\mathbb{E}$ to denote the expected values of random variables. Further, we denote with $(X|E)$ the conditional probability distribution of X given event E.

2.1 Almost Normally Distributed Loop Summations

We consider stochastic processes induced by the probabilistic program of Fig. 2, which uses the random variable series $\{S_n\}_{n \in \mathbb{N}}$ and $\{Z_n\}_{n \in \mathbb{N}}$ corresponding to the values of s and z in the n-th iteration. We respectively denote by F_{S_n} and F_{Z_n} the cumulative distribution function (cdf) of S_n and Z_n. The notation $z \sim \mathcal{N}_{c_0}^{\delta_1, C_1}(0, \sigma_n)$ indicates that z is drawn from a distribution that is almost equivalent to a Normal distribution $\mathcal{N}(0, \sigma_n)$ with variance σ_n^2, in a sense, that c_0 bounds the absolute deviation of cdf of Z_n from the cdf of the Normal variable. Additionally, δ_1 bounds the shift of the sub-Gaussian tail bound and C_1 is the multiplicative deviation of that bound's variance (see Lemma 1). The initial value of s is also drawn from such an almost Normal distribution with variance σ_0^2. In each loop iteration n, a sample is drawn from an almost Normal distribution with variance σ_n^2 and then added to s. The loop is exited, once s is within the smallest ϵ-fraction of S_n.

Recall that the stochastic processes induced by $\{S_n\}$ and $\{Z_n\}$ represent a Markov chain [7]. Based on the semantics of Fig. 2, the series S_n sums variables $Z_0, \ldots, Z_n$, as follows:

- $S_0 = Z_0$;
- $S_{n+1} = (S_n \mid S_n \geq F_{S_n}^{-1}(\epsilon)) + Z_{n+1}$, where $F_{S_n}^{-1}$ is the inverse of the cdf of S_n.

Note that for S_n only the paths are considered, for which the program has not yet terminated. To reason about termination of Fig. 2, we rely on a right tail bound of the random variable S_n, as it gives an upper limit on the probability of a random variable exceeding a certain value [11]. In other words, the right tail bound of S_n quantifies how likely it is to observe values of S_n in the extreme right (or upper) tail of a distribution. While such extreme cases might yield to the non-termination of Fig. 2, in Sect. 2.2 we show that a tail bound for S_n can be derived from the tail bound of Z_n, which ensures that this behavior is unlikely. Moreover, by adjusting standard properties of distributions [7], we also bound the behavior of Z_n, as listed below.

Lemma 1 (CDF deviation and bound). *Let $\{Z_n\}_{n \in \mathbb{N}}$ be a sequence of random variables, as defined in Fig. 2. Recall that Z_n follows an almost Normal distribution $\mathcal{N}_{c_0}^{\delta_1, C_1}(0, \sigma_n)$. The following holds:*

1. *The cdf of Z_n, denoted as F_{Z_n} deviates from a Normal distribution only by at most c_0. That is, $|F_{Z_n}(z) - \Phi(\frac{z}{\sigma_n})| \leq c_0$, where Φ denotes the cumulative distribution function of the Normal distribution.*

2. The variable Z_n admits a sub-Gaussian tail bound on its right tail, which is offset by $\delta_1 \sigma_n$ and has variance of $\frac{\sigma_n^2}{C_1}$. That is,

$$\forall a \in \mathbb{R}^+ : \mathbb{P}(Z_n \geq a + \delta_1 \sigma_n) \leq \exp\left(C_1 \frac{-a^2}{2\sigma_n^2}\right).$$

2.2 Tail Bound for S_n

Lemma 1 limits the behavior of variable z in Fig. 2. We next show that, in addition to z, also variable s does not grow in an uncontrolled way. Similarly to Lemma 1, we reason about the right tail of S_n and prove that it admits a sub-Gaussian tail bound (Lemma 2); as such, the probability of S_n being significantly greater than its expected value is limited.

Lemma 2 (Preservation of sub-gaussian tail bound). *Consider the random variable series $\{S_n\}_{n \in \mathbb{N}}$ induced by s in Fig. 2. Assume that the variance σ_n^2 grows such that the inequality $\sigma_{n+1}^2 \geq d(\sigma_1^2 + \cdots + \sigma_n^2)$ holds for some $d \in \mathbb{R}^+ \setminus \{0\}$. Then, S_n admits a sub-Gaussian (but not centered) tail bound:*

$$\forall a \in \mathbb{R}^+ : \mathbb{P}\left(S_n \geq a + \left(\frac{\sqrt{1+d}}{\sqrt{1+d}-1}\delta_1 + b\right)\sqrt{\sum_{i=0}^n \sigma_i^2}\right) \leq \exp\left(C_1 \frac{-a^2}{2\sum_{i=0}^n \sigma_i}\right),$$

where $b \geq \frac{\sqrt{2 \ln \frac{1}{1-\epsilon}}}{\sqrt{C_1}(\sqrt{(1+d)}-1)}$.

2.3 Inductive Bound Set for S_n

While Lemma 2 gives an upper bound for the right tail of S_n, the sub-Gaussian tail bound of S_n is not very sharp for values close to the mean of S_n. In this section we extend Lemma 2 with a *union-bound compositional approach* to tighten cdf bounds, as follows. We (i) split the cdf into $m \in \mathbb{N}$ pieces, (ii) provide lower bounds $B(S_n)$ for each piece of the cdf, and (iii) combine the lower bounds inductively into a tighter upper bound for the cdf of S_n.

Our compositional framework is inductive over the lower bounds of cdf pieces: S_0 satisfies the bound $B(S_0)$ (base case) and, if S_n satisfies the bound $B(S_n)$, then S_{n+1} satisfies $B(S_{n+1})$ (induction step). Our bound $B(S_n)$ is uniquely defined by a set of inequalities using two vectors $\boldsymbol{a}_B, \boldsymbol{b}_B$ of bounding values, where elements of $\boldsymbol{a}_B, \boldsymbol{b}_B$ provide the location of and lower bounds on the cdf pieces of S_n. As such, we set:

$$B(S_n) = \left\{ \mathbb{P}\left(S_n \leq \boldsymbol{a}_{B,1}\sqrt{\sum_{i=0}^n \sigma_i^2}\right) \geq \boldsymbol{b}_{B,1}, \quad \ldots, \quad \mathbb{P}\left(S_n \leq \boldsymbol{a}_{B,m}\sqrt{\sum_{i=0}^n \sigma_i^2}\right) \geq \boldsymbol{b}_{B,m} \right\}. \quad (1)$$

For simplicity, we assume $\boldsymbol{a}_1 \leq 0$ and $\boldsymbol{b}_1 \geq \epsilon$, in order to ensure that only negative values of s exit the loop therefore enforcing $F_{S_n}^{-1}(\epsilon) \leq 0$. If each inequality in $B(S_n)$ is valid, we say that the bound $B(S_n)$ holds. By simple arithmetic reasoning, we state the following property over bounds.

Lemma 3 (Partial order of bounds). *The bounds $B(S_n)$ admit an ordering whenever they describe the same intervals and probabilities are ordered. That is:*

$$B'(\cdot) \leq B(\cdot) \iff \boldsymbol{a}_{B'} = \boldsymbol{a}_B \wedge \forall_{1 \leq i \leq m} : b_{B',i} \geq b_{B,i}$$

Inductive Computations of Bound Set for S_n. With regard to the partial order, we provide our inductive computation for bounds of S_n: from a bound $B(\cdot)$ that holds for S_n, we compute a bound $B'(\cdot)$ that holds for S_{n+1}. If $B'(\cdot) \leq B(\cdot)$, then $B(S_m)$ holds for all $m \geq n$, as our computation of new bounds ensures monotonicity of bounds. Doing so and starting with $B(S_n)$, (i) the left tail with a cdf of ϵ is cut away from S_n, yielding S'_n. Then, using Lemma 1, we (ii) add an almost Normal variable with variance $\sigma^2_{n+1} \geq d \sum_{i=0}^n \sigma_i^2$ to the resulting distribution of S'_n, and compute a new bound $B'(S_{n+1})$. Our bound computation uses a union-bound approach for deriving interval boundaries. In addition to the bounds from B, we use tail bounds from Lemma 2, as otherwise obtaining an inductive bound for S_n with $\epsilon > 0$ is not possible.

The next example illustrates our inductive bound set computation for S_n.

Example 1. Consider an instance of the stochastic process of Fig. 2, by setting $\epsilon = 0.1$ and $d = 3$ and using a set of almost Normal variables $\{Z_n\}_{n \in \mathbb{N}}$ with parameters $C_1 = 1$, $\delta_1 = 0$, and $c_0 = 10^{-3}$. For this instance of Fig. 2, we define the bound set:

$$B(S_n) = \left\{ \mathbb{P}(S_n \leq 0) \geq 0.1, \quad \mathbb{P}\left(S_n \leq \sqrt{\sum\nolimits_{i=0}^n \sigma_i^2}\right) \geq 0.4 \right\}.$$

For the tail bounds of Lemma 2, we (arbitrarily) pick the value $3\sqrt{\sum_{i=0}^n \sigma_i^2}$ and derive:

$$\mathbb{P}\left(S_n \geq 3\sqrt{\sum\nolimits_{i=0}^n \sigma_i^2}\right) \leq \exp\left(\frac{-(2.54)^2}{2}\right) \leq 0.04.$$

Therefore, $\mathbb{P}(S_n \leq 3\sqrt{\sum_{i=0}^n \sigma_i^2}) \geq 0.96$. Using these inequalities, we compute new bounds S'_n. Here, $S'_n = (S_n \geq F_{S_n}^{-1}(\epsilon))$ is the variable obtained from S_n by cutting away the left tail with weight ϵ. For readability, we denote the summation of variances up to S_n through $\sigma^2_{S_n} := \sum_{i=0}^n \sigma_i^2$, and similarly $\sigma^2_{S_{n+1}} := \sum_{i=0}^{n+1} \sigma_i^2$. With this notation at hand, we have:

$$\mathbb{P}(S_{n+1} \leq 0) \geq \mathbb{P}(S'_n \leq \sigma_{S_n}) \cdot \mathbb{P}(Z_{n+1} \leq -\sigma_{S_n}) +$$
$$\mathbb{P}(\sigma_{S_n} \leq S'_n \leq 3\sigma_{S_n}) \cdot \mathbb{P}(Z_{n+1} \leq -3\sigma_{S_n})$$
$$\mathbb{P}(S_{n+1} \leq \sigma_{S_{n+1}}) \geq \mathbb{P}(S'_n \leq \sigma_{S_n}) \cdot \mathbb{P}(Z_{n+1} \leq \sigma_{S_{n+1}} - \sigma_{S_n}) +$$
$$\mathbb{P}(\sigma_{S_n} \leq S'_n \leq 3\sigma_{S_n}) \cdot \mathbb{P}(Z_{n+1} \leq \sigma_{S_{n+1}} - 3\sigma_{S_n})$$

Note that $\sqrt{\sum_{i=0}^{n+1} \sigma_i^2} \geq \sqrt{1+d}\sqrt{\sum_{i=0}^n \sigma_i^2}$ and Z_{n+1} has standard deviation $\sqrt{3\sum_{i=0}^n \sigma_i^2}$. Using the bound set $B(S_n)$ and Lemma 1, we get:

$$\mathbb{P}\left(S_{n+1} \leq 0\right) \geq \frac{0.4 - \epsilon}{1-\epsilon}\left(\Phi\left(\frac{-1}{\sqrt{3}}\right) - c_0\right) + \frac{0.56}{1-\epsilon}\left(\Phi\left(\frac{-3}{\sqrt{3}}\right) - c_0\right) \approx 0.1188$$

Similarly, for the second bound $\mathbb{P}(S_{n+1} \leq \sqrt{\sum_{i=0}^{n+1} \sigma_i^2}) \gtrsim 0.4138$. The bound B is inductive as the new bound which we computed for S_{n+1} is smaller than the initial bound B which is assumed to hold for S_n. △

On the Existence of Inductive Bounds. Our approach to inductively computing bound sets for S_n relies on an union-bound argument to improve the cdf bound of Lemma 2. Recall that the sub-Gaussian tail bound of S_n only depends on $d \in \mathbb{R}^+ \setminus 0$ and $\epsilon \in \mathbb{R}^+ \setminus 0$. We next show that the existence of an inductive bound set is conditioned only by such a d. Namely, Theorem 1 ensure that, for every d there is an $\epsilon \in \mathbb{R}^+ \setminus 0$, such that an inductive bound exists for the corresponding series $\{S_n\}_{n \in \mathbb{N}}$, with $F_{S_n}^{-1}(\epsilon) \leq 0$. The probability of S_n being smaller than zero is therefore *lower bounded* by some nonzero percentage.

Theorem 1 (Inductive bound set). *For every $d \in \mathbb{R}^+ \setminus 0$ there exists an $\epsilon \in \mathbb{R}^+ \setminus 0$ such that an inductive bound set $B(S_n)$ holds, with $F_{S_n}^{-1}(\epsilon) \leq 0$, given that c_0 converges to 0 and can be chosen arbitrarily small.*

3 Polynomial Random Walks

Section 2 showed that stochastic processes defined by Fig. 2 have bounded behavior, allowing us to lower bound the termination probability via sub-Gaussian tail bounds and inductive bound sets. In this section we map the termination analysis of certain polynomial programs, called polynomial random walks, to the framework of Sect. 2. Importantly, *we reduce the problem of verifying PAST of polynomial random walks to the problem of ensuring existence of inductive bounds* (Theorem 2). Our recipe consists of transforming a polynomial random walk program $\mathcal{P}$ to a program that (i) bounds PAST of $\mathcal{P}$ and (ii) is equivalent to the stochastic process of Fig. 2.

$n \leftarrow 0$
$y \leftarrow y_0$
while $y > 0$ **do**
$\quad n \leftarrow n + 1$
$\quad x \leftarrow q_1[n] \oplus_p q_2[n]$
$\quad y \leftarrow y + x$
end while

Fig. 3. Polynomial random walk $\mathcal{P}$

3.1 Programming Model

We define the class of *polynomial random walks* via the programming model of Fig. 3, where $q_1[n], q_2[n] \in \mathbb{R}[n]$ are arbitrary polynomial expressions in the loop counter n. The *degree of a polynomial random walk program* $\mathcal{P}$, written as $\deg(\mathcal{P})$, is given by the maximum degree of its polynomials, that is $\deg(\mathcal{P}) = \max\{\deg(q_1[n]), \deg(q_2[n])\}$. The series $\{X_n\}_{n \in \mathbb{N}}, \{Y_n\}_{n \in \mathbb{N}}$ induced by the random loop variables x, y are next defined.

Definition 1 (Random walk variables). *The random walk variable X_n corresponding to the loop variable x at iteration n in Fig. 3 is*

$$X_n = \begin{cases} q_1[n] & \text{with probability } p \\ q_2[n] & \text{with probability } 1-p. \end{cases}$$

The random walk variable Y_n captures the distribution of y after iteration n, as:

$$Y_{n+1} = (Y_n | Y_n > 0) + X_{n+1}.$$

The second-order moment of a random variable X_n is written as $Var(X_n)$. For Fig. 3, we have $\mathbb{E}(X_n) = q_1[n]p + q_2[n](1-p)$ and $Var(X_n) = q_1[n]^2 p + q_2[n]^2(1-p)$, capturing the mean (first moment) and variance (second moment) of X_n; note that both moments of X_n are also polynomials in n.

To prove PAST of Fig. 3 we need to prove that the expected value of its stopping time is finite [12]. Based on the semantics of Fig. 3, it is easy to see that the stopping time of Fig. 3 is given by the first iteration n in which Y_n becomes negative.

Definition 2 (Expected stopping time). *Let T be $\inf\{n \geq 0 : Y_n \leq 0\}$, where T denotes the stopping time of the stochastic process induced by the polynomial random walk of Fig. 3. The expected stopping time of Fig. 3 is defined as $\mathbb{E}(T) = \sum_{n=1}^{\infty} \mathbb{P}(T \geq n)$.*

We exploit Definition 2 to show that Fig. 3 is PAST under additional conditions. Namely, we translate Fig. 3 into Fig. 4 and ensure that the stopping time of Fig. 4 becomes finite above a certain threshold; this threshold depends only on the maximum polynomial degree of Fig. 3 and the variable k. We then show that finiteness of the stopping time of Fig. 4 implies PAST of Fig. 3 (Lemma 4).

Program Transformation. We translate Fig. 3 into the stochastic process of Fig. 4. This program transformation is defined through the parameters n_0, k and g. The loop body of Fig. 3 is initially executed several times, accumulating n_0 steps. In every iteration of the outer loop k times as many steps as before are summed up, before the loop guard is checked again.

```
n ← 0
y ← y₀
while n ≤ n₀ do
    n ← n + 1
    x ← q₁[n] ⊕_p q₂[n]
    y ← y + x
end while
while y > g do            ▷ where g ≤ 0
    z ← 0
    n' ← n
    while n ≤ n' · k do
        n ← n + 1
        x ← q₁[n] ⊕_p q₂[n]
        z ← z + x
    end while
    y ← y + z
end while
```

Fig. 4. Transformed random walk

Furthermore, the loop guard of Fig. 4 might be relaxed, as $g \leq 0$. We highlight similarities between Fig. 4 and the summation of almost Normal variables with conditioning in Fig. 2: the inner loop of

Fig. 4 computes the value for z by summing up X_i. As argued in Sect. 3.2, this is similar to drawing z from an almost Normal distribution as in Fig. 2.

We have that the expected stopping time of Fig. 4 is larger than of Fig. 3.

Lemma 4 (Stopping Time Inequality). *Let T' be $\inf\{n \geq n_0 : Y_n \leq g\}$, denoting the stopping time of Fig. 4. Then, $\mathbb{E}(T) \leq \mathbb{E}(T')$.*

We denote with p_{term} a lower bound for the probability of the outer loop terminating each iteration. In what follows, we will ensure the stopping time of the program in Fig. 4 is finite when the probability p_{term} is high enough and k is small (Lemma 8). The existence of a nonzero lower bound for p_{term} is implied by Theorem 1; we note that p_{term} depends on the probability of choosing a branch p and grows as $k^{2\deg(\mathcal{P})} + 1$ increases. By setting k to its maximum value, we derive a threshold $d_{\min}(p)$ for the degree $\deg(\mathcal{P})$ of the polynomial random walk of Fig. 4 and prove that the stopping time of Fig. 3 above this threshold $d_{\min}(p)$ is finite (Theorem 2). We thus use $d_{\min}(p)$ to provide a sufficient condition for verifying PAST of the polynomial random walks in Fig. 3.

3.2 Loop Summations of Polynomial Random Walk Increments

We now establish the formal connection between the polynomial random walks of Fig. 4 and the stochastic processes of Fig. 2. We prove that the loop summation (defined below) of the increments of the random walk in Fig. 4 is almost normally distributed as given in Lemma 1, when an inequality over the degrees of the expected value of the step and its variance is true. This inequality holds, whenever the leading terms of the steps cancel out.[2]

Definition 3 (Random walk loop summation). *The random variables $U_0 = y_0 + X_0 + \cdots + X_{n_0}$ and $U_{n'} = X_{n'} + \cdots + X_{\lceil n' \cdot k \rceil}$ are (loop) summations of the random variables X_i of Fig. 4.*

Lemma 5 then shows that the absolute deviation c_0 from the cdf of the Normal distribution converges to 0. Further, Lemma 6 conjectures that the summation of random walk increments admits a sub-Gaussian tail bound with $C_1 = 4p(1-p)$ and δ_1 converging to 0, thus establishing, that the loop summation follows an almost Normal distribution $\mathcal{N}_{c_0}^{\delta_1, 4p(1-p)}$.

Lemma 5 (Convergence of cdf deviation). *Assume that $\deg(Var(X_i)) > 2\deg(\mathbb{E}(X_i)) + 1$ holds for Fig. 4. Then, the normalizations of the loop summations U_0 and $U_{n'}$ follow a Normal distribution up to a constant error c_0, with c_0 converging to 0 with increasing n_0:*

$$\forall n' \geq n_0 : \left| F_{U_{n'}}(z) - \Phi\left(\frac{z}{\sqrt{\sum_{i=n_0}^{\lceil n' \cdot k \rceil} Var(X_i)}}\right) \right| \leq c_0 \qquad (2)$$

---

[2] in case the inequality does not hold but the expected value of the step is asymptotically negative, the polynomial random walk can be transformed into a program with asymptotically larger stopping time for which the inequality holds.

where n_0 is as given in Fig. 4 and $F_{U_n'}$ denotes the cdf of $U_{n'}$.

Using Lemma 5, we derive that the loop summations of polynomial random walks follow an almost Normal distribution, similarly to the stochastic process of Fig. 2 in Sect. 2.

Lemma 6 (Tail bound for $U_{n'}$). *Let $\sigma_{U_{n'}}$ be the standard deviation of $U_{n'}$ and assume $\deg(Var(X_i)) > 2\deg(\mathbb{E}(X_i)) + 1$. Then, the right tail probability is bounded, with δ_1 converging to 0, as follows:*

$$\mathbb{P}(U_{n'} \geq \lambda \sigma_{U_{n'}} + \delta_1 \sigma_{U_{n'}}) \leq \exp\left(4(1-p)p\frac{-\lambda^2}{2}\right), \text{ and}$$
$$\mathbb{P}(U_0 \geq \lambda \sigma_{U_0} + \delta_1 \sigma_{U_0}) \leq \exp\left(4(1-p)p\frac{-\lambda^2}{2}\right).$$

Example 2. Consider our motivating example from Fig. 1. In order to ensure that its loop summations follow an almost Normal distribution, with c_0 and δ_1 converging to zero, we need to ensure that $\deg(Var(X_i)) > 2\deg(\mathbb{E}(X_i)) + 1$. This inequality is true, since $Var(X_i) = (i+1)^2$ and $\mathbb{E}(X_i) = 4$, hence $\deg(Var(X_i)) = 2$ and $\deg(\mathbb{E}(X_i)) = 0$. Consequently, Lemmas 5–6 can be used. □

In the remaining, we define the random variable series $\{Z_n\}_{n \in \mathbb{N}}$ corresponding to the loop summation of the inner loop of Fig. 4. That is, Z_n captures the program variable z at the end of every iteration of the outer loop of Fig. 4, with Z_0 being the variable corresponding to z after its first loop. As such,

- $Z_0 = U_0$, and
- $Z_n = U_{n'_{(n)}}$, where $n'_{(n)}$ is the value of n' in the n-th iteration of the outer loop of Fig. 4.

Further, Y_n is induced by the program variable y of Fig. 4, capturing the loop summation of Z_n with repeated conditioning. In order to use inductive bound sets as in Theorem 1, the variance of $\{Z_n\}_{n \in \mathbb{N}}$ must grow consistently and exponentially. This is however clearly ensured by choosing $k > 1$ in Fig. 4, implying the following result.[3]

Lemma 7 (Growth of variance). *The variance $\{\sigma_n^2\}_{n \in \mathbb{N}}$ of $\{Z_n\}_{n \in \mathbb{N}}$ grows exponentially, with δ' converging to 1:*

$$\sigma_{n+1}^2 \geq \left(\delta' k^{2\deg(\mathcal{P})+1} - 1)\right) \sum_{i=0}^{n} \sigma_i^2$$

Lemmas 5 and 6 establish that Z_n follows an almost Normal distribution as in Lemma 1. Together with Lemma 7, this ensures that the right tail of Y_n can be bounded (Lemma 2), and therefore inductive bounds can be used. Based on this bounds, Sect. 3.3 introduces conditions on the stopping time T of Fig. 4 being finite, implying thus PAST of Fig. 3.

---

[3] see Appendix A.2 in [27] for proof.

3.3 Bounding the Stopping Time and PAST

Recall that, using Definition 2, the expected stopping time $\mathbb{E}(T)$ of Fig. 4 is determined by the loop summation variables Y_n and is set to:

$$\mathbb{E}(T) = \sum_{n=1}^{\infty} \mathbb{P}(T \geq n).$$

Using Lemma 7, we obtain the following bound on $\mathbb{P}(T \geq n)$, and hence on $\mathbb{E}(T)$.

Lemma 8 (Bounding the stopping time). *Assume that the outer loop of Fig. 4 terminates with probability p_{term} after some n_0. Then,*

$$\mathbb{P}(T \geq n) \leq \min\left\{1, Bn^{\frac{\ln(1-p_{\text{term}})}{\ln(k+\frac{1}{n_0})}}\right\}$$

where $B = \dfrac{1}{(1-p_{\text{term}})^{\log_{k+\frac{1}{n_0}}(n_0)+2}}$. *Therefore, if* $\ln(1-p_{\text{term}}) < -\ln(k+\frac{1}{n_0})$ *holds, then the expected stopping time $\mathbb{E}(T)$ is finite.*

On the Finiteness of Stopping Times. Lemma 8 formulates conditions under which Fig. 4 has finite stopping time. These conditions effectively only depend on the probability p_{term} and k, as n_0 can be chosen arbitrarily. As such, *finiteness of $\mathbb{E}(T)$ and PAST of Fig. 4 is reduced to finding an inductive bound*, with $d = \delta' k^{2\deg(\mathcal{P})+1} - 1$, $C = p(1-p)$ and ϵ so large, that the inequality in Lemma 8 is satisfies. The terms δ_1, δ' and c_0 can be computed from a finite, arbitrary n_0.

To this end, let $p_{\text{i.b.}}(d, n_0, p)$ be a to-be-determined function that returns the largest ϵ for which an inductive bound exists, which is a lower bound for p_{term}. Then,

$$\mathbb{P}(T \geq n) \leq \inf_{1 < k}\left\{\inf_{0 \leq n_0}\left\{\min\left\{Bn^{\frac{\ln(1-p_{\text{i.b.}}(\delta' k^{2\deg\{\mathcal{P}+1\}}-1, n_0, p))}{\ln(k+\frac{1}{n_0})}}\right\}\right\}\right\} \quad (3)$$

with $B = (1 - p_{\text{i.b.}}(\delta' k^{2\deg\{\mathcal{P}+1\}} - 1, n_0, p))^{-(\log_{k+\frac{1}{n_0}}(n_0)+2)}$. Enforcing (3) requires however *solving a non-trivial optimization problem*: we need to approximate the function $p_{\text{i.b.}}(d, n_0, p)$. While in Sect. 4 we show that this approximation can be done using linear programming and a genetic algorithm, the statement of (3) has theoretical consequences. The existence of an inductive bound set from Theorem 1 implies that an ϵ for $p_{\text{i.b.}}(d, n_0, p)$ always exist, allowing us to state a PAST condition over polynomial random walks $\mathcal{P}$ from Fig. 3.

Theorem 2 (PAST of polynomial random walks). *Let $\mathcal{P}$ be a polynomial random walk program of Fig. 3. For every probabilistic choice p in $\mathcal{P}$ there exists a threshold $d_{\min}(p)$ such that $\mathcal{P}$ has finite expected stopping time, when $\deg(\mathcal{P}) > d_{\min}(p)$ and $\deg(\mathcal{P}) > \deg(p(q_1[n]) + (1-p)q_2[n])$.*

Example 3 (PAST of Fig. 1). There exists an inductive bound with $\epsilon \leq 0.1128$ and $d \geq 0.4102$ for $p = 0.5$. These constants are chosen so that convergence is guaranteed; see Appendix C in [27] for details. As $d = \delta' k^{2\deg(\mathcal{P})+1} - 1$ and δ' converges to 1 (Lemma 7), $k \geq 1.1214$ ensures that d is large enough when the degree of a polynomial random walk program is at least 1 (that is, at least linear updates).

Through this bound and Lemma 8, the stopping time of a polynomial random walk program $\mathcal{P}$ with $\deg(\mathcal{P}) \geq 1$ is bounded: $\mathbb{E}(T) \leq \sum_{n=1}^{\infty} Bn^{\frac{\ln(0.8872)}{\ln(1.1214)+\tau}}$, where τ converges to 0. This stopping time bound has an exponent which is smaller than -1.04; therefore, the loop summation of the respective Fig. 4 is finite and $d_{\min}(0.5) \leq 1$.

Using Theorem 2 we conclude that polynomial random walks with linearly (or faster) increasing step size and branching probability 0.5 have finite expected stopping time and are PAST, given that $\deg(\mathcal{P}) > \deg(p(q_1[n]) + (1-p)q_2[n])$. In particular, this is true for Fig. 1, as shown in Example 2, hence it is PAST. □

Higher Moments of the Stopping Time. We conclude this section by noting that solving (3) and applying Theorem 2 allows us to derive not only PAST, but also higher moments of the stopping times of polynomial random walks $\mathcal{P}$. That is, the bound we compute for $\mathbb{P}(T \geq n)$ by solving (3) is of the form Bn^m and this bound can be used to bound higher moments N of the stopping time. In particular, $\mathbb{E}(T^N) = \sum_{n=1}^{\infty} \mathbb{P}(T^N \geq n) = \sum_{n=1}^{\infty} \mathbb{P}(T \geq \sqrt[N]{n}) \leq \sum_{n=0}^{\infty} Bn^{\frac{m}{N}}$. Therefore, when (3) is solved using a bound with $m < -N$, then $\mathbb{E}(T^N)$ is finite.

4 Implementation and Experiments

Theorem 2 states sufficient conditions under which the polynomial random walk programs $\mathcal{P}$ of Fig. 3 are PAST. These sufficient conditions can be checked by solving inequalities among random walk updates and, importantly, by finding solutions to the optimization problem of (3). In this section, we detail our implementation to find tight bounds to (3), allowing us to conclude PAST of $\mathcal{P}$. Our implementation involves heuristic optimization techniques to find provably correct solutions. Our experiments provide practical evidence on the tightness of computed stopping time bounds and give evidence of the reliability of our approach[4], despite the absence of convergence guarantees.

4.1 Computing Tight Bounds on Stopping Times

We solve (3) in extension of the Polar program analyzer [22]. We use Polar to compute closed form expressions for the loop-guard changes of probabilistic branches, allowing us to support programs $\mathcal{P}$ that are even more general than Fig. 3. We combine Polar with linear programming through OR-Tools [24] and derive inductive bounds for fixed program transformation parameters. To find

---
[4] see Appendix C.3 in [27].

the best values for these parameters, we rely on genetic algorithms, such that the fitness functions of these genetic algorithms are controlled by our linear solver. Doing so, we use the Gurobi-solver [10] to solve linear models. By integrating algebraic reasoning, linear programming and genetic algorithms, our implementation in Polar minimizes the exponent in the bound of $\mathbb{P}(T \geq n)$ in (3), which is sufficient to prove PAST and finiteness of further higher moments of $\mathcal{P}$ (Theorem 2). By changing the objective function, our implementation can also minimize an *explicit* bound for the expected stopping time $\mathbb{E}(T)$.

Inferring Inductive Bound Sets. To compute an inductive bound set B in (1), the parameters $\epsilon, d, c_0, C_1, \delta_1$ must be fixed. Additionally, we require vectors $\boldsymbol{a}_B$ and $\boldsymbol{c}_B$, specifying respectively which m lower bounds for the inductive bound-set are computed and which k tail bounds are used. We compute values for the bounds by solving the linear inequality:

$$b_i \geq \sum_{j=1}^{m} d_j \left(\Phi \left(\frac{a_i \sqrt{1+d} - a_j}{\sqrt{d}} \right) - c_0 \right) + \sum_{j=1}^{k} d_{m+j} \left(\Phi \left(\frac{a_i \sqrt{1+d} - c_j}{\sqrt{d}} \right) - c_0 \right) \quad (4)$$

for $i = 1, \ldots, m$. The vector $\boldsymbol{d}$ denotes auxiliary variables, which describe the difference of neighbouring bounds. Additionally, we enforce[5] that the initial, almost Normal, distribution of Z_0 satisfies the bound set B.

Our implementation invokes linear programming over the linear model (4) in the form of an indicator function. This function returns 1, when an inductive bound set B is found for the given parameters $\epsilon, d, c_0, C_1, \delta_1$; and 0 otherwise.

Genetic Agorithm. We use a genetic algorithm to solve the optimization problem (3) and find the best parameter values in (4), for which an inductive bound set B exists. Our genetic algorithm repeatedly modifies a collection of individual solutions: we select individuals from the current set of solutions and use them to produce next individuals/solutions. An individual has (i) the properties d, ϵ, and n_0 to capture the program transformation of Fig. 4 and (ii) the parameters g, s, c to specify the vectors $\boldsymbol{a}$ and $\boldsymbol{c}$ of (4) for the inductive bound B. Specifically, we set $\boldsymbol{a}_1 = 0, \boldsymbol{a}_2 = \frac{s}{g-1}, \ldots, \boldsymbol{a}_g = s$, and $\boldsymbol{c} = \begin{pmatrix} c \end{pmatrix}$.

The fitness of an individual is calculated by first calculating the exponent of the bound. In case an explicit bound should be computed, the error-terms c_0 and δ_1 are inferred from n_0. Otherwise, we choose very small values, such as $c_0 = \delta_1 = 10^{-8}$, $\delta' = 1 - 10^{-8}$. Next, we solve our linear model (4). If no solution is found, we set $m = 0$; otherwise, we take $m = \frac{\ln(1-\epsilon)}{\ln(k+\frac{1}{n_0})}$ with $k = \left(\frac{(d+1)}{\delta'} \right)^{\frac{1}{2 \deg(\mathcal{P})+1}}$. If $m < -1$ and an explicit bound is sought, we compute the summation $\mathbb{E}(T) = \sum_{n=1}^{\infty} \mathbb{P}(T \geq n)$ using the Hurwitz ζ-function[6]. The fitness of an individual is further expressed via the tuple $(\mathbb{E}(T), m)$, which is minimized/optimized with respect to the usual lexicographical ordering.

From a given solution set (generation), a new solution set (population) is generated using random mutations of the parameters. The property d is biased to decrease, while ϵ is biased to increase. Additionally, n_0 is biased to increase

---

[5] see Appendix B in [27].
[6] (25.11)NIST:DLMF.

when the exponent is not smaller than −1, and biased to decrease otherwise. Furthermore, new individuals are generated by randomly selecting the properties of two parent individuals.

Table 1. Derived bounds on stopping times for polynomial random walk programs $\mathcal{P}$, with increasing maximal degree deg P and different probabilistic choices p. The program in the 3rd line of the table corresponds to Fig. 1.

deg($\mathcal{P}$)	p	measured exponent	tightest bound
0.25	0.5	−0.744	−0.5589
0.5	0.5	−0.997	−0.7436
1	0.5	−1.508	−1.1189
2	0.5	−2.442	−1.8639
5	0.5	−4.588	−4.0843
3	0.5	−3.448	−2.5971
3	0.9	−3.334	−2.4453
3	0.1	−3.57	−2.4453
3	0.99	−3.152	−1.9321
3	0.01	−3.516	−1.9321
3	0.999	−3.144	−1.359
3	0.001	−3.562	−1.359

4.2 Experimental Results

We evaluated our approach for computing stopping time bounds, and hence, inferring PAST, using various polynomial random walk programs $\mathcal{P}$. To this end, we took instances of Fig. 3 with different random walk degrees deg P and various values of the probabilistic choice p. The PAST analysis of such programs is out of reach of existing tools (see Sect. 5), notably Amber [21], eco-imp [2], KoAT [15], LexRsm [1], and LazyLexRsm [25].

Table 1 summarizes our experiments, with the third line of Table 1 being our motivating example from Fig. 1. Column 3 of Table 1 reports the empirical exponents of $\mathbb{P}(T \leq n)$, further detailed in Fig. 5a. Column 4 of Table 1 states the smallest (tightest) exponent obtained through our approach using inductive bounds. Our experiments were run on a machine with 2x AMD EPYC 7502 32-Core processor with one task per core and hyperthreading disabled.

Experimental Analysis. Figure 5a displays empirically measured rates of $\mathbb{P}(T \geq n)$ for symmetric random walks with varying degree. These probabilities appear to converge towards a line in the log-log plot, which suggests, that $\mathbb{P}(T \geq n)$ eventually is of form Bn^m, coinciding with the form of our bound. The observed exponent of this probability is the slope of the robust log-log regression lines [26], displayed as dashed lines and displayed in Column 3 of Table 1.

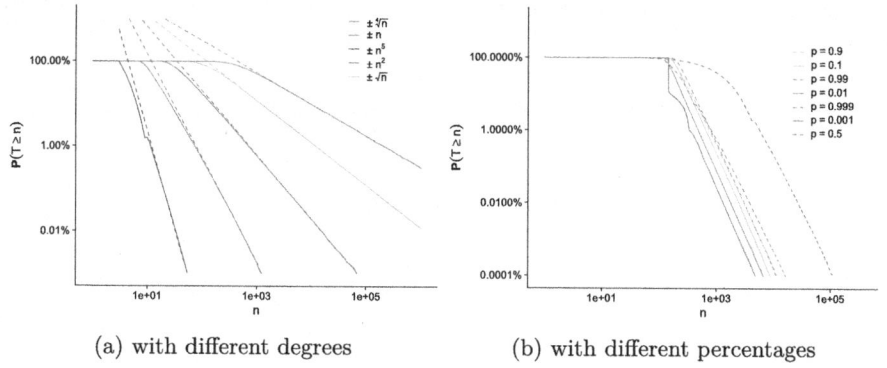

Fig. 5. Empirical results on stopping times of polynomial random walks

In Fig. 5b we display the stopping times $\mathbb{P}(T \geq n)$ of zero-mean polynomial random walks with different values of p and degree 3. The approximated values of the exponent, as well as the tightest bound found by our method can be found in Table 1. The increasing unsharpness for small (or large) values of p stem from the use of Hoeffding's lemma in the proof of Lemma 6. While for individual bounds of centered X_i this bound is sharp, for the product this no longer is the case and the plot suggests, that a tighter bound might be found.

Explicit Bound Analysis. Our genetic algorithms can be used to compute explicit bounds on the running times. Figure 6 shows the tightest explicit bound found for some random walk programs. The explicit bound is off by several orders of magnitude, as listed in Table 2. One of the main reasons for the explicit bound being unsharp stems from the fact that we choose a specific n_0 in (3), from which the parameters for the inductive bounds B are computed. This could be improved by computing a bound with multiple segments, and therefore inferring multiple exponents that decrease with growing n.

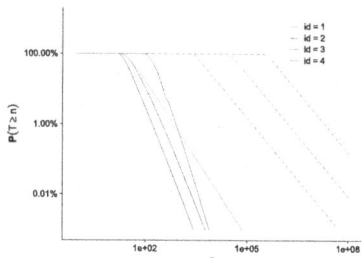

Fig. 6. Examples of explicit bounds

Table 2. Explicit bounds and empirical means of stopping time, with various polynomial updates q_1, q_2 and probabilistic choice p of polynomial random walks from Fig. 3

id	q_1	p	q_2	y_0	empiric $\mathbb{E}(T)$	explicit bound
1	n	0.5	$-n$	100	89.62	9562887
2	n^2	0.5	$-n^2$	100	36.4	17708
3	$n^2 + 2n + 20$	0.5	$-n^2 + 2n + 20$	1000	59.47	213570
4	$\frac{n^3}{0.99}$	0.99	$-\frac{n^3}{0.01}$	10^8	212.8	2671328

5 Related Work

Reasoning about probabilistic program termination is much harder than for deterministic programs [12], turning the automated analysis of probabilistic loops into a challenging problem. Most approaches rely on proof rules for proving (positive) almost-sure termination [4,6,16,20], which in turn require additional expressions, notably loop invariants and martingale variants, for the applicability in the proof rules. As the generation of invariants and martingales is undecidable in general, automation of these approaches requires user-provided invariants/martingales. Our work is limited to polynomial random walks with the benefit of providing sufficient conditions under which PAST can automatically be inferred. While restrictive, Fig. 1 shows advantages of our approach: proving PAST of this program using proof rules from [4,6,16] requires an auxiliary ranking super-martingale, whose computability is still an open question.

Alternative approaches to automating termination analysis have been proposed by focusing on restrictive classes of probabilistic loops, whose (P)AST analysis becomes (semi-)decidable [9,21]. Notably, constant probability loops [9] limit probabilistic loop updates to constant increments over random variable and their (P)AST is decidable. A more expressive class of programs is given in [21], with (P)AST analysis shown to be semi-decidable and automated. Key to automation is the ability to inferring (ranking) super-martingales from loop guards and relaxing proof rules to "eventual" reasoning over polynomial loop updates. Our approach complements these works by using arbitrary polynomial updates in polynomial random walks. Such loops cannot be analyzed by [9,21]; in particular, PAST of Fig. 1 cannot be inferred.

The analysis of probabilistic programs with arbitrary polynomial updates and control flow is shown to be difficult, especially due to the lack of compositionality [13]. By adjustments of the weakest precondition [18,19] calculus, runtime bounds are inferred as sufficient conditions for proving PAST in [2,23]. Control-flow refinement methods are also advocated in [15] to derive runtime bounds on probabilistic loops. Further, lexicographical extensions of synthesizing ranking super-martingales are presented in [1,25] for the purpose of PAST inference. While powerful, automation of these works depend on the suitable martingales. Unlike our technique, proving PAST of polynomial random walks, in particular of Fig. 1, cannot yet be achieved by other works.

6   Conclusions

We study the positive almost-sure termination (PAST) problem of polynomial probabilistic programs implementing random walks with increasing increments. We show that PAST can be proven for polynomial random walks by checking conditions via solving linear inequalities over the polynomial program updates, without requiring additional user input in the form of invariants and/or martingales. Our experiments demonstrate that our approach determines PAST of non-trivial probabilistic programs. Notably, we show PAST for programs beyond the scope of existing methods: for such programs, state-of-the-art works would require ranking super-martingales whose computation is undecidable in general. For such loops, we prove PAST by finding bounds on the probability of termination, depending on the degrees and the branching probability of the polynomial updates. Future work includes the extension of our results to (i) deriving hardness results on PAST decidability for polynomial random walks, and (ii) dealing with probabilistic programs with nondeterminism and more complex updates.

Acknowledgments. This research was partially supported by the European Research Council Consolidator Grant ARTIST 101002685 and the Vienna Science and Technology Fund WWTF 10.47379/ICT19018 grant ProbInG. We thank Marcel Moosbrugger for valuable discussions and providing guidance on the Polar analyzer.

Data Availability Statement. The models, scripts, and tools to reproduce our experimental evaluation are archived and publicly available at DOI 10.5281/zenodo.15257958.

Disclosure of Interests. The authors have no competing interests to declare that are relevant to the content of this article.

References

1. Agrawal, S., Chatterjee, K., Novotný, P.: Lexicographic ranking supermartingales: an efficient approach to termination of probabilistic programs. Proc. ACM Program. Lang. **2**(POPL), 34:1–34:32 (2018)
2. Avanzini, M., Moser, G., Schaper, M.: A modular cost analysis for probabilistic programs. Proc. ACM Program. Lang. **4**(OOPSLA) (2020)
3. Bournez, O., Garnier, F.: Proving positive almost-sure termination. In: RTA, pp. 323–337 (2005)
4. Chakarov, A., Sankaranarayanan, S.: Probabilistic program analysis with martingales. In: CAV, pp. 511–526 (2013)
5. Chatterjee, K., Fu, H., Goharshady, A.K.: Termination analysis of probabilistic programs through positivstellensatz's. In: CAV, pp. 3–22 (2016)
6. Chatterjee, K., Novotný, P., Žikelić, D.: Stochastic invariants for probabilistic termination. In: POPL, pp. 145–160 (2017)
7. Durrett, R.: Probability: Theory and Examples. Cambridge University Press (2019)
8. Fioriti, L.M.F., Hermanns, H.: Probabilistic termination: soundness, completeness, and compositionality. In: POPL, pp. 489–501. ACM (2015)

9. Giesl, J., Giesl, P., Hark, M.: Computing expected runtimes for constant probability programs. In: Fontaine, P. (ed.) Automated Deduction - CADE 27, pp. 269–286. Springer International Publishing, Cham (2019)
10. Gurobi Optimization, LLC: Gurobi Optimizer Reference Manual (2025). https://www.gurobi.com
11. Harchol-Balter, M.: Introduction to Probability for Computing. Cambridge University Press (2023)
12. Hark, M., Kaminski, B.L., Giesl, J., Katoen, J.: Aiming low is harder: induction for lower bounds in probabilistic program verification. Proc. ACM Program. Lang. **4**(POPL), 37:1–37:28 (2020)
13. Kaminski, B.L., Katoen, J.P., Matheja, C., Olmedo, F.: Weakest precondition reasoning for expected runtimes of randomized algorithms. J. ACM **65**(5) (2018)
14. Kozen, D.: A probabilistic PDL. J. Comput. Syst, Sci (1985)
15. Lommen, N., Meyer, É., Giesl, J.: Control-flow refinement for complexity analysis of probabilistic programs in KoAT (short paper). In: IJCAR, pp. 233–243 (2024)
16. Majumdar, R., Sathiyanarayana, V.R.: Positive almost-sure termination: Complexity and proof rules. Proc. ACM Program. Lang. **8**(POPL) (2024)
17. Majumdar, R., Sathiyanarayana, V.R.: Sound and complete proof rules for probabilistic termination. Proc. ACM Program. Lang. **9**(POPL), 1871–1902 (2025)
18. McIver, A., Morgan, C.: Abstraction. Springer, Refinement and Proof for Probabilistic Systems (2005)
19. McIver, A., Morgan, C., Kaminski, B.L., Katoen, J.: A new proof rule for almost-sure termination. Proc. ACM Program. Lang. **2**(POPL), 33:1–33:28 (2018)
20. Meyer, F., Hark, M., Giesl, J.: Inferring expected runtimes of probabilistic integer programs using expected sizes. In: TACAS, pp. 250–269 (2021)
21. Moosbrugger, M., Bartocci, E., Katoen, J.P., Kovács, L.: The probabilistic termination tool Amber. In: FM, pp. 667–675 (2021)
22. Moosbrugger, M., Stankovic, M., Bartocci, E., Kovács, L.: This is the moment for probabilistic loops. Proc. ACM Program. Lang. **6**(OOPSLA2), 1497–1525 (2022)
23. Ngo, V.C., Carbonneaux, Q., Hoffmann, J.: Bounded expectations: resource analysis for probabilistic programs. In: PLDI, pp. 496–512 (2018)
24. Perron, L., Furnon, V.: Or-tools. https://developers.google.com/optimization/
25. Takisaka, T., Zhang, L., Wang, C., Liu, J.: Lexicographic ranking supermartingales with lazy lower bounds. In: CAV, pp. 420–442 (2024)
26. Venables, W.N., Ripley, B.D.: Modern Applied Statistics with S. Springer, New York, fourth edn. (2002). https://www.stats.ox.ac.uk/pub/MASS4/
27. Winkler, L., Kovács, L.: Positive Almost-Sure Termination of Polynomial Random Walks (2025). https://arxiv.org/abs/2504.19575

Noninterference Analysis of Deterministically Timed Reversible Systems

Andrea Esposito, Alessandro Aldini, and Marco Bernardo(✉)

Dipartimento di Scienze Pure e Applicate, Università di Urbino, Urbino, Italy
marco.bernardo@uniurb.it

Abstract. Information flow theory aims at guaranteeing the absence of covert channels among different security levels. As for the verification of noninterference via equivalence checking, in nondeterministic and probabilistic settings weak bisimilarity is adequate only for forward-computing systems, while branching bisimilarity has turned out to be appropriate for reversible systems too. In this paper we investigate noninterference for deterministically timed systems based on the model of Moller and Tofts. After recasting a selection of noninterference properties via timed variants of weak and branching bisimilarities, we analyze their preservation and compositionality aspects, establish their taxonomy, and compare it with the nondeterministic taxonomy for (ir)reversible systems. We illustrate the adequacy of our proposal on real-time database transactions.

1 Introduction

The notion of noninterference was introduced in [34] to reason about the way in which illegitimate information flows can occur in multi-level security systems due to covert channels from high-level agents to low-level ones. Since the first definition, conceived for deterministic systems, a lot of work has been done to extend the approach to a variety of more expressive domains, such as nondeterministic systems, systems in which quantitative aspects like time and probability play a central role, and reversible systems; see, e.g., [26, 3, 44, 35, 65, 58, 9, 6, 4, 37, 25, 23] and the references therein. Likewise, to verify information-flow security properties based on noninterference, several different approaches have been proposed ranging from the application of type theory [70] and abstract interpretation [30] to control flow and equivalence or model checking [27, 45, 5].

Noninterference guarantees that low-level agents cannot infer from their observations what high-level ones are doing. Regardless of its specific definition, noninterference is closely tied to the notion of behavioral equivalence [32] because, given a multi-level security system, the idea is to compare the system behavior with high-level actions being prevented and the system behavior with the same actions being hidden. A natural framework in which to study system behavior is given by process algebra [46]. In this setting, weak bisimilarity has been employed in [26] to reason formally about covert channels and illegitimate information flows as well as to study a classification of noninterference properties for nondeterministic forward-computing systems.

Noninterference analysis has been recently extended to reversible systems – which feature forward and backward computations – both in the nondeterministic setting [25] and in the probabilistic one [23]. Reversibility has started to gain attention in computing since it has been shown that it may achieve lower levels of energy consumption [40, 10]. Its applications range from biochemical reaction modeling [54, 55] and parallel discrete-event simulation [52, 60] to robotics [43], wireless communications [61], fault-tolerant systems [19, 66, 41, 64], program debugging [29, 42], and distributed algorithms [68, 13].

As shown in [25, 23], noninterference properties based on weak bisimilarity are not adequate in a reversible context because they fail to detect information flows emerging when backward computations are triggered. A more appropriate semantics turns out to be branching bisimilarity [33] because it coincides with weak back-and-forth bisimilarity [21]. The latter behavioral equivalence requires systems to be able to mimic each other's behavior stepwise not only when performing actions in the standard forward direction, but also when undoing those actions in the backward direction. Formally, weak back-and-forth bisimilarity is defined on computation paths instead of states thus preserving not only causality but also history, as backward moves are constrained to take place along the same path followed in the forward direction even in the presence of concurrency.

In this paper we extend the approach of [25, 23] to a deterministically timed setting, in which delays are fixed (as opposed to being subject to stochastic fluctuations), so as to address noninterference properties in a framework featuring nondeterminism, time, and reversibility. To accomplish this we move to a model combining nondeterminism and time inspired by [47, 48], in which transitions are divided into action transitions, each labeled with an action, and timed transitions, each labeled with a positive natural number that expresses a delay. The reason for choosing – in the vast realm of timed process calculi [57, 47, 69, 8, 15, 2, 36, 50, 56, 17, 49, 63] – this model in which time passing is orthogonal to action execution instead of a model in which action execution and time passing are integrated (see [11] for encodings between integrated-time and orthogonal-time calculi) is that the former naturally supports the definition of behavioral equivalences abstracting from unobservable actions [48] – which are necessary for noninterference analysis – whereas this is not the case in the latter.

Following [47] we build a process calculus featuring action prefix separated from delay prefix. As for behavioral equivalences, we adopt the weak timed bisimilarity of [48] and introduce a novel timed branching bisimilarity. By using these two equivalences we recast the noninterference properties of [26, 28] for irreversible systems and the noninterference properties of [25] for reversible systems, respectively, to study their preservation and compositionality aspects as well as to provide a taxonomy similar to those in [26, 25, 23]. Reversibility comes into play by extending one of the results of [21] to our orthogonal-time model;

we show that a timed variant of weak back-and-forth bisimilarity coincides with our timed branching bisimilarity.

This paper is organized as follows. In Sect. 2 we recall the orthogonal-time model of [47] along with various definitions of strong and weak bisimilarities for it and a process calculus interpreted on it. In Sect. 3 we recast in our timed framework a selection of noninterference properties taken from [26, 28, 25]. In Sect. 4 we study their preservation and compositionality characteristics as well as their taxonomy, which in Sect. 5 we relate to the nondeterministic taxonomy of [25]. In Sect. 6 we establish a connection with reversibility by introducing a weak timed back-and-forth bisimilarity and proving that it coincides with timed branching bisimilarity. In Sect. 7 we present a real-time database management system example to show the adequacy of our approach when dealing with information flows in reversible systems featuring nondeterminism and time. Finally, in Sect. 8 we provide some concluding remarks.

2 Background Definitions and Results

In this section we recall the timed model of [47] (Sect. 2.1) along with weak timed bisimilarity [48] and define timed branching bisimilarity (Sect. 2.2). Then we introduce a timed process language inspired by [47] through which we will express bisimulation-based information-flow security properties accounting for nondeterminism and time (Sect. 2.3).

2.1 Timed Labeled Transition Systems

To represent the behavior of a process featuring nondeterminism and time, we use a timed labeled transition system. This is a variant of a labeled transition system [39] whose transitions are labeled with actions or positive natural numbers expressing delays [47]. We assume that the action set $\mathcal{A}_\tau$ contains a set $\mathcal{A}$ of observable actions and a single action $\tau \notin \mathcal{A}$ representing unobservable actions.

Definition 1. *A timed labeled transition system (TLTS) is a triple* $(\mathcal{S}, \mathcal{A}_\tau, \longrightarrow)$ *where* $\mathcal{S} \neq \emptyset$ *is an at most countable set of states,* $\mathcal{A}_\tau = \mathcal{A} \cup \{\tau\}$ *is a countable set of actions, and* $\longrightarrow = \longrightarrow_a \cup \longrightarrow_t$ *is the transition relation, with* $\longrightarrow_a \subseteq \mathcal{S} \times \mathcal{A}_\tau \times \mathcal{S}$ *being the action transition relation whilst* $\longrightarrow_t \subseteq \mathcal{S} \times \mathbb{N}_{>0} \times \mathcal{S}$ *being the timed transition relation.* ∎

An action transition (s, a, s') is written $s \xrightarrow{a}_a s'$ while a timed transition (s, t, s') is written $s \xrightarrow{t}_t s'$, where s is the source state and s' is the target state. We say that s' is reachable from s, written $s' \in \mathit{reach}(s)$, iff $s' = s$ or there exists a sequence of finitely many transitions such that the target state of each of them coincides with the source state of the subsequent one, with the source of the first transition being s and the target of the last one being s'.

Following [47] we assume that timed transitions are subject to *time determinism*, i.e., every state has at most one outgoing timed transition, and *time additivity*, i.e., a timed transition can be split into a sequence of timed transitions whose overall duration is equal to the duration of the original transition, as well as a sequence of timed transitions can be merged into a single timed transition whose duration is equal to the sum of the durations of the original transitions. As for the interplay between action transitions and timed ones, we assume *eagerness*, i.e., actions must be performed as soon as they become enabled without any delay, thereby implying that their execution is urgent. Moreover τ-transitions, which cannot be disabled by the environment where the system executes, take precedence over timed ones; this property is called *maximal progress*.

2.2 Bisimulation Equivalences

Bisimilarity [51, 46] identifies processes mimicking each other's behavior stepwise, i.e., having the same branching structure. In our setting this extends to timed behavior [47]. Due to maximal progress, timed transitions are compared only in states s with no outgoing τ-transitions, which is denoted by $s \not\xrightarrow{\tau}_a$.

Definition 2. *Let $(S, A_\tau, \longrightarrow)$ be a TLTS. We say that $s_1, s_2 \in S$ are strongly timed bisimilar, written $s_1 \sim_t s_2$, iff $(s_1, s_2) \in B$ for some strong timed bisimulation B. A symmetric relation B over S is a strong timed bisimulation iff, whenever $(s_1, s_2) \in B$, then:*

- *For each $s_1 \xrightarrow{a}_a s_1'$ there exists $s_2 \xrightarrow{a}_a s_2'$ such that $(s_1', s_2') \in B$.*
- *If $s_1 \not\xrightarrow{\tau}_a$, for each $s_1 \xrightarrow{t}_t s_1'$ there exists $s_2 \xrightarrow{t}_t s_2'$ such that $(s_1', s_2') \in B$.* ∎

Weak bisimilarity [46] is additionally capable of abstracting from unobservable actions. Let $\Longrightarrow_a$ be the reflexive and transitive closure of $\xrightarrow{\tau}_a$. Moreover let $\xRightarrow{\hat{a}}_a$ stand for $\Longrightarrow_a$ if $a = \tau$ or $\Longrightarrow_a \xrightarrow{a}_a \Longrightarrow_a$ if $a \neq \tau$, while $\xRightarrow{t}_t$ stands for $\Longrightarrow_a \xrightarrow{t_1}_t \Longrightarrow_a \ldots \Longrightarrow_a \xrightarrow{t_n}_t \Longrightarrow_a$ where $\sum_{1 \leq i \leq n} t_i = t$ and every t_i-transition departs from a state with no outgoing τ-transitions. The weak timed bisimilarity below is taken from [48].

Definition 3. *Let $(S, A_\tau, \longrightarrow)$ be a TLTS. We say that $s_1, s_2 \in S$ are weakly timed bisimilar, written $s_1 \approx_{tw} s_2$, iff $(s_1, s_2) \in B$ for some weak timed bisimulation B. A symmetric relation B over S is a weak timed bisimulation iff, whenever $(s_1, s_2) \in B$, then:*

- *For each $s_1 \xrightarrow{a}_a s_1'$ there exists $s_2 \xRightarrow{\hat{a}}_a s_2'$ such that $(s_1', s_2') \in B$.*
- *If $s_1 \not\xrightarrow{\tau}_a$ then there exists $s_2 \Longrightarrow_a \bar{s}_2$ such that $\bar{s}_2 \not\xrightarrow{\tau}_a$, $(s_1, \bar{s}_2) \in B$, and for each $s_1 \xrightarrow{t}_t s_1'$ there exists $\bar{s}_2 \xRightarrow{t}_t s_2'$ such that $(s_1', s_2') \in B$.* ∎

Branching bisimilarity [33] is finer than weak bisimilarity as it preserves the branching structure of processes even when abstracting from τ-actions – see condition $(s_1, \bar{s}_2) \in B$ in the action transitions matching of the definition below. We adapt it to the timed setting as follows.

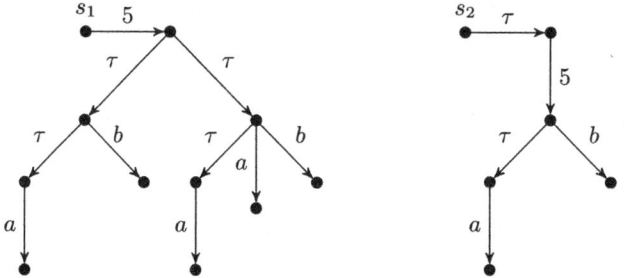

Fig. 1. States s_1 and s_2 are related by $\approx_{\text{tw}}$ but distinguished by $\approx_{\text{tb}}$.

Definition 4. *Let* $(\mathcal{S}, \mathcal{A}_\tau, \longrightarrow)$ *be a TLTS. We say that* $s_1, s_2 \in \mathcal{S}$ *are timed branching bisimilar, written* $s_1 \approx_{\text{tb}} s_2$, *iff* $(s_1, s_2) \in \mathcal{B}$ *for some timed branching bisimulation* $\mathcal{B}$. *A symmetric relation* $\mathcal{B}$ *over* $\mathcal{S}$ *is a timed branching bisimulation iff, whenever* $(s_1, s_2) \in \mathcal{B}$, *then:*

- *For each* $s_1 \xrightarrow{a}_a s_1'$:
  - *either* $a = \tau$ *and* $(s_1', s_2) \in \mathcal{B}$;
  - *or there exists* $s_2 \Longrightarrow_a \bar{s}_2 \xrightarrow{a}_a s_2'$ *such that* $(s_1, \bar{s}_2) \in \mathcal{B}$ *and* $(s_1', s_2') \in \mathcal{B}$.
- *If* $s_1 \not\xrightarrow{\tau}_a$ *then there exists* $s_2 \Longrightarrow_a \bar{s}_2$ *such that* $\bar{s}_2 \not\xrightarrow{\tau}_a$, $(s_1, \bar{s}_2) \in \mathcal{B}$, *and for each* $s_1 \xrightarrow{t}_t s_1'$ *there exists* $\bar{s}_2 \Longrightarrow_t s_2'$ *such that* $(s_1', s_2') \in \mathcal{B}$. ∎

It is worth noting that the clause for timed transitions in the two definitions above implies that the states along $\bar{s}_2 \Longrightarrow_t s_2'$ that are connected by τ-transitions, for which time is not progressing, belong to the same equivalence class. This feature, which is a piecewise variant of the *stuttering property* for τ-computations of [33], is established by the following proposition.

Proposition 1. *Let* $s_1, s_2 \in \mathcal{S}$ *and* $\approx \in \{\approx_{\text{tw}}, \approx_{\text{tb}}\}$. *Suppose that* $s_1 \approx s_2$, $s_1 \not\xrightarrow{\tau}_a$, $s_2 \not\xrightarrow{\tau}_a$, $s_1 \xrightarrow{t}_t s_1'$, $s_2 \xrightarrow{t_1}_t s_{2,1} \Longrightarrow_a \ldots \Longrightarrow_a s_{2,n-1}' \xrightarrow{t_n}_t s_{2,n} \Longrightarrow_a s_2'$, $\sum_{1 \leq i \leq n} t_i = t$, *and* $s_1' \approx s_2'$. *Then* $s_{2,i} \approx s_{2,i}'$ *for all* $s_{2,i} \Longrightarrow_a s_{2,i}'$. ∎

It may be argued that the weak bisimilarity of Definition 3 is already very close to branching bisimilarity, because maximal progress forces a check on the branching structure of the considered processes. We show that our novel Definition 4, which sticks to the original one of [33], is more discriminating. Consider Fig. 1, where every TLTS is depicted as a directed graph in which vertices represent states and action- or delay-labeled edges represent transitions. The initial states s_1 and s_2 of the two TLTSs are weakly timed bisimilar but not timed branching bisimilar. On the one hand, each of the two states reachable from s_1 after 5 time units and a τ-transition and the state reachable from s_2 after a τ-transition and 5 time units are all weakly timed bisimilar. On the other hand, the two states reachable from s_1 are not timed branching bisimilar, because if the one on the right performs a then the one on the left cannot respond by

Table 1. Operational semantic rules for action transitions

Prefix	$a\,.\,P \xrightarrow{a}_a P$	
Choice	$\dfrac{P_1 \xrightarrow{a}_a P_1'}{P_1 + P_2 \xrightarrow{a}_a P_1'}$	$\dfrac{P_2 \xrightarrow{a}_a P_2'}{P_1 + P_2 \xrightarrow{a}_a P_2'}$
Parallel	$\dfrac{P_1 \xrightarrow{a}_a P_1' \quad a \notin L}{P_1 \,\|_L\, P_2 \xrightarrow{a}_a P_1' \,\|_L\, P_2}$	$\dfrac{P_2 \xrightarrow{a}_a P_2' \quad a \notin L}{P_1 \,\|_L\, P_2 \xrightarrow{a}_a P_1 \,\|_L\, P_2'}$
Synch	$\dfrac{P_1 \xrightarrow{a}_a P_1' \quad P_2 \xrightarrow{a}_a P_2' \quad a \in L}{P_1 \,\|_L\, P_2 \xrightarrow{a}_a P_1' \,\|_L\, P_2'}$	
Restriction	$\dfrac{P \xrightarrow{a}_a P' \quad a \notin L}{P \setminus L \xrightarrow{a}_a P' \setminus L}$	
Hiding	$\dfrac{P \xrightarrow{a}_a P' \quad a \in L}{P / L \xrightarrow{\tau}_a P' / L}$	$\dfrac{P \xrightarrow{a}_a P' \quad a \notin L}{P / L \xrightarrow{a}_a P' / L}$

performing τ followed by a because the state reached after τ no longer enables b. Thus, with respect to timed branching bisimilarity, s_1 reaches two inequivalent states, while s_2 reaches only one of them.

2.3 A Timed Process Calculus with High and Low Actions

We now introduce a timed process calculus to formalize the security properties of interest. To address two security levels, we partition the set $\mathcal{A}$ of observable actions into $\mathcal{A}_\mathcal{H} \cup \mathcal{A}_\mathcal{L}$, with $\mathcal{A}_\mathcal{H} \cap \mathcal{A}_\mathcal{L} = \emptyset$, where $\mathcal{A}_\mathcal{H}$ is the set of high-level actions, ranged over by h, and $\mathcal{A}_\mathcal{L}$ is the set of low-level actions, ranged over by l. Note that $\tau \notin \mathcal{A}_\mathcal{H} \cup \mathcal{A}_\mathcal{L}$.

The set $\mathbb{P}$ of process terms is obtained by considering typical operators from CCS [46] and CSP [16] together with delay prefix from [47]. In addition to prefix, choice, and parallel composition – taken from CSP so as not to turn synchronizations among high-level actions into τ as would happen with the CCS parallel composition – we include restriction and hiding as they are necessary to formalize noninterference properties. The syntax for $\mathbb{P}$ is:

$$P ::= \underline{0} \mid a\,.\,P \mid (t)\,.\,P \mid P + P \mid P \,\|_L\, P \mid P \setminus L \mid P / L$$

where:

- $\underline{0}$ is the terminated process.
- $a\,.\,\_$, for $a \in \mathcal{A}_\tau$, is the action prefix operator describing a process that can initially perform action a.
- $(t)\,.\,\_$, for $t \in \mathbb{N}_{>0}$, is the delay prefix operator describing a process that can initially let t time units pass.
- $\_ + \_$ is the alternative composition operator expressing a choice between two processes, which is nondeterministic in case of actions, governed by time determinism in case of delays [47], or subject to maximal progress otherwise.

Table 2. Operational semantic rules for timed transitions

$TimedPrefix$	$(t).P \xrightarrow{t}_t P$
$TimedSplit$	$\dfrac{t = t_1 + t_2 \quad t_1, t_2 \in \mathbb{N}_{>0}}{(t).P \xrightarrow{t_1}_t (t_2).P}$
$TimedMerge$	$\dfrac{P \xrightarrow{t_2}_t P' \quad t = t_1 + t_2}{(t_1).P \xrightarrow{t}_t P'}$
$TimedChoice$	$\dfrac{P_1 \xrightarrow{t}_t P_1' \quad P_2 \xrightarrow{t}_t P_2'}{P_1 + P_2 \xrightarrow{t}_t P_1' + P_2'}$
$TimedSynch$	$\dfrac{P_1 \xrightarrow{t}_t P_1' \quad P_2 \xrightarrow{t}_t P_2'}{P_1 \parallel_L P_2 \xrightarrow{t}_t P_1' \parallel_L P_2'}$
$TimedRestriction$	$\dfrac{P \xrightarrow{t}_t P'}{P \setminus L \xrightarrow{t}_t P' \setminus L}$
$TimedHiding$	$\dfrac{P \xrightarrow{t}_t P'}{P / L \xrightarrow{t}_t P' / L}$

- $\_\parallel_L\_$, for $L \subseteq \mathcal{A}$, is the parallel composition operator allowing two processes to proceed independently on any action not in L and forcing them to synchronize on every action in L and on delays due to time determinism [47].
- $\_\setminus L$, for $L \subseteq \mathcal{A}$, is the restriction operator, which prevents the execution of all actions belonging to L.
- $\_/ L$, for $L \subseteq \mathcal{A}$, is the hiding operator, which turns all the executed actions belonging to L into the unobservable action τ.

The operational semantic rules for the process language are shown in Tables 1 and 2 for action and timed transitions respectively. Together they produce the TLTS $(\mathbb{P}, \mathcal{A}_\tau, \longrightarrow)$ where $\longrightarrow = \longrightarrow_a \cup \longrightarrow_t$, with $\longrightarrow_a \subseteq \mathbb{P} \times \mathcal{A}_\tau \times \mathbb{P}$ and $\longrightarrow_t \subseteq \mathbb{P} \times \mathbb{N}_{>0} \times \mathbb{P}$, to which the bisimulation equivalences defined in Sect. 2.2 are applicable. Following [47], rules $TimedSplit$ and $TimedMerge$ implement time additivity, while rules $TimedChoice$ and $TimedSynch$ implement time determinism, according to which time does not solve choices and does not decide which subprocess advances in a parallel composition.

3 Timed Information-Flow Security Properties

The intuition behind noninterference in a two-level security system is that, if a group of agents at the high level performs some actions, the effect of those actions should not be seen by any agent at the low level. To formalize this, the restriction and hiding operators play a central role.

In this section we recast the noninteference properties defined in [26, 28, 25] – *Nondeterministic Non-Interference* (NNI) and *Non-Deducibility on Composition*

(NDC) – by taking as behavioral equivalence the weak or branching bisimilarity of Sect. 2.2. In the acronyms of the following variants of NNI and NDC properties, B stands for bisimulation-based, S stands for strong, and P stands for persistent.

Definition 5. *Let $P \in \mathbb{P}$ and $\approx \,\in \{\approx_{tw}, \approx_{tb}\}$:*

- $P \in \text{BSNNI}_\approx \iff P \setminus \mathcal{A}_\mathcal{H} \approx P / \mathcal{A}_\mathcal{H}$.
- $P \in \text{BNDC}_\approx \iff$ *for all $Q \in \mathbb{P}$ such that all of its action prefixes belongs to $\mathcal{A}_\mathcal{H}$ whilst its timed prefixes match the ones in P and for all $L \subseteq \mathcal{A}_\mathcal{H}$, $P \setminus \mathcal{A}_\mathcal{H} \approx ((P \|_L Q) / L) \setminus \mathcal{A}_\mathcal{H}$.*
- $P \in \text{SBSNNI}_\approx \iff$ *for all $P' \in reach(P)$, $P' \in \text{BSNNI}_\approx$.*
- $P \in \text{P\_BNDC}_\approx \iff$ *for all $P' \in reach(P)$, $P' \in \text{BNDC}_\approx$.*
- $P \in \text{SBNDC}_\approx \iff$ *for all $P', P'' \in reach(P)$ such that $P' \xrightarrow{h}_a P''$, $P' \setminus \mathcal{A}_\mathcal{H} \approx P'' \setminus \mathcal{A}_\mathcal{H}$.* ■

Bisimulation-based Strong Nondeterministic Non-Interference (BSNNI) has been one of the first and most intuitive proposals. Basically, it is satisfied by any process P that behaves the same when its high-level actions are prevented (as modeled by $P \setminus \mathcal{A}_\mathcal{H}$) or when they are considered as hidden, unobservable actions (as modeled by $P / \mathcal{A}_\mathcal{H}$). The equivalence between these two low-level views of P states that a low-level agent cannot deduce the high-level behavior of the system. For instance, in our timed setting, a low-level agent that observes the execution of l in $(t) \, . \, 1 \, . \, \underline{0} + h \, . \, h \, . \, (t) \, . \, l \, . \, \underline{0}$. cannot infer anything about the execution of h. Indeed, a low-level user always observes the execution of l after a delay of t units of time. Formally, $P \setminus \{h\} \approx P / \{h\}$ because $(t) \, . \, l \, . \, \underline{0} \approx (t) \, . \, l \, . \, \underline{0} + \tau \, . \, \tau \, . \, (t) \, . \, l \, . \, \underline{0}$.

$\text{BSNNI}_\approx$ is not powerful enough to capture covert channels that derive from the behavior of a high-level agent interacting with the system. For instance, $(t) \, . \, l \, . \, \underline{0} + h_1 \, . \, h_2 \, . \, (t) \, . \, l \, . \, \underline{0}$ is $\text{BSNNI}_\approx$ for the same reason discussed above. However, a high-level agent could decide to enable only h_1, thus yielding the low-level view of the system $(t) \, . \, l \, . \, \underline{0} + \tau \, . \, \underline{0}$, which is clearly distinguishable from $(t) \, . \, l \, . \, \underline{0}$ as the former is forced, due to maximal progress, to perform τ and reach a terminal state, while the latter can let t units of time pass and then perform l. To avoid such a limitation, the most obvious solution consists of checking explicitly the interaction on any action set $L \subseteq \mathcal{A}_\mathcal{H}$ between the system and every possible high-level agent Q. The resulting property is the *Bisimulation-based Non-Deducibility on Composition* (BNDC), which features a universal quantification over Q containing only high-level actions.

Note that in our timed setting the high-level agent Q must allow the same amount of time as P to pass, otherwise the property BNDC would never be satisfied. To see why, consider the trivially safe process $(1) \, . \, l \, . \, \underline{0}$ and the high-level agent $h \, . \, \underline{0}$. The processes $((1) \, . \, l \, . \, \underline{0}) \setminus \mathcal{A}_\mathcal{H}$ and $(((1) \, . \, l \, . \, \underline{0} \|_L h \, . \, \underline{0}) / L) \setminus \mathcal{A}_\mathcal{H}$ are not equivalent, regardless of the specific $L \subseteq \mathcal{A}_\mathcal{H}$ chosen, because the former can let time pass, while the latter cannot, as it is blocked by the process $h \, . \, \underline{0}$.

To overcome the verification problems related to the quantification over Q, several properties have been proposed that are stronger than BNDC. They all express some persistency conditions, stating that the security checks have to be extended to all the processes reachable from a secure one. Three of the most representative ones among such properties are the variant of BSNNI that requires every reachable process to satisfy BSNNI itself, called *Strong* BSNNI (SBSNNI), the variant of BNDC that requires every reachable process to satisfy BNDC itself, called *Persistent* BNDC (P_BNDC), and *Strong* BNDC (SBNDC), which requires the low-level view of every reachable process to be the same before and after the execution of any high-level action, meaning that the execution of high-level actions must be completely transparent to low-level agents. In the nondeterministic and probabilistic settings, P_BNDC and SBSNNI have been proven to coincide in the case of both weak bisimilarity and branching bisimilarity [28, 25, 23].

4 Characteristics of Timed Security Properties

In this section we investigate preservation and compositionality characteristics of the noninterference properties introduced in the previous section (Sect. 4.1) as well as the inclusion relationships between the ones based on $\approx_{\mathrm{tw}}$ and the ones based on $\approx_{\mathrm{tb}}$ (Sect. 4.2).

4.1 Preservation and Compositionality

All the timed noninterference properties of Definition 5 turn out to be preserved by the bisimilarity employed in their definition. This means that if a process P_1 is secure under any of such properties, then every other equivalent process P_2 is secure too according to the same property. This is very useful for automated property verification, as it allows us to work with the process with the smallest state space among the equivalent ones.

These results immediately follow from the next lemma, which states that $\approx_{\mathrm{tw}}$ and $\approx_{\mathrm{tb}}$ are congruences with respect to action prefix, delay prefix, parallel composition, restriction, and hiding. Some of these results were already proven in [48] for weak timed bisimilarity. Here we extend those results to the operators of our calculus as well as to timed branching bisimilarity.

Lemma 1. *Let* $P_1, P_2 \in \mathbb{P}$ *and* $\approx \in \{\approx_{\mathrm{tw}}, \approx_{\mathrm{tb}}\}$. *If* $P_1 \approx P_2$ *then:*

1. $a . P_1 \approx a . P_2$ *for all* $a \in \mathcal{A}_\tau$.
2. $(t) . P_1 \approx (t) . P_2$ *for all* $t \in \mathbb{N}_{>0}$.
3. $P_1 \|_L P \approx P_2 \|_L P$ *and* $P \|_L P_1 \approx P \|_L P_2$ *for all* $L \subseteq \mathcal{A}$ *and* $P \in \mathbb{P}$.
4. $P_1 \setminus L \approx P_2 \setminus L$ *for all* $L \subseteq \mathcal{A}$.
5. $P_1 / L \approx P_2 / L$ *for all* $L \subseteq \mathcal{A}$. ∎

Theorem 1. *Let* $P_1, P_2 \in \mathbb{P}$, $\approx \in \{\approx_{\mathrm{tw}}, \approx_{\mathrm{tb}}\}$, *and* $\mathcal{P} \in \{\mathrm{BSNNI}_\approx, \mathrm{BNDC}_\approx, \mathrm{SBSNNI}_\approx, \mathrm{P\_BNDC}_\approx, \mathrm{SBNDC}_\approx\}$. *If* $P_1 \approx P_2$ *then* $P_1 \in \mathcal{P} \iff P_2 \in \mathcal{P}$. ∎

As far as modular verification is concerned, like in the nondeterministic and probabilistic settings [26, 25, 23] only the local properties $\text{SBSNNI}_\approx$, $\text{P\_BNDC}_\approx$, and $\text{SBNDC}_\approx$ are compositional, i.e., are preserved by some operators of the calculus in certain circumstances. Moreover, similar to [25, 23] compositionality with respect to parallel composition is limited, for $\text{SBSNNI}_{\approx_{tb}}$ and $\text{P\_BNDC}_{\approx_{tb}}$, to the case in which synchronizations can take place only among low-level actions, i.e., $L \subseteq \mathcal{A}_\mathcal{L}$. A limitation to low-level actions applies to action prefix and hiding as well, whilst this is not the case for restriction. Another analogy with the nondeterministic and probabilistic settings [26, 25, 23] is that none of the considered noninterference properties is compositional with respect to alternative composition. As an example, let us examine processes $P_1 = l.\underline{0}$ and $P_2 = h.\underline{0}$. Both processes are $\text{BSNNI}_\approx$, as $(l.\underline{0}) \setminus \{h\} \approx (l.\underline{0}) / \{h\}$ and $(h.\underline{0}) \setminus \{h\} \approx (h.\underline{0}) / \{h\}$, but $P_1 + P_2 \notin \text{BSNNI}_\approx$, because $(l.\underline{0} + h.\underline{0}) \setminus \{h\} \approx l.\underline{0} \not\approx l.\underline{0} + \tau.\underline{0} \approx (l.\underline{0} + h.\underline{0}) / \{h\}$. It is easy to check that $P_1 + P_2 \notin \mathcal{P}$ also for $\mathcal{P} \in \{\text{BNDC}_\approx, \text{SBSNNI}_\approx, \text{SBNDC}_\approx\}$.

Theorem 2. *Let $P, P_1, P_2 \in \mathbb{P}$, $\approx \, \in \{\approx_{tw}, \approx_{tb}\}$, $\mathcal{P} \in \{\text{SBSNNI}_\approx, \text{P\_BNDC}_\approx, \text{SBNDC}_\approx\}$. Then:*

1. $P \in \mathcal{P} \Longrightarrow a.P \in \mathcal{P}$ *for all* $a \in \mathcal{A}_\mathcal{L} \cup \{\tau\}$.
2. $P \in \mathcal{P} \Longrightarrow (t).P \in \mathcal{P}$ *for all* $t \in \mathbb{N}_{>0}$.
3. $P_1, P_2 \in \mathcal{P} \Longrightarrow P_1 \|_L P_2 \in \mathcal{P}$ *for all* $L \subseteq \mathcal{A}_\mathcal{L}$ *if* $\mathcal{P} \in \{\text{SBSNNI}_{\approx_{tb}}, \text{P\_BNDC}_{\approx_{tb}}\}$ *or for all* $L \subseteq \mathcal{A}$ *if* $\mathcal{P} \in \{\text{SBSNNI}_{\approx_{tw}}, \text{P\_BNDC}_{\approx_{tw}}, \text{SBNDC}_{\approx_{tw}}, \text{SBNDC}_{\approx_{tb}}\}$.
4. $P \in \mathcal{P} \Longrightarrow P \setminus L \in \mathcal{P}$ *for all* $L \subseteq \mathcal{A}$.
5. $P \in \mathcal{P} \Longrightarrow P / L \in \mathcal{P}$ *for all* $L \subseteq \mathcal{A}_\mathcal{L}$. ∎

As for the limitation to $L \subseteq \mathcal{A}_\mathcal{L}$ for parallel composition under $\text{SBSNNI}_{\approx_{tb}}$, for example both $P_1 = h.\underline{0} + l_1.\underline{0} + \tau.\underline{0}$ and $P_2 = h.\underline{0} + l_2.\underline{0} + \tau.\underline{0}$ are $\text{SBSNNI}_{\approx_{tb}}$, but $P_1 \|_{\{h\}} P_2$ is not because the transition $(P_1 \|_{\{h\}} P_2) / \mathcal{A}_\mathcal{H} \xrightarrow{\tau}_a (\underline{0} \|_{\{h\}} \underline{0}) / \mathcal{A}_\mathcal{H}$ arising from the synchronization between the two h-actions cannot be matched by $(P_1 \|_{\{h\}} P_2) \setminus \mathcal{A}_\mathcal{H}$ in the timed branching bisimulation game. Indeed, the only two possibilities are $(P_1 \|_{\{h\}} P_2) \setminus \mathcal{A}_\mathcal{H} \Longrightarrow_a (P_1 \|_{\{h\}} P_2) \setminus \mathcal{A}_\mathcal{H} \xrightarrow{\tau}_a (\underline{0} \|_{\{h\}} P_2) \setminus \mathcal{A}_\mathcal{H} \xrightarrow{\tau}_a (\underline{0} \|_{\{h\}} \underline{0}) \setminus \mathcal{A}_\mathcal{H}$ and $(P_1 \|_{\{h\}} P_2) \setminus \mathcal{A}_\mathcal{H} \Longrightarrow_a (P_1 \|_{\{h\}} P_2) \setminus \mathcal{A}_\mathcal{H} \xrightarrow{\tau}_a (P_1 \|_{\{h\}} \underline{0}) \setminus \mathcal{A}_\mathcal{H} \xrightarrow{\tau}_a (\underline{0} \|_{\{h\}} \underline{0}) \setminus \mathcal{A}_\mathcal{H}$. However, neither $(\underline{0} \|_{\{h\}} P_2) \setminus \mathcal{A}_\mathcal{H}$ nor $(P_1 \|_{\{h\}} \underline{0}) \setminus \mathcal{A}_\mathcal{H}$ is timed branching bisimilar to $(P_1 \|_{\{h\}} P_2) \setminus \mathcal{A}_\mathcal{H}$ when $l_1 \neq l_2$. Note that $(P_1 \|_{\{h\}} P_2) / \mathcal{A}_\mathcal{H} \approx (P_1 \|_{\{h\}} P_2) \setminus \mathcal{A}_\mathcal{H}$ because $(P_1 \|_{\{h\}} P_2) / \mathcal{A}_\mathcal{H} \xrightarrow{\tau}_a (\underline{0} \|_{\{h\}} \underline{0}) / \mathcal{A}_\mathcal{H}$ is matched by $(P_1 \|_{\{h\}} P_2) \setminus \mathcal{A}_\mathcal{H} \Longrightarrow_a (\underline{0} \|_{\{h\}} \underline{0}) \setminus \mathcal{A}_\mathcal{H}$. Similar to [25, 23], it is not only a matter of the higher discriminating power of $\approx_{tb}$ with respect to $\approx_{tw}$. If we used the CCS parallel composition operator [46], which turns into τ the synchronization of two actions thus combining communication with hiding, then the parallel composition of P_1 and P_2 with restriction on $\mathcal{A}_\mathcal{H}$ would be able to respond with a single τ-transition reaching the parallel composition of $\underline{0}$ and $\underline{0}$ with restriction on $\mathcal{A}_\mathcal{H}$.

4.2 Taxonomy of Security Properties

First of all, similar to the nondeterministic and probabilistic settings [26, 25, 23] the properties in Definition 5 turn out to be increasingly finer. This result holds for both those based on $\approx_{\text{tw}}$ and those based on $\approx_{\text{tb}}$.

Theorem 3. *Let* $\approx \, \in \{\approx_{\text{tw}}, \approx_{\text{tb}}\}$. *Then:*
$$\text{SBNDC}_\approx \subsetneq \text{SBSNNI}_\approx = \text{P\_BNDC}_\approx \subsetneq \text{BNDC}_\approx \subsetneq \text{BSNNI}_\approx \qquad \blacksquare$$

All the inclusions are strict as shown by the following counterexamples:

- The process $\tau.l.\underline{0} + l.l.\underline{0} + h.l.\underline{0}$ is SBSNNI$_\approx$ (resp. P_BDNC$_\approx$) because $(\tau.l.\underline{0} + l.l.\underline{0} + h.l.\underline{0}) \setminus \{h\} \approx (\tau.l.\underline{0} + l.l.\underline{0} + h.l.\underline{0})/\{h\}$ and action h is enabled only by the initial process so every derivative is BSNNI$_\approx$ (resp. BNDC$_\approx$). It is not SBNDC$_\approx$ because the low-level view of the process reached after action h, i.e., $(l.\underline{0}) \setminus \{h\}$, is not $\approx$-equivalent to $(\tau.l.\underline{0} + l.l.\underline{0} + h.l.\underline{0}) \setminus \{h\}$.
- The process $l.\underline{0} + l.l.\underline{0} + l.h.l.\underline{0}$ is BNDC$_\approx$ because, whether there are synchronizations with high-level actions or not, the overall process can always perform either an l-action or a sequence of two l-actions. It is not SBSNNI$_\approx$ (resp. P_BNDC$_\approx$) because the reachable process $h.l.\underline{0}$ is not BSNNI$_\approx$ (resp. BNDC$_\approx$).
- The process $l.\underline{0} + h.h.l.\underline{0}$ is BSNNI$_\approx$ as $(l.\underline{0} + h.h.l.\underline{0}) \setminus \{h\} \approx (l.\underline{0} + h.h.l.\underline{0})/\{h\}$. It is not BNDC$_\approx$ as $(((l.\underline{0} + h.h.l.\underline{0}) \|_{\{h\}} (h.\underline{0}))/\{h\}) \setminus \{h\} \not\approx (l.\underline{0} + h.h.l.\underline{0}) \setminus \{h\}$ in that the former behaves as $l.\underline{0} + \tau.\underline{0}$ while the latter behaves as $l.\underline{0}$.

Secondly, we observe that all the $\approx_{\text{tb}}$-based noninterference properties imply the corresponding $\approx_{\text{tw}}$-based ones, due to the fact that $\approx_{\text{tb}}$ is finer than $\approx_{\text{tw}}$.

Theorem 4. *The following inclusions hold:*

1. $\text{BSNNI}_{\approx_{\text{tb}}} \subsetneq \text{BSNNI}_{\approx_{\text{tw}}}$.
2. $\text{BNDC}_{\approx_{\text{tb}}} \subsetneq \text{BNDC}_{\approx_{\text{tw}}}$.
3. $\text{SBSNNI}_{\approx_{\text{tb}}} \subsetneq \text{SBSNNI}_{\approx_{\text{tw}}}$.
4. $\text{P\_BNDC}_{\approx_{\text{tb}}} \subsetneq \text{P\_BNDC}_{\approx_{\text{tw}}}$.
5. $\text{SBNDC}_{\approx_{\text{tb}}} \subsetneq \text{SBNDC}_{\approx_{\text{tw}}}$. $\qquad \blacksquare$

All the inclusions above are strict by virtue of the following result; for an example of P_1 and P_2 below, see Fig. 1.

Theorem 5. *Let* $P_1, P_2 \in \mathbb{P}$ *be such that* $P_1 \approx_{\text{tw}} P_2$ *but* $P_1 \not\approx_{\text{tb}} P_2$. *If no high-level actions occur in* P_1 *and* P_2, *then* $Q \in \{P_1 + h.P_2, P_2 + h.P_1\}$ *is such that:*

1. $Q \in \text{BSNNI}_{\approx_{\text{tw}}}$ *but* $Q \notin \text{BSNNI}_{\approx_{\text{tb}}}$.
2. $Q \in \text{BNDC}_{\approx_{\text{tw}}}$ *but* $Q \notin \text{BNDC}_{\approx_{\text{tb}}}$.

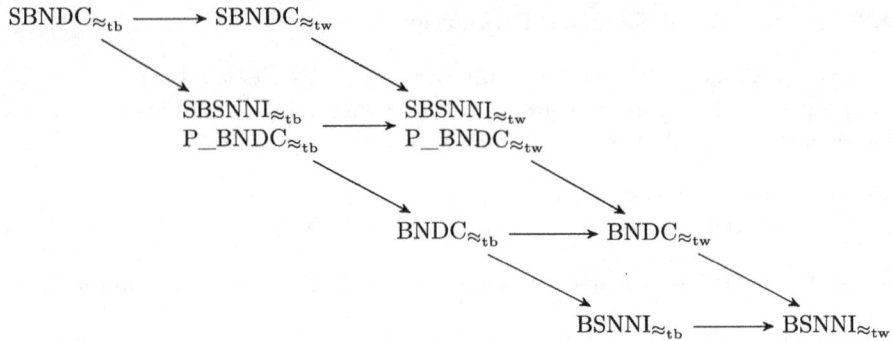

Fig. 2. Taxonomy of security properties based on timed bisimilarities.

3. $Q \in \text{SBSNNI}_{\approx_{tw}}$ but $Q \notin \text{SBSNNI}_{\approx_{tb}}$.
4. $Q \in \text{P\_BNDC}_{\approx_{tw}}$ but $Q \notin \text{P\_BNDC}_{\approx_{tb}}$.
5. $Q \in \text{SBNDC}_{\approx_{tw}}$ but $Q \notin \text{SBNDC}_{\approx_{tb}}$. ∎

The diagram in Fig. 2, which follows the same pattern as the nondeterministic and probabilistic settings [25, 23], summarizes the relationships among the various noninterference properties based on the results in Theorems 3 and 4. In the diagram, $\mathcal{P} \rightarrow \mathcal{Q}$ means that $\mathcal{P}$ is strictly included in $\mathcal{Q}$, while missing arrows express incomparability and are justified by the following counterexamples:

- $\text{SBNDC}_{\approx_{tw}}$ vs. $\text{SBSNNI}_{\approx_{tb}}$. The process $\tau . l . \underline{0} + l . l . \underline{0} + h . l . \underline{0}$ is $\text{BSNNI}_{\approx_{tb}}$ as $\tau . l . \underline{0} + l . l . \underline{0} \approx_{tb} \tau . l . \underline{0} + l . l . \underline{0} + \tau . l . \underline{0}$. It is also $\text{SBSNNI}_{\approx_{tb}}$ because every reachable process does not enable any more high-level actions. However, it is not $\text{SBNDC}_{\approx_{tw}}$, because after executing the high-level action h it can perform a single action l, while the original process with the restriction on high-level actions can go along a path where it performs two l-actions. On the other hand, the process Q mentioned in Theorem 5 is $\text{SBNDC}_{\approx_{tw}}$ but neither $\text{BSNNI}_{\approx_{tb}}$ nor $\text{SBSNNI}_{\approx_{tb}}$.
- $\text{SBSNNI}_{\approx_{tw}}$ vs. $\text{BNDC}_{\approx_{tb}}$. The process $l . h . l . \underline{0} + l . \underline{0} + l . l . \underline{0}$ is $\text{BSNNI}_{\approx_{tb}}$ as $l . \underline{0} + l . \underline{0} + l . l . \underline{0} \approx_{tb} l . \tau . l . \underline{0} + l . \underline{0} + l . l . \underline{0}$. The same process is $\text{BNDC}_{\approx_{tb}}$ too as it includes only one high-level action, hence the only possible high-level strategy coincides with the check conducted by $\text{BSNNI}_{\approx_{tb}}$. However, the process is not $\text{SBSNNI}_{\approx_{tw}}$ because of the reachable process $h . l . \underline{0}$, which is not $\text{BSNNI}_{\approx_{tw}}$. On the other hand, the process Q mentioned in Theorem 5 is $\text{SBSNNI}_{\approx_{tw}}$ but not $\text{BSNNI}_{\approx_{tb}}$ and, therefore, cannot be $\text{BNDC}_{\approx_{tb}}$.
- $\text{BNDC}_{\approx_{tw}}$ vs. $\text{BSNNI}_{\approx_{tb}}$. The process $(t) . l . \underline{0} + h_1 . h_2 . (t) . l . \underline{0}$ is $\text{BSNNI}_{\approx_{tb}}$ as discussed in Sect. 3, but it is not $\text{BNDC}_{\approx_{tw}}$. In contrast, the process Q mentioned in Theorem 5 is both $\text{BSNNI}_{\approx_{tw}}$ and $\text{BNDC}_{\approx_{tw}}$, but not $\text{BSNNI}_{\approx_{mb}}$.

5 Relating Nondeterministic and Timed Taxonomies

Let us compare our timed taxonomy with the nondeterministic one of [25]. In the following, we assume that $\approx_w$ denotes the weak nondeterministic bisimilarity of [46] and $\approx_b$ denotes the nondeterministic branching bisimilarity of [33]. These can also be obtained from the corresponding definitions in Sect. 2.2 by ignoring the clause about timed transitions. Since we are abstracting from delays, given a process $P \in \mathbb{P}$ we can obtain its nondeterministic variant, denoted by $nd(P)$, by replacing every occurrence of $(t).P'$ with $\tau.P'$. However, to respect maximal progress, first we have to eliminate every subprocess starting with a delay prefix that is alternative to a subprocess starting with a τ-prefix. To accomplish this transformation syntactically, we focus on the set $\mathbb{P}_{\mathrm{seq}}$ of sequential processes, i.e., without parallel composition; this is not too restrictive because, in the absence of recursion, parallel composition can be eliminated by repeatedly applying a timed variant of the expansion law [47].

The next proposition states that if two sequential processes are equivalent according to any of the weak bisimilarities in Sect. 2.2, then their nondeterministic variants are equivalent according to the corresponding nondeterministic weak bisimilarity. The inverse does not hold; e.g., processes $P_1 = (1).a.\underline{0}$ and $P_2 = (2).a.\underline{0}$ are such that $P_1 \not\approx_{tw} P_2$ and $P_1 \not\approx_{tb} P_2$, but their nondeterministic counterparts coincide as both of them are equal to $\tau.a.\underline{0}$.

Proposition 2. *Let $P_1, P_2 \in \mathbb{P}_{\mathrm{seq}}$. Then:*

- $P_1 \approx_{tw} P_2 \implies nd(P_1) \approx_w nd(P_2)$.
- $P_1 \approx_{tb} P_2 \implies nd(P_1) \approx_b nd(P_2)$. ∎

An immediate consequence is that if a sequential process is secure under any of the timed noninterference properties of Sect. 3, then its nondeterministic variant is secure under the corresponding nondeterministic property. The taxonomy of Fig. 2 thus extends to the left the one in [25], as each of the properties of Sect. 3 is finer than its nondeterministic counterpart.

Corollary 1. *Let $\mathcal{P}_{tm} \in \{\mathrm{BSNNI}_{\approx_{tm}}, \mathrm{BNDC}_{\approx_{tm}}, \mathrm{SBSNNI}_{\approx_{tm}}, \mathrm{P\_BNDC}_{\approx_{tm}},$ $\mathrm{SBNDC}_{\approx_{tm}}\}$ and $\mathcal{P}_{nd} \in \{\mathrm{BSNNI}_{\approx_{nd}}, \mathrm{BNDC}_{\approx_{nd}}, \mathrm{SBSNNI}_{\approx_{nd}}, \mathrm{P\_BNDC}_{\approx_{nd}},$ $\mathrm{SBNDC}_{\approx_{nd}}\}$ for $\approx_{tm} \in \{\approx_{tw}, \approx_{tb}\}$ and $\approx_{nd} \in \{\approx_w, \approx_b\}$, where $\mathcal{P}_{nd}$ is the nondeterministic variant of $\mathcal{P}_{tm}$. Then $P \in \mathcal{P}_{tm} \implies nd(P) \in \mathcal{P}_{nd}$ for all $P \in \mathbb{P}_{\mathrm{seq}}$.* ∎

6 Reversibility via Timed Back-and-Forth Bisimilarity

In [21] it was shown that, for nondeterministic processes, weak back-and-forth bisimilarity coincides with branching bisimilarity. We now extend that result so that timed branching bisimilarity can be employed in the noninterference analysis of reversible processes featuring nondeterminism and time.

A TLTS $(\mathcal{S}, \mathcal{A}_\tau, \longrightarrow)$ represents a reversible process if each of its transitions is seen as bidirectional. When going backward, it is of paramount importance to respect causality, i.e., the last performed transition must be the first one to be undone. Following [21] we set up an equivalence that enforces not only causality but also history preservation. This means that, when going backward, a process can only move along the path representing the history that brought the process to the current state even in the presence of concurrency. To accomplish this, the equivalence has to be defined over computations, not over states, and the notion of transition has to be revised so that it has source and target paths instead of states. We start by adapting the notation of the nondeterministic setting of [21] to our nondeterministic and timed setting. We use ℓ for a label in $\mathcal{A}_\tau \cup \mathbb{N}_{>0}$.

Definition 6. *A sequence* $\xi = (s_0, \ell_1, s_1)(s_1, \ell_2, s_2)\ldots(s_{n-1}, \ell_n, s_n) \in \longrightarrow^*$ *is a path of length n from state s_0. We let first$(\xi) = s_0$ and last$(\xi) = s_n$; the empty path is indicated with ε. We denote by path(s) the set of paths from s.* ■

Definition 7. *A pair $\rho = (s, \xi)$ is called a* run *from state s iff $\xi \in$ path(s), in which case we let path$(\rho) = \xi$, first$(\rho) =$ first$(\xi) = s$, last$(\rho) =$ last(ξ), with first$(\rho) =$ last$(\rho) = s$ when $\xi = \varepsilon$. We denote by run(s) the set of runs from state s. Given $\rho = (s, \xi) \in$ run(s) and $\rho' = (s', \xi') \in$ run(s'), their composition $\rho\rho' = (s, \xi\xi') \in$ run(s) is defined iff last$(\rho) =$ first$(\rho') = s'$. We write $\rho \xrightarrow{\ell} \rho'$ iff there exists $\rho'' = (\bar{s}, (\bar{s}, \ell, s'))$ with $\bar{s} =$ last(ρ) such that $\rho' = \rho\rho''$; note that first$(\rho) =$ first(ρ').* ■

In the considered TLTS we work with the set $\mathcal{U}$ of runs in lieu of $\mathcal{S}$. Following [21], given a run ρ, we distinguish between *outgoing* and *incoming* transitions of ρ during the weak bisimulation game, both for action transitions and for timed ones, depending on whether we examine the forward or backward direction.

Definition 8. *Let $(\mathcal{S}, \mathcal{A}_\tau, \longrightarrow)$ be a TLTS. We say that $s_1, s_2 \in \mathcal{S}$ are* weakly timed back-and-forth bisimilar, *written $s_1 \approx_{\text{tbf}} s_2$, iff $((s_1, \varepsilon), (s_2, \varepsilon)) \in \mathcal{B}$ for some* weak timed back-and-forth bisimulation $\mathcal{B}$*. A symmetric relation $\mathcal{B}$ over $\mathcal{U}$ is a* weak timed back-and-forth bisimulation *iff, whenever $(\rho_1, \rho_2) \in \mathcal{B}$, then:*

- *For each $\rho_1 \xrightarrow{a}_a \rho'_1$ there exists $\rho_2 \xRightarrow{\hat{a}}_a \rho'_2$ such that $(\rho'_1, \rho'_2) \in \mathcal{B}$.*
- *For each $\rho'_1 \xrightarrow{a}_a \rho_1$ there exists $\rho'_2 \xRightarrow{\hat{a}}_a \rho_2$ such that $(\rho'_1, \rho'_2) \in \mathcal{B}$.*
- *For each $\rho_1 \Longrightarrow_a \rho'_1$ with $\rho'_1 \not\xrightarrow{\tau}_a$ there exists $\rho_2 \Longrightarrow_a \rho'_2$ with $\rho'_2 \not\xrightarrow{\tau}_a$ such that $(\rho'_1, \rho'_2) \in \mathcal{B}$ and for each $\rho'_1 \xrightarrow{t}_t \rho''_1$ there exists $\rho'_2 \xRightarrow{t}_t \rho''_2$ such that $(\rho''_1, \rho''_2) \in \mathcal{B}$.*
- *For each $\rho'_1 \xrightarrow{t}_t \rho_1$ with $\rho'_1 \not\xrightarrow{\tau}_a$ there exists $\rho'_2 \xRightarrow{t}_t \rho_2$ with $\rho'_2 \not\xrightarrow{\tau}_a$ such that $(\rho'_1, \rho'_2) \in \mathcal{B}$.* ■

We show that weak timed back-and-forth bisimilarity over runs coincides with $\approx_{\text{tb}}$, the forward-only timed branching bisimilarity over states. We proceed by adopting the proof strategy followed in [21] to show that their weak back-and-forth bisimilarity over runs coincides with the forward-only branching bisimilarity over states of [33]. Therefore we start by proving that $\approx_{\text{tbf}}$ satisfies the *cross property*. This means that, whenever two runs of two $\approx_{\text{tbf}}$-equivalent states can perform a sequence of finitely many τ-transitions, such that each of the two target runs is $\approx_{\text{tbf}}$-equivalent to the source run of the other sequence, then the two target runs are $\approx_{\text{tbf}}$-equivalent to each other as well.

Lemma 2. *Let $s_1, s_2 \in \mathcal{S}$ with $s_1 \approx_{\text{tbf}} s_2$. For all $\rho'_1, \rho''_1 \in \text{run}(s_1)$ such that $\rho'_1 \Longrightarrow_a \rho''_1$ and for all $\rho'_2, \rho''_2 \in \text{run}(s_2)$ such that $\rho'_2 \Longrightarrow_a \rho''_2$, if $\rho'_1 \approx_{\text{tbf}} \rho'_2$ and $\rho''_1 \approx_{\text{tbf}} \rho'_2$ then $\rho''_1 \approx_{\text{tbf}} \rho''_2$.* ∎

Theorem 6. *Let $s_1, s_2 \in \mathcal{S}$. Then $s_1 \approx_{\text{tbf}} s_2 \iff s_1 \approx_{\text{tb}} s_2$.* ∎

In conclusion, the properties $\text{BSNNI}_{\approx_{\text{tb}}}$, $\text{BNDC}_{\approx_{\text{tb}}}$, $\text{SBSNNI}_{\approx_{\text{tb}}}$, $\text{P\_BNDC}_{\approx_{\text{tb}}}$, and $\text{SBNDC}_{\approx_{\text{tb}}}$ do not change if $\approx_{\text{tb}}$ is replaced by $\approx_{\text{tbf}}$. This allows us to study noninterference properties for reversible systems featuring nondeterminism and time by using $\approx_{\text{tb}}$ in a process calculus like the one of Sect. 2.3, without having to resort to external memories [18], communication keys [53], or executed action decorations [14, 12] like in reversible process calculi.

7 Use Case: Real-Time Database Transactions

Integrating security in real-time systems is a critical issue in several application domains, ranging from database management systems (DBMS) [1] to cyberphysical [7] and embedded [67] systems. In particular, the processing of concurrent transactions in real-time, multi-level secure database systems has to respect noninterference security properties about *values* (i.e., data read by low-level users cannot be affected by actions performed by high-level transactions), *delays* (i.e., the delay experienced by low-level transactions cannot depend on the execution of high-level transactions), and *recovery* (the abort of low-level transactions, as well as the actions taken to recover, cannot be influenced by the presence of high-level transactions) [38,62]. The satisfaction of these conditions is even more complicated in systems where transactions with real-time requirements are assigned priorities, as they are served according to their priorities rather than on a first-come-first-served basis [1].

Let us first explain through some examples inspired by [62] the subtleties of potential covert channels in such a complex scenario. To this aim, consider a sequence of three transactions, each with its own security level, priority, arrival time for scheduling, and execution time. Depending on these parameters, we will show that covert channels may or may not arise. In the following, we assume that the first transaction, arriving at time 1, is the high-level transaction HT_1. Then, we have the two low-level transactions LT_2 and LT_3, arriving at time 8 and 11 respectively, such that the priority of LT_3 is higher than the priority of LT_2. Moreover, HT_1 requests read access to variable x, while the two low-level transactions request write access to that variable. The three transactions follow a classical two-phase locking (2PL) mechanism based on the acquisition and release of a read/write lock before and after the requested operation. By abstracting from the lock operations and denoting by hr_1, lw_2, and lw_3 the three access operations, we obtain the following process:

$$DBMS = (1).(\tau.(7).lw_2.(3).lw_3.\underline{0} + hr_1.(7).lw_2.(3).lw_3.\underline{0})$$

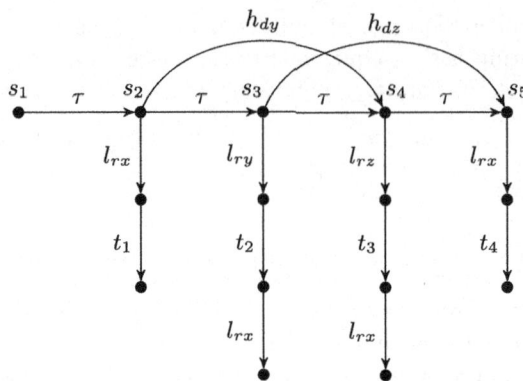

Fig. 3. Interleaving of two concurrent transactions (r for read, d for delete)

which represents the case in which HT_1, if scheduled, terminates before the arrival of LT_2, which in turn terminates before the arrival of LT_3. This means that every low-level transaction does not experience any delay due to the other transactions. Both *value* and *delay* security hold and, in particular, it can be easily verified that $DBMS \setminus \{hr_1\}$ and $DBMS / \{hr_1\}$ enable weakly/branching bisimilar behaviors in this nondeterministic and timed setting.

Now, consider the following variant:
$$DBMS' \;=\; (1) \,.\, (\tau \,.\, (7) \,.\, lw_2 \,.\, (3) \,.\, lw_3 \,.\, \underline{0} \;+\; hr_1 \,.\, (9) \,.\, lw_2 \,.\, (1) \,.\, lw_3 \,.\, \underline{0})$$
expressing that if HT_1 is scheduled then LT_2 is delayed by 2 time units with respect to its arrival time (e.g., because the duration of HT_1 exceeds 7 time units, thus blocking the lock acquisition for LT_2), while LT_3 does not experience any delay. *Delay* security does not hold anymore and, in particular, this can be verified via $DBMS' \setminus \{hr_1\}$ and $DBMS' / \{hr_1\}$ not being timed weakly/branching bisimilar. Note that, in contrast, their nondeterministic versions are still nondeterministic weakly/branching bisimilar as in the previous case.

Finally, consider the following further variant, where the execution of HT_1 requires 11 time units:
$$DBMS'' \;=\; (1) \,.\, (\tau \,.\, (7) \,.\, lw_2 \,.\, (3) \,.\, lw_3 \,.\, \underline{0} \;+\; hr_1 \,.\, (11) \,.\, lw_3 \,.\, (1) \,.\, lw_2 \,.\, \underline{0})$$
Note that, because of the latency due to HT_1, LT_3 arrives when LT_2 is still waiting, thus preempting it because of the higher priority level. Hence, in this case, both *value* and *delay* security do not hold anymore, as confirmed by the fact that $DBMS'' \setminus \{hr_1\}$ and $DBMS'' / \{hr_1\}$ are not nondeterministic/timed weakly/branching bisimilar.

Several approaches have been proposed to make 2PL robust with respect to these kinds of interferences. Here we consider a solution in which conflicting lock operations are not delayed thanks to the use of *virtual locks*. From the user viewpoint, such operations are transparent and the variables are accessed immediately on demand (see, e.g., [62, 20] for details). In this scenario, we describe an example that emphasizes how the branching semantics helps to capture violations of *recovery* security whenever the DBMS supports reversible transactions [22,25]. Consider a low-level transaction LT and a high-level transaction HT accessing three variables x, y, and z. Their interleaving is shown in Fig. 3, where actions of the form l_{rv} denote a read access on variable v by LT, while actions of the form h_{dv} denote the deletion of variable v by HT. Since the virtual lock operations are transparent from the user viewpoint, they are not modeled. The execution starts in s_1 with LT and HT that are activated. Then, in s_2, LT has to choose between some internal activity and the reading of variable x, after which a delay of t_1 time units follows, while at the same time HT may access variable y to delete it. The interpretation of the subsequent branches is analogous. Note that the high-level transaction departing from s_2 skips s_3, because variable y is deleted and, consequently, LT could not access it.

It is worth noting that the system in Fig. 3, call it $DBMS_{vl}$, satisfies the SBSNNI$_{\approx_{tw}}$ property. In particular, the most interesting case is given by the transition $s_2 \xrightarrow{\tau}_a s_4$ in $DBMS_{vl} / \mathcal{A}_\mathcal{H}$, which is simulated by the sequence $s_2 \xrightarrow{\tau}_a s_3 \xrightarrow{\tau}_a s_4$ in $DBMS_{vl} \setminus \mathcal{A}_\mathcal{H}$ (we can reason analogously for the transition $s_3 \xrightarrow{\tau}_a s_5$). However, this does not hold when considering the $\approx_{tb}$-based semantics, because the intermediate state s_3 is not timed branching bisimilar to the departing state s_2. From the back-and-forth perspective, consider executing the run $\tau.\tau.l_{rz}$ of $DBMS_{vl} / \mathcal{A}_\mathcal{H}$, which can be matched by the run $\tau.\tau.\tau.l_{rz}$ of $DBMS_{vl} \setminus \mathcal{A}_\mathcal{H}$. By undoing the actions of the former run (e.g., due to the recovery following an abort on the reading operation), it is not possible to go back to a state enabling action l_{ry}. Instead, this is possible by undoing the latter run. This is enough to distinguish the two versions of the system in the setting of reversible transactions. As a consequence, it turns out that the BSNNI$_{\approx_{tb}}$ property is not satisfied, thus revealing that the $\approx_{tb}$-based semantics is adequate to verify *recovery* security.

8 Conclusions

In this paper we have extended to a deterministically timed setting our previous compositionality, preservation, and classification results about a selection of noninterfence properties for irreversible or reversible systems developed in a nondeterministic setting [25] and in a probabilistic one [23] (stochastic time has been recently addressed in [24]). To represent the passing of time, we have assumed time determinism and time additivity. The two behavioral equivalences – designed to comply with the further assumption of maximal progress – for those noninterference properties are weak timed bisimilarity [48] and a newly defined timed branching bisimilarity. Since we have shown that timed branching bisimilarity coincides with a timed variant of the weak back-and-forth bisimilarity of [21], noninterference properties based on this equivalence can be applied to reversible timed systems, thus extending the results in [25, 23] for nondeterministic and probabilistic systems.

As for future work, we would like to include recursion in the considered process language, thus allowing one to model systems that may not terminate. This requires identifying adequate timed variants of the up-to technique for weak [59] and branching [31] bisimilarities, to be used in the proof of some results where we can now proceed by induction on the depth of the tree-like TLTS underlying the considered process term. Another direction that we want to pursue is addressing dense time [69].

Acknowledgment. This research has been supported by the PRIN 2020 project *NiRvAna – Noninterference and Reversibility Analysis in Private Blockchains*.

References

1. Abbott, R.K., Garcia-Molina, H.: Scheduling real-time transactions: a performance evaluation. ACM Trans. Database Syst. **17**, 513–560 (1992)
2. Aceto, L., Murphy, D.: Timing and causality in process algebra. Acta Informatica **33**, 317–350 (1996)
3. Aldini, A.: Classification of security properties in a Linda-like process algebra. Sci. Comput. Program. **63**, 16–38 (2006)
4. Aldini, A., Bernardo, M.: A general framework for nondeterministic, probabilistic, and stochastic noninterference. In: Proc. of the 1st Joint Workshop on Automated Reasoning for Security Protocol Analysis and Issues in the Theory of Security (ARSPA/WITS 2009). LNCS, vol. 5511, pp. 18–33. Springer (2009)
5. Aldini, A., Bernardo, M.: Component-oriented verification of noninterference. J. Syst. Architect. **57**, 282–293 (2011)
6. Aldini, A., Bravetti, M., Gorrieri, R.: A process-algebraic approach for the analysis of probabilistic noninterference. J. Comput. Secur. **12**, 191–245 (2004)
7. Baek, H., Lee, J., Lee, Y., Yoon, H.: Preemptive real-time scheduling incorporating security constraint for cyber physical systems. IEICE Trans. Inf. Syst. **99-D**, 2121–2130 (2016)
8. Baeten, J.C.M., Bergstra, J.A.: Real time process algebra. Formal Aspects Comput. **3**, 142–188 (1991)
9. Barbuti, R., Tesei, L.: A decidable notion of timed non-interference. Fund. Inform. **54**, 137–150 (2003)
10. Bennett, C.H.: Logical reversibility of computation. IBM J. Res. Dev. **17**, 525–532 (1973)
11. Bernardo, M., Corradini, F., Tesei, L.: Timed process calculi with deterministic or stochastic delays: commuting between durational and durationless actions. Theoret. Comput. Sci. **629**, 2–39 (2016)
12. Bernardo, M., Esposito, A., Mezzina, C.A.: Alternative characterizations of hereditary history-preserving bisimilarity via backward ready multisets. In: Proc. of the 28th Int. Conf. on Foundations of Software Science and Computation Structures (FOSSACS 2025). LNCS, vol. 15691, pp. 67–87. Springer (2025)
13. Bernardo, M., Lanese, I., Marin, A., Mezzina, C.A., Rossi, S., Sacerdoti Coen, C.: Causal reversibility implies time reversibility. In: Proc. of the 20th Int. Conf. on Quantitative Evaluation of Systems (QEST 2023). LNCS, vol. 14287, pp. 270–287. Springer (2023)
14. Bernardo, M., Rossi, S.: Reverse bisimilarity vs. forward bisimilarity. In: Proc. of the 26th Int. Conf. on Foundations of Software Science and Computation Structures (FOSSACS 2023). LNCS, vol. 13992, pp. 265–284. Springer (2023)
15. Bolognesi, T., Lucidi, F.: LOTOS-like process algebras with urgent or timed interactions. In: Proc. of the 4th Int. Conf. on Formal Description Techniques for Distributed Systems and Communication Protocols (FORTE 1991). IFIP Transactions, vol. C-2, pp. 249–264 (1991)

16. Brookes, S.D., Hoare, C.A.R., Roscoe, A.W.: A theory of communicating sequential processes. J. ACM **31**, 560–599 (1984)
17. Corradini, F., Vogler, W., Jenner, L.: Comparing the worst-case efficiency of asynchronous systems with PAFAS. Acta Informatica **38**, 735–792 (2002)
18. Danos, V., Krivine, J.: Reversible communicating systems. In: Gardner, .P., Yoshida, N. (eds.) CONCUR 2004. LNCS, vol. 3170, pp. 292–307. Springer, Heidelberg (2004). https://doi.org/10.1007/978-3-540-28644-8_19
19. Danos, V., Krivine, J.: Transactions in RCCS. In: Abadi, M., de Alfaro, L. (eds.) CONCUR 2005. LNCS, vol. 3653, pp. 398–412. Springer, Heidelberg (2005). https://doi.org/10.1007/11539452_31
20. David, R., Son, S.H.: A secure two phase locking protocol. In: Proc. of the 12th IEEE Symp. on Reliable Distributed Systems (SRDS 1993), pp. 126–135. IEEE-CS Press (1993)
21. De Nicola, R., Montanari, U., Vaandrager, F.: Back and forth bisimulations. In: Baeten, J.C.M., Klop, J.W. (eds.) CONCUR 1990. LNCS, vol. 458, pp. 152–165. Springer, Heidelberg (1990). https://doi.org/10.1007/BFb0039058
22. Engblom, J.: A review of reverse debugging. In: Proc. of the 4th System, Software, SoC and Silicon Debug Conf. (S4D 2012), pp. 1–6. IEEE-CS Press (2012)
23. Esposito, A., Aldini, A., Bernardo, M.: Noninterference analysis of reversible probabilistic systems. In: Proc. of the 44th Int. Conf. on Formal Techniques for Distributed Objects, Components, and Systems (FORTE 2024). LNCS, vol. 14678, pp. 39–59. Springer (2024)
24. Esposito, A., Aldini, A., Bernardo, M.: Noninterference analysis of stochastically timed reversible systems. In: Proc. of the 45th Int. Conf. on Formal Techniques for Distributed Objects, Components, and Systems (FORTE 2025). LNCS, vol. 15732, pp. 75–95. Springer (2025)
25. Esposito, A., Aldini, A., Bernardo, M., Rossi, S.: Noninterference analysis of reversible systems: an approach based on branching bisimilarity. Logical Methods Comput. Sci. **21(1)**, 6:1–6:28 (2025)
26. Focardi, R., Gorrieri, R.: Classification of security properties. In: Proc. of the 1st Int. School on Foundations of Security Analysis and Design (FOSAD 2000). LNCS, vol. 2171, pp. 331–396. Springer (2001)
27. Focardi, R., Piazza, C., Rossi, S.: Proofs methods for bisimulation based information flow security. In: Cortesi, A. (ed.) VMCAI 2002. LNCS, vol. 2294, pp. 16–31. Springer, Heidelberg (2002). https://doi.org/10.1007/3-540-47813-2_2
28. Focardi, R., Rossi, S.: Information flow security in dynamic contexts. J. Comput. Secur. **14**, 65–110 (2006)
29. Giachino, E., Lanese, I., Mezzina, C.A.: Causal-Consistent Reversible Debugging. In: Gnesi, S., Rensink, A. (eds.) FASE 2014. LNCS, vol. 8411, pp. 370–384. Springer, Heidelberg (2014). https://doi.org/10.1007/978-3-642-54804-8_26
30. Giacobazzi, R., Mastroeni, I.: Abstract non-interference: a unifying framework for weakening information-flow. ACM Trans. on Privacy and Security **21(2)**, 9:1–9:31 (2018)
31. van Glabbeek, R.J.: A complete axiomatization for branching bisimulation congruence of finite-state behaviours. In: Borzyszkowski, A.M., Sokołowski, S. (eds.) MFCS 1993. LNCS, vol. 711, pp. 473–484. Springer, Heidelberg (1993). https://doi.org/10.1007/3-540-57182-5_39
32. van Glabbeek, R.J.: The linear time – branching time spectrum I. In: Handbook of Process Algebra, pp. 3–99. Elsevier (2001)
33. van Glabbeek, R.J., Weijland, W.P.: Branching time and abstraction in bisimulation semantics. J. ACM **43**, 555–600 (1996)

34. Goguen, J.A., Meseguer, J.: Security policies and security models. In: Proc. of the 2nd IEEE Symp. on Security and Privacy (SSP 1982), pp. 11–20. IEEE-CS Press (1982)
35. Hedin, D., Sabelfeld, A.: A perspective on information-flow control. In: Software Safety and Security – Tools for Analysis and Verification, pp. 319–347. IOS Press (2012)
36. Hennessy, M., Regan, T.: A process algebra for timed systems. Inf. Comput. **117**, 221–239 (1995)
37. Hillston, J., Marin, A., Piazza, C., Rossi, S.: Persistent stochastic non-interference. Fund. Inform. **181**, 1–35 (2021)
38. Keefe, T.F., Tsai, W.T., Srivastava, J.: Multilevel secure database concurrency control. In: Proc. of the 6th Int. Conf. on Data Engineering (ICDE 1990), pp. 337–344. IEEE-CS Press (1990)
39. Keller, R.M.: Formal verification of parallel programs. Commun. ACM **19**, 371–384 (1976)
40. Landauer, R.: Irreversibility and heat generation in the computing process. IBM J. Res. Dev. **5**, 183–191 (1961)
41. Lanese, I., Lienhardt, M., Mezzina, C.A., Schmitt, A., Stefani, J.-B.: Concurrent flexible reversibility. In: Felleisen, M., Gardner, P. (eds.) ESOP 2013. LNCS, vol. 7792, pp. 370–390. Springer, Heidelberg (2013). https://doi.org/10.1007/978-3-642-37036-6_21
42. Lanese, I., Nishida, N., Palacios, A., Vidal, G.: CauDEr: A Causal-Consistent Reversible Debugger for Erlang. In: Gallagher, J.P., Sulzmann, M. (eds.) FLOPS 2018. LNCS, vol. 10818, pp. 247–263. Springer, Cham (2018). https://doi.org/10.1007/978-3-319-90686-7_16
43. Laursen, J.S., Ellekilde, L.P., Schultz, U.P.: Modelling reversible execution of robotic assembly. Robotica **36**, 625–654 (2018)
44. Mantel, H.: Information flow and noninterference. In: Encyclopedia of Cryptography and Security, pp. 605–607. Springer (2011)
45. Martinelli, F.: Analysis of security protocols as open systems. Theoret. Comput. Sci. **290**, 1057–1106 (2003)
46. Milner, R.: Communication and Concurrency. Prentice Hall (1989)
47. Moller, F., Tofts, C.: A temporal calculus of communicating systems. In: Baeten, J.C.M., Klop, J.W. (eds.) CONCUR 1990. LNCS, vol. 458, pp. 401–415. Springer, Heidelberg (1990). https://doi.org/10.1007/BFb0039073
48. Moller, F., Tofts, C.: Behavioural abstraction in TCCS. In: Kuich, W. (ed.) ICALP 1992. LNCS, vol. 623, pp. 559–570. Springer, Heidelberg (1992). https://doi.org/10.1007/3-540-55719-9_104
49. Nicollin, X., Sifakis, J.: An overview and synthesis on timed process algebras. In: Proc. of the REX Workshop on Real Time: Theory in Practice. LNCS, vol. 600, pp. 526–548. Springer (1991)
50. Nicollin, X., Sifakis, J.: The algebra of timed processes ATP: theory and application. Inf. Comput. **114**, 131–178 (1994)
51. Park, D.: Concurrency and automata on infinite sequences. In: Deussen, P. (ed.) GI-TCS 1981. LNCS, vol. 104, pp. 167–183. Springer, Heidelberg (1981). https://doi.org/10.1007/BFb0017309
52. Perumalla, K.S., Park, A.J.: Reverse computation for rollback-based fault tolerance in large parallel systems - Evaluating the potential gains and systems effects. Clust. Comput. **17**, 303–313 (2014)
53. Phillips, I., Ulidowski, I.: Reversing algebraic process calculi. J. Logic Algebraic Program. **73**, 70–96 (2007)

54. Phillips, I., Ulidowski, I., Yuen, S.: A reversible process calculus and the modelling of the ERK signalling pathway. In: Proc. of the 4th Int. Workshop on Reversible Computation (RC 2012). LNCS, vol. 7581, pp. 218–232. Springer (2012)
55. Pinna, G.M.: Reversing steps in membrane systems computations. In: Proc. of the 18th Int. Conf. on Membrane Computing (CMC 2017). LNCS, vol. 10725, pp. 245–261. Springer (2017)
56. Quemada, J., Frutos, D., Azcorra, A.: TIC: a timed calculus. Formal Aspects Comput. **5**, 224–252 (1993)
57. Reed, G.M., Roscoe, A.W.: A timed model for communicating sequential processes. Theoret. Comput. Sci. **58**, 249–261 (1988)
58. Sabelfeld, A., Sands, D.: Probabilistic noninterference for multi-threaded programs. In: Proc. of the 13th IEEE Computer Security Foundations Workshop (CSFW 2000), pp. 200–214. IEEE-CS Press (2000)
59. Sangiorgi, D., Milner, R.: The problem of "weak bisimulation up to". In: Proc. of the 3rd Int. Conf. on Concurrency Theory (CONCUR 1992). LNCS, vol. 630, pp. 32–46. Springer (1992)
60. Schordan, M., Oppelstrup, T., Jefferson, D.R., Barnes, P.D., Jr.: Generation of reversible C++ code for optimistic parallel discrete event simulation. N. Gener. Comput. **36**, 257–280 (2018)
61. Siljak, H., Psara, K., Philippou, A.: Distributed antenna selection for massive MIMO using reversing Petri nets. IEEE Wireless Commun. Lett. **8**, 1427–1430 (2019)
62. Son, S.H., Mukkamala, R., David, R.: Integrating security and real-time requirements using covert channel capacity. IEEE Trans. on Knowledge Data Eng. **12**, 865–879 (2000)
63. Ulidowski, I., Yuen, S.: Extending process languages with time. In: Johnson, M. (ed.) AMAST 1997. LNCS, vol. 1349, pp. 524–538. Springer, Heidelberg (1997). https://doi.org/10.1007/BFb0000494
64. Vassor, M., Stefani, J.B.: Checkpoint/rollback vs causally-consistent reversibility. In: Proc. of the 10th Int. Conf. on Reversible Computation (RC 2018). LNCS, vol. 11106, pp. 286–303. Springer (2018)
65. Volpano, D., Smith, G.: Probabilistic noninterference in a concurrent language. In: Proc. of the 11th IEEE Computer Security Foundations Workshop (CSFW 1998), pp. 34–43. IEEE-CS Press (1998)
66. Vries, E., Koutavas, V., Hennessy, M.: Communicating transactions. In: Gastin, P., Laroussinie, F. (eds.) CONCUR 2010. LNCS, vol. 6269, pp. 569–583. Springer, Heidelberg (2010). https://doi.org/10.1007/978-3-642-15375-4_39
67. Xie, T., Qin, X.: Improving security for periodic tasks in embedded systems through scheduling. ACM Trans. in Embedded Comput. Syst. **6**(3), 20:1–20:19 (2007)
68. Ycart, B.: The philosophers' process: an ergodic reversible nearest particle system. Ann. Appl. Probability **3**, 356–363 (1993)
69. Yi, W.: CCS + time = an interleaving model for real time systems. In: Proc. of the 18th Int. Coll. on Automata, Languages and Programming (ICALP 1991). LNCS, vol. 510, pp. 217–228. Springer (1991)
70. Zheng, L., Myers, A.: Dynamic security labels and noninterference. In: Proc. of the 2nd IFIP Workshop on Formal Aspects in Security and Trust (FAST 2004). IFIP AICT, vol. 173, pp. 27–40. Springer (2004)

Controller Synthesis for Parametric Timed Games

Mikael Bisgaard Dahlsen-Jensen[1], Baptiste Fievet[2], Laure Petrucci[2(✉)], and Jaco van de Pol[1]

[1] Aarhus University, Aarhus, Denmark
{mikael,jaco}@cs.au.dk

[2] CNRS, Université Sorbonne Paris Nord, LIPN, 93430 Villetaneuse, France
{Baptiste.Fievet,Laure.Petrucci}@lipn.univ-paris13.fr

Abstract. We present a (semi)-algorithm to compute winning strategies for parametric timed games. Previous algorithms only synthesized constraints on the clock parameters for which the game is winning. A new definition of (winning) strategies is proposed, and ways to compute them. A transformation of these strategies to (parametric) timed automata allows for building a controller enforcing them. The feasibility of the method is demonstrated by an implementation and experiments for the Production Cell case study.

1 Introduction

Timed Games (TG) [17] extend Timed Automata (TA) [1] by distinguishing controllable and uncontrollable transitions and introducing a reachability goal. The game is won by the controller, if he can play controllable actions, such that no matter which uncontrollable actions are taken by the environment, the goal is reached. An on-the-fly algorithm was introduced in [7], to decide whether a timed game is won by the controller. This algorithm forms the basis of UPPAAL Tiga [5]. Parametric Timed Automata (PTA) [2] enable the study of an infinite family of TA by replacing concrete constraints by parameters. Given a desired property, the problem is to find all parameter valuations that satisfies it. Unlike TA, most problems are undecidable for PTA [3], although several fragments are known to be decidable, e.g., L/U PTA [13] and PTA with only one clock [6].

Parametric Timed Games (PTG) combine the ideas of PTA and TG and were introduced along with a semi-algorithm for parameter synthesis for reachability objectives [14,15], extended and implemented in [10]. The semi-algorithm in that work enumerates all constraints on clock parameters, for which the game is won.

We extend the previous work in several directions. Our main goal is to generate a *concrete strategy*, that tells the controller how to win the game. Thus, we

---

This work was partially supported by CNRS international PhD programme, the CNRS International Research Network CLoVe and Innovationsfonden Danmark's DIREC project SIoT (Secure Internet of Things).

© The Author(s), under exclusive license to Springer Nature Switzerland AG 2026
P. Prabhakar and A. Vandin (Eds.): QEST+FORMATS 2025, LNCS 16143, pp. 314–332, 2026.
https://doi.org/10.1007/978-3-032-05792-1_17

automatically synthesize a controller that is correct by construction. This strategy is represented itself by a parametric timed automaton, which can be put in parallel to the original system. In the synchronous product of the controller and the system, every run will eventually reach the goal location.

UPPAAL Tiga [5] allows generating winning strategies for TG and [12] proposes a translation from these to controller timed automata. However, these strategies have the "implicit semantics" that their instructions must be taken *as soon as possible*, causing ill-defined behavior in some cases. We improve the situation by a new type of strategies, whose instructions explicitly define the interval in which a transition must be taken, and extend this to PTG. This new notion of strategy allows for expressing the controller as a Parametric Timed Automaton.

We also reconciled the algorithm from [7] with the actual implementation in [5]. In [7] (and also in its extension [10]), a strategy can only rely on controllable actions, while UPPAAL Tiga takes the more realistic approach, in which the strategy can also consider uncontrollable actions, as long as they cannot be postponed forever. In this paper, we develop the theory of these *forced uncontrollable transitions*, and adapt the algorithms accordingly.

1.1 Motivating Example

Let us first recall how [7], its parametric extension [14,15] as well as the tool UPPAAL Tiga itself defines strategies: they are memoryless, i.e. provide the next action based solely on the current state. This action can be either a discrete transition or a *Wait* instruction. In case of a *Wait*, the controller must wait until a state is reached where the strategy provides a discrete transition, which should then be taken *as soon as possible*. However, the "first" future state where the strategy outputs a discrete transition may not exist.

Example 1. Consider the example in Fig. 1a. It is a Timed Game with a clock x and a reachability objective with Win as the only target location. Controllable (resp. uncontrollable) transitions are drawn by solid (resp. dashed) arrows, have action labels, optional guards and clock resets.

The environment can force a transition from L_0 to L_1 for any value $0 \leq x \leq 2$. When in L_1 there are three cases: If $1 < x < 2$, we can immediately reach Win by taking c_2 and c_3. If $x \leq 1$, we can safely wait to reach the interval $1 < x < 2$, and apply the previous case. If $x = 2$, we should still take c_2 immediately, else we would be stuck in L_1 forever. However, we might not be able to take c_3 from L_2 before the environment sends us back to L_1 with $x = 0$, where the previous case applies. Since we have a winning strategy from L_0 when the environment moves with $x \leq 2$, we can just wait if it happens, or else win with action c_1 when x reaches value 2. The strategy synthesized by UPPAAL is illustrated in Fig. 1b. Notice that c_2 winning in L_1 when $x = 2$ is merged with applying c_2 in L_1 when $1 < x < 2$.

Let us focus on how to win from state $(L_1, x = 0)$. From (3), we wait until $x > 1$ where (2) can be applied, taking action c_2 as soon as possible. But such a moment does not exist: when $x = 1$, it is too early to take c_2; when $x = t > 1$, it is too late as we could have taken c_2 earlier at a time t' such that $1 < t' < t$. Therefore, no run going through $(L_1, x = 0)$ complies with the strategy provided by UPPAAL. Yet, we need to be able to react if the environment moves from L_0 to L_1 when $x = 0$.

The way we handle this in our strategy is by adding additional information on how long to wait before acting. In Fig. 1b (bottom), we show how we modify (2). This new specification requires the player to commit to an action in advance and wait until the specified time interval before executing it. In L_1 when applying (2a), the action c_2 can be taken at any point in time within $1 < x < 2$. This avoids the issue in UPPAAL's strategy, where an action is always fired at the first available moment (even if it might not exist).

A key subtlety arises: unlike UPPAAL's synthesis, we cannot simply merge (2a) and (2b). If we did, it would be possible to follow the strategy but end up looping infinitely between L_1 and L_2. Instead, we keep them separate and ensure they remain disjoint. With this change, at most one iteration of the loop can happen, no matter how well the environment plays the uncontrollable actions. We will discuss the exact procedure for synthesizing such a strategy later.

1.2 Outline

In Sect. 2, we recall the basic definitions of the model for Parametric Timed Games. Section 3 introduces our new representation of a controller strategy, the strategy specification. Modifications to the strategy synthesis algorithm provide a strategy specification representing a winning strategy for all parameter values found winning. In Sect. 4, we adapt the model to account for forced transitions. We also introduce the notion of temporal bounds used in an updated version of our algorithm to solve Parametric Timed Games with forced transitions. Section 5 presents an algorithm that synthesizes a controller from a strategy specification such that the controller, when synchronized with the parametric timed

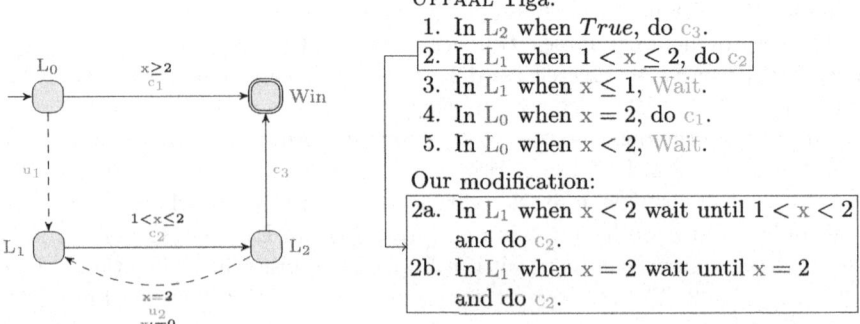

(a) Example of a Timed Game.   (b) Specification of winning strategy.

Fig. 1. Timed Game, strategy from UPPAAL Tiga, and a modified fix.

game, limits its possible runs to a subset of the runs that are coherent with the strategy specification, without restricting the environment actions. Then, Sect. 6 presents our implementation and the experiments conducted. Finally, conclusion and perspectives are drawn in Sect. 7.

2 Model of Parametric Timed Games

A Parametric Timed Game (PTG) is a structure based on timed automata (TA). Like classical automata, it has locations connected by discrete transitions. It also has clocks. Locations are associated a condition on clock valuations (invariant) that must be satisfied while staying in the location. An action in a timed automaton is either to take a discrete transition or to let some time pass. Discrete transitions have a guard that must be satisfied in order to take the transition. In a parametric setting, these conditions use linear terms over clocks and parameters. Parameters are unspecified but constant during a run. A discrete transition has a subset of clocks which are reset when the transition is taken.

A *clock valuation* is a function $v_X \in \mathbb{R}_{\geq 0}^X$ assigning a non-negative real value to each clock. A *parameter valuation* $v_P \in \mathbb{Q}_{\geq 0}^P$ assigns a non-negative rational value to each parameter. A *valuation of a game* is a pair $v = (v_X, v_P)$. The set of all valuations of the game is denoted $V = \mathbb{R}_{\geq 0}^X \times \mathbb{Q}_{\geq 0}^P$. A *linear term* over P is a term defined by the following grammar: $plt := k \mid kp \mid plt + plt$ where $k \in \mathbb{Q}$ and $p \in P$. *Zones* allow for capturing a set of valuations in a game. The set of *parametric zones* $\mathcal{Z}(X, P)$ is the set of formulas defined inductively by the following grammar: $\phi := \top \mid \phi \wedge \phi \mid x \sim plt \mid x - y \sim plt \mid plt' \sim plt$ where $x, y \in X$, $\sim \in \{<; \leq; =; \geq; >\}$ and plt and plt' are linear terms over P.

In a two-player timed game, discrete transitions are partitioned between controllable transitions and uncontrollable (environment) transitions.

Definition 1 (PTG). *A Parametric Timed Game is a tuple of the form* $G = (L, X, P, Act, T_c, T_u, \ell_0, Inv)$ *such that*

- L, X, P, Act *are sets of* locations, clocks, parameters, transition labels.
- $T = T_c \sqcup T_u$ *is partitioned into sets of* controllable *and* uncontrollable *transitions.* $T \subseteq L \times \mathcal{Z}(X, P) \times Act \times \mathcal{P}(X) \times L$ *is the set of transitions of the form* (ℓ, g, a, Y, ℓ') *where:* ℓ, ℓ' *are source and target locations, g is the guard,[1] a the label, Y the set of clocks to reset.*
- ℓ_0 *is the initial location.*
- $Inv : L \to \mathcal{Z}(X, P)$ *associates an* invariant *with each location.*

Function v_P is naturally extended to linear terms on parameters, by replacing each parameter in the term with its valuation. With $v \models \phi$, we denote that valuation $v = (v_X, v_P)$ *satisfies* a zone ϕ. Zones can also be seen as a convex set in the space of valuations by considering those satisfying the condition.

---
[1] Note that we extend the definition in [10] by allowing diagonal constraints in guards. The main reason is that the synthesized controller will need diagonal constraints.

2.1 Semantics of Parametric Timed Games

A *state* of a PTG consists of a location and a valuation of clocks and parameters. Transitions modify clock valuations by letting time pass or resetting clocks.

Let $v = (v_X, v_P)$ be a valuation of the game and $\delta \geq 0$ a delay. $\forall x \in X : (v_X + \delta)(x) = v_X(x) + \delta$ and $v + \delta = (v_X + \delta, v_P)$. Let $Y \subseteq X$. $v_X[Y := 0]$ is the valuation obtained by *resetting the clocks* in Y, i.e.: $\forall x \in Y : v_X[Y := 0](x) = 0$ and $\forall x \in X \setminus Y : v_X[Y := 0](x) = v_X(x)$, and $v[Y := 0] = (v_X[Y := 0], v_P)$.

The semantics of a Parametric Timed Game is defined as a timed transition system on states, with timed and discrete transitions, and a winning condition. A *state* of a PTG is a pair (ℓ, v) where ℓ is a location and v a valuation of the game satisfying its invariant: $v \models Inv(\ell)$. The *state space* is then $\mathbb{S} = \{(\ell, v) \in L \times V \mid v \models Inv(\ell)\} = \bigcup_{\ell \in L} \{\ell\} \times Inv(\ell)$. Let $\delta \in \mathbb{R}_{\geq 0}$ be a time delay. A *timed transition* is a relation $\to^\delta \in \mathbb{S} \times \mathbb{S}$ s.t. $\forall (\ell, v), (\ell', v') \in \mathbb{S} : (\ell, v) \to^\delta (\ell', v')$ iff $\ell = \ell'$ and $v' = v + \delta$. Let $t = (\ell, g, a, Y, \ell') \in T$ be a transition. A *discrete transition* is a relation $\to^t \in \mathbb{S} \times \mathbb{S}$ s.t. $\forall (\ell, v), (\ell', v') \in \mathbb{S} : (\ell, v) \to^t (\ell', v')$ iff $v \models g$ and $v' = v[Y := 0]$.

States are grouped in symbolic states, similar to valuations grouped in zones.

Definition 2 (symbolic state). *A symbolic state of a PTG is a pair* $\xi = (\ell, Z)$ *where* ℓ *is a location and* Z *a zone of the game satisfying its invariant:* $Z \models Inv(\ell)$. *The symbolic state* (ℓ, Z) *represents the subset of states* $\{\ell\} \times Z \subseteq \mathbb{S}$. *The set of* temporal successors *of* ξ *is defined as* $\xi^\nearrow = \{s' \mid \exists \delta, \exists s \in \xi : s \to^\delta s'\}$. *Similarly, its* temporal predecessors *are defined* $\xi^\swarrow = \{s \mid \exists \delta, \exists s' \in \xi : s \to^\delta s'\}$.

Let $\vec{0}$ be the clock valuation where all clocks have value 0. The set of possible initial states of the PTG is $\xi_0 = \{(\ell_0, (\vec{0}, v_P)) \mid v_P \in \mathbb{Q}^P_{\geq 0} : (\vec{0}, v_P) \models Inv(\ell_0)\}$.

A *run* of the PTG G is a finite or infinite sequence of states $s_0 s_1 s_2 \ldots$ s.t. $s_0 \in \xi_0$ and $\forall i \in \mathbb{N}, \exists \delta \in \mathbb{R}_{\geq 0}, \exists s \in \mathbb{S}, s_i \to^\delta s \to^t s_{i+1}$.

A *history* is a finite run. Given a history h, we define $ls(h)$ to be the last state of h. Given a game G, we denote its set of runs by $\mathcal{R}$ and the histories by $\mathcal{H}$.

Since we deal with reachability games, the objective is specified by a set of locations in the PTG. A run is winning if it reaches one of these locations. Let $R \subseteq L$ be a *reachability objective*. The set of *winning runs* $\Omega_{Reach}(R)$ is the subset of runs that visit R: $\Omega_{Reach}(R) = \{r \in \mathcal{R} \mid \exists \ell \in R, \exists v \in V : (\ell, v) \in r\}$.

2.2 Strategies in Parametric Timed Games

A controller strategy σ_c models decision-making. Building upon the strategy definition of [8], it is a function taking a history and deciding to either wait indefinitely, or to wait for a finite delay and apply a discrete transition. In the latter case, we will require the transition to be available after the delay (i.e. the invariant holds and the guard of the transition is enabled).

Section 1.1 exhibited a critical delay interval to select an action. We now introduce controller strategies that, given a history, return a non-empty set of valid decisions, one of which will be chosen non-deterministically.

Definition 3 (controller strategy). *A controller strategy σ is a function $\sigma : \mathcal{H} \to \mathcal{P}((\mathbb{R}_{\geq 0} \times T_c) \cup \{\infty\}) \setminus \{\emptyset\}$, s.t. for all $h \in \mathcal{H}$, if $(\delta, t_c) \in \sigma(h)$, then for some $s_\delta, s_{t_c} \in \mathbb{S}$, $ls(h) \to^\delta s_\delta \to^{t_c} s_{t_c}$.*
The delay of a decision is: $delay(\delta, t) = \delta$ and $delay(\infty) = \infty$.

Note ∞ denotes that we "give up our turn", not necessarily that we stay in this state forever. So we do not require that the invariant stays true in this case.

A controller strategy σ induces a set of runs coherent with the strategy. Either a decision of the controller strategy is applied, or a decision leading to an uncontrollable action occurs with a delay lower or equal to the delay of a controller decision of the strategy. This corresponds to the case where an uncontrollable action intercepts a controller decision of the strategy. If both the controller and the environment select the same delay, we consider that the controller cannot guarantee that its transition will be taken, thus the environment can intercept.

Definition 4 (run coherent with controller strategy). *Run $r = s_0 s_1 s_2 \ldots$ is coherent with the controller strategy σ if and only if $\forall i \in \mathbb{N}$:*

- *Either the run ends in s_i, and $\infty \in \sigma(s_0 s_1 \ldots s_i)$*
- *Or there exists $(\delta, t_c) \in \sigma(s_0 s_1 \ldots s_i)$ and $s' \in \mathbb{S}$ s.t. $s_i \to^\delta s' \to^{t_c} s_{i+1}$.*
- *Or there exists $d \in \sigma(s_0 s_1 \ldots s_i)$, $\delta' \leq delay(d) \in \mathbb{R}_{\geq 0}^\infty$, $t_u \in T_u$ and $s' \in \mathbb{S}$ such that $s_i \to^{\delta'} s' \to^{t_u} s_{i+1}$.*

Definition 5 (winning strategy). *A controller strategy σ is winning w.r.t. a reachability objective R iff all runs coherent with σ are winning w.r.t. R.*

Previous work [10] introduced an algorithm aimed at solving the following question: Given a Parametric Timed Game G and a Reachability Objective R, for which parameter valuations is there a winning controller strategy from the initial state? We now aim to solve the following questions: *How can we produce a controller strategy that guarantees to win from those parameters valuations? Can we express the controller as a PTA, to be synchronized with the system?*

3 Strategy Specification and Synthesis

We now introduce strategy specifications, as a means to specify infinite strategies by finite objects. A strategy specification is a list of instructions. Instructions can be of two types: First, $(\xi, \bot, Wait)$ where ξ is a symbolic state and *Wait* is a signal to wait. This instruction represents the instruction "When in ξ, wait indefinitely". Second, (ξ, ξ', t_c) where ξ and ξ' are symbolic states and t_c is a controllable transition. This represents the instruction "When in ξ, wait until ξ' is reached, and apply the discrete transition t_c".

Definition 6. *A strategy specification L is a list of instructions i of the form: $i = (\xi, \xi', t_c)$ or $i = (\xi, \bot, Wait)$, where ξ and ξ' are symbolic states, t_c is a controllable transition such that t_c is applicable from every state s' of ξ' and $\xi \subseteq \xi'^{\checkmark}$, $\bot$ is undefined and Wait is a signal to wait.*

The source of an instruction i is denoted by $src(i) = \xi$.

A state $s \in \mathbb{S}$ matches an instruction i of some strategy specification L when $s \in src(i)$. If the current state does not match any instruction of L, the default action is to wait indefinitely, as with an instruction of the form $(\xi, \bot, \textit{Wait})$. If the state matches an instruction of the form (ξ, ξ', t_c), a delay leading to ξ' is selected non-deterministically. From there, transition t_c is taken.

Definition 7 (decisions). *The set of possible decisions specified by instructions from a strategy specification L, given current state $s \in \mathbb{S}$, is defined as follows:*

$$dec((\xi, \bot, \textit{Wait}), s) := \{\infty\}$$
$$dec((\xi, \xi', t), s) := \{\delta \in \mathbb{R}_{\geq 0} \mid \exists s_\delta \in \xi' \text{ s.t. } s \to^\delta s_\delta\} \times \{t\}$$

The controller strategy σ_L specified by L on history $h \in \mathcal{H}$ is defined as follows:

$$\sigma_L(h) = \bigcup \{dec(i, ls(h)) \mid i \in L \text{ and } ls(h) \text{ matches } i\} \quad \text{if not empty,}$$
$$\sigma_L(h) = \{\infty\} \quad \text{otherwise.}$$

3.1 Synthesis of Strategy Specification

An algorithm to compute valuations of the parameters for which a winning strategy exists, is presented in [10]. We quickly recall how this algorithm works, before introducing some modifications to synthesize a winning strategy specification. Algorithm 1 presents the algorithm, where our modifications are highlighted in green boxes. The blue boxes will be addressed in Sect. 4. It features two main parts: the exploration (EXPLORE) and the back-propagation of winning states (UPDATE).

Original Algorithm from [10]. The overall idea is to compute reachable states while propagating winning conditions backward based on controllable and uncontrollable transitions. Starting from the initial symbolic state ξ_0, as long as some states remain to be explored or updated, they are handled by the appropriate function (ll. 6 and 7).

Procedure EXPLORE chooses a symbolic state ξ in the set of states to explore *WaitingExplore* and removes it from this set (l. 10). It then considers one by one all discrete transitions that can be taken from ξ (ll. 11–14). The successor symbolic states ξ' are generated by computing the states reachable from ξ via the transition (*Succ*) and letting time pass. *Depends* records ξ as a predecessor of ξ'. ξ' is added to the set of states to explore, if not already explored yet. If the location of ξ is a target, it is considered winning, and its predecessors added to the set waiting for a back-propagation update (l. 15–17). In any case, ξ is also added to that set (l. 23), and declared explored (l. 23).

Procedure UPDATE performs the back-propagation of winning zones. It operates on a symbolic state ξ picked from *WaitingUpdate*. First, the predecessors (*Pred*) of successors of ξ by an uncontrollable transition that are not already known as winning are stored in a set *UnCtrl* (l. 26). The predecessors of the

Controller Synthesis for Parametric Timed Games 321

Algorithm 1 For PTG $G = (L, X, P, Act, T_c, T_u, \ell_0, Inv)$ and reachability objective R, returns the set of all parameter valuations that win the game.

1: *Explored, WaitingUpdate, WaitingExplore* $\leftarrow \emptyset, \emptyset, \{\xi_0^\nearrow\}$ ▷ *Symbolic state sets*
2: $Win[\,], Depends[\,]$, $ForcedMoves[\,]$ $\leftarrow \emptyset, \emptyset, \emptyset$ ▷ *Maps from symbolic states*
3: $InstrList := [\,]$ ▷ *List of triplets (Symbolic state, Symbolic state, action)*
4: $WinningParam := \textbf{False}$
5: **function** SOLVEPTG
6:    **while** *WaitingExplore* $\neq \emptyset \lor$ *WaitingUpdate* $\neq \emptyset$ **do**
7:       Choose either EXPLORE() or UPDATE()
8:    **return** ($WinningParam$, $InstrList$)
9: **procedure** EXPLORE
10:    Pick ξ from *WaitingExplore*
11:    **for** t transition from ξ **do**
12:       $\xi' := Succ(t, \xi)^\nearrow$
13:       $Depends[\xi'] \leftarrow Depends[\xi'] \cup \{\xi\}$
14:       **if** $\xi' \notin Explored$ **then** *WaitingExplore* $\leftarrow$ *WaitingExplore* $\cup \{\xi'\}$
15:    **if** $\xi.\ell \in R$ **then**
16:       $Win[\xi] \leftarrow \xi$, *WaitingUpdate* $\leftarrow$ *WaitingUpdate* $\cup$ $Depends[\xi]$
17:       $InstrList \leftarrow InstrList \cup \{(\xi, \bot, Wait)\}$
18:    $UnCtrl := \bigcup\{guard(t_u) \mid t_u \in T_u \land t_u \text{ transition from } \xi\}$
19:    $Ctrl := \bigcup\{guard(t_c) \mid t_c \in T_c \land t_c \text{ transition from } \xi\}$
20:    $(InvBound_{In}, InvBound_{Out}) := UTempSplit(Inv(\xi.\ell))$
21:    $ForcedMoves[\xi] \leftarrow (InvBound_{In} \cap UnCtrl) \setminus Ctrl$
22:    $ForcedMoves[\xi] \leftarrow ForcedMoves[\xi] \cup (InvBound_{Out} \cap \overline{UnCtrl}) \setminus \overline{Ctrl}$
23:    *WaitingUpdate* $\leftarrow$ *WaitingUpdate* $\cup \{\xi\}$, *Explored* $\leftarrow$ *Explored* $\cup \{\xi\}$
24: **procedure** UPDATE
25:    Pick ξ from *WaitingUpdate*
26:    $UnCtrl := \bigcup\{Pred(t_u, \xi' \setminus Win[\xi']) \mid t_u \in T_u \land \xi' = Succ(t_u, \xi)^\nearrow\}$
27:    **for** t controllable transition from ξ **do**
28:       $WinningMove := Pred(t, Win[Succ(t, \xi)^\nearrow])$ $\setminus UnCtrl$ ▷ *Union of zones*
29:       **for** $\xi_i \in WinningMove$ **do**
30:          $NewWin_i := SafePred(\xi_i, UnCtrl) \cap \xi$
31:          **for** $(\xi_{solved}, \_, \_) \in InstrList$ **do**
32:             $NewWin_i \leftarrow NewWin_i \setminus \xi_{solved}$
33:          $InstrList \leftarrow InstrList \cup (NewWin_i, \xi_i, t)$
34:          $NewWin \leftarrow NewWin \cup NewWin_i$
35:    **for** ξ_i zone in the union of zones $ForcedMoves[\xi]$ **do**
36:       $NewWin_i := SafePred(\xi_i, UnCtrl) \cap \xi$
37:       **for** $(\xi_{solved}, \_, \_) \in InstrList$ **do**
38:          $NewWin_i \leftarrow NewWin_i \setminus \xi_{solved}$
39:       $NewWin \leftarrow NewWin \cup NewWin_i$
40:       $InstrList \leftarrow InstrList \cup (NewWin_i, \bot, Wait)$
41:    **if** $NewWin \not\subseteq Win[\xi]$ **then**
42:       *WaitingUpdate* $\leftarrow$ *WaitingUpdate* $\cup Depends[\xi]$, $Win[\xi] \leftarrow Win[\xi] \cup NewWin$
43:    $WinningParam \leftarrow (Win[\xi] \cap \xi_0)\!\downarrow_P$

Green boxes are additions to the original algorithm [10] for generating a strategy specification. Blue boxes denote additions for forced transition semantics, explained in Sec. 4.

winning part of the successors of ξ via controllable actions outside *UnCtrl* are considered as winning (l. 28). For such states, the safe temporal predecessors (*SafePred*) are computed, i.e. the temporal predecessors that avoid reaching a state in *UnCtrl* (l. 30). These are the states from which the controller can win in ξ. Finally, if the found winning part (*NewWin*) is different from the one stored in *Win*[ξ], it is updated with the new value and a back propagation is initiated by adding the predecessors of ξ to *WaitingUpdate* (ll. 41–42). The projection of the zone on its parameters valid in the initial state is reported (l. 43).

Modifications for Generating the Strategy Specification. In this paper, we not only report the parameters but also construct the winning strategy. [14] and [15] also discuss how to do this, however their constructed strategies are equivalent to those of UPPAAL (with the addition of parameters) that we compare with in Sect. 1. Procedure EXPLORE constructs at l. 17 the list of instructions in the strategy, as defined in Def. 6, in the case of a target state reached, and thus time can elapse. The other case, where a controllable transition must be taken from a state, is handled by the UPDATE procedure at ll. 31–32: states that are already present in the instructions list cannot be considered as newly winning. The remaining lead to an instruction in the strategy specification (l. 33).

We now state the correctness of the algorithm, as proved in [11, App. A].

Theorem 1 (correct strategy). *For all winning initial states, the strategy associated with the specification synthesized by Algorithm 1 is winning.*

Example 2. Applying the algorithm on the game of Fig. 2a returns $\text{p} \geq 0$ as well as the strategy depicted in Fig. 2b. The strategy is exactly as expected: the player should take c_1 to L_1 when $\text{x} \geq \text{p}$ (as well as when $\text{x} > 1$, imposed by the guard) and then c_2 to Win immediately.

4 Forced Transition Semantic

In [7], the environment may remain in a location with an invariant when an action is available the moment the invariant breaks—effectively allowing the environment to violate the invariant by idling. While this is a valid choice, it is not very natural and disallowed in practice; e.g., UPPAAL Tiga [5] enforces that the environment takes an available action (if no controllable action exists) when

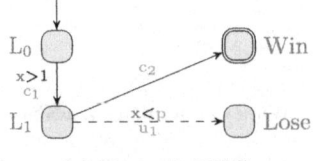

Location	Condition	Wait Until	Action
L_0	$0 \leq p$	$0 \leq p \leq x \wedge 1 < x$	c_1 to reach L_1.
L_1	$0 \leq p \leq x \wedge 1 < x$	$0 \leq p \leq x \wedge 1 < x$	c_2 to reach Win

(a) Example PTG

(b) Strategy to Ensure Controller Win

Fig. 2. Example PTG and a corresponding strategy generated by the algorithm

an invariant is about to break. We introduce a forced transition semantics to formalize this behavior, thus justifying the practice in [5].

Definition 8 (forced transition constraint). *A run r satisfies the* forced transition constraint *if and only if one of the following conditions holds:*

- *r is an infinite sequence.*
- *r is a finite sequence and for all $\delta \in \mathbb{R}_{\geq 0}^{\infty}$, there exists $s_\delta \in \mathbb{S}$ such that $ls(r) \to^\delta s_\delta$ (No further discrete transition and no invariant break).*
- *r is a finite sequence and there exists a delay $\delta \in \mathbb{R}_{\geq 0}^{\infty}$ such that for all $\delta' \geq \delta$ and $s_{\delta'} \in \mathbb{S}$ such that $ls(r) \to^{\delta'} s_{\delta'}$, a controllable action is available in $s_{\delta'}$ or no uncontrollable action is available in $s_{\delta'}$.*

A strategy is winning under the forced transition constraint if all its coherent runs satisfying the forced transition constraint are winning. A state is winning under the forced transition constraint if it has a winning controller strategy.

4.1 Temporal Bounds

To solve a PTG with forced transitions, we describe when a transition can be forced. This happens at the temporal upper bound of the invariant. We also distinguish if this bound intersects the invariant or not. To compute the temporal upper bound of a zone, we distinguish how upper and lower temporal constraints restrict time elapsing. Diagonal constraints do not restrict time elapse.

Definition 9. *Upper (resp. lower) temporal constraints ϕ of a zone $Z = \bigwedge \phi_i$ are constraints ϕ_i of type $x \sim plt$ with $\sim \in \{<, \leq, =\}$ (resp. $\sim \in \{>, \geq, =\}$). Furthermore, all zones have implicit temporal lower constraints $x \geq 0$ for all $x \in X$. The set of upper (resp. lower) temporal constraints of Z is denoted by $U_{Temp}(Z)$ (resp. $L_{Temp}(Z)$). All other constraints are called* diagonal *constraints.*

Note that the temporal upper bound of a zone may not be included into the zone itself: In Fig. 3, both invariant $x \leq p$ and $x < p$ have the same temporal upper bound $x = p$ while only $x \leq p$ contains it.

Definition 10. *Let zone Z consist of the set of constraints ϕ. The upper (resp. lower)* temporal closure *of Z, denoted by $\overline{Z^\uparrow}$ (resp. $\overline{Z^\downarrow}$) keeps all diagonal constraints unchanged, and applies the following transformations to ϕ:*

- *Make all upper (resp. lower) temporal constraints of ϕ non-strict.*
- *Make all lower (resp. upper) temporal constraints of ϕ strict.*

Let Z be a zone with an upper temporal constraint $\phi = x \sim plt$. Then Z gets invalidated due to ϕ after time $x = plt$.

Definition 11. *Let Z be a zone and $\phi = \text{x} \sim plt$ an upper (resp. lower) temporal constraint of Z. The* temporal upper (resp. lower) bound *associated with ϕ and denoted $\partial Z_\phi^\uparrow$ (resp. $\partial Z_\phi^\downarrow$) is the zone obtained by replacing $\text{x} \leq plt$ (resp. $\text{x} \geq plt$) by $\text{x} = plt$ in Z if ϕ is non-strict, or in $\overline{Z^\uparrow}$ otherwise.*

Note that the bound associated with a strict constraint $\text{x} < plt$ is necessarily outside the zone Z, since it satisfies $\text{x} = plt$, whereas for a non-strict constraint $\text{x} \sim plt$ it is necessarily within Z. The temporal bounds of a zone are the union of the temporal bounds of its constraints. The proof of Theorem 2 is in [11, App. B].

Definition 12. *Let Z be a zone. The* upper (resp. lower) temporal bound *of Z is the union of the upper (resp. lower) temporal bounds of its temporal upper (resp. lower) constraints ϕ.*

$$\partial Z^\uparrow = \bigcup_{\phi \in U_{Temp}(Z)} \partial Z_\phi^\uparrow. \qquad \partial Z^\downarrow = \bigcup_{\phi \in L_{Temp}(Z)} \partial Z_\phi^\downarrow.$$

Theorem 2. *The upper (lower) temporal bound of zone Z corresponds to:*

- $\partial Z^\uparrow = \{v + \delta_{sup} \mid v \in Z \text{ such that } \delta_{sup} = \sup\{\delta \in \mathbb{R}_{\geq 0} \mid v + \delta \in Z\} \in \mathbb{R}_{\geq 0}\}.$
- $\partial Z^\downarrow = \{v - \delta_{sup} \mid v \in Z \text{ such that } \delta_{sup} = \sup\{\delta \in \mathbb{R}_{\geq 0} \mid v - \delta \in Z\} \in \mathbb{R}_{\geq 0}\}.$

4.2 Algorithm Handling Forced Transitions

To solve the PTG, we add so-called *ForcedMoves* from states where the invariant breaks and only uncontrollable transitions are available. The controller can use *ForcedMoves* to claim the victory, since the environment didn't respect the forced-move semantics. We use the temporal upper bounds to make this precise.

Example 3. Figure 3 shows two PTGs with initial state $(L_0, \text{x} = \text{y} = 0)$. In both cases, the temporal upper bound of the invariant is $\text{x} = \text{p}$. In Fig. 3a, this bound can be reached, so the uncontrollable transition u_1 can be forced, and the game is winning. In Fig. 3b, this bound cannot be reached, so u_1 cannot be forced, and the game is losing. In Fig. 3a, since the constraint $\text{x} \leq \text{p}$ is non-strict, we intersect the bound directly with the guard of u_1. We get *ForcedMoves* equals $\text{x} = \text{p} \land \text{y} \geq \text{p}$. In Fig. 3b, since the constraint $\text{x} < \text{p}$ is strict, we intersect the bound of the invariant $\text{x} = \text{p}$ with the *temporal upper closure* $\text{y} > \text{p}$ of the guard of u_1. We get *ForcedMoves* equals $\text{x} = \text{p} \land \text{y} > \text{p}$ (which cannot be reached).

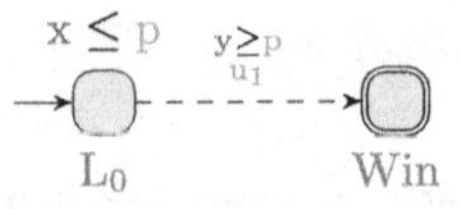

(a) PTG with a forced transition 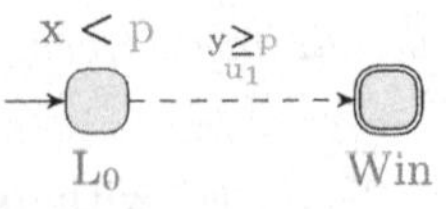   (b) PTG without forced transitions

Fig. 3. Examples of PTG with and without forced transitions

The blue boxes in Algorithm 1 handle the forced transitions. During the exploration (EXPLORE in Algorithm 1), *ForcedMoves* are collected for the processed symbolic state ξ. First, at ll. 18 and 19, the guards associated with controllable and uncontrollable transitions available in ξ are collected. Then, function $UTempSplit(Inv(\xi.\ell))$ (l. 20) computes the upper temporal bounds, grouped as being inside the zone or outside it. For both of these sets, we save the part that has available uncontrollable transitions and no controllable transitions in *ForcedMoves*$[\xi]$ (l. 21, l. 22). The computation with the temporal bounds outside the zone is done with the closed (un)controllable sets (l. 22). In procedure UPDATE, at ll. 35–40, the forced moves do not correspond to controllable transitions. Therefore, the strategy of the controller must be to wait until the environment is forced to move.

5 Controller Synthesis

We now synthesize a controller automaton that enforces a given winning strategy, without restricting uncontrollable actions. A method for constructing a controller from a strategy specification was previously proposed in [12]. However, it is based on the UPPAAL strategy semantics, so it exhibits the same issues described in Sect. 1. We take inspiration from the proposed structure adapting it to our strategy specification, using a new concept of ϵ-bounds. We assume that for all $s \in S$ there exists at most one transition labelled $a \in Act$.

5.1 Controller Synchronization

A controller is a PTA with its own locations, clocks and parameters. It may read the values of the clocks and parameters with the PTG it controls, but it may not change them (i.e. not reset clocks). When synchronised with the original PTG, it will thus restrict its behaviour. The controller has its own discrete transitions that may share a label with transitions of the controlled PTA.

Definition 13 (controller). *A controller $\mathcal{C}$ for a Parametric Timed Game $G = (L^G, X^G, P^G, Act^G, T_c^G, T_u^G, \ell_0^G, Inv^G)$, with $T^G = T_c^G \sqcup T_u^G$, is a PTA $(X^\mathcal{C}, P^\mathcal{C}, L^\mathcal{C}, Act^\mathcal{C}, T^\mathcal{C}, \ell_0^\mathcal{C})$ such that $\mathcal{C}$ has no clock reset on $X^\mathcal{C} \cap X^G$.*

In the parallel composition of the controller PTA and the controlled PTG, a transition with a shared label a occurs simultaneously in both automata. Transitions with a non-shared label occur independently.

Definition 14. *The parallel composition $(\mathcal{C} \parallel G)$ of a controller and its PTG is a labelled transition system where:*

- *The set of states $\mathbb{S}^{(\mathcal{C} \parallel G)}$ is the subset of $L^\mathcal{C} \times L^G \times V(X^\mathcal{C} \cup X^G, P^\mathcal{C} \cup P^G)$ such that, for all $(\ell^\mathcal{C}, \ell^G, v) \in \mathbb{S}^{(\mathcal{C} \parallel G)}$, $v \models Inv(\ell^G) \wedge Inv(\ell^\mathcal{C})$.*
- *It is equipped with 3 types of transitions:*
  - *Timed transitions: for $\delta \in \mathbb{R}_{\geq 0}^\infty$, $(\ell^\mathcal{C}, \ell^G, v) \to^\delta (\ell^\mathcal{C}, \ell^G, v')$ iff $v' = v + \delta$.*

- Internal discrete transitions: For $a \in Act^C \setminus Act^G$, $(\ell_0^C, \ell_0^G, v) \to^a (\ell_1^C, \ell_0^G, v')$ iff $\exists t^C = (\ell_0^C, g, a, Y, \ell_1^C) \in T^C$ such that $v \models g$ and $v' = v[Y := 0]$. For $a \in Act^G \setminus Act^C$, $(\ell_0^C, \ell_0^G, v) \to^a (\ell_0^C, \ell_1^G, v')$ iff $\exists t^G = (\ell_0^G, g, a, Y, \ell_1^G) \in T^G$ such that $v \models g$ and $v' = v[Y := 0]$.
- Parallel discrete transitions: For $a \in Act^C \cap Act^G$, $(\ell_0^C, \ell_0^G, v) \to^a (\ell_1^C, \ell_1^G, v')$ iff $\exists t^C = (\ell_0^C, g^C, a, Y^C, \ell_1^C) \in T^C \wedge \exists t^G = (\ell_0^G, g^G, a, Y^G, \ell_1^G) \in T^G$ such that $v \models g^C \wedge g^G$, and $v' = v[Y^C \cup Y^G := 0]$.

5.2 ϵ-Lower Temporal Bound

For an instruction (ξ_1, ξ_2, t) the controller should force the transition t to occur in ξ_2. We use an upper bound of ξ_2 as an invariant to force the transition to happen before leaving ξ_2. In case ξ_2 does not have an upper bound, we create one, modelled using a new parameter ϵ: the transition should then be taken within an ϵ delay after ξ_2 is reached.

Definition 15. *Let Z be a zone in $\mathcal{Z}(X, P)$ represented by the formula ϕ and $\phi = x \sim plt$ a lower temporal constraint of Z. The ϵ-temporal lower bound $\partial Z_\phi^{\downarrow \epsilon}$ is the zone in $\mathcal{Z}(X, P \cup \{\epsilon\})$ defined as $\phi \wedge x \leq plt + \epsilon$.*

Function ADDEPSILONBOUNDS in Algorithm 2 describes the modification of the strategy by adding instructions for zones without upper temporal constraints. For an instruction (ξ_1, ξ_2, t_c) such that ξ_2 has no upper temporal constraint (condition on l. 4), the instruction is split between the ϵ-lower temporal bounds of the zone of ξ_2. For every ϵ-lower temporal bound $\partial Z_\phi^{\downarrow \epsilon}$ (their union is computed by *EpsilonLTempBound* at l. 5), a new instruction is created from the intersection of its predecessors with ξ_1 (l. 6). When ξ_2 can be entered directly in a state located after a delay ϵ from its lower bound, an instruction to take the transition immediately (if not interrupted by the environment) is added (l. 8).

5.3 Synthesizing the Controller

Function CONTROLLERSYNTHESIS in Algorithm 2 considers all strategy instructions one by one from the augmented list. All actions are initially part of the

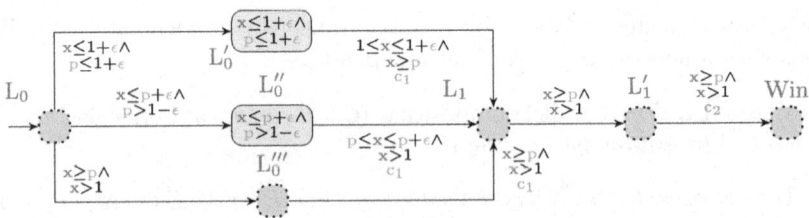

Fig. 4. Controller generated for Fig. 2a. Yellow circles indicate urgent locations. (Color figure online)

controller (if not created by the algorithm, their synchronisation is not possible and thus they are deactivated in the game). In l. 14, if the location of the instruction does not exist in the controller yet, it is created. This location is made urgent, i.e. a location where a discrete action should be taken without letting time elapse. Then a location q_i is created that can only be reached from ξ_1 (ll. 15–16). When d is a controllable action, the invariant of the new state q_i matches the predecessors of the symbolic state that should be reached according to the strategy, and the target location is created if necessary, as well as the transition for applying the strategy (ll. 19–21). If ξ_1 is included in ξ_2 and ξ_2 has no temporal upper bound, then the instruction was added by the pre-processing (l. 8) where the associated transition should be taken immediately. Thus, location q_i is made urgent (ll. 22–23). Finally, all uncontrollable transitions from $\xi.\ell$ are inserted in the controller to allow environment moves (ll. 24–26).

Algorithm 2 Algorithm for controller synthesis from a strategy list. Contains the method to force transition with no upper temporal bound as well

1: **function** ADDEPSILONBOUNDS(*InstrList*)
2:     $InstrList' := \emptyset$
3:     **for** $(\xi_1, \xi_2, d) \in InstrList$ **do**
4:         **if** $(d \neq Wait)$ and $\neg HasUpperTempCons(\xi_2.Z)$ **then**
5:             **for** ξ_2' in the union of zones $EpsilonLTempBound(\xi_2)$ **do**
6:                 $InstrList' \leftarrow InstrList' \cup \{(\xi_1 \cap (\xi_2'^{\swarrow}), \xi_2', d)\}$
7:             **if** $\xi_1 \cap \xi_2 \neq \emptyset$ **then**
8:                 $InstrList' \leftarrow InstrList' \cup \{(\xi_1 \cap \xi_2, \xi_2, d)\}$
9:         **else** $InstrList' \leftarrow InstrList' \cup \{(\xi_1, \xi_2, d)\}$
10:     **return** $InstrList'$

11: **function** CONTROLLERSYNTHESIS(*InstrList*)
12:     $Act^\mathcal{C} := Act^G$
13:     **for** $i = (\xi_1, \xi_2, d) \in$ ADDEPSILONBOUNDS(*InstrList*) **do**
14:         **if** $q_{\xi_1.\ell}$ does not exist **then** create urgent $q_{\xi_1.\ell}$
15:         Create q_i ▷ *New location for the instruction*
16:         Add transition from $q_{\xi_1.\ell}$ to q_i with guard $\xi_1.Z$, no clock reset
17:         **if** $d \neq Wait$ **then**   ▷ *The decision is a controllable transition*
18:             Match d with $(\ell, \_, a, \_, \ell')$
19:             Set q_i invariant to $\xi_2^{\swarrow}$
20:             **if** $q_{\ell'}$ does not exist **then** create urgent $q_{\ell'}$
21:             Add transition from q_i to $q_{\ell'}$ with guard $\xi_2.Z$, no clock reset, label a.
22:             **if** $\xi_1 \subseteq \xi_2$ and $\neg HasUpperTempCons(\xi_2.Z)$ **then**
23:                 Set q_i as urgent
24:         **for** $t_u = (\xi_1.\ell, \_, a, \_, \ell')$ uncontrollable transition from $\xi_1.\ell$ **do**
25:             **if** $q_{\ell'}$ does not exist **then** create urgent $q_{\ell'}$
26:             Add transition from q_i to $q_{\ell'}$ with guard $\top$, no clock reset, label a.
27:     **return** the controller

Example 4. The most interesting aspect of the game in Fig. 2a happens in the controller generation. Notice that in the strategy we can wait as long as we want in L_0 and L_1. Thus, a controller must at some point force us to move, which is where the epsilon parameter operates. See Fig. 4 for the generated controller.

From L_0 there are three choices generated from the corresponding strategy entry in Fig. 2 due to the pre-processing step described by Algorithm 2. The transition to L_0' can only be taken if $x \geq 1$ happens after $x \geq p$. The opposite is true for L_0''. L_0''' is added by the pre-processing in case we arrive in L_0 after we have passed the two boundaries in which case we can continue immediately. Hence, L_0''' is also made urgent. The guards of the transitions from L_0' and L_0'' along with their invariants ensure that the action c_1 is fired at the appropriate time according to the strategy. After arriving in the mirror location L_1, no pre-processing is actually necessary because of a small optimization in the implementation: for a strategy entry (ξ_1, ξ_2, d) if $\xi_1 \subseteq \xi_2$ then it is left untouched. Since we can take the action and stay winning, we do not need to worry about finding bounds.

5.4 Properties of the Controller

The controller $\mathcal{C}$ generated by Algorithm 2 with a strategy specification L as input satisfies the following properties, as long as all states of a history h match an instruction of L and for $\epsilon > 0$: $\mathcal{C}$ does not cause the run to end prematurely; $\mathcal{C}$ does not prevent uncontrollable transitions from happening; $\mathcal{C}$ restricts controllable discrete transitions to match the strategy L. Thus, as proved in [11, App. C]:

Theorem 3 (Controller Correctness). *Let G be a PTG with a reachability objective. Let L be the strategy generated by Algorithm 1. Let $\mathcal{C}$ be the controller synthesized by Algorithm 2 from L. Then all runs of $(\mathcal{C} \parallel G)$ starting in an initial state with a winning parameter valuation and $\epsilon > 0$ eventually reach a target location.*

6 Implementation and Experimental Evaluation

The algorithm proposed by [7] was extended with parameters and implemented by [10] along with some optimizations. The implementation is an extension to IMITATOR model checker [4], supporting a wide set of PTA synthesis algorithms.

IMITATOR allows the user to specify a model which consists of parameters, clocks, a network of parametric timed automata and (for PTG) a partitioning of actions into controllable and uncontrollable. The winning parameters for a PTG can then be synthesized by querying with property Win to select AlgoPTG. Thus, property := #synth Win(state_predicate) synthesizes parameters for a PTG, where state_predicate defines winning states. Predicate accepting, captures all states with an accepting location.

We extended the PTG parameter synthesis algorithm (AlgoPTG) of [10] and the source code can be found on GitHub[2]. Our extensions are threefold:

---

[2] https://github.com/imitator-model-checker/imitator, branch: develop.

1. Forced uncontrollable action semantics (Sect. 4)
2. Strategy synthesis (Sect. 3)
3. Controller synthesis (Sect. 5)

6.1 Verification

In order to verify that a controller actually works, we can use IMITATOR again. The input game (Fig. 2a) and the generated controller (Fig. 4) are combined in a single IMITATOR model with synchronized actions c_1 and c_2. By querying with `AF Accepting`, we synthesize the parameters for which all runs of the composition reach an accepting state. If the controller is indeed correct, this should reproduce the results of the winning parameter synthesis algorithm (plus $\epsilon > 0$).

While IMITATOR has a tailored algorithm for unavoidability synthesis (`AF`), this query can also be solved by viewing the parallel composition as a game, in which all actions are uncontrollable. This game is won by the controller if and only if the goal is unavoidable. We have found that the algorithm proposed in this paper is faster than the tailored AF-synthesis. This is used in our experiments to confirm the generated controllers are correct.

6.2 Experiments

We evaluate the scalability on the *Production Cell* case study [16]. It has two conveyor belts (incoming/ exiting), a press and a robot with arms A/B (Fig. 5 shows a visualization). The arms are coupled, moving in unison between the incoming conveyor belt, the press and the exiting conveyor belt. The incoming belt transports unprocessed plates to a buffer zone below arm A. Once the preceding plate has been processed at the press, arm A transfers the unprocessed plate to the press, while arm B transfers the processed plate to the exiting belt.

We model this in IMITATOR with 1–4 plates. The reachability goal for the n plate experiment is that all n plates make it to the exiting conveyor belt. More precisely, no two plates must be at the buffer zone simultaneously. If this happens the game is immediately lost via a trap location. There are several good candidates for parameters: the min/max waiting time between consecutive plates, the min/max time it takes for the robot to rotate and the processing time of a plate. For our experiments, we have one parameter for the min waiting time, while all other quantities are constant. The exact timing of plates arriving at the incoming conveyor belt and the robot arms movement is uncontrollable, although the environment must respect the constants and min waiting time parameter.

The PTG model is largely inspired by the one in [10]. The only key difference is that with the new forced uncontrollable semantics, the modelling becomes simpler - there is no need for auxiliary controllable actions to simulate the environment respecting invariants. See [11, App. E] for the 1-plate IMITATOR model.

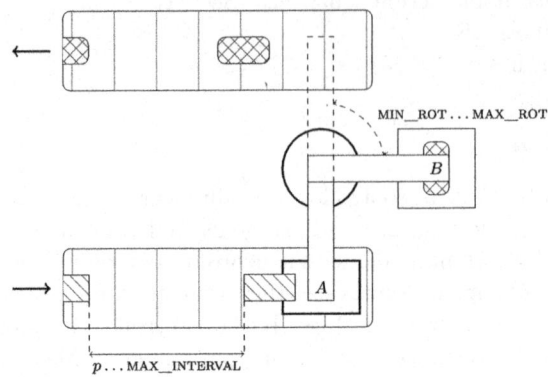

Fig. 5. Production Cell case study.

Table 1. Experimental results for 1–4 plates. "P" refers to parameter synthesis (original algorithm), while "P+C" also includes strategy construction with controller synthesis (our contribution). Values in parentheses indicate the percentage w.r.t. the "P" row.

		Solving Game			Controller Verification	
Plates	Version	P. Synth.	C. Synth.	Size	P. Synth (%)	Size (%)
1	P+C	0.080s	0.029s	73	0.074s (158%)	95 (130%)
1	P	0.047s		73		
2	P+C	0.68s	0.20s	496	0.49s (127%)	378 (76%)
2	P	0.39s		496		
3	P+C	2.46s	0.34s	1167	0.46s (26%)	241 (21%)
3	P	1.76s		1167		
4	P+C	149s	2.89s	7477	3.33s (5.6%)	703 (9.4%)
4	P	60s		7477		

6.3 Experimental Results

Results are in Table 1 and experimental details in [11, App. D]. Strategy construction has a 250% slowdown in the 4-plate experiment, revealing an overhead in strategy construction. Controller synthesis is very fast, indicating that the majority of the computational effort occurs during the main algorithm for strategy construction, while controller translation requires minimal additional resources.

Verification speeds up relatively as the model grows. Due to the way the controller is constructed, it always has more discrete locations than the input model, but when exploring the parallel composition, the symbolic state space quickly becomes smaller. This is expected, as the controller restricts the availability of controllable actions, directing the exploration more efficiently.

Synthesis times are 4000% smaller than those in [10] (P column). Beyond hardware/library/implementation differences, the key conceptual difference is that forced uncontrollable transitions make the model smaller and avoid recomputations of winning moves.

7 Conclusion

This paper proposes a method to compute a controller for a parametric timed reachability game. The controller is represented by a parametric timed automaton, to be synchronized with the game. To this end, we introduced a new notion of strategy specifications, which allow for planned actions to happen in a specified time interval. We also updated the semantics of timed games, to properly take care of forced environment transitions. We demonstrate experimentally that controller synthesis is feasible. Most extra time is spent on generating the strategy specification while solving the game on-the-fly. Generating the controller from the strategy is fast. The experiment also demonstrates that the composition of the game with the controller is guaranteed to reach the goal.

Data Availability Statement. The models, scripts, and tools to reproduce our experimental evaluation are archived and publicly available at [9].

References

1. Alur, R., Dill, D.L.: A theory of timed automata. Theoret. Comput. Sci. **126**(2), 183–235 (1994)
2. Alur, R., Henzinger, T.A., Vardi, M.Y.: Parametric real-time reasoning. In: STOC, pp. 592–601. ACM (1993)
3. André, É.: What's decidable about parametric timed automata? Int. J. Softw. Tools Technol. Transf. **21**(2), 203–219 (2019)
4. André, É.: IMITATOR 3: synthesis of timing parameters beyond decidability. In: Silva, A., Leino, K.R.M. (eds.) CAV 2021. LNCS, vol. 12759, pp. 552–565. Springer, Cham (2021). https://doi.org/10.1007/978-3-030-81685-8_26
5. Behrmann, G., Cougnard, A., David, A., Fleury, E., Larsen, K.G., Lime, D.: UPPAAL-Tiga: time for playing games! In: Damm, W., Hermanns, H. (eds.) CAV 2007. LNCS, vol. 4590, pp. 121–125. Springer, Heidelberg (2007). https://doi.org/10.1007/978-3-540-73368-3_14
6. Bundala, D., Ouaknine, J.: On parametric timed automata and one-counter machines. Inf. Comput. **253**, 272–303 (2017). GandALF 2014
7. Cassez, F., David, A., Fleury, E., Larsen, K.G., Lime, D.: Efficient on-the-fly algorithms for the analysis of timed games. In: Abadi, M., de Alfaro, L. (eds.) CONCUR 2005. LNCS, vol. 3653, pp. 66–80. Springer, Heidelberg (2005). https://doi.org/10.1007/11539452_9
8. Chatain, T., David, A., Larsen, K.G.: Playing games with timed games. In: ADHS. IFAC Proceedings Volumes, vol. 42, pp. 238–243. Elsevier (2009)
9. Dahlsen-Jensen, M.B., van de Pol, J.C., Petrucci, L., Fievet, B.: Artifact for "controller synthesis for parametric timed games". Zenodo (2025). https://doi.org/10.5281/zenodo.15284734

10. Dahlsen-Jensen, M.B., Fievet, B., Petrucci, L., van de Pol, J.: On-the-fly algorithm for reachability in parametric timed games (extended version) (2024)
11. Dahlsen-Jensen, M.B., Fievet, B., Petrucci, L., van de Pol, J.: Controller synthesis for parametric timed games. arXiv (2025). https://doi.org/10.48550/arxiv.2506.15532
12. David, A., Fang, H., Larsen, K.G., Zhang, Z.: Verification and performance evaluation of timed game strategies. In: Legay, A., Bozga, M. (eds.) FORMATS 2014. LNCS, vol. 8711, pp. 100–114. Springer, Cham (2014). https://doi.org/10.1007/978-3-319-10512-3_8
13. Hune, T., Romijn, J., Stoelinga, M., Vaandrager, F.W.: Linear parametric model checking of timed automata. J. Log. Algebraic Methods Program. **52–53**, 183–220 (2002)
14. Jovanovic, A., Faucou, S., Lime, D., Roux, O.H.: Real-time control with parametric timed reachability games. In: IFAC WODES, pp. 323–330. Elsevier (2012)
15. Jovanovic, A., Lime, D., Roux, O.H.: A game approach to the parametric control of real-time systems. Int. J. Control **92**(9), 2025–2036 (2019)
16. Lewerentz, C., Lindner, T.: Case study "production cell": a comparative study in formal specification and verification. In: Broy, M., Jähnichen, S. (eds.) KORSO: Methods, Languages, and Tools for the Construction of Correct Software. LNCS, vol. 1009, pp. 388–416. Springer, Heidelberg (1995). https://doi.org/10.1007/BFb0015473
17. Maler, O., Pnueli, A., Sifakis, J.: On the synthesis of discrete controllers for timed systems. In: Mayr, E.W., Puech, C. (eds.) STACS 1995. LNCS, vol. 900, pp. 229–242. Springer, Heidelberg (1995). https://doi.org/10.1007/3-540-59042-0_76

Symbolic Reduction for Formal Synthesis of Global Lyapunov Functions

Jun Liu(✉) and Maxwell Fitzsimmons

University of Waterloo, Waterloo, Ontario N2L 3G1, Canada
{j.liu,mfitzsimmons}@uwaterloo.ca

Abstract. We investigate the formal synthesis of global polynomial Lyapunov functions for polynomial vector fields. We establish that a sign-definite polynomial must satisfy specific algebraic constraints, which we leverage to develop a set of straightforward symbolic reduction rules. These rules can be recursively applied to symbolically simplify the Lyapunov candidate, enabling more efficient and robust discovery of Lyapunov functions via optimization or satisfiability modulo theories (SMT) solving. In many cases, without such simplification, finding a valid Lyapunov function is often infeasible. When strict Lyapunov functions are unavailable, we design synthesis procedures for finding weak Lyapunov functions to verify global asymptotic stability using LaSalle's invariance principle. Finally, we encode instability conditions for Lyapunov functions and develop SMT procedures to disprove global asymptotic stability. Through a series of examples, we demonstrate that the proposed symbolic reduction, LaSalle-type conditions, and instability tests allow us to efficiently solve many cases that would otherwise be challenging.

Keywords: Nonlinear dynamical systems · Lyapunov functions · Formal verification

1 Introduction

Stability analysis is a fundamental problem in the formal modelling and verification of dynamical systems. Lyapunov methods [16,21,33] remain among the most effective tools for establishing stability, particularly in determining whether a system will converge to a desired equilibrium point from a region beyond a neighborhood of the equilibrium. By constructing a Lyapunov function—a scalar function that decreases along system trajectories—one can certify stability without explicitly solving the system equations. Furthermore, any level set of a Lyapunov function provides a certified region of attraction. If a Lyapunov function is valid across the entire state space and its level sets are unbounded in all directions (a property known as radial unboundedness), then we can establish global attraction. Despite their effectiveness, finding suitable Lyapunov functions, especially global ones, is often challenging. Consequently, the construction of Lyapunov functions has received considerable attention [11,12].

© The Author(s), under exclusive license to Springer Nature Switzerland AG 2026
P. Prabhakar and A. Vandin (Eds.): QEST+FORMATS 2025, LNCS 16143, pp. 333–352, 2026.
https://doi.org/10.1007/978-3-032-05792-1_18

Sum-of-squares (SOS) programming has been the state-of-the-art approach for computing Lyapunov functions for polynomial systems, both globally [23] and locally [30]. It allows for verifying the negative or positive (semi)-definiteness of polynomials by expressing them as a sum of squares of other polynomials, which can be efficiently managed using semidefinite programming (SDP). SOSTOOLS [22] is a widely used toolbox for SOS optimization, including Lyapunov function synthesis. While SDP provides a candidate solution in floating-point form, obtaining an exact symbolic representation requires postprocessing, typically by either filtering out smaller terms or retaining them, often in an ad hoc manner.

On the other hand, satisfiability modulo theories (SMT) solvers [5,8] offer an alternative approach with rigorous handling of exact arithmetic. SMT solvers enable direct verification and synthesis of Lyapunov functions by encoding stability conditions as logical formulas over real numbers. This approach allows for precise verification without numerical errors, making it particularly suitable for systems where symbolic precision is critical and SMT solving is feasible. SMT-based methods have recently gained traction as a promising tool for the formal synthesis of Lyapunov functions and controllers [1,3,6,9,19,20,24,25,32], especially in cases where SDP-based methods face limitations due to floating-point precision or the non-convex nature of the problems.

Linear programming (LP) is another framework that has been effectively used for the synthesis of Lyapunov functions. In [26], linear constraints are derived using interval evaluation of the polynomial form and "Handelman representations" for positive polynomials over polyhedral sets. Linear constraints can also be generated from simulations for a given template Lyapunov function, with counterexamples provided by an SMT solver [14]. This is often done through a counterexample-guided iterative synthesis (CEGIS) procedure. The works in [24,25] were among the first to employ the CEGIS procedure for Lyapunov function synthesis. The study in [3] also investigated the sound synthesis of Lyapunov functions and developed both LP- and SMT-based CEGIS implementations aimed at being simple and effective. One can also formulate the problem of identifying a polynomial Lyapunov function as a semi-algebraic problem and solve it using a computer algebra system (CAS). Notably, the works in [27,28] employ this approach to discover local Lyapunov functions and perform local stability analysis.

In this work, we aim to extend the use of SMT and LP solvers for the synthesis of global Lyapunov functions, addressing several gaps in the existing literature. First, for a candidate Lyapunov function to be a valid global Lyapunov function, both the Lyapunov function and its Lie derivative, i.e., the derivative along the vector field, must be globally sign-definite. For a polynomial vector field and a polynomial Lyapunov function candidate, simple examples reveal that such global sign-definiteness imposes specific restrictions on the Lie derivative and, as a result, on the Lyapunov function candidate, which can be checked and resolved algebraically. While one might expect a CEGIS procedure to identify these constraints via counterexamples, this is not always feasible if the template is initially unrestricted (see our motivating example in Sect. 3.2 and Claim 5.4 in

Sect. 5.4). We take a first step in this direction by proposing an explicit symbolic reduction procedure that can be applied prior to optimization and SMT solving.

Second, not all globally asymptotically stable dynamical systems admit readily available strict Lyapunov functions, particularly if we restrict the search to polynomial functions of a given degree. A well-known result in dynamical systems, LaSalle's invariance principle [15,16], allows global asymptotic stability to be established using a weak Lyapunov function. However, applying LaSalle's invariance principle often requires verifying that the largest invariant set within the set where the Lie derivative vanishes is, in fact, the origin, which is generally infeasible because it seemingly requires exact computation of a semi-algebraic set. We address this by developing an alternative sufficient condition that can be encoded as a satisfiability problem over polynomials.

Finally, stability analysis for polynomial vector fields remains an open question. Currently, it is unknown if the problem is decidable, with conjectures suggesting it is not [4]. Additionally, it is unclear whether a locally asymptotically stable polynomial vector field, even a homogeneous polynomial one, necessarily admits a polynomial Lyapunov function [2]. Practically, this implies that when a Lyapunov function of a specific form cannot be found, no definitive conclusion can be made about the system's stability. Developing Lyapunov certificates for instability would thus be beneficial in providing information in the opposite direction. While instability results are well known in dynamical systems stability analysis, they are not yet implemented within SMT frameworks.

To bridge these gaps, this paper develops automated symbolic reduction procedures that enable efficient and robust synthesis of global Lyapunov functions using optimization and SMT solvers. We also present a synthesis procedure that leverages LaSalle's principle to prove global asymptotic stability using weak Lyapunov functions, as well as a method for conducting instability tests. Through a range of examples, from low- to high-dimensional systems, we demonstrate that symbolic reduction significantly improves synthesis efficiency and mitigates numerical challenges in both optimization- and SMT-based synthesis. The instability tests enable highly efficient proofs of instability for randomly generated dynamical systems with up to ten state dimensions.

2 Preliminaries

Consider a nonlinear system described by

$$\dot{x} = f(x), \tag{1}$$

where $f : \mathbb{R}^n \to \mathbb{R}^n$ defines the vector field, which is assumed to be a multivariate polynomial function in this paper. We also assume that $f(0) = 0$. We denote the unique solution to (1) from the initial condition $x(0) = x_0$ by $\phi(t, x_0)$ for t in the maximal interval of existence for ϕ from x_0.

We are concerned with the stability (or instability) of the equilibrium $x = 0$ of system (1). Of particular interest is the construction of Lyapunov functions that can either prove or disprove the global asymptotic stability of the origin.

To this end, we provide the basic definition of global asymptotic stability and cite three well-known results from the literature on Lyapunov functions.

Definition 1. *The equilibrium $x = 0$ of (1) is said to be*

- *stable if, for every $\varepsilon > 0$, there exists a $\delta > 0$ such that $|x_0| < \delta$ implies $|\phi(t, x_0)| < \varepsilon$ for all $t \geq 0$;*
- *asymptotically stable if it is stable and there exists some $\rho > 0$ such that $|x_0| < \rho$ implies $\phi(t, x_0) \to 0$ as $t \to \infty$;*
- *globally asymptotically stable if it is stable and $\phi(t, x_0) \to 0$ as $t \to \infty$ for all $x_0 \in \mathbb{R}^n$.*

A function $V : \mathbb{R}^n \to \mathbb{R}$ is said to be *positive definite* if it satisfies $V(0) = 0$ and $V(x) > 0$ for all $x \in \mathbb{R}^n \setminus \{0\}$. It is *positive semi-definite* if $V(0) = 0$ and $V(x) \geq 0$ for all $x \in \mathbb{R}^n$. We say V is *negative definite* (or *negative semi-definite*) if $-V$ is positive definite (or positive semi-definite). Finally, V is *radially unbounded* if $V(x) \to \infty$ as $|x| \to \infty$.

We start with the Lyapunov theorem for global stability.

Theorem 1. *[15, Theorem 4.2] Suppose that there exists a continuously differentiable function $V : \mathbb{R}^n \to \mathbb{R}$ that satisfies the following:*

1. *V is positive definite and radially unbounded; and*
2. *$\dot{V} = \nabla V \cdot f$ is negative definite.*

Then $x = 0$ is globally asymptotically stable (GAS) for (1).

With a slight abuse of terminology, we sometimes simply say (1) is GAS if the origin is GAS for (1).

The previous theorem requires a strict Lyapunov function, in the sense that $\dot{V}$ is negative definite, to conclude global asymptotic stability. LaSalle's invariance principle provides an alternative to prove GAS with a weak Lyapunov function, where $\dot{V}$ is only required to be negative semi-definite.

Theorem 2. *[15, Corollary 4.2] Suppose that there exists a continuously differentiable function $V : \mathbb{R}^n \to \mathbb{R}$ satisfying:*

1. *V is positive definite and radially unbounded;*
2. *$\dot{V} = \nabla V \cdot f$ is negative semi-definite;*
3. *the largest invariant set in $\left\{ x \in \mathbb{R}^n : \dot{V}(x) = 0 \right\}$ is $\{0\}$.*

Then $x = 0$ is GAS for (1).

In fact, LaSalle's invariance principle [16] (see also [15, Theorem 4.4]) is more general than the theorem cited above and states that any bounded solution of (1) converges to the largest invariant set in $\{x \in \mathbb{R}^n : \nabla V(x) = 0\}$, which implies Theorem 2 as a special case. Furthermore, Theorem 2 implies Theorem 1 as a special case where $\{x \in \mathbb{R}^n : \nabla V(x) = 0\} = \{0\}$.

Finally, a Lyapunov function can also be used to infer instability.

Theorem 3. *[16, p. 38] If there exists a continuously differentiable function $V : \mathbb{R}^n \to \mathbb{R}$ such that $\dot{V}$ is negative semi-definite (or $\dot{V}(x) \leq 0$ whenever $V(x) \leq 0$) and some $z \in \mathbb{R}^n$ such that $V(z) < V(0)$, then $x = 0$ is not GAS for (1).*

There are many variants of instability theorems [16]. We only discuss instability for GAS here. The proof of Theorem 3 is straightforward. By the conditions in the theorem, we have $V(\phi(t,z)) \leq V(z) < V(0)$ for all $t \geq 0$, which prevents $\phi(t,z) \to 0$.

We adopt the multi-index notation to write polynomials and monomials. Let $\alpha = (\alpha_1, \alpha_2, \ldots, \alpha_n) \in \mathbb{N}^n$ be a multi-index representing the exponents of each variable. Here, $\mathbb{N}^n$ denotes the set of all n-tuples of non-negative integers, allowing each α_i to take any non-negative integer value. For example, we can write a polynomial p of degree k without a constant term as

$$p(x) = \sum_{1 \leq |\alpha| \leq k} c_\alpha x^\alpha = \sum_{1 \leq \alpha_1 + \alpha_2 + \cdots + \alpha_n \leq k} c_\alpha x_1^{\alpha_1} x_2^{\alpha_2} \ldots x_n^{\alpha_n} \quad (2)$$

where $|\alpha| = \alpha_1 + \alpha_2 + \cdots + \alpha_n$ denotes the total degree of the monomial term x^α, and c_α represents the corresponding coefficient.

3 Necessary Conditions on Non-positivity of Polynomials

3.1 A Motivating Example

Searching for Lyapunov functions is usually done by fixing a candidate Lyapunov template and then determining parameters in this template to make it a valid Lyapunov function.

Example 1. Consider

$$\dot{x} = f(x) = \begin{bmatrix} -x_1^3 + x_1^5 x_2 \\ -x_2^3 - x_1^6 \end{bmatrix}. \quad (3)$$

We construct a quadratic Lyapunov function template of the form $V(x) = c_0 x_1^2 + c_1 x_2^2 + c_2 x_1 x_2$. It can be easily computed that

$$\dot{V}(x) = \nabla V(x) \cdot f(x) = (2c_0 - 2c_1)x_1^6 x_2 - c_2 x_1^7 + c_2 x_1^5 x_2^2 \\ - 2c_0 x_1^4 - 2c_1 x_2^4 - c_2 x_1^3 x_2 - c_2 x_1 x_2^3.$$

As we shall discuss in more detail in later sections, there are different approaches one can take to determine these coefficients through SMT solvers or optimization (e.g., gradient descent or LP). One of the necessary conditions we shall establish in Sect. 3.2 is that all highest and lowest odd monomials in $\dot{V}$ should vanish for it to be negative semi-definite. As a result, we can symbolically derive $c_0 = c_1$ and $c_2 = 0$, which means the only possible quadratic Lyapunov function for this system should be of the form

$$V(x) = c_1(x_1^2 + x_2^2) \quad (4)$$

On the one hand, if we use a complete SMT procedure to determine the coefficients such that $V(x) > 0$ and $\dot{V}(x)$ hold for all $x \in \mathbb{R}^n \setminus \{0\}$, we should be able to recover this template (4) and conclude that any $c_1 > 0$ provides a valid Lyapunov function. However, a complete SMT procedure can be slow for more than three decision variables, based on the evaluations we conduct. Therefore, being able to reduce the number of decision variables is crucial.

On the other hand, if we were to use an optimization-based approach to solve for the parameters, any optimization procedure not uncovering (or retaining) the *exact* template (4) would not yield a valid Lyapunov function for this system. For example, one could potentially use gradient descent on the parameters with a loss function that encourages

$$V(y_j) > 0, \quad \dot{V}(y_j) < 0, \tag{5}$$

for all y_j in a finite set of sample points $\{y_j\}_{j=1}^N$. Alternatively, one can solve for (c_0, c_1, c_2) from (5) using LP, since these are linear constraints in the coefficients. Similarly, an incomplete SMT procedure can be used to solve for (c_0, c_1, c_2) such that (5) holds for a finite number of samples, potentially aided by a CEGIS procedure [3] that can add any counterexamples to the list of samples to improve the chance of discovering (4).

Regardless of the procedure employed, as long as it is sample-based or requires numerical updates to the parameters (c_0, c_1, c_2), it will face the difficulty, if not impossibility, of discovering the exact form (4). This limitation applies to several synthesis procedures we present in Sect. 5 (see Claim 5.4), and we will discuss it further in Sect. 6 with numerical examples. This motivates the technical result in the next subsection, which can be used to symbolically reduce algebraic constraints such as $c_2 = 0$ and $c_0 = c_1$ in this example.

3.2 Necessary Conditions on Non-positivity of Polynomials

We will focus on necessary conditions for the non-positivity of polynomials.

Proposition 1. *Consider a polynomial p of the form*

$$p(x) = \sum_{l \leq |\alpha| \leq k} c_\alpha x^\alpha,$$

where $l \geq 1$ and $k \geq l$ indicate the lowest and highest degrees of p, i.e., the lowest l (or highest k) such that there exists some $\alpha \in \mathbb{N}^n$ with $|\alpha| = l$ (or k) with $c_\alpha \neq 0$. Assume that $p(x) \leq 0$ for all $x \in \mathbb{R}^n$. Then the following statements are true:

1. *The degrees l and k are even. Both $p_l(x) = \sum_{|\alpha|=l} c_\alpha x^\alpha$ and $p_k(x) = \sum_{|\alpha|=k} c_\alpha x^\alpha$ are negative semi-definite.*
2. *If any monomial $c_\alpha x^\alpha$ in $p(x)$ contains a factor $x_i^{\alpha_i}$ whose degree is the highest among all monomials in $p(x)$, then α_i is even.*

3. If any monomial x^α that appears in $p_l(x)$ (or $p_k(x)$) contains a factor $x_i^{\alpha_i}$ whose degree is the highest among all monomials in $p_l(x)$ (or $p_k(x)$), then α_i is even.
4. If any monomial in p takes the form $c x_i^d$ (i.e., with a single factor x_i^d and coefficient c) and d is either the lowest or highest degree among terms of the same form, then d is even and $c \leq 0$.

Proof. (1) Suppose that k is odd. Pick x such that $p_k(x) = \sum_{|\alpha|=k} c_\alpha x^\alpha$ is nonzero. Consider any $\lambda \in \mathbb{R}$. Then

$$p(\lambda x) = \lambda^k p_k(x) + \sum_{l \leq |\alpha| < k} c_\alpha (\lambda x)^\alpha.$$

This is an odd-degree polynomial in λ (note that the leading term coefficient $p_k(x) \neq 0$). As a result we cannot have $p(\lambda x)$ to be sign-definite for arbitrary $\lambda \in \mathbb{R}$. Hence k must be even. We also must have $p_k(x) \leq 0$ in order for $p(\lambda x) \leq 0$ for all $\lambda \in \mathbb{R}$. Since x is chosen arbitrarily, $p_k(x)$ must be negative semi-definite.

Now suppose that l is odd. For $y \neq 0$, consider

$$p\left(\frac{y}{|y|^2}\right) = p_l\left(\frac{y}{|y|^2}\right) + \sum_{l < |\alpha| \leq k} c_\alpha \left(\frac{y}{|y|^2}\right)^\alpha.$$

Multiply both sides by $|y|^{2k}$. We obtain

$$\hat{p}(y) := |y|^{2k} p\left(\frac{y}{|y|^2}\right) = \underbrace{|y|^{2k} p_l\left(\frac{y}{|y|^2}\right)}_{:=\hat{p}_l(y)} + \underbrace{|y|^{2k} \sum_{l < |\alpha| \leq k} c_\alpha \left(\frac{y}{|y|^2}\right)^\alpha}_{:=\hat{R}(y)}.$$

It follows that $\hat{p}_l(y)$ is a polynomial of degree $2k - l$ and $\hat{R}(y)$ is a polynomial of degree strictly less than $2k - l$. Note that $\hat{p}$ share the same sign with p. By the same argument above, where we showed the highest degree k for p cannot be odd, $2k - l$ cannot be odd because it is the highest degree for $\hat{p}$. Therefore, l must be even and $\hat{p}_l(x)$ is negative semi-definite. Since $\hat{p}_l$ share the same sign with p_l, p_l is also negative semi-definite.

(2) Let $\{c_\alpha x^\alpha\}_{\alpha \in I}$ denote the set of monomials that contain a factor $x_i^{\alpha_i}$ whose degree α_i is the highest among all monomials in $p(x)$. We can write $p(x) = \sum_{\alpha \in I} c_\alpha x^\alpha + \sum_{\alpha \notin I} c_\alpha x^\alpha$. Let $x_\lambda = (x_1, \ldots, \lambda x_i, \ldots, x_n)$. Then $p(x_\lambda) = \lambda^{\alpha_i} \sum_{\alpha \in I} c_\alpha x^\alpha + \sum_{\alpha \notin I} c_\alpha (x_\lambda)^\alpha$. Given any x such that $\sum_{\alpha \in I} c_\alpha x^\alpha \neq 0$, $p(x_\lambda)$ can be viewed as a polynomial of λ, whose highest degree is α_i. Since we must have $p(x_\lambda) \leq 0$ for all $\lambda \in \mathbb{R}$, α_i must be even.

(3) Since $p_k(x)$ and $p_k(x)$ must be negative semi-definite as shown in part (1), we can apply the argument in part (2) to each of them to obtain the conclusion.

(4) Pick $x = (0, \ldots, x_i, \ldots, 0)$. We have $p(x) = \sum_{l \leq d \leq k} c_d x_i^d$, which is a polynomial in x_i. By part (1), the leading terms are negative semi-definite with respect to x_i. This implies that d is even and $c_d \leq 0$ for both $d = l$ and $d = k$.

Remark 1. *Clearly, if $p(x) \geq 0$ for all $x \in \mathbb{R}^n$, we can draw the same conclusions as that of Proposition 1 with "negative semi-definiteness" replaced by "positive semi-definiteness" for $p_l(x)$ and $p_k(x)$ in part (1) and $c \geq 0$ replacing $c \leq 0$ in part (4).*

Remark 2. *Determining the positivity of polynomials is an NP-hard problem. For this reason, the necessary conditions and the reduction rules derived from them are definitely not exhaustive. Furthermore, when combined with other incomplete synthesis procedures (e.g., linear programming or SMT solving based on inequality constraints evaluated at sample points), we can apply additional heuristic reductions to the parameters in the Lyapunov candidate.*

4 A Sufficient Condition for LaSalle's Invariance Principle

We use the following example to show that, although LaSalle's principle is a well-known result, a reformulation of its conditions may be necessary for it to be verifiable by an SMT solver in an automated manner.

Example 2. Consider

$$\dot{x} = f(x) = \begin{bmatrix} x_2 \\ -x_1^3 - x_2^3 \end{bmatrix}. \tag{6}$$

Consider the Lyapunov function $V(x) = x_1^4 + 2x_2^2$. We can compute

$$\dot{V}(x) = \nabla V(x) \cdot f(x) = -4x_2^4. \tag{7}$$

To apply LaSalle's invariance principle, we aim to show that the largest invariant set contained in $\left\{ x \in \mathbb{R}^2 : \dot{V}(x) = 0 \right\} = \{-4x_2^4 = 0\}$ is $\{0\}$. While it is straightforward to see in this particular example that the set $\left\{ x \in \mathbb{R}^2 : \dot{V}(x) = 0 \right\}$ is simply $\{x_2 = 0\}$, we cannot generally expect to solve for this set symbolically. Therefore, we proceed with the analysis using the set $\{-4x_2^4 = 0\}$. Differentiating $\dot{V} = -4x_2^4$ along solutions of (6), we obtain

$$\ddot{V}(x) = \nabla(\dot{V}) \cdot f(x) = -16x_2^3(-x_1^3 - x_2^3). \tag{8}$$

Intuitively, if

$$\dot{V}(x) = 0 \quad \text{and} \quad x \neq 0 \Longrightarrow \ddot{V}(x) \neq 0, \tag{9}$$

then the largest invariant set in $\left\{ \dot{V} = 0 \right\}$ must be $\{0\}$. This is because a solution $\phi(t,x)$ of (6) with $\ddot{V}(x) \neq 0$ will necessarily leave the set $\left\{ \dot{V} = 0 \right\}$ instantly. In view of (7) and (8), we do not have (9) hold. Nonetheless, we can continue with the same reasoning and seek to verify that there exists some integer $r \geq 1$ such that

$$L_f^k(x) = 0, \quad \forall k < r, \quad \text{and} \quad x \neq 0 \Longrightarrow L_f^r(x) \neq 0, \tag{10}$$

where $L_f^k(x)$ denotes the k-th order derivatives of V along solutions (formally defined in (11) below). If this can be verified for some $r \geq 1$, then we know the largest invariant set in $\{\dot{V} = 0\}$ must be $\{0\}$ (see Proposition 2 for a formal proof), and we can use LaSalle's invariance principle to conclude global asymptotic stability. For the example above, one can see that $L_f^5 V$ will expose a term that only involves x_1^3, while the rest of the terms all have a factor of x_2, which would imply (10) holds. In fact, we can verify that $L_f^1(x) = \dot{V}(x) = 0$ and $x \neq 0$ imply $x_2 = 0$ and $x_1 \neq 0$, which in turn imply $L_f^5 V(x) \neq 0$. This provides the idea for the technical result below.

Consider system (1) and a sufficiently smooth function $V : \mathbb{R}^n \to \mathbb{R}$. Define the various order Lie derivatives of V with respect to (1) as

$$L_f^1 V(x) = L_f^1 V(x) = \nabla V(x) \cdot f(x),$$
$$L_f^{k+1} V(x) = \nabla(L_f^k V(x)) \cdot f(x), \quad k \geq 1. \tag{11}$$

Introduce the sets

$$\mathcal{C}^1 = \{x \in \mathbb{R}^n : L_f V(x) = 0\},$$
$$\mathcal{C}^{k+1} = \left\{x \in \mathbb{R}^n : L_f^{k+1} V(x) = 0\right\} \cap \mathcal{C}^k, \quad k \geq 1. \tag{12}$$

The following result provides a sufficient condition for applying LaSalle's invariance principle to prove global asymptotic stability.

Proposition 2. *Suppose that there exists a sufficiently smooth, positive definite, and radially unbounded function $V : \mathbb{R}^n \to \mathbb{R}$ such that $\dot{V}(x) \leq 0$ for all $x \in \mathbb{R}^n$. Furthermore, there exists some integer $r \geq 1$ such that the sets defined in (12) satisfy*

$$\bigcap_{k=1}^r \mathcal{C}^k = \{0\}, \tag{13}$$

then the origin is globally asymptotically stable for (1).

Proof. If $\mathcal{C}^1 = \{0\}$, GAS follows from Theorem 1. Now suppose that $\mathcal{C}^1 \neq \{0\}$. To apply LaSalle's invariance principle, we aim to show that, for any $x \in \mathcal{C}^1 \setminus \{0\}$, the solution $\phi(t, x)$ cannot stay indefinitely in $\mathcal{C}^1$. Let $l \in \{2, \ldots, r\}$ be the smallest integer such that $x \notin \mathcal{C}^l$. Such an l exists because $x \neq 0$ and (13) holds. Then we have $L_f^k V(x) = 0$ for $k = 1, \ldots, l-1$ and $L_f^l V(x) \neq 0$. As a result, there exists a small $\tau_1 > 0$ such that $L_f^{l-1} V(\phi(t,x)) \neq 0$ for all $t \in (0, \tau_1]$. Continuing this argument leads to the existence of a small $\tau_{l-1} > 0$ such that $\dot{V}(\phi(t,x)) \neq 0$ for all $t \in (0, \tau_{l-1}]$, which implies $\phi(t,x)$ must leave $\mathcal{C}^1$. Hence, the largest invariant set within the set $\mathcal{C}^1$ is $\{0\}$. By LaSalle's invariance principle (Theorem 2), the origin is GAS.

Remark 3. *We note that similar conditions for computing the limit invariant set and synthesizing relaxed (weak) Lyapunov functions using LaSalle's principle have been derived in [10,17]. In particular, [10] (Proposition 1 and the remarks following Proposition 3) notes that the finite termination of a condition analogous to (13) can be established using the descending chain condition for algebraic varieties, as discussed in [7].*

5 Synthesis Algorithms

In this section, we discuss several algorithmic procedures that can be used to efficiently synthesize a polynomial Lyapunov function with the aid of SMT and optimization solvers.

5.1 Symbolic Reduction

Based on Proposition 1, we can formulate the following symbolic reduction procedure. Consider a Lyapunov function template

$$V(x) = \sum_{\alpha \in I} c_\alpha x^\alpha. \tag{14}$$

The set of multi-indices, I, specifies which monomial terms are included in the Lyapunov function candidate. For example, one may wish to include all monomial terms in a quadratic form. For higher degrees, it is common practice to retain only even-degree monomials in V. For a polynomial vector field f, we then compute the Lie derivative as

$$\dot{V}(x) = \nabla V(x) \cdot f(x) = \sum_{\alpha \in \hat{I}} \hat{c}_\alpha x^\alpha. \tag{15}$$

The possible terms that appear in $\hat{I}$ are obviously determined by that of f and V. For V to be a valid global Lyapunov function candidate (for both Lyapunov and LaSalle conditions), we require at least $\dot{V}(x) \leq 0$ for all $x \in \mathbb{R}^n$. Therefore, according to Proposition 1, certain terms cannot appear. We can use the result of Proposition 1 to introduce symbolic equality and inequality constraints on $\{\hat{c}_\alpha\}_{\alpha \in}$ and solve them to reduce the number of terms in V.

We revisit Example 1 to illustrate the procedure.

Example 3 (Example 1 revisited). The Lyapunov candidate is

$$V(x) = c_0 x_1^2 + c_1 x_2^2 + c_2 x_1 x_2.$$

and the computed Lie derivative $\dot{V}$ is

$$(2c_0 - 2c_1)x_1^6 x_2 - c_2 x_1^7 + c_2 x_1^5 x_2^2 - 2c_0 x_1^4 - 2c_1 x_2^4 - c_2 x_1^3 x_2 - c_2 x_1 x_2^3.$$

By Proposition 1(1), for $\dot{V}$ to be negative semi-definite, the highest-degree terms must be even. Setting $2c_0 - 2c_1 = 0$, $-c_2 = 0$, and $c_2 = 0$ yields $c_0 = c_1$ and $c_2 = 0$. This results in the only valid quadratic Lyapunov function candidate, $V(x) = c_1(x_1^2 + x_2^2)$.

5.2 Complete Synthesis

Given a Lyapunov function candidate $V(x) = \sum_{\alpha \in I} c_\alpha x^\alpha$, possibly after symbolic reduction outlined in Sect. 5.1, we propose the following SMT procedures to synthesize global Lyapunov functions for GAS either using a strict Lyapunov function or a weak Lyapunov function by LaSalle's invariance principle. We also discuss how to synthesize a Lyapunov function to prove instability. The procedure is said to be complete in the sense that if the formulated SMT conditions encoding the Lyapunov conditions on a particular Lyapunov function template are satisfiable, then a Lyapunov function can be found. If they are not satisfiable, then a Lyapunov function of this particular form does not exist to satisfy the given SMT conditions.

Strict Lyapunov Function: The satisfiability problem for finding a strict Lyapunov function can be formulated as follows:

$$\exists \{c_\alpha\}((\forall x(x \neq 0 \Rightarrow (V(x) > 0 \land \nabla V(x) \cdot f(x) < 0)))). \tag{16}$$

For a polynomial Lyapunov function V and a polynomial vector field f, the SMT solver Z3 [8] can readily solve this problem. If the satisfiability check is successful, a "model" — an assignment of values to the coefficients $\{c_i\}$ that satisfies all constraints — can be extracted, defining a valid Lyapunov function. By Theorem 1, this proves GAS of (1), provided that V is also radially unbounded.

Weak Lyapunov Function and GAS Using LaSalle's Principle: If a strict Lyapunov function cannot be found, GAS can still be established using LaSalle's invariance principle. In this case, the SMT procedure relaxes the conditions to search for a weak Lyapunov function V where:

$$\exists \{c_\alpha\}((\forall x(x \neq 0 \Rightarrow (V(x) > 0 \land \nabla V(x) \cdot f(x) \leq 0)))). \tag{17}$$

Additionally, we use Proposition 2 to formulate the following LaSalle's condition:

$$\forall x((\nabla V \cdot f(x) = 0 \land x \neq 0) \Rightarrow \vee_{2 \leq k \leq r}(L_f^k(x) \neq 0)). \tag{18}$$

Here, r defines the number of Lie derivatives we are checking LaSalle's condition with. Note that (18) is equivalent to (13). A simpler sufficient condition for (18) is

$$\forall x((\nabla V \cdot f(x) = 0 \land x \neq 0) \Rightarrow L_f^r(x) \neq 0), \tag{19}$$

for some $r \geq 2$. Both (18) and (19) can be easily coded in an iterative loop to check if a Lyapunov function of a particular form can be found using LaSalle's condition up to a degree-r Lie derivative.

Remark 4. *There is a choice to be made regarding whether to code the LaSalle conditions directly in the synthesis part, or to first synthesize a weak Lyapunov function satisfying (17), and then verify the LaSalle condition (18) or (19). Verification is presumably cheaper than synthesis, but synthesis can be comprehensive in determining whether a weak Lyapunov function of a particular exists that satisfies a LaSalle-type condition, whereas a weak Lyapunov function synthesized without the LaSalle condition may not always verify the LaSalle condition.*

Lyapunov Function for Instability. According to Theorem 3, if the following condition:

$$\exists \{c_\alpha\}((\forall x(V(x) \leq 0 \Rightarrow \nabla V(x) \cdot f(x) \leq 0) \wedge \exists z(V(z) < V(0))))$$

is satisfiable, then we can use V to prove that the origin is not globally asymptotically stable for (1). Even though the condition does not fully require $\dot{V}$ to be negative semi-definite, we can still use the symbolic procedure outlined in Sect. 5.1 to potentially reduce the number of decision variables in SMT synthesis. This turns out to be highly effective, and we demonstrate in Sect. 6 that we can efficiently disprove GAS of non-trivial examples up to 10 dimensions in under a second for each example.

5.3 Sampling-Based Partial Synthesis

In contrast to the complete synthesis method discussed in the previous section, one can also use a sampling-based partial synthesis approach to compute a candidate Lyapunov function and then formally verify it with respect to the Lyapunov conditions. This approach is especially useful when exact SMT synthesis may be computationally prohibitive. Sampling-based approaches are usually combined with a counterexample-guided iterative synthesis (CEGIS) approach [3,25] to synthesize valid Lyapunov functions.

LP-CEGIS. The Lyapunov condition (16) depends linearly on the coefficients $\{c_\alpha\}$. Instead of solving $\{c_\alpha\}$ such that (16) holds for all $x \in \mathbb{R}^n$, we can sample a number of points $\{y_j\}_1^N \in \mathbb{R}^n \setminus \{0\}$ and use them formulate $2N$ linear constraints in $\{c_\alpha\}$:

$$\sum_{\alpha \in I} c_\alpha (y_j)^\alpha > 0, \quad j = 1, \ldots, N,$$

$$\sum_{i=1}^{n} \sum_{\alpha \in I} c_\alpha \alpha_i y_j^{\alpha - e_i} f_i(y_j) < 0, \quad j = 1, \ldots, N, \quad (20)$$

where $e_i = (0, 0, \ldots, 1, \ldots, 0)$. The purpose of writing this down is simply to illustrate that these constraints are indeed linear in $\{c_\alpha\}$. In practice, the evaluation and formulation of these constraints can be automated. However, solving (20) directly with linear programming (LP) will very likely return $c_\alpha = 0$ for all $\alpha \in I$. To obtain nontrivial solutions, we need to reformulate the problem. We implemented the following as a benchmark for solving the examples presented in this paper:

$$\sum_{\alpha \in I} c_\alpha (y_j)^\alpha \geq \mu \min(|y_j|^{l_V}, |y_j|^{k_V}), \quad j = 1, \ldots, N,$$

$$\sum_{i=1}^{n} \sum_{\alpha \in I} c_\alpha \alpha_i y_j^{\alpha - e_i} f_i(y_j) \leq -\mu \min(|y_j|^{l_{\dot{V}}}, |y_j|^{k_{\dot{V}}}), \quad j = 1, \ldots, N, \quad (21)$$

where $\mu > 0$ is a small constant, l_V and $l_{\dot{V}}$ are the lowest degrees of monomials in V and $\dot{V}$, respectively, and k_V and $k_{\dot{V}}$ are the highest degrees. We can solve (21) to obtain a Lyapunov candidate. While this may not yield a valid Lyapunov function, one can use an SMT solver to further verify the strict Lyapunov condition (16) or the weak Lyapunov condition (17), along with the LaSalle condition (18) and or (19). If a counterexample is found during verification of V, this counterexample can be added to the set of samples to re-solve for $\{c_\alpha\}$. This procedure is referred to as LP-CEGIS.

Z3-CEGIS. Instead of using an LP solver, one can also directly use an SMT solver to solve for $\{c_\alpha\}$ from the constraints (20) provided by a finite set of samples, although this approach is potentially more expensive than solving an LP. When the SMT solver is Z3, we refer to this as Z3-CEGIS. Notably, both LP-CEGIS and Z3-CEGIS were previously proposed in [3] for the synthesis of Lyapunov functions. We implement these as baselines for the numerical case studies in this paper.

5.4 LP-CEGIS-SR and Z3-CEGIS-SR

We revisit Example 1 (and Example 3) to illustrate the necessity of using symbolic reduction when applying a CEGIS procedure. We prove the following claim.

Claim. Consider Example 1 and a quadratic Lyapunov function candidate $V(x) = c_0 x_1^2 + c_1 x_2^2 + c_2 x_1 x_2$. For any finite set of samples $\{y_j\}_1^N$, there exist $\{c_0, c_1, c_2\}$ such that $c_0 \neq c_1$ and V satisfies both (20) and (21).

Proof. Recall that the Lie derivative of V is

$$(2c_0 - 2c_1)x_1^6 x_2 - c_2 x_1^7 + c_2 x_1^5 x_2^2 - 2c_0 x_1^4 - 2c_1 x_2^4 - c_2 x_1^3 x_2 - c_2 x_1 x_2^3.$$

We can simply let $c_2 = 0$. The positive definiteness constraints are always satisfied, provided that c_0 and c_1 are sufficiently large positive values relative to μ. We have

$$\dot{V}(x) = (2c_0 - 2c_1)x_1^6 x_2 - 2c_0 x_1^4 - 2c_1 x_2^4.$$

For any sample $y_j = (y_{j1}, y_{j2}) = 0$, the constraints (20) and (21) are trivially satisfied. Hence, assume, without loss of generality, that $y_j \neq 0$ for all $j = 1, \ldots, N$. For sufficiently large c_0 and c_1, each constraint from (20) and (21) takes the form $(2c_0 - 2c_1)b_j \leq d_j$, where $d_j > 0$. As a result, we can choose c_0 to be sufficiently close, but not equal, to c_1 to satisfy all the constraints.

By Proposition 1 in Sect. 5.1, any $c_0 \neq c_1$ makes V an invalid Lyapunov function. Adding further counterexamples will not help either, due to the claim above. This simple analysis reveals that symbolic reduction, as discussed in Sect. 5.1, can play a crucial role in identifying and eliminating algebraic constraints before optimization and satisfiability solvers are effectively deployed for Lyapunov function synthesis. We observe in our implementations that CEGIS (without symbolic reduction and with exact arithmetic) fails to converge on this example.

Remark 5. *We note, however, that rounding the solutions returned by LP or SMT solvers can mitigate the issue raised in Claim 5.4 because, in the analysis above, we would need to choose c_0 and c_1 sufficiently close with increasing number of samples. Nonetheless, this approach would be ad hoc and akin to how we filter out small coefficients returned by SOS programming. Hence, we believe that symbolic reduction is a more principled way of handling these constraints. Furthermore, an obvious benefit of using symbolic reduction is to reduce the number of decision variables in optimization or SMT synthesis.*

Remark 6. *Intuitively, symbolic reduction aims (albeit without guarantees) to reduce V to the form $V(x) = \sum_{i=1}^{N} c_i p_i(x)$, where each $p_i(x)$ is a polynomial, such that the set*

$$\Theta = \{c = (c_1, \ldots, c_N) : V \text{ satisfies } (17)\}$$

has a positive measure in $\mathbb{R}^N$. If this condition holds, it is expected that computing c can be achieved more robustly.

In the next section, we refer to LP-CEGIS and Z3-CEGIS with symbolic reduction as **LP-CEGIS-SR** and **Z3-CEGIS-SR**, respectively. We refer to complete synthesis with symbolic reduction as Z3-Complete-SR. We encode the LaSalle condition (19) for synthesis in Z3-Complete-SR-LaSalle and for verification of weak Lyapunov functions in **LP-CEGIS-SR-LaSalle** and **Z3-CEGIS-SR-LaSalle**.

Remark 7. *We omit the discussion of training polynomial Lyapunov functions using gradient descent in this paper. To do so, one can use a quadratic activation function in a multi-layer neural network and train it with a loss function that encourages the inequality constraints (20) or (21) to be satisfied; see, e.g., [9, 19]. A typical gradient descent (GD) optimization algorithm can be called to minimize the loss. This approach can also be combined with a CEGIS procedure as implemented in [9]. Such an approach can be called GD-CEGIS or NN-CEGIS (NN stands for neural networks). We implemented this, but the results are not comparable with LP-CEGIS, and it suffers from the same drawbacks as LP-CEGIS and Z3-CEGIS, where counterexamples cannot be exhausted due to certain underlying algebraic constraints (stipulated by Proposition 1 and illustrated by Claim 5.4) that are not eliminated before optimization/SMT solvers are deployed.*

6 Case Studies and Experiments

In this section, we revisit a suite of examples from related work [3, 23, 26] and introduce new examples to illustrate the main results of this paper. In particular, we contrast different approaches: Z3-Complete-SR, LP-CEGIS-SR, Z3-CEGIS-SR, and the baseline LP-CEGIS and Z3-CEGIS without symbolic reduction. We also include results obtained using SOSTOOLS [22] for comparison. All computations were performed on an Intel Xeon Gold 6326 CPU @ 2.90 GHz with 32 cores and 16 GB of RAM on a computing cluster. Implementations are done using the LyZNet toolbox [19].

Data Availability. The models, scripts, and tools to reproduce our experimental evaluation are archived and publicly available at DOI 10.5281/zenodo.15272621.

A more detailed description of the examples, along with the Lyapunov functions successfully synthesized using different approaches, is provided in the Appendix of the arXiv version of this paper [18].

We run experiments on a total of 10 examples, which are polynomial vector fields of dimensions 2 to 6, labeled as (E1)–(E10) and listed below:

$$(\mathbf{E1})\ \dot{x} = \begin{bmatrix} -x_1^3 + x_1^5 x_2 \\ -x_2^3 - x_1^6 \end{bmatrix},\ (\mathbf{E2})\ \dot{x} = \begin{bmatrix} -x_1^7 + x_1 x_2 \\ -x_2^3 - x_1^2 \end{bmatrix},\ (\mathbf{E3})\ \dot{x} = \begin{bmatrix} -x_1 - 1.5 x_1^2 x_2^3 \\ -x_2^3 + 0.5 x_1^3 x_2^2 \end{bmatrix},$$

$$(\mathbf{E4})\ \dot{x} = \begin{bmatrix} -x_1^3 + x_2 \\ -x_1 - x_2 \end{bmatrix},\ (\mathbf{E5})\ \dot{x} = \begin{bmatrix} -\sigma x_1 + \sigma x_2 \\ r x_1 - x_2 - x_1 x_3 \\ -b x_3 + x_1 x_2 \end{bmatrix},\ \sigma = 10,\ r = 0.9999,\ b = \frac{8}{3},$$

$$(\mathbf{E6})\ \dot{x} = \begin{bmatrix} -x_1 + x_2^3 - 3 x_3 x_4 \\ -x_1 - x_2^3 \\ x_1 x_4 - x_3 \\ x_1 x_3 - x_4^3 \end{bmatrix},\ (\mathbf{E7})\ \dot{x} = \begin{bmatrix} -x_1^3 + 4 x_2^3 - 6 x_3 x_4 \\ -x_1 - x_2 + x_5^3 \\ x_1 x_4 - x_3 + x_4 x_6 \\ x_1 x_3 + x_3 x_6 - x_4^3 \\ -2 x_2^3 - x_5 + x_6 \\ -3 x_3 x_4 - x_5^3 - x_6 \end{bmatrix}.$$

$$(\mathbf{E8})\ \dot{x} = \begin{bmatrix} x_2 \\ -x_1^3 - x_2^3 \end{bmatrix},\ (\mathbf{E9})\ \dot{x} = \begin{bmatrix} x_2 \\ -x_1^5 - 3 x_2 \end{bmatrix},\ (\mathbf{E10})\ \dot{x} = \begin{bmatrix} x_2 \\ -x_1 - 7 x_2^5 \end{bmatrix}.$$

The main objective is to assess whether symbolic reduction can improve the likelihood of discovering a valid Lyapunov function. Results, including computational times, for (E1)–(E7) are reported in Table 1, and for (E8)–(E10) in Table 2.

For sampling-based approaches, the initial domain for all examples was set to $[-10, 10]^n$, but subsequent samples (counterexamples) are drawn from $\mathbb{R}^n$. We selected 3,000 initial samples for LP-CEGIS/LP-CEGIS-SR and 300 for Z3-CEGIS/Z3-CEGIS-SR, as the latter is considerably slower and significantly impacted by the number of constraints and decision variables. We do a total of 10 CEGIS steps for all iterative synthesis approaches. While LP[1] is solved with floating-point arithmetic, we convert it[2] to exact rational arithmetic when extracting the Lyapunov function expression for verification. For solving with SOSTOOLS [22], we use the build-in function `findlyap` with the default solver, SeDuMi [29].

---

[1] For the examples in this paper, we solved LP using `scipy.optimize.linprog` [31] with the HiGHS solver [13].

[2] We use `symp.Rational` on the solution returned by LP, rounding it to a specified precision, and apply the same process to the solution returned by Z3. We observed that rounding has two benefits: first, it mitigates the issue of unsolved constraints (see Remark 5), and second, it provides a more interpretable expression of the Lyapunov functions.

Results in Table 1 show that incorporating symbolic reduction clearly increases the likelihood of discovering a valid Lyapunov function in most cases. Furthermore, Table 2 shows that with symbolic reduction, all approaches using LaSalle's principle successfully find a Lyapunov function, whereas approaches without LaSalle fail to do so. Finally, we also conducted extensive tests to evaluate the effect of symbolic reduction on instability detection using Z3. As shown in Table 3, symbolic reduction significantly improves performance, enabling successful verification even for high-dimensional systems, while the baseline approach without reduction fails (100% timeout) even in low dimensions.

Table 1. Runtimes (in seconds) of Lyapunov function synthesis methods for the systems in Examples (E1)–(E7). Symbolic reduction clearly plays a crucial role in increasing the likelihood of discovering a valid Lyapunov function in most cases.

	Dim	Z3-Complete-SR	LP-CEGIS	LP-CEGIS-SR	Z3-CEGIS	Z3-CEGIS-SR	SOS
E1	2	0.226	0.193	0.075	4.863	1.402	3.200
E2	2	0.202	0.168	0.071	-	1.479	1.566
E3	2	0.245	0.193	0.101	-	1.483	-
E4	2	0.206	0.167	0.077	-	1.429	-
E5	3	0.357	-	0.170	-	19.090	1.629
E6	4	-	-	0.113	-	-	-
E7	6	-	-	0.349	-	-	-

Table 2. Runtimes (in seconds) of Lyapunov and LaSalle synthesis methods for the three systems in (E8)–(E10). Synthesis without the LaSalle module failed to find a valid Lyapunov function in all cases.

	Dim	Z3-Complete-SR-LaSalle	LP-CEGIS-SR-LaSalle	Z3-CEGIS-SR-LaSalle
E8	2	0.412	0.291	1.449
E9	2	0.097	0.135	1.382
E10	2	1.032	0.609	1.771

Table 3. Experiment results for "speedy" instability tests using the SMT solver Z3: A total of 100 systems is randomly generated for each case, with a prescribed maximum degree for the monomials. The number of monomial terms for each dimension is set to 3. Systems that are trivially unstable, where the linearization has an eigenvalue with a positive real part, are eliminated from the trials. We also exclude vector fields that contain a zero component, as they cannot be asymptotically stable. A timeout (t.o.) of 1 s is applied to each call of Z3-complete-SR for stability or instability tests. The table includes the average synthesis time for stable and unstable cases, encompassing both SMT solver time and the overhead for symbolic reduction. Notably, a considerable number of systems can be efficiently proven to be not globally asymptotically stable in under 1 s, even for systems up to dimension 10.

Dim (n)	Deg	Unstable Unstable %	t.o. (%)	Time (s)	Stable Stable %	t.o. (%)	Time (s)	No. Systems
2	2	61.0	1.0	0.033	3.0	0.0	0.045	100
2	3	34.0	7.0	0.033	1.0	21.0	0.050	100
3	2	57.0	4.0	0.038	0.0	3.0	0.000	100
3	3	38.0	20.0	0.068	0.0	48.0	0.000	100
4	2	58.0	8.0	0.061	0.0	6.0	0.000	100
4	3	25.0	21.0	0.075	0.0	60.0	0.000	100
5	2	53.0	8.0	0.066	0.0	6.0	0.000	100
5	3	27.0	18.0	0.102	0.0	58.0	0.000	100
6	2	58.0	2.0	0.103	0.0	1.0	0.000	100
6	3	23.0	24.0	0.158	0.0	65.0	0.000	100
7	2	50.0	4.0	0.174	0.0	3.0	0.000	100
7	3	23.0	24.0	0.272	0.0	70.0	0.000	100
8	2	58.0	4.0	0.305	0.0	0.0	0.000	100
8	3	25.0	9.0	0.379	0.0	53.0	0.000	100
9	2	60.0	1.0	0.443	0.0	0.0	0.000	100
9	3	24.0	24.0	0.524	0.0	61.0	0.000	100
10	2	51.0	2.0	0.608	0.0	2.0	0.000	100
10	3	12.0	21.0	0.718	0.0	90.0	0.000	100

7 Conclusions

In this paper, we investigated automated synthesis of global Lyapunov functions and proposed a simple set of algebraic checks that can be used to symbolically solve for parameters in a polynomial Lyapunov function candidate, before an optimization or SMT-based synthesis procedure is used to compute a global Lyapunov function. We also proposed sufficient LaSalle-type conditions that can be implemented in an SMT procedure, either as verification or synthesis, for global Lyapunov functions using a weak Lyapunov function candidate. Additionally, we encoded instability conditions and designed SMT-based procedures for disproving global asymptotic stability. Through a suite of examples, we demonstrated that the proposed symbolic reduction, LaSalle-type conditions, and instability tests allow us to efficiently solve many examples that would be otherwise unsolvable.

For future work, one could further build upon the symbolic reduction rules and encode additional algebraic necessary conditions for sign-definite polynomials. It would be interesting to investigate how the symbolic reduction procedure could be developed for non-polynomial vector fields. We expect that if the vector field is locally smooth, parts of the reduction rules may still apply. Similarly, if the vector field behaves well at infinity, similar reductions could be performed. While this paper focused on global asymptotic stability, one could also specialize to local stability, for which symbolic reduction, LaSalle's principle, and instability tests could be developed accordingly.

Acknowledgements. This work was supported in part by the Natural Sciences and Engineering Research Council of Canada and the Canada Research Chairs Program. Jun Liu would like to thank Alessandro Abate, Alec Edwards, and Andrea Peruffo for helpful discussions.

References

1. Abate, A., et al.: Automated formal synthesis of digital controllers for state-space physical plants. In: Proceedings of International Conference on Computer Aided Verification, pp. 462–482. Springer (2017)
2. Ahmadi, A.A., Parrilo, P.A.: Stability of polynomial differential equations: complexity and converse Lyapunov questions. arXiv preprint arXiv:1308.6833 (2013)
3. Ahmed, D., Peruffo, A., Abate, A.: Automated and sound synthesis of Lyapunov functions with SMT solvers. In: Proceedings of International Conference on Tools and Algorithms for the Construction and Analysis of Systems, pp. 97–114 (2020)
4. Arnold, V.: Problems of present day mathematics, xvii (dynamical systems and differential equations). In: Proc. Symp. Pure Math. vol. 28 (1976)
5. Barrett, C., Tinelli, C.: Satisfiability modulo theories. Handbook of model checking, pp. 305–343 (2018)
6. Chang, Y.C., Roohi, N., Gao, S.: Neural lyapunov control. Adv. Neural Inf. Proc. Syst. **32** (2019)
7. Cox, D.A., Little, J., O'Shea, D.: Ideals, varieties, and algorithms: An Introduction to Computational Algebraic Geometry and Commutative Algebra. Springer, 4 edn. (2015)
8. De Moura, L., Bjørner, N.: Z3: an efficient SMT solver. In: International conference on Tools and Algorithms for the Construction and Analysis of Systems, pp. 337–340. Springer (2008)
9. Edwards, A., Peruffo, A., Abate, A.: Fossil 2.0: formal certificate synthesis for the verification and control of dynamical models. In: Proceedings of the 27th ACM International Conference on Hybrid Systems: Computation and Control, pp. 1–10 (2024)

10. Gerbet, D., Röbenack, K.: Proving asymptotic stability with LaSalle's invariance principle: on the automatic computation of invariant sets using quantifier elimination. In: 2020 7th International Conference on Control, Decision and Information Technologies (CoDIT). vol. 1, pp. 306–311. IEEE (2020)
11. Giesl, P.: Construction of global lyapunov functions using radial basis functions. Springer (2007)
12. Giesl, P., Hafstein, S.: Review on computational methods for lyapunov functions. Discrete and Continuous Dynamical Systems-B **20**(8), 2291–2331 (2015)
13. Huangfu, Q., Galabova, I., Feldmeier, M., Hall, J.A.J.: HiGHS - high performance software for linear optimization. https://highs.dev/ (2023)
14. Kapinski, J., Deshmukh, J.V., Sankaranarayanan, S., Arechiga, N.: Simulation-guided lyapunov analysis for hybrid dynamical systems. In: Proceedings of the 17th International Conference On Hybrid Systems: Computation And Control, pp. 133–142 (2014)
15. Khalil, H.K.: Nonlinear Systems. Prentice-Hall (2002)
16. La Salle, J., Lefschetz, S.: Stability by Liapunov's direct method with applications. Academic Press (1961)
17. Liu, J., Zhan, N., Zhao, H.: Automatically discovering relaxed lyapunov functions for polynomial dynamical systems. Math. Comput. Sci. **6**(4), 395–408 (2012)
18. Liu, J., Fitzsimmons, M.: Symbolic reduction for formal synthesis of global Lyapunov functions. arXiv preprint (2025)
19. Liu, J., Meng, Y., Fitzsimmons, M., Zhou, R.: LyZNet: a lightweight python tool for learning and verifying neural Lyapunov functions and regions of attraction. In: Proceedings of the 27th ACM International Conference on Hybrid Systems: Computation and Control, pp. 1–8 (2024)
20. Liu, J., Meng, Y., Fitzsimmons, M., Zhou, R.: Physics-informed neural network lyapunov functions: pde characterization, learning, and verification. Automatica **175**, 112193 (2025)
21. Lyapunov, A.M.: The general problem of the stability of motion. Int. J. Control **55**(3), 531–534 (1992)
22. Papachristodoulou, A., et al.: SOSTOOLS: sum of squares optimization toolbox for MATLAB (2013)
23. Papachristodoulou, A., Prajna, S.: On the construction of Lyapunov functions using the sum of squares decomposition. In: Proceedings of the 41st IEEE Conference on Decision and Control, 2002. vol. 3, pp. 3482–3487. IEEE (2002)
24. Ravanbakhsh, H., Sankaranarayanan, S.: Counter-example guided synthesis of control Lyapunov functions for switched systems. In: 2015 54th IEEE conference on decision and control (CDC), pp. 4232–4239. IEEE (2015)
25. Ravanbakhsh, H., Sankaranarayanan, S.: Learning control lyapunov functions from counterexamples and demonstrations. Auton. Robot. **43**, 275–307 (2019)
26. Sankaranarayanan, S., et al.: Lyapunov function synthesis using handelman representations. IFAC Proc. Vol. **46**(23), 576–581 (2013)
27. She, Z., Li, H., Xue, B., Zheng, Z., Xia, B.: Discovering polynomial lyapunov functions for continuous dynamical systems. J. Symb. Comput. **58**, 41–63 (2013)
28. She, Z., Xia, B., Xiao, R., Zheng, Z.: A semi-algebraic approach for asymptotic stability analysis. Nonlinear Anal. Hybrid Syst **3**(4), 588–596 (2009)
29. Sturm, J.F.: Using SeDuMi 1.02, a MATLAB toolbox for optimization over symmetric cones. Optim. Meth. Softw. **11**(1-4), 625–653 (1999)
30. Topcu, U., Packard, A., Seiler, P.: Local stability analysis using simulations and sum-of-squares programming. Automatica **44**(10), 2669–2675 (2008)

31. Virtanen, P., et al.: SciPy 1.0: fundamental algorithms for scientific computing in Python. Nature Meth. **17**(3), 261–272 (2020)
32. Zhou, R., Quartz, T., Sterck, H., Liu, J.: Neural lyapunov control of unknown nonlinear systems with stability guarantees. Adv. Neural. Inf. Process. Syst. **35**, 29113–29125 (2022)
33. Zubov, V.I.: Methods of AM Lyapunov and their application, vol. 4439. US Atomic Energy Commission (1961)

Using Communication to Bound Clock Drift in Local-Timed Negotiations

Abhinav Garg[2], Madhavan Mukund[1,3], Adwitee Roy[1], B. Srivathsan[1,3(✉)], and Gautham Viswanathan[1]

[1] Chennai Mathematical Institute, Chennai, India
sri@cmi.ac.in
[2] Indian Institute of Technology Kanpur, Kanpur, India
[3] CNRS, ReLaX, IRL 2000, Siruseri, India

Abstract. In the model of negotiations introduced by Esparza et al. (2013), a distributed set of agents strive to achieve a final outcome through a sequence of negotiations within potentially smaller groups. Mukund et al. (2023) extended this model to include timing constraints between negotiations. In this model of *local-timed negotiations* (LTNs), each agent has its own local reference clock that evolves independently of the others, and represents its local-time. Apart from their reference clocks, agents can have other local clocks that can be checked for guards and updated using resets (like in a timed automaton), when they move from one negotiation to another. LTNs come with a special feature where negotiations can be specified to be *time-synchronizing* or not. Agents participating in a time-synchronizing negotiation are required to synchronize their local reference clocks before deciding on an outcome. Due to this feature, agents which have drifted far apart in their local times are forced to resynchronize their local-times again. This makes reachability undecidable for LTNs.

In this paper, we consider LTNs which ensure a *connected communication*: in every cyclic behaviour, every agent participates in a time-synchronization with every other agent, either directly or indirectly. We show that connected communication makes reachability in LTNs decidable. A counterpart of this notion of connected communication has been used in message sequence graphs by Alur et al. (1999), and in distributed controller synthesis by Thiagarajan et al. (2005) to gain decidability.

Keywords: concurrency · timed systems · negotiations

1 Introduction

Concurrent computation can be modeled in many different ways—some form of communicating automata, Petri Nets, negotiations, message sequence graphs, etc. In the communicating automata perspective, local behaviours are described using automata and the way the global system behaves is determined by a specified parallel composition operation. The global system computed as the product is exponential in the size of the local automata. In the *negotiations* perspective [3], the computation is carried out by a set of agents. The joint interactions of the agents, called *atomic negotiations*, are specified first. Each atomic negotiation results in an *outcome*. Every agent has a "next" function that determines for each outcome the set of next negotiations that an agent is ready to engage in. The negotiations model is amenable to efficient static analysis and several problems on negotiations have been shown to have polynomial-time algorithms [4,5].

Motivated by the attractive algorithmic properties of negotiations, an extension of negotiations to the real-time setting was proposed in [10]. Each agent has its own *local-time*. Local-time preserves independence properties: if agent p does an action a when its local-time is 3, and agent q does an action b when its local-time is 2, the actions a and b are still independent and "commutable", since they deal with disjoint agents, who have their own local-time. Whereas if all agents proceed via a single global time scale, then b happens-before a, destroying independence. Local-time semantics has been closely studied in the context of verification of networks of timed automata [2,6,7]. In local-timed negotiations (LTNs) of [10], a unique feature was introduced: by default, agents participating in an atomic negotiation are not required to synchronize their local-times when deciding on an outcome—unlike in the local-time semantics for networks of timed automata, where all joint actions are taken when the local times of all participating agents are equal; instead, the model can explicitly specify which of the atomic negotiations are *time-synchronizing*—in such atomic negotiations, the participating agents should necessarily synchronize their local times before deciding on an outcome (see Section 2.1 of [10] for examples applying this feature). Surprisingly, this feature gives immense computing power and makes reachability undecidable for LTNs. The core issue behind the undecidability is the possibility for two agents to arbitrarily drift apart in their local-times and then come together for a later time-synchronization.

Contributions. In this work, we consider a *bounded-drift* property for an LTN, similar to the notion considered in [8] for networks of timed automata. An LTN is said to have drift bounded by K if in every behaviour, the local-time delays can be *tightened* so that the drift between the local-times of every pair of processes is always bounded by a constant K. Similar to [8], we conclude that reachability is decidable for bounded-drift negotiations, by building a region automaton for the set of untimed behaviours.

Subsequently, we consider a syntactic fragment of LTNs, called *connectedly-communicating* negotiations. An LTN is said to be connectedly-communicating, if in every cycle of a behaviour, each agent has a time-synchronizing negotiation with every other agent, either directly or indirectly. This idea is formalized by looking at the so-called *marking graph* of the underlying untimed negotiation, and by enforcing that a certain *communication graph* of each loop to be connected. The inspiration for this definition comes from the work of [1] on hierarchical message sequence graphs, where a similar idea makes the emptiness problem decidable. The terminology "connectedly-communicating" comes from the work of [9] on the distributed controller synthesis problem, which is unde-

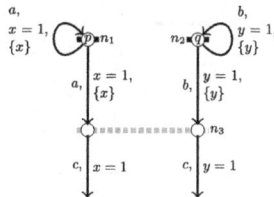

Fig. 1. An LTN exhibiting non-regular behaviours.

cidable in general. Once again the connectedness restriction makes the problem decidable. The setting of [9] is slightly different and weaker than our definition: in their fragment, if process p does not hear from q either directly or indirectly, they never participate in a synchronizing action in the future. Towards the end of this manuscript, we comment on how our techniques can be extended to this weaker setting as well.

Related Work. Local-timed negotiations were introduced in [10]. It was shown that reachability is decidable in two cases: either when all nodes are time-synchronizing (called the fully synchronizing fragment), or none of the nodes is time-synchronizing (called the synchronization-free fragment). The idea of drifting the bound between processes has been studied for networks of timed automata in [7,8]. For networks of timed automata, local-time semantics has been studied in the context of efficient zone-based verification methods. Due to independence, the zone graph computed using the local-time semantics contains diamonds, and hence avoids one dimension of exponential blowup caused due to interleaving of independent actions. In this context, bounded-drift is helpful in coming up with simulation techniques to make the zone graph exploration finite. Related works from the concurrency literature have been mentioned above.

Organization of the Paper. In Sect. 2, we recall the definition of local-timed negotiations and present a new view of its behaviours as traces. Section 3 discusses reachability for drift-bounded negotiations. Our main technical contribution starts in Sect. 4 where we show that connectedly-communicating negotiations have a bounded drift. Section 5 summarizes the work and discusses some future work.

2 Local-Timed Negotiations (LTNs)

We start with some informal examples to intuitively describe local-timed negotiations. Figure 1 of [10] gives an example of a transaction between a customer, ATM and a bank as an LTN. Here, we present some abstract examples to illustrate interesting mechanics of LTNs, and then recall the formal syntax and semantics of LTNs from [10].

Example 1. Figure 1 gives a two agent negotiation from [10] that exhibits non-regular behaviours. A similar example also appears in the study of local-time

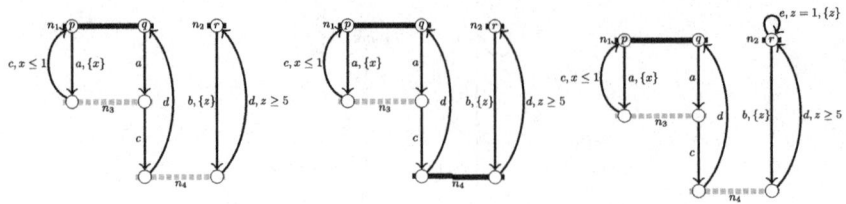

Fig. 2. Examples of Local-Timed Negotiations (LTNs).

semantics for timed automata [7]. Agent p starts at atomic negotiation (or node) n_1, and q starts at node n_2. Both of them independently do a and b, and can non-deterministically reach node n_3. Each a incurs a local-time of 1 unit for p due to the guard and reset of x. Similarly, each b incurs a local-time of 1 unit for q. Node n_3 is a time-synchronizing negotiation (denoted by the dashed line). When the outcome c happens, both p and q need to be at the same time. This imposes a condition that the number of a and the number of b seen before outcome c need to be equal.

Example 2. Figure 2a describes an LTN between three agents p, q, r. The atomic negotiations (or nodes) are $\{n_1, n_2, n_3, n_4\}$. Agents p and q participate in n_1 and n_3, agents q and r participate in n_4 and node n_2 is local to agent r. Assume that p, q, r start at n_1, n_1, n_2 respectively. The initial values of their local-times are $t_p = 0, t_q = 0, t_r = 0$. At node n_1, p and q decide the outcome a and move to n_3. When a is executed, say the local-times are $t_p = 1, t_q = 2$ (they need not be equal). At n_3, p and q execute c. Notice the reset of clock x on a and a guard $x \leq 1$ on c. Moreover, n_3 is a time-synchronizing node (denoted by dashed lines). When outcome c is executed, both the local-times need to be the same, say c is executed at $t_p = 2, t_q = 2$ (observe that the guard is also satisfied). If p had elapsed less than 1 time unit while executing a, then it would not be possible to do c because the combination of the guard and the time-synchronizing constraint cannot be met simultaneously in that case.

Another remark is that, as the LTN continues executing, the time that p spends between c and the next a, is also governed by r. Notice that the bd loop done by r requires at least 5 time units. When q time-synchronizes with r at n_4, its local-time catches up with that of r. Agent q "shares this information" with p during the atomic negotiation n_1. Since p cannot spend more than 1 time unit between a and the next c, agent p has to plan ahead and elapse enough time at n_1 before executing a.

In Fig. 2b, node n_4 is no longer time-synchronizing. Here agent r has no influence over the local-time of agent p. In Fig. 2c, agent r has a loop at node n_2, with each execution of the loop increasing its local-time by 1 unit. Now, n_4 is still time-synchronizing. So, each time r talks to q at n_4, agent q catches up with the time of r, depending on how many times the loop e was executed by r. This way, r can force an unbounded drift between q and r at n_4, and for p and q at negotiation n_1.

2.1 Formal Syntax and Semantics

Let $\mathbb{R}_{\geq 0}$ and $\mathbb{N}$ denote the set of non-negative reals and natural numbers, respectively. A *clock* is a variable ranging over $\mathbb{R}_{\geq 0}$ used for enforcing timing constraints. Let X be a set of clocks. A *guard* over X is a conjunction of clock constraints of the form $x \bowtie c$ where $x \in X$, $\bowtie \in \{<, \leq, =, >, \geq\}$ and $c \in \mathbb{N}$. Let $\Phi(X)$ denote the set of guards over X.

Fix a finite set of agents $\mathcal{A}$, a finite alphabet Σ of outcomes, and a finite set X of clocks. The clocks are partitioned as $\{X_p\}_{p \in \mathcal{A}}$, with X_p being the *local clocks* of agent p. Apart from local clocks X_p, every agent p is assumed to have a special *reference clock* t_p, denoting the local-time of p. The clocks t_p are never reset, nor used in guards.

Definition 1 (Local-timed negotiation (LTN)). *A local-timed negotiation (LTN) over the set of agents $\mathcal{A}$, is a tuple $\mathcal{N} = (N, dom, \delta, \mathsf{TSync})$ where:*

- *N is a finite set of nodes (also called atomic negotiations),*
- *$dom : N \to \mathcal{P}(\mathcal{A})$ maps each node to a non-empty subset of agents; for $p \in \mathcal{A}$, we let $N_p := \{n \in N \mid p \in dom(n)\}$,*
- *for each agent p, we mark a special node $n_p^{init} \in N_p$ as the initial node for the agent,*
- *the transition function $\delta = \{\delta_p\}_{p \in \mathcal{A}}$ is a tuple, one for each agent, where $\delta_p : N_p \times \Sigma \to \Phi(X) \times \mathcal{P}(N_p) \times \mathcal{P}(X_p)$ maps each node-outcome pair (n, a) to a guard $g \in \Phi(X)$, a set of nodes $S_p \subseteq N_p$ that p becomes ready to engage in after this outcome, and a set $Y \subseteq X_p$ of clocks that get reset,*
- *$\mathsf{TSync} \subseteq N$ is a subset of* time-synchronizing *nodes.*

Formally, the semantics of a negotiation is described using *markings* and *valuations*. A marking C is a function assigning each agent p to a subset of N_p, giving the set of nodes that each agent is ready to engage in. A valuation $v : X \cup T \to \mathbb{R}_{\geq 0}$ maps every clock (including reference clocks) to a non-negative real such that $v(x) \leq v(t_p)$ for all $x \in X_p$, and all agents $p \in \mathcal{A}$. The interpretation is that clocks in X_p move at the same pace as t_p, the local reference clock. Since t_p is never reset, it gives the local time at agent p and is greater than or equal to every other local clock of p. For a constraint $x \bowtie c$ we say $v \models x \bowtie c$ if $v(x) \bowtie c$. We say v satisfies guard $g \in \Phi(X)$, written as $v \models g$, if v satisfies every atomic constraint appearing in g.

A *local-delay* $\Delta \in \mathbb{R}_{\geq 0}^{|\mathcal{A}|}$ is a vector of non-negative reals, giving a time elapse for each agent. Given a valuation v and a local-delay Δ, we write $v + \Delta$ for the valuation obtained as follows: for agent $p \in \mathcal{A}$, we have $(v + \Delta)(y) = v(y) + \Delta(p)$ for every $y \in \{t_p\} \cup X_p$. For a set of clocks $Y \subseteq X$, we denote by $v[Y]$ the valuation satisfying $v[Y](y) = 0$ if $y \in Y$ and $v[Y](y) = v(y)$ otherwise.

A configuration is a pair (C, v) consisting of a marking C and a valuation v. The *initial configuration* (C_0, v_0) contains a marking C_0 which maps every agent to its initial node and valuation v_0 maps all clocks to 0. We write $(C, v) \xrightarrow{\Delta} (C, v + \Delta)$ for the local-delay transition Δ at configuration (C, v).

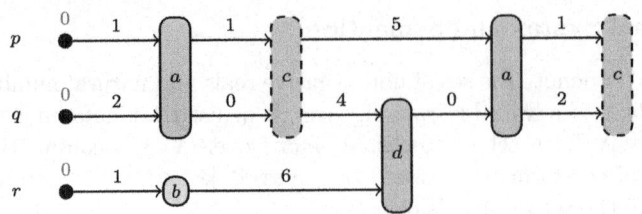

Fig. 3. Trace of a run.

An *event* is a pair consisting of a node and an outcome. An event $e = (n, a)$ can be executed at a configuration (C, v) leading to a configuration (C', v'), written as $(C, v) \xrightarrow{e} (C', v')$, provided there is an entry $\delta_p(n, a) = (g_p, S_p, Y_p)$ for all $p \in dom(n)$ such that:

- *current marking enables the negotiation:* $n \in C(p)$ for all $p \in dom(n)$,
- *synchronization condition is met:* if $n \in \mathsf{TSync}$, then $v(t_p) = v(t_q)$ for all $p, q \in dom(n)$,
- *guard is satisfied:* $v \models g_p$ for all $p \in dom(n)$,
- *target marking is correct:* $C'(p) = S_p$ for all $p \in dom(n)$, $C'(p) = C(p)$ for $p \notin dom(n)$,
- *resets are performed:* $v'(y) = 0$ for $y \in \bigcup_{p \in dom(n)} Y_p$

We call $(C, v) \xrightarrow{\Delta} (C, v + \Delta) \xrightarrow{e} (C', v')$ a *step* and write this as $(C, v) \xrightarrow{\Delta, e} (C', v')$ for conciseness. A *run* is a sequence of steps starting from the initial configuration. We say that an event $e = (n, a)$ is *reachable* (in the negotiation) if there is a run starting from the initial configuration, containing a step that executes e. Reachability for LTNs is known to be undecidable in general [10]. When there are no time-synchronizing nodes, or when every node is time-synchronizing, reachability becomes decidable. In this paper, we present a new decidable class.

2.2 Trace of a Run

In this work, we will often view a run of an LTN as a *trace* over the events of the run. Figure 3 gives an example of a trace for the negotiation depicted in Fig. 2 and described in Example 2. Fix an LTN $\mathcal{N}$. Suppose we are given a run $\rho := (C_1, v_1) \xrightarrow{\Delta_1, e_1} (C_2, v_2) \xrightarrow{\Delta_2, e_2} \cdots (C_k, v_k)$, starting from some arbitrary configuration (C_1, v_1).

From Run to Trace. The trace $\mathrm{Tr}(\rho)$ is a directed graph $(E_\rho, \{\to_p\}_{p \in \mathcal{A}})$ equipped with an assignment $\{\tau_p^0\}_{p \in \mathcal{A}}$ which gives an initial local-time for each agent and:

- the vertex set E_ρ is given by the set of events $\{e_1, e_2, \ldots, e_{k-1}\}$ along with special events $\{\perp_p\}_{p \in \mathcal{A}}$ denoting an initial event for each agent; in Fig. 3, the initial events $\perp_p$ are denoted by dots.

- the edge relation is partitioned as $\{\to_p\}_{p\in\mathcal{A}}$; there is an edge $e_i \xrightarrow{\theta}_p e_j$ with label θ if e_j is the next event that p participates after e_i, and θ is the time spent by p in between. Formally, we have $i < j$, $p \in dom(e_i) \cap dom(e_j)$, and there is no $i < i' < j$ such that $p \in dom(e_{i'})$; moreover, $\theta = \Delta_i(p) + \cdots + \Delta_{j-1}(p)$, which is the time spent by agent p between the events e_i and e_j.

Given a trace with its initial assignment of local-times, and an event e, we can compute a local-time at e for every agent in $dom(e)$, by summing up the initial local time and the delays along the $\to_p$ edges starting from $\perp_p$ to e. For an event e and an agent $p \in dom(e)$, we will write $\tau_p(e)$ for this timestamp. For example, in Fig. 3, all initial timestamps are 0, and for the first a and c events we have $\tau_p(a) = 1$, $\tau_p(c) = \tau_q(c) = 2$. For a time-synchronizing event s, we sometimes write $\tau(s)$ for $\tau_p(s)$ with $p \in dom(s)$. We can do this since for a time-synchronizing event, all the time stamps of the participating processes are the same.

From Trace to Run. We have seen how to get a trace $\text{Tr}(\rho)$ from a run ρ. For the reverse, given a trace Tr and an initial configuration (C_0, v_0) that matches with the initial timestamps τ_p^0 of Tr, i.e. $v_0(t_p) = \tau_p^0$ for all agents p, we can derive run $\rho[(C_0, v_0), \text{Tr}]$ as follows: to each event e a step Δ_e can be associated, where for each $p \in dom(e)$, $\Delta_e(p)$ is the label of the incoming $\to_p$ edge to e, and for $p \notin dom(e)$, we set $\Delta_e(p)$ to 0; then we get a run by starting with (C_0, v_0), linearizing the events of Tr, executing the corresponding steps, and computing intermediate configurations of the LTN. The sequence $\rho[(C_0, v_0), \text{Tr}]$ gives a run of the LTN if all the events of the trace can be executed this way.

3 Drift-Bounded LTNs

A configuration (C, v) of an LTN $\mathcal{N}$ is said to be k-drift-bounded for a constant $k \in \mathbb{N}$ if $|v(t_p) - v(t_q)| \leq k$ for all agents $p, q \in \mathcal{A}$. Extending this notion, a run $(C_1, v_1) \xrightarrow{(\Delta_1, e_1)} (C_2, v_2) \cdots$ is k-drift-bounded, if every configuration (C_i, v_i) is drift-bounded. Finally, an LTN $\mathcal{N}$ is said to be k-drift-bounded if for every run $(C_1, v_1) \xrightarrow{\Delta_1, e_1} (C_2, v_2) \cdots$ there is a way to change the local-delays to get a run $(C_1, v_1) = (C_1, v_1') \xrightarrow{\Delta_1', e_1} (C_2, v_2') \xrightarrow{\Delta_2', e_2} \cdots$ that is k-drift bounded.

Example 3. Figure 4 shows an example of an LTN $\mathcal{N}_1$ with unbounded drift and a 1-drift bounded LTN $\mathcal{N}_2$. In $\mathcal{N}_1$, there are two self-loops in the nodes n_1 and n_2. For any k, the loop on n_1 can be taken $k+1$ times and the loop on n_2 may not be taken at all, making $|v(t_p) - v(t_q)| \geq k$ at a configuration where n_3 is ready to be fired. Note that there is no way to reduce the drift in this case: if a local-time of more than 1 time unit is elapsed for q while p keeps doing the a loops, agent q will not be able to perform the self-loop b. Therefore, in order to execute the specific run $a^{k+1}b^{k+1}$, the drift needs to be kept at least k.

In $\mathcal{N}_2$, each time the outcome on the time-synchronizing node n_3 is taken, both the agents have the same time, and hence drift at that point is 0. Then, a

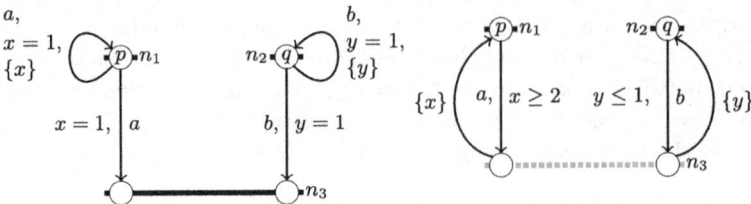

Fig. 4. LTN $\mathcal{N}_1$ with unbounded drift (left) and 1-drift bounded LTN $\mathcal{N}_2$ (right).

and b can be done with local-time 2 and 1 for p and q respectively. This maintains a drift of 1 between them.

Given a maximum constant M, a region equivalence $\equiv_M^k$ has been defined over k-drift-bounded configurations in [8]. There are two minor differences: [8] studies the equivalence in the context of networks of timed automata; and secondly, it makes use of an *offset* representation for valuations where only the reference clocks are assumed to elapse, and each local clock maintains the local time at which it was last reset. The equivalence can be adapted to our setting in a straightforward manner. The equivalence $\equiv_M^k$ has been shown to be a time-abstract bisimulation on the space of k-drift-bounded configurations. In other words, suppose $(C, v) \equiv_M^k (C, v')$ for two k-drift-bounded configurations (C, v) and (C, v')—the equivalence on configurations imposes the markings (control state) to be the same in both configurations—then, (1) for every local delay Δ, there is a local delay Δ' such that $(C, v + \Delta) \equiv_M^k (C, v' + \Delta')$, (2) v satisfies a guard with constant atmost M iff v' satisfies it, and (3) $v(t_p) = v(t_q)$ iff $v'(t_p) = v'(t_q)$ for every pair of agents p, q. All these constraints ensure that for every k-drift-bounded run that can be executed from (C, v), there is an equivalent k-drift-bounded run from (C, v'). Using the equivalence classes of $\equiv_M^k$ as states, a region automaton can be constructed. The region automaton accepts all untimed behaviours (sequences of outcomes) for which there is a k-drift-bounded timed run of the LTN. When the LTN is k-drift-bounded, this set simply equals all untimed behaviours of the LTN. This makes reachability decidable for k-drift-bounded LTNs: given such an LTN, construct the region graph and check for reachability of a certain edge.

The question now becomes whether we can test a given LTN to be k-drift bounded: given an LTN $\mathcal{N}$ and a $k \in \mathbb{N}$, is it k-drift-bounded. The decidability of this problem is open, even in the timed automaton setting. Sufficient conditions for a network of timed automata to be k-drift-bounded have been presented in [7]. In the next section, we present a structural restriction on local-timed negotiations that makes it drift bounded.

4 Connectedly Communicating LTNs

For an LTN $\mathcal{N}$, we write $\mathcal{N}_{\text{untime}}$ for the (untimed) negotiation obtained by syntactically removing all guards and resets from transitions. Notice that $\mathcal{N}_{\text{untime}}$

over-approximates the set of untimed behaviours of $\mathcal{N}$. For $\mathcal{N}_{\text{untime}}$ we can define a marking graph $\text{MG}(\mathcal{N}_{\text{untime}})$ as follows: vertices of the marking graph are markings C which map each agent to a set of nodes that it participates in; there is an edge $C \xrightarrow{n} C'$, if $n \in C(p)$ for all agents $p \in dom(n)$. This is a finite graph, of size $2^{|N|} \cdot |\mathcal{A}|$ where $|N|$ denotes the number of nodes of $\mathcal{N}$. A loop in $\text{MG}(\mathcal{N}_{\text{untime}})$ is a cyclic sequence $C = C_1 \xrightarrow{n_1} C_2 \xrightarrow{n_2} \cdots \xrightarrow{n_k} C_{k+1} = C$. The communication graph of a loop contains the set of agents as its vertices; there is an edge (p,q) if p and q participate in a time-synchronizing node in the loop, in other words, there is some n_i such that $p, q \in dom(n_i)$ and $n_i \in \mathsf{TSync}$.

An LTN $\mathcal{N}$ is said to be connectedly communicating if the communication graph is connected for every loop of the marking graph $\text{MG}(\mathcal{N}_{\text{untime}})$. For example, the LTN in Fig. 2a is connectedly communicating, whereas the other two are not. In Fig. 2b, in the loop $(\{n_1\}, \{n_1\}, \{n_2\}) \xrightarrow{abdc} (\{n_1\}, \{n_1\}, \{n_2\})$, the node r does not participate in a time-synchronizing node. Hence it will be isolated in the communication graph. In Fig. 2c, the self-loop in node n_2 on r creates a loop which is not connectedly communicating. In the Conclusion, we briefly explain how we can weaken the definition of connectedly communicating to include the LTN of Fig. 2b, and still apply the technique we explain in this section.

Our goal in this section is to prove that connectedly-communicating LTNs have a bounded drift.

Theorem 1. *Every connectedly communicating LTN $\mathcal{N}$ is k-drift bounded with $k \in \mathcal{O}(|N| \cdot |\mathcal{A}|^3 \cdot (M+1))$, where N is the set of nodes of $\mathcal{N}$, $\mathcal{A}$ the set of agents and M the maximum constant appearing in $\mathcal{N}$.*

We will prove the theorem in two steps: (1) every run of a connectedly communicating LTN can be *tightened*, and (2) tight runs have bounded drift. Fix a connectedly communicating LTN $\mathcal{N}$ for the rest of the section.

4.1 Tightening a Run

We first explain the idea of tightening through examples. Figure 5 shows a trace for a single agent negotiation, with maximum constant $M = 10$. The time elapsed between a_1 and a_2 is 15, and the time between a_3 and a_4 is 20. These time delays are greater than $M + 1 = 11$. Suppose we want to reduce these delays to 11. This gives the trace $\overline{\text{Tr}}_1$. This will still give us a valid trace that corresponds to a run of the negotiation, because of the following property: while executing an event a_i, every clock that was previously smaller than M, retains the same value, and every clock that was previously greater than M continues to remain so. Since every guard has a constant at most M, the reduced delays enable the same set of guards. The trace $\overline{\text{Tr}}_1$ is said to be *tightened*.

Now consider the trace Tr_2 shown in Fig. 6.

Assume both p and q were at local-time τ before the start of this trace. Agent p has large values 15 and 20, greater than $M + 1$. However, reducing any of them would make the time-synchronization at s_1 impossible. Therefore, this trace is tight.

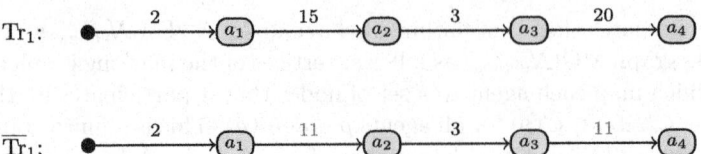

Fig. 5. A trace Tr_1 from a single agent negotiation. We assume $M = 10$. Run $\overline{Tr}_1$ is obtained by tightening Tr_1.

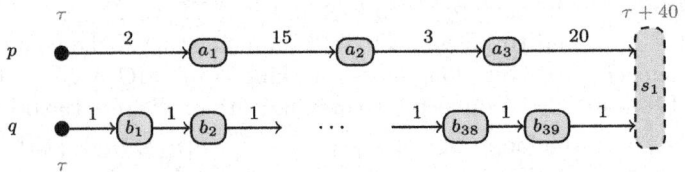

Fig. 6. A trace Tr_2 for a two agent negotiation, with $M = 10$.

Trace Tr_3 of Fig. 7 is not tight, in other words, there is still some *slack*. Agent p starts at τ_0, and q starts at τ_1, and the time-synchronizing event s_1 occurs at $\tau_2 = \tau_0 + 40 = \tau_1 + 30$.

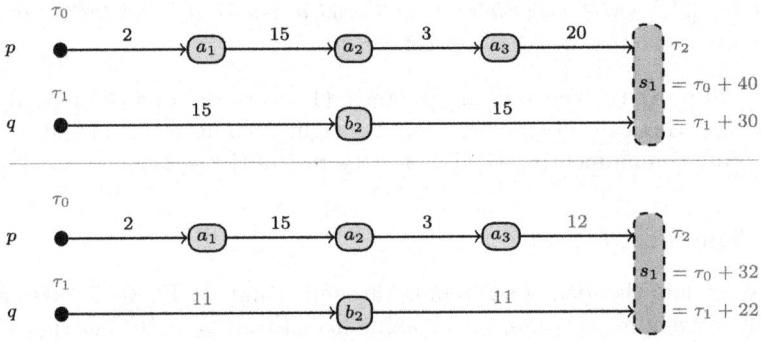

Fig. 7. Above: Trace Tr_3, with $M = 10$. Here, the slack is determined by the agent q. Below: Tightened trace $\overline{Tr}_3$, with a local-time 8 shaved off from both p and q.

When considered independently, agent q has two large values 15, and 15, amounting to a slack of 8 $((15 - 11) + (15 - 11))$, whereas agent p has a slack of $(15 - 11) + (20 - 11) = 13$. The final tightening is controlled by the smaller slack: reducing the total delays of p by more than 8 would make the time-synchronization at s_1 impossible. The tightened trace $\overline{Tr}_3$ is shown below in the figure.

As a final example, consider trace Tr_4 in Fig. 8. This exhibits an interesting aspect where the slack is determined not just by the *past* events, but also by *future* time-synchronizations. Trace Tr_4 contains trace Tr_3 as a prefix. However, agent p time-synchronizes with agent r at event s_2. Notice that there is no slack for agent r until s_2. If the local-delays of p are reduced, the local-time stamp of p at s_2 will reduce. Since it is a time-synchronizing node, the local-time of r needs to be reduced by the same amount. However, there are no large delays for r, and hence no slack. Therefore, the trace Tr_4 is already tight, even though there are some large values for p and q.

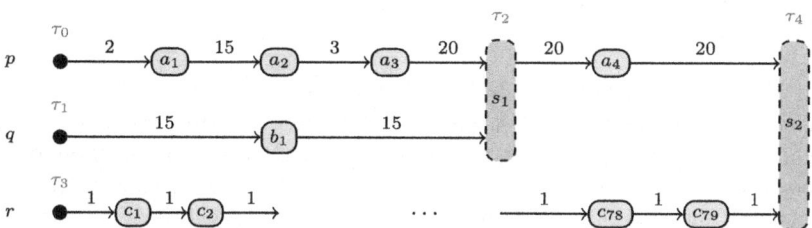

Fig. 8. Trace Tr_4 over three agents, and $M = 10$. The trace is tight since there is no slack for agent r up to the time-synchronizing event s_2.

The Slack Graph of a Trace. We will now formalize the ideas of "slack" and "tightening of a trace". Given a trace Tr, its slack graph Slack-Graph(Tr) is obtained from Tr as follows:

- for every edge $e \xrightarrow{\theta} e'$ in Tr, if $\theta \leq M+1$, then replace θ by 0; if $\theta > M+1$, replace θ by $(\theta - (M+1))$—these new values are called the *slack* in those edges
- for every agent p, and every pair of time-synchronizing events e_i and e_j that p participates in, if $e_i(\rightarrow_p)^+ e_j$ (that is, e_i happens before e_j), then add a *back edge* $e_j \xrightarrow{0}_p e_i$ with weight 0.

For example, in Slack-Graph(Tr_4), we have $a_1 \xrightarrow{4}_p a_2$, $a_3 \xrightarrow{9}_p s_1$, $s_1 \xrightarrow{9}_p a_4$, $a_4 \xrightarrow{9}_p s_2$, $\bullet \xrightarrow{4}_q b_1$, $b_1 \xrightarrow{4}_q s_1$, and every other edge has weight 0. Additionally, there is a back edge $s_2 \xrightarrow{0}_p s_1$. Back edges help propagate slack constraints from future synchronizations.

Definition 2 (Tight traces). *A trace Tr is said to be* tight *if for every time-synchronizing event s, there is a path of weight 0 in* Slack-Graph(Tr) *from an initial node $\perp_p$ to s, for some arbitary agent $p \in \mathcal{A}$.*

Trace Tr_4 is tight since there is a zero weight path from $\perp_r$ to s_1 and s_2. Trace Tr_3 is not tight. The shortest paths between the initial nodes and the time-synchronizing nodes for traces Tr_2, Tr_3 and Tr_4 are depicted in Fig. 9.

A Method to Tighten a Trace. Suppose we are given a trace Tr. The following notations and discussion are over this fixed trace Tr. For a time-synchronizing

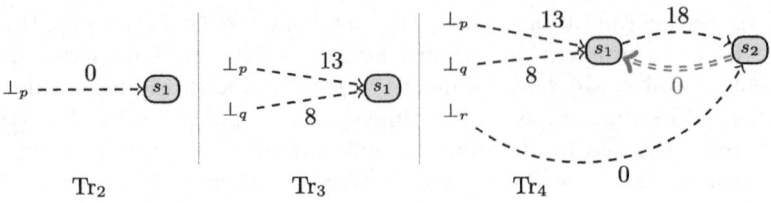

Fig. 9. The shortest weighted paths to time-synchronizing nodes in slack graphs. The red edge denotes a back edge. (Color figure online)

event s, let $\sigma_p(s)$ denote the weight of the shortest path from $\perp_p$ to s, and let $\sigma(s) := \min_{p \in \mathcal{A}}\{\sigma_p(s)\}$. For instance, for trace Tr_4 and event s_1, we have $\sigma_p(s_1) = 13$, $\sigma_q(s_1) = 8$ and $\sigma_r(s_1) = \sigma(s) = 0$. Here is a useful monotonicity property of this σ function. Let $p \in \mathcal{A}$ and s, s' be time-synchronizing events for p such that $s(\rightarrow_p)^+ s'$ and there is no other time-synchronizing event for p between s and s'. Then:

- $\sigma(s) \leq \sigma(s')$ (since there is a back-edge $s' \xrightarrow{0}_p s$ in Slack-Graph(Tr), the value $\sigma(s)$ cannot be strictly greater than $\sigma(s')$),
- $\sigma(s') \leq \sigma(s) + \theta_p$ where θ_p is the sum of the slacks for p along the path of $\rightarrow_p$ edges from s to s'. This simply follows by definition of σ.

We expand on the first item above, which also illustrates the need for back edges. Consider the trace Tr_4 and the graph representing shortest paths in Fig. 9. If we did not have the back edge depicted in red, then $\sigma(s_1) = 8$ and $\sigma(s_2) = 0$. Therefore, even though there is a path from s_1 to s_2 over edges representing agents p and q, the value $\sigma(s_1)$ would be greater than $\sigma(s_2)$ when there are no back edges. Item 1 above will no longer be true. However, adding the back edge communicates the slack of s_2 due to r on to s_1 too. Now, $\sigma(s_1)$ becomes 0 as well and item 1 is true.

The consequence of these two observations on the properties of $\sigma(s)$ noted above is that we can always reduce the timestamp of a time-synchronizing event s by $\sigma(s)$: the difference can be compensated by reducing the local-times contributing to the non-zero slack. For example, in trace Tr_3, we have $\sigma(s_1) = 8$. This value 8 can be compensated by reducing the large values as shown in Fig. 7. We summarize this discussion in the following lemma.

Lemma 1. *From every trace* Tr, *we can obtain a trace* $\overline{\text{Tr}}$ *where*

- *for every time-synchronizing event* s, *we have* $\overline{\tau}(s) = \tau(s) - \sigma(s)$, *where* $\tau(s)$ *and* $\overline{\tau}(s)$ *denote the time stamp of* s *in* Tr *and* $\overline{\text{Tr}}$ *respectively,*
- *for every edge* $a \xrightarrow{t}_p b$ *in* Tr *with* $t \leq M + 1$, *we have* $a \xrightarrow{t}_p b$ *in* $\overline{\text{Tr}}$ *too (in other words, delays less than* $M + 1$ *are preserved).*

Furthermore, for any configuration (C, v) *that matches the initial timestamps of* Tr, *there is a run* $\rho[(C, v), \text{Tr}]$ *of the LTN iff there is a run* $\rho[(C, v), \overline{\text{Tr}}]$.

Proof. The two items follow from the discussion above. The last point follows by noting that at every event, each clock that had a value $\leq M$ continues to have the same value, and a clock that had a value $> M$ will have a value $> M$. Furthermore, by construction, time-synchronization constraints are met. □

We next state a crucial property of tight traces.

Lemma 2. *Let Tr be a tight trace. Let τ_m be the maximum time-stamp among $\tau_p(\bot_p)$, the time-stamps of the initial events. Then, for every time-synchronizing event s of Tr, we have $\tau(s) \leq \tau_m + k \cdot (M+1)$, where k is the length of a shortest (weighted) path from some $\bot_p$ to s in* Slack-Graph(Tr).

Proof. Let s be a time-synchronizing event of Tr. Since Tr is tight, there is some agent p such that the shortest path from $\bot_p$ to s has weight 0. Let this path π be: $\bot_p = e_0 \to_{p_0} e_1 \to_{p_1} e_2 \cdots \to_{p_k} e_k = s$. As the path has weight 0, no edge in π has non-zero slack. Therefore, the time-delays in Tr corresponding to these edges are at most $M + 1$. This proves that $\tau(s) \leq \tau_p(\bot_p) + (k) \cdot (M + 1)$, as required. □

4.2 Tight Runs of a Connectedly Communicating LTN are Drift-Bounded

Section 4.1 gave a generic framework to reduce delays of runs in an LTN: any run can be reduced to a tight run. Moreover, this tight run has timestamps bounded by the size of the run, as shown in Lemma 2. To show an LTN to be drift-bounded, we need a uniform bound across all the runs. Lemma 2 does not immediately provide an answer, since the bound depends on the size of the run. In this section, we will closely examine the property of connected communication, and derive a uniform bound. Fix a connectedly communicating LTN $\mathcal{N}$ for the rest of this section.

Consider a segment of a tight run: $(C_1, v_1) \xrightarrow{(\Delta_1, e_1)} (C_2, v_2) \cdots (C_k, v_k)$ such that $C_1 = C_k$. This behaviour corresponds to a loop in the marking graph. If we pick two agents p and q that time-synchronize somewhere in the loop, then at the end of the loop, we can imagine a bound between t_p and t_q given by Lemma 2. What if p and q do not directly time-synchronize in this loop? For simplicity, let us consider $|\mathcal{A}|$ iterations of the same loop of the marking graph. Let p—p_1—$p_2 \cdots p_\ell$—q be the path between p and q in the communication graph of this loop (which by hypothesis is connected). Then, after $|\mathcal{A}|$ iterations, we can deduce some bound between t_p and t_{p_1}, between t_{p_1} and t_{p_2} and so on, until t_{p_ℓ} and t_q. Summing them up gives a bound on the difference $t_p - t_q$. However, there are two challenges:

- What about differences in the local-times in the intermediate part of the run? We need to ensure that they do not drift far and then later get closer.
- In general, there are multiple loops each with its own communication graph, and they can be executed in any order. How do we take care of combinations of different loops?

To tackle these two challenges, here is the proof template that we adopt. Given a run ρ, we break it into as suitable chunks $\rho_1 \rho_2 \ldots \rho_k$, and tighten each chunk independently and stitch them together. Each chunk is big enough that there is some loop of the marking graph in each of them—different chunks may contain different loops; it only matters that there is some loop, thereby handling challenge 2 above. Then we show that for each chunk ρ_i there is a bound on the difference between the maximum timestamp and the minimum timestamp occurring in ρ_i. We consider $\text{Tr}(\rho_i)$. Clearly, the maximum timestamp occurs in a maximal node of $\text{Tr}(\rho_i)$ and the minimal timestamp appears in an inital node $\perp_p$.

Suppose we are given a run $\rho := (C_0, v_0) \xrightarrow{(\Delta_1, e_1)} (C_1, v_1) \cdots (C_k, v_k)$. For technical convenience, let us assume that the final event e_{k-1} is a global synchronization for all agents. If this was not the case, we can always add a dummy node at the end, and then apply the tightening algorithm to this extended run. So, let us assume ρ is a tight run with a global synchronization at the end.

We will first break ρ into chunks. To do this, we make use of the size of the marking graph $\text{MG}(\mathcal{N}_{\text{untime}})$. Suppose $\text{MG}(\mathcal{N}_{\text{untime}})$ has $(N^* - 1)$ nodes. Let ρ_1 be the prefix of ρ consisting of the first N^* events, ρ_2 the next N^* events and so on, with ρ_ℓ being the last chunk consisting of at most N^* elements:

$$\rho_1 := (C_0, v_0) \cdots (C_{N^*}, v_{N^*})$$
$$\rho_2 := (C_{N^*}, v_{N^*}) \cdots (C_{2N^*}, v_{2N^*})$$
$$\ldots$$

Each ρ_i is a segment of a tight run ρ, but ρ_i by itself may not be tight: recall the tight trace Tr_4 of Fig. 8 which contains Tr_3 of Fig. 7 as prefix; trace Tr_3 as a standalone is not tight, however, when extended to Tr_4, the resulting trace is tight.

Let us now define some traces that we will work with. For the tight global run ρ, let Tr be its corresponding trace. For the segments $\rho_1, \rho_2, \ldots$, let Tr_1, $\text{Tr}_2, \ldots$ be the corresponding traces. We will denote the initial events of Tr_i by $\perp_p^i$. Notice that the timestamp associated to $\perp_p^0, \perp_p^1, \perp_p^2 \ldots$ are respectively $v_0(t_p), v_{N^*}(t_p), v_{2N^*}(t_p)$, and so on. Moreover for each initial node $\perp_p^i$ the maximal event of p in Tr_{i-1} has the same time stamp as $\perp_p^i$ in Tr_i. For an agent p, let e_p^i be the maximal event in Tr_i in which p participates. The event e_p^i need not be time-synchronizing.

Our goal is to reason about the maximum and minimum timestamps within each trace Tr_i. We do this in a series of claims.

Claim 1. Consider an index i. Let f be an arbitrary event in Tr_i. Let f' be an arbitrary event in $\text{Tr}_{i+|\mathcal{A}|+1}$. Then there is a path of forward edges from f to f' in Slack-Graph(Tr).

Proof. Let $j = i + |\mathcal{A}| + 1$. Suppose $p_1 \in dom(f')$. If p_1 is also in $dom(f)$, then there is a path of $\to_{p_1}$ edges from f to f'. Since there are N^* (non-initial) events in ρ_{j-1}, some marking repeats in the run ρ_{j-1} and hence, in Tr_{j-1}, agent p

synchronizes with at least one other process p_2 in some event f^{j-1}. If either p_1 or p_2 is in the domain of f, then there is a path of forward edges from f to f^{j-1}, and a path of forward edges from f^{j-1} to f'; this is because $f \in \text{Tr}_i, f^{j-1} \in \text{Tr}_{j-1}$, and $j - 1 > i$. Otherwise, we continue. Due to the connectedly communicating hypothesis, there is at least one process p_3 different from p_1 and p_2, that time-synchronizes with one of p_1 or p_2 in Tr_{j-2}. Let the time-synchronizing event be f^{j-2}—again, since $j - 2 > i$, there is a path of forward edges from f to f^{j-2}, and a path of forward edges from f^{j-2} to f' (possibly via f^{j-1}). At some point, we will reach an $f^{j-i'}$ such that $dom(f) \cap dom(f^{j-i'})$ is non-empty, with the index $i' \leq |\mathcal{A}|$. When we reach this event, we argue as above to get a path from f to f'. □

Claim 2. Consider an index i. Let s be an arbitrary time-synchronizing event in Tr_i, and let s' be an arbitrary time-synchronizing event in $\text{Tr}_{i+|\mathcal{A}|+1}$. There is a path of back edges from s' to s in Slack-Graph(Tr).

Proof. From Claim 1 there is a forward path from s to s'. Let $s_1, s_2, \ldots, s_\ell$ be the sequence of time-sychronizing events in this path. Then, by definition there is a back edge from s' to s_ℓ, from s_ℓ to $s_{\ell-1}$ and so on until s. □

Let τ_i^{max} be the largest timestamp present in Tr_i: there is an event e^i and an agent p_i participating in e^i such that $\tau_{p_i}(e_i) = \tau_i^{max}$. Let τ_i^{min} be the smallest timestamp occurring in Tr_i: there is an initial event $\perp_{q_i}$ of an agent q_i with timestamp τ_i^{min}.

Claim 3. Consider an index $i > |\mathcal{A}|$. Let $i' = i - |\mathcal{A}|$. Then: $\tau_{i'}^{max} \leq \tau_i^{min}$.

Proof. Let $\theta = \tau_{i'}^{max}$. Assume θ is witnessed by an agent p_1. Since there is a loop in $\rho_{i'+1}$, at least one agent p_2 synchronizes with p_1. Hence, by the end of $\rho_{i'+1}$, the times of $\{p_1, p_2\}$ are at least θ. Similarly, in $\rho_{i'+2}$, there is at least one agent p_3 that time-synchronizes with either of p_1 or p_2, due to the connectedness hypothesis. Hence the times of $\{p_1, p_2, p_3\}$ are greater than θ by the end of $\rho_{i'+2}$. Applying this argument iteratively shows that by the end of $\rho_{i'+|\mathcal{A}|-1}$, all agents have local-times greater than θ. This means, $\theta \leq \tau_p(\perp_p^{i'+|\mathcal{A}|})$ for all agents p. This proves the claim, since $i = i' + |\mathcal{A}|$, and τ_i^{min} equals $\tau_q(\perp_q^i)$ for some agent q. □

Claim 4. Let f^i be a maximal node of Tr_i, and q any process in $dom(f^i)$. There exists a minimal node $\perp_{p_i}^i$ of Tr_i such that: $\tau_q(f^i) \leq \tau_{p_i}(\perp_{p_i}^i) + \mathcal{O}(N^* \cdot (|\mathcal{A}|)^2 \cdot (M+1))$.

Proof. Pick a time-synchronizing node s^{i+1} in Tr_{i+1} whose domain intersects with the domain of f^i. If i is the last index, then by our assumption, f^i in itself will be a time-synchronizing node and we pick it. Otherwise, such a node exists because Tr_{i+1} has N^* events and our negotiation is connectedly communicating. Since $f^i \in \text{Tr}_i$, we have $\tau_q(f^i) \leq \tau(s^{i+1})$. Now, since Tr is tight, there is a zero slack path starting from some initial node of Tr up to s^{i+1} in Slack-Graph(Tr).

This path could zig-zag along Tr_i going far left and far right of events in i. We need to bound the amount by which this path goes to the right of i.

Let $\perp_{p_i}^i$ be the final occurrence of a minimal element of Tr_i on this path. That is, after this event, the path only passes through events which are in Tr_l for $l \geq i$. If the path visits any time-synchronizing event in $i + |\mathcal{A}| + 2$, we can take a sequence of back edges to come to s^{i+1}, thanks to Claim 2. Therefore, we can assume that the path does not pass through any event beyond $\text{Tr}_{i+|\mathcal{A}|+2}$ (recall that every back edge has slack zero). Hence, the total number of edges present in this path after $\perp_{p_i}^i$ is of the order $\mathcal{O}(N^* \cdot |\mathcal{A}|^2)$. We now apply Lemma 2 on the suffix of Tr beginning with Tr_i, noting that we have just produced a shortest weighted path of length $\mathcal{O}(N^* \cdot |\mathcal{A}|^2)$ to infer:

$$\tau(s^{i+1}) \leq \tau_{p_i}(\perp_{p_i}^i) + \mathcal{O}(N^* \cdot |\mathcal{A}|^2 \cdot (M+1)).$$

Since $\tau_q(f^i) \leq \tau(s^{i+1})$, we conclude the claim. □

Now, let $i' = i - |\mathcal{A}| - 1$. Note that, by construction, every minimal element $\perp_{p_i}^i$ of Tr_i corresponds to a maximal element of Tr_{i-1}. This allows us to iteratively apply Claim 4 to deduce that there is a minimal event $\perp_{p'}^{i'}$ in $\text{Tr}_{i'}$ such that

$$\tau_q(f^i) \leq \tau_{p'}(\perp_{p'}^{i'}) + (|\mathcal{A}| + 1) \cdot \mathcal{O}(N^* \cdot |\mathcal{A}|^2 \cdot (M+1)),$$

for any $q \in dom(f^i)$. By definition, $\tau_{p'}(\perp_{p'}^{i'}) \leq \tau_{i'}^{max}$. Secondly, taking f^i to be e^i, the event giving rise to τ_i^{max}, we infer:

$$\tau_i^{max} \leq \tau_{i'}^{max} + \mathcal{O}(N^* \cdot |\mathcal{A}|^3 \cdot (M+1))$$

Combining this inequality with Claim 3 gives $\tau_i^{max} - \tau_i^{min} \leq \mathcal{O}(N^* \cdot |\mathcal{A}|^3 \cdot (M+1))$ for all indices greater than $|\mathcal{A}|+1$. When $i < |\mathcal{A}|+1$, we apply Claim 4 iteratively as above to reach an initial node $\perp_p^0$ of Tr and conclude the bound.

Finally, to show the required bound on the drift of the run, notice that in any configuration (C, v) obtained by executing an event of Tr_i, the difference between two agents $v(t_p) - v(t_q)$ is at most $\tau_i^{max} - \tau_i^{min}$.

5 Conclusion

We have considered the notion of connectedly communicating negotiations, where in every cycle, every agent communicates with every other agent either directly or indirectly. Firstly, we have shown in Sect. 4.1 how a behaviour can be tightened in terms of reducing some unnecessary time elapses above the maximum constant. The tightening becomes challenging due to the intricate interactions among the agents. We have formalized a notion of slack, a slack graph of a trace and a way to determine the time reductions at each time-synchronizing node. Following this, we have shown that tight runs in a connectedly communicating LTN have bounded drift. The challenge in the proof lies in finding good chunks of traces where the bound between maximum and minimum timestamps lies within a bound.

The definition of connectedly communicating in [9] is weaker. A counterpart of that in our case would be: if agent p and q do not participate in a time-synchronization in a cycle of $\mathrm{MG}(\mathcal{N}_{\mathsf{untime}})$, they will never time-synchronize anytime in the future. At every marking, the set of agents can be partitioned as $\pi_1, \pi_2, \ldots, \pi_\ell$ such that no agent from an element π_i ever time-synchronizes with an agent in π_j for $i \neq j$. Notice that if $C \xrightarrow{(n,a)} C'$ in $\mathrm{MG}(\mathcal{N}_{\mathsf{untime}})$, then the partition at C' can only refine the partition at C. Hence, in any loop the partition does not change. We can now ask that for each element π_i in the partition, the communication graph should be connected. A similar change can be incorporated in the definition of drift-boundedness: at a configuration (C, v) we require that for every p, q within an element π_i of the partition corresponding to C, we need $|v(t_p) - v(t_q)|$ to be bounded. With these book-keeping overheads, our techniques can be extended to this weaker setting.

We now have a non-trivial class of local-timed negotiations (LTNs) with a decidable reachability problem. LTNs are attractive since they allow for a good amount of concurrency and restrict time-synchronizations to a select few nodes specified by the user. This suggests that LTNs should be amenable to efficient verification methods. However, as we know, an unrestricted use of time-synchronizations leads to undecidability. We hope that our work, showing decidability for a non-trivial fragment, motivates a further study of efficient algorithmics and use-cases for LTNs.

References

1. Alur, R., Yannakakis, M.: Model checking of message sequence charts. In: Baeten, J.C.M., Mauw, S. (eds.) CONCUR 1999. LNCS, vol. 1664, pp. 114–129. Springer, Heidelberg (1999). https://doi.org/10.1007/3-540-48320-9_10
2. Bengtsson, J., Jonsson, B., Lilius, J., Yi, W.: Partial order reductions for timed systems. In: Sangiorgi, D., de Simone, R. (eds.) CONCUR 1998. LNCS, vol. 1466, pp. 485–500. Springer, Heidelberg (1998). https://doi.org/10.1007/BFb0055643
3. Desel, J., Esparza, J., Hoffmann, P.: Negotiation as concurrency primitive. Acta Informatica **56**(2), 93–159 (2019)
4. Esparza, J., Kuperberg, D., Muscholl, A., Walukiewicz, I.: Soundness in negotiations. Log. Methods Comput. Sci. **14**(1) (2018)
5. Esparza, J., Muscholl, A., Walukiewicz, I.: Static analysis of deterministic negotiations. In: 32nd Annual ACM/IEEE Symposium on Logic in Computer Science, LICS 2017, Reykjavik, Iceland, 20–23 June 2017, pp. 1–12. IEEE Computer Society (2017)
6. Govind, R., Herbreteau, F., Srivathsan, B., Walukiewicz, I.: Revisiting local time semantics for networks of timed automata. In: Fokkink, W.J., van Glabbeek, R. (eds.) 30th International Conference on Concurrency Theory, CONCUR 2019, 27–30 August 2019, Amsterdam, The Netherlands. LIPIcs, vol. 140, pp. 16:1–16:15. Schloss Dagstuhl - Leibniz-Zentrum für Informatik (2019)
7. Govind, R., Herbreteau, F., Srivathsan, B., Walukiewicz, I.: Abstractions for the local-time semantics of timed automata: a foundation for partial-order methods. In: Baier, C., Fisman, D. (eds.) LICS 2022: 37th Annual ACM/IEEE Symposium on Logic in Computer Science, Haifa, Israel, 2–5 August 2022, pp. 24:1–24:14. ACM (2022)

8. Herbreteau, F., Srivathsan, B., Walukiewicz, I.: Checking timed büchi automata emptiness using the local-time semantics. In: Klin, B., Lasota, S., Muscholl, A. (eds.) 33rd International Conference on Concurrency Theory, CONCUR 2022, Warsaw, Poland, 12–16 September 2022. LIPIcs, vol. 243, pp. 12:1–12:24. Schloss Dagstuhl - Leibniz-Zentrum für Informatik (2022)
9. Madhusudan, P., Thiagarajan, P.S., Yang, S.: The MSO theory of connectedly communicating processes. In: Sarukkai, S., Sen, S. (eds.) FSTTCS 2005. LNCS, vol. 3821, pp. 201–212. Springer, Heidelberg (2005). https://doi.org/10.1007/11590156_16
10. Mukund, M., Roy, A., Srivathsan, B.: A local-time semantics for negotiations. In: Petrucci, L., Sproston, J. (eds.) FORMATS 2023. LNCS, vol. 14138, pp. 105–121. Springer, Cham (2023). https://doi.org/10.1007/978-3-031-42626-1_7

Numerical Errors in Quantitative System Analysis With Decision Diagrams

Sebastiaan Brand^(✉), Arend-Jan Quist, Richard M. K. van Dijk, and Alfons Laarman

Leiden Institute of Advanced Computer Science,
Leiden University, Leiden, The Netherlands
s.o.brand@liacs.leidenuniv.nl

Abstract. Decision diagrams (DDs) are a powerful data structure that is used to tackle the state-space explosion problem, not only for discrete systems, but for probabilistic and quantum systems as well. While many of the DDs used in the probabilistic and quantum domains make use of floating-point numbers, this is not without challenges. Floating-point computations are subject to small rounding errors, which can affect both the correctness of the result and the effectiveness of the DD's compression. In this paper, we investigate the numerical stability, i.e. the robustness of an algorithm to small numerical errors, of matrix-vector multiplication with multi-terminal binary decision diagrams (MTBDDs). Matrix-vector multiplication is of particular interest because it is the function that computes successor states for both probabilistic and quantum systems. We prove that the MTBDD matrix-vector multiplication algorithm can be made numerically stable under certain conditions, although in many practical implementations of MTBDDs these conditions are not met. Additionally, we provide a case study of the numerical errors in the simulation of quantum circuits, which shows that the extent of numerical errors in practice varies greatly between instances.

Keywords: Decision diagrams · numerical stability · floating-point numbers · quantum circuit simulation

1 Introduction

The analysis and verification of discrete, probabilistic, or quantum systems involves exploring a space of possible behaviors that is exponential in the system's description. Decision diagrams (DDs) are a data structure that can be used to symbolically represent huge subspaces of such systems [10,15,16,31], lifting the computation to this representation.

More specifically, DDs are directed graphs that represent pseudo-Boolean functions of n variables with a co-domain of real or complex numbers in the probabilistic and quantum settings. Every node in a DD represents a sub-function, and DDs achieve compactness by merging nodes that represent equivalent sub-functions. While it is possible to represent values in the co-domain exactly, using for example rational [10] or other algebraic representations [21,23], this is often not practical. In the probabilistic domain, the commonly-used value iteration algorithm becomes inefficient when using rational numbers [10], while in the quantum domain, algorithms such as the ubiquitous quantum Fourier transform cannot be represented exactly with algebraic representations such as presented in [21].

Many DD packages therefore include floating-point representation for their values [5,16,25,28,35]. However, floating-point representations come with their issues: they are often not exact, and operations on them can introduce small rounding errors. These errors not only affect the correctness of the output but can also prevent equivalent sub-functions from being recognized, thus undermining the DD's compression. Issues with floating-point errors have been reported for various types of DDs [21,25,27,34]. In the implementation of many DDs that make use of floating-point values, authors mitigate this problem by introducing a small merging threshold [25,26,28,34], which we will call δ. Two floating-point values a and a' are considered equivalent if $|a - a'| \leq \delta$. However, as the framework we introduce will show, this value merging is now a new source of potential errors, which we will call "merging errors".

To gain a better insight into the effects of floating-point errors that occur during the analysis of probabilistic and quantum systems using DDs, we perform error analysis on the matrix-vector multiplication algorithm for multi-terminal binary decision diagrams (MTBDDs). We pick matrix-vector multiplication specifically because it is ubiquitous in the verification and analysis of discrete, probabilistic, and quantum systems, where it is used to exhaustively explore the system's behavior by computing successor states under the system's transition relation [13,15,20].

When analyzing how numerical errors grow during the analysis of such systems, a metric for the size of the system is needed. Two options present themselves here. We can either consider the size of the original system description —for instance, a Bayesian network or a quantum circuit— or the size of a symbolic representation of the system's behavior —such as the states and transition relations recorded in a decision diagram. The latter is important in practice because it is often used in efficient symbolic analysis methods [15,24,35]. In the context of the analysis of combinatorial systems, previous studies on the numerical stability of linear algebra effectively consider this measure, and the relevant vectors and matrices are of exponential length in the size of the original system. The former, however, represents the more strict viewpoint, as the system's description can be exponentially more succinct than the DDs computed when exhaustively exploring its behavior. This viewpoint is therefore considered in the complexity analysis of such symbolic methods [7]. Additionally, in practice, it is arguably more relevant to understand how the error grows with the size of the system under analysis. For example, in the numerical analysis of a quantum system, it is interesting to understand how the numerical errors grow with the number of quantum bits that make up the system, or in probabilistic inference, we might consider the size of the Bayesian network. *For this reason, we study numerical stability from the stringent perspective of the original system description.*

We obtain bounds on the worst-case errors, stated in Theorem 1 for arbitrary matrices and vectors, and in Corollary 1 for probabilistic and quantum systems specifically. The bounds are parameterized in three variables: the size of the underlying system, expressed in the number of Boolean variables n used to represent its states, the maximum rounding error ε which depends on the floating-point precision, and the merging threshold δ which allows values that are δ-close to be merged.

In line with error analyses of other linear algebra algorithms [3,4], we analyze the *forward error*, which assumes the input to be exact and measures how floating-point errors accumulate in the output. Additionally, we consider *componentwise errors*, i.e. errors on the individual elements of the output vector. We consider the algorithm *forward stable* if the forward error grows at most polynomially with the size n of the underlying system.

Our theoretical findings can be summarized as follows.

1. The MTBDD matrix-vector multiplication algorithm is not componentwise forward stable in n for arbitrary vectors and matrices.
2. The MTBDD matrix-vector multiplication algorithm is componentwise forward stable in n for probabilistic and quantum systems when the merging threshold $\delta \in O(2^{-n})$, or simply $\delta = 0$.
3. The MTBDD matrix-vector multiplication algorithm is not componentwise forward stable in n for quantum or probabilistic systems if δ is constant and non-zero.

Finally, we also provide a case study of numerical errors when using MTBDDs to simulate quantum circuits, where we evaluate the effect of the choice of δ on both the DD size as well as the errors on the output. For this evaluation, we make use of an MTBDD implementation that allows for arbitrary-precision floating-point values.[1] This case study shows that, in practice, the effects of numerical errors can greatly vary between different types of instances.

2 Preliminaries

In this section, we briefly explain MTBDDs, how floating-point errors can cause issues with a DD's compactness, and the basics of floating-point error analysis.

2.1 MTBDDs

MTBDDs, also referred to as algebraic decision diagrams (ADDs) [8,9], are rooted, directed, acyclic graphs, and can be used to represent pseudo-Boolean functions $f : \{0,1\}^n \to \mathbb{D}$, where the domain $\mathbb{D}$ can for example be real or complex numbers. The function $f(x)$ can also be seen as a 2^n-dimensional vector, where x specifies an index in the vector and $f(x)$ gives the corresponding value.

---

[1] This implementation is part of the Q-Sylvan DD package and can be found online here: https://github.com/System-Verification-Lab/Q-Sylvan.

MTBDDs have two types of nodes: leaf nodes, which have a value $\mathsf{val}(\cdot) \in \mathbb{D}$, and internal nodes, which each have a variable $\mathsf{var}(\cdot) \in \{x_0, \ldots x_{n-1}\}$. MTBDDs are ordered; that is, on every path from the root to a leaf, the variables are encountered in the same order, $x_j \prec x_{j+1}$, although variables may be skipped.

An internal MTBDD node v can be understood as the Shannon decomposition of the function f^v it encodes:

$$f^v = \overline{x}_j \cdot f^v_{|0} + x_j \cdot f^v_{|1}$$

This decomposition can be read as "if $x_j = 0$ then $v[0]$, else $v[1]$", where $v[0]$ and $v[1]$ correspond to the sub-functions of f^v where x_j has been set to 0 and 1 respectively. In general we let $f_{|0}$ ($f_{|1}$) denote the sub-function of f where the first variable of f is set to 0 (1). Every path through the MTBDD corresponds to a single value of the vector/function it represents. For example, the value $f(x_0, x_1, x_2)_{|100} = f(100) = 3$ can be retrieved from the MTBDD in Fig. 1a by following the 1 (solid) edge for x_0, and the 0 (dashed) edge for x_1 and x_2.

Decision diagram algorithms are typically defined recursively, using the Shannon decomposition described above. As an example, the addition of two vectors a and b can be split up recursively as $a + b = (a_{|0} \; a_{|1})^\mathsf{T} + (b_{|0} \; b_{|1})^\mathsf{T} = (a_{|0} + b_{|0} \; a_{|1} + b_{|1})^\mathsf{T} = (c_{|0} \; c_{|1})^\mathsf{T} = c$. The corresponding MTBDD algorithm is given in Algorithm 1. Caching of previously computed values is used to avoid redundant recursive calls. If A and B are leaves, a leaf with the sum of their values is returned. The MAKELEAF function ensures that leaves are canonical, i.e. for every unique value there is only one corresponding leaf node. If a leaf with value $\mathsf{val}(A) + \mathsf{val}(B)$ already exists, that leaf is returned. Otherwise, a new leaf is created. The MAKENODE function does the same for internal nodes. This can be done efficiently since the DD is built from the bottom up, thus allowing MAKENODE to rely on the already canonical representation of child nodes.

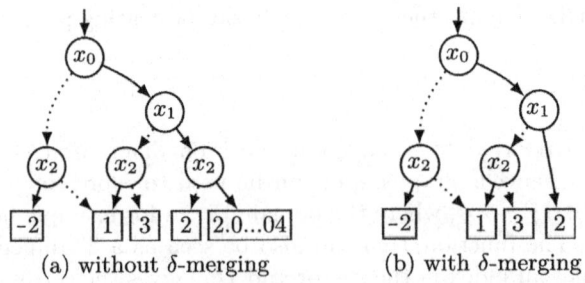

(a) without δ-merging   (b) with δ-merging

Fig. 1. An MTBDD that encodes the vector $(1 \; 1 \; \text{-}2 \; \text{-}2 \; 3 \; 1 \; 2 \; \hat{2})^\mathsf{T}$, with $\hat{2} = 2.0\ldots04$ in (a) and $\hat{2} = 2$ in (b).

Algorithm 1: Vector addition using MTBDDs.

```
1  def PLUS(MTBDD A, MTBDD B)                    ▷ assuming var(A) = var(B)
2    if A and B are leaves then                  ▷ terminal case
3      └ return MAKELEAF(val(A) + val(B))
4    if exists (R ← cache[PLUS,A,B]) then return R
5    R₀ ← PLUS(A[0], B[0])
6    R₁ ← PLUS(A[1], B[1])
7    R ← MAKENODE(var(A), R₀, R₁)
8    cache[PLUS,A,B] ← R
9    return R
```

Similarly to vector addition, matrix-vector multiplication can be split up as

$$Mv = \begin{pmatrix} M_{|00} & M_{|01} \\ M_{|10} & M_{|11} \end{pmatrix} \begin{pmatrix} v_{|0} \\ v_{|1} \end{pmatrix} = \begin{pmatrix} M_{|00}v_{|0} + M_{|01}v_{|1} \\ M_{|10}v_{|0} + M_{|11}v_{|1} \end{pmatrix}.$$

The corresponding pseudo-code is given in Algorithm 2. To efficiently access the submatrices $M_{|i,j}$, variables representing column and row indices are interleaved.

2.2 Floating-Point Values in DDs

Floating-point operations are typically not exact. For example, $0.1 + 0.2 - 0.1$ might yield 0.20000000000000004 instead of 0.2. For an MTBDD that has floating-point values in its leaves this creates a problem: new leaves are introduced that prevent the DD from staying as compact as it might have been with exact computations. To allow nodes to merge despite possible floating-point errors, the merging rule for leaves might be slightly relaxed. Specifically, we can allow two leaf nodes u and v to merge when $|\text{val}(u) - \text{val}(v)| \leq \delta$, for some choice of δ. An example is given in Figs. 1a and 1b. The allowance of approximate equality between floating-point values is used in implementations of various decision diagrams [25,28,34], and is not only used to overcome floating-point errors, but can also be explicitly used as a trade-off between errors and DD compactness [26,29].

The actual "merging" of floating-point values can be handled in different ways. Typically, MAKELEAF(a) returns a value a' that satisfies $|a' - a| \leq \delta$, if there is such an a' among existing leaves. The value a is then completely forgotten. When multiple values are δ-close to a, for example one slightly greater and one slightly less than a, which value is returned depends on the specific DD implementation. To keep our analysis agnostic of such implementation details, the only assumption we make is that MAKELEAF(a) returns a leaf node with value a', such that $|a - a'| \leq \delta$.

While floating-point rounding errors are multiplicative (i.e. the error is relative to the magnitude of the value), errors introduced by δ-merging are additive.

Algorithm 2: Matrix-vector multiplication using MTBDDs.

```
1  def MULTIPLY(MTBDD M, MTBDD V)
       ▷ M: MTBDD over 2n variables representing a 2ⁿ × 2ⁿ matrix
       ▷ V: MTBDD over n variables representing a 2ⁿ-sized vector
2      if M and V are leaves then
3          return MAKELEAF(val(M) · val(V))
4      if exists (R ← cache[MULTIPLY,M,V]) then return R
5      R₀₀ ←MULTIPLY(M[00], V[0])
6      R₁₀ ←MULTIPLY(M[10], V[0])
7      R₀₁ ←MULTIPLY(M[01], V[1])
8      R₁₁ ←MULTIPLY(M[11], V[1])
9      R₀ ←PLUS(R₀₀, R₁₀)
10     R₁ ←PLUS(R₀₁, R₁₁)
11     R ← MAKENODE(var(v), R₀, R₁)
12     cache[MULTIPLY,M,V] ← R
13     return R
```

These additive errors have a significantly greater impact than multiplicative errors, as illustrated in Table 1. Despite this, having a relative merging rule (e.g. merge a and a' when $|\frac{a}{a'} - 1| \leq \delta$) is not desirable either, since this will prevent values from merging with 0, taking away a potentially larger source of DD space reduction.

2.3 Error Analysis and Numerical Stability

The analysis of floating-point errors in algorithms is a well-established research domain. In this section, we recap the terms and definitions that are relevant to this work, for which we take [11] as a basis.

The general assumption for analyzing the error of floating-point operations is that every operation $\text{OP}(\cdot, \cdot) \in \{+, -, \times\}$ on two real or complex[2] numbers outputs the correct value up to a rounding error, i.e. the computed value $\text{OP}(a,b)_{\text{comp}}$ is equal to $\text{OP}(a,b)(1+\theta)$, for some θ with $|\theta| \leq \varepsilon$. Here, ε depends on the numerical precision of floating-point numbers [11, Eq.(2.4)] and can be assumed to scale as $\varepsilon \approx 2^{-b}$, where b is the number of bits representing the mantissa [4]. The maximum rounding error ε is typically of the order $2^{-53} \approx 10^{-16}$ ($2^{-24} \approx 10^{-7}$) for 64-bit (32-bit) floating-point arithmetic.

For a computation f, where f describes a concrete algorithm, on an input x we denote $\hat{y} = f_{\text{comp}}(x)$ as the computed output, and $y = f(x)$ as the exact

---

[2] Rounding errors for complex arithmetic can be slightly greater than for real numbers since complex multiplication requires multiple operations. However, this only affects the value of ε ($\varepsilon_{\text{complex}} \approx 2\varepsilon_{\text{real}}$) and does not affect the analysis.

output. The quality of a computed $\hat{y}$ can be measured by either its forward or backward error. The forward error is the difference $|y-\hat{y}|$ between the computed value and the exact value, while the backward error is the smallest $|\Delta x|$ such that $f(x + \Delta x) = \hat{y}$. An algorithm is called forward (backward) stable if the forward (backward) error is small, where the definition of "small" is context dependent. Alternatively put, an algorithm is forward stable if $\hat{y}$ is only slightly wrong for the correct input, and backward stable if $\hat{y}$ is the correct output for a slightly wrong (or perturbed) input.

In the context of natural sciences, where calculations are performed on empirical data, the backward error is natural to consider since the perturbations on the input due to measurement imprecision might be much larger than the numerical imprecision of the computation. In a setting where one assumes the input to be exact (up to machine precision), the forward error is arguably a more natural point of view.

Independent of whether one considers the forward or backward error, different error metrics can be used. For example, in perturbation theory of Markov chains, ℓ_1 and ℓ_2 norms are considered [1,18,19,33], although the maximum error on individual components is also used [1,33]. In the analysis of numerical errors of matrix multiplication algorithms, both errors on the individual components [11, 17] and errors on the norm [4] have been used. A componentwise (normwise) stable algorithm is an algorithm that has a small error bound on the components (norm) of the output.

2.4 Numerical Stability of Matrix Multiplication

Error analyses have been performed for various numerical algorithms, including matrix-multiplication algorithms. The traditional $O(N^3)$ algorithm is known to be componentwise stable [11, Eq.(3.13)], while, on the other hand, theoretically faster matrix-multiplication algorithms, such as the $O(N^{2.81})$ Strassen algorithm [30], cannot be componentwise stable [17] but are normwise stable [4].

3 Error Analysis

In this section, we give bounds on the worst-case error for matrix-vector multiplication with MTBDDs, expressed in the size of the underlying system they represent. We also show how errors that approach this bound can be obtained when simulating quantum circuits. Additionally, based on these bounds, we provide suggestions for the merging threshold δ and the floating-point precision such that the error remains low.

3.1 Framework

We lay out the following framework, containing the error model and a parameterization of the problem, as a foundation not just for the error analysis when systems and their subspaces are represented as MTBDDs, but also for the future error analysis when making use of different types of decision diagrams. Furthermore, the framework applies to real and complex numbers.

1. Error model:
   (a) We assume that the computed outcome of every numerical operation $\text{OP}_{\text{comp}}(\cdot,\cdot) \in \{+,-,\times,\div\}$ on two real or complex floating-point numbers yields a value equal to $\text{OP}(a,b)(1+\theta)$ with $|\theta| \leq \varepsilon$, which is the textbook assumption on floating-point errors [11]. Section 2.3 provides a more detailed discussion.
   (b) We assume an absolute merging threshold δ, i.e. MAKELEAF merges two values a and a' only if $|a - a'| \leq \delta$, thus that MAKELEAF(a) yields a leaf with value $a + \lambda$ with $|\lambda| \leq \delta$. A more detailed discussion is given in Sect. 2.2.
2. Parameterization:
   (a) n: the size of the underlying system, expressed in the number of DD variables required to encode its states.
   (b) ε: an upper bound on the floating-point rounding errors resulting from a single primitive operation $\{+,-,\times,\div\}$.
   (c) δ: an absolute threshold such that two floating-point values a and a' in the DD are considered equivalent if $|a - a'| \leq \delta$.

3.2 Error Bounds

We present componentwise forward error bounds for matrix-vector multiplication with MTBDDs, specifically for Algorithm 2, as a function of the size of the underlying system that generates the matrices and vectors. Theorem 1 gives a bound for matrices and vectors without restrictions, while Corollar 1 gives a bound for when the input is restricted to a probabilistic or quantum system. The corresponding proofs are given in [2, App. A]. Sect. 3.3 shows how the exponential terms in the error bounds can be realized.

Theorem 1. *Given a weighted transition system that has states of size n (giving rise to a 2^n-sized state space and $2^n \times 2^n$ sized transition relations), the MTBDD matrix-vector multiplication algorithm that computes successor states for a transition relation (matrix) M and a weighted set of states (vector) V yields an output such that every entry of the resulting vector has an error of at most $(n+1)\varepsilon C + \delta 2^{n+1} + O(\varepsilon^2) + O(\delta\varepsilon 2^n)$, where $C = \max_i \sum_{j=0}^{2^n-1} |M_{i,j} V_j| \leq 2^n c_M c_V$, with c_M and c_V the largest absolute values of the elements of M and V.*

Theorem 1 can be specified to a probabilistic or quantum case, where V is either a probability distribution (unit ℓ_1-norm) and M is a column stochastic matrix (preserving ℓ_1-norm), or V is a quantum state vector (unit ℓ_2-norm) and M is a unitary matrix (preserving ℓ_2-norm).

Corollary 1 (probabilistic/quantum case of Theorem 1). *When M and V are restricted to a probabilistic setting (M a stochastic matrix and V a probability distribution) or a quantum setting (M a unitary matrix and V a quantum state), every entry of the resulting vector has an error of at most $(n+1)\varepsilon + \delta 2^{n+1} + O(\varepsilon^2) + O(\delta\varepsilon 2^n)$.*

Table 1. Worst-case componentwise errors on the output of Algorithm 2, for a $2^n \times 2^n$-sized unitary matrix and a 2^n-sized vector with unit norm represented by MTBDDs, with machine precision $\varepsilon = 1.11 \cdot 10^{-16}$ (smallest epsilon for 64-bit floats), merging threshold $\delta = 10^{-15}$. Without δ-merging, i.e. $\delta = 0$, the error 2 is zero.

n	error 1: $(n+1)\varepsilon$	error 2: $\delta 2^{n+1}$
10	$1.221 \cdot 10^{-15}$	$2.048 \cdot 10^{-12}$
20	$2.331 \cdot 10^{-15}$	$2.097 \cdot 10^{-9}$
30	$3.441 \cdot 10^{-15}$	$2.147 \cdot 10^{-6}$
40	$4.551 \cdot 10^{-15}$	$2.199 \cdot 10^{-3}$
50	$5.661 \cdot 10^{-15}$	$2.252 \cdot 10^{0}$
60	$6.771 \cdot 10^{-15}$	$2.306 \cdot 10^{3}$

Theorem 1 shows an error bound that grows exponentially in n, even when $\delta = 0$. As we show in Sect. 3.3, this exponential growth can be realized under adversarial assumptions. Like [4], we call an algorithm numerically stable if the error grows at most polynomially. When choosing n as our size parameter, we thus find that for general matrices and vectors, the MTBDD matrix-vector multiplication algorithm is not componentwise stable. This instability is due to the choice of size parameter. For $N \times N$ matrices, the traditional $O(N^2)$ matrix-vector multiplication is componentwise stable in N [11, Eq.(3.12)], i.e. the error only grows polynomially in N. However, transition systems where states are of size $O(n)$ give rise to state-spaces (and thus also vectors) of size $O(2^n)$. Our analysis considers the errors to grow exponentially, not due to the algorithm itself, but due to expressing the error in terms of the underlying system size.

In probabilistic and quantum settings, MTBDD matrix-vector multiplication can be made componentwise stable in n if the merging threshold δ is chosen appropriately. This is because both of these settings restrict individual values and inner products (if computed exactly) to have a magnitude of at most 1, which makes the term $C \leq 1$. This numerical stability is achieved when setting δ exponentially small in n. When δ is kept constant, the bound grows exponentially in n, as also shown in Table 1. The reason that merging errors (the δ terms) contribute much more than floating-point errors (the ε terms) in this worst-case bound is that merging errors are additive (absolute) while floating-point errors are multiplicative (relative). As the example in the next subsection will show, when summing over exponentially many exponentially small terms, the multiplicative errors remain small when accumulated because the terms and thus also the errors are exponentially small. The size of the additive errors, however, is independent of the size of the values and will thus sum together to a term that grows linearly with the number of terms in the summation.

3.3 Adversarial Instance That Realizes Exponential Error

While the bounds presented in the previous section are worst-case bounds, i.e. they assume that all operations yield errors and that all errors accumulate without canceling out, we show here that the largest term in the quantum and probabilistic bounds, $\delta 2^{n+1}$, is tight, at least up to a factor 2. We also show that without the quantum or probabilistic restrictions on the input, exponential errors can be achieved even when $\delta = 0$.

Take the following $2^n \times 2^n$-sized matrix $H^{\otimes n}$ and 2^n-sized vector $\boldsymbol{x}$.

$$H^{\otimes n} = \frac{1}{\sqrt{2^n}} \begin{pmatrix} 1 & 1 & 1 & 1 & \cdots \\ 1 & -1 & 1 & -1 & \cdots \\ 1 & 1 & -1 & -1 & \cdots \\ 1 & -1 & -1 & 1 & \cdots \\ \vdots & \vdots & \vdots & \vdots & \ddots \end{pmatrix} \qquad \boldsymbol{x} = \frac{1}{\sqrt{2^n}} \begin{pmatrix} 1 \\ 1 \\ 1 \\ 1 \\ \vdots \end{pmatrix}$$

The matrix $H^{\otimes n}$ represents the parallel application of n of Hadamard quantum gates to n quantum bits (qubits), and $\boldsymbol{x}$ represents a quantum state (specifically a so-called uniform superposition) made up of n qubits. Both are a common occurrence in quantum computing [20]. The effect of a gate on a register of qubits can be computed through matrix-vector multiplication.

When computing $H^{\otimes n}\boldsymbol{x} = \boldsymbol{y}$, the first entry of $\boldsymbol{y}$ should equal

$$y_0 = \sum^{2^n} \frac{1}{\sqrt{2^n}} \cdot \frac{1}{\sqrt{2^n}} = \sum^{2^n} \frac{1}{2^n} = 1.$$

However, let us assume that the multiplication $\frac{1}{\sqrt{2^n}} \cdot \frac{1}{\sqrt{2^n}}$ yields both a floating point rounding error $|\varepsilon'| \leq \varepsilon$ and a merging error $|\delta'| \leq \delta$ (and that all additions happen without errors). The computed value will be

$$\hat{y}_0 = \sum^{2^n} \left(\frac{1}{\sqrt{2^n}} \cdot \frac{1}{\sqrt{2^n}}\right)_{\text{comp}} = \sum^{2^n} \left(\frac{1}{2^n}(1+\varepsilon') + \delta'\right) = 1 + \varepsilon' + \delta' 2^n,$$

which yields an error of $|y_0 - \hat{y}_0| = |\varepsilon + \delta 2^n|$. The only adversarial assumption required here is that there exists some other leaf (for example from a previous computation) with value $\frac{1}{2^n}(1+\varepsilon') + \delta'$. This results in the computed value $\frac{1}{2^n}(1+\varepsilon')$ being merged with $\frac{1}{2^n}(1+\varepsilon') + \delta'$, and thus picking up a δ' error. Since both floating-point rounding and leaf merging are deterministic, all computed values pick up the same error, ultimately accumulating in an error $\delta' 2^n$.

We can use a modified version of the example above to show exponentially large errors for the general setting (Theorem 1), even when $\delta = 0$. This modification consists of replacing the $1/\sqrt{2^n}$ terms from $H^{\otimes n}$ and $\boldsymbol{x}$ with some value a that is independent of n. By doing so, we are also not in a quantum or probabilistic setting anymore. This modification would yield an error of $|y_0 - \hat{y}_0| = |a^2 \cdot \varepsilon 2^n + \delta 2^n|$, which grows as $O(2^n \varepsilon)$ when $\delta = 0$. An error of $O(n\varepsilon 2^n)$ can be obtained by also assuming errors on the additions.

3.4 Parameter Suggestions

Next, we discuss suggestions for the merging threshold δ and the required floating-point precision in the probabilistic and quantum settings. These values

Table 2. Values for δ such that the error is bounded by 10^{-3} and 10^{-6}, assuming 64-bit floating-point values, giving $\varepsilon = 1.11 \cdot 10^{-16}$

n	δ s.t. error $\leq 10^{-3}$	δ s.t. error $\leq 10^{-6}$
10	$4.883 \cdot 10^{-7}$	$4.883 \cdot 10^{-10}$
20	$4.768 \cdot 10^{-10}$	$4.768 \cdot 10^{-13}$
30	$4.657 \cdot 10^{-13}$	$4.657 \cdot 10^{-16}$
40	$4.547 \cdot 10^{-16}$	$4.547 \cdot 10^{-19}$
50	$4.441 \cdot 10^{-19}$	$4.441 \cdot 10^{-22}$

affect both the error as well as the number of nodes in the DD. The suggested parameter values provide theoretical guarantees on the maximum error, but only serve as a heuristic for the number of DD nodes.

The bound $|y_i - \hat{y}_i| \leq (n+1)\varepsilon + \delta 2^{n+1} + O(\varepsilon^2) + O(\delta\varepsilon 2^n)$ can be rewritten to obtain the following suggestion for the merging threshold δ, given a certain amount of allowed error $\geq (n+1)\varepsilon$.

$$\delta < \frac{\text{allowed-error} - (n+1)\varepsilon}{2^{n+1}} \quad (1)$$

If we want the errors on the output to grow at most polynomially with n, δ must thus be set as $\delta \in O(2^{-n})$. Table 2 shows concrete values of δ for different errors allowed.

However, since the purpose of δ-merging is to keep the DD compact despite floating-point rounding errors, we wish that at least individual rounding errors are smaller than δ. As mentioned in Sect. 2.3, floating-point values with b significant bits give a relative rounding error $\varepsilon \approx 2^{-b}$. This means that for values with magnitude in the order of 2^k, the absolute errors resulting from floating-point imprecision are of the order of $2^k \cdot 2^{-b} = 2^{k-b}$. For δ-merging to achieve its goal, we need $\delta > 2^{k-b}$. In the probabilistic and quantum settings, where the magnitude of values never exceeds 1, we can set $k = 0$, which leaves $\delta > 2^{-b}$, or

$$b > \log_2(\delta^{-1}). \quad (2)$$

4 Case Study

In this section, we perform a case study of the effect of different values of δ on both the numerical error and the size of the DDs during the simulation of several quantum circuits.

4.1 Experimental Setup

Using the decision diagram package Q-Sylvan, we run quantum circuit simulation on a selection of quantum circuits from the MQT Bench [22] benchmark set.[3]

---

[3] Our setup, including instructions on how to reproduce the results, can be found online at https://github.com/System-Verification-Lab/mtbdd-benchmarks, and has been permanently archived at https://doi.org/10.5281/zenodo.15322337.

Algorithm 3: Quantum circuit simulation using MTBDDs.
---
1 **def** SIMULATE(Circuit U)
     ▷ U: n-qubit quantum circuit with gates $U_1,\ldots,U_m$
2      $S \leftarrow$ CREATEINITIALMTBDD(n)   ▷ create MTBDD for $(1\ 0\ 0\ \cdots\ 0)^\mathsf{T}$
3   **for** $k = 1$ **to** m **do**
4          $M \leftarrow$ CREATEMTBDD(U_k)   ▷ create MTBDD for matrix U_k
5          $S \leftarrow$ MULTIPLY(M, S)
6   **return** S

A quantum circuit is a sequence of gates, defined by matrices, that act on an initial state, described by a vector. As a convention, the initial state for an n-qubit quantum circuit is the 2^n-sized vector $(1\ 0\ 0\ \cdots\ 0)^\mathsf{T}$. If a circuit U needs to run on a different input state $\boldsymbol{x}$, the circuit that creates $\boldsymbol{x}$ can simply be prepended to U. For a quantum circuit with m gates $U_1,\ldots,U_m$, and an initial state $\boldsymbol{x}$, simulating the quantum circuit can be done by computing $U_m\ldots U_1\boldsymbol{x}$. This is typically computed as $U_m(\ldots(U_1\boldsymbol{x}))$, since computing $(U_m\ldots U_1)\boldsymbol{x}$ tends to be much harder computationally, including when using decision diagrams [14].

Since computing the effect of a single quantum gate is not a typical use case in quantum computing, and because computing the matrix that represents the whole quantum circuit is computationally infeasible, we choose to evaluate the errors when simulating the quantum circuit gate-by-gate. The corresponding pseudo-code is given in Algorithm 3. While this yields results that show accumulated errors over multiple matrix-vector multiplications, rather than the errors on individual matrix-vector multiplications that our bounds apply to, such an evaluation is more representative of the actual use cases of DDs in quantum computing.

We pick three types of circuits from the MQT Bench set where MTBDDs are able to achieve good compression for the final state: the Deutsch-Jozsa (DJ) algorithm, exact quantum phase estimation (QPE), and W-state preparation. For each of these three types of circuits, the matrices that make up the circuits are categorically different, which results in different behaviors of the errors. An overview of the matrix elements for each type of circuit is given in Table 3.

As a proxy for the ground truth, we use the result from a floating-point computation that uses 128 bits to represent the mantissa (with $\delta = 0$) by making use of the GNU MPC library [6], which allows for arbitrary precision floating-point complex numbers. For the non-ground-truth computations, we use MPC values that have a 24 and 53-bit mantissa, equivalent to IEEE 754 single- and double-precision floats, which gives $\varepsilon \approx 10^{-7}$ and $\varepsilon \approx 10^{-16}$ respectively. The error metric is the largest absolute deviation from the 128-bit output, i.e. the reported error $= \max_j |y_j^{(\text{128-bit})} - \hat{y}_j|$.

Table 3. Overview of matrix elements that occur in each type of quantum circuit.

circuit	types of matrix elements
Deutsch-Jozsa	$\{0, \pm 1, \pm 1/\sqrt{2}, \pm \cos(\pi/4)\}$
Quantum phase estimation (QPE)	$\{0, \pm 1, \pm 1/\sqrt{2}, e^{i\theta}\}$, with most $\theta = -\pi/2^k$ with $k \in \{1, 2, \ldots, n-2\}$ and some random θ's
W-state preparation	$\{0, \pm 1, \cos(\theta), \sin(\theta)\}$, for a great variety of θ's given as finite-precision decimal numbers

4.2 Results

We focus on the results that use a 53-bit mantissa (equivalent to 64-bit doubles), shown in Fig. 2. We first discuss the results for each circuit category separately, after which we discuss general observations.

First, for the Deutsch-Jozsa circuits we find that the numerical errors are very small and mostly independent of δ and the number of qubits. This is no great surprise, as the circuits and resulting states are effectively discrete, i.e. discrete up powers of $1/\sqrt{2}$. The errors for $\delta = 10^{-3}$ and $n \geq 19$ are likely due to terms $(1/\sqrt{2})^n$ and $(1/\sqrt{2})^{n-1}$ becoming δ-close, and thus incorrectly merge. It is interesting to note that the small errors for $\delta > 0$ are not errors at all, but rather the result of a slight imprecision in the 128-bit ground truth computation: while $\cos(\pi/4)$ should equal $1/\sqrt{2}$, the computed values differ slightly from each other. When $\delta = 0$, as it is for the ground-truth computation, occurrences of $1/\sqrt{2}$ and $-\cos(\pi/4)$ fail to perfectly cancel each other out, leaving a result with several very small values that should instead equal 0. Setting $\delta > 0$ allows for the occurrences of $\cos(\pi/4)$ to merge with $1/\sqrt{2}$ and very small values to be merged with 0. This shows that for such "almost discrete" instances, having a small non-zero δ not only benefits the DD size, but can also correct small errors in the computation.

Next, for the QPE circuits, the $e^{i\theta}$ terms complicate the computation in two ways: the random θ's increase the number of unique values that do not have an exact floating-point representation, and θ's of the form $\theta = -\pi/2^k$ introduce values that are exponentially small in n. When either δ or n becomes too big, the exponentially small values are merged with 0, which causes significant errors in the result. The coincidence between erroneous instances and large DDs is caused by the errors destroying the symmetry that allows the DDs to stay compact.

Finally, the W-state circuits contain many different values that do not have exact floating-point representations. Under an exact computation, these values should all multiply and sum together such that the resulting vector only contains values in $\{0, 1/\sqrt{n}\}$. However, the complexity with which this happens leaves a lot of opportunity for errors to be introduced. As δ and n increase, the errors for the W-state circuit increase more gradually than for the QPE circuit. This is likely due to the W-state circuit producing more unique values during the computation, resulting in merging errors occurring more frequently but affecting fewer values when they do.

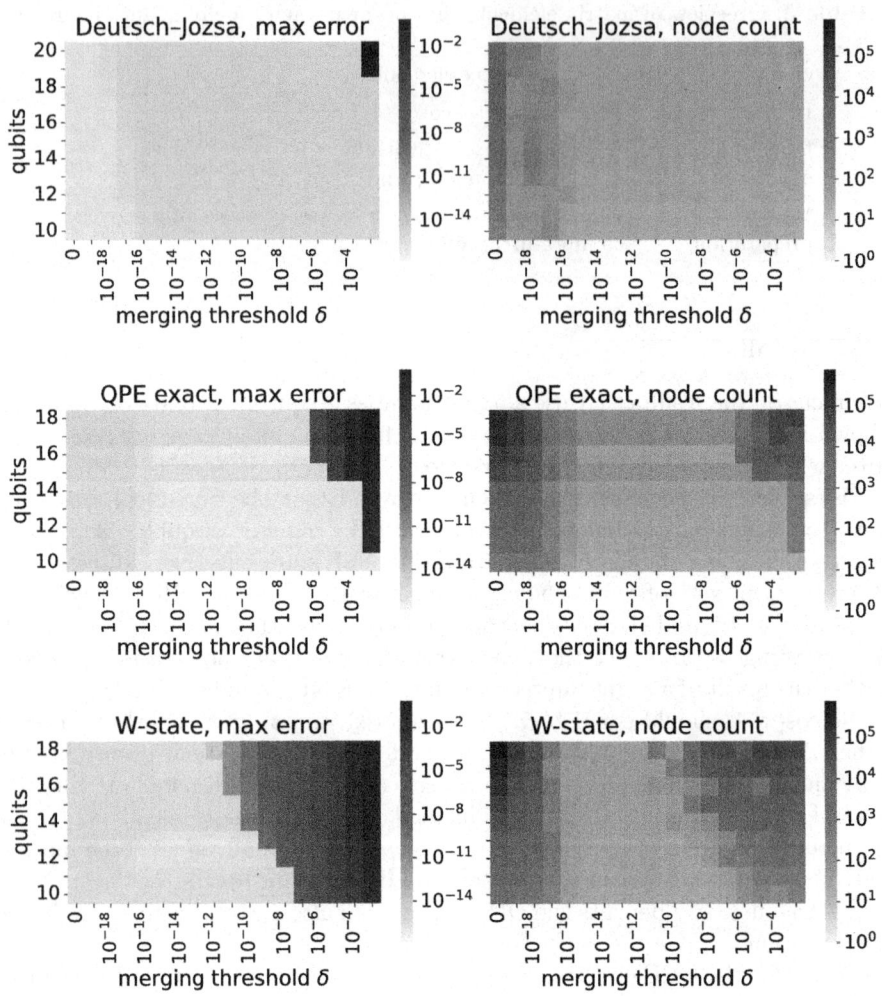

Fig. 2. Errors and number of DD nodes of the final state for different circuits, with $b = 53$ mantissa bits, equivalent to double precision floats.

For all three circuits, we find that when $\delta = 0$ the errors are small (as also discussed in Sect. 3.2), but the number of DD nodes is also significantly greater than it should be if the calculation was exact. As discussed in Sect. 2.2, this is caused by sub-functions that should theoretically be equal no longer being recognized as such due to small floating-point errors. This illustrates that in order to benefit from the compression DDs provide, some sort of δ-merging must be used when working with floating-point values.

The issue with $1/\sqrt{2}$ and $\cos(\pi/4)$ not being recognized as equal also highlights that the exact choice of gates matters. For example, it is likely that decomposing a Toffoli gate (a 3-qubit gate made up of 1's and 0's) into single- and two-qubit

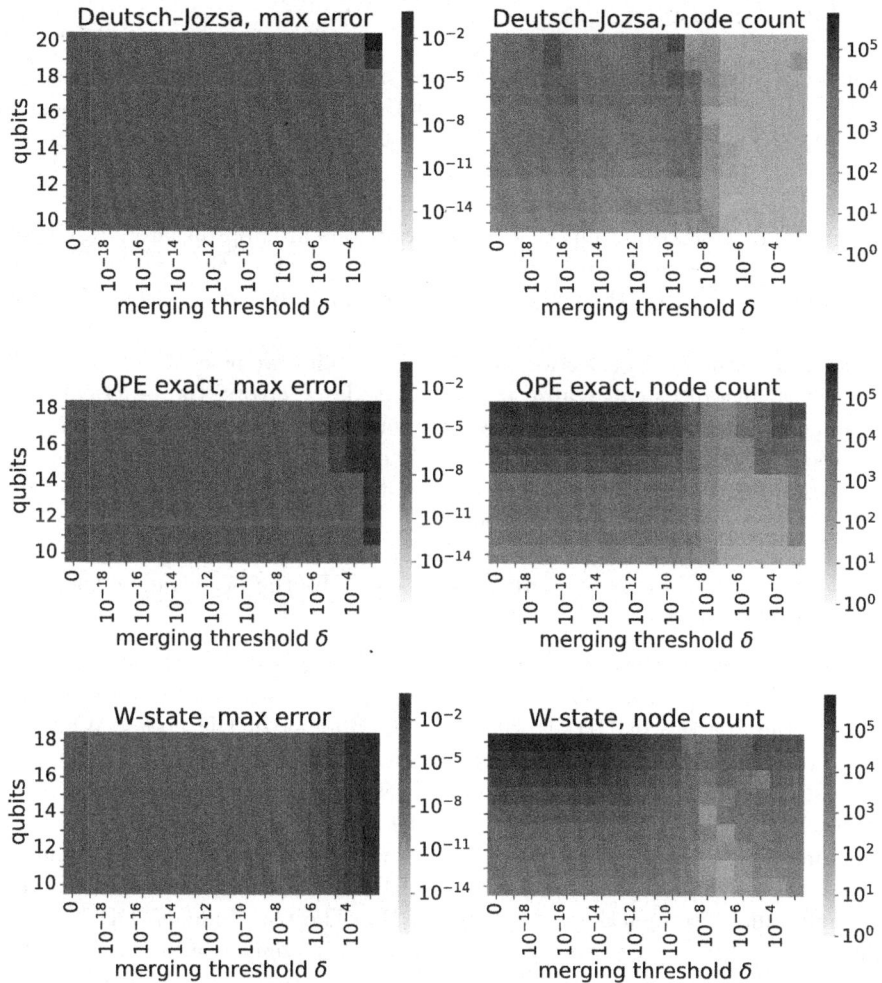

Fig. 3. Errors and number of DD nodes of the final state for different circuits, with $b = 24$ mantissa bits, equivalent to single precision floats.

gates (which contain terms $1/\sqrt{2}$ and $e^{i\pi/4}$) before simulation yields significantly worse numerical errors.

Figure 3 shows errors and DD sizes for MPC values that use a 24-bit mantissa (equivalent to 32-bit floats). It is unsurprising that a lower precision results in greater errors. However, considering the node counts in Figs. 2 and 3, we find that the range of δ that yields compact DDs gets wider (narrower) when the number of mantissa bits increases (decreases). Additionally, the node counts for the QPE and W-state circuits show that this range decreases as n increases. This advocates for DD implementations that can increase the precision of the floating-point numbers as n increases, to maximize their compactness.

5 Conclusion

Summary. In this paper, we analyzed the numerical stability of the matrix-vector multiplication algorithm for MTBDDs, which is a crucial component in the exhaustive exploration of discrete, probabilistic, and quantum systems. We found that the numerical errors that can accumulate during the execution of this algorithm can be kept polynomially small relative to the size n of the system under investigation, only if certain parameters are appropriately restricted. Specifically, for quantum and probabilistic systems, to keep the errors from growing exponentially in n, the merging threshold should be set exponentially small in n.

In a case study on a selection of quantum circuits, we have shown that while adversarial instances can realize exponentially large errors, the practical effect of numerical errors can greatly vary between different types of instances. The empirical results also suggest that at least in quantum use cases, a good heuristic for setting the merging threshold δ is to set it as small as possible while keeping $\delta > 2^{-b}$, where b is the number of bits in the mantissa of the floating-point representation. Ideally, b can grow with the system size n, however, when restricted to 64-bit doubles, setting δ to 10^{-15} appears to be the best trade-off between errors and DD size.

Future Work. For our error bounds, we have picked the size of (the states of) the underlying system as the size parameter, however, another natural size parameter for the analysis of DD algorithms is the number of DD nodes. While our error analysis is completely independent of the number of nodes, it would be interesting to investigate if a different analysis can be made, either for MTBDDs or other DDs, that relates the error and the number of DD nodes.

While our results show both theoretical and empirical insights into the numerical errors in computations done with MTBDDs, other types of DDs exist that can offer more compactness than MTBDDs. These include EVDDs [16,35], CFLOBDDs [27], and LIMDDs [12,32]. Our error analysis and the characterization of the problem in terms of the number of variables n, the floating-point errors ε, and the merging threshold δ can serve as a foundation for the analysis of other DD algorithms as well as other types of DDs.

References

1. Abbas, K., Berkhout, J., Heidergott, B.: A critical account of perturbation analysis of Markov chains. Markov Process. Related Fields **22**, 227–265 (2016)
2. Brand, S., Quist, A.J., van Dijk, R.M., Laarman, A.: Numerical errors in quantitative system analysis with decision diagrams. arXiv preprint (2025)
3. Demmel, J., Dumitriu, I., Holtz, O.: Fast linear algebra is stable. Numer. Math. **108**(1), 59–91 (2007)
4. Demmel, J., Dumitriu, I., Holtz, O., Kleinberg, R.: Fast matrix multiplication is stable. Numer. Math. **106**(2), 199–224 (2007)
5. van Dijk, T., Van de Pol, J.: Sylvan: multi-core framework for decision diagrams. Int. J. Softw. Tools Technol. Transfer **19**, 675–696 (2017)

6. Enge, A., Gastineau, M., Théveny, P., Zimmermann, P.: GNU MPC (2022). https://www.multiprecision.org/
7. Feigenbaum, J., Kannan, S., Vardi, M.Y., Viswanathan, M.: Complexity of problems on graphs represented as OBDDs. In: Morvan, M., Meinel, C., Krob, D. (eds.) STACS 1998. LNCS, vol. 1373, pp. 216–226. Springer, Heidelberg (1998). https://doi.org/10.1007/BFb0028563
8. Frohm, R.I.B.E.A., Hachtel, C.M.G.G.D.: Algebraic decision diagrams and their applications. In: International Conference on Computer Aided Design (1993)
9. Fujita, M., McGeer, P.C., Yang, J.Y.: Multi-terminal binary decision diagrams: an efficient data structure for matrix representation. Formal Methods Syst. Des. **10**, 149–169 (1997)
10. Hensel, C., Junges, S., Katoen, J.P., Quatmann, T., Volk, M.: The probabilistic model checker Storm. Int. J. Softw. Tools Technol. Transfer 1–22 (2022)
11. Higham, N.J.: Accuracy and Stability of Numerical Algorithms. SIAM (2002)
12. Hong, X., Dai, A., Gao, D., Li, S., Ji, Z., Ying, M.: LimTDD: a compact decision diagram integrating tensor and local invertible map representations. arXiv preprint arXiv:2504.01168 (2025)
13. Katoen, J.P.: The probabilistic model checking landscape. In: LICS, pp. 31–45 (2016)
14. Matsunaga, Y., McGeer, P.C., Brayton, R.K.: On computing the transitive closure of a state transition relation. In: Proceedings of the 30th International Design Automation Conference, pp. 260–265 (1993)
15. McMillan, K.L.: Symbolic model checking: an approach to the state explosion problem. Ph.D. thesis, Carnegie Mellon University (1992)
16. Miller, D.M., Thornton, M.A.: QMDD: a decision diagram structure for reversible and quantum circuits. In: 36th International Symposium on Multiple-Valued Logic (ISMVL 2006), p. 30. IEEE (2006)
17. Miller, W.: Computational complexity and numerical stability. In: Proceedings of the Sixth Annual ACM Symposium on Theory of Computing, pp. 317–322 (1974)
18. Mitrophanov, A.Y.: Stability and exponential convergence of continuous-time Markov chains. J. Appl. Probab. **40**(4), 970–979 (2003)
19. Negrea, J., Rosenthal, J.S.: Approximations of geometrically ergodic reversible Markov chains. Adv. Appl. Probab. **53**(4), 981–1022 (2021)
20. Nielsen, M.A., Chuang, I.L.: Quantum Computation and Quantum Information. Cambridge University Press, Cambridge (2010)
21. Niemann, P., Zulehner, A., Drechsler, R., Wille, R.: Overcoming the tradeoff between accuracy and compactness in decision diagrams for quantum computation. IEEE Trans. Comput. Aided Des. Integr. Circuits Syst. **39**(12), 4657–4668 (2020)
22. Quetschlich, N., Burgholzer, L., Wille, R.: MQT Bench: benchmarking software and design automation tools for quantum computing. Quantum (2023). MQT Bench is available at https://www.cda.cit.tum.de/mqtbench/
23. Quist, A.J., Coopmans, T., Laarman, A.: Exact quantum decision diagrams scaling in the T-count (2025, in preparation)
24. Sang, T., Beame, P., Kautz, H.A.: Performing Bayesian inference by weighted model counting. In: AAAI, vol. 5, pp. 475–481 (2005)
25. Sanner, S., McAllester, D.: Affine algebraic decision diagrams (AADDs) and their application to structured probabilistic inference. In: IJCAI, vol. 2005, pp. 1384–1390 (2005)
26. Sanner, S., Uther, W.T., Delgado, K.V., et al.: Approximate dynamic programming with affine ADDs. In: AAMAS, pp. 1349–1356 (2010)

27. Sistla, M., Chaudhuri, S., Reps, T.: Symbolic quantum simulation with Quasimodo. In: CAV, pp. 213–225. Springer, Cham (2023)
28. Somenzi, F.: CUDD: CU decision diagram package release 3.0.0 (2015). https://web.archive.org/web/20171208230728/http://vlsi.colorado.edu/~fabio/CUDD/cudd.pdf
29. St-Aubin, R., Hoey, J., Boutilier, C.: APRICODD: approximate policy construction using decision diagrams. In: Advances in Neural Information Processing Systems, vol. 13 (2000)
30. Strassen, V.: Gaussian elimination is not optimal. Numer. Math. **13**(4), 354–356 (1969)
31. Viamontes, G.F., Markov, I.L., Hayes, J.P.: High-performance QuIDD-based simulation of quantum circuits. In: Proceedings Design, Automation and Test in Europe Conference and Exhibition, vol. 2, pp. 1354–1355. IEEE (2004)
32. Vinkhuijzen, L., Coopmans, T., Elkouss, D., Dunjko, V., Laarman, A.: LIMDD: a decision diagram for simulation of quantum computing including stabilizer states. Quantum **7**, 1108 (2023)
33. Yin, G., Zhang, H.: Singularly perturbed Markov chains: limit results and applications. Ann. Appl. Probab. **17**, 207–229 (2007)
34. Zulehner, A., Hillmich, S., Wille, R.: How to efficiently handle complex values? Implementing decision diagrams for quantum computing. In: 2019 IEEE/ACM International Conference on Computer-Aided Design (ICCAD), pp. 1–7. IEEE (2019)
35. Zulehner, A., Wille, R.: Advanced simulation of quantum computations. IEEE Trans. Comput. Aided Des. Integr. Circuits Syst. **38**(5), 848–859 (2018)

Modeling Uncertainty: From Simulink to Stochastic Hybrid Automata

Pauline Blohm[1]($\boxtimes$), Felix Schulz[1], Lisa Willemsen[2], Anne Remke[1,2], and Paula Herber[1,2]

[1] University of Münster, Münster, Germany
{pauline.blohm,anne.remke,paula.herber}@uni-muenster.de,
felix.s.schulz@gmx.de
[2] University of Twente, Enschede, The Netherlands
l.c.willemsen@utwente.nl

Abstract. Simulink is widely used in industrial design processes to model increasingly complex embedded control systems. Thus, their formal analysis is highly desirable. However, this comes with two major challenges: First, Simulink models often provide an idealized view of real-life systems and omit uncertainties such as, aging, sensor noise or failures. Second, the semantics of Simulink is only informally defined. In this paper, we present an approach to formally analyze safety and performance of embedded control systems modeled in Simulink in the presence of uncertainty. To achieve this, we 1) model different types of uncertainties as stochastic Simulink subsystems and 2) extend an existing formalization of the Simulink semantics based on stochastic hybrid automata (SHA) by providing transformation rules for the stochastic subsystems. Our approach gives us access to established quantitative analysis techniques, like statistical model checking and reachability analysis. We demonstrate the applicability of our approach by analyzing safety and performance in the presence of uncertainty for two smaller case studies.

Keywords: Simulink · Stochastic Hybrid Automata · Uncertainty

1 Introduction

Embedded control systems require high functionality and flexibility, especially since they are increasingly used in safety-critical environments, such as cars, airplanes, or energy control systems. Thus, formal verification is desirable to ensure their safety, performance and resilience. Model-driven development tools such as MATLAB Simulink allow to graphically model and simulate complex hybrid control systems, i.e. systems that combine discrete and continuous behavior. One aspect that is often omitted when modeling real-life systems is their inherent uncertainty, e.g. caused by aging, sensor noise or failures.

This research is partly funded by the DFG project RealySt (471367371).

Furthermore, simulation executes the system only for selected inputs and sound statistical methods like statistical model checking (SMC) are required to provide stochastic guarantees [11]. However, existing approaches for reachability analysis and SMC of Simulink models either rely on a transformation of the Simulink model into a formal representation or perform SMC directly on the Simulink model. While the former approaches enable formal verification and even provide formal guarantees about crucial properties, they mostly disregard uncertainties and probabilistic behavior, thus, the results become useless in the presence of real-life effects such as aging, noise or failures. As Simulink does neither offer specification techniques for more complex path properties, nor hypothesis testing, it cannot be directly used for SMC. Existing approaches, e.g. [39], apply SMC directly on the Simulink model, or include SMC methods in Simulink [1]. However, statistical evaluations in Simulink are very costly due to the high overhead of simulations in Simulink. Furthermore, this does not provide a formal model which would be amenable to formal verification, like reachability analysis.

In this paper, we present an approach to model uncertainties in Simulink and formalize them using stochastic hybrid automata (SHA). We build on previous work, where we proposed a modular and extensible transformation from Simulink to SHA, which gives us access to quantitative analysis methods. Previously, the transformation was only amenable for simplified failure-repair models which illustrates the potential of introducing stochasticity into Simulink models. Our contribution in this paper is twofold: 1) We provide a library of Simulink subsystems for different types of stochastic behavior to capture uncertainties. 2) We provide a formalization of the presented subsystems via a transformation into a dedicated SHA formalism that accommodates the stochastic extensions. Then, we seamlessly integrate the new transformation rules into our existing transformation from Simulink to SHA. The transformation into SHA makes the whole approach amenable to both formal reachability analysis techniques and SMC. We demonstrate the feasibility of our approach by using the SMC tool modes [10] on two small case studies, namely a temperature control system with sensor losses, and a simple energy measurement unit with stochastic switching.

The rest of this paper is structured as follows: In Sect. 2, we introduce the necessary background. In Sect. 3, we propose Simulink subsystems that are designed to model different types of uncertainties. We present their respective formalization via SHA templates in Sect. 4 and evaluate our approach in Sect. 5. Finally, we summarize related work in Sect. 6 and conclude in Sect. 7.

2 Background

This section introduces the necessary background for the remainder of this paper, namely Simulink and stochastic hybrid automata (SHA) and our transformation from Simulink to SHA.

2.1 Simulink

Simulink [33] is an industrially well established graphical modeling language for hybrid systems. It comes with a tool suite for simulation and automated code

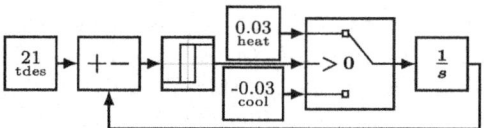

Fig. 1. Simulink model of a temperature control system.

generation. Simulink models consist of blocks that are connected by discrete or continuous signals via ports. The Simulink block library provides a large set of predefined blocks, from arithmetics over control flow blocks to integrators and complex transformations. Together with the MATLAB library, linear and non-linear differential equations can be modeled and simulated. Furthermore, the Simulink library provides random blocks to sample values from a probability distribution. Simulink also provides the user with the option to define custom masks for subsystems, effectively allowing the user to create subsystems that can be parameterized and used like regular Simulink blocks.

Example. Figure 1 shows a Simulink model of a temperature control system, which aims to keep the temperature in the room close to the desired temperature *tdes*. Heating and cooling rates are modeled as constant blocks *heat* and *cool*. The system switches to heating if the temperature is below a specified lower threshold, and to cooling if it's above and upper threshold. A *relay* block is used to prevent rapid switching, i.e. the system only switches if the temperature deviation is above a given tolerance.

2.2 Stochastic Hybrid Automata

SHA are an extension of hybrid automata (HA) with stochastic behavior. HA [4] allow to capture the interaction of discrete and continuous behavior. Formally, they are defined in [23] as follows:

Definition 1 (Hybrid Automata). *A* hybrid automaton (HA) *is a tuple* $\mathcal{H} = (Loc, Var, Flow, Inv, Lab, Edge, Init)$ *with components:*

- *Loc is a non-empty finite set of* locations *or* control modes.
- $Var = \{x_1, \ldots, x_d\}$ *is a finite ordered set of* variables. *We call* $\nu \in \mathcal{V}$ *a* valuation, *and* $\sigma = (l, \nu) \in Loc \times \mathcal{V} = \Sigma$ *a* state *of* $\mathcal{H}$.
- $Flow : Loc \to (\mathcal{V} \to \mathcal{V})$ *specifies for each location its* flow *or* dynamics.
- $Inv : Loc \to 2^{\mathcal{V}}$ *specifies an* invariant *for each location.*
- $Lab = \{a_1, \ldots, a_k\}$ *is a non-empty finite ordered set of* labels.
- $Edge \subseteq Loc \times Lab \times 2^{\mathcal{V}} \times (\mathcal{V} \to \mathcal{V}) \times Loc$ *is a finite set of* edges. *For an edge* $(l, a, g, r, l') \in Edge$, *l and l' are its* source *resp.* target locations, *a its label, g its* guard, *and r its* reset. *Guards need to be disjoint for each pair of edges with identical source location and label.*
- $Init : Loc \to 2^{\mathcal{V}}$ *defines* initial *valuations for each location. We call a state* $(l, \nu) \in \Sigma$ initial *if* $\nu \in Inv(l) \cap Init(l)$.

Different formalisms exist to integrate stochastic behavior into HA. Here, we extend the definition of *decomposed HA with eager non-predictive specification* (DHA) from [34,35] with a *reset kernel* that allows to stochastically set the valuation of a continuous variable according to the current state of the automaton. For the required preliminaries from probability theory we refer to [9].

Definition 2 (HA with stochastic kernels). *A hybrid automaton with stochastic kernels (HAwK) is a tuple* $\mathcal{A} = (\mathcal{H}, \Psi, \Psi^R)$ *with* $\Psi = (\Psi_1, \ldots, \Psi_k)$, *where:*

- $\mathcal{H} = (Loc, Var, Flow, Inv, Lab, Edge, Init)$ *a HA with* $|Var| = d$.
- $\Psi_i : \mathcal{B}(\mathbb{R}_{\geq 0}) \times \Sigma \to [0,1]$, $i = 1, \ldots, k$, *where* $k = |Lab|$, *are continuous stochastic kernels from* $(\Sigma, \mathcal{B}(\Sigma))$ *to* $(\mathbb{R}_{\geq 0}, \mathcal{B}(\mathbb{R}_{\geq 0}))$, *called* delay *kernels*.
- $\Psi^R : \mathcal{B}(\mathbb{R}^d) \times (\Sigma \times Lab) \to [0,1]$ *is a continuous stochastic kernel from* $((\Sigma \times Lab), \mathcal{B}((\Sigma \times Lab)))$ *to* $(\mathbb{R}^d, \mathcal{B}(\mathbb{R}^d))$, *called* reset *kernel*.

The execution semantics of a HAwK follows the semantics of a DHA [35]. Similarly to DHA, a HAwK $\mathcal{A} = (\mathcal{H}, (\Psi_1, \ldots, \Psi_k), \Psi^R)$ extends the underlying HA $\mathcal{H}$ with k so-called *random clocks* $c_1, \ldots, c_k$. In each state, the i-th random clock evolves with rate 1 if an edge associated with label a_i is enabled and with rate 0 otherwise. The random clock c_i is reset to 0 if an edge associated with label a_i is scheduled. During the execution of a HAwK, the expiration time of the random clock c_i is sampled based on the delay kernel Ψ_i for the associated edge and stored in a vector $\mathcal{R}$. An edge is taken if one random clock reaches its indicated expiration time, i.e. if $c_j = \mathcal{R}[j]$, for $1 \leq j \leq k$. For completeness, we provide a summary of the formal construction of DHA as defined in [35] in [9]. As an extension to DHA, HAwK include an additional *reset step*, which directly follows each discrete step and immediately resets the continuous state of $\mathcal{H}$ according to the probability distribution given by the reset kernel Ψ^R.

In the following, we assume that the function $w(i)$ specifies the index of the random clock which reached its indicated delay for the i-th step of the execution. We denote the valuation of the continuous variables from $\mathcal{H}$ as $\sigma.\nu_\mathcal{H}$. For a *probability density function (PDF)* $f : \mathbb{R}_{\geq 0} \to \mathbb{R}_{\geq 0}$ we define its *support* as $supp(f) = \{\omega \in \mathbb{R} \mid f(\omega) > 0\}$.

Definition 3 (Semantics of HAwK). *A path* π *of a given HAwK* $\mathcal{A} = (\mathcal{H}, (\Psi_1, \ldots, \Psi_k), \Psi^R)$, *has the form* $\pi = (\sigma_0, \mathcal{R}_0) \xrightarrow{t_0} (\sigma'_0, \mathcal{R}_0) \xrightarrow{a_{w(0)}} (\sigma'_1, \mathcal{R}_1) \xrightarrow{r_0} (\sigma_1, \mathcal{R}_1) \xrightarrow{t_1} \ldots$ *such that*

- $\sigma_i \xrightarrow{t} \sigma'_i \xrightarrow{a_{w(0)}} \sigma_{i+1}$ *is governed by* $\mathcal{H}$,
- $\mathcal{R}_i \in \mathbb{R}^k_{\geq 0}$ *for all* $0 \leq i \leq len(\pi)$,
- $\mathcal{R}_0[j] \in supp(\mathrm{Dist}_{\sigma_0}^{\Psi_j})$ *for all* $j \in \{1, \ldots, k\}$,
- $\nu'_i(c_{w(i)}) = \mathcal{R}_i[w(i)]$, $\nu'_i(c_j) \leq \mathcal{R}_i[j]$, $\mathcal{R}_{i+1}[w(i)] \in supp(\mathrm{Dist}_{\sigma_{i+1}}^{\Psi_{w(i)}})$ *and* $\mathcal{R}_{i+1}[j] = \mathcal{R}_i[j]$ *for all* $0 \leq i < len(\pi)$ *and* $j \in \{1, \ldots, k\} \setminus \{w(i)\}$,
- $\sigma_i.\nu_\mathcal{H} \in supp(\mathrm{Dist}_{(\sigma_i \times w(i))}^{\Psi^R})$, *and*

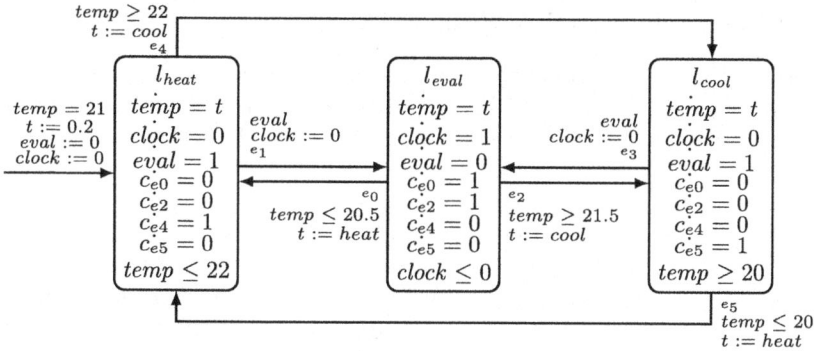

Fig. 2. Simple Temperature Control Unit given as a HAwK.

– if π is finite and it ends with a time step $(\sigma_i, \mathcal{R}_i) \xrightarrow{t_i} (\sigma'_i, \mathcal{R}_i)$ then $\sigma'_i(c_j) \leq \mathcal{R}_i[j]$ for all $j \in \{1, \ldots, k\}$.

To ease notation, we write $\Psi^R_{var_i}(\sigma, a)$ to indicate the probability measure specified by Ψ^R for the continuous variable $var_i \in Var$ in state $\sigma \in \Sigma$ at edges with label $a \in Lab$.

Example. The HAwK shown in Fig. 2 models a simple temperature control unit that can either heat (l_{heat}) or cool (l_{cool}) the room with rate t that depending on the state of the model is either chosen from $\mathcal{U}(0.1, 0.3)$ or from $\mathcal{U}(-0.1, -0.3)$. If the temperature $temp$ is too high the system switches to cooling, and to heating if it gets to cold. After a $\mathcal{N}(8, 1)$-distributed delay, we move to l_{eval} where the temperature is compared to given bounds. Then, if we move to l_{heat} or l_{cool}, the rate t is resampled according to the reset kernel. For each edge, we specify a delay kernel, which characterizes the distribution of the expiration time of the random clock associated with the edge. Here, the delay kernels for the edges e_i for $i \in \{0, 2, 4, 5\}$ can be specified as $\Psi_{e_i} \sim \mathcal{D}(d_i)$ and for $i \in \{1, 3\}$ as $\Psi_{e_i} \sim \mathcal{N}(8, 1)$. For $i \in \{0, 2\}$, i.e. for the urgent edges, $d_i = 0$. For $i \in \{4, 5\}$, the delays d_i are resolved such that $\int_0^{d_i} t\, dx + \nu(temp) = \theta_i$, where θ_i is the bound of the temperature at the discrete edge, i.e. $\theta_4 = 22$ for e_4 and $\theta_5 = 20$ for e_5 in our example. The random clocks $eval$, c_{e0}, c_{e2}, c_{e4} and c_{e5} track the enabling time of the corresponding edge. Further, the reset kernels for each variable at each edge can be defined for all states $\sigma \in \Sigma$ as follows: $\Psi^R_{temp}(\sigma, e_i) \sim \mathcal{D}(temp)$, $\Psi^R_{clock}(\sigma, e_i) \sim \mathcal{D}(clock)$ for $i \in \{0, \ldots, 5\}$, $\Psi^R_t(\sigma, e_i) \sim \mathcal{D}(t)$ for $i \in \{1, 3\}$, $\Psi^R_t(\sigma, e_i) \sim \mathcal{U}(0.1, 0.3)$ for $i \in \{0, 5\}$, and $\Psi^R_t(\sigma, e_i) \sim \mathcal{U}(-0.1, -0.3)$ for $i \in \{2, 4\}$. To ease notation, we do not explicitly state the stochastic kernels that follow a Dirac distribution in the remainder of this paper, as their definition directly follows from the specification of the underlying HA, as illustrated in this example. Thus, we omit (i) random clocks (c_i for $i \in \{0, 2, 4, 5\}$ in Fig. 2), (ii) kernels specifying the stochastic delays following a Dirac distribution and (iii) the definition of the stochastic reset kernel for Dirac distributed resets.

2.3 Formalizing Simulink Using SHA

To formally analyze Simulink models, we have previously proposed a modular transformation from Simulink to a subclass of SHA, namely *linear hybrid automata with random clocks* (*LHAC*) [8]. Note that HAwK are a conservative extension of *LHAC* with stochastic kernels, so each *LHAC* can easily be translated into a HAwK by specifying the corresponding delay kernel for each random clock. The key idea of our transformation from Simulink to *LHAC* is as follows: The Simulink model is separated into the singular blocks and the signal flow. Each block is transformed independently using transformation rules defined by so-called *SHA templates*. SHA templates are given as *LHAC*sync, which extend *LHAC* by introducing synchronization labels as well as distinguishing between *input* and *output* variables and also relax the definition of the *LHAC* s' flow and initial state. While output variables are used to model the signal driven by the corresponding block and thus have a known flow and initial value, input variables represent the signal lines connected to the inport. Therefore, they do not have a given flow or initial value. As a result, the SHA templates do not have a defined execution semantics. To maintain the execution order and correctly map the output variables to their corresponding input variables, a discrete-event synchronization via *synchronization mappings* is derived from the signal lines. Then, the SHA templates together with the synchronization mappings are composed using a modified parallel composition which results in a monolithic SHA following the *LHAC* formalism. This automaton can then be analyzed with established tools for quantitative analysis, e.g. REALYST [19] or MODESTTOOLSET [10].

3 Modeling Uncertainty in Simulink

Modeling real-life systems enables us to simulate component interaction and system behavior. However, models often portray an idealized view of the real-life system and omit uncertainties like aging, sensor noise or failures. To bridge this gap, we design Simulink subsystems that model different kinds of uncertainties using probability distributions. The subsystems are masked, i.e. they can be used in the same way as standard Simulink blocks, and different parameters and settings can be used to conveniently adjust them in a simple user interface.

In [7], we have identified the following sources of uncertainties in cyber-physical systems: measurement errors, noise, component failures, failed memory accesses, bit flips, clock errors and skews, and uncertain effects of chosen control values as well as uncertain physical effects. Conceptually, all of these uncertainties can be modeled via stochastic sampling of a signal or clock value. Measurement errors like noise can be modeled by adding a random value to a base signal. Failures can be modeled via a stochastic timeout after which a component fails. Bit flips and failed memory accesses can be modeled using stochastic switching between correct and failed bit or memory accesses. Clock errors or clock skews can be modeled using stochastic sampling, where sampling times are randomly chosen. Uncertain effects of control values as well as uncertain physical effects

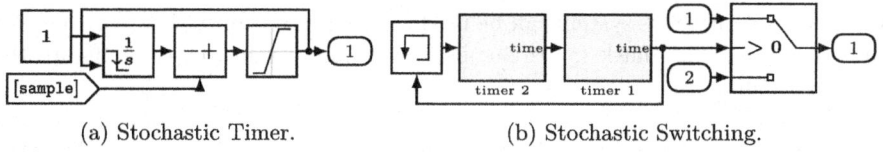

(a) Stochastic Timer.   (b) Stochastic Switching.

Fig. 3. Simulink Subsystems for Timer and Switching.

can be similarly modeled as noise and failures using stochastic noise or stochastic timeouts. Overall, to accommodate these uncertainties, we provide Simulink subsystems for *Stochastic Timers*, *Stochastic Sampling*, *Stochastic Noise*, and *Stochastic Switching*. Furthermore, as many of these effects are worsening over time, we provide subsystems for *Discrete Aging* and *Continuous Aging*.[1]

Stochastic Timer. The subsystem shown in Fig. 3a models a timer whose expiration time is sampled from a probability distribution. The stochastic timer is used in most of the stochastic subsystems presented in the following to model a randomly distributed time delay. It functions as follows: Upon a trigger, an expiration time is sampled from a probability distribution and an integrator block is reset to zero. The integrator block functions as a clock by integrating over the value 1. Subtracting the clock value from the expiration time gives the current value of the timer, provided as the output. Once the timer reaches zero, the falling edge re-triggers this subsystem. The stochastic timer supports selecting between a uniform distribution, with configurable minimum and maximum parameters, or a folded-normal distribution with $mean = 0$ and a configurable variance. The seed for the random number generators can be set in the mask.

Stochastic Switching. The subsystem shown in Fig. 3b models a stochastic extension of the switch block provided by Simulink. In contrast to the regular switch block, the switching condition for the stochastic switch does not depend on a third signal but rather on stochastic delays provided by two stochastic timers. While the value of the first timer is greater than zero, i.e. it has not yet expired, the signal provided at the first inport is passed to the outport. Similarly, while the value of the second timer is greater than zero, the signal provided at the second inport is passed to the outport. Once a timer expires, an expiration time for the other timer is sampled according to its distribution, which is specified in the mask. The stochastic switch will initially sample an expiration time for the first timer and therefore output the signal at the first inport.

Stochastic Sampling. This subsystem shown in Fig. 4a models a random sampling of a signal. A stochastic timer is used to provide randomly distributed sampling times. When the timer expires, the value of the output signal is updated to the current value of the input signal and the timer is triggered to sample a new expiration time. While the timer is running, the output signal holds the latest

---

[1] The .slx-files of the subsystems are provided in the artifact.

value. Exemplary, this system can be used to model a sensor with a non-constant sampling rate. The mask for this subsystems allows the user to configure the stochastic timer.

Stochastic Noise. The subsystem shown in Fig. 4b models a signal distorted by noise sampled from a normal distribution. This subsystem uses Simulinks' random number generator to sample random numbers from a normal distribution at specified discrete intervals. The random number is then either multiplied with a constant noise factor or with the input signal to calculate the amount of distortion which is then added to the input signal. The subsystems mask allows the designer to enable and disable the noise, specify the noise factor, mean and variance for the distribution as well as the sample time and a seed.

Stochastic Discrete Aging. The subsystem shown in Fig. 5a models a signal that is degraded by repeatedly reducing the input signal by an aging factor until a lower bound is reached that triggers a repair of the signal. The aging factor is increased every time a stochastic timer expires. The lower bound is derived from the maximum number of times the factor can be decreased. Once this bound is reached, a repair is triggered whose duration is provided by a stochastic timer. During the repair, either an explicit repair signal is passed to the outport or a specified percentage of the input signal. The mask allows the designer to specify a multitude of options such as the underlying distributions for the reduction time steps and the repair time, the maximum number of reduction steps or whether an explicit repair signal is used.

Stochastic Continuous Aging. The subsystem shown in Fig. 5b also models signal degradation up to a specified lower bound. However, in this case the signal is degraded linearly based on the time passed since the last repair. Similarly to the discrete aging, once a lower bound is reached a repair of the signal is triggered and during the repair the designer can choose whether an explicit repair signal or a specified percentage of the input signal is passed to the outport. Additionally, the signal degradation can also be paused and resumed based on a stochastic timer. The mask enables the designer to customize the distributions of the different timers, whether an external repair signal is used and the rate of degradation.

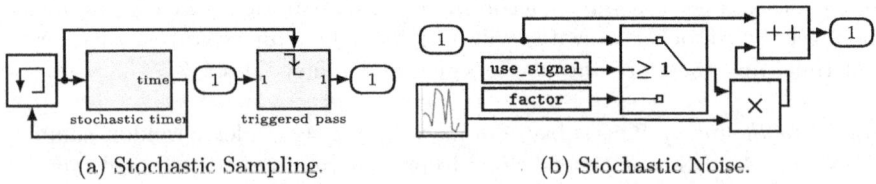

(a) Stochastic Sampling. (b) Stochastic Noise.

Fig. 4. Simulink Subsystems for Sampling and Noise.

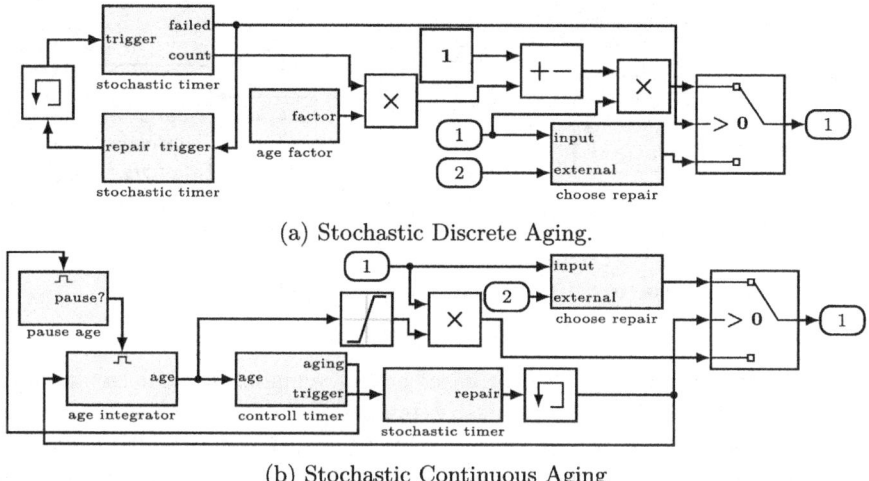

Fig. 5. Simulink Subystems for Discrete and Continuous Aging.

4 Formalizing Stochastic Subsystems Using SHA

To formalize the presented Simulink subsystems, we use SHA and build upon an existing modular transformation from Simulink to SHA presented in Sect. 2.3. In particular, we provide individual transformation rules that translate each of the Simulink subsystems into a SHA template. To enable seamless integration of these transformation rules into our existing transformation, we first lift the parallel composition of SHA templates from *LHAC*sync to *HAwK*sync to account for the extended stochastic behavior. In the following, we first explain this lifting and then the individual transformation rules for each Simulink subsystem.

4.1 Lifting SHA Template Composition to HAwK

While the concept of our transformation from Simulink to SHA is generally applicable to a wide range of automata classes, some adjustments are necessary to correctly compose the more complex stochastic behavior of HAwK. First, to define SHA templates we introduce *HAwK*sync analogously to the *LHAC*sync used in [8]. Intuitively, *HAwK*sync extend the definition of HAwK by introducing synchronization labels to indicate sending and receiving of variables and splitting the variable set into input and output variables, and relax it by only requiring flow and initial value for output variables. For *HAwK*sync, the delay and reset kernels are also only defined for output variables. A formal definition is provided in [9]. Note, that analogous to *LHAC*sync, *HAwK*sync also do not have an execution semantics.

To correctly compose the *HAwK*sync using the previously proposed transformation, the rules for synchronized and non-synchronized edges presented in [8] need to be lifted to include the stochastic kernels. Intuitively, the rules are

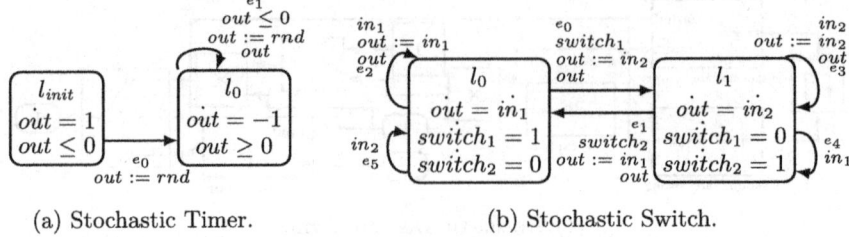

(a) Stochastic Timer.   (b) Stochastic Switch.

Fig. 6. SHA Templates for Timer and Switch.

extended as follows: For non-synchronized edges we maintain the delay and reset kernels defined for this edge in the SHA templates. For synchronized edges, we assign the continuous kernel from the sending edge for the resulting synchronized edge. When assigning the reset kernel to the new edge we combine all reset kernels from the individual edges. This is possible as the reset kernel of a *HAwK* sync only considers the output variables which are unique for each SHA template. For a more formal definition please see [9]. The result of the composition is a monolithic SHA which follows the definition of a HAwK.

4.2 Individual Transformation Rules for Each Stochastic Subsystem

We present SHA templates that formalize the semantics of the stochastic Simulink subsystem presented in Sect. 3. The graphical illustrations indicate random clocks or sampling from non-Dirac distributions in pink, receiving of the corresponding variable in light blue and sending of the corresponding variable in green. Parameters provided by the Simulink block are `represented` accordingly. If not stated otherwise, $Init(l) = \{\emptyset, \texttt{false}\}$ for any location l.

Stochastic Timer. To formalize the stochastic timer subsystem (see Fig. 3a), we define the SHA template shown in Fig. 6a. The variable *out* represents the value of the output signal which is initially sampled from the distribution specified in the mask of the subsystem. To realize this initial sampling, the urgent location l_{init} is added where no time can pass and the immediate edge to l_0 assigns the value of *out*. This value then decreases with a rate of -1 until it reaches zero, which results in taking the self-loop where a new random value is assigned to *out* and the discrete update of *out* is sent. As the Simulink block does not have any input, the template also does not have any input variables. The reset kernel is defined as $\Psi^R_{out}(\sigma, e_i) \sim \texttt{Dist}$ for $i \in \{0,1\}$ where $\texttt{Dist} = \mathcal{U}(\texttt{low}, \texttt{high}) \mid \mathcal{N}_{\geq 0}(\texttt{var})$ for all states σ and $Init(l_{init}) = (\{out = 0\}, \texttt{true})$.

Stochastic Switch. To formalize the stochastic switch subsystem (see Fig. 3b), we define the SHA template shown in Fig. 6b. Each location represents one of the two cases for the switch, i.e. either the variable *out* has the same value as in_1 or as in_2. Switching between these locations depends on the random clocks *switch1*

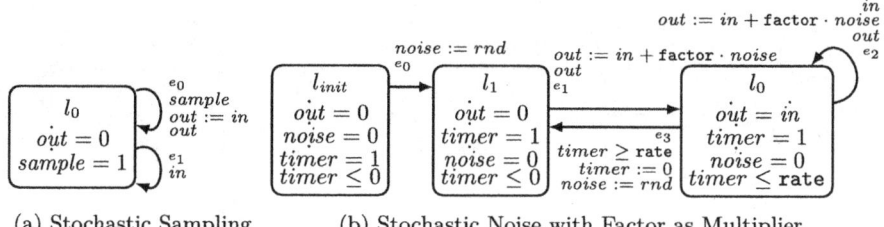

(a) Stochastic Sampling   (b) Stochastic Noise with Factor as Multiplier.

Fig. 7. SHA Templates for Sampling and Noise.

and *switch2*. Initially, the value of in_1 is assigned to *out*. Once the expiration time for *switch1* is reached, the edge to l_1 is taken immediately upon which the value of *out* is updated to in_2. Similarly, once the expiration time for *switch2* is reached, the edge back to l_0 is taken. To catch discrete updates of the two input variables, self-loops are added that process this information. The delay kernel is defined as $\Psi_{e_0}(\sigma) \sim \text{Dist}_1, \Psi_{e_1}(\sigma) \sim \text{Dist}_2$ with $\text{Dist}_i = \mathcal{U}(\text{low}, \text{high}) \mid \mathcal{N}_{\geq 0}(\text{var})$ for $i \in \{1, 2\}$ and all states $\sigma \in \Sigma$. The initial state is given by $Init(l_0) = (\{out = in_1, switch1 := 0, switch2 := 0\}, \text{true})$.

Stochastic Sampling. To formalize the stochastic sampling subsystem (see. Fig. 4a), we define the SHA template shown in Fig. 7a. The variable *out* represents the output of the subsystems, the random clock *sample* effectively models the random timer. The expiration time of *sample* is sampled from the distribution specified in the mask. Once the expiration time is reached the stochastic edge is taken which causes *out* to be updated to the current value of *in*. This discrete update is sent and a new expiration time for *sample* is sampled. As the subsystem has one input, discrete updates of the corresponding variable *in* have to be received, however, this does not affect the value of *out* or *sample* and does not result in a location change. The delay kernel is defined as $\Psi_{e_0}(\sigma) \sim \text{Dist}$ with $\text{Dist} = \mathcal{U}(\text{low}, \text{high}) \mid \mathcal{N}_{\geq 0}(\text{var})$ for all states $\sigma \in \Sigma$. The initial state is given by $Init(l_{init}) = (\{out = in, sample := 0\}, \text{true})$.

Stochastic Noise. To formalize the stochastic noise subsystem (see Fig. 4b), we define two SHA template: one where the noise is multiplied with a constant factor and one where the rate of distortion depends on the input signal. Exemplary, we show the SHA template for the former in Fig. 7b. The value of the variable representing the output, i.e. *out*, is initially set to the initial value of the input plus a randomly-distributed noise. Again, we use an initial location to avoid sampled values in the initial state. *out* then evolves with the same rate as the input and every **rate** time units, a new value for the variable *noise* is sampled according to a normal distribution. As the resets on transitions are non-deterministic, we use an urgent location l_1 to ensure that the reset of *out*, where the noise is added to the current value of the input signal *in*, is executed after sampling the noise. Discrete changes of the input variable *in* are handled at the

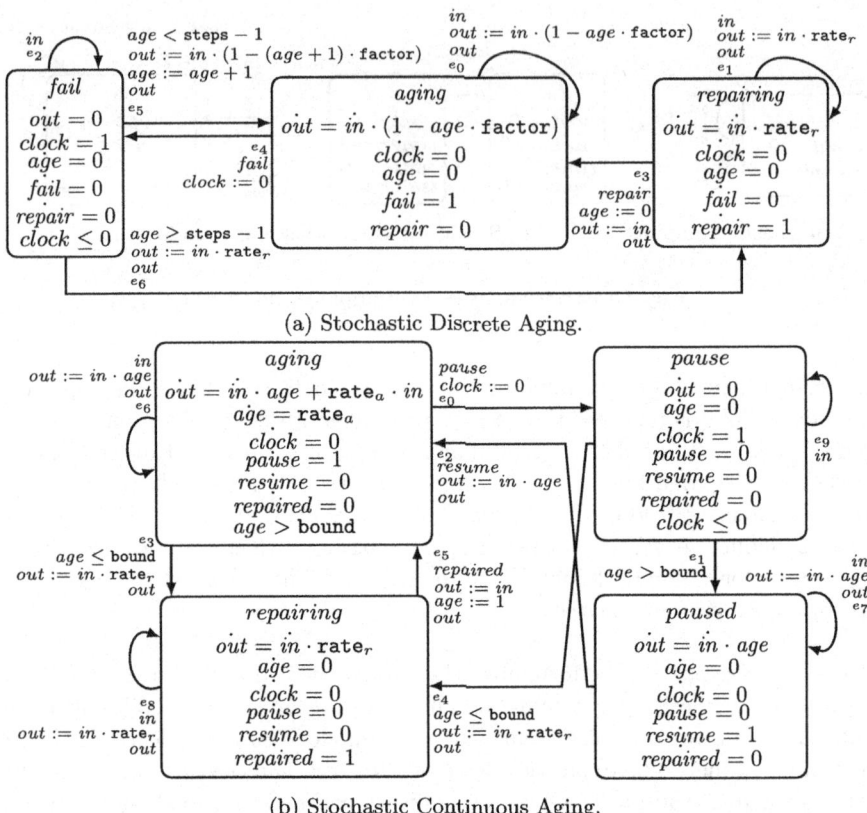

Fig. 8. SHA Templates for Aging without Explicit Repair Signal.

self-loop of l_0 by updating the value of *out* accordingly. The reset kernel is given by $\Psi^R_{noise}(\sigma, e_j) \sim \mathcal{N}(\text{mean}, \text{var})$ for $j \in \{0, 3\}$ and all states σ. The initial state is given by $Init(l_{init}) = (\{out = 0, timer := 0, noise := 0\}, \text{true})$. For the SHA template where the rate depends on the input signal we refer to [9].

Stochastic Discrete Aging. To formalize the stochastic discrete aging subsystem (see Fig. 5a), we define two SHA templates, one for the case without an explicit repair signal and one with the repair signal. Exemplary, we show the SHA template without using an explicit repair signal in Fig. 8a. The variable *out* represents the output of the subsystem which initially equals the value of *in*, i.e. the value of the input signal. Once the random clock *fail* reaches its sampled expiration time, the variable *age* is increased which controls the rate of the degradation. The value of *out* is then updated to *in* multiplied by the current aging rate. Once max degradation is reached, the expiration of *fail* triggers the repair of the signal whose duration depends on the expiration time of *repair*. Once the signal is repaired, *out* is assigned the value of *in* without degradation. Discrete changes of the input signal are handled by the self-loops. The delay kernel is defined as $\Psi_{e_3}(\sigma) \sim \text{Dist}_1, \Psi_{e_4}(\sigma) \sim \text{Dist}_2$ with $\text{Dist}_i = \mathcal{U}(\text{low}, \text{high}) \mid \mathcal{N}_{\geq 0}(\text{var})$ for $i \in \{1, 2\}$ and all states σ. The initial state is given by $Init(l_{aging}) = (\{out = in, fail := 0, repair := 0, age = 0, clock = 0\}, \text{true})$. The SHA template using an explicit repair signal is shown in [9].

Stochastic Continuous Aging. To formalize the stochastic continuous aging subsystem (see Fig. 5b), we again define two SHA templates. Exemplary, we show the SHA template for the case without using an explicit repair signal shown in Fig. 8b. Similarly to the discrete aging, the variable *out* represents the output of the subsystems which is degraded over time until a lower bound is reached. In contrast to the discrete aging, the rate of the degradation is not increased at discrete time points but reduces continuously, represented by the variable *age* that evolves with a specified $rate_a$. Additionally, the aging can be paused and resumed, which is controlled via the two random clocks. Again, discrete updates of *in* are received via the self-loops in all locations and handled via updating *out*. The delay kernel is defined as $\Psi_{e_0}(\sigma) \sim \text{Dist}_1, \Psi_{e_2}(\sigma) \sim \text{Dist}_2, \Psi_{e_5}(\sigma) \sim \text{Dist}_3$ with $\text{Dist}_i = \mathcal{U}(\text{low}, \text{high}) \mid \mathcal{N}_{\geq 0}(\text{var})$ for $i \in \{1, 2, 3\}$ and all states σ. The initial state is given by $Init(l_{aging}) = (\{out = in \cdot age, pause := 0, resume := 0, repair := 0, age = 1, clock = 0\}, \text{true})$. The SHA template using an explicit repair signal is shown in [9].

5 Evaluation

To demonstrate the feasibility of our approach, we provide quantitative results for two small case studies. The first is a modified version of the temperature control system shown in Fig. 1, where sensor losses are modeled as delayed sampling. The second is a simplified energy measurement unit, which stochastically switches between low and high loads. To formally analyze both systems, we have used the transformation presented in [8] extended as outlined in Sect. 4.

Then, we have used the tool modes [10] from the MODESTTOOLSET to apply statistical model checking, which provides us with statistical guarantees in the form of confidence intervals (CI). We compare the CI provided by modes with a simulation-based evaluation of the Simulink model. While Simulink does not provide CI as-is, we use the implementation presented in [1] to compute them based on the Wilson score [36]. For both tools we use a confidence level of $\lambda = 0.95$ and the Wilson score to compute the CI. We use a time horizon of

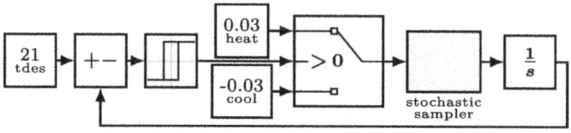

Fig. 9. Simulink Model for the Temperature Control System with Loss.

Table 1. Results for Temperature Control System with Sensor Loss.

Tool		$P(\Diamond\ tmp \leq 20)$	$P(\Diamond\ tmp \leq 20.2)$	$P(\Diamond\ tmp \leq 20.4)$	$P(\Diamond\ tmp \leq 20.5)$
Simulink	CI	[0.0652, 0.0795]	[0.4454, 0.4730]	[0.7262, 0.7506]	[0.9992, 1]
	midpoint	0.0723	0.4592	0.7384	0.9996
modes	CI	[0.0497, 0.05885]	[0.4463, 0.4663]	[0.7322, 0.7498]	[0.9997, 1]
	midpoint	0.0541	0.4562	0.7409	1

$t = 100$ and perform 5000 runs in the Simulink model and 9704 for modes, as the significantly faster runtime of modes allow us to perform more runs. In Simulink a fixed-step solver with a step size $s = 0.05$ is used and for modes either a uniform scheduler (temperature control) or a ASAP scheduler (energy consumption) is used.

Temperature Control System. The temperature control system with stochastic sensor loss is shown in Fig. 9. It is similar to the example we have used in the introduction to Simulink (cf. Fig. 1). To model a sensor with stochastic loss, i.e. that the sensor can only successfully read the temperature at stochastically chosen time points, we use a stochastic sampling block. The corresponding HAwK is shown in Fig. 10. Please note that we have applied some optimizations to eliminate redundant or unused variables, locations and edges. The delay kernel is defined as $\Psi_{e_2} = \Psi_{e_3}(\sigma) \sim \mathcal{U}(10, 20)$ for all states σ and $Init(l_{cool}) = \{temp = 21, sample := 0, rate = -0.03\}$. We use a PCTL-like notation to express the properties that a temperature stays below a given threshold, e.g., $P(\Diamond\ temp \leq 20)$ gives the probability that a $temp$ of 20 or lower is reached during the observed time frame. Table 1 shows that the CI provided by modes are tighter and lie within the CI computed with the Simulink model. As expected, a temperature of 20.5 is reached almost certainly, whereas, a temperature of 20 is quite unlikely, as the controller aims to keep the temperature at 21°. On average, the analysis performed with modes was significantly faster with only 0.3 s, whereas the Simulink evaluation exceeded 30 min (1802 s).

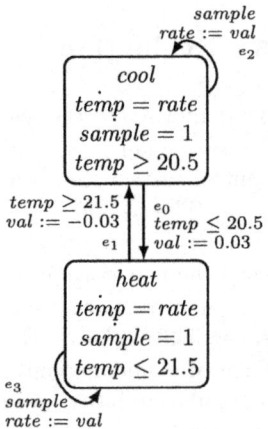

Fig. 10. Temperature Control as HAwK.

Energy Measurement Unit. The second case study is inspired by an energy measurement unit. The Simulink model shown in Fig. 11 uses two stochastic switches that both model a unit that can stochastically switch between a high or low load. Two integrators are used to measure the total energy consumed and the amount

Table 2. Results for the Energy Measurement Unit.

Tool		$P(\lozenge\ load \geq 10)$	$P(\lozenge\ load \geq 30)$	$P(\lozenge\ total \geq 13000)$	$P(\lozenge\ total \geq 16000)$
Simulink	CI	[0.8934, 0.9099]	[0.0000, 0.0008]	[0.8895, 0.9063]	[0.0050, 0.0097]
	midpoint	0.9017	0.0004	0.8979	0.0074
modes	CI	[0.8975, 0.9094]	[0, 0.0005]	[0.8914, 0.9036]	[0.0043, 0.0074]
	midpoint	0.9034	0	0.8975	0.0057

of time passed with a maximum load, i.e. with both consumers in high load mode. The corresponding HAwK is shown in Fig. 12. The four different locations reflect the four different states of energy consumption: 1) both units have a low load(l_0) 2) only one unit has high load (l_1 and l_2) or 3) both units have high load (l_3). The delay kernel is defined as $\Psi_{e_0} = \Psi_{e_2}(\sigma) \sim \mathcal{U}(10, 20)$, $\Psi_{e_1} = \Psi_{e_3} = \Psi_{e_4} = \Psi_{e_6}(\sigma) \sim \mathcal{U}(5, 10)$, and $\Psi_{e_5} = \Psi_{e_7}(\sigma) \sim \mathcal{U}(5, 15)$ for all states σ and $Init(l_0 = \{total = 0, max = 0, switch_{1,1} = 0, switch_{1,2} = 0, switch_{2,1} = 0, switch_{2,2} = 0\}$.

We analyzed whether the total energy consumption and the time spent at max load exceed certain thresholds. Table 2 shows that the CI provided by modes are tighter and lie within the CI provided by Simulink for the first three properties. For the fourth property $P(\lozenge total \geq 16000)$, the CI provided by modes is also tighter but there is only a large overlap. As the probabilities for this property are very low, we suspect that slight differences in the sampling process might be the cause. Further investigation is needed to better understand the causes. The average runtime using modes was significantly faster with only 0.3 s while Simulink exceeded 12 min (736 s).

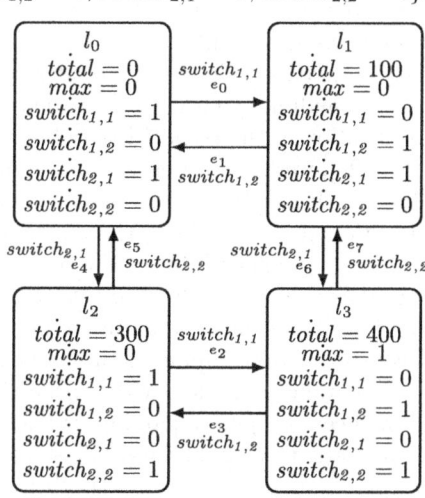

Fig. 12. Energy Measurement Unit as a HAwK.

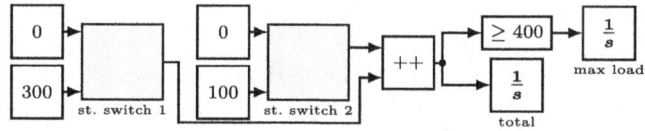

Fig. 11. Simulink Model for the Energy Measurement System.

6 Related Work

Different works have investigated hybrid systems under uncertainty modeled in Simulink, e.g. [13,14,25,38]. In [25], a stochastic timer is used to model uncertain operation times. [14] presents a power plant with a failure-repair model, where they also consider degradation over time. [38] presents a model of a controller for an air-craft elevator system and introduce random failures to three hydraulic circuits based on Poisson processes. [13] presents a Simulink model of a human heart with stochastic delay between subsequent heartbeats. While these approaches model and analyze relevant uncertainties, they are rather specific to the used case studies. Additionally, all of these works perform simulation within Simulink and are not amenable for a sophisticated quantitative analysis.

There have been quite some efforts to enable the formal verification of hybrid systems modeled in Simulink, e.g. [12,15,27,28,30,37]. In [15], the authors propose the tool CheckMate to model HA in Simulink, which can then be formally verified via reachability analysis. Similarly, in [30] the authors present a transformation from a subset of Simulink to the HA dialect SpaceEx [21]. However, they focus on techniques for a special class of systems and do not provide general transformation rules for a broader set of blocks. In [12,37], the authors transform Simulink models with Stateflow parts into Hybrid CSP and enable the verification in the Hybrid Hoare Logic Prover. Finally, in our own previous work [2,27,28], we have presented a transformation from Simulink into the differential dynamic logic [31], which enables deductive verification using KeYmaera X [22]. Additionally, in [5] the authors propose a modular correct-by-construction approach for embedded systems where HA are first modeled and verified and then translated and embedded into Simulink/Stateflow models. However, all of these methods focus on the qualitative analysis of safety properties, and none of them take stochastic components into consideration.

To evaluate models with stochastic behavior, SMC approaches for Simulink have been proposed [26,39]. While [39] is based on Bayesian statistics and hypothesis testing, [26] uses Plasma lab with Monte Carlo simulation for probability estimation. However, both model uncertainties in an ad hoc manner for a specific scenario and rely on expensive simulations in Simulink. Furthermore, their approach is not amenable to quantitative analysis beyond SMC. In [20], the authors propose a transformation from Simulink into stochastic timed automata that can be analyzed with UPPAAL SMC [16]. However, they do not consider stochastic blocks and transform the Simulink models into a deterministic STA model where all probabilities are one.

Our transformation from Simulink into SHA requires a well-defined formalism that can express the hybrid and stochastic behavior present in Simulink. HA [3, 24] have been extended with stochastic components as e.g. stochastic timed automata [6], or singular and rectangular automata with random events [17,18, 32]. However, these approaches do not allow us to model the complex continuous behavior present in Simulink. More general classes for stochastic hybrid models are considered in [29], which even allow for stochastic differential equations to express stochastic noise. However, to the best of our knowledge, this class is not directly amenable for the computation of reachability probabilities.

7 Conclusion

In this paper, we have presented an approach to model uncertainties in Simulink by providing a library of stochastic subsystems. This library allows us to capture uncertainties like noise or aging and model real-life systems more accurately. We have also presented transformation rules for these subsystems to formally capture their behavior via SHA templates and seamlessly integrated these templates into our previously proposed transformation from Simulink to SHA. This allows us to formally argue about safety and performance under uncertainty. To demonstrate the feasibility of our approach, we have presented two small case studies using our newly presented stochastic subsystems. We have analyzed different properties using SMC on the transformed SHA with the tool modes as well as an evaluation in Simulink. Our results show that the analysis using SHA provides tighter CI with only a fraction of the computational effort required for the Simulink evaluation. Additionally, the SHA are not limited to SMC and could be analyzed with other quantitative analysis techniques in the future.

In future work, we plan to apply more sophisticated quantitative analysis techniques like reachability analysis, e.g. using REALYST [18,19]. Furthermore, we plan to investigate how the existing tools can be extended to analyze more complex, non-linear systems, for example by combining them with deductive verification, which can not cope well with stochasticity but might be useful to analyze non-linear differential equations.

Data Availability Statement. The models, scripts, and tools to reproduce our experimental evaluation, the Simulink files for the stochastic subsystems and an extended version of the paper including an appendix are archived and publicly available at DOI 10.5281/zenodo.15273669.

References

1. Adelt, J., Bruch, S., Herber, P., Niehage, M., Remke, A.: Shielded learning for resilience and performance based on statistical model checking in simulink. In: Steffen, B. (ed.) Bridging the Gap Between AI and Reality, pp. 94–118. Springer, Cham (2024). https://doi.org/10.1007/978-3-031-46002-9_6
2. Adelt, J., Liebrenz, T., Herber, P.: Formal verification of intelligent hybrid systems that are modeled with simulink and the reinforcement learning toolbox. In: Huisman, M., Păsăreanu, C., Zhan, N. (eds.) FM 2021. LNCS, vol. 13047, pp. 349–366. Springer, Cham (2021). https://doi.org/10.1007/978-3-030-90870-6_19
3. Alur, R., et al.: The algorithmic analysis of hybrid systems. Theoret. Comput. Sci. **138**, 3–34 (1995). https://doi.org/10.1016/0304-3975(94)00202-T
4. Alur, R., Courcoubetis, C., Henzinger, T.A., Ho, P.-H.: Hybrid automata: an algorithmic approach to the specification and verification of hybrid systems. In: Grossman, R.L., Nerode, A., Ravn, A.P., Rischel, H. (eds.) HS 1991-1992. LNCS,

vol. 736, pp. 209–229. Springer, Heidelberg (1993). https://doi.org/10.1007/3-540-57318-6_30
5. Bak, S., Beg, O.A., Bogomolov, S., Johnson, T.T., Nguyen, L.V., Schilling, C.: Hybrid automata: from verification to implementation. Int. J. Softw. Tools Technol. Transfer **21**(1), 87–104 (2019)
6. Bertrand, N., et al.: Stochastic timed automata. Logical Methods in Comput. Sci. **10** (2014). https://doi.org/10.2168/LMCS-10(4:6)2014
7. Blohm, P., Fränzle, M., Herber, P., Kröger, P., Remke, A.: Towards probabilistic contracts for intelligent cyber-physical systems. In: Margaria, T., Steffen, B. (eds.) ISoLA 2024, pp. 26–47. Springer, Cham (2025). https://doi.org/10.1007/978-3-031-75380-0_3
8. Blohm, P., Herber, P., Remke, A.: Towards quantitative analysis of simulink models using stochastic hybrid automata. In: Kosmatov, N., Kovács, L. (eds.) IFM 2024, pp. 172–193. Springer, Cham (2025). https://doi.org/10.1007/978-3-031-76554-4_10
9. Blohm, P., Schulz, F., Willemsen, L., Remke, A., Herber, P.: Modeling uncertainty: from simulink to stochastic hybrid automata (2025). https://arxiv.org/abs/2506.14581
10. Budde, C.E., D'Argenio, P.R., Hartmanns, A., Sedwards, S.: A statistical model checker for nondeterminism and rare events. In: Beyer, D., Huisman, M. (eds.) TACAS 2018. LNCS, vol. 10806, pp. 340–358. Springer, Cham (2018). https://doi.org/10.1007/978-3-319-89963-3_20
11. Budde, C.E., Hartmanns, A., Meggendorfer, T., Weininger, M., Wienhöft, P.: Sound statistical model checking for probabilities and expected rewards. CoRR (2024). https://doi.org/10.48550/ARXIV.2411.00559
12. Chen, M., et al.: MARS: a toolchain for modelling, analysis and verification of hybrid systems. In: Hinchey, M.G., Bowen, J.P., Olderog, E.-R. (eds.) Provably Correct Systems. NMSSE, pp. 39–58. Springer, Cham (2017). https://doi.org/10.1007/978-3-319-48628-4_3
13. Chen, T., Diciolla, M., Kwiatkowska, M., Mereacre, A.: A simulink hybrid heart model for quantitative verification of cardiac pacemakers. In: Proceedings of the 16th International Conference on Hybrid Systems: Computation and Control, pp. 131–136 (2013). https://doi.org/10.1145/2461328.2461351
14. Chiacchio, F., Famoso, F., D'Urso, D., Brusca, S., Aizpurua, J.I., Cedola, L.: Dynamic performance evaluation of photovoltaic power plant by stochastic hybrid fault tree automaton model. Energies **11**, 306 (2018). https://doi.org/10.3390/en11020306
15. Chutinan, A., Krogh, B.H.: Computational techniques for hybrid system verification. IEEE Trans. Automatic Control **48**(1), 64–75 (2003). https://doi.org/10.1109/TAC.2002.806655
16. David, A., Larsen, K.G., Legay, A., Mikučionis, M., Poulsen, D.B.: UPPAAL SMC tutorial. Int. J. Softw. Tools Technol. Transfer **17**(4), 397–415 (2015). https://doi.org/10.1007/s10009-014-0361-y
17. Delicaris, J., Remke, A., Ábrahám, E., Schupp, S., Stübbe, J.: Maximizing reachability probabilities in rectangular automata with random events. Sci. Comput. Program. **240**, 103213 (2025). https://doi.org/10.1016/J.SCICO.2024.103213
18. Delicaris, J., Schupp, S., Ábrahám, E., Remke, A.: Maximizing reachability probabilities in rectangular automata with random clocks. In: David, C., Sun, M. (eds.) TASE 2023. LNCS, vol. 13931, pp. 164–182. Springer, Cham (2023). https://doi.org/10.1007/978-3-031-35257-7_10

19. Delicaris, J., Stübbe, J., Schupp, S., Remke, A.: RealySt: a C++ tool for optimizing reachability probabilities in stochastic hybrid systems. In: Kalyvianaki, E., Paolieri, M. (eds.) VALUETOOLS 2023, vol. 539, pp. 170–182. Springer, Cham (2023). https://doi.org/10.1007/978-3-031-48885-6_11
20. Filipovikj, P., Mahmud, N., Marinescu, R., Seceleanu, C., Ljungkrantz, O., Lönn, H.: Simulink to UPPAAL statistical model checker: analyzing automotive industrial systems. In: Fitzgerald, J., Heitmeyer, C., Gnesi, S., Philippou, A. (eds.) FM 2016. LNCS, vol. 9995, pp. 748–756. Springer, Cham (2016). https://doi.org/10.1007/978-3-319-48989-6_46
21. Frehse, G., Kateja, R., Le Guernic, C.: Flowpipe approximation and clustering in space-time. In: Proceedings of the 16th International Conference on Hybrid Systems: Computation and Control, pp. 203–212. ACM (2013). https://doi.org/10.1145/2461328.2461361
22. Fulton, N., Mitsch, S., Quesel, J.-D., Völp, M., Platzer, A.: KeYmaera X: an axiomatic tactical theorem prover for hybrid systems. In: Felty, A.P., Middeldorp, A. (eds.) CADE 2015. LNCS (LNAI), vol. 9195, pp. 527–538. Springer, Cham (2015). https://doi.org/10.1007/978-3-319-21401-6_36
23. Henzinger, T.A.: The theory of hybrid automata. In: Inan, M.K., Kurshan, R.P. (eds.) Verification of Digital and Hybrid Systems, vol. 170, pp. 265–292. Springer, Heidelberg (2000). https://doi.org/10.1007/978-3-642-59615-5_13
24. Henzinger, T.A., Kopke, P.W., Puri, A., Varaiya, P.: What's decidable about hybrid automata? J. Comput. Syst. Sci. **57**(1), 94–124 (1998)
25. Kuriakose, R.B., Vermaak, H.J.: Customized mixed model stochastic assembly line modelling using simulink. Int. J. Simul. Syst. Sci. Technol. **20**(1), 61–69 (2019). https://doi.org/10.5013/IJSSST.a.20.S1.06
26. Legay, A., Traonouez, L.-M.: Statistical model checking of simulink models with plasma lab. In: Artho, C., Ölveczky, P.C. (eds.) FTSCS 2015. CCIS, vol. 596, pp. 259–264. Springer, Cham (2016). https://doi.org/10.1007/978-3-319-29510-7_15
27. Liebrenz, T., Herber, P., Glesner, S.: Deductive verification of hybrid control systems modeled in simulink with KeYmaera X. In: Sun, J., Sun, M. (eds.) ICFEM 2018. LNCS, vol. 11232, pp. 89–105. Springer, Cham (2018). https://doi.org/10.1007/978-3-030-02450-5_6
28. Liebrenz, T., Herber, P., Glesner, S.: A service-oriented approach for decomposing and verifying hybrid system models. In: Arbab, F., Jongmans, S.-S. (eds.) FACS 2019. LNCS, vol. 12018, pp. 127–146. Springer, Cham (2020). https://doi.org/10.1007/978-3-030-40914-2_7
29. Lygeros, J., Prandini, M.: Stochastic hybrid systems: a powerful framework for complex, large scale applications. Eur. J. Control. **16**(6), 583–594 (2010). https://doi.org/10.3166/ejc.16.583-594
30. Minopoli, S., Frehse, G.: SL2SX translator: from Simulink to SpaceEx models. In: International Conference on Hybrid Systems: Computation and Control, pp. 93–98. ACM (2016). https://doi.org/10.1145/2883817.2883826
31. Platzer, A.: Differential dynamic logic for hybrid systems. J. Autom. Reason. **41**(2), 143–189 (2008). https://doi.org/10.1007/s10817-008-9103-8
32. da Silva, C., Schupp, S., Remke, A.: Optimizing reachability probabilities for a restricted class of stochastic hybrid automata via flowpipe-construction. ACM Trans. Model. Comput. Simul. **33**(4) (2023). https://doi.org/10.1145/3607197
33. The MathWorks: Simulink. https://de.mathworks.com/products/simulink.html
34. Willemsen, L., Remke, A., Ábrahám, E.: Comparing two approaches to include stochasticity in hybrid automata. In: Jansen, N., Tribastone, M. (eds.) QEST 2023.

LNCS, vol. 14287, pp. 238–254. Springer, Cham (2023). https://doi.org/10.1007/978-3-031-43835-6_17

35. Willemsen, L., Remke, A., Ábrahám, E.: (de-)Composed and more: eager and lazy specifications (CAMELS) for stochastic hybrid systems. In: Jansen, N., et al. (eds.) Principles of Verification: Cycling the Probabilistic Landscape, Part III. LNCS, vol. 15262, pp. 309–337. Springer, Cham (2025). https://doi.org/10.1007/978-3-031-75778-5_15

36. Wilson, E.: Probable inference, the law of succession, and statistical inference. J. Am. Stat. Assoc. **22**(158), 209–212 (1927). https://doi.org/10.2307/2276774

37. Zou, L., Zhan, N., Wang, S., Fränzle, M.: Formal verification of simulink/stateflow diagrams. In: Finkbeiner, B., Pu, G., Zhang, L. (eds.) ATVA 2015. LNCS, vol. 9364, pp. 464–481. Springer, Cham (2015). https://doi.org/10.1007/978-3-319-24953-7_33

38. Zuliani, P., Baier, C., Clarke, E.M.: Rare-event verification for stochastic hybrid systems. In: Proceedings of the 15th ACM International Conference on Hybrid Systems: Computation and Control, pp. 217–226. ACM (2012). https://doi.org/10.1145/2185632.2185665

39. Zuliani, P., Platzer, A., Clarke, E.M.: Bayesian statistical model checking with application to stateflow/simulink verification. Formal Methods Syst. Des. 338–367 (2013). https://doi.org/10.1007/s10703-013-0195-3

Statistical Bayesian Inference for Stochastic Process Discovery

Pierre Cry[1], Paolo Ballarini[1(✉)], András Horváth[2], and Pascale Le Gall[1]

[1] Université Paris Saclay, CentraleSupélec, MICS, Gif-sur-Yvette, France
{pierre.cry,paolo.ballarini,pascale.legall}@centralesupelec.fr
[2] Università di Torino, Turin, Italy
horvath@di.unito.it

Abstract. Stochastic process discovery is concerned with deriving a model capable of reproducing the stochastic character of observed executions of a given process, stored in a log. This leads to an optimisation problem in which the model's parameter space is searched for, driven by the resemblance between the log's and the model's stochastic languages. The bottleneck of such optimisation problem lay in the determination of the model's stochastic language which existing approaches deal with through, hardly scalable, exact computation approaches. In this paper we introduce a novel framework in which we combine a simulation-based Bayesian parameter inference scheme, used to search for the "optimal" instance of a stochastic model, with an expressive statistical model checking engine, used (during inference) to approximate the language of the considered model's instance. Because of its simulation-based nature, the payoff is that, the runtime for discovering of the optimal instance of a model can be easily traded in for accuracy, hence allowing to treat large models which would result in a prohibitive runtime with non-simulation based alternatives. We validate our approach on several popular event logs concerning real-life systems.

Keywords: Statistical model checking · Hybrid automata · Stochastic process mining · Stochastic languages · Earth Movers Distance

1 Introduction

The Process Mining Problem. The primary goal of *process mining* [25] is *discovering* of formal models that adequately mimic the dynamics of a business process. Discovery relies on observations of the considered process stored as *traces*, in an *event log*. A *trace* consists of a sequence of activities that represent one observed execution of the process. Classic process discovery algorithms [2,14,24,27,29,30], aim at capturing the workflow aspects of the considered process, that is, extracting a so-called *workflow* model, normally in the form of a structured Petri net, capable of reproducing the *log's language*. The quality of discovery algorithms is assessed by means of *conformance checking* indicators such as, e.g., *fitness*, which measures how much of the language (i.e., the set of unique traces) of the log is reproduced by the discovered model, and *precision*, which, conversely, quantifies how much of the language of the model (the traces issued by the model) is contained in the log.

The Stochastic Process Mining Problem. By taking into account how often each unique sequence of actions has been detected while observing the process (i.e., trace multiplicity) leads to the stochastic extension of the process discovery problem, whose aim is to devise a stochastic model whose *stochastic language* resembles that of the log. The resemblance is assessed via *stochastic conformance* indicators, such as those based on adaptations of the Earth Movers Distance (EMD) [16,17] or those based on entropy measures [1]. Within the still relatively emergent literature [9–11,13,21,28], mining of a stochastic model is mainly achieved *indirectly*, that is, first a non-stochastic, workflow model is extracted from the log through a standard discovery algorithm (hence disregarding trace multiplicity), then the obtained model is converted into a stochastic one by associating *weight* parameters to each event. This leads to a *parameter optimization* problem whose goal is to identify adequate parameter (weight) values so that the corresponding stochastic language is *as close as possible* to that of the log.

Our Contribution. Discovery of optimal parameters requires an effective procedure to determine the stochastic language emitted by the model. Existing approaches [11,15] rely on exact computation of a model's stochastic language and, although effective, suffer from poor scalability, making it impractical for complex logs. As a remedy to such bottleneck, in this paper, we propose an alternative optimization framework, which relies on a *simulation-based* engine to obtain an (arbitrarily precise) approximation of the model's stochastic language, through a statistical model checking tool [3]. The search for optimal model's parameters is then achieved through an adaptation of the Approximate Bayesian Computation (ABC) [19,23] scheme, a *likelihood-free* Bayesian parameter inference method, in which the model's language approximation engine is plugged in. Experimental evidence shows that depending on the complexity of the considered event log (hence of the mined model), the simulation-based discovery of optimal stochastic models we propose here may be better than numerical-based ones, resulting in the discovery of more conformant models in less time.

The paper is organized as follows: Sect. 2 introduces preliminary notions used throughout the manuscript and succinctly overview background material including the HASL model checking framework and the ABC parameter inference scheme. In Sect. 3, the novel simulation-based framework for discovery of optimal instances of sWN models. In Sect. 4, we demonstrate the novel framework through several experiments on popular real-life event logs. We wrap up the manuscript with conclusive remarks and future perspectives in Sect. 5.

Related Work. Several contributions have been proposed within the still thin yet growing literature on stochastic process discovery. In their seminal work [21] authors introduced a framework to discover generalized stochastic Petri net (GSPN) models extended with generally distributed timed transitions so to allow for performance analysis of the mined process. In close relationship with the problem we face in this paper, i.e. discovery of untimed, probabilistic models, Burke et al.'s [9] instead addressed the problem of converting a workflow Petri net, mined through a conventional discovery algorithm, into an adequate (untimed) stochastic workflow Petri net through weight estimation. Specifically, [9] introduced six weight estimators that combine summary statistics computed on the log with statistics computed on the model while also taking into account structural relationships between the Petri net nodes (e.g., *transitions causality*). These estimators enjoy being computationally light as, by definition, they do not need to assess the language of the Petri net model. However, the price for such simplicity is paid in terms of conformance, as the distance between the resulting model's and the log's stochastic languages appears to be far from optimal in many cases. In a follow-up work [10], the same authors introduced a framework to directly discover an *untimed* GSPN model from a log based on traces' frequency.

The problem of discovering stochastic models through optimization has been the subject of a few recent research works. The main difficulty in this respect is that finding the optimal parameters for the stochastic process requires computing the probability of the traces issued by the model, which is algorithmically non-trivial due to the size and the likely infiniteness of the model language.

In [8], authors introduce an approach focused on optimizing the earth mover's stochastic conformance score of a discovered stochastic Petri net through subgradient ascent. Their approach consists of two main steps: first, they derive the stochastic language of the net by analyzing its structure to assess conformance with the event log. In the second step, they perform subgradient optimization on the loss function with respect to the model's stochastic language. This subgradient is then propagated to update the weights in the Petri net, progressively improving its alignment with the log's observed behavior. The approach has been implemented by the WAWE tool [22].

In [15], authors propose an optimization scheme that requires extracting the analytical expression for the probability of each trace issued by the model. These expressions are obtained as solutions of n (n being the traces in the log) absorbing state probability problems for n different discrete-time Markov chains (given by the cross-product of the stochastic reachability graph underlying the considered stochastic Petri net with a deterministic finite automaton corresponding to a trace of the log). Although effective, this approach faces scalability issues: the size of the analytical expressions extracted from the model increases dramatically with the length of the traces, making the approach unfeasible even for relatively simple models. The approach has been implemented by the SLPN miner tool [18].

2 Preliminaries

Alphabet, Trace, Language, Stochastic Language. We let Σ denote the alphabet of an event log's activities (we use letters to denote activities of an alphabet, e.g., $\Sigma = \{a, b, c\}$) and Σ^* the set of traces (words) composed of activities in Σ where

$\varepsilon \in \Sigma^*$ represents the empty trace. A stochastic language over an alphabet Σ is a function $L : \Sigma^* \to [0, 1]$ that provides the probability of the traces such that $\sum_{t \in \Sigma^*} L(t) = 1$.

Event Log. An event log E is a multi-set of traces built on an alphabet that we denote Σ_E, i.e., $E \in Bag(\Sigma_E^*)$. Given a trace $t \in E$ we denote by $f(E, t)$ its multiplicity (i.e., its frequency in E). The stochastic language induced by an event log E is straightforwardly obtained by computing the probability of the traces as $p(E, t) = f(E, t) / \sum_{t \in Supp(E)} f(E, t)$ where $Supp(E)$ denotes the set of unique traces in E, i.e., its support. The stochastic language of an event log E will be denoted by L_E. For E an event log over an n-letters alphabet $\Sigma_E = \{a_1, \ldots, a_n\}$ we denote l_i $(1 \leq i \leq n)$ the maximum number of occurrence of letter a_i in any trace of E and c_m the maximal length of any trace in Σ_E.

Petri Net. A labeled Petri net model is a tuple $N = (P, T, F, \Sigma, \lambda, M_0)$, where P is the set of places, T the set of transitions, $F : (P \times T) \cup (T \times P) \to \mathbb{N}$ gives the arcs' multiplicity (0 meaning absence of arc), $\lambda : T \to (\Sigma \cup \tau)$ associates each transition with an action (τ being the silent action) and $M_0 : P \to \mathbb{N}$ is the initial marking. For N a PN, we denote $RG(N) = (S, A)$, where $S = RS(N)$ is the reachability set (the set of markings reachable from the initial one, and $A \subseteq RS(N) \times RS(N) \times T$ is the set of arcs whose elements $(M, M', t) \in A$ are such that $M[t\rangle M'$, i.e. M' is reached from M by firing of t.

Workflow Net. A workflow net is a 1-safe[1] Petri net with the following structural constraints: 1) there exists a unique place, denoted *source* with no incoming transition and a unique place denoted *sink* with no outgoing transitions; 2) the initial marking is $M_0(source) = 1$ and $M(p) = 0$ for any place p different from *source* and 3) the net graph can be turned into a strongly connected one by adding a single transition outgoing place *sink* and ingoing place *source*.

Stochastic Workflow Net. In the remainder, we consider the stochastic, untimed extension of workflow nets, which we refer to as stochastic workflow nets (sWN). In practice a sWN $N_s = (P, T, F, W, \Pi, \Sigma, \lambda, M_0)$ is a generalized stochastic Petri net (GSPN) [20] consisting uniquely of *immediate* transitions, i.e., $T = T_i \cup T_t$, with $T_t = \emptyset$ (T_t, resp. T_i, being the subset of timed transitions, resp. immediate transitions), each of which is associated with a non-negative weight $W : T \to \mathbb{R}_{>0}$ while priorities are all equal ($\Pi(t) = 1, \forall t \in T$), therefore in the remainder we omit Π from the characterization of a sWN. In a marking M transition t of a sWN fires with a probability that is a function of their weights, i.e., $P(t|M) = W(t) / \sum_{t' \in en(M)} W(t')$. The probability of firing a transition induces a probability over sequences of transition firings. Therefore, if marking M' is reachable from M through the sequence $M = M_1[t_1\rangle M_2[t_2\rangle \ldots [t_n\rangle M_{n+1} = M'$, then the probability of the sequence of transitions $\langle t_1, \ldots, t_n \rangle$ (hence of the corresponding trace $\langle \lambda(t_1), \lambda(t_2), \ldots, \lambda(t_n) \rangle$) starting from marking $M = M_1$ is $P(\langle t_1, \ldots, t_n \rangle | M) = \prod_{i=1}^{n} P(t_i | M_i)$. We denote by L_{N_s} the stochastic language

---

[1] In any marking each place may contain at most one token.

associated with the sWN N_s. In order to make explicit the role of the weights, $N_s(\overline{W})$ will denote the sWN N_s with weights given in the vector $\overline{W}$.

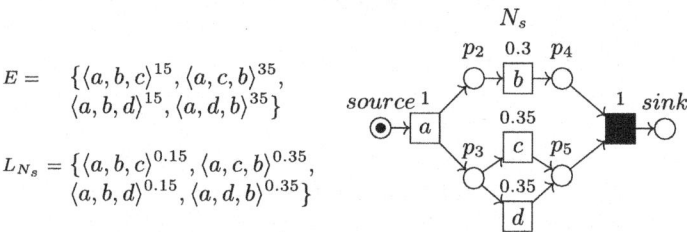

$$E = \{\langle a,b,c\rangle^{15}, \langle a,c,b\rangle^{35},$$
$$\langle a,b,d\rangle^{15}, \langle a,d,b\rangle^{35}\}$$

$$L_{N_s} = \{\langle a,b,c\rangle^{0.15}, \langle a,c,b\rangle^{0.35},$$
$$\langle a,b,d\rangle^{0.15}, \langle a,d,b\rangle^{0.35}\}$$

Fig. 1. An log E with its corresponding sWN (N_s): transition weights are depicted above each transition, while silent transitions are depicted as black rectangles. Notice that the stochastic language L_{N_s} of the sWN conforms that of the log E.

Example 1. Figure 1 shows an example of the event log and corresponding stochastic workflow net together with its stochastic language.

Earth Mover's Distance (EMD). EMD [17] evaluates the resemblance between the stochastic languages of logs and (sWN) models by measuring the cost of transforming the distribution of the log's traces into the distribution of the traces issued by the model. It depends on the distance between traces which, in turns, we determine by the Levenshtein distance[2]. In order to deal with possible infiniteness of the model's language, in the remainder we use the so-called restricted EMD (rEMD), which consists in applying EMD to compare the log's traces with the subset of the model's traces that belong to the log.

HASL Model Checking. In Sect. 3 we introduce a formal approach to approximate the stochastic language of a sWN which is based on the HASL statistical model checking approach (HASL-SMC) [3]. HASL-SMC (Fig. 2) allows for assessing sophisticated performance indicator of a (timed or untimed) GSPN model N_s, via a property $\varphi \equiv (\mathcal{A}, Z)$ formally encoded by a combination of linear hybrid automaton $\mathcal{A}$ and a target expression Z. The functioning of the framework can be summarised as follows: a sufficiently large number of (finite) traces are sampled (via simulation) from N_s and synchronised (*on-the-fly*) with $\mathcal{A}$ and those that meet the acceptance condition(s) of $\mathcal{A}$ are used (together with the statistics collected in the variables of $\mathcal{A}$) to build an $\epsilon\%$-confidence level estimate (with confidence interval width δ) of the measure of interest Z. In the reminder we use $CI(N_s, \varphi, \epsilon, \delta)$ to denote such confidence interval.

---

[2] Which measures the distance between two traces as the minimum *alignment* [26], i.e., the minimum number of single-character edits (insertion, deletion or substitution of an action) needed to change one trace into the other.

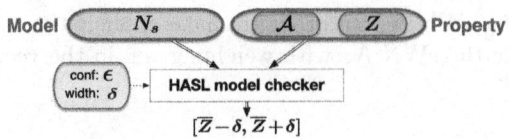

Fig. 2. The HASL statistical model checking scheme.

Within HASL a lynear hybrid automata (LHA) associated to a GSPN[3] $N_s = (P, T, F, W, \Pi, \Sigma, \lambda, M_0)$ is a tuple $\mathcal{A}_{N_s} = \langle Ev, L, I, F, X, \text{flow}, \Lambda, \rightarrow \rangle$, where $Ev = \Sigma \cup \{\tau\}$ is the alphabet of observed events, L is a finite set of locations, with $I \subset L$, and $F \subset L$, the initial, respectively, the final (accepting) locations, $X = \{x_1, \ldots x_n\}$ a finite set of real-valued variables, $\text{flow} : L \rightarrow (RS(N_s) \rightarrow \mathbb{R}^n)$ specifies (for each location) the rate (i.e. first derivative) with which each variable x_i evolves depending on the current marking of N_s, $\Lambda : L \rightarrow (RS(N_s) \rightarrow \mathbb{B})$ are the location *invariants* (i.e. boolean evaluated propositions built based depending on the current marking of N_s) and $\rightarrow$ is a set of transitions $l \xrightarrow{Ev', \gamma, U} l'$ where γ is an enabling guard (an inequality built on top of variables X), Ev' is either a set of events names (i.e. the transition is *synchronously* traversed on the occurrence of any reaction in Ev' occurring in the path being sampled) or ♯ (i.e. the transition is *autonomously* traversed without synchronization) and U are the variable updates.

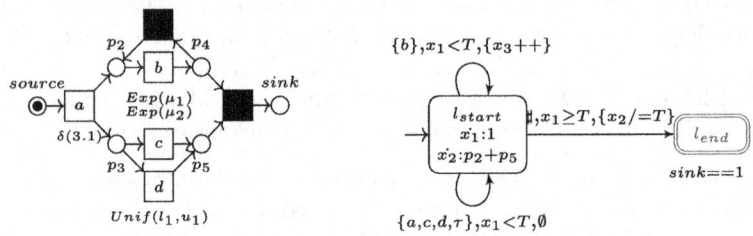

Fig. 3. An LHA synchronisying with a (timed) GSPN.

Example 2. Figure 3 shows an example of GSPN model N_s (left) and a corresponding synchronising LHA $\mathcal{A}_{N_s}$ (right). Model N_s is a timed extension of that in Fig. 1 (right) such that transitions labeled with actions a, b, c and d are timed and associated to different kind of delay distribution (Dirac, Exponential and Uniform) while silent transitions are assumed to be immediate. The LHA has an initial (l_{start}) and a final (l_{end}) location and uses three variables $X = \{x_1, x_2, x_3\}$ whose flows (rates) are $\text{flow}(l_{start}) = (1, p_2 + p_5, 0)$ and $\text{flow}(l_{end}) = (0, 0, 0)$ which means that x_1 is used as clock (rate $\dot{x}_1 = 1$), x_2 is used to hold the integral of

---

[3] Whose timed transitions may be associated with non-exponential distributions.

the sum of tokens in places p_2 and p_5 ($\dot{x}_2 = p_2 + p_5$) while x_3 is not evolving as $\mathcal{A}_{N_s}$ spends time in l_{start}. Finally $\mathcal{A}_{N_s}$ has two self-loop *synchronised* transitions $l_{start} \xrightarrow{\{b\},x_1<T,x_3+=1} l_{start}$, $l_{start} \xrightarrow{\{a,c,d,\tau\},x_1<T,\emptyset} l_{start}$, traversed by synchronisation with occurrences of transitions of model N_s and an *autonomous* transition $l_{start} \xrightarrow{\sharp,x_1=T,x_2/=T} l_{end}$, traversed autonomously as soon as its guard ($x_1 \geq T$) is satisfied, on condition that the invariant $\Lambda(l_{end}) : sink == 1$ of destination location l_{end} is satisfied. The traces accepted by $\mathcal{A}_{N_s}$ are then used to assess relevant performance measures formally given by an expression Z, such as, for example, $Z = PDF(last(x_2), 0.01, 0, 2)$ that represents the distribution of the *number of tokens contained in places p_2 and p_5 within time at least T* approximated using $[0,2]$ as support set and discretizing $[0,2]$ with buckets. Notice that differently from the example in Fig. 3 within this paper HASL is referred exclusively to the sWN subclass of GSPN (i.e., untimed models).

Approximate Bayesian Computation. The optimization framework we introduce is an adaptation of Approximate Bayesian Computation (ABC) [19,23], a family of *likelihood-free* methods, that given a model with parameters θ, an initial belief on their value expressed by means of a prior distribution $\pi(\theta)$ and based on some observed data y_{obs}, is concerned with obtaining an arbitrarily precise estimate of the posterior distribution $\pi(\theta|y_{obs})$ expressed by the Bayes theorem as:

$$\pi(\theta|y_{obs}) = \frac{p(y_{obs}|\theta)\pi(\theta)}{\int_{\theta'} p(y_{obs}|\theta')\pi(\theta')\,d\theta'} \quad (1)$$

In most models, the likelihood functions $p(y_{obs}|\theta)$ is too expensive to compute or even intractable, hindering the determination of the posterior distribution. In these cases ABC algorithms still allow one to obtain an arbitrarily precise approximation, denoted $\pi_{ABC,\epsilon}$ ($\epsilon \in \mathbb{R}_+^*$ being a tolerance value), of the posterior $\pi(\theta|y_{obs})$. In its simplest, *rejection sampling*, form (Algorithm 1) ABC consists of a simple iterative procedure where n parameters value θ' are selected from the parameter space through sampling from the prior distribution $\pi()$, up until the distance between the traces issued by the corresponding model instance ($y' \sim p(.|\theta')$) and the observations y_{obs} is below tolerance ϵ (i.e. $\rho(\eta(y'), \eta(y_{obs})) \leq \epsilon^4$). Though effective, ABC rejection sampling suffers of slow convergence, particularly for small tolerance values ϵ. Therefore, the sequential Monte Carlo extension of ABC, named ABC-SMC [5], has been introduced to speed up the search of acceptable parameters, essentially by means of a multilevel parameter search procedure through which parameters are progressively accepted using decreasing tolerance levels.

Since ABC algorithms rely on a distance measure (ρ), they can be adapted to calibrating models w.r.t. specific behavioral characteristics (rather than w.r.t. observations y_{obs}) as long as an adequate distance can be defined to measure how far a model instance (issued by a parameter vector θ_i) is from exhibiting the desired characteristic (e.g., see [4,6,7]).

---

[4] Where $\eta : \mathcal{Y} \to \mathcal{S} \subset \mathbb{R}^{k_1}$ is a function that computes summary statistics on the observations and $\rho : \mathcal{S} \times \mathcal{S} \to \mathbb{R}^+$ is a distance in the space of summary statistics.

Algorithm 1. ABC rejection sampling

Require: y_{obs} (observations), ϵ (tolerance), ρ (distance metric), η (summary statistics)
Ensure: $(\theta_i)_{0 \leq i \leq n}$ drawn from $\pi_{ABC,\epsilon}$
    for $i = 1 : n$ **do**
        repeat
            $\theta' \sim \pi(.)$
            $y' \sim p(.|\theta')$
        until $\rho(\eta(y'), \eta(y_{obs})) \leq \epsilon$
        $\theta_i \leftarrow \theta'$
    end for

In Sect. 3, we introduce a novel version of the ABC-SMC algorithm which, based on the restricted Earth Movers Distance (rEMD), allows for inferring the weights of a sWN model N_s (mined from a given event log E) so to minimize the distance between the stochastic language L_{N_s}, issued by the sWN, and that of the corresponding event log L_E.

3 Method

We introduce a simulation-based method for discovery of optimal weight parameters of a sWN model N_s. The method combines the HASL-SMC procedure for approximating the stochastic language L_{N_s} issued by N_s, with an adaptation of the ABC-SMC scheme to identify the "best" transitions weights for N_s. Figure 4 outlines the functioning of the method. Taking from an event log, a workflow net is discovered (using the Inductive Miner method [14], which guarantees that the model reproduces all traces in the log) and fed in, enriched with a weights parameter vector $\overline{W}$, to the ABC-SMC inference engine together with a number of hyper-parameters needed to control both the inference scheme as well as the HASL-based approximation of L_{N_s}. The framework outputs the estimates of the marginal distributions of optimal weights parameters.

3.1 HASL-Based Approximation of a sWN Stochastic Language

We present an HASL formula (Definition 3), that relying on a dedicated *stochastic language detector* (hybrid) automaton $\mathcal{A}_{sld(E)}$ (Definition 2) allows one to obtain an arbitrarily precise approximation of the stochastic language L_{N_s}.

To this aim we first define a mapping through which words over an alphabet Σ are mapped to unique integer values.

Definition 1 (Word mapping). *Given an alphabet Σ consisting of n letters ($|\Sigma| = n$) and an injective function $f_l : \Sigma \to \mathbb{N}$ we define the word mapping function $w_m : \Sigma^* \to \mathbb{N}$ as:*

$$w_m(\sigma) = \sum_{i=1}^{|\sigma|} f_l(\sigma[i]) \cdot n^{i-1} \qquad (2)$$

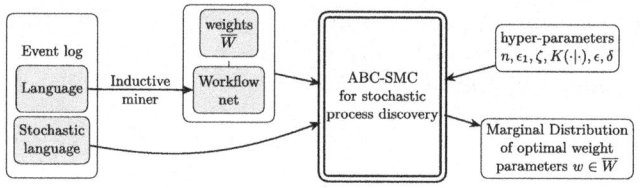

Fig. 4. Stochastic process discovery based on an ABC-SMC procedure.

Notice that, trivially, mapping w_m is injective as, by hypothesis, the letters mapping f_l is injective.

Example 3. Let $\Sigma = \{a, b, c\}$ be an alphabet, and let us assume $f_l(a) = 1$, $f_l(b) = 2$ and $f_l(c) = 3$ as letters' mapping, then the following are examples of mapping of words in Σ^*: $w_m(aa) = 1 \cdot 3^0 + 1 \cdot 3^1 = 4$, $w_m(cb) = 3 \cdot 3^0 + 2 \cdot 3^1 = 9$, $w_m(abc) = 1 \cdot 3^0 + 2 \cdot 3^1 + 3 \cdot 3^2 = 34$, $w_m(cba) = 3 \cdot 3^0 + 2 \cdot 3^1 + 1 \cdot 3^2 = 18$.

In the remainder, given a finite event log $E \in Bag(\Sigma_E^*)$ and a word mapping w_m (defined on alphabet Σ_E), we denote $w_m(E) \subset \mathbb{N}$ the set of naturals to which the words of $Supp(E)$ are mapped.

Convex Remapping. As mapping (2) commonly yields a non-convex (sparse) support set with large[5] supremum, which would negatively impact the HASL-based confidence interval estimation of the corresponding probability density function (PDF) expression[6], we use a convex re-mapping scheme through which each word of a finite log E is mapped back over the convex interval $\{0, 1, 2, ..., |Supp(E)|-1\}$. To this aim we first compute (offline) $w_m(E)$ and then re-map each element in $w_m(E)$ to a corresponding value in $\{0, 1, 2, ..., |Supp(E)|-1\}$. Therefore in the remainder for a word $\sigma \in Supp(E)$ we denote its convex remapping $w_{cm}(w_m(\sigma)) \in \{0, 1, 2, ..., |Supp(E)|-1\}$.

Definition 2 (Stochastic language detector automaton). *Given N_s, a sWN discovered from an event log E with alphabet $\Sigma_E = \{a_1, \ldots, a_n\}$ and given an injective mapping $f_l : \Sigma_E \to \mathbb{N}$, the stochastic language estimator LHA $\mathcal{A}_{sld(E)} = \langle Ev, L, I, F, X, flow, \Lambda, \to \rangle$ is defined as follows: event set $Ev = \Sigma_E \cup \{\tau\}$, locations $L = I \cup F$ with $I = \{l_{start}\}$ and $F = \{l_{end}\}$, variables set $X = \{w, w_c, c, c_1, \ldots, c_n\}$ consisting of the following $n + 3$ (integer) variables,*

- *w: mapping of detected word*
- *w_c: convex mapping of detected word*
- *c: length of detected word*
- *c_i $(1 \leq i \leq n)$: number of occurrences of letter a_i in the detected word*

---

[5] Exponential in the size of the word.
[6] In this context, what we deal with are probability mass functions (PMF). We still use PDF because it is a keyword in Cosmos covering both PDFs and PMFs.

each with constant rate of evolution in every location flow(l) = $(0,\ldots,0)$ ($\forall l \in L$), $\Lambda(l_{start}) = \mathbf{true}$, location invariants $\Lambda(l_{end}) = (sink == 1)$ and transition set $\rightarrow$ consisting the following $n+1$ transitions:

- n self-loop synchronised *transitions* (with $1 \leq i \leq n$) defined as follows:

$$l_{start} \xrightarrow{\{a_i\},c_i \leq l_i \wedge (\sum_i c_i) \leq c_m, \{w += f_l(a_i) \cdot n^c; c+=1; c_i+=1;\}} l_{start}$$

where n, l_i and c_m are constants referred to the log E (n number of letters, l_i maximum number of the i-th letter in any word of E, c_m length of the longest word in E).

- $|Supp(E)|$ autonomous *transitions* (with $1 \leq i \leq |Supp(E)|$) defined as

$$l_{start} \xrightarrow{\sharp, w == w_m(\sigma_i), \{w_c = w_{cm}(w_m(\sigma_i))\}} l_{end}$$

where $\sigma_i \in Supp(E)$ is a unique trace of E, $w_m(\sigma_i) \in \mathbb{N}$ is its mapping and $w_{cm}(w_m(\sigma_i)) \in \mathbb{N}$ its convex re-mapping.

The goal of automaton $\mathcal{A}_{sld(E)}$, depicted in Fig. 5[7]), is to detect, among the traces issued by model N_s it synchronies with, those that belong to the log E by retrieving their corresponding mapping. To this aim $\mathcal{A}_{sld(E)}$ is equipped with $n+3$ integer variables whose goal is 1) to store (variable w) the mapping (as per Definition 1) of the trace currently being scanned, 2) to count the occurrences of each activity a_i (variables c_i, $1 \leq i \leq n$) as it is observed on the scanned trace, 3) to store the length of the word (in terms of total number of observed actions a_i (variable c) and, finally, 4) to store the convex remapping (variable w_c) when the scanned trace yield a word of E. Each observed activity is detected by traversal of the corresponding synchronized self-loop arc on the initial location (l_{start}). Notice that on traversal of the self-loop arc corresponding to activity a_i, the value of w is added up with $f_l(a_i) \cdot n^c$ while the a_i activity counter c_i is incremented.

Scanning of the traces issued by a sWN model is guaranteed to terminate by either accepting or rejecting the currently observed trace. A trace $\sigma \in \Sigma_E^*$ is accepted if and only if 1) it is generated by a sequence of transitions $\langle t_1, \ldots, t_m \rangle$ whose last transition t_m reaches the deadlock marking by adding a token in place *sink* (i.e., invariant $sink == 1$ of final location l_{end}) and 2) it belongs to the event log $\sigma \in supp(E)$ (that is, if its mapping w is in $w_m(E)$, which is captured by the guard $w \in w_m(E)$ on $l_{start} \rightarrow l_{end}$). Any other trace $\sigma \in \Sigma_E^* \setminus supp(E)$ results in the automaton to block and, hence, it is rejected (thus its mapping value w is discarded). Notice that in order to rule out infinite traces (that may result from a net that contains loops) and more generally to shorten the trace detection process, each self-loop, synchronous, a_i arc is enabled only on condition that

---

[7] For the sake of space we slightly abuse the LHA syntax and subsume the $|Supp(E)|$ transitions $l_{start} \rightarrow l_{end}$ described in Definition 2 by a single, semantically equivalent, transition with guard $w \in w_m(E)$.

1) the number of a_i observed on the current trace is not above the maximum number l_i computed for the traces in E (captured by constraint $c_i \leq l_i$) and 2) the length of the current trace does not trespasses c_m, that is, the length of the longest trace in E (captured by constraint $\sum_i c_i \leq c_m$).

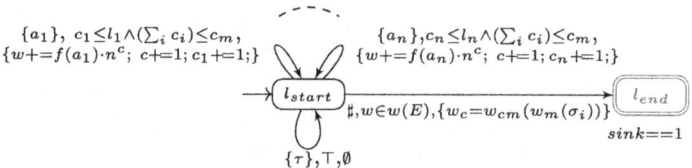

Fig. 5. The stochastic language detector automaton $\mathcal{A}_{sld(E)}$.

By definition, automaton $\mathcal{A}_{sld(E)}$ enjoys the property that the language $L_{N_s \times \mathcal{A}_{sld(E)}}$, issued by the product process $N_s \times \mathcal{A}_{sld(E)}$, is contained in the event log E from which the model N_s has been discovered. We state this property in Theorem 1.

Theorem 1. *Let N_s be a sWN discovered with a discovery algorithm Alg from an event log $E \in Bag(\Sigma_E^*)$ then $L_{N_s \times \mathcal{A}_{sld(E)}} \subseteq supp(E)$.*

Proof. Straightforward consequence of HASL operational semantics.

Lemma 1. *If algorithm Alg used to discover N_s has fitness 1 then $L_{N_s \times \mathcal{A}_{sld(E)}} = supp(E)$.*

As a consequence of Theorem 1 by combining automaton $\mathcal{A}_{sld(E)}$ with target expression $PDF(last(w), s, l, h)$ (with adequate buckets and support set parameters s, l and h) we obtain a confidence interval estimator of the probability distribution with which model N_s generates the traces of log E.

Definition 3 (HASL stochastic language estimator formula). *Given N_s a sWN discovered from an event log E we define the HASL stochastic language estimator formula $\varphi_{sle(E)} \equiv (\mathcal{A}_{sld(E)}, PDF(last(w_c), 1, 0, |Supp(E)|-1))$, where $\mathcal{A}_{sld(E)}$ is the stochastic language detector automaton (Definition 2), w its corresponding (word mapping) variable. Notice that since w_c is a discrete random variable in the target expression $PDF(last(w_c), 1, 0, |Supp(E)|-1))$ we use buckets of size $s = 1$, while we use $[l, h] = [0, |Supp(E)|-1]$ as support set for the sought approximation of the PDF of w_c.*

3.2 An ABC Framework for Optimised Stochastic Process Discovery

To discover optimal weights $\overline{W}$ for a sWN model N_s (mined from an event log E with alphabet Σ_E), we introduce an adaptation of the ABC sequential Monte

Carlo (ABC-SMC) parameter inference approach [5] in which we plug in the HASL-based stochastic language approximation method (Sect. 3.1) as a mean to retrieve the stochastic language issued by the model's instance $N_s(\overline{W})$ that corresponds to a specific parameter vector $\overline{W}$. In our context, the parameter space is k-dimensional ($k = |T|$ being the number of transitions of N_s) and its elements are vectors $\overline{W} \in [0,1]^k$ corresponding to the weights of the transitions. The optimization objective is to minimize the rEMD distance between the stochastic language $L_{N_s(\overline{W})}$ issued by model $N_s(\overline{W})$ and the stochastic language L_E of the log E. Following the ABC-SMC scheme, the novel algorithm we introduce (Algorithm 2) operates over a sequence of $m \in \mathbb{N}$ telescopic layers i ($1 \leq i \leq m$) with decreasing tolerances ϵ_i, such that $\epsilon_1 > \epsilon_2 > \ldots > \epsilon_m$.

At each layer i, a user-defined number (n) of *particles* (in our case, weight vectors $\overline{W}$) are iteratively sampled from a given probability distribution until the measured distance (in this case, the rEMD between the event log language and the language generated by the model with weights $\overline{W}$) falls within the tolerance level ϵ_i. At the first layer ($i = 1$), the initial n particles ($\overline{w}_1^{(1)}, \ldots, \overline{w}_N^{(1)}$) are selected using a stochastic process discovery adaptation of the simple ABC rejection sampling method (Algorithm 1). In this step, weight vectors are sampled from a k-dimensional uniform prior distribution $\mathcal{U}(0,1)^k$, with a sufficiently permissive user-defined tolerance ϵ_1 to ensure a satisfactory acceptance rate.

At each successive level ($i \geq 2$), the n particles are selected through a two-step probabilistic procedure. First, a particle from the set accepted at the previous level, $(\overline{w}_j^{(i-1)})_{1 \leq j \leq n}$, is randomly selected based on its probabilistic weight, $(\delta_j)_{1 \leq j \leq n}$, where δ_j represents the probability of selecting particle j. Next, a new particle for the current level is sampled using a kernel distribution $K(.|\overline{w}_j')$ centered around the previously selected particle $\overline{w}_j'$. As the weights of a sWN must be non-negative, we employed a k-dimensional normal distribution $\mathcal{N}(a,b)$, truncated to $[0,1]$. The mean a and covariance matrix b at the i-th layer are derived from the collection of n particles accepted at the previous ($i-1$)-th layer [12]. We stress that in Algorithm 2 we apply an adaptive tolerance scheme, that is: the tolerance ϵ_i for each layer $i \geq 2$, rather then being a required input constant of the algorithm (as in the original ABC-SMC method [5]), is computed as the median of the rEMD distances of the particles accepted at the previous layer $i-1$. This dynamic approach allows for more flexibility and a better control of the parameter search process as the required m threshold levels (ϵ_i) of the original ABC-SMC method are replaced by a single *improvement threshold* parameter (ζ) used to establish when to terminate the search process. Specifically, at the end of each layer, the algorithm determines whether to proceed to the next layer or terminate the procedure by comparing the prospective tolerance of the next layer with the current one (i.e., $\epsilon_i - Median(rEMD(L_E, L_{N_s(\overline{w}_j^{(i)})})_{1 \leq j \leq n}) < \zeta$).

If the improvement in tolerance is not significant (specifically, if it falls below the user-defined improvement threshold ζ), the algorithm has likely reached a local minimum at the current layer and terminates as continuing to the next layer would not result in significantly better particles. Finally, notice that such adaptive tolerance scheme ensures convergence, as the tolerance for the next layer is

Algorithm 2. ABC-SMC for stochastic process discovery

Require: E (event log), Σ_E (alphabet), $f_l(\Sigma_E)$ (activity mapping), $N_s(\overline{W})$ (sWN mined from E with weights $\overline{W}$), n (particles), ϵ_1 (first layer threshold), ζ (improvement threshold), $K(\cdot|\cdot)$ (transition Kernel), ϵ (confidence-level), δ (interval-width)
Ensure: $(\overline{w}_j)_{1 \leq j \leq n}$ drawn from π_{ABC, ϵ_M}
▷ We use $L_{N_s(\overline{w})} = CI(N_s(\overline{w}), \varphi_{slc(E)}, \epsilon, \delta)$ to denote the stochastic language of $N_s(\overline{w})$ computed by simulation.
Iteration $i = 1$: Find $(\overline{w}_j^{(1)})_{1 \leq j \leq n}$ using the sWN adaptation of ABC rejection sampling (Algorithm 1), which employs the threshold ϵ_1 as the particle acceptance parameter.
$(\delta_j)_{1 \leq j \leq n} \leftarrow \frac{1}{n}$
repeat
    $i \leftarrow i + 1$
    $\epsilon_i \leftarrow Median\left(rEMD\left(L_E, L_{N_s\left(\overline{w}_j^{(i-1)}\right)}\right)_{1 \leq j \leq n}\right)$
    for $j = 1 : n$ **do**
        repeat
            Take $\overline{w}_j'$ from $(\overline{w}_j^{(i-1)})_{1 \leq j \leq n}$ with probabilities $(\delta_j)_{1 \leq j \leq n}$
            $\overline{w}_j^{(i)} \sim K(\cdot|\overline{w}_j')$
        until $rEMD\left(L_E, L_{N_s\left(\overline{w}_j^{(i)}\right)}\right) < \epsilon_i$
        $\delta_j \leftarrow \pi(\overline{w}_j^{(i)}) \times \left(\sum_{j'=1}^n \delta_{j'}^{(i-1)} K(\overline{w}_j^{(i)}|\overline{w}_{j'}^{(i-1)})\right)^{-1}$
    end for
    Normalize $(\delta_j)_{1 \leq j \leq n}$
until $\epsilon_i - Median\left(rEMD\left(L_E, L_{N_s\left(\overline{w}_j^{(i)}\right)}\right)_{1 \leq j \leq n}\right) < \zeta$

---

strictly lower than that of the preceding layer ($\epsilon_{i-1} > \epsilon_i$). This property holds because the median is calculated solely from particles whose rEMD distances are strictly less than the previous tolerance ϵ_{i-1}.

4 Experiments and Results

We developed a prototype implementation of the HASL-based ABC-SMC parameter inference scheme described in Sect. 3. To validate it, we conducted two types of experiments. The first (Sect. 4.1) evaluates the precision and cost of the HASL-based stochastic language estimator (Definition 3) in isolation. The second (Sect. 4.2) tests the ability of ABC-SMC to infer weights such that the stochastic language of the sWN aligns with that of the log. All experiments were conducted on real-life event logs of varying complexity[8], with corresponding WN models discovered using the inductive miner algorithm [14], guaranteeing fitness equal to 1, that is, models that reproduces all traces of the log.

Prototype Tool. The tool, written in Python, uses the simulator generated by Cosmos for a GSPN instance to compute the approximate stochastic language and integrates it with a Python-based ABC-SMC implementation for parameter inference. Source code and results are publicly available in the Git repository at https://github.com/DocPierro/modelchecking_spd.git. All experiments were conducted on an Ubuntu machine with a 2.60 GHz CPU.

---

[8] From the Business Process Intelligence challenge https://data.4tu.nl/.

4.1 Accuracy of HASL-Based Stochastic Language Estimates

To assess the accuracy of the HASL-based estimation of the stochastic language of a given sWN model, we run a number of experiments using the Cosmos model checker. For each log, we discovered the corresponding WN N and then obtained several instances $N_s(\overline{W}_i)$ of the corresponding sWN by randomly generating weight vectors $\overline{W}_i \in [0,1]^{|T|}$ for the transitions T of N. For each model instance $N_s(\overline{W}_i)$, we approximated its stochastic language, evaluating the formula $\varphi_{sle(E)}$ on Cosmos, using $\epsilon = 99\%$ as confidence level and varying the length δ of the confidence interval (note that ϵ and δ determine the number of necessary simulation runs). The obtained approximation was compared, in terms of rEMD, to the exact stochastic language of the model, computed via the reachability graph unfolding method introduced in [11]. The results are shown on the left in Fig. 6. The plot indicates an increasing precision of the HASL approximations of $L_{N_s(\overline{W})}$ as the length of confidence interval δ is decreased, with the approximated and the exact language getting essentially indistinguishable (rEMD ≈ 0 with $\delta = 10^{-3}$) for all logs.

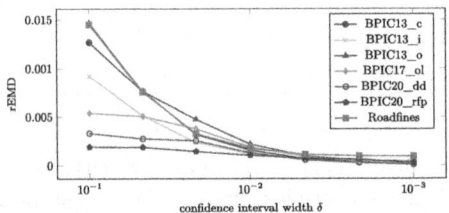

Fig. 6. HASL-based stochastic language approximation: accuracy and runtime.

The table in Fig. 6 details the computational costs of the experiments, both in terms of the number of simulations and the corresponding runtime for different values of δ. Combined with the plot in Fig. 6 the values allow one to quantify the trade-off between precision and computational costs. Notice that even with a relatively large interval size ($\delta = 10^{-1}$) we obtain precise estimates of the language (rEMD < 0.015) in a few tenths of a second (except for one log). Clearly, an increased precision is paid in terms of longer runtime which yet remains below one second for 5 out of 7 of the considered logs when an excellent precision is chosen (i.e., $\delta = 10^{-2}$ resulting in a rEMD < 0.001 for all logs).

4.2 HASL Based ABC-SMC Stochastic Process Discovery

Table 1 compares the outcomes of sWN parameter estimation resulting from the ABC-SMC procedure outlined in Sect. 3 (column "ABC-SMC opt") with (1) the reachability graph unfolding optimization method [11] (column "Unfolding opt"), (2) multiple weight estimators derived from statistic activity relations [9] (column "Burke's WE"), (3) the Wasserstein weight estimator [8] (column "WAWE") and (4) the stochastic labeled Petri net miner [15] (column "SLPN Miner"). Column "$|Tr|$" indicates the number of unique traces in the log and "$|T|$" the number of transitions in the corresponding mined WN (i.e., the dimensionality of the parameter space). The "rEMD" columns report the distance between the stochastic language of the optimized model and that of the log.

Table 1. Comparison of rEMD distances obtained with ABC-HASL parameter inference against alternative approaches: best measured distances are in bold.

Event log	$|Tr|$	$|T|$	ABC-SMC opt				Unfolding opt [11]		Burke's WE [9]		WAWE [8]	SLPN Miner [15]
			ζ	m	rEMD	runtime	rEMD	runtime	name	rEMD	rEMD	rEMD
BPIC13_c	183	19	0.01	12	0.082	563s	**0.04**	603s	fork	0.63	0.2	T/O
			0.005	17	0.081	1326s						
			0.0025	24	0.062	2770s						
BPIC13_i	1511	19	0.01	17	0.23	9401s	0.18	87086s	lhpair	0.7	0.69	T/O
			0.005	24	0.19	24427s						
			0.0025	32	**0.18**	108585s						
BPIC13_o	108	20	0.01	6	0.19	61s	**0.076**	137s	pairs	0.24	0.26	T/O
			0.005	15	0.14	340s						
			0.0025	20	0.09	1352s						
BPIC17_ol	19	11	0.01	8	0.09	197s	0.09	1.27s	freq	0.09	0.08	0.25
			0.005	19	0.07	434s						
			0.0025	29	**0.04**	4583s						
BPIC20_dd	99	43	0.01	23	0.086	2799s	**0.02**	3946s	fork	0.93	0.62	T/O
			0.005	28	0.082	5007s						
			0.0025	29	0.072	4583s						
BPIC20_rfp	89	51	0.01	14	0.58	2261s	0.41	59819s	freq	0.99	0.98	T/O
			0.005	28	0.28	7692s						
			0.0025	76	**0.08**	57634s						
Roadfines	231	34	0.01	12	0.29	1180s	0.08	1276s	pairs	0.27	0.48	T/O
			0.005	20	0.11	2312s						
			0.0025	24	**0.04**	5201s						

For "Burke's WE", among the six estimators introduced in [9], only the one yielding the lowest rEMD is reported. Experiments with the WAWE tool require setting of five hyper-parameters that may strongly affect the result quality and in this respect we resorted to reference settings given in [8]. Further, following the experimental procedure in [8] and due to variability in the output, we ran each WAWE experiment 30 times, computed the rEMD for each run, and reported the median rEMD. Except for BPIC17_ol, optimization with SLPN Miner consistently failed to terminate due to the symbolic representation of trace probabilities becoming computationally infeasible in complex scenarios. In the "ABC-SMC opt" column, ζ indicates the improvement threshold used during parameter inference and m denotes the number of explored layers. As ζ decreases, rEMD consistently improves for every log, reflecting better model fitting. This comes at the cost of longer runtimes due to the larger number of layers. For example, for the BPIC13_c log, reducing ζ from 0.01 to 0.0025 improves rEMD from 0.082 to 0.062 but increases runtime from 563 seconds to 2770 seconds.

The results in Table 1 were obtained with $N = 100$ particles, using $\epsilon = 99\%$ confidence level and confidence interval width $\delta = 0.1$ which, as illustrated in Sect. 4.1, provides one with sufficiently good accuracy.

ABC-SMC consistently discovers more accurate parameters (i.e., with lower rEMD) than "Burke's WE", "WAWE" and "SLPN Miner", and delivers competitive and sometimes superior performance compared to "Unfolding opt". While "Unfolding opt" can occasionally yield slightly better rEMD values, ABC-SMC remains competitive, especially at finer thresholds (e.g., $\zeta = 0.0025$), and is generally faster at coarser ones (e.g., $\zeta = 0.01$). This trade-off between accuracy and efficiency makes ABC-SMC a practical option when both are important. It performs robustly across logs of varying complexity, from BPIC17_ol to large-scale cases like BPIC13_i and BPIC20_rfp. For instance, on BPIC20_rfp (51 transitions), ABC-SMC achieves a better rEMD (0.28) than "Unfolding opt" (0.41) in under one-seventh of the time. While runtime increases with log complexity, especially for logs with many unique traces like BPIC13_i, ABC-SMC remains effective. This overhead stems from the extended PDF support used in Cosmos for estimating the stochastic language, which increases computational cost.

The particle search in ABC-SMC is parallelizable, offering substantial potential to reduce runtime. All reported runtimes were obtained using 16 parallel jobs. On a supercomputer, parallelization could match the number of particles per layer, bringing ABC-SMC runtimes closer to those of other methods.

Additional experiments varying the number of particles showed that fewer particles reduced computation time without significantly affecting rEMD values. Increasing the number of particles led to longer runtimes with minimal rEMD improvement. While more particles occasionally enabled faster convergence, the accuracy gains did not offset the higher computational cost. However, using more particles yields a more precise posterior over transition weights, offering better insights into their influence on the stochastic behavior of the net.

4.3 ABC Estimates of the Marginal Posterior Distributions

The n particles provided by the ABC procedure are samples from the target posterior distribution. Here we present the resulting marginal distributions over transition weights for one of the considered logs, namely, BPIC17_ol, which describes a loan application process from a Dutch financial institution and involves eight distinct activities. The corresponding WN, mined using the inductive miner algorithm, is shown in Fig. 7. It consists of 11 transitions: 8 labeled with log activities and 3 silent. Since the net contains no loops, its language L_N is finite.

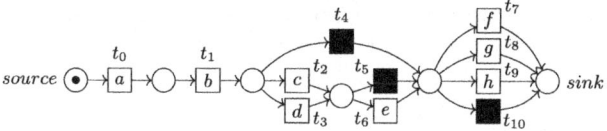

Fig. 7. WN model discovered from the BPIC17_ol log.

Figure 8 illustrates the approximation of the marginal posterior distributions for the weights of the 11 transitions t_i ($0 \leq i \leq 10$) of the sWN (Fig. 7) corresponding to log BPIC17_ol resulting from $n = 10000$ particles sampled through Algorithm 2 and using the interval $[0, 1]$ as support for each weight.

The non-informative (uniform-like) distribution of certain transitions weights (i.e., t_0, t_1) indicates their little relevance w.r.t. the conformance between the net and the log stochastic languages. Conversely, transitions whose marginal concentrates the probability mass close to the supremum of the support "dominate" those whose probability mass is close to the infimum. For example, t_2 dominates both t_3 and t_4 with which it is in conflict and, similarly, t_9 and, to a less extent, t_7 dominate over t_8 and t_{10} which they are in conflict with.

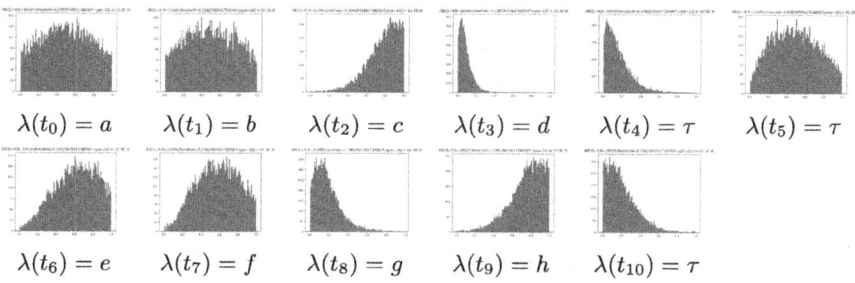

Fig. 8. Marginal posterior distributions of the weights for the 11 transitions of the BPIC17_ol model: obtained with Algorithm 2 using $n = 10000$ particles.

These dominance relationships reflect statistical characteristics of the 16 traces in the BPIC17_ol log. For instance, the conflict among transitions t_2 (labeled c), t_3 (labeled d), and t_4 (silent) arises because some traces contain only c, others only d, some neither, and none both. In the log traces with c occur with probability 0.924, those with d with 0.047, and those with neither c nor d with 0.029, summing to 1. This indicates that activity c is far more likely than d or neither. In the sWN, this distribution must be reflected in the transition weights: t_2 should have a much higher probability than t_3 and t_4. The marginal posterior distributions confirm this: all three are approximately a truncated Gaussian, with t_2 centered near 1 and t_3, t_4 near 0. This shows that t_2 must carry significantly more weight to accurately capture the observed trace probabilities. A similar pattern emerges in the conflict between t_7, t_8, t_9, and

t_{10}. Here, t_9 labeled h, which is the most probable event in the log, has a weight close to 1 to reflect this feature of the log.

5 Conclusion

We introduced a novel, simulation-based framework for stochastic process discovery. The framework replaces the computationally heavy, exact computation of the discovered model's stochastic language (of [11,15]) by an arbitrarily precise approximation obtained via synchronization with a stochastic language detector automaton. Optimal model parameters are obtained via an adaptation of *likelihood free* Bayesian inference scheme where the stochastic language approximation engine is plugged in. Through experiments on real-life event logs, we have demonstrated that, in terms of conformance, our approach 1) outperforms the weight estimators [9] as well as the WAWE framework, 2) it manages to optimize models that cannot be treated with the SLPN tool [15] and 3) it may result in a better conformance w.r.t. unfolding based optimisation [11]. Therefore we argue that simulation-based discovery of optimal models is a valuable alternative for scaling to real-world process mining applications. Finally, it is worth pointing out that posterior distribution analysis (e.g., Fig. 8), supported by our framework (but not by alternative ones), provides the user with useful added insights. We envisage two directions as future developments: first the integration, within the framework, of entropy relevance [1] as an alternative to rEMD based objective function as this could reduce the runtime of optimization and secondly to consider the extension to timed stochastic models.

References

1. Alkhammash, H., Polyvyanyy, A., Moffat, A., García-Bañuelos, L.: Entropic relevance: a mechanism for measuring stochastic process models discovered from event data. Inf. Syst. **107**, 101922 (2022)
2. Augusto, A., Conforti, R., Dumas, M., Rosa, M., Polyvyanyy, A.: Split miner: automated discovery of accurate and simple business process models from event logs. Knowl. Inf. Syst. **59**(2), 251–284 (2019)
3. Ballarini, P., Barbot, B., Duflot, M., Haddad, S., Pekergin, N.: HASL: a new approach for performance evaluation and model checking from concepts to experimentation. Perform. Eval. **90**, 53–77 (2015)
4. Ballarini, P., Bentriou, M., Cournède, P.H.: A formal approach for tuning stochastic oscillators. In: Pang, J., Niehren, J. (eds) CMSB 2023. LNCS, vol. 14137, pp. 1–17. Springer, Cham (2023). https://doi.org/10.1007/978-3-031-42697-1_1
5. Beaumont, M.A., Cornuet, J.-M., Marin, J.-M., Robert, C.P.: Adaptive approximate Bayesian computation. Biometrika **96**(4), 983–990 (2009)
6. Bentriou, M., Ballarini, P., Cournède, P.-H.: Reachability design through approximate Bayesian computation. In: Bortolussi, L., Sanguinetti, G. (eds.) CMSB 2019. LNCS, vol. 11773, pp. 207–223. Springer, Cham (2019). https://doi.org/10.1007/978-3-030-31304-3_11

7. Bentriou, M., Ballarini, P., Cournède, P.-H.: Automaton-ABC: a statistical method to estimate the probability of spatio-temporal properties for parametric Markov population models. Theor. Comput. Sci. **893**, 191–219 (2021)
8. Brockhoff, T., Uysal, M.S., Van Der Aalst, W.M.P.: Wasserstein weight estimation for stochastic petri nets. In: 2024 6th International Conference on Process Mining (ICPM), pp. 81–88 (2024)
9. Burke, A., Leemans, S.J.J., Wynn, M.T.: Stochastic process discovery by weight estimation. In: Leemans, S., Leopold, H. (eds.) ICPM 2020. LNBIP, vol. 406, pp. 260–272. Springer, Cham (2021). https://doi.org/10.1007/978-3-030-72693-5_20
10. Burke, A., Leemans, S.J.J., Wynn, M.T.: Discovering stochastic process models by reduction and abstraction. In: Buchs, D., Carmona, J. (eds.) PETRI NETS 2021. LNCS, vol. 12734, pp. 312–336. Springer, Cham (2021). https://doi.org/10.1007/978-3-030-76983-3_16
11. Cry, P., Horváth, A., Ballarini, P., Gall, P.: A framework for optimisation based stochastic process discovery. In: Hillston, J., Soudjani, S., Waga, M. (eds.) QEST+FORMATS 2024. LNCS, vol. 14996, pp. 34–51. Springer, Cham (2024). https://doi.org/10.1007/978-3-031-68416-6_3
12. Filippi, S., Barnes, C., Cornebise, J., Stumpf, M.P.H.: On optimality of kernels for approximate Bayesian computation using sequential Monte Carlo (2012)
13. Horváth, A., Ballarini, P., Cry, P.: Probabilistic process discovery with stochastic process trees. ValueTools 2024, preprint available at arXiv
14. Leemans, S.J.J., Fahland, D., Aalst, W.M.P.: Discovering block-structured process models from event logs - a constructive approach. In: Colom, J.-M., Desel, J. (eds.) PETRI NETS 2013. LNCS, vol. 7927, pp. 311–329. Springer, Heidelberg (2013). https://doi.org/10.1007/978-3-642-38697-8_17
15. Leemans, S.J.J., Li, T., Montali, M., Polyvyanyy, A.: Stochastic process discovery: can it be done optimally? In: Guizzardi, G., Santoro, F., Mouratidis, H., Soffer, P. (eds.) CAiSE 2024. LNCS, vol. 14663. Springer, Cham (2024). https://doi.org/10.1007/978-3-031-61057-8_3
16. Leemans, S.J.J., Syring, A.F., Aalst, W.M.P.: Earth movers' stochastic conformance checking. In: Hildebrandt, T., van Dongen, B.F., Röglinger, M., Mendling, J. (eds.) BPM 2019. LNBIP, vol. 360, pp. 127–143. Springer, Cham (2019). https://doi.org/10.1007/978-3-030-26643-1_8
17. Leemans, S.J.J., Aalst, W.M.P., Brockhoff, T., Polyvyanyy, A.: Stochastic process mining: earth movers' stochastic conformance. Inf. Syst. **102**, 101724 (2021)
18. Lit, B.W.: SLPN-Miner: mining stochastic labelled petri nets from event logs (2023). https://github.com/brucelit/slpn-miner
19. Marin, J.-M., Pudlo, P., Robert, C.P., Ryder, R.J.: Approximate Bayesian computational methods. Stat. Comput. **22**(6), 1167–1180 (2012)
20. Ajmone Marsan, M., Balbo, G., Conte, G., Donatelli, S., Franceschinis, G.: Modelling with generalized stochastic Petri nets. SIGMETRICS Perform. Eval. Rev. **26**(2), 2 (1998)
21. Rogge-Solti, A., Aalst, W.M.P., Weske, M.: Discovering stochastic Petri nets with arbitrary delay distributions from event logs. In: Lohmann, N., Song, M., Wohed, P. (eds.) BPM 2013. LNBIP, vol. 171, pp. 15–27. Springer, Cham (2014). https://doi.org/10.1007/978-3-319-06257-0_2
22. Rütschlin, T.: Wasserstein-spn-weight-estimation (2021). https://github.com/tbr-git/wasserstein-spn-weight-estimation
23. Sisson, S.A., Fan, Y., Beaumont, M.: Handbook of approximate Bayesian computation. Chapman and Hall/CRC (2018)

24. van der Aalst, W., Weijters, T., Maruster, L.: Workflow mining: discovering process models from event logs. IEEE Trans. Knowl. Data Eng. **16**(9), 1128–1142 (2004)
25. Aalst, W.: Process Mining: Data Science in Action, 2nd edn. Springer, Cham (2016)
26. Aalst, W., Adriansyah, A., Dongen, B.: Replaying history on process models for conformance checking and performance analysis. WIREs Data Min. Knowl. Discov. **2**(2), 182–192 (2012)
27. Werf, J.M.E.M., Dongen, B.F., Hurkens, C.A.J., Serebrenik, A.: Process discovery using integer linear programming. In: van Hee, K.M., Valk, R. (eds.) PETRI NETS 2008. LNCS, vol. 5062, pp. 368–387. Springer, Heidelberg (2008). https://doi.org/10.1007/978-3-540-68746-7_24
28. Dongen, B.F., Medeiros, A.K.A., Verbeek, H.M.W., Weijters, A.J.M.M., Aalst, W.M.P.: The ProM framework: a new era in process mining tool support. In: Ciardo, G., Darondeau, P. (eds.) ICATPN 2005. LNCS, vol. 3536, pp. 444–454. Springer, Heidelberg (2005). https://doi.org/10.1007/11494744_25
29. Broucke, S.K.L.M., Weerdt, J.: Fodina: a robust and flexible heuristic process discovery technique. Decis. Support Syst. **100**, 109–118 (2017). Smart Business Process Management
30. Weijters, A.J.M.M., van der Aalst, W.M.P., Alves de Medeiros, A.K.: Process Mining with the Heuristics Miner-algorithm. BETA Working Paper Series, WP 166, Eindhoven University of Technology, Eindhoven (2006)

Conservation Analysis and Discrete Probabilistic Approximations for Parameter Estimation of Biochemical Networks

Olivier Bouët-Willaumez, Adrien Le Coënt[✉],
Benoît Barbot, and Nihal Pekergin

Univ Paris Est Creteil, LACL, F-94010 Créteil, France
{olivier.bouet-willaumez,adrien.le-coent,benoit.barbot,
nihal.pekergin}@u-pec.fr

Abstract. In this paper, we aim to efficiently estimate parameters of biological systems. Biological systems, and more specifically biochemical networks, are usually modeled by Ordinary Differential Equations (ODEs). Instead of working directly on the ODE models, our approach relies on approximating biological systems with discrete probabilistic models, more specifically Dynamic Bayesian Networks (DBNs). The discrete approximation is used to efficiently estimate the parameters of the system by relying on Bayesian inference instead of classical optimization methods that fail due to the high dimensionality and number of parameters encountered in this type of systems. While more efficient than classical approaches, our method is also limited by dimensionality. To mitigate the dimensionality issue as much as possible, we tackle the problem of model reduction of biological systems, and the implications it has on our parameter estimation method. More specifically, we discuss an exact model reduction method that allows us to simplify the system without losing information, and we adapt the discrete models and parameter estimation methods to this reduction method. We illustrate our methods on some concrete case studies.

Keywords: Parameter estimation · Markov chains · Bayesian networks · Biochemical networks · Model reduction

1 Introduction

Ordinary Differential Equations (ODEs) are a widely used formalism for modeling dynamical systems, including biological and biochemical networks. In such systems, estimating parameters from experimental data is a central challenge, with applications in drug discovery, clinical diagnostics, and system biology.

Several ODE-based approaches have been proposed for analyzing such systems, each with specific strengths and limitations, as discussed in [16]. Classical estimation methods rely on repeatedly simulating ODEs and minimizing a cost function, but this becomes computationally prohibitive for stiff or high-dimensional models. Probabilistic approaches, particularly those based on Markov Chain Monte Carlo (MCMC), have been explored to tackle uncertainty and improve inference robustness [46], but they also suffer from high computational costs when the parameter space is large or when model stiffness is significant.

In prior work [26], we proposed a method for parameter estimation that transforms ODE models into discrete probabilistic models, specifically, Dynamic Bayesian Networks (DBNs), to enable faster cost evaluation and avoid the need for repeated ODE simulations. This approach has been shown to outperform classical ODE-based optimization for moderately sized systems (approximatively 10 variables) [27]. However, it still faces scalability limitations for larger models (approximatively 100 variables) due to the exponential growth of the state space and the increasing complexity of computing and storing transition probabilities within the DBN framework.

To address this challenge, we explore the use of model reduction techniques, which aim to simplify the system while preserving key dynamical properties. A comprehensive overview of such methods for biochemical networks is provided in [44]. These approaches can generally be divided into two categories. The first includes **structural simplification** methods, which reformulate the model to reveal intrinsic properties such as conservation laws. In particular, conservation analysis [39,47] exploits algebraic relationships among species to reduce the system's dimensionality without altering its dynamic behavior. The second category consists of **component reduction** methods, which simplify the system by eliminating or aggregating species or reactions. This includes classical lumping techniques [13] as well as more recent exact approaches based on symbolic computation [4,21,31], species lumping in stochastic models [9], and equivalence-based reductions using backward bisimulation principles [10,11]. Additionally, timescale separation methods [23,40] fall under this category, treating fast and slow subsystems separately to facilitate simplification. While such methods are promising for handling large models, their integration into parameter estimation pipelines remains nontrivial and often requires careful adaptation.

In this paper, we extend our previous discrete probabilistic framework for parameter estimation [26] by integrating conservation analysis as a model reduction technique.

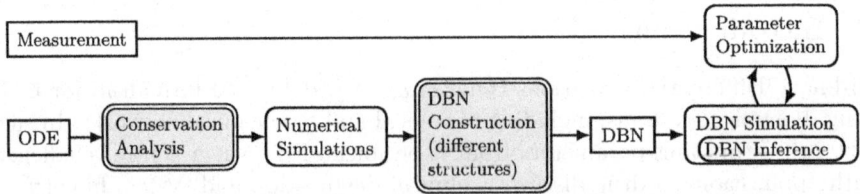

Fig. 1. Workflow of the method. Square nodes represent data structures, while rounded nodes indicate processing phases. Nodes with a gray background and double border correspond to novel contributions introduced in this paper.

The goal is to reduce the dimensionality of the biochemical system prior to discretization, thereby improving scalability and computational efficiency. To better illustrate the overall approach, Fig. 1 summarizes the proposed parameter estimation pipeline. The procedure, including the discretization and Bayesian inference steps, will be detailed in Sect. 2.

Once the reduced system is obtained through conservation analysis, we propose several alternative structures of DBNs to represent its dynamics. These different structures reflect distinct interpretations of the reduced system in terms of node selection and dependency modeling, offering various trade-offs between accuracy and computational cost. We compare their performance in terms of parameter estimation quality and runtime on two biochemical case studies.

A full version of this paper, including additional appendices, is available as a preprint on HAL [2].

The paper is structured as follows: Sect. 2 introduces the problem, the case studies, and the parameter estimation framework. Section 3 presents the conservation analysis method. Section 4 details the integration of this last reduction step into the parameter estimation pipeline. Section 5 presents the numerical results. We conclude in Sect. 6.

2 Problem Setting

2.1 Biochemical Networks

We consider the following set of differential equations as the base model for biochemical networks:

$$\dot{x}_i(t) = f_i(\boldsymbol{x}(t), \boldsymbol{k}), \tag{1}$$

the variables $\boldsymbol{x}(t) = (x_1(t), \ldots, x_n(t))$ are real-valued concentrations of species at time t. The real-valued vector $\boldsymbol{k} = (k_1, \ldots, k_m)$ contains parameter values, which are supposed to be constant over time, but unknown.

The ODE can be written in matrix form as follows:

$$\dot{\boldsymbol{x}}(t) = \boldsymbol{N} \cdot \boldsymbol{v}(\boldsymbol{x}(t), \boldsymbol{k}) \tag{2}$$

where $\boldsymbol{v}(\boldsymbol{x}(t), \boldsymbol{k}) = (v_1(\boldsymbol{x}(t), \boldsymbol{k}), \ldots, v_r(\boldsymbol{x}(t), \boldsymbol{k}))^T$ is the rate vector that contains the rate equations of the r reactions. Note that $v_1(\boldsymbol{x}(t), \boldsymbol{k}), \ldots, v_r(\boldsymbol{x}(t), \boldsymbol{k})$ are rational functions of the state variables x_i and parameters k_j. The Stoichiometric matrix $\boldsymbol{N}$ is $n \times r$ dimensional and is relating the species to the reactions they participate in [16,28]. We now present the running examples used in this paper.

Enzyme-Catalyzed Reaction Network. is a typical biochemical enzyme-catalyzed reaction with $n = 4$ variables and $m = 3$ parameters. Its continuous dynamics is given by the system of equations (3).

Let us write the rate vector $\boldsymbol{v}$ with $r = 3$ reactions: $v_1(\boldsymbol{x}(t), \boldsymbol{k}) = k_1 x_1 x_2$, $v_2(\boldsymbol{x}(t), \boldsymbol{k}) = k_2 x_3$, $v_3(\boldsymbol{x}(t), \boldsymbol{k}) = k_3 x_3$. Then the system of equations (3) can be written in matrix form (2) with the stoichiometric matrix $\boldsymbol{N}$ as defined in (4).

$$\begin{cases} \dot{x}_1 = -k_1 x_1 x_2 + k_2 x_3 \\ \dot{x}_2 = -k_1 x_1 x_2 + (k_2 + k_3) x_3 \\ \dot{x}_3 = k_1 x_1 x_2 - (k_2 + k_3) x_3 \\ \dot{x}_4 = k_3 x_3 \end{cases} \quad (3) \qquad N = \begin{bmatrix} -1 & 1 & 0 \\ -1 & 1 & 1 \\ 1 & -1 & -1 \\ 0 & 0 & 1 \end{bmatrix} \quad (4)$$

The EGF-NGF signaling pathway that is considered in this paper as a case study is a biochemical network which comprises two interacting sub-networks, namely the Epidermal Growth Factor (EGF) signaling pathway and the Nerve Growth Factor (NGF) signaling pathway. Both of EGF and NGF, and their receptors were discovered by Nobel Laureates Rita Levi-Montalcini and Stanley Cohen [12] through the intense experimental research on the regulation of chemical reactions related to cell growth. The EGF-NGF signaling pathway model was extensively studied in [5,27] to describe the effects of EGF and NGF in rat pheochromocytoma (PC12) cells. The system's dynamics and network structure are depicted in the appendices of [2]. The original model has 32 variables, 26 reactions and 48 parameters, 28 of which are known, the remaining ones needing to be learned.

2.2 Discrete Approximations of Biochemical Networks

To enable the dynamic Bayesian network formalizing, we assume that both the time and state space are discretized. We now present the Discrete-Time Markov Chain (DTMC) and Bayesian Network (BN) models used to construct approximate probabilistic representations of biochemical networks.

DTMC Approximation. Let us suppose that the ODE state intervals $\mathcal{X} = [x_1^{min}, x_1^{max}] \times \cdots \times [x_n^{min}, x_n^{max}]$ are divided into L_i sub-intervals: $[x_i^{min}, x_i^1)$, $[x_i^1, x_i^2), \ldots, [x_i^{L_i-1}, x_i^{max}]$, and that the parameter intervals $\mathcal{K} = [k_1^{min}, k_1^{max}] \times \cdots \times [k_m^{min}, k_m^{max}]$ are divided into M_j sub-intervals: $[k_j^{min}, k_j^1), [k_j^1, k_j^2), \ldots, [k_j^{M_j-1}, k_j^{max}]$. Suppose furthermore that time is discretized with time-step δ to specify an increasing sequence of time instants $(t_j)_{j=0}^T$, where $t_0 = 0$ and $\delta \cdot T = \tau$ is the observation horizon. The concentration level of each species is given by a vector of random variables:

$$\boldsymbol{X}^t = (X_1^t, X_2^t, \ldots, X_n^t)$$

where X_i^t denotes the sub-interval (discrete state of) the ith species at time t. We assume that the model satisfies the Markov property:

$$p(\boldsymbol{X}^{t+1} | \boldsymbol{X}^t, \boldsymbol{X}^{t-1}, \ldots, \boldsymbol{X}^0) = p(\boldsymbol{X}^{t+1} | \boldsymbol{X}^t)$$

Thus, the joint probability distribution from time step 0 to T can be computed as follows:

$$p(\boldsymbol{X}^{0:T}) = p(\boldsymbol{X}^0) p(\boldsymbol{X}^1 | \boldsymbol{X}^0) \cdots p(\boldsymbol{X}^T | \boldsymbol{X}^{T-1}) = p(\boldsymbol{X}^0) \prod_{t=1}^{T} p(\boldsymbol{X}^t | \boldsymbol{X}^{t-1})$$

When the ODE is time-homogeneous (time-invariant), the transition probabilities do not depend on the time step and remain constant, *i.e.* $\forall t \geq 1$:

$$p(\boldsymbol{X}^t \mid \boldsymbol{X}^{t-1}) = p(\boldsymbol{X}' \mid \boldsymbol{X})$$

This model is thus a (time-homogeneous) DTMC. The transition probabilities of a DTMC are typically stored in a transition probability matrix, $\boldsymbol{P}$ such that

$$\boldsymbol{P}[i,j] = p(\boldsymbol{X}^t = j \mid \boldsymbol{X}^{t-1} = i) = p(\boldsymbol{X}' = j \mid \boldsymbol{X} = i)$$

where i, j are any two (vector) states of the underlying stochastic process.

However, the state space of the DTMC approximations grows exponentially with the dimension of the original system, causing the transition matrix $\boldsymbol{P}$ to become prohibitively large to store explicitly for high-dimensional models. To address this limitation, we employ (dynamic) Bayesian networks, which provide a compact representation of transition probabilities.

BN Approximation of the DTMC. In essence, a Dynamic Bayesian Network (DBN) [25] allows to model conditional (in)dependencies in a probabilistic transition system. In our case, it is indeed reasonable to assume that the evolution of a given species does not depend on the entire system (at least in the short term), but rather on a subset of the species. One could for example consider that a species' dynamic depends only on those that appear on the right hand side of the ODE. While in practice this would lead to reasonable numerical results, there are some subtleties associated with these independencies hypotheses. These are beyond the scope of this paper but are discussed in detail in [26]. To summarize, we can always make some independence hypotheses that lead to significant transition probability matrix size reduction. However, the validity and impact of these assumptions depend on how the underlying probabilistic models (DTMCs or BNs) are constructed. The formal definition a (non dynamic) Bayesian network is the following.

Definition 1. *A Bayesian network is a pair* $\mathcal{B} = (\mathcal{G}, P_B)$. $\mathcal{G}$ *is a Directed, Acyclic Graph (DAG) whose nodes are random variables* $\boldsymbol{X} = X_1, \ldots, X_n$. *The graph structure represent the dependency relations among random variables: there is an incoming arc to node* X_i *from node* X_j, *if and only if the value of* X_i *depends on* X_j. *Let* $Pa^{\mathcal{G}}_{X_i}$ *denote the set of parents of* X_i *in* $\mathcal{G}$. *The distribution over* $\boldsymbol{X} = X_1, \ldots, X_n$ *can be factorized as a product:*

$$p(\boldsymbol{X}) = \prod_{i=1}^{n} p(X_i \mid Pa^{\mathcal{G}}_{X_i})$$

$p(X_i \mid Pa^{\mathcal{G}}_{X_i})$ *is called the Conditional Probability Distribution for node* X_i *specified as a Conditional Probability Table (CPT) at node* X_i. P_B *is the set of CPTs associated with* $\mathcal{G}$'s *nodes.*

The independence assumptions encoded by a DAG for Bayesian networks allows to specify the joint distribution with fewer parameters compared to unstructured models, especially when the DAG structure is sparse.

Bayesian networks are generalized to temporal models. Under Markovian and time-homogeneity assumptions, the probability distribution over infinite trajectories can be specified compactly by the initial state distribution and the conditional probabilities for the transition model in one time step, $p(\boldsymbol{X}'|\boldsymbol{X})$.

Definition 2. *A 2-Time-slice Bayesian Network (2-TBN) for a process over $\boldsymbol{X}$ is a Bayesian network specifying the evolution of random variables at time $t+1$, $\boldsymbol{X}^{t+1}$, given the random variables at time t, $\boldsymbol{X}^t$. $\boldsymbol{X}_I \subseteq \boldsymbol{X}$ is a set of input variables having no parents in $\mathcal{G}$. $\boldsymbol{X}'$ is the set of output random variables. The transition model, for any time step t, is given by conditional probabilities $p(\boldsymbol{X}' | \boldsymbol{X}_I)$. Therefore, for each output random variable X_i, the CPT $p(X'_i | Pa^{\mathcal{G}}_{X'_i})$ is given once to specify the transition probabilities from $\boldsymbol{X}^t$ at time t to $\boldsymbol{X}^{t+1}$ at time $t+1$ with $\boldsymbol{X}^t \subseteq \boldsymbol{X}_I$ for all t.*

Dynamic Bayesian Networks, represented as 2-TBNs, are well suited to the time-homogeneous models we consider. For this reason, we adopt them as approximate representations of such ODE systems. An example of a 2-TBN graph is shown in Fig. 2(a), illustrating the enzyme-catalyzed reaction case. Before proceeding with parameter estimation, the CPTs must be computed. Since exact computation is infeasible, we approximate them using the following procedure.

2.3 Computation of a BN Approximation and Parameter Estimation of ODE-Based Models

To approximate a given system with a DBN model, the transition probabilities need to be estimated for the said system. To do so, we perform a number of ODE simulations, and, for each simulation, we keep count of the visited discrete states. By doing that, we can estimate the probability of transitioning from a starting state to an arrival state as the sum of the number of trajectories that have led to transition from the starting state to the arrival state, divided by the sum of the number of trajectories that have departed from the starting state. This counting procedure also needs to keep track of the variables and parameters on which a variable depends. The details of the procedure and the associated formal definitions are given in [26].

The main task of parameter estimation consists in finding the parameter values of a system for which we have some measurements. The measurements are possibly incomplete, in the sense that they are scattered within time, or that some species are not observed, but we assume that at least some initial conditions are given. Once a DBN is computed, parameter estimation is performed using the fact that efficient inference algorithms allow to compute distributions of species concentrations over time for given parameters. More precisely, for given parameter values, and initial conditions (which can be given as distributions), the distributions of the species concentration over time are obtained by "unrolling the DBN" (inferring the distributions at a time step given the distributions at the previous time step). From this distribution, we can reconstruct some point values that can be compared to the measurements. Then, an exploration of the parameter space allows to find the parameter values that best reproduce the measurements.

Formally, given a DBN approximating a biochemical network, and a set of measurements $(t^k, x^k)_{1 \leq k \leq M}$ taken from an instance of this network. We assume that $\forall k \in \{1, \ldots, M\}$, there exists $j \in \{1, \ldots, T\}$ such that $t^k = \delta j$. For a given parameter set $\boldsymbol{K}$ and initial condition $\boldsymbol{X^0}$, species concentrations over time are computed from the posterior by unrolling the DBN:

$$p(\boldsymbol{X}^{0:T} | \boldsymbol{K}, \boldsymbol{X^0}) \tag{5}$$

Recall that the concentration of species i, X_i is a discrete value representing an interval: $X_i = l$, $1 \leq l \leq L_i$ (see Subsect. 2.2). Let $\bar{x}_i^l$ denote the midpoint of interval l: $\bar{x}_i^l = \frac{(x_i^{l-1} + x_i^l)}{2}$. For each time instant $j \in \{1, \ldots, T\}$, the real-valued concentration of species i, $\tilde{x}_i^j$, is derived through the posterior distribution as follows:

$$\tilde{x}_i^j = \sum_{l \in \{1, \ldots, L_i\}} p(X_i^j = l \mid \boldsymbol{K}, \boldsymbol{X^0}) \bar{x}_i^l \tag{6}$$

The parameters are then estimated by minimizing the sum of squared errors:

$$\varepsilon(\boldsymbol{K}) = \sum_{\{j | \exists k \text{ s.t. } t^k = \delta j\}} (x^k - \tilde{x}^j)^2 \tag{7}$$

The optimization problem: $\underset{\boldsymbol{K}}{\mathrm{argmin}}\ \varepsilon(\boldsymbol{K})$ can be solved by brute-force if the search space is small enough (e.g. for the enzyme-catalyzed reaction example with moderate state and parameter discretization), or with more advanced strategies when it cannot be explored exhaustively (e.g. for the EGF-NGF example). Many hybrid optimization strategies combining global exploration and local refinement have been proposed for discrete problems lacking derivative information [6,20,24]. When brute-force is infeasible, we use a two-phase approach: (1) global search via Halton sequence sampling [17], where the objective is evaluated at each point, and (2) local optimization of the top n_{hj} candidates using a Hooke and Jeeves pattern search [18], with up to n_{iter} iterations per run. This strategy follows similar hybrid methods found in the literature [22,36].

Limitations. The approach we propose for parameter estimation works well in practice for moderately sized case studies. It is more efficient than standard ODE-based optimization methods whenever the dimension and number of parameters gets too large. However, we are also limited by the dimension of the system since the size of the discrete state space grows exponentially with the dimension: $(n + m)^L$ if each species and parameter state is divided in L

sub-intervals. And ideally, one would want to explore the discrete state space as much as possible when building the DBN approximation. From the numerous experiments we carried out, we concluded that a reasonable trade-off between accuracy and computation time regardless of the dimension is obtained with $L \approx 5$. Consequently, in our framework, dimensionality is the primary focus of system simplification methods. Another limiting factor is the CPTs sizes. They do not depend directly on the dimension, but on the number of parents a node depends on (referred to as the *fan-in F* of a node, the CPT size of a node with *fan-in F* is $(F+1)^L$ if every node is divided in L sub-intervals).

A common drawback that we identified in usual biological systems reduction methods, such as classical lumping methods [44], is the increase in the variables dependencies in the system. Differential equivalence methods [8,9], and more specifically backward differential equivalence, are another type of lumping reduction which is less subject to the increase in variables dependencies, but they still require keeping track of initial conditions. This leads to an unavoidable increase in the *fan-in*, and consequently a significant increase in the CPTs sizes for the reduced system, canceling any computation time gain made with the dimension reduction. Conservation analysis allows to remove variables that can be expressed as linear combinations of others. When implemented correctly, *i.e.*, using a reduced system DBN graph structure that avoids a drastic increase in *fan-in*, it is possible to significantly reduce the system's dimensionality while maintaining a manageable *fan-in*. This, in turn, results in substantial computational time savings for both DBN construction and parameter estimation.

3 Model Reduction by Conservation Analysis

In the original system (2), species dynamics are determined by reaction rates. However, certain network structures induce intrinsic conservation relations [39, 44,47], leading to linear dependencies among species and among the rows of the stoichiometric matrix $\boldsymbol{N}$. Consequently, the rank of $\boldsymbol{N}$ is reduced compared to the case without conservation relations. We denote this rank by n_0.

The conservation analysis method aims to partition species into independent and dependent subsets [47]. This is achieved by separating linearly independent and dependent rows of the stoichiometric matrix $\boldsymbol{N}$. Let n_0 denote the number of independent species $\boldsymbol{x}_i(t)$, with the remaining $n - n_0$ being dependent species $\boldsymbol{x}_d(t)$. By rearranging rows, $\boldsymbol{N}$ can be written as:

$$\boldsymbol{N} = \begin{bmatrix} \boldsymbol{N_I} \\ \boldsymbol{N_D} \end{bmatrix}, \tag{8}$$

where $\boldsymbol{N_I} \in \mathbb{R}^{n_0 \times r}$ is full rank and contains the independent rows of $\boldsymbol{N}$, and $\boldsymbol{N_D} \in \mathbb{R}^{(n-n_0) \times r}$ contains the dependent rows of $\boldsymbol{N}$.

Note that in certain networks, such a partition of the stoichiometric matrix $\boldsymbol{N}$ may not be possible [39], as all species may be independent, *i.e.*, $n = n_0$ and $\boldsymbol{N} = \boldsymbol{N_I}$. We propose a theoretical upper bound on n_0 in the appendices of [2].

3.1 Link-Zero Matrix L_0 and System Partitioning

We begin by analyzing the relationship between the submatrix of dependent species, N_D and the submatrix of independent species, N_I.

Proposition 1 (Link-Zero Matrix [35,39,47]). *Let N_D and N_I denote the dependent and independent row blocks of the stoichiometric matrix N, respectively. If the rows of N_D are linear combinations of those in N_I, then there exists a matrix L_0 such that*

$$N_D = L_0 \cdot N_I \tag{9}$$

This result enables the reformulation of the system dynamics in terms of independent and dependent species via the partitioned stoichiometric matrix.

Corollary 1 (Partitioning of the System and Species Dynamics). *Given the stoichiometric matrix partition (8), the dynamics of the full system (2) can be rewritten as*

$$\dot{x}(t) = \begin{bmatrix} N_I \\ L_0 \cdot N_I \end{bmatrix} \cdot v(t) \tag{10}$$

This form separates the dynamics of the independent and dependent species. More explicitly, the system can be expressed as:

$$\dot{x}_i(t) = N_I \cdot v(t) \tag{11}$$
$$\dot{x}_d(t) = L_0 \cdot N_I \cdot v(t) \tag{12}$$

where $x_i(t)$ and $x_d(t)$ denote the concentrations of independent and dependent species, respectively.

Building on this formulation, we next simplify the dynamics of the dependent species by expressing them directly in terms of the independent species:

Corollary 2 (Simplified Dependent Species Dynamics). *From the partitioned equations (11) and (12), the dynamics of the dependent species reduce to*

$$\dot{x}_d(t) = L_0 \cdot \dot{x}_i(t) \tag{13}$$

Integrating this equation gives the dependent species' concentrations as a function of the independent species:

$$x_d(t) = x_d(0) + L_0 \cdot [x_i(t) - x_i(0)] \tag{14}$$

where $x_d(0)$ is the initial concentration of dependent species, and $x_i(t)$ and $x_i(0)$ are the concentrations of the independent species at time t and at $t = 0$, respectively.

3.2 Computing the Link-Zero Matrix L_0

A variety of approaches were proposed to find the link-zero matrix L_0, depending on the size of the network considered as well as the nature of the equations [39,44]. These include computing the null space of the stoichiometric matrix N, using Gaussian elimination or SVD [44,47]. However, such methods are often numerically unstable for very large systems. In [47], Vallabhajosyula et al. uses full pivoting QR decomposition with Householder reflections, a more numerically stable method, even for systems with large condition number. The definition of the full pivoting QR Decomposition is the following:

Definition 3 (Full Pivoting QR Decomposition [47]). *Let $A \in \mathbb{R}^{n \times r}$ be a non-symmetric matrix. A **full pivoting QR decomposition** of A is a factorization of the form*

$$A \cdot P = Q \cdot R \tag{15}$$

where $Q \in \mathbb{R}^{n \times n}$ is an orthogonal matrix ($Q^\top Q = I_n$), $P \in \mathbb{R}^{r \times r}$ is a permutation matrix representing column exchanges in A, and $R \in \mathbb{R}^{n \times r}$ is an upper trapezoidal matrix (i.e., lower triangular elements are zero). The permutation matrix P enhances numerical stability by reordering the columns of A during the QR factorization process, typically using Householder reflections [19], as refined by Golub [15]. Full pivoting improves the decomposition's robustness by selecting optimal pivots at each step [45].

Application to N^T Matrix. To compute the link-zero matrix L_0, Vallabhajosyula et al. [47] use the transpose of the stoichiometric matrix N^T and apply a full pivoting QR decomposition. This is because the permutation matrix P reorders the columns of N^T (equivalent to the rows of N), which is essential for generating Q and R. The full pivoting QR decomposition of N^T is:

$$N^T \cdot P_* = Q_* \cdot R_* \tag{16}$$

where the index $*$ denotes matrices derived from the QR decomposition of N^T [47]. Multiplying both sides of (16) by Q_*^T yields:

$$Q_*^T \cdot N^T \cdot P_* = R_* \tag{17}$$

Since R_* is upper trapezoidal, if the rows of N are linearly dependent, R_* can be partitioned into 4 sub-matrices [47]. After performing a Gaussian elimination on R_*, each non-zero row will have a unity on the main diagonal and zeros above the diagonal, leading to:

$$R_* = \left[\begin{array}{c|c} I & M \\ \hline 0 & 0 \end{array}\right] \tag{18}$$

Finally, the link-zero matrix is given by $L_0 = M^T$ [35,39,47].

3.3 Application on the Running Examples

For the enzyme-catalyzed reaction, the method reduces the original system (3) of 4 species to a subsystem with only 2 species:

$$\begin{cases} \dot{x}_2 = -k_1 x_1 x_2 + (k_2 + k_3)\left(-x_2 + x_3(0) + x_2(0)\right) \\ \dot{x}_1 = -k_1 x_1 x_2 + k_2 \left(-x_2 + x_3(0) + x_2(0)\right) \end{cases} \quad (19)$$

The reduction process and resulting system are detailed in the appendices of [2].

In the case of the EGF-NGF model, the original system comprising 32 species is reduced to a subsystem with only 13 independent species, as shown in the appendices of [2]. This reduction is especially beneficial, as it significantly lowers the computational cost of numerical integration. For instance, running 100,000 simulations with 11 time steps using SciPy's `solve_ivp` LSODA integrator [32,42] took 561 s for the original system, whereas the same setup applied to the reduced system required only 394 s. This numerical integration phase is approximately reduced by a factor of 1.42.

4 BN-Based Parameter Estimation Relying on Conservation Analysis

The expression of the dependent species in (14) can be substituted into the independent species equations (11), allowing simulation of the system using only the independent variables. This reduced system is used for simulations which significantly decreases the computation time spent in the ODE solver during DBN construction as shown in Subsect. 3.3. However, since our parameter estimation procedure relies on measurements that may include dependent species, we must still reconstruct the full set of species concentrations over time although they are rarely required in the parameter estimation procedure.

DBN Construction. A key challenge is defining an appropriate DBN structure for the reduced system. As shown in Equation (14), species' initial conditions can be substituted into the system dynamics and retained in the graph. We propose four specific methods to incorporate this information. It is important to note that explicitly representing the independent variables while approximating the dependent variables through injected initial conditions may result in some loss of information.

In the *no-var* approach, the species' initial conditions are entirely omitted from the DBN, resulting in the most compact graph in terms of CPT size but at the cost of significant information loss. In contrast, the *unstacked* method retains each relevant initial condition as a distinct parameter node in the graph, preserving the full dynamics of the original system but increasing the *fan-in*. The *stacked* approach mitigates this increase by aggregating frequently recurring algebraic expressions of initial conditions (often sums, as seen in (14) and (19)) into a single parameter node. While this reduces complexity, it introduces information loss due to discretizing only the aggregated value. Finally, in the *merged* method, suitable when the DBN simulation horizon is limited to a single time step (*i.e.* $\tau = \delta$), the initial condition of a dependent species is identical to its parent node's value, and only the latter is retained. This ensures good coverage of the discrete state space under tight simulation budgets. For longer time horizons, this method requires reconstructing intermediate values, and we propose an *ad hoc* procedure to reconstruct the initial condition distribution during unrolled inference.

In Fig. 2, we illustrate these proposed methods on the enzyme-catalyzed reaction for which the corresponding reduced system (19) is given in Subsect. 3.3.

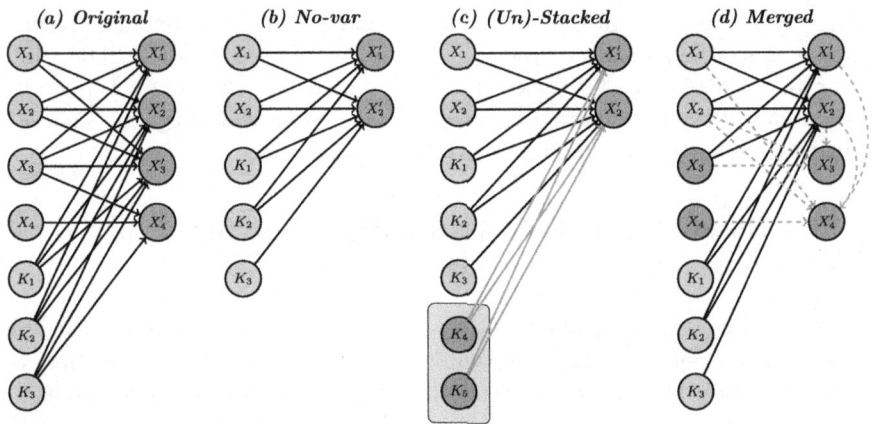

Fig. 2. 2-TBN structures for the original and reduced models in the enzyme-catalyzed case. (a) *Original* model as depicted in [27]. (b) New parameters are treated as constants. (c) New parameters are grouped (*Stacked*) or treated individually (*Unstacked*). (d) Dependent species are computed via conservation laws at each step.

Reconstruction Procedure for *Merged* Graph Inference. We discuss here how the inference may be performed from the proposed DBN structures letting us take into account the species initial conditions. For the *stacked* and *unstacked* graphs, unrolling inference is straightforward, as only the distribution of the species at time 0 is required to estimate the distributions of the independent species at any time t. However, *merged* graphs require a distribution at time t to estimate the distribution at time $t + 1$.

More precisely, let us denote by $\boldsymbol{X}_i$ the random variables associated to the independent variables, and by $\boldsymbol{X}_d$ the random variables associated to the dependent variables, we add an exponent t to denote their values at (discrete) time t. In the *stacked* and *unstacked* graphs, we compute the distributions of species over time using:
$$p(\boldsymbol{X}_i^{t+1}|\boldsymbol{X}_i^t, \boldsymbol{K}, \boldsymbol{X}^0)$$

where the values of the species at time 0 can be viewed as parameters. We can see in Fig. 2 that X_2^0 and X_3^0 are injected as parameters K_4 and K_5.

In the *merged* graph, we estimate the distributions of species by iterating:

$$p(\boldsymbol{X}_i^{t+1}|\boldsymbol{X}_i^t, \boldsymbol{X}_d^t, \boldsymbol{K}, \boldsymbol{X}^0).$$

Since CPTs are computed on the interval $[0, \delta]$, $\boldsymbol{X}_d^t = \boldsymbol{X}_d^0$, this is thus the same as computing

$$p(\boldsymbol{X}_i^{t+1}|\boldsymbol{X}^t, \boldsymbol{K})$$

For a one step computation, there is no approximation. But to compute $\boldsymbol{X}_i^{t+1}$ for $t \geq 2$, we need to reconstruct a value for $\boldsymbol{X}_d^t$, we do this with a reconstruction function $\boldsymbol{X}_d^{t+1} = h(\boldsymbol{X}^t, \boldsymbol{X}_i^{t+1})$ which implements a heuristics that computes a distribution for $\boldsymbol{X}_d^{t+1}$ by first computing a continuous value from $\boldsymbol{X}^t$, $\boldsymbol{X}_i^{t+1}$ and then computing a random variable with realizations that fit this continuous value. The reconstruction is illustrated by the red dotted lines in Fig. 2.

Fan-In Analysis. The maximum *fan-in* is generally the most critical, bottleneck inducing value, as it reflects the size of the largest CPT in the DBN. This section outlines how different reduced model graphs affect the fan-in of a node.

The *no-var* graph always reduces fan-in when conservation analysis identifies dependent species, as their removal leads to fewer incoming arcs to the remaining nodes. The *unstacked* graph significantly increases *fan-in*, often counter-balancing any dimensionality reduction. Although dependent variables are removed, a separate parameter node is introduced per species' initial condition, adding numerous new arcs. The *stacked* graph leads to a null or moderate increase in *fan-in*. While dependent variables are removed, they are replaced by a new node representing the species' initial conditions sums (see (14)). As a result, some independent species may depend on more variables than in the original model. The *merged* graph preserves the original *fan-in*, and all the information about species' initial conditions is kept for inference. Overall, the *stacked* graph is the most broadly applicable, as it requires no hypothesis on the DBN construction, whereas the *merged* graph offers the best balance between reduction and inference accuracy.

5 Numerical Results

Implementation. The approach was implemented in Python and is available online as a prototype called BayeSBML [1]. Biochemical networks were specified using SBML files via SimpleSBML [7,38]. CPTs were constructed with pgmpy [33], and Bayesian network modeling and inference were performed using pyAgrum [34]. To improve efficiency, we applied Just-In-Time (JIT) compilation with Numba [29], achieving near-C performance.

All computations were performed on a standard laptop[1], except for CPTs generation in the EGF-NGF case, which required a more powerful machine[2].

We now compare the reduced models presented in Sect. 4 with the original model using our running examples.

Enzyme-Catalyzed Reaction Network. The CPTs are generated with 78,125 ODE simulations via SciPy's `solve_ivp` LSODA integrator [32,42], over 11 time steps.

Table 1. Comparison of the original and reduced models for the enzyme-catalyzed reaction network example.

Model	Original	No-var	Stacked	Unstacked	Merged
$\operatorname*{argmin}_{K} \varepsilon(K) := \tilde{K}$	(3,4,1)	(3,2,2)	(3,2,1)	(1,1,1)	(2,2,1)
$\operatorname*{min}_{K} \varepsilon(K) := \varepsilon(\tilde{K})$	0.0538	0.0604	0.0502	0.0431	0.0481
State-space Size	78125	3125	31250	78125	15625
Total CPTs Rows	172500	18750	187500	468750	93750
fan-in (Cum., Max)	(20,6)	(9,5)	(11,6)	(13,7)	(11,6)

Due to the system's low dimensionality, the increased *fan-in* in the *unstacked* model is not substantial. As reported in Table 1, this is the only model with a higher maximum *fan-in*, despite a reduced number of species. All other reduced models achieve lower *fan-in* values, both cumulative and maximal, compared to the original. As expected, the reduction in state space size is most significant for the *no-var* model, with an approximate 20-fold decrease. The state space is reduced by a factor of 5 in the *merged* model and by a factor of 2.5 in the *stacked* model, while no reduction is observed in the *unstacked* model.

We also observe that the total CPTs size (summed across all nodes) increases significantly for the *unstacked* model. In the *no-var* and *merged* models, some nodes are entirely removed, resulting in smaller CPTs. For the *stacked* model, although two variables are eliminated, a new variable representing species' initial conditions sum is introduced. This variable, discretized 10 times (5 per original variable), appears in all reduced equations, slightly increasing the overall CPTs size. A similar mechanism occurs in the *unstacked* model, but since a separate variable is introduced for each initial condition, the increase in CPTs size is more pronounced.

To evaluate the predictive quality of the models and the accuracy of their DBN approximations, we carry out a parameter estimation procedure identical to that described in Subsect. 2.3. The continuous parameter space is discretized

---

[1] Asus Zenbook 15 equipped with an 8-Core AMD Ryzen^{TM} 7 7735U processor and 32 GB of RAM.
[2] 16-Core AMD EPYC 7281 processor with 128 GB of RAM.

into 5 intervals per parameter as described in Subsect. 2.2, resulting in a grid of $5^3 = 125$ possible configurations. Next, we generate a reference trajectory with a known initial condition $x_0 = (5, 1, 7.5, 1.5)$ and a continuous parameter vector $k = (0.3, 0.3, 0.3)$, $K = (1, 1, 1)$ if discretized, that we aim to recover. A brute-force search is then performed over this discrete space to identify the optimal parameter configuration, $\tilde{K}$ that minimizes the objective function defined in (7). Once the optimal discrete combination $\tilde{K}$ is found, it is mapped back to a continuous estimate $\tilde{k}$ by taking the midpoints of the corresponding intervals. This estimated parameter vector is then used to simulate the model and assess its predictive performance.

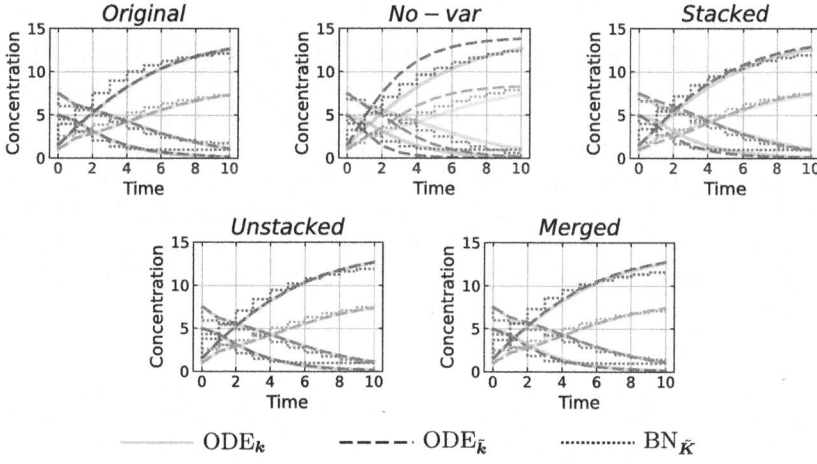

Fig. 3. Species concentrations in the enzyme-catalyzed reaction example. ODE_k: simulation with nominal parameters; $ODE_{\tilde{k}}$: simulation with learned parameters through LSODA scheme and 5 discrete states; $BN_{\tilde{K}}$: the expected values for the proposed BN-based approximation with learned parameters.

Figure 3 compares the original ODE trajectory (ODE_k), the ODE trajectory using the estimated parameters ($ODE_{\tilde{k}}$), and the BN-derived trajectory based on $\tilde{K}$. The latter is indeed the expected value of the state at each time step using the midpoint-weighted distributions as described in (6).

As shown in Table 1, all reduced models, except *no-var*, match or outperform the original in predictive accuracy. However, the estimated parameter configurations $\tilde{K}$ differ notably from the reference $K = (1, 1, 1)$ for most models, highlighting a well-known issue in biochemical modeling: ***parameter identifiability*** [14,43,48,49]. That is, distinct parameter sets can yield similar system behavior, complicating reliable parameter inference.

The EGF-NGF Model. The CPTs are generated using 1,000,000 ODE simulations over 2 time steps. The high dimensionality makes the *unstacked* version impractical, with a state space size of 5^{84} (vs. 5^{52} for the original), resulting in massive CPTs of size 5.9 Go (vs 10.6 Mo for the original) and infeasible inference times, due to the addition of a parameter node per species' initial condition.

Table 2. Comparison of the original and reduced models for the EGF-NGF case study.

Model	Original	No-var	Stacked	Unstacked	Merged
Total CPTs - Rows	1.01e6	3.31e4	3.96e7	5.69e8	4.66e5
Total CPTs - Size (Mo)	10.62	0.35	415.68	5967.24	4.88
fan-in (Cum., Max)	(206,11)	(77,10)	(108,13)	(131,16)	(94,11)

The *merged* and *no-var* models are efficient with significantly smaller CPTs, as shown in Table 2. The *stacked* model is more complex but still usable. We focus on the *original*, *merged*, *stacked*, and *no-var* models, excluding the *unstacked* model due to its prohibitive computational cost.

Regarding structural complexity, the cumulative *fan-in* in the reduced models is up to half that of the original (see Table 2). The *stacked* model has a higher maximum *fan-in* than the original, leading to prohibitive inference costs, while the *merged* and *no-var* models maintain a maximum *fan-in* equal to or smaller than the original.

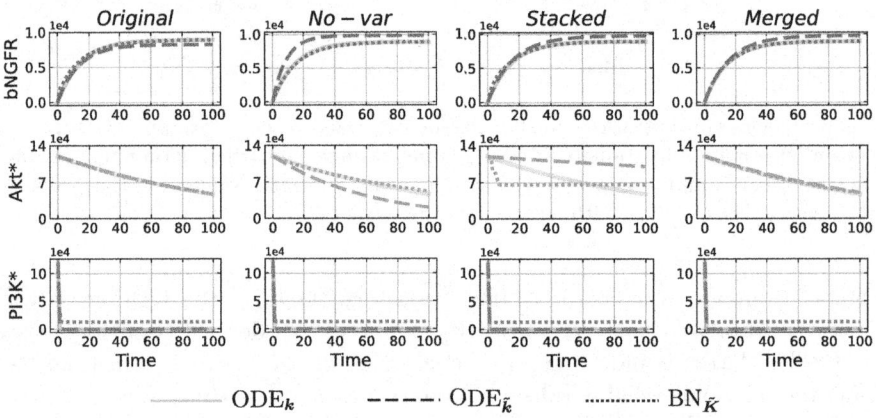

Fig. 4. boundNGFR(x_6), inactive Akt(x_{24}) and inactive PI3K(x_{22}) concentrations in the EGF-NGF reaction example. ODE$_k$ simulation with nominal parameters; ODE$_{\tilde{k}}$: simulation with learned parameters through LSODA scheme and 5 discrete states; BN$_{\tilde{K}}$: the expected values for the proposed BN-based approximation with learned parameters.

For parameter estimation, we generated a reference trajectory with x_0 and k specified in the appendices of [2]. The continuous parameter space is discretized into 5 intervals, resulting in a grid of 5^{20} possible configurations. Due to the large size, we employ the exploration/exploitation method outlined in Sect. 2.3, using 250 Halton sequence samples, $n_{hj} = 3$, and $n_{iter} = 25$.

We selected key signaling species from the MAPK/ERK and PI3K-Akt pathways, which are the primary cascades activated by EGF and NGF stimulation [27,30,37,41]. The analysis focuses on the dynamics of boundNGFR (x_6), and the inactive forms of PI3K (x_{22}) and Akt (x_{24}), due to their central roles in signal transduction and feedback regulation.

Similarly to the enzyme-catalyzed case, Fig. 4 compares the original ODE trajectory (ODE$_k$), the ODE trajectory using the estimated parameters (ODE$_{\tilde{k}}$), and the BN-derived trajectory based on $\tilde{K}$.

For both running examples, we observe comparable behaviors, except for the *stacked* model that struggles due to its large state space in the EGF-NGF case (see Akt* in Fig. 4).

6 Conclusion

The proposed approach leverages a model simplification technique that directly impacts the structure of our probabilistic approximation. Though previously unexplored, it yields significant performance gains, enhancing the method's applicability to high-dimensional case studies.

In future work, we aim to address the parameter identifiability problem to ensure robust and reliable estimation. We also plan to collaborate with biologists to apply our methods to real-world data, further validating their practical applicability. Other possible directions include parameter identification of more complex models, such as partial differential equations, and better sampling techniques.

Regarding implementation, we intend to integrate efficient CPU parallelization and batching into our existing framework BayeSBML to accelerate numerical simulations (and thus reduce the CPTs construction time) as well as the parameter optimization procedure.

Funding Information. This work has been partly supported by ANR projects MAVeriQ (ANR-20-CE25-0012) and the joint ANR-JST project CyPhAI, France.

Data Availibility Statement. All models, scripts, and tools used to reproduce our experimental evaluations, including the resulting tables and figures, are archived and publicly available at DOI https://doi.org/10.5281/zenodo.15284362 [3].

References

1. bayesbml. https://git.lacl.fr/barbot/bayesbml.git
2. Bouët-Willaumez, O., Le Coënt, A., Barbot, B., Pekergin, N.: Conservation Analysis and Discrete Probabilistic Approximations for Parameter Estimation of Biochemical Networks (2025). https://hal.science/hal-05104966. Working paper or preprint

3. Bouët-Willaumez, O., Barbot, B., Le Coënt, A., Pekergin, N.: Artifact: conservation analysis and discrete probabilistic approximations for parameter estimation of biochemical networks (2025). https://doi.org/10.5281/zenodo.15284362
4. Brenig, L.: Reducing nonlinear dynamical systems to canonical forms. Philos. Trans. Roy. Soc. A Math. Phys. Eng. Sci. **376**(2124), 20170384 (2018)
5. Brown, K.S., et al.: The statistical mechanics of complex signaling networks: nerve growth factor signaling. Phys. Biol. **1**(3), 184 (2004)
6. Caflisch, R.E., Morokoff, W., Owen, A.B.: Valuation of mortgage-backed securities using brownian bridges to reduce effective dimension. J. Comput. Finance **1**(1), 27–46 (1997)
7. Cannistra, C., Medley, K., Sauro, H.M.: SimpleSBML: a python package for creating and editing SBML models. bioRxiv (2015). https://api.semanticscholar.org/CorpusID:61794957
8. Cardelli, L., Tribastone, M., Tschaikowski, M., Vandin, A.: Symbolic computation of differential equivalences. ACM SIGPLAN Not. **51**(1), 137–150 (2016)
9. Cardelli, L., Tribastone, M., Tschaikowski, M., Vandin, A.: ERODE: a tool for the evaluation and reduction of ordinary differential equations. In: Legay, A., Margaria, T. (eds.) TACAS 2017. LNCS, vol. 10206, pp. 310–328. Springer, Heidelberg (2017). https://doi.org/10.1007/978-3-662-54580-5_19
10. Cardelli, L., Tribastone, M., Tschaikowski, M., Vandin, A.: Maximal aggregation of polynomial dynamical systems. Proc. Natl. Acad. Sci. **114**(38), 10029–10034 (2017). https://doi.org/10.1073/pnas.1702697114. https://www.pnas.org/doi/abs/10.1073/pnas.1702697114
11. Cardelli, L., Tribastone, M., Tschaikowski, M., Vandin, A.: Symbolic computation of differential equivalences. Theor. Comput. Sci. **777**, 132–154 (2019). https://doi.org/10.1016/j.tcs.2019.03.018. https://www.sciencedirect.com/science/article/pii/S0304397519301719
12. Cohen, S.: Origins of growth factors: NGF and EGF. J. Biol. Chem. **283**(49), 33793–33797 (2008). https://doi.org/10.1074/jbc.X800008200
13. Dokoumetzidis, A., Aarons, L.: Proper lumping in systems biology models. IET Syst. Biol. **3**(1), 40–51 (2009)
14. Eisenberg, M.C., Hayashi, M.A.: Determining identifiable parameter combinations using subset profiling. Math. Biosci. **256**, 116–126 (2014). https://doi.org/10.1016/j.mbs.2014.08.008. https://www.sciencedirect.com/science/article/pii/S0025556414001631
15. Golub, G.H., Van Loan, C.F.: Matrix Computations. The Johns Hopkins University Press (1996)
16. Gratie, D.E., Iancu, B., Petre, I.: ODE analysis of biological systems, pp. 29–62. Springer, Heidelberg (2013)
17. Halton, J.: On the efficiency of certain quasi-random sequences of points in evaluating multi-dimensional integrals. Numerische Mathematik **2**, 84–90 (1960). http://eudml.org/doc/131448
18. Hooke, R., Jeeves, T.A.: Direct search solution of numerical and statistical problems. J. ACM (JACM) **8**(2), 212–229 (1961)
19. Householder, A.S.: Unitary triangularization of a nonsymmetric matrix. J. ACM **5**(4), 339–342 (1958)
20. Huyer, W., Neumaier, A.: Global optimization by multilevel coordinate search. J. Global Optim. **14**(4), 331–355 (1999)
21. Jiménez-Pastor, A., Jacob, J.P., Pogudin, G.: Exact linear reduction for rational dynamical systems. In: International Conference on Computational Methods in Systems Biology, pp. 198–216. Springer (2022)

22. Kang, F., Li, J., Li, H.: Artificial bee colony algorithm and pattern search hybridized for global optimization. Appl. Soft Comput. **13**(4), 1781–1791 (2013). https://doi.org/10.1016/j.asoc.2012.12.025. https://www.sciencedirect.com/science/article/pii/S1568494612005698
23. Klonowski, W.: Simplifying principles for chemical and enzyme reaction kinetics. Biophys. Chem. **18**(2), 73–87 (1983)
24. Kolda, T.G., Lewis, R.M., Torczon, V.: Optimization by direct search: new perspectives on some classical and modern methods. SIAM Rev. **45**(3), 385–482 (2003)
25. Koller, D., Friedman, N.: Probabilistic Graphical Models - Principles and Techniques. MIT Press (2009)
26. Le Coënt, A., Barbot, B., Pekergin, N.: Efficient probabilistic inference in biochemical networks. Comput. Biol. Med. **183**, 109280 (2024). https://doi.org/10.1016/j.compbiomed.2024.109280. https://www.sciencedirect.com/science/article/pii/S0010482524013659
27. Liu, B., Hsu, D., Thiagarajan, P.S.: Probabilistic approximations of odes-based bio-pathway dynamics. Theoret. Comput. Sci. **412**(21), 2188–2206 (2011)
28. Maarleveld, T., Khandelwal, R., Olivier, B., Teusink, B., Bruggeman, F.: Basic concepts and principles of stoichiometric modeling of metabolic networks. Biotechnol. J. **8**(9), 997–1008 (2013)
29. numba. https://numba.pydata.org/
30. Okada-Hatakeyama, M., et al.: A computational model on the modulation of mitogen-activated protein kinase (MAPK) and AKT pathways in Heregulin-induced erbb signalling. Biochem. J. **373**, 451–63 (2003). https://doi.org/10.1042/BJ20021824
31. Ovchinnikov, A., Pérez Verona, I., Pogudin, G., Tribastone, M.: Clue: exact maximal reduction of kinetic models by constrained lumping of differential equations. Bioinformatics **37**(12), 1732–1738 (2021)
32. Petzold, L.: Automatic selection of methods for solving stiff and nonstiff systems of ordinary differential equations. SIAM J. Sci. Stat. Comput. **4** (1983)
33. pgmpy. https://pgmpy.org/
34. pyagrum. https://agrum.gitlab.io/
35. Reder, C.: Metabolic control theory: a structural approach. J. Theor. Biol. **135**(2), 175–201 (1988)
36. Rios-Coelho, A., Sacco, W., Henderson, N.: A metropolis algorithm combined with Hooke-Jeeves local search method applied to global optimization. Appl. Math. Comput. **217**(2), 843–853 (2010). https://doi.org/10.1016/j.amc.2010.06.027. https://www.sciencedirect.com/science/article/pii/S0096300310007125
37. Santos, S.D.M., Verveer, P.J., Bastiaens, P.I.H.: Growth factor-induced MAPK network topology shapes ERK response determining PC-12 cell fate. Nat. Cell Biol. **9**(3), 324–330 (2007)
38. Sauro, H.M.: Simplesbml: a python package for creating, editing, and interrogating SBML models: version 2.0 (2021). https://arxiv.org/abs/2009.01969
39. Sauro, H.M., Ingalls, B.: Conservation analysis in biochemical networks: computational issues for software writers. Biophys. Chem. **109**(1), 1–15 (2004)
40. Schauer, M., Heinrich, R.: Quasi-steady-state approximation in the mathematical modeling of biochemical reaction networks. Math. Biosci. **65**(2), 155–170 (1983)
41. Schoeberl, B., Eichler-Jonsson, C., Gilles, E., Müller, G.: Computational modeling of the dynamics of the map kinase cascade activated by surface and internalized EGF receptors. Nat. Biotechnol. **20**, 370–375 (2002). https://doi.org/10.1038/nbt0402-370

42. scipy.integrate. https://docs.scipy.org/doc/scipy/reference/integrate.html
43. Simpson, M.J., Baker, R.E.: Parameter identifiability, parameter estimation and model prediction for differential equation models (2025). https://arxiv.org/abs/2405.08177
44. Snowden, T.J., van der Graaf, P.H., Tindall, M.J.: Methods of model reduction for large-scale biological systems: a survey of current methods and trends. Bull. Math. Biol. **79**(7), 1449–1486 (2017)
45. Strang, G.: Linear Algebra and Learning from Data. Cambridge University Press (2019)
46. Valderrama-Bahamóndez, G.I., Fröhlich, H.: MCMC techniques for parameter estimation of ode based models in systems biology. Front. Appl. Math. Stat. **5**, 55 (2019)
47. Vallabhajosyula, R.R., Chickarmane, V., Sauro, H.M.: Conservation analysis of large biochemical networks. Bioinformatics **22**(3), 346–353 (2006)
48. Villaverde, A., Barreiro, A.: Identifiability of large nonlinear biochemical networks. MATCH Commun. Math. Comput. Chem. **76**, 259–296 (2016)
49. Wieland, F.G., Hauber, A.L., Rosenblatt, M., Tönsing, C., Timmer, J.: On structural and practical identifiability. Curr. Opinion Syst. Biol. **25**, 60–69 (2021). https://doi.org/10.1016/j.coisb.2021.03.005

Formal Approximations of the Transient Distributions of the M/G/1 Workload Process

Fabian Michel(✉) [ID] and Markus Siegle [ID]

Universität der Bundeswehr München,
Werner-Heisenberg-Weg 39, 85579 Neubiberg, Germany
{fabian.michel,markus.siegle}@unibw.de

Abstract. This paper calculates transient distributions of a special class of Markov processes with continuous state space and in continuous time, up to an explicit error bound. We approximate specific queues on $\mathbb{R}$ with one-sided Lévy input, such as the M/G/1 workload process, with a finite-state Markov chain. The transient distribution of the original process is approximated by a distribution with a density which is piecewise constant on the state space. Easy-to-calculate error bounds for the difference between the approximated and actual transient distributions are provided in the Wasserstein distance. Our method is fast: to achieve a practically useful error bound, it usually requires only a few seconds or at most minutes of computation time.

Keywords: Formal error bounds · Lévy-driven queues · Markov chain approximation · Transient distributions

1 Introduction

Most of the theory in formal methods for stochastic systems is restricted to systems where either the state space or the time is discrete. In contrast, we consider systems where both the state space and the time are continuous. In particular, we would like to calculate transient distributions of a Markov process in continuous time and with continuous state space. As exact computations are typically infeasible, we approximate the transient distributions using a discretization approach, and we provide formal error bounds for the difference between the actual and approximated transient distribution. In this paper, we focus on queues with one-sided Lévy input, as the case of a general Markov process seems to be very difficult to analyze.

The queues covered by our method include the M/G/1 workload process, in which jobs with arbitrarily distributed sizes arrive according to a Poisson process with rate λ and are served by a single server. As a motivating example, consider a server setup where the capacity was chosen such that all jobs can be dealt with reasonably quickly under the average expected load. Now, assume that a higher than usual job arrival rate is expected during a short time period (e.g. due to new events becoming available in a booking system, which many users try to book at once). Then, we can look at the transient workload distribution of the server with the higher job arrival rate to assess how congested the server will become in the short heavy load period. Using our error bounds, we can provide guarantees that, with high probability, the server's workload at a given time will e.g. not exceed a given amount. In practice, if the probability of a catastrophic congestion in the heavy load period is too high, the system administrator could decide to temporarily increase the system's capacity.

While many results on particular properties of the transient distributions are available for these types of processes (e.g. moments [4], probability of being idle [7, p. 24]), calculating the transient distributions themselves up to some controllable formal error has received little attention. Our method can be used to do exactly that (compute these distributions together with formal error bounds), and, with the transient distribution available, a variety of questions about the underlying process can be answered. The error bounds are explicit and easy to calculate, and the computational cost to meet a predetermined accuracy is reasonably small – more precisely, the cost grows cubically with the inverse of the desired accuracy. Other approaches, such as the numerical inversion of Laplace transforms of the transient distributions, usually do not offer error bounds or only at an unreasonable amount of computational cost (see [3, p. 10], [5, Section 5]). Compared to the Laplace transform inversion algorithms which are most widely available in libraries, our method is both faster and more accurate in our numerical experiments (see Sect. 4).

1.1 Literature Review

Formal error bounds for approximations of Markov processes have been considered in various settings, but mostly for models where either space or time are discrete. An exception is [14], but it only looks at pure jump processes.

Next to [14], one of the most similar papers to the present work is [13], which considers a Markov process with general state space in discrete time. The transition kernel as well as the initial distribution are assumed to be expressible with a (Lipschitz-continuous) probability density, the process is approximated with a finite-state Markov chain and the densities of the transient distributions are approximated by piecewise constant densities. This is the same approach that we follow below. However, we look at continuous-time models and do not assume that the transient distributions of the original model admit a density. We therefore use a different metric to measure the error: the Wasserstein distance instead of the $\|\cdot\|_\infty$-norm applied to densities as in [13]. On the other hand, we restrict ourselves to the state space $\mathbb{R}$, unlike [13].

There are some works, such as [2], on approximate model checking for stochastic hybrid systems, which usually have a continuous component in the state space. For more literature in that direction, also consult the reference lists from [2,13]. There is also a large body of work on models with discrete state space and continuous time, i.e., continuous-time Markov chains. For example, [1,11] provide error bounds for an approximation of discrete- and continuous-time Markov chains via state space reduction.

While there seems to be no literature on formal error bounds for approximating the transient distribution of general Markov processes, some models with continuous time and continuous state space have received considerable attention, in particular in the analysis of queueing systems. In [4], some transient characteristics of the M/G/1 workload process are considered, in particular its moments. The authros of [7] give a good overview on the theory behind a more general class of queues with continuous state space and in continuous time: so-called Lévy-driven queues. They are defined using a Lévy process (a special type of Markov process with stationary and independent increments) whose state space is then restricted to the non-negative reals. This is the setting we focus on, and we will give more details in Sect. 2.1.

As it turns out, the transient distribution of such queues can often be characterized by explicit expressions for their Laplace transforms. Thus, another approach for calculating transient distributions is the numerical inversion of these transforms. One should note that the characterizations are often only given in terms of double or triple transforms, where next to the Laplace transform of the distribution, additional transforms in the time variable or in the initial state are considered. [6] follows this approach for the distribution of the running maximum of a Lévy process, and reports promising results using the inversion technique from [8], albeit without formal error bounds. [3,5] propose to use different inversion techniques and compare the results to estimate the error, as the computational cost for meeting a pre-defined formal error bound is often excessive. We will show that a Markov chain approximation can solve this problem for transient distributions of Lévy-driven queues.

1.2 Our Contribution

We present an easy-to-implement method which approximates specific queues with one-sided Lévy input by a finite-state Markov chain, and which provides explicit and easy-to-calculate error bounds for the transient distributions in the Wasserstein distance. In particular, the transient distribution of the original Lévy-driven queue at time points which are multiples of the discretization parameter Δ will be approximated by a density which is piecewise constant on the intervals $(k\Delta, (k+1)\Delta]$ for $k \in \mathbb{N}$. This density is obtained by lifting the discrete distribution of the approximate model to the original, continuous state space.

2 Preliminaries

Consider a general Markov process X_t with continuous state space and in continuous time. Assume we want to calculate the transient distribution at a given point in time. One of the main issues when approximating the transient distributions is the famous butterfly effect – small deviations can result in a completely different future behavior [12]. Therefore, we consider Lévy-driven queues which offer the advantage that the process behavior is basically the same everywhere in the state space, making it easier to control approximation errors.

2.1 Lévy-Driven Queues

We will restrict ourselves to a subclass of Markov processes: the workload processes of M/G/1 queues, and queues fed by spectrally negative compound Poisson processes. The state Q_t of the M/G/1 workload process at time t is the current workload which needs to be processed by the single server. New jobs arrive at rate λ. Their sizes are iid with the distribution of some arbitrary random variable on $\mathbb{R}$ which we will call B. When a job arrives, Q_t jumps upwards by the size of the new job. Otherwise, Q_t decreases at constant rate r, corresponding to a constant processing speed, if $Q_t > 0$. Both these M/G/1 queues and their spectrally negative counterpart belong to the class of Lévy-driven queues, for which the theory is already well developed. We follow [7] to present the most important concepts in this context.

A Lévy process on $\mathbb{R}$ is a Markov process X_t with stationary and independent increments, and we also require $X_0 = 0$. Lévy processes can be described by three components: a deterministic drift r (the process moves with constant speed r upwards or downwards), a Brownian motion part, and a jump part. For a more concise presentation, we only consider processes without a Brownian motion part, and which allow only finitely many jumps in a finite time interval. In addition, we consider so-called spectrally one-sided Lévy processes, which either only jump upwards or only jump downwards. Our method can be generalized to some extent, but this would lead to unnecessarily complex formulas.

For such Lévy processes (no Brownian motion part, finite jump intensity, spectrally one-sided), the description is simpler than for general Lévy processes. The jump rate of X_t into the set $X_t + A$ is defined as $\Pi(A)$, where Π is the so-called Lévy measure on $\mathbb{R} \setminus \{0\}$, and where $A \subseteq \mathbb{R}$ is a Borel set. The measure Π is finite in our case, and either supported on $(0, \infty)$ for upward jumps, or on $(-\infty, 0)$ for downward jumps. The term spectrally positive process is used for the former and spectrally negative for the latter.

Given a spectrally positive or negative Lévy process X_t with $X_0 = 0$, we want to define a queue $Q_t \geq 0$ with net input X_t by letting Q_t be a modified version of X_t which can never go below 0. Q_t, started at Q_0, is formally defined by setting $Q_t = X_t + \max\{Q_0, -\inf_{0 \leq s \leq t} X_s\}$. Q_t behaves like a shifted version of X_t, except if $Q_t = 0$ and X_t moves down – in this case, Q_t stays at 0. Q_t is no Lévy process, but it is a Markov process.

Q_t is called the workload of the queue at time t, and X_t is called the net input process, the latter incorporating both the arrivals and the processing of jobs. X_t being spectrally positive corresponds to jobs with varying workloads arriving (according to the measure Π), and then being processed at a constant rate r by the server, given by the deterministic speed of X_t. This type of process is also called a compound Poisson process (see also [7, page 12, item (2)]).

We will use the following notation for compound Poisson processes, both of spectrally positive and negative type: jumps occur at rate $\lambda = \Pi(\mathbb{R} \setminus \{0\})$, and we let the random variable $\widetilde{B}$ have law $\lambda^{-1}\Pi$. The jump distances are then an iid sequence with the distribution of the random variable $B = |\widetilde{B}|$. In many typical examples, the deterministic speed of the Lévy process X_t is in the direction opposite to the jump directions. We thus denote by $r > 0$ the constant speed at which X_t decreases in the spectrally positive case, while we use $r > 0$ for the speed at which X_t increases in the spectrally negative case.

As mentioned before, the queue arising from a spectrally positive compound Poisson process can be seen as the workload process of an M/G/1 queue with server speed r, job arrival rate λ and job size distribution given by B. The spectrally negative case could for example be used to model an insurance company which receives premium at rate r per time unit and which has to pay claims with size distribution B, arriving at rate λ.

Stationary and transient distributions of Lévy-driven queues can be computed by numerically inverting (single, double, or even triple) Laplace transforms, as explained in [7, Sections 3.1, 3.2, 4.1, and 4.2]. A brief overview is also given in Appendix 1 of [10]. The discretization approach which we use below has the advantage of providing better error bounds at a lower computational cost.

2.2 The Wasserstein Distance

In our approach to approximate the transient distribution of Q_t, we will use the Wasserstein distance to formally bound the distance between the actual transient distribution and its approximation. The choice of the Wasserstein distance is deliberate. Other distance measures such as the total variation distance often assign the maximal distance to two probability measures which are orthogonal/singular, which is the case for a Dirac measure and a measure with a density w.r.t. the Lebesgue measure. This would be problematic since we discretize the original process and approximate its transient distribution by combinations of uniform distributions over small intervals (see below). For example, a process started with $Q_0 = x > 0$, i.e., a Dirac measure, would already cause the maximal possible error in the initial approximation if we used the total variation distance. Even if the initial distribution is not an issue, jump distributions with atoms, among others, will not work well in conjunction with such distances.

The Wasserstein distance is better suited to our approach. For two probability measures μ and ν on $\mathbb{R}$ (with the Borel σ-algebra), it is defined as

$$\mathrm{WD}(\mu,\nu) = \inf_{\gamma} \int_{\mathbb{R}^2} |x-y| \, \mathrm{d}\gamma(x,y) \stackrel{[15]}{=} \min_{\gamma} \int_{\mathbb{R}^2} |x-y| \, \mathrm{d}\gamma(x,y) \qquad (1)$$

where γ ranges over all couplings of μ and ν, i.e., we have $\gamma(A \times \mathbb{R}) = \mu(A)$ and $\gamma(\mathbb{R} \times A) = \nu(A)$ for measurable A (the marginal distributions of γ are μ and ν, respectively). The coupling minimizing the above expression describes how to shift the probability mass of one distribution along the real line in an optimal way to obtain the second distribution. We note that, by [15], as μ and ν are distributions on $\mathbb{R}$, this definition is equivalent to

$$\text{WD}(\mu, \nu) = \int_{\mathbb{R}} |F_\mu(x) - F_\nu(x)|\, dx \tag{2}$$

where F_μ and F_ν are the cumulative distribution functions (CDFs) of μ and ν.

3 Discretization with Formal Error Bounds

We will start by approximating the evolution of the workload process Q_t of an M/G/1 queue with a discrete-time Markov chain. This will allow us to obtain approximations of the transient distributions of the process. To simplify notation, we will assume that the service speed of the M/G/1 queue is fixed at $r = 1$. As we still allow an arbitrary job arrival rate $\lambda > 0$, this is no real restriction.

We discretize the model in space and time, and we truncate the state space to $[0, M]$ with $M > 0$. The precision of the approximation is controlled via the discretization parameter Δ, and we choose M to be a multiple of Δ: $M = M_\Delta \cdot \Delta$ with $M_\Delta \in \mathbb{N}$. We approximate Q_t with a discrete-time Markov chain $\widetilde{Q}_k$ on the state space $\{0, 1, \ldots, M_\Delta\}$. The state $\widetilde{Q}_k = 0$ approximates the state $Q_{k\Delta} = 0$ in the original model, while $\widetilde{Q}_k = i \geq 1$ should hold (approximately) when $Q_{k\Delta} \in ((i-1)\Delta, i\Delta]$. We discretize space and time with precisely the same step size Δ due to the service speed being 1. This will be important later.

If μ_t is the law of Q_t, and if $p_k \in \mathbb{R}^{M_\Delta + 1}$ is the distribution of $\widetilde{Q}_k$, given by $p_k^\mathsf{T} = p_0^\mathsf{T} P^k$ (with p_0 and P still to be defined), then we approximate $\mu_{k\Delta}$ with

$$\begin{aligned}\widetilde{\mu}_k &:= \mathbb{P}\left[\widetilde{Q}_k = 0\right] \cdot \delta_0 + \sum_{i=1}^{M_\Delta} \mathbb{P}\left[\widetilde{Q}_k = i\right] \cdot U\big((i-1)\Delta, i\Delta\big) \\ &= p_k(0) \cdot \delta_0 + \sum_{i=1}^{M_\Delta} p_k(i) \cdot U\big((i-1)\Delta, i\Delta\big)\end{aligned} \tag{3}$$

where δ_0 is the Dirac measure in 0 and $U(a,b)$ is a uniform distribution over the interval $[a, b]$. We later provide a formal bound on $\text{WD}(\mu_{k\Delta}, \widetilde{\mu}_k)$, so that $\widetilde{\mu}_k$ (which we can calculate easily) and this bound can then be used in practice to verify properties of the actual transient distribution $\mu_{k\Delta}$. Note that $\widetilde{\mu}_k$ is supported on $[0, M]$, while $\mu_{k\Delta}$ is supported on the entire positive half-line.

The most reasonable choice for p_0 is the following:

$$p_0(0) := \mathbb{P}[Q_0 = 0], \qquad p_0(i) := \mathbb{P}\left[Q_0 \in ((i-1)\Delta, i\Delta]\right] \ (i \geq 1) \tag{4}$$

If M is chosen large enough such that the initial distribution μ_0 of Q_0 is supported on $[0, M]$, then this choice of p_0 ensures that $\text{WD}(\mu_0, \widetilde{\mu}_0) \leq \Delta$: the probability mass contained in every interval of length Δ is correctly represented in $\widetilde{\mu}_0$ by definition (see (3) and (4)), and, in consequence, the mass in $\widetilde{\mu}_0$ has to be shifted by a distance of at most Δ to obtain μ_0. Hence, there exists a coupling γ for which the integral in (1) is $\leq \Delta$. We now proceed to explain how P should be calculated, and how to derive error bounds.

3.1 Transition Matrix of Discretized M/G/1 Model

We want to choose P such that

$$P(i,j) \approx \mathbb{P}\left[Q_\Delta \in ((j-1)\Delta, j\Delta] \,|\, Q_0 \sim U((i-1)\Delta, i\Delta)\right]$$

for $i,j \geq 1$ (we have to adapt the expression for the special state 0). This ensures that (approximately) the right amount of probability mass is transferred from the interval $((i-1)\Delta, i\Delta]$ to the interval $((j-1)\Delta, j\Delta]$ in the discrete model if the starting distribution is uniform over the discretization intervals. The distribution of Q_Δ will in general not be uniform over these intervals. This incurs a discretization error at every time step, as we replace the actual distribution of Q_Δ with a combination of uniform distributions in the discretized model.

We can calculate $P(i,j)$ for the M/G/1 queue explicitly, up to a controllable error. Recall that jobs whose sizes are iid arrive at rate λ and are served at constant speed $r = 1$. We write F_B for the CDF of B, a random variable having the job size distribution. With probability $e^{-\lambda \Delta}$, no new job arrives within time Δ and the probability mass is simply shifted by Δ downwards in the state space. With probability $\lambda \Delta e^{-\lambda \Delta}$, one new job arrives in that same time interval. We will ignore two or more jobs arriving within the interval $[0, \Delta]$. The reason will become apparent later – basically, it is enough to consider only one job arriving in order to obtain error bounds which decrease with decreasing Δ (for a given fixed time point at which the transient distribution should be calculated). This is discussed in more detail in Sect. 3.4.

Conditional One-Jump CDFs. Let

$$F_{oj}^{(s)}(y) := \mathbb{P}\left[Q_\Delta \leq y \,|\, Q_0 = s, 1 \text{ job arrival in } [0, \Delta]\right]$$

be the CDF of Q_Δ, conditioned on one jump (oj) in the time interval $[0, \Delta]$ and started with $Q_0 = s$. We further write

$$F_{oj}^{[i]}(y) := \mathbb{P}\left[Q_\Delta \leq y \,|\, Q_0 \sim U((i-1)\Delta, i\Delta), 1 \text{ job arrival in } [0, \Delta]\right] \quad (i \geq 1)$$

$$F_{oj}^{[0]}(y) := \mathbb{P}\left[Q_\Delta \leq y \,|\, Q_0 = 0, 1 \text{ job arrival in } [0, \Delta]\right] = F_{oj}^{(0)}(y)$$

We will proceed by deriving expressions for $F_{oj}^{(s)}(y)$ and $F_{oj}^{[i]}(y)$, which we can then use to calculate $P(i,j)$. We have to distinguish two cases with respect to s.

Case $s \geq \Delta$. If $s \geq \Delta$, then the server will not idle within time Δ and

$$\begin{aligned} F_{oj}^{(s)}(y) &= \mathbb{P}[Q_\Delta \leq y \,|\, Q_0 = s, 1 \text{ job arrival in } [0, \Delta]] \\ &= \mathbb{P}[s + B - \Delta \leq y] = \mathbb{P}[B \leq y + \Delta - s] \quad & (s \geq 1) \\ &= F_B(y + \Delta - s) \end{aligned}$$

This holds because for Q_Δ to be $\leq y$, we need that the starting workload s plus the new job size B minus the processed workload within the time interval $[0, \Delta]$

(that is, Δ, due to $r = 1$) is $\leq y$. In consequence,

$$F_{\text{oj}}^{[i]}(y) = \frac{1}{\Delta}\int_{(i-1)\Delta}^{i\Delta} F_{\text{oj}}^{(s)}(y)\,\mathrm{d}s = \frac{1}{\Delta}\int_{y-(i-1)\Delta}^{y-(i-2)\Delta} F_B(s)\,\mathrm{d}s \qquad (i \geq 2)$$

Here, we averaged with respect to the uniform distribution over the interval $[(i-1)\Delta, i\Delta]$, which is the starting distribution of Q_0 in the definition of $F_{\text{oj}}^{[i]}$.

Case $s < \Delta$. For $s < \Delta$, we need to consider that the server might idle some of the time within the interval $[0, \Delta]$. To simplify calculations, we will define the idle time as the time spent at 0 before the new job arrives (we are still conditioning on one job arrival). It is possible that Q_t first reaches 0, then a very small job arrives, and Q_t reaches 0 again before time Δ. However, it will be easier to consider only the time spent at 0 before the arrival as the idle time. In fact, for the following calculations, we will let Q_t take negative values instead of being absorbed in 0, continuing to decrease at constant speed 1, but *only if* Q_t reaches 0 *after* the new job has already arrived. Before the new job arrives, Q_t will be held at 0 as before in case the workload s present at time 0 has already been processed. $F_{\text{oj}}^{(s)}(y)$ will thus be positive for $y > -\Delta$, and we still have

$$F_{\text{oj}}^{(s)}(y) = \mathbb{P}\left[Q_\Delta \leq y \mid Q_0 = s, 1 \text{ job arrival in } [0, \Delta]\right] \qquad (y \geq 0)$$

both in the original setting and if we let Q_t take negative values after the job arrival, the equality just does not hold for $y < 0$.

The idle time of the server before the new job arrival within $[0, \Delta]$, if started with workload s at time 0, is distributed as $\frac{\Delta-s}{\Delta}U(0, \Delta - s) + \frac{s}{\Delta}\delta_0$: the time of the new job arrival is distributed uniformly over $[0, \Delta]$ (when conditioning on one arrival), and thus, with probability $\frac{s}{\Delta}$, the new job arrives before the old workload is processed (which would happen at time s) and the server does not idle. With probability $\frac{\Delta-s}{\Delta}$, the job arrives after 0 has been reached; then, the idle time is uniformly distributed between 0 and $\Delta - s$. The processing time is distributed as Δ minus the idle time, i.e., its distribution is $\frac{\Delta-s}{\Delta}U(s, \Delta) + \frac{s}{\Delta}\delta_\Delta$.

We can now write down the equation for $F_{\text{oj}}^{(s)}(y)$:

$$F_{\text{oj}}^{(s)}(y) = \frac{\Delta - s}{\Delta} \cdot \frac{1}{\Delta - s}\int_s^\Delta F_B(y + t - s)\,\mathrm{d}t + \frac{s}{\Delta}F_B(y + \Delta - s) \qquad (y \geq -\Delta)$$

$$= \frac{1}{\Delta}\int_s^\Delta F_B(y + t - s)\,\mathrm{d}t + \frac{s}{\Delta}F_B(y + \Delta - s)$$

Note: we averaged over the possible processing times of the server, and the factor $\frac{1}{\Delta-s}$ in the first line originates from the density of the distribution $U(s, \Delta)$.

The above expression directly yields $F_{\text{oj}}^{[0]}(y)$:

$$F_{\text{oj}}^{[0]}(y) = F_{\text{oj}}^{(0)}(y) = \frac{1}{\Delta}\int_0^\Delta F_B(y + t)\,\mathrm{d}t = \frac{1}{\Delta}\int_y^{y+\Delta} F_B(s)\,\mathrm{d}s$$

Furthermore, we have

$$F_{oj}^{[1]}(y) = \frac{1}{\Delta}\int_0^{\Delta} F_{oj}^{(s)}(y)\,ds$$

$$= \frac{1}{\Delta^2}\int_0^{\Delta}\left(\int_s^{\Delta} F_B(y+t-s)\,dt + sF_B(y+\Delta-s)\right)ds$$

$$= \ldots = \frac{2}{\Delta^2}\int_y^{y+\Delta}(y+\Delta-s)F_B(s)\,ds$$

The final expression can be obtained by exchanging the order of the inner and the outer integral, as well as by a linear substitution in the integration variables.

Calculating $P(i,j)$. We can use the CDFs from above for a first approximation

$$\check{P}(i,j) := e^{-\lambda\Delta}\left(\mathbb{1}_{\{j=i-1 \vee i=j=0\}} + \lambda\Delta\left(F_{oj}^{[i]}(j\Delta) - F_{oj}^{[i]}((j-1)\Delta)\right)\right)$$

The indicator function corresponds to the case that no jobs arrive (in which case the probability mass simply shifts one discrete state to the left), and the second summand to the case with one job arrival – more job arrivals are ignored in this approximation. As we ignore more jumps and as we cut off jumps out of the truncated state space, $\check{P}$ will be a substochastic matrix. We define P by $P = \check{P} + D$ where D is a diagonal matrix with non-negative entries such that P is stochastic.

We can make the above expression for $\check{P}(i,j)$ more explicit: for $i \geq 2$,

$$F_{oj}^{[i]}(j\Delta) - F_{oj}^{[i]}((j-1)\Delta) = \frac{1}{\Delta}\int_{(j-i+1)\Delta}^{(j-i+2)\Delta} F_B(s)\,ds - \frac{1}{\Delta}\int_{(j-i)\Delta}^{(j-i+1)\Delta} F_B(s)\,ds$$

$$= \frac{1}{\Delta}\left(\int_{(j-i)\Delta}^{(j-i+1)\Delta}(F_B(s+\Delta) - F_B(s))\,ds\right)$$

and hence (equivalent calculations can be done for $i=0, i=1$)

$$\check{P}(i,j) = e^{-\lambda\Delta}\left(\mathbb{1}_{\{j=i-1\}} + \lambda\int_{(j-i)\Delta}^{(j-i+1)\Delta}(F_B(s+\Delta) - F_B(s))\,ds\right) \quad (i \geq 2)$$

$$\check{P}(0,j) = e^{-\lambda\Delta}\left(\mathbb{1}_{\{j=0\}} + \lambda\int_{(j-1)\Delta}^{j\Delta}(F_B(s+\Delta) - F_B(s))\,ds\right)$$

$$\check{P}(1,j) = e^{-\lambda\Delta}\left(\mathbb{1}_{\{j=0\}} + \frac{2\lambda}{\Delta}\int_{(j-1)\Delta}^{j\Delta}(j\Delta - s)(F_B(s+\Delta) - F_B(s))\,ds\right)$$

To find $\check{P}$, we thus need to integrate the function $s \mapsto F_B(s+\Delta) - F_B(s) = \mathbb{P}[s < B \leq s+\Delta]$ (for $\check{P}(1,j)$, we actually calculate a convolution with a piecewise linear triangle function and not just a simple integral). Depending on the distribution of B, we might be able to derive exact expressions for these integrals, otherwise we use numerical integration.

3.2 Transition Matrix of Discretized Spectrally Negative Model

Assume now that Q_t is the Lévy-driven queue fed by a spectrally negative Lévy process X_t. X_t is a compound Poisson process with constant upwards speed $r = 1$ and with downward jumps occurring at rate λ, the jump sizes being iid with the distribution of the random variable B.

We discretize the state space exactly as in the M/G/1 case, described at the beginning of Sect. 3. The discretized state 0 can be dropped in the spectrally negative case, as 0 will be left immediately if a jump down to 0 occurs, due to the constant positive speed of 1. However, in some situations, it might make sense to make the state 0 absorbing in the spectrally negative case, corresponding e.g. to an insurance company going bankrupt. In such a case, we would keep the discretized state 0 (and we would of course also have to adapt the transition probabilities of the discrete model).

The calculations here are simpler than in the M/G/1 case, and can be found in Appendix 2 of [10]. We also end up with a transition matrix P of the discrete model (indexed by indices 1 through M_Δ, if we drop state 0), defined as $P = \check{P} + D$ where D is a diagonal matrix with non-negative entries ensuring stochasticity and

$$\check{P}(i,j) = \begin{cases} e^{-\lambda \Delta}\left(\mathbb{1}_{\{j=i+1\}} + \lambda \int_{(i-j)\Delta}^{(i-j+1)\Delta} \left(F_B(s+\Delta) - F_B(s)\right) ds\right) & \text{if } j \geq 2 \\ e^{-\lambda \Delta} \cdot \lambda \left(\Delta - \int_{(i-1)\Delta}^{i\Delta} F_B(s) \, ds\right) & \text{if } j = 1 \end{cases}$$

3.3 Error Bounds

We now derive an error bound for every step in the discrete model – a bound on how much the difference between the actual transient distribution and the approximated distribution can increase per step in the Wasserstein distance. Assume that the process starts with initial law μ_0, i.e., $Q_0 \sim \mu_0$. We are given an approximation $\tilde{\mu}_0$ of μ_0 via the distribution p_0 of $\tilde{Q}_0$ over the aggregates/intervals as in (3). We do *not* assume that p_0 satisfies (4) because we want to apply the analysis below to all time steps and not just the initial one. Instead, we assume that we have a bound b_0 on the Wasserstein distance $\text{WD}(\mu_0, \tilde{\mu}_0)$.

We calculate the distribution of $\tilde{Q}_1$ via the matrix P, and we want to bound $\text{WD}(\mu_\Delta, \tilde{\mu}_1)$, where μ_Δ is the distribution of Q_Δ, which we want to approximate with $\tilde{\mu}_1$, obtained from the distribution of $\tilde{Q}_1$. This bound can be applied iteratively to upper bound the Wasserstein distance $\text{WD}(\mu_{k\Delta}, \tilde{\mu}_k)$ for any k and therefore give a formal error estimate. We use the strategy depicted in Fig. 1:

- First, we look at how the error which is already present in the initial approximation evolves over the time interval $[0, \Delta]$. Consider Markov processes Q and Q', started with initial distributions $Q_0 \sim \mu_0$ and $Q'_0 \sim \tilde{\mu}_0$, both evolving according to the original dynamics of the Lévy-driven queue. Given the bound $\text{WD}(\mu_0, \tilde{\mu}_0) \leq b_0$, we will derive a bound b_1 on $\text{WD}(\mu_\Delta, \text{Law}(Q'_\Delta))$.

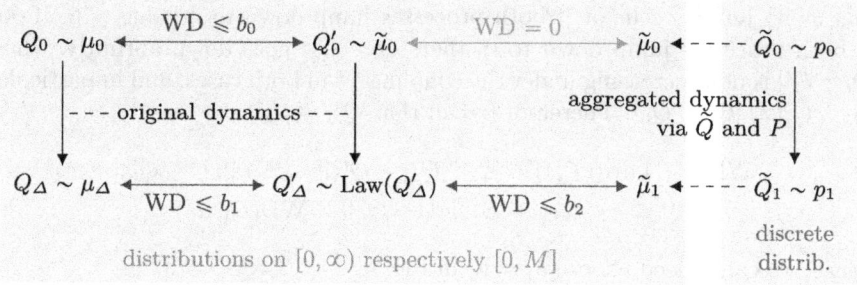

Fig. 1. Bounding the Wasserstein distance

- Next, we look at the error caused by approximating the dynamics (averaging over the intervals and truncation). We will derive a bound b_2 on $\mathrm{WD}(\mathrm{Law}(Q'_\Delta), \widetilde{\mu}_1)$ where $\widetilde{\mu}_1$ is the distribution as given by (3) for $k = 1$. We can calculate the distribution of $\widetilde{Q}_1$ easily via the matrix P.
- By the triangle inequality for the Wasserstein distance, we can then conclude that $\mathrm{WD}(\mu_\Delta, \widetilde{\mu}_1) \leq \mathrm{WD}(\mu_\Delta, \mathrm{Law}(Q'_\Delta)) + \mathrm{WD}(\mathrm{Law}(Q'_\Delta), \widetilde{\mu}_1) \leq b_1 + b_2$.

Error Caused by Initial Approximation. Here, we show that we can choose $b_1 = b_0$ in Fig. 1. That is, if we consider processes started with $Q_0 \sim \mu_0$ and $Q'_0 \sim \widetilde{\mu}_0$, both evolving according to the same dynamics of the Lévy-driven queue, then the Wasserstein distance of their transient distributions is bounded by the initial distance $\mathrm{WD}(\mu_0, \widetilde{\mu}_0)$. The proof uses couplings. We can find a coupling γ of Q_0 and Q'_0 with $\mathbb{E}_\gamma[|Q_0 - Q'_0|] = \mathrm{WD}(\mu_0, \widetilde{\mu}_0)$. We will extend γ to a coupling of the two entire processes (and not just their initial states).

Let $t_1, t_2, \ldots$ be the sequence of times at which the process Q_t (with $Q_0 \sim \mu_0$) jumps, and let $h_1, h_2, \ldots$ be the corresponding jump heights. Note: t_1 as well as $t_j - t_{j-1}$ for $j \geq 2$ are iid with distribution $\mathrm{Exp}(\lambda)$, independently of the particular value of Q_0, and they are in addition independent of $h_1, h_2, \ldots$. The sequence $h_1, h_2, \ldots$ is itself also an iid sequence of jump heights with the distribution of B. The jump times and heights of Q'_t (with $Q'_0 \sim \widetilde{\mu}_0$) follow the same distribution, for both the M/G/1 and the spectrally negative case.

We can therefore extend the coupling γ from the pair (Q_0, Q'_0) to a coupling γ^* of the pair $((Q_t)_{t \geq 0}, (Q'_t)_{t \geq 0})$ by simply letting Q'_t jump with the same height whenever Q_t jumps. The remaining behavior of both processes is determined by the constant speed $r = 1$. We look at how the distance $|Q_t - Q'_t|$ evolves with t under this extended coupling γ^*: both processes will perform synchronous jumps, and they will both move downwards with speed 1 (or upwards in the spectrally negative case), as long as they are not in 0. Hence, for the M/G/1 queue, the distance $|Q_t - Q'_t|$ will stay constant as long as the processes are either both > 0 or both in 0. When only one process is in 0, then the distance $|Q_t - Q'_t|$ decreases with speed 1, as the other process will approach 0 with speed 1. In the spectrally negative case, the behavior is similar: $|Q_t - Q'_t|$ will stay constant as long as no jumps occur or if both processes jump down to a value > 0. If one or both processes jump down to 0, then $|Q_t - Q'_t|$ goes down abruptly. Thus, $|Q_t - Q'_t|$ is non-increasing under the coupling γ^* in both cases, and in particular $|Q_t - Q'_t| \leq |Q_0 - Q'_0|$. Therefore (recall that $Q_\Delta \sim \mu_\Delta$),

$$\mathrm{WD}(\mu_\Delta, \mathrm{Law}(Q'_\Delta)) \leq \mathbb{E}_{\gamma^*}\left[|Q_\Delta - Q'_\Delta|\right] \leq \mathbb{E}_{\gamma^*}\left[|Q_0 - Q'_0|\right]$$
$$= \mathbb{E}_\gamma\left[|Q_0 - Q'_0|\right] = \mathrm{WD}(\mu_0, \widetilde{\mu}_0)$$

Hence, we can indeed choose $b_1 = b_0$ in Fig. 1.

Error Caused by Aggregated Dynamics. Here, we derive a bound b_2 for Fig. 1. That is, we let Q'_0 start with distribution $\widetilde{\mu}_0$ as obtained from a given p_0 using (3), and we then want to bound the distance between the law of Q'_Δ (where Q'_t evolves according to the original process dynamics) and $\widetilde{\mu}_1$ as obtained from p_1, where $p_1^\mathsf{T} = p_0^\mathsf{T} P$ (with P as defined in Sect. 3.1 or Sect. 3.2). We thus consider the error caused by approximating the density of Q'_Δ with a density which is piecewise constant over the aggregation intervals, and by approximating the transition probabilities between the aggregates by P.

There is no error when zero jumps occur in the time interval $[0, \Delta]$, except in the spectrally negative case for the topmost discrete state: if no jump occurs, the probability mass in that state would move out of the truncated state space. However, we will consider the error caused by truncation separately below. In principle, if we ignore truncation effects, the approximation of the density of Q'_Δ (started with $\widetilde{\mu}_0$ and conditioned on no jump) is exact.

In contrast, there is an approximation error in the one-jump densities: the total probability mass in every aggregate is correct, as we defined P this way, but assuming that it is uniformly distributed over the intervals is an approximation. The Wasserstein distance of the piecewise uniform one-jump approximation and the actual distribution of Q'_Δ, conditioned on one jump, is bounded by Δ, as we only have to redistribute probability mass within distance Δ (within one interval) to go from the approximation to the actual distribution. The probability of one jump occurring is $\lambda \Delta e^{-\lambda \Delta}$, so the error per step is at most

$$e_{\mathrm{jmpagg}}(\lambda, \Delta) = \Delta \cdot \lambda \Delta e^{-\lambda \Delta} = \lambda \Delta^2 \cdot \frac{1}{e^{\lambda \Delta}} \leq \lambda \Delta^2 \tag{5}$$

where the first factor Δ is the distance by which we have to shift the probability mass at most to go from one distribution to the other, and where the second factor $\lambda \Delta e^{-\lambda \Delta}$ corresponds to the amount of mass we might have to shift. We can further improve the error bound, as the Wasserstein distance between the piecewise uniform one-jump approximation and the actual distribution of Q'_Δ, conditioned on one jump, will often be lower than Δ. See Sect. 4 for details.

We have a second error source: ignoring more than one jump per time step of length Δ. The probability mass moving due to two or more jumps in the original model stays where it is in the discretized version. Here, the analysis for the M/G/1 and the spectrally negative queue differ. We focus on the M/G/1 queue first. As we allow general jump height distributions, we might also have to ignore large single jumps in the M/G/1 case, in particular if arbitrarily large jumps are possible. The error introduced by these two types of cut-off can be bounded by:

$$\mathbb{P}\left[1 \text{ jump in } [0, \Delta]\right] \cdot \mathbb{E}\left[(\text{jump height})\mathbb{1}_{\{\text{jmp. hgt.}>M-i\Delta\}}\right]$$
$$+ \sum_{j=2}^{\infty} \mathbb{P}\left[j \text{ jumps in } [0, \Delta]\right] \cdot \mathbb{E}\left[\text{total jump height} \mid j \text{ jumps in } [0, \Delta]\right] \qquad (6)$$

where i is the index of the starting interval in the discrete model. This follows from the definition of the Wasserstein distance via couplings. Informally speaking, we can couple the part of $\text{Law}(Q'_\Delta)$ where two or more jumps occurred in $[0, \Delta]$ or where a single jump led out of the truncated state space with the equal-sized part of $\tilde{\mu}_1$ resulting from the amount we added to the diagonal of P to make $\check{P}$ stochastic. The expectation in the above expression is the integral of the distance of two points w.r.t. (a part of) the coupled measures, as in the definition of the Wasserstein distance. In fact, we could subtract Δ from the jump height within the expectation in most cases because of the constant processing speed 1. However, we will not do so as the above expression also gives an upper bound on the contribution to the Wasserstein distance if the initial distribution is concentrated on $[0, \Delta]$, where the processing time within time $[0, \Delta]$ is not necessarily Δ. Rewriting (6) in terms of B, we get (for the M/G/1 case)

$$\lambda \Delta e^{-\lambda \Delta} \cdot \mathbb{E}\left[B\mathbb{1}_{\{B>M-i\Delta\}}\right] + \sum_{j=2}^{\infty} \frac{(\lambda \Delta)^j}{j!} e^{-\lambda \Delta} \cdot j \cdot \mathbb{E}[B]$$
$$= \lambda \Delta e^{-\lambda \Delta} \cdot \mathbb{E}\left[B\mathbb{1}_{\{B>M-i\Delta\}}\right] + \lambda \Delta e^{-\lambda \Delta} \cdot \mathbb{E}[B] \cdot \sum_{j=2}^{\infty} \frac{(\lambda \Delta)^{j-1}}{(j-1)!}$$
$$= \underbrace{\lambda \Delta e^{-\lambda \Delta} \cdot \int_{(M-i\Delta, \infty)} x \, dF_B(x)}_{=: \, e_{\text{trunc}}^{(\text{spos})}(\lambda, \Delta, i)} + \underbrace{\lambda \Delta \left(1 - e^{-\lambda \Delta}\right) \cdot \mathbb{E}[B]}_{=: \, e_{\text{jmpcut}}^{(\text{spos})}(\lambda, \Delta)}$$

Note that the Wasserstein error bound only works if $\mathbb{E}[B]$ exists.

For the spectrally negative case, we also ignore two or more jumps per time interval, but large single jumps are not an issue as jumps cannot go below 0. Instead, as previously mentioned, an error occurs when the probability mass in the topmost discrete space should move out of the truncated state space due to no jump occurring. For the error caused by ignoring two or more jumps, we can almost use the same bound as in the M/G/1 case, but we can take additional advantage of the fact that jumps are stopped in 0. As the distribution of Q'_0 is supported on $[0, M]$, no jumps of size larger than $M + \Delta$ can occur within time

$[0, \Delta]$. Therefore, the error caused by ignoring two or more jumps is bounded by

$$\sum_{j=2}^{\infty} \frac{(\lambda\Delta)^j}{j!} e^{-\lambda\Delta} \cdot \min\{j \cdot \mathbb{E}[B], M + \Delta\}$$

$$\leqslant \min\left\{\lambda\Delta(1 - e^{-\lambda\Delta})\mathbb{E}[B],\ (1 - (1+\lambda\Delta)e^{-\lambda\Delta})(M+\Delta)\right\} =: e_{\text{jmpcut}}^{(\text{sneg})}(\lambda, \Delta)$$

In fact, we do not need to require that the expectation of B exists in this case. For the truncation error with respect to the starting interval i, we get

$$e_{\text{trunc}}^{(\text{sneg})}(\lambda, \Delta, i) = 0 \text{ if } i < M_\Delta, \qquad e_{\text{trunc}}^{(\text{sneg})}(\lambda, \Delta, i) = \Delta \cdot e^{-\lambda\Delta} \leqslant \Delta \text{ if } i = M_\Delta$$

This is because in the topmost interval (index M_Δ), the mass which should move upwards by Δ in case of no jump is $e^{-\lambda\Delta}$ (the probability of no jump).

Putting everything together, we can bound the error per step in the discrete model by choosing the following b_2 in Fig. 1, for both the M/G/1 and the spectrally negative case (but with different expressions for e_{jmpcut} and e_{trunc}):

$$b_2 := \sum_{i=0}^{M_\Delta} p_0(i) \cdot \left(e_{\text{jmpagg}}(\lambda, \Delta) + e_{\text{jmpcut}}(\lambda, \Delta) + e_{\text{trunc}}(\lambda, \Delta, i)\right)$$

where p_0 is the distribution of the discrete model before the current time step.

Overall Error Bound. We have now proved the following central result:

Theorem 1. *Consider a Lévy-driven queue fed by a spectrally one-sided Lévy process with finite jump intensity and no Brownian motion part (a compound Poisson process). Assume that the deterministic drift is 1 in the direction opposite to the jump directions. If we approximate the transient distribution μ_t of the original queue with the measures $\widetilde{\mu}_k$ as described in Sect. 3.1 and Sect. 3.2, and if we assume that the initial law μ_0 is supported on $[0, M]$, then:*

$$\text{WD}(\mu_{k\Delta}, \widetilde{\mu}_k)$$
$$\leqslant \Delta + \sum_{j=0}^{k-1} \sum_{i=0}^{M_\Delta} p_j(i) \cdot \left(e_{\text{jmpagg}}(\lambda, \Delta) + e_{\text{jmpcut}}(\lambda, \Delta) + e_{\text{trunc}}(\lambda, \Delta, i)\right)$$

where $e_{\text{jmpagg}}(\lambda, \Delta) = \lambda\Delta^2 e^{-\lambda\Delta}$ and, in the spectrally positive case,

$$e_{\text{jmpcut}}(\lambda, \Delta) = \lambda\Delta(1 - e^{-\lambda\Delta}) \cdot \mathbb{E}[B]$$
$$e_{\text{trunc}}(\lambda, \Delta, i) = \lambda\Delta e^{-\lambda\Delta} \cdot \mathbb{E}\left[B\mathbb{1}_{\{B > M - i\Delta\}}\right]$$

while for the spectrally negative case,

$$e_{\text{jmpcut}}(\lambda, \Delta) = \min\left\{\lambda\Delta(1 - e^{-\lambda\Delta})\mathbb{E}[B],\ (1 - (1+\lambda\Delta)e^{-\lambda\Delta})(M+\Delta)\right\}$$
$$e_{\text{trunc}}(\lambda, \Delta, i) = \Delta e^{-\lambda\Delta} \cdot \mathbb{1}_{\{i = M_\Delta\}}$$

The runtime of our approach is $\mathcal{O}(\frac{1}{\Delta^3})$ for $\Delta \to 0$ because P has $\mathcal{O}(\frac{1}{\Delta^2})$ entries (if we assume that M is fixed) and we need to take $\mathcal{O}(\frac{1}{\Delta})$ discrete steps to arrive at the transient distribution of a fixed time point in the original model.

3.4 Asymptotical Error Bound Behavior

We want to conclude with an analysis of the behavior of the accumulated error at time 1 in the original model (after $\frac{1}{\Delta}$ steps in the discrete model). For $\Delta \to 0$, the accumulated error should approach 0 as well, such that we can actually gain precision by making the aggregation intervals smaller. If we ignore the truncation part $e_{\text{trunc}}(\lambda, \Delta, i)$, then two remaining parts $e_{\text{jmpcut}}(\lambda, \Delta)$ and $e_{\text{jmpagg}}(\lambda, \Delta)$ are both of order $\mathcal{O}(\Delta^2)$. This is clear for $e_{\text{jmpagg}}(\lambda, \Delta)$, and we have:

$$e_{\text{jmpcut}}(\lambda, \Delta) \leqslant \lambda\Delta\left(1 - e^{-\lambda\Delta}\right) \cdot \mathbb{E}[B] = \lambda\Delta\left(\lambda\Delta + \mathcal{O}(\Delta^2)\right) \cdot \mathbb{E}[B] = \mathcal{O}(\Delta^2)$$

(For the spectrally negative case, we also have $e_{\text{jmpcut}}(\lambda, \Delta) = \mathcal{O}(\Delta^2)$ if $\mathbb{E}[B]$ does not exist). $e_{\text{trunc}}(\lambda, \Delta, i)$ is of order $\mathcal{O}(\Delta)$ (for fixed M). The only requirement for the Wasserstein bound to be usable in practice is that the error made in the approximation of the densities of Q'_t per step in the discrete model is

- $\mathcal{O}(\Delta^2)$ for the density approximations conditioned on zero jumps (which is true if there is no error in the zero-jump approximation as in our case)
- $\mathcal{O}(\Delta)$ for the density approximations conditioned on one jump (which is true if the probability per aggregate is correct in the one-jump approximation as in our case)

As a jump only occurs with a probability of $\mathcal{O}(\Delta)$ within time $[0, \Delta]$, this implies that the total error per time step is at most $\mathcal{O}(\Delta^2)$. This, in turn, implies that the error at original time 1 (after $\frac{1}{\Delta}$ steps in the discretized model) is $\mathcal{O}(\Delta)$, i.e., it does get smaller if we decrease Δ. Note that this would *not* be the case anymore if we used different discretization parameters for space and time. The analysis ignores the error due to truncation, which is a valid approximation in practical settings if the truncation point is chosen large enough such that only a small part of the probability mass would have exited the truncated state space within the considered time horizon. In fact, $e_{\text{trunc}}(\lambda, \Delta, i)$ accumulates to an error of $\mathcal{O}(1)$ after $\frac{1}{\Delta}$ steps (for $\Delta \to 0$ and M fixed), but we can make it arbitrarily small by letting $M \to \infty$.

4 Numerical Example

We conclude with a demonstration of the practical applicability of the presented techniques and error bounds using a numerical example.

The error bounds reported below actually use an improved version of e_{jmpagg} from (5): we can calculate the exact CDF of Q'_Δ in Fig. 1 (conditioned on one jump) with the help of the CDFs $F_{\text{oj}}^{[i]}$ obtained in Sect. 3.1 (or Appendix 2 of [10] for the spectrally negative case). We can then use (2) to calculate the Wasserstein distance between the exact distribution of Q'_Δ (conditioned on one jump) and the piecewise uniform approximation, and replace $e_{\text{jmpagg}}(\lambda, \Delta)$ by $\lambda\Delta e^{-\lambda\Delta}$ times the calculated Wasserstein distance.

Consider the M/G/1 queue started at $Q_0 = 1$ with job arrival rate $\lambda = \frac{1}{4}$ and B having a uniform distribution over $[1, 5]$. This ensures that the process

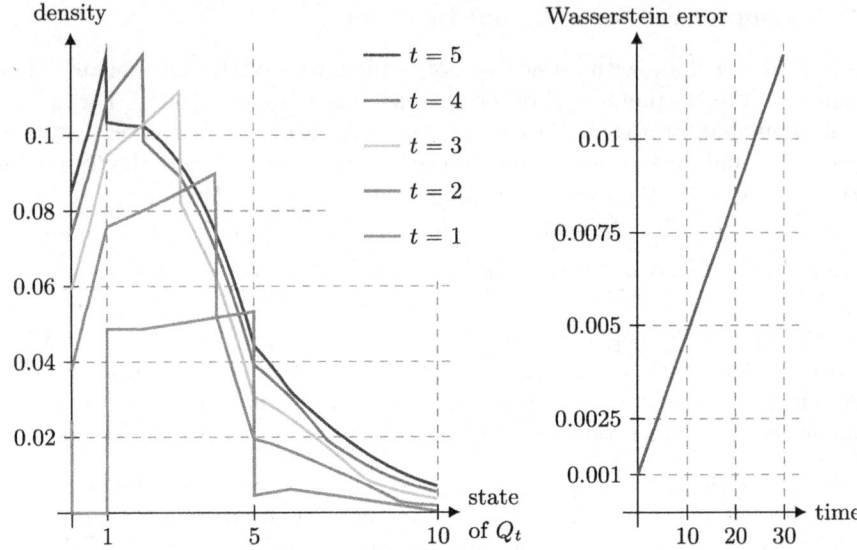

Fig. 2. Transient densities of the M/G/1 workload process started with $Q_0 = 1$ at different times t. The service speed is 1, the job arrival rate is $\lambda = \frac{1}{4}$, job sizes are distributed uniformly over the interval $[1, 5]$. Discretization parameter $\Delta = \frac{1}{500}$, truncation parameter $M = 50$. On the right: the corresponding Wasserstein error bounds.

always returns to 0. Figure 2 shows how the density of Q_t evolves (the non-negligible atoms at 0 are not shown; note that the depicted densities integrate to a value smaller than 1). For example, at time 1, the density is the sum of the densities conditioned on a fixed number of jumps, scaled with the probability of the respective number of jumps ($1 \gg \Delta$, so our discrete model allows more than one jump up to time 1). The 1-jump part is the uniform distribution over $[1, 5]$ which is very prominent in for $t = 1$. The 2-jump part (for $t = 1$) has a triangle shape starting at 2 and going back down to zero at 10, which is less prominent.

The plot on the right of Fig. 2 shows how the error bounds from Sect. 3.3 evolve. Here, we used the more precise version of e_{jmpagg} mentioned above. The initial error $\frac{\Delta}{2} = 0.001$ is the Wasserstein distance of the Dirac measure at 1 to a uniform distribution on the neighboring interval $(1 - \Delta, 1]$ of length Δ. The error increases almost linearly as the truncation error is comparatively small.

In Fig. 3, we compare setting $\Delta = \frac{1}{500}$ to $\Delta = \frac{1}{10}$ for $t = 1$. The density obtained with $\Delta = \frac{1}{10}$ is already quite close to the approximation obtained with $\Delta = \frac{1}{500}$, which shows that even coarse discretizations can yield good approximations. We also compare with the result obtained with a double inverse Laplace transform as explained in the appendix of [10]. That result, obtained by Mathematica [16], is similar to our results, although without any associated formal error bounds, and there are oscillatory artifacts near the discontinuities.

We want to give a short informal account to show that our method is also attractive in terms of runtime. Calculating the transient density approxima-

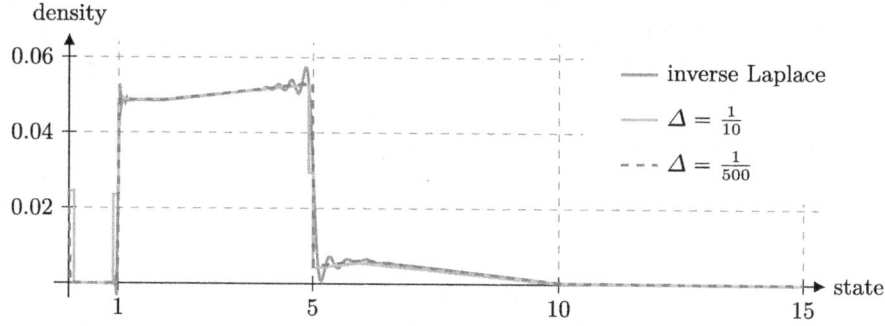

Fig. 3. Transient densities of the M/G/1 workload process started with $Q_0 = 1$ at time $t = 1$. The parameters are the same as in Fig. 2. Two different discretization parameters as well as the inverse Laplace transform approach are shown.

tion with $\Delta = \frac{1}{10}$ (and the corresponding error bounds) took less than one second on our test machine (single-threaded, Intel Core i7-1260P CPU at 4.7 GHz), while Mathematica needs around eight minutes. For $\Delta = \frac{1}{500}$, the runtime for the discretization approach was around two minutes, and a common Python library for Laplace transform inversion, mpmath [9], did not manage to compute the double inverse at all in a reasonable amount of time. Here, a more in-depth comparison, e.g. with the inversion technique from [8] would be interesting.

The positive density on the interval $(0, 0.1]$ for $\Delta = \frac{1}{10}$ in Fig. 3 is a discretization error resulting from ignoring more than one jump per time step. In Appendix 3 of [10], we also give an example of an M/G/1 queue under heavy load and an example of a spectrally negative queue.

5 Conclusion

We calculated transient distributions of (a subclass of) queues with one-sided Lévy input by approximation with a finite Markov chain, together with explicit error bounds in the Wasserstein distance. Within a few seconds or minutes of computation time, the proposed approach can deliver good approximations with error bounds which are useful in practice. The method is both faster and more accurate than common inverse Laplace transform approaches, and does not only compute the transient distribution at a fixed time point like the Laplace approach does.

5.1 Future Work

A more in-depth comparison with some Laplace inversion techniques which are not widely supported by tools like Python and Mathematica, such as [8], is still missing. As a next step, we would also like to extend our approach to a wider class of processes, e.g. queues with a Brownian motion part, two-sided input processes, or an M/G/1 queue with two distinct server speeds depending on the current load. However, there seem to be fundamental issues when looking beyond Lévy processes and queues with Lévy input. New methods are required for these cases.

Data Availibility Statement. The models, scripts, and tools to reproduce our experimental evaluation are archived and publicly available at DOI 10.5281/zenodo.15199111.

Disclosure of Interests. The authors have no competing interests to declare that are relevant to the content of this article.

References

1. Abate, A., Andriushchenko, R., Češka, M., Kwiatkowska, M.: Adaptive formal approximations of Markov chains. Perform. Eval. **148**(102207) (2021). https://doi.org/10.1016/j.peva.2021.102207. https://www.sciencedirect.com/science/article/pii/S0166531621000249
2. Abate, A., Katoen, J.P., Lygeros, J., Prandini, M.: Approximate model checking of stochastic hybrid systems. Eur. J. Control **16**(6), 624–641 (2010). https://doi.org/10.3166/ejc.16.624-641. https://www.sciencedirect.com/science/article/pii/S0947358010706919
3. Abate, J., Whitt, W.: The fourier-series method for inverting transforms of probability distributions. Queueing Syst. **10**(1), 5–87 (1992). https://doi.org/10.1007/BF01158520. https://www.columbia.edu/~ww2040/FourierSeries1992.pdf
4. Abate, J., Whitt, W.: Transient behavior of the M/G/1 workload process. Oper. Res. **42**(4), 750–764 (1994). https://doi.org/10.1287/opre.42.4.750. https://www.columbia.edu/~ww2040/transientworkloadOR94.pdf
5. Abate, J., Whitt, W.: Numerical inversion of Laplace transforms of probability distributions. ORSA J. Comput. **7**(1), 36–43 (1995). https://doi.org/10.1287/ijoc.7.1.36. https://www.columbia.edu/~ww2040/LaplaceInversionJoC95.pdf
6. Asghari, N.M., den Iseger, P., Mandjes, M.: Numerical techniques in Lévy fluctuation theory. Methodol. Comput. Appl. Probab. **16**(1), 31–52 (2014). https://doi.org/10.1007/s11009-012-9296-5
7. Dębicki, K., Mandjes, M.: Queues and Lévy Fluctuation Theory. Springer (2015). https://doi.org/10.1007/978-3-319-20693-6_1
8. den Iseger, P.: Numerical transform inversion using Gaussian quadrature. Probab. Eng. Informational Sci. **20**(1), 1–44 (2006). https://doi.org/10.1017/S0269964806060013. https://papers.ssrn.com/sol3/papers.cfm?abstract_id=1013507
9. Johansson, F., et al.: mpmath: a Python library for arbitrary-precision floating-point arithmetic (version 1.3.0) (2023). https://github.com/mpmath/mpmath
10. Michel, F., Siegle, M.: Formal approximations of the transient distributions of the M/G/1 workload process. arXiv preprint (2025). https://doi.org/10.48550/arXiv.2504.01193
11. Michel, F., Siegle, M.: Formal error bounds for the state space reduction of Markov chains. Perform. Eval. **167**, 102464 (2025). https://doi.org/10.1016/j.peva.2024.102464. https://www.sciencedirect.com/science/article/pii/S0166531624000695

12. Motter, A.E., Campbell, D.K.: Chaos at fifty. Phys. Today **66**(5), 27–33 (2013). https://doi.org/10.1063/PT.3.1977. https://pubs.aip.org/physicstoday/article/66/5/27/615055/Chaos-at-fiftyIn-1963-an-MIT-meteorologist
13. Soudjani, S.E.Z., Abate, A.: Precise approximations of the probability distribution of a Markov process in time: an application to probabilistic invariance. In: Ábrahám, E., Havelund, K. (eds.) Tools and Algorithms for the Construction and Analysis of Systems, pp. 547–561. Springer (2014). https://doi.org/10.1007/978-3-642-54862-8_45
14. Soudjani, S.E.Z., Majumdar, R., Abate, A.: Safety verification of continuous-space pure jump Markov processes. In: Chechik, M., Raskin, J.F. (eds.) Tools and Algorithms for the Construction and Analysis of Systems, pp. 147–163. Springer, Heidelberg (2016). https://doi.org/10.1007/978-3-662-49674-9_9. https://www.cs.ox.ac.uk/people/alessandro.abate/publications/bcSMA16.pdf
15. Vallender, S.S.: Calculation of the Wasserstein distance between probability distributions on the line. Theory Probab. Appl. **18**(4), 784–786 (1974). https://doi.org/10.1137/1118101
16. Wolfram Research, Inc.: Mathematica, Version 14.0, Champaign, Illinois (2024). https://www.wolfram.com/mathematica

On Choice of Loss Functions for Neural Control Barrier Certificates

Alireza Nadali[✉][iD], Ashutosh Trivedi[iD], and Majid Zamani[iD]

University of Colorado Boulder, Boulder, CO 80303, USA
a_nadali@colorado.edu

Abstract. The design of controllers with correctness guarantees is a primary concern for safety-critical control systems. A Control Barrier Certificate (CBC) is a real-valued function over the state space of the system that provides an inductive proof of the existence of a safe controller. Recently, neural networks have been successfully deployed for data-driven learning of control barrier certificates. These approaches encode the conditions for the existence of a CBC using a rectified linear unit (ReLU) loss function. The resulting encoding, while sound, tends to be conservative, which results in slower training and limits scalability to large, complex systems. Can altering the loss function alleviate some of the problems associated with ReLU loss and lead to faster learning?

This paper proposes a novel encoding with a Mean Squared Error loss function, which allows for more scalable and efficient training, while addressing some of the theoretical limitations of previous methods. We also encode one of the main conditions of CBC in a non-conservative way, enabling us to derive CBC where existing methods have failed. The proposed approach derives a validity condition based on Lipschitz continuity to formally characterize safety guarantees, eliminating the need for a post-hoc verification. The effectiveness of the proposed loss functions is demonstrated through six case studies curated from the existing literature. Our results provide a strong argument for exploring alternative loss function choices as a novel approach to optimizing the design of CBCs.

1 Introduction

Recent advances in deep learning have accelerated the integration of autonomous systems into various safety-critical areas of everyday life, including self-driving cars, robotic manipulators, and personalized implantable medical devices. Consequently, even a minor fault in the control logic of these systems can lead to catastrophic consequences, such as loss of human life, severe financial losses, legal liabilities, and damage to infrastructure. In response to this grand challenge, the development of formally certified control methods for autonomous systems has received considerable research interest in recent years [43,52,53,58]. Control Barrier Certificates (CBCs) [3,41]—and their neural network representations [5,15,16,19,29,53,55]—have emerged as a leading approach to designing a safety controller along with an inductive proof of correctness. This paper focuses on the crucial role that the choice of loss functions plays in the scalable design of safety controllers and their corresponding CBCs.

Neural Control Barrier Certificates. The key idea behind *control barrier certificates* (CBCs) is as follows: if one can construct a real-valued function defined over the state space of a dynamical system such that (1) the function is negative in the initial states, (2) positive in the unsafe states, and (3) for every state where the function value is non-positive, there exists a control signal that enables a transition to another state with a non-positive value, then a feedback control exists that ensures the system remains safe indefinitely. Traditionally, Sum-of-Squares (SOS) optimization has been employed to synthesize such certificates and their corresponding controllers [41,44,57]. However, this approach often requires human expertise to identify an appropriate template and suffers from poor scalability.

In contrast, CBCs parameterized by neural networks—commonly referred to as Neural Control Barrier Certificates (NCBCs)—have recently gained significant attention. This is due to their universal approximation capabilities, ease of automation, and the growing availability of robust tool support [5,16,29,53]. Due to their data-centric nature, NCBCs only provide guarantees over the finite set of data points used during training. As a result, the resulting controller requires formal verification to ensure rigorous safety guarantees across the entire continuous state space. This verification is typically formulated as a constraint satisfaction problem and solved using Satisfiability Modulo Theories (SMT) solvers, such as Z3 [17,18]. However, the necessity for such post-hoc verification introduces an additional bottleneck that limits scalability.

Choice of Loss Functions. The success of deep-learning-based approximation hinges on a well-designed loss function, which ensures that the model learns the correct objective, converges efficiently, and generalizes well to unseen data [26,30]. Following Zhao et al. [54], most work on NCBCs [1,5,19,55,60] encodes the control barrier conditions using a ReLU function ($x \in \mathbb{R} \mapsto \max(x,0)$). The ReLU loss function is straightforward to implement and provides a natural termination condition for training, as training stops when the loss reaches zero. However, suffers from several fundamental drawbacks, such as having a zero Hessian everywhere (which limits interpretability [47]) and multiple sub-gradients near the global minimum (which negatively impacts convergence and robustness).

Furthermore, prior work [5] has shown that using this loss function results in large Lipschitz constants for the trained networks (and consequently for the resulting controllers). In practice, small Lipschitz constants are desirable for controllers to ensure robust control [10]. Similarly, barrier certificates with small Lipschitz constants are preferable due to their robustness to small perturbations in the dynamical system model (caused by factors such as mechanical wear and tear or changes in operating conditions), thereby enhancing the applicability of the resulting guarantees. A natural question arises: *Can modifying the loss function mitigate issues associated with the loss and lead to faster learning?*

Mean Squared Error (MSE) Loss. Mean Squared Error (MSE) is a popular choice [21] for loss functions in regression problems due to its strong convergence guarantees [2,11]. We investigate the suitability of MSE loss functions for NCBCs by posing the following research questions:

RQ1 Can MSE effectively encode the conditions of neural control barrier certificates?

RQ1 Can an MSE-based loss function support intuitive termination checks?

RQ1 Current methods typically fail to scale to more parameterized neural networks and high-dimensional systems. To what extent do MSE loss functions alleviate this drawback?

RQ1 Small Lipschitz constants are desirable for 1) interpretability, 2) robustness of training, 3) robustness of the resulting controller, and 4) transferability of the resulting guarantees. How do MSE-based NCBCs compare to ReLU-based NCBCs in this regard?

Contributions. Our contributions in addressing these research questions are summarized below.

1. We begin by introducing a more relaxed condition for barrier certificates compared to state-of-the-art methods (**RQ1**). Subsequently, in Sect. 4, we reformulate the CBC conditions using the MSE loss function. By leveraging the MSE loss, we facilitate smoother gradients, which enhances the stability and convergence of neural network training.
2. In Sect. 5, we utilize mild Lipschitz continuity assumptions on the system to derive specific *validity conditions* (Theorem 2) for the resulting network. When these conditions are satisfied, they enable us to terminate training and provide safety guarantees across the entire state space. This approach removes the need for post-hoc verification (**RQ2**), significantly enhancing the scalability of the overall method.
3. In Sect. 6, we address **RQ3-4** experimentally by applying our approach to six case studies drawn from state-of-the-art literature [5,19,53,57]. Our results show that, compared to existing methods, our approach achieves significant improvements in scalability with respect to system dimensions and neural network architectures, while also enabling the more efficient discovery of formally correct NCBCs. Furthermore, our experiments indicate that our approach produces barrier certificates and controllers with smaller Lipschitz constants, which, in turn, streamline the correctness establishment process compared to state-of-the-art methods using ReLU loss.

Organization. We begin the paper with a discussion of related work in Sect. 2. In Sect. 3, we formally define the problem of learning a neural control barrier certificate by introducing the underlying system model and relevant concepts. Section 4 presents the neural control barrier certificates with MSE loss and outlines an algorithm for training them. The key correctness proof is provided in Sect. 5. Experimental evaluation results are discussed in Sect. 6.

2 Related Work

Prajna and Jadbabaie [40] first introduced the notion of barrier certificates (BCs) as functions whose level sets provide over-approximations of the reachable sets of systems. Control barrier certificates (CBCs) emerged as a promising approach to synthesize safe controllers [3,12,14,24,51]. Traditionally, Sum-of-Squares (SOS) optimization is deployed to synthesize such controllers [41,44,57].

Zhao et al. [54] first introduced a notion of neural network based barrier certificates. They consider a simple, one hidden layer neural network to represent a barrier certificate, and employ Mixed-Integer Linear Programming (MILP) to verify its correctness. Later, Peruffo et al. [38] utilized SMT solvers to find counterexamples to a candidate neural barrier certificate and used those counterexamples to train their neural networks. Neural barrier certificates have also been employed for safety verification of hybrid [56] and stochastic [31] systems.

To tackle the controller synthesis, neural control barrier certificates (NCBCs) have been proposed recently [15,16,27–29,42]. Existing work utilizes methods such as SMT solvers [1,19,54], reachable set verification [50], polynomial approximation [45], Lipschitz continuity [5], and ReLU networks verification [25], to formally verify the correctness of NCBCs. More recently, Zhang et al. [53] proposed a novel algorithm for exact verification of NCBCs, by considering a given barrier certificateand synthesizing a controller if that barrier is correct. Most of the aforementioned algorithms require an exact model of a system (with the exception of [5]), and utilize SMT solvers; these solvers cannot deal with deep neural networks efficiently, as the computational complexity grows exponentially with respect to the number of parameters, which restricts the architecture of neural networks.

3 Problem Formulation

We denote the set of reals, non-negative reals, and positive reals by $\mathbb{R}$, $\mathbb{R}_{\geq 0}$, and $\mathbb{R}_{>0}$, respectively. For sets A and B, we write $A \setminus B$ and $A \times B$ for their difference and Cartesian product, respectively. We write $|A|$ for the cardinality of the set A. We consider n-dimensional Euclidean space $\mathbb{R}^n$ equipped with infinity norm $\|\cdot\|$, defined as $\|x - y\| := \max_{1 \leq i \leq n} |x_i - y_i|$ and Euclidean norm as $\|x - y\|_2 := \sqrt{\sum_{i=1}^{n}(x_i - y_i)^2}$, where $x=(x_1, x_2, \ldots, x_n), y=(y_1, y_2, \ldots, y_n)$ belong to $\mathbb{R}^n$. We denote the rectified linear unit function by $\mathsf{ReLU}(x) := \max(x, 0)$, and mean squared error function by $\mathsf{MSE}(x, y) := \frac{1}{n}\sum_{i}^{n}(x_i - y_i)^2$.

3.1 Control Barrier Certificates

Systems studied in this paper are modeled as a discrete-time control system (dtCS), defined as follows.

Definition 1 (Discrete-Time Control System). *A discrete-time control system (dtCS) is a tuple $\mathfrak{S} := (\mathcal{X}, \mathcal{X}_0, U, f)$, where $\mathcal{X} \subseteq \mathbb{R}^n$ represents the continuous state set, $\mathcal{X}_0 \subseteq \mathcal{X}$ is the initial state set, and $U \subseteq \mathbb{R}^m$ is the set of inputs.*

Furthermore, $f : \mathcal{X} \times U \to \mathcal{X}$ is the state transition function. The evolution of the system under an input sequence $u = \langle u(1), u(2), \ldots \rangle$ is given by

$$\mathfrak{S} : x(t+1) = f(x(t), u(t)). \tag{1}$$

We assume that sets $\mathcal{X}$, and U are bounded, and the map f is unknown but can be simulated via a black-box model, and f is Lipschitz continuous, as stated in the following assumption.

Assumption 1 (Lipschitz Continuity). *For a given dtCS $\mathfrak{S}=(\mathcal{X}, \mathcal{X}_0, U, f)$, we assume that f is Lipschitz continuous, i.e., there exists (Lipschitz) constants $\mathcal{L}_u, \mathcal{L}_x \in \mathbb{R}_{\geq 0}$ such that for all $x, x' \in \mathcal{X}$, and $u, u' \in U$, we have*

$$\|f(x, u) - f(x', u')\| \leq \mathcal{L}_x \|x - x'\| + \mathcal{L}_u \|u - u'\|. \tag{2}$$

A discrete-time control system (dtCS) $\mathfrak{S} = (\mathcal{X}, \mathcal{X}_0, U, f)$ equipped with a feedback controller $k(x) : \mathcal{X} \to U$ is *safe* with respect to a set of unsafe states $\mathcal{X}_u \subseteq \mathcal{X}$ if, for every state trajectory (a.k.a. trace) of the system originating from $\mathcal{X}_0$ under the control inputs provided by k, the trajectory never reaches $\mathcal{X}_u$. The primary safety problem addressed in this work is formalized as follows.

Problem 1 (Safety Problem). Given a discrete-time control system (dtCS) $\mathfrak{S} = (\mathcal{X}, \mathcal{X}_0, U, f)$, the goal is to design a feedback controller $k : \mathcal{X} \to U$ such that $\mathfrak{S}$ is safe with respect to the initial set of states $\mathcal{X}_0 \subseteq \mathcal{X}$ and the unsafe set $\mathcal{X}_u \subseteq \mathcal{X}$. Specifically, for every trace $\langle x(0), x(1), \ldots \rangle$, where $x(t+1) = f(x(t), k(x(t)))$ and $x(0) \in \mathcal{X}_0$, it must hold that $x(t) \notin \mathcal{X}_u$ for all $t \in \mathbb{N}$.

We employ the following notion of control barrier certificates (CBCs) [4] which provides sufficient conditions for ensuring safety.

Definition 2 (Control Barrier Certificates). *Consider a dtCS $\mathfrak{S} = (\mathcal{X}, \mathcal{X}_0, U, f)$. A function $B : \mathcal{X} \to \mathbb{R}$ is called a control barrier certificate (CBC) for $\mathfrak{S}$ with respect to initial set of states $\mathcal{X}_0 \subseteq \mathcal{X}$ and unsafe set $\mathcal{X}_u \subseteq \mathcal{X}$ if there exists a controller $k : \mathcal{X} \to U$ such that, for some $\eta \in \mathbb{R}_{\geq 0}$:*

$$B(x) \leq -\eta, \text{ for all } x \in \mathcal{X}_0, \tag{3}$$
$$B(x) > \eta, \text{ for all } x \in \mathcal{X}_u, \text{ and} \tag{4}$$
$$(B(x) \leq 0) \implies (B(f(x, k(x))) \leq 0), \text{ for all } x \in \mathcal{X}. \tag{5}$$

We adapt the following theoretical result from [4], which demonstrates the effectiveness of CBCs.

Theorem 1 (CBCs Imply Safety). *Consider a dtCS $\mathfrak{S} = (\mathcal{X}, \mathcal{X}_0, U, f)$ and an unsafe set of states $\mathcal{X}_u \subseteq \mathcal{X}$. A control barrier certificate satisfying conditions (3)–(5) guarantees that the system $\mathfrak{S}$, when equipped with the CBC's controller, will never reach $\mathcal{X}_u$ from any initial state $x \in \mathcal{X}_0$.*

3.2 Neural Control Barrier Certificates

Neural networks, as universal approximators [23], are capable of representing any Borel-measurable function based on input-output data. Consider a neural network F with k fully connected layers, where each layer i is characterized by a weight matrix W_i and a bias vector b_i of appropriate dimensions, followed by an activation function. Such a network can be represented as a function $F : \mathbb{R}^{n_i} \to \mathbb{R}^{n_o}$. Given $y_0 \in \mathbb{R}^{n_i}$, a network computes its output $y_k \in \mathbb{R}^{n_o}$ as:

$$y_1 = \sigma(W_1 y_0 + b_1), \cdots, y_k = \sigma(W_k y_{k-1} + b_k).$$

We denote y_{i-1} and y_i, for $i \in \{1, \ldots, k\}$, as the input and output of the i-th layer, respectively, where σ represents the activation function. It is observed that neural networks with ReLU activations ($\sigma(x) = \max(0, x)$) describe locally Lipschitz continuous functions with a Lipschitz constant $\mathcal{L}_F \in \mathbb{R}_{\geq 0}$. Specifically, for all $x_1', x_2' \in \mathbb{R}^{n_i}$, the following condition holds:

$$\|F(x_1') - F(x_2')\| \leq \mathcal{L}_F \|x_1' - x_2'\|. \tag{6}$$

Moreover, an upper bound on the Lipschitz constant of a neural network with ReLU activations can be computed using the spectral norm [13]. Although tighter Lipschitz upper bounds for neural networks have been extensively investigated [7,20,32,37,39,48,59], our experiments reveal that these methods are either overly restrictive or introduce significant computational complexity during the training process. In contrast, the spectral norm approach achieves a favorable balance: it offers a considerably tighter bound than the trivial upper bound while maintaining computational efficiency.

We focus on training neural networks to serve as control barrier certificates. To achieve this, we first outline the construction of the training set. Specifically, we partition the set $\mathcal{X}$ into finitely many disjoint hypercubes $X_1, X_2, \ldots, X_M$ by selecting a *discretization parameter* $\epsilon > 0$ such that:

$$\|x - x_i\| \leq \frac{\epsilon}{2}, \text{ for all } x \in X_i, \tag{7}$$

where x_i denotes the center of the hypercube X_i containing x, with $i \in \{1, \ldots, M\}$. Accordingly, we pick the centers of these hypercubes as sample points, and denote the set of all sample points by $\mathcal{X}_d := \{x_1, \ldots, x_M\}$. Now we have all the ingredients to propose our notion of NCBCs.

Definition 3 (Neural Control Barrier Certificates). *Consider a dtCS $\mathfrak{S} = (\mathcal{X}, \mathcal{X}_0, U, f)$, constants $\epsilon, \eta, \gamma \in \mathbb{R}_{>0}$ such that $\gamma \leq \eta$, an unsafe set $\mathcal{X}_u \subseteq \mathcal{X}$, and neural networks $B : \mathcal{X} \to \mathbb{R}$ and $k : \mathcal{X} \to U$. We assert that B and k form a neural control barrier certificate (NCBC) if the following conditions are satisfied:*

$$B(x) \leq -\eta, \text{ for all } x \in \mathcal{X}_0 \cap \mathcal{X}_d, \tag{8}$$

$$B(x) > \eta, \text{ for all } x \in \mathcal{X}_u \cap \mathcal{X}_d, \text{ and} \tag{9}$$

$$(B(x) \leq \gamma) \implies (B(f(x, k(x))) \leq -\eta), \text{ for all } x \in \mathcal{X} \cap \mathcal{X}_d, \tag{10}$$

where $\mathcal{X}_d$ is constructed according to (7), with discretization parameter ϵ.

In previous works, condition (5) is often replaced with

$$B(f(x,k(x))) - B(x) \leq -\eta,$$

which enforces that the barrier certificate must decrease as the system evolves. While this is a more conservative condition, it simplifies the verification process [5,35,36]. Moreover, it is commonly assumed that this decreasing condition must hold over the entire state space, a restrictive requirement since some states may not even be reachable, yet are still obligated to satisfy this condition. To address this, we adopt an implication-based approach, particularly by setting $\gamma = \mathcal{L}_B \frac{\epsilon}{2}$, where ϵ is the discretization parameter.

Current methods for training neural networks to act as control barrier certificates for a dtCS $\mathfrak{S}$ utilize the loss function $L_{\text{ReLU}} := L_1 + L_2 + L_3$, where

$$L_1 := \text{ReLU}(B(x), -\eta), \quad \text{for all } x \in \mathcal{X}_d \cap \mathcal{X}_0,$$
$$L_2 := \text{ReLU}(B(x), \eta), \quad \text{for all } x \in \mathcal{X}_d \cap \mathcal{X}_u,$$
$$L_3 := \text{ReLU}(B(f(x,k(x))) - B(x), -\eta), \text{ for all } x \in \mathcal{X}_d \setminus \mathcal{X}_u,$$

and L_1, L_2, and L_3 correspond to conditions (3) to (5), respectively (with a slight modification in (5) by replacing the implication with an inequality).

The advantage of using ReLU is that training can be stopped once the loss reaches zero. However, from both theoretical and practical perspectives, this loss leads to unstable training. As a result, algorithms relying on ReLU do not scale with respect to the system's dimension or the number of neural network parameters. To address this, we employ the mean squared error (MSE) loss, which provides convergence guarantees for over-parameterized networks [2,11].

4 Neural Barrier Certificates with MSE Loss

We propose an alternative approach by replacing the ReLU activation function in the loss function with an MSE-based formulation for constructing Neural Control Barrier Certificates. This substitution is motivated by the smooth and continuous nature of MSE, which facilitates more efficient gradient-based optimization and enhances the overall performance and robustness of the system equipped with the designed controller.

We train $B(x)$ and $k(x)$ using the following loss, $L_{\text{MSE}} = L_1 + L_2 + L_3$, where:

$$L_1 := \text{MSE}(B(x), -\eta), \quad \text{for all } x \in \mathcal{X}_d \cap \mathcal{X}_0, \tag{11}$$
$$L_2 := \text{MSE}(B(x), \eta), \quad \text{for all } x \in \mathcal{X}_d \cap \mathcal{X}_u, \tag{12}$$
$$L_3 := \text{MSE}(B(f(x,k(x))), -\eta), \quad \text{for all } x \in \mathcal{X}_d \setminus \mathcal{X}_u, \text{such that } B(x) \leq \gamma, \tag{13}$$

where $\eta \in \mathbb{R}_{>0}$ is a design parameter. Specifically, L_1, L_2, and L_3 encode the conditions (3) to (5) of the control barrier certificate, respectively. Additionally, we train $k(x)$ using L_3. Note that this loss function depends on both networks, so training $B(x)$ and $k(x)$ involves addressing the moving target problem [33]. To mitigate this, we fix B for a predefined number of iterations while optimizing L_3 in (13).

Algorithm 1. Training Neural Control Barrier Certificates

Input: Sets $\mathcal{X}_0, \mathcal{X}, U$ for a dtCS $\mathfrak{S}$, respectively, as in Definition (1); discretization parameters ϵ for the set $\mathcal{X}$ as in (7); robustness parameters $\eta \in \mathbb{R}_{>0}$ as in Definition (3); $\mathcal{L}_x, \mathcal{L}_u$ as introduced in Assumption (1); the number of iterations N for fixing network B; the architecture of the neural networks B and k; and number of iterations $N_{\max}$.
Output: B and k.

1: Construct the training data set $\mathcal{X}_d$ according to 7.
2: Initialize networks B and k [21].
3: $\mathcal{L}_B \leftarrow$ Upper bound of Lipschitz constant of B [13].
4: $\mathcal{L}_k \leftarrow$ Upper bound of Lipschitz constant of k [13].
5: $i \leftarrow 0$
6: **while** Conditions (8)–(10) and conditions (14)–(15) are not met and $i \leq N_{\max}$ **do**
7:    **if** i=nN **then**
8:       $B_3 = B$.
9:    **end if**
10:    Train B with loss $L_{\mathsf{MSE}} = L_1 + L_2 + L_3$, with L_1, L_2, and L_3 as in 11-13, respectively.
11:    Train k via loss L_3 generated from B_3.
12:    $i \leftarrow i + 1$
13:    $\mathcal{L}_B \leftarrow$ Upper bound of Lipschitz constant of B [13].
14:    $\mathcal{L}_k \leftarrow$ Upper bound of Lipschitz constant of k [13].
15: **end while**
16: Return B, k

To theoretically motivate the use of MSE, we present the following simple example. Consider a scalar system $\mathfrak{S} = (\mathcal{X}, \mathcal{X}_0, U, f)$, where the dynamics are given by $f(x) = \frac{x}{2}$, with $\mathcal{X} = [-10, 10]$, the initial set $\mathcal{X}_0 = [3, 4]$, and the unsafe set $\mathcal{X}_u = [-10, 0)$. Since $x(t) \geq 0$ for all $t \in \mathbb{N}$, the system is trivially safe. Now, consider a barrier certificate represented by a linear neural network $B(x) = Mx$, with a Lipschitz constant $\mathcal{L}_B = |M|$. When using ReLU loss, any non-positive value of M results in a loss of 0. In contrast, with MSE loss, a non-positive M with a large absolute value—corresponding to a larger Lipschitz constant for the barrier certificate—produces a higher loss. In this sense, MSE inherently favors barrier certificates with smaller Lipschitz constants, encouraging the construction of smoother and more effective barrier functions.

Algorithm 1 summarizes our training framework. First, the training dataset $\mathcal{X}_d$ is constructed, and the networks are initialized. Training then begins using the loss function L_{MSE}. During training, we iteratively check for the smallest value of η that satisfies conditions (8)–(10) and conditions (14)–(15). If an admissible η is found, training concludes; otherwise, the process continues. Additionally, we include small regularizers for both networks B and k to encourage them to maintain small Lipschitz constants, as suggested in [21].

Note that a neural control barrier certificate is not necessarily a valid control barrier certificate as in Definition (2), since the training is performed only over a finite set of data. To address this issue, we propose the following validity conditions, which will be utilized to prove that a neural control barrier certificate satisfies conditions of Definition (2), *i.e.*, extend guarantees from training samples to unseen samples.

Assumption 2 (Validity Conditions). *Consider a dtCS* $\mathfrak{S} = (\mathcal{X}, \mathcal{X}_0, U, f)$ *and two neural networks* $B : \mathcal{X} \to \mathbb{R}$ *and* $k : \mathcal{X} \to U$, *with* ReLU *activations that satisfy conditions (8) to (10) for* $\mathcal{X}_d$ *constructed according to (7). We assume the following validity conditions:*

$$\mathcal{L}_B(\mathcal{L}_x + \mathcal{L}_u \mathcal{L}_k)\frac{\epsilon}{2} - \eta \leq 0, \tag{14}$$

$$\mathcal{L}_B \frac{\epsilon}{2} - \eta \leq 0, \tag{15}$$

where $\mathcal{L}_B$ *and* $\mathcal{L}_k$ *are the Lipschitz constants of the networks* B *and* k, *respectively,* $\mathcal{L}_x$ *and* $\mathcal{L}_u$ *are the Lipschitz constants of* $\mathfrak{S}$ *as defined in (1),* ϵ *is the discretization parameter, and* $\eta \in \mathbb{R}_{>0}$ *is a robustness hyper-parameter.*

Lipschitz continuity allows us to extend guarantees from a finite set of training data to the entire state set. Assumption 2 provides a critical condition that facilitates this extension. Specifically, it ensures that if a sample point satisfies the control barrier certificate conditions, then all points within a neighborhood centered at the sample point, with a radius of $\frac{\epsilon}{2}$, also satisfy those conditions. This approach establishes the theoretical foundation required to bridge the gap between finite training data and overall correctness across the entire state set.

Although η is user-defined, a control barrier certificate (CBC) does not necessarily need to satisfy conditions (8)–(10) and conditions (14)–(15) with that specific value of η. Any positive value of η that satisfies these conditions provides a formal guarantee of safety.

5 Proof of Correctness

In this section, we present the main theoretical result of our paper and formally prove that a neural control barrier certificate, synthesized according to Algorithm 1 (provided it terminates), is indeed a valid control barrier certificate. In other words, it satisfies conditions (3)–(5) and provides a solution to Problem 1.

Theorem 2 (Validity Conditions Imply Formal Correctness). *Consider a dtCS* $\mathfrak{S} = (\mathcal{X}, \mathcal{X}_0, U, f)$ *with Lipschitz constants* $\mathcal{L}_x$ *and* $\mathcal{L}_u$ *as defined in Assumption 1, and a constant* $\epsilon \in \mathbb{R}_{>0}$ *used to construct* $\mathcal{X}_d$ *as in (7). Let* $B : \mathcal{X} \to \mathbb{R}$ *and* $k : \mathcal{X} \to U$ *be neural networks with Lipschitz constants* $\mathcal{L}_B$ *and* $\mathcal{L}_k$, *respectively, trained according to Algorithm 1 and representing a neural control barrier certificate. Then,* $\mathfrak{S}$ *is safe with respect to the unsafe set* $\mathcal{X}_u \subseteq \mathcal{X}$ *under the controller* k.

Proof. We first prove that condition (5) is satisfied. Consider any $x \in \mathcal{X}$. If $B(x) > 0$, then implication in (5) is trivially satisfied. From now on, we just consider the case that $B(x) \leq 0$. By construction of $\mathcal{X}_d$ as in (7), there exists

$x_i \in \mathcal{X}_d$ such that $\|x - x_i\| \leq \frac{\epsilon}{2}$. To obtain an upper bound for $B(x_i)$, we employ Lipschitz continuity as follows:

$$B(x_i) = B(x_i) - B(x) + B(x) \leq \mathcal{L}_B \|x - x_i\| + B(x) \leq \mathcal{L}_B \frac{\epsilon}{2} = \gamma.$$

Based on (10), for any $x_i \in \mathcal{X}_d$ with $B(x_i) \leq \gamma$, we have:

$$B(f(x_i, k(x_i))) \leq -\eta.$$

It follows that,

$$\begin{aligned}B(f(x, k(x))) &= B(f(x, k(x))) - B(f(x_i, k(x_i))) + B(f(x_i, k(x_i))) \\ &\leq B(f(x, k(x))) - B(f(x_i, k(x_i))) - \eta \\ &\leq \mathcal{L}_B \|f(x, k(x)) - f(x_i, k(x_i))\| - \eta,\end{aligned}$$

where the last inequality follows from Lipschitz continuity of B. Moreover:

$$\begin{aligned}\mathcal{L}_B \|f(x, k(x)) - f(x_i, k(x_i))\| - \eta &\leq \mathcal{L}_B(\mathcal{L}_x \|x - x_i\| + \mathcal{L}_u \|k(x) - k(x_i)\|) - \eta \\ &\leq \mathcal{L}_B(\mathcal{L}_x + \mathcal{L}_u \mathcal{L}_k)\|x - x_i\| - \eta,\end{aligned}$$

which is followed by Assumption 1 and Lipschitz continuity of k. According to Algorithm 1, validity condition (14) holds, thus:

$$B(f(x, k(x))) \leq \mathcal{L}_B(\mathcal{L}_x + \mathcal{L}_u \mathcal{L}_k)\frac{\epsilon}{2} - \eta \leq 0.$$

Therefore, condition (10) combined with the validity condition (14) implies condition (5). Similar arguments can be used to show that conditions (3) and (4) also hold; however, these proofs are omitted here for brevity. Consequently, a neural control barrier certificate synthesized using Algorithm 1 satisfies the conditions of a control barrier certificate as stated in Definition 2, thereby guaranteeing the safety of $\mathfrak{S}$ under the controller k, as established in Theorem 1.

6 Experimental Evaluation

Thus far, we have answered **RQ1** and **RQ2** in previous sections, and here, we aim to address **RQ3** and **RQ4**. We demonstrate the efficacy of our Algorithm with six case studies. Table 1 shows a detailed comparison between our method and other state-of-the-art algorithms. We considered methods that 1) provide formal guarantee and 2) train a feedback controller. Among these methods, [5] is model-free, rest require closed-form mathematical expression of map f. Moreover, some methods such as [53] are for continuous time systems only, however, we discretize systems with forward Euler method [22] for the sake of comparison.

6.1 Case Study Descriptions and System Models

We evaluate our approach on a diverse set of nonlinear dynamical systems. For each system, we describe the state-space representation, transition dynamics, objectives, and relevant parameters.

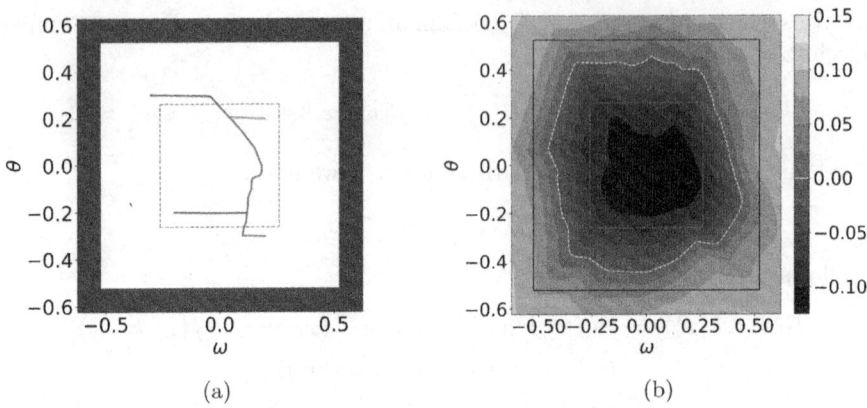

Fig. 1. Four state sequences of the inverted pendulum are depicted in Fig. 1a, starting from different initial conditions; Dotted blue lines indicate the initial set, and red areas depict the unsafe set. Level set of NCBC for the inverted pendulum are depicted in Fig. 1b, dotted white, blue, and black lines show the zero-level, the initial set, and the unsafe set of states, respectively. (Color figure online)

Inverted Pendulum (IP). We consider a dtCS $\mathfrak{S} = (\mathcal{X}, \mathcal{X}_0, U, f)$ modeling an inverted pendulum. The state space is defined as $\mathcal{X} = [\frac{-\pi}{4}, \frac{\pi}{4}] \times [\frac{-\pi}{4}, \frac{\pi}{4}]$, with initial set $\mathcal{X}_0 = [\frac{-\pi}{12}, \frac{\pi}{12}] \times [\frac{-\pi}{12}, \frac{\pi}{12}]$, and unsafe set $\mathcal{X}_u = \mathcal{X} \setminus [\frac{-\pi}{6}, \frac{\pi}{6}] \times [\frac{-\pi}{6}, \frac{\pi}{6}]$. The system dynamics are given by:

$$\begin{bmatrix}\theta(t+1)\\ \omega(t+1)\end{bmatrix} = \begin{bmatrix}\theta(t) + \tau\omega(t) \\ \omega + \frac{g\tau}{l}\sin(\theta(t)) + \frac{10\tau}{ml^2}k(x(t))\end{bmatrix},$$

where $x(t) := [\theta(t), \omega(t)]$, and θ and ω are the angular position and velocity, respectively. Moreover, $g = 9.8$ is the gravitational acceleration, and $l=1$ and $m=1$ are the length and mass of the pendulum, respectively. Constant $\tau = 0.01$ is the sampling rate, and Lipschitz constants $\mathcal{L}_x = 1.098$, $\mathcal{L}_u = 0.1$, based on Assumption (1). The discretization parameter and input set are $\epsilon = 1.2 * 10^{-3}$, and $U=[-2.5, 2.5]$, respectively. Our method converged with the following parameters: $\mathcal{L}_B = 0.48, \mathcal{L}_k = 2.3$, and $\eta = 0.0037$. [5] report a Lipschitz constant of $\mathcal{L}_B=21$ for barrier certificate and $\mathcal{L}_K=20$ for its controller. Some state sequences and level sets of CBC are depicted in Fig. 1a and Fig. 1b, respectively.

Double Inverted Pendulum (DIP). For our next case study, we consider a double inverted pendulum modeled as a dtCS $\mathfrak{S} = (\mathcal{X}, \mathcal{X}_0, U, f)$, where the transition function f is defined as:

$$\begin{bmatrix}\theta_1(t+1)\\ \omega_1(t+1)\\ \theta_2(t+1)\\ \omega_2(t+1)\end{bmatrix} = \begin{bmatrix}\theta_1(t)+\tau\omega_1(t)\\ \omega_1(t)+\tau(g\sin(\theta_1(t))-\sin(\theta_1(t)-\theta_2(t))\omega_1^2(t))\\ \theta_2(t)+\tau\omega_2(t)\\ \omega_2(t)+\tau(g\sin(\theta_2(t))+\sin(\theta_1(t)-\theta_2(t))\omega_2^2(t))\end{bmatrix} + \tau\begin{bmatrix}0 & 0\\ 30 & 0\\ 0 & 0\\ 0 & 39\end{bmatrix}u(t),$$

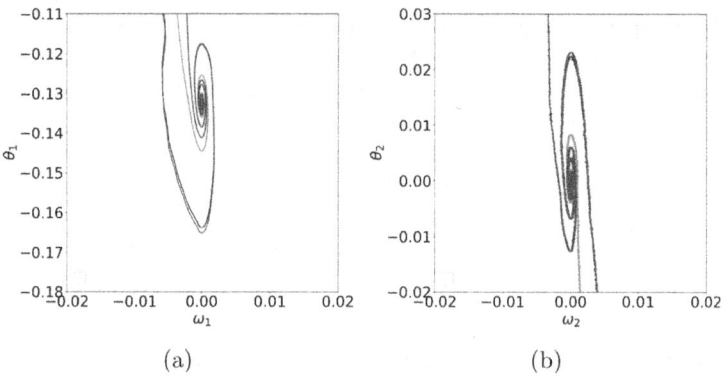

Fig. 2. Some trajectories of the double inverted pendulum, starting from different initial conditions. Figure 2a and Fig. 2b depict the trajectories for the first and second joint, respectively.

where $x(t) := [\theta_1(t); \omega_1(t); \theta_2(t); \omega_2(t)] \in [\frac{-\pi}{4}, \frac{\pi}{4}]^4$, θ_1 and θ_2 represent the angular position of the first and the second joint, respectively, and ω_1 and ω_2 are the angular velocity of the first and the second joint, respectively, and $U = [-3.5, 3.5]^2$ are the inputs applied to the first and second joint, respectively. Constant $g=9.8$ is the gravitational acceleration, and Lipschitz constants $\mathcal{L}_x=1.098, \mathcal{L}_u=0.39$, based on Assumption (1). The initial and unsafe set of states are $\mathcal{X}_0 = [\frac{-\pi}{20}, \frac{\pi}{20}]^4, \mathcal{X}_u = \mathcal{X} \setminus [\frac{-\pi}{6}, \frac{\pi}{6}]^4$, respectively, and $\epsilon = 10^{-2}$. Our algorithm converged with the following parameters: $\mathcal{L}_B=0.17, \mathcal{L}_K=1.8$, and $\eta=0.00326$. Some trajectories of the system are depicted in Fig. 2a and Fig. 2b.

Spacecraft Rendezvous (SR). This case study is adapted from [53]. It models a spacecraft rendezvous scenario, where a chaser spacecraft approaches a target in orbit. The relative motion of the chaser with respect to the target is described using the linearized Clohessy–Wiltshire–Hill (CWH) equations. The system state is defined as: $x(t) := [p_x(t); p_y(t); p_z(t); v_x(t); v_y(t); v_z(t)]$, where (p_x, p_y, p_z) denote the relative position and (v_x, v_y, v_z) denote the relative velocity components. The control input is given by $u(t) = [u_x(t); u_y(t); u_z(t)]$ and dynamics defined as follows.

$$\begin{bmatrix} p_x(t+1) \\ p_y(t+1) \\ p_z(t+1) \\ v_x(t+1) \\ v_y(t+1) \\ v_z(t+1) \end{bmatrix} = \begin{bmatrix} p_x(t) \\ p_y(t) \\ p_z(t) \\ v_x(t) \\ v_y(t) \\ v_z(t) \end{bmatrix} + \tau \begin{bmatrix} 1 & 0 & 0 & 0 & 0 & 0 \\ 0 & 1 & 0 & 0 & 0 & 0 \\ 0 & 0 & 1 & 0 & 0 & 0 \\ 3n^2 & 0 & 0 & 0 & 2n & 0 \\ 0 & 0 & 0 & -2n & 0 & 0 \\ 0 & 0 & -n^2 & 0 & 0 & 0 \end{bmatrix} \begin{bmatrix} p_x(t) \\ p_y(t) \\ p_z(t) \\ v_x(t) \\ v_y(t) \\ v_z(t) \end{bmatrix} + \tau \begin{bmatrix} 0 & 0 & 0 \\ 0 & 0 & 0 \\ 0 & 0 & 0 \\ 1 & 0 & 0 \\ 0 & 1 & 0 \\ 0 & 0 & 1 \end{bmatrix} \begin{bmatrix} u_x(t) \\ u_y(t) \\ u_z(t) \end{bmatrix}.$$

The objective is to steer the chaser to the origin, representing perfect rendezvous with the target, while ensuring that the trajectory remains within a prescribed safety region.

Obstacle Avoidance (OA). This experiment is adapted from [53]. The system models a planar motion of an aircraft, where the state consists of the 2-D position and the yaw angle: $x(t) := [x_1(t); x_2(t); \psi(t)]$. Here, (x_1, x_2) denotes the position of the aircraft in the plane, and ψ represents its yaw (heading) angle. The control input $u(t)$ directly influences the yaw rate of the aircraft. The discrete-time dynamics with time step τ are given by:

$$\begin{bmatrix} x_1(t+1) \\ x_2(t+1) \\ \psi(t+1) \end{bmatrix} = \begin{bmatrix} x_1(t) \\ x_2(t) \\ \psi(t) \end{bmatrix} + \tau \begin{bmatrix} v\sin(\psi(t)) \\ v\cos(\psi(t)) \\ 0 \end{bmatrix} + \tau \begin{bmatrix} 0 \\ 0 \\ u \end{bmatrix},$$

where v is the constant forward velocity of the aircraft. The goal is to control the yaw rate to guide the aircraft as it navigates through a field of obstacles.

Darboux (DB). This experiment is adapted from [53], and concerns the verification of an autonomous nonlinear system known as the *Darboux system*, a planar benchmark commonly used in formal verification literature to evaluate reachability and safety analysis techniques. The system is uncontrolled (i.e., there are no control inputs), and the state is defined as $x(t) := [x_1(t); x_2(t)]$ and the dynamics are given by:

$$\begin{bmatrix} x_1(t+1) \\ x_2(t+1) \end{bmatrix} = \begin{bmatrix} x_1(t) \\ x_2(t) \end{bmatrix} + \tau \begin{bmatrix} x_2(t) + 2x_1(t)x_2(t) \\ -x_1(t) + 2x_1^2(t) - x_2^2(t) \end{bmatrix}.$$

Since the system is autonomous (i.e., there are no control inputs), the goal is to verify whether its natural evolution satisfies certain safety properties.

Bicycle Steering (BS). This experiment is adapted from [55]. The control objective is to stabilize a bicycle by regulating its tilt through handlebar steering. The system state is defined as: $x(t) := [x_1(t); x_2(t); x_3(t)]$ where $x_1(t)$ denotes the tilt angle of the bicycle, $x_2(t)$ is the angular velocity of the tilt, and $x_3(t)$ represents the handlebar angle relative to the bicycle body. The control input $u(t)$ manipulates the handlebar torque. The discrete-time dynamics are:

$$\begin{bmatrix} x_1(t+1) \\ x_2(t+1) \\ x_3(t+1) \end{bmatrix} =$$

$$\begin{bmatrix} x_1(t) \\ x_2(t) \\ x_3(t) \end{bmatrix} + \tau \begin{bmatrix} x_2(t) \\ c_1(g\sin x_1(t) + \frac{v^2}{b}\cos x_1(t)\tan x_3(t)) \\ 0 \end{bmatrix} + \tau \begin{bmatrix} 0 \\ c_2 \cdot \frac{\cos x_1(t)}{\cos^2 x_3(t)} \\ 1 \end{bmatrix} u(t).$$

The task is to design a controller that stabilizes the bicycle in an upright position while maintaining feasible handlebar motion.

Table 1. Comparison of our proposed method and state-of-the-art. Results showcase our algorithm's independence from the architecture of control barrier certificate, since we do not utilize SMT solvers. We denote the runtime by "NA" when an algorithm fails to converge. Each number, in the architecture column (same architecture for both B and k), represents number of neurons for each hidden layer (*i.e.*,10-10-10 refers to a neural network with 3 hidden layers, each consist of 10 neurons), and all networks have ReLU activations. Last two columns depicts the training time for our method. Last column uses [13] to estimate Lipschitz contants of neural networks, while second to last column utilizes the state-of-the-art neural network verifier $\alpha, \beta-$ CROWN [59].

Benchmark	Architecture	[19]	[5]	[55]	[53]	Ours α,β-crown	Lipschitz
IP (2d)	10-10-10	250 s	45 min	40 min	450 s	750 s	**130 s**
IP (2d)	200-200-200-200	NA	NA	NA	NA	80 min	**120 s**
DIP (4d)	200-200-200-200	NA	NA	NA	NA	12 h	**800 s**
SR (6d)	10-10-10-10	130 s	NA	95 min	300 s	30 min	**110 s**
SR (6d)	200-200-200-200	NA	NA	NA	NA	60 min	**92 s**
OA (3d)	10	130 s	60	1.2 h	**7 s**	600 s	120 s
OA (3d)	200-200-200-200	NA	NA	NA	NA	90 min	**70 s**
DB (2d)	10	50 s	600 s	450 s	8 s	400 s	**5.8 s**
DB (2d)	200-200-200-200	NA	NA	NA	NA	30 min	**3.5 s**
BS (3d)	10		300 s	2800 s	2100 s	**20 s**	300 s	45 s
BS (3d)	200-200-200-200	NA	NA	NA	NA	25 min	**42 s**

6.2 Discussion

The results in [53] assume that a candidate NCBC is already provided, and after verification, they synthesize an admissible controller. Consequently, their method performs well on shallow networks. Moreover, they assume access to the exact model of the system, similar to [19,55]. In contrast, our approach trains an NCBC from scratch and assumes access only to a black-box representation of the system. The only system information required is the Lipschitz constants $\mathcal{L}_x$ and $\mathcal{L}_u$, as stated in Assumption 1. If these constants are unknown, sampling-based methods can be utilized to estimate them [8,46,49].

As demonstrated earlier, even under milder assumptions, our method outperforms existing approaches, scaling effectively to higher-dimensional and more complex systems. Additionally, our method can leverage over-parameterized networks to capitalize on their representational capacity. Beyond scalability, our synthesized controller also exhibits a smaller Lipschitz constant compared to prior work. This advantage arises from encoding CBC conditions using MSE loss, which is differentiable at its global minimum (unlike ReLU) and comes with convergence guarantees [2,11].

We acknowledge that other approaches in the literature provide tighter bounds on the Lipschitz constant of neural networks compared to [13], such as those proposed in [7,20,37,48]. However, these methods are computationally expensive. Our numerical experiments indicate that approximately 99% of the training time is spent calculating the Lipschitz constants, with only 1% dedicated to the actual training process (see Table 1).

The spectral norm approach strikes a favorable balance: it provides a much tighter bound than the trivial upper bound while remaining computationally efficient. We conducted an *a posteriori* comparison between the spectral norm approach and the method proposed in [20]. Our results demonstrate that the spectral norm approach is an order of magnitude faster than [20], while the upper bound it provides is only 40% to 50% larger than that obtained by [20]. This slight increase in the upper bound can be compensated for by appropriately tuning ϵ and η. Notably, this finding aligns with the results reported in [20]. In next subsections, we have primarily highlighted the pendulum examples, as they present the greatest challenges due to the nonlinearity.

7 Conclusion

This paper presents advancements in the synthesis and verification of Neural Control Barrier Certificates (NCBCs) by addressing key limitations in prior works. First, by introducing a more relaxed condition and reformulating the CBC conditions using MSE loss functions, we achieved smoother gradients, leading to more stable and efficient neural network training. Second, by leveraging Lipschitz continuity assumptions, we established training termination conditions that ensure guaranteed safety across the entire state space, eliminating the need for post-hoc verification and enhancing scalability. Finally, through experimental validation on six state-of-the-art case studies, we demonstrated that our method improves scalability in terms of system dimensions and network architecture. Additionally, our approach produces synthesized barrier certificates and controllers with smaller Lipschitz constants, simplifying the verification process while enhancing robustness and transferability. A possible future direction is to encode NCBC conditions using alternative loss functions and to explore the effects of MSE on other neural certificates, such as Lyapunov functions [9] and Closure certificates [34]. Further research could also focus on reducing sample complexity by incorporating system properties, such as monotonicity [6].

Acknowledgments. This research was supported in part by the National Science Foundation (NSF) under grant CNS-2111688 and by NSF CAREER Awards CNS-2145184 and CCF-2146563. Ashutosh Trivedi holds the position of Royal Society Wolfson Visiting Fellow and gratefully acknowledges the support of the Wolfson Foundation and the Royal Society for this fellowship.

References

1. Abate, A., Ahmed, D., Giacobbe, M., Peruffo, A.: Formal synthesis of lyapunov neural networks. IEEE Control Syst. Lett. **5**(3), 773–778 (2020)

2. Allen-Zhu, Z., Li, Y., Song, Z.: A convergence theory for deep learning via overparameterization. In: International Conference on Machine Learning, pp. 242–252. PMLR (2019)
3. Ames, A.D., Coogan, S., Egerstedt, M., Notomista, G., Sreenath, K., Tabuada, P.: Control barrier functions: theory and applications. In: 18th European Control Conference (ECC), pp. 3420–3431. IEEE (2019)
4. Anand, M., Murali, V., Trivedi, A., Zamani, M.: K-inductive barrier certificates for stochastic systems. In: Proceedings of the 25th ACM International Conference on Hybrid Systems: Computation and Control, pp. 1–11 (2022)
5. Anand, M., Zamani, M.: Formally verified neural network control barrier certificates for unknown systems. IFAC-PapersOnLine **56**(2), 2431–2436 (2023)
6. Angeli, D., Sontag, E.D.: Monotone control systems. IEEE Trans. Autom. Control **48**(10), 1684–1698 (2003)
7. Araujo, A., Havens, A., Delattre, B., Allauzen, A., Hu, B.: A unified algebraic perspective on Lipschitz neural networks. In: ICLR (2023)
8. Calliess, J.P.: Lipschitz optimisation for Lipschitz interpolation. In: 2017 American Control Conference (ACC), pp. 3141–3146. IEEE (2017)
9. Chang, Y.C., Roohi, N., Gao, S.: Neural Lyapunov control. In: Advances in Neural Information Processing Systems, vol. 32 (2019)
10. Chen, B.M.: Robust and H_∞ Control. Springer (2013)
11. Cheridito, P., Jentzen, A., Riekert, A., Rossmannek, F.: A proof of convergence for gradient descent in the training of artificial neural networks for constant target functions. J. Complex. **72**, 101646 (2022)
12. Clark, A.: Verification and synthesis of control barrier functions. In: 60th IEEE Conference on Decision and Control (CDC), pp. 6105–6112. IEEE (2021)
13. Combettes, P.L., Pesquet, J.C.: Lipschitz certificates for layered network structures driven by averaged activation operators. SIAM J. Math. Data Sci. **2**(2), 529–557 (2020)
14. Dai, H., Permenter, F.: Convex synthesis and verification of control-Lyapunov and barrier functions with input constraints. In: American Control Conference (ACC), pp. 4116–4123. IEEE (2023)
15. Dawson, C., Gao, S., Fan, C.: Safe control with learned certificates: a survey of neural lyapunov, barrier, and contraction methods for robotics and control. IEEE Trans. Rob. **39**(3), 1749–1767 (2023)
16. Dawson, C., Qin, Z., Gao, S., Fan, C.: Safe nonlinear control using robust neural Lyapunov-barrier functions. In: Conference on Robot Learning, pp. 1724–1735. PMLR (2022)
17. De Moura, L., Bjørner, N.: Z3: an efficient SMT solver. In: International Conference on Tools and Algorithms for the Construction and Analysis of Systems, pp. 337–340. Springer (2008)
18. De Moura, L., Bjørner, N.: Satisfiability modulo theories: introduction and applications. Commun. ACM **54**(9), 69–77 (2011)
19. Edwards, A., Peruffo, A., Abate, A.: Fossil 2.0: formal certificate synthesis for the verification and control of dynamical models. In: Proceedings of the 27th ACM International Conference on Hybrid Systems: Computation and Control, pp. 1–10 (2024)
20. Fazlyab, M., Robey, A., Hassani, H., Morari, M., Pappas, G.: Efficient and accurate estimation of Lipschitz constants for deep neural networks. In: Advances in Neural Information Processing Systems, vol. 32 (2019)
21. Goodfellow, I., Bengio, Y., Courville, A.: Deep Learning, vol. 1. MIT Press (2016)

22. Gottlieb, S., Shu, C.W., Tadmor, E.: Strong stability-preserving high-order time discretization methods. SIAM Rev. **43**(1), 89–112 (2001)
23. Hornik, K., Stinchcombe, M., White, H.: Multilayer feedforward networks are universal approximators. Neural Netw. **2**(5), 359–366 (1989)
24. Jagtap, P., Soudjani, S., Zamani, M.: Formal synthesis of stochastic systems via control barrier certificates. IEEE Trans. Autom. Control **66**(7), 3097–3110 (2020)
25. Katz, G., Barrett, C., Dill, D.L., Julian, K., Kochenderfer, M.J.: Reluplex: an efficient SMT solver for verifying deep neural networks. In: Majumdar, R., Kunčak, V. (eds.) CAV 2017. LNCS, vol. 10426, pp. 97–117. Springer, Cham (2017). https://doi.org/10.1007/978-3-319-63387-9_5
26. Li, H., Xu, Z., Taylor, G., Studer, C., Goldstein, T.: Visualizing the loss landscape of neural nets. In: Advances in Neural Information Processing Systems, vol. 31 (2018)
27. Lindemann, L., et al.: Learning hybrid control barrier functions from data. In: Conference on Robot Learning, pp. 1351–1370. PMLR (2021)
28. Lindemann, L., Robey, A., Jiang, L., Das, S., Tu, S., Matni, N.: Learning robust output control barrier functions from safe expert demonstrations. IEEE Open J. Control Syst. (2024)
29. Liu, S., Liu, C., Dolan, J.: Safe control under input limits with neural control barrier functions. In: Conference on Robot Learning, pp. 1970–1980. PMLR (2023)
30. Ma, J., et al.: Loss odyssey in medical image segmentation. Med. Image Anal. **71**, 102035 (2021)
31. Mathiesen, F.B., Calvert, S.C., Laurenti, L.: Safety certification for stochastic systems via neural barrier functions. IEEE Control Syst. Lett. **7**, 973–978 (2022)
32. Meunier, L., Delattre, B.J., Araujo, A., Allauzen, A.: A dynamical system perspective for Lipschitz neural networks. In: International Conference on Machine Learning, pp. 15484–15500. PMLR (2022)
33. Mnih, V., et al.: Human-level control through deep reinforcement learning. Nature **518**(7540), 529–533 (2015)
34. Nadali, A., Murali, V., Trivedi, A., Zamani, M.: Neural closure certificates. In: Proceedings of the AAAI Conference on Artificial Intelligence, vol. 38, pp. 21446–21453 (2024)
35. Nejati, A., Lavaei, A., Jagtap, P., Soudjani, S., Zamani, M.: Formal verification of unknown discrete-and continuous-time systems: a data-driven approach. IEEE Trans. Autom. Control **68**(5), 3011–3024 (2023)
36. Nejati, A., Zamani, M.: Data-driven synthesis of safety controllers via multiple control barrier certificates. IEEE Control Syst. Lett. **7**, 2497–2502 (2023)
37. Pauli, P., Koch, A., Berberich, J., Kohler, P., Allgöwer, F.: Training robust neural networks using lipschitz bounds. IEEE Control Syst. Lett. **6**, 121–126 (2021)
38. Peruffo, A., Ahmed, D., Abate, A.: Automated and formal synthesis of neural barrier certificates for dynamical models. In: International Conference on Tools and Algorithms for the Construction and Analysis of Systems, pp. 370–388. Springer (2021)
39. Prach, B., Lampert, C.H.: Almost-orthogonal layers for efficient general-purpose Lipschitz networks. In: European Conference on Computer Vision, pp. 350–365. Springer (2022)
40. Prajna, S., Jadbabaie, A.: Safety verification of hybrid systems using barrier certificates. In: Hybrid Systems: Computation and Control, pp. 477–492. Springer, Heidelberg (2004)

41. Prajna, S., Jadbabaie, A., Pappas, G.J.: A framework for worst-case and stochastic safety verification using barrier certificates. IEEE Trans. Autom. Control **52**(8), 1415–1428 (2007)
42. Robey, A., et al.: Learning control barrier functions from expert demonstrations. In: 2020 59th IEEE Conference on Decision and Control (CDC), pp. 3717–3724. IEEE (2020)
43. Salamati, A., Lavaei, A., Soudjani, S., Zamani, M.: Data-driven verification and synthesis of stochastic systems via barrier certificates. Automatica **159**, 111323 (2024)
44. Schneeberger, M., Dörfler, F., Mastellone, S.: SOS construction of compatible control lyapunov and barrier functions. IFAC-PapersOnLine **56**(2), 10428–10434 (2023)
45. Sha, M., et al.: Synthesizing barrier certificates of neural network controlled continuous systems via approximations. In: 58th ACM/IEEE Design Automation Conference (DAC), pp. 631–636. IEEE (2021)
46. Strongin, R., Barkalov, K., Bevzuk, S.: Acceleration of global search by implementing dual estimates for Lipschitz constant. In: International Conference on Numerical Computations: Theory and Algorithms, pp. 478–486. Springer (2019)
47. Torop, M., Masoomi, A., Hill, D., Kose, K., Ioannidis, S., Dy, J.: Smoothhess: relu network feature interactions via stein's lemma. In: Proceedings of the 37th International Conference on Neural Information Processing Systems. NIPS 2023. Curran Associates Inc., Red Hook (2024)
48. Wang, Z., et al.: On the scalability and memory efficiency of semidefinite programs for Lipschitz constant estimation of neural networks. In: The Twelfth International Conference on Learning Representations (2024)
49. Wood, G.R., Zhang, B.: Estimation of the lipschitz constant of a function. J. Global Optim. **8**, 91–103 (1996)
50. Xiang, W., Tran, H.D., Johnson, T.T.: Output reachable set estimation and verification for multilayer neural networks. IEEE Trans. Neural Netw. Learn. Syst. **29**(11), 5777–5783 (2018)
51. Xiao, W., Belta, C.: Control barrier functions for systems with high relative degree. In: IEEE 58th Conference on Decision and Control (CDC), pp. 474–479. IEEE (2019)
52. Xu, X., Grizzle, J.W., Tabuada, P., Ames, A.D.: Correctness guarantees for the composition of lane keeping and adaptive cruise control. IEEE Trans. Autom. Sci. Eng. **15**(3), 1216–1229 (2017)
53. Zhang, H., Wu, J., Vorobeychik, Y., Clark, A.: Exact verification of relu neural control barrier functions. In: Advances in Neural Information Processing Systems, vol. 36 (2024)
54. Zhao, H., Zeng, X., Chen, T., Liu, Z.: Synthesizing barrier certificates using neural networks. In: Proceedings of the 23rd International Conference on Hybrid Systems: Computation and Control, pp. 1–11 (2020)
55. Zhao, H., Zeng, X., Chen, T., Liu, Z., Woodcock, J.: Learning safe neural network controllers with barrier certificates. Formal Aspects Comput. **33**(3), 437–455 (2021). https://doi.org/10.1007/s00165-021-00544-5
56. Zhao, Q., et al.: Synthesizing relu neural networks with two hidden layers as barrier certificates for hybrid systems. In: Proceedings of the 24th International Conference on Hybrid Systems: Computation and Control, pp. 1–11 (2021)
57. Zhao, W., He, T., Wei, T., Liu, S., Liu, C.: Safety index synthesis via sum-of-squares programming. In: American Control Conference (ACC), pp. 732–737. IEEE (2023)

58. Zhong, B., Cao, H., Zamani, M., Caccamo, M.: Towards safe AI: sandboxing DNNs-based controllers in stochastic games. In: Proceedings of the AAAI Conference on Artificial Intelligence, vol. 37, pp. 15340–15349 (2023)
59. Zhou, D., Brix, C., Hanasusanto, G.A., Zhang, H.: Scalable neural network verification with branch-and-bound inferred cutting planes. arXiv preprint arXiv:2501.00200 (2024)
60. Žikelić, D., Lechner, M., Verma, A., Chatterjee, K., Henzinger, T.: Compositional policy learning in stochastic control systems with formal guarantees. In: Advances in Neural Information Processing Systems, vol. 36 (2024)

Computing the Congestion Phases of Dynamical Systems with Priorities and Application to Emergency Departments

Xavier Allamigeon, Pascal Capetillo(✉), and Stéphane Gaubert

INRIA and CMAP, École polytechnique,
Institut polytechnique de Paris, CNRS, Palaiseau, France
{Xavier.Allamigeon,Pascal.Capetillo,
Stephane.Gaubert}@inria.fr

Abstract. Medical emergency departments are complex systems in which patients must be treated according to priority rules based on the severity of their condition. We develop a model of emergency departments using Petri nets with priorities, described by nonmonotone piecewise linear dynamical systems. The collection of stationary solutions of such systems forms a "phase diagram", in which each phase corresponds to a subset of bottleneck resources (like senior doctors, interns, nurses, consultation rooms, etc.). Since the number of phases is generally exponential in the number of resources, developing automated methods is essential to tackle realistic models. We develop a general method to compute congestion diagrams. A key ingredient is a polynomial time algorithm to test whether a given "policy" (configuration of bottleneck tasks) is achievable by a choice of resources. This is done by reduction to a feasibility problem for an unusual class of lexicographic polyhedra. Furthermore, we show that each policy uniquely determines the system's throughput. We apply our approach to a case study, analyzing a simplified model of an emergency department from Assistance Publique – Hôpitaux de Paris.

Keywords: Performance evaluation · Emergency departments · Piecewise-linear Dynamics · Petri nets with priorities · Polyhedral Computation

1 Introduction

Context: Piecewise Linear Models of Timed Discrete Event Systems. The "max-plus approach" to discrete event systems allows the analysis of synchronization and concurrency phenomena by means of piecewise linear dynamics. Initially developed to model synchronization phenomena (i.e., the subclass of timed Petri nets called timed-event graphs [6,11,18,23]), it was later extended to *monotone* systems [7,12,16], to account for concurrency (resource sharing) via arbitration by preselection rules. More recently, the framework was adapted to include arbitration by priority rules, first to a special case of road traffic in [15], later to a general setting [1], with an application to the staffing of emergency call centers that was subsequently developed in [2].

The general form of the piecewise dynamical systems arising from the max-plus approach involves *counter functions* $z_i \colon \mathbb{R}_{\geq 0} \to \mathbb{R}_{\geq 0}$; each associated with a type of discrete events in the system, such as the firing of a given transition in a timed Petri net. Every z_i is an increasing function of time t, where the value $z_i(t)$ represents the cumulative number of times the event of type i has occurred up to and including time t. The dynamical system over the functions z_i then writes as follows:

$$z_i(t) = \min_{a \in \mathcal{A}_i} \left(r_i^a + \sum_{\tau \in \mathcal{T}} \sum_{j \in [n]} (P_\tau^a)_{ij} z_j(t - \tau) \right), \; i \in [n], \; t \geq \max \mathcal{T}. \qquad (D)$$

Here, the finite set $\mathcal{T} \subset \{0, 1, 2, \dots\}$ represents the time delays in the system. For every $i \in [n]$,[1] $\mathcal{A}_i$ is a finite set that corresponds to the possible limiting prerequisites for event i. The parameters r_i^a and $(P_\tau^a)_{ij}$ take real values. Notably, the parameters r_i^a are used to model the resources of systems (e.g., staffing, equipment, facilities), and play a central role in its behavior.

Performance evaluation can be addressed by seeking *stationary solutions* of the dynamics (D), i.e., solutions of the form $z_i(t) = u_i + \rho_i t$ for all $i \in [n]$, where $u, \rho \in \mathbb{R}^n$. Then, the vector ρ represents the system's *throughput* with each entry corresponding to the rate of events occurring for each type. When the system is monotone (e.g., when $(P_\tau^a)_{ij}$ are nonnegative) and has a stoichiometric invariant, the system (D) can be reformulated as the dynamic programming equation of a semi-Markov decision process. Then, ρ is uniquely determined by the parameters of (D), and a pair (ρ, u) can be computed in polynomial time [2,12].

The study of systems governed by priority rules breaks the monotonicity of the dynamics (i.e., $(P_\tau^a)_{ij}$ may take negative values) and poses an additional challenge for computing stationary solutions. In fact, understanding the conditions of existence and the properties of stationary solutions of nonmonotone systems was already stated as Problem #34 in the list of open problems in control theory collected in 1999, see [25]. Progress was made for the explicit dynamics (D) in [4], where we identified sufficient conditions, satisfied by a large family of models, that guarantee stationary solutions to exist, independent of the resource allocation (i.e., the values of the parameters r_i^a). In contrast, the computation of the stationary solutions has only been done by hand on a few specific and small-size models [2,10,15] and has remained unresolved in general.

Contributions. We develop an algorithm that computes the stationary solutions of a timed discrete event system governed by a dynamics of the form (D). The algorithm iterates over the set of *policies*, each specifying a selection $a \in \mathcal{A}_i$ for every $i \in [n]$ in the minimum in (D). It relies on two key ingredients. First,

---

[1] Throughout the paper, we use the notation $[k] := \{1, \dots, k\}$.

we develop a polynomial time procedure that checks whether a policy is strictly feasible, meaning that it is associated with a stationary regime with a nonidentically zero throughput vector (Theorem 2). This boils down to the study of lexicographic polyhedra, i.e., sets defined by finitely many linear inequalities in the lexicographic sense. Lexicographic polyhedra encompass other already studied generalizations of polyhedral models, such as lexicographic linear programming [19] or polyhedra defined by mixed systems of strict and non-strict linear inequalities [8]. We show that the feasibility problem of such polyhedra can be decided in polynomial time by reduction to successive linear programs. As a second ingredient, we show that, for generic time delays, the throughput associated with a given policy is uniquely determined by the resource vector, and we provide an explicit formula for it (Theorem 3) under a condition that 1 is a semisimple eigenvalue of the matrix associated with this policy. The semisimplicity condition holds for a broad class of systems [4].

This work is motivated by a real case study: the dimensioning of a medical emergency department (ED), carried out as part of an ongoing project with Assistance Publique – Hôpitaux de Paris. We build a timed Petri net model of such an ED and apply our algorithm to generate a congestion phase diagram— a collection of polyhedral regions that, as a function of the resources, identify the congested components of the system, as well as an explicit expression of the throughput. In practice, this diagram can help ED physician coordinator measure the effect of limited resources on the system, give quantitative thresholds for staff allocation in order to remove critical bottlenecks and get into a "fluid" phase.

The paper is organized as follows. In Sect. 2, we present the Petri net model of the ED and the phase diagram returned by our algorithm. Section 3 deals with a general overview of the algorithm, and the two ingredients, namely checking strict feasibility of a policy in polynomial time, and determining the throughput vector, are presented in Sect. 4 and 5. We conclude with further experimental results and implementation details in Sect. 6. Proofs and extra material can be found in the appendix of the full version of the paper.[2]

Related Work. As mentioned, we build on a series of works on max-plus and piecewise-linear approaches to discrete event systems [6,12,16,18]. More recent developments of this approach include invariant space methods [5,14,21] and models of resource sharing [17]. We also refer the reader to [23] for a survey. We point out that the Petri net model that we use is a continuous relaxation of classical discrete Petri nets, where fractional tokens are allowed. This model differs from the continuous timed Petri net model of [13,26], which is based on a system of piecewise-linear ODEs. However, as shown in [3], the two models share the same stationary regimes.

The idea of computing phase diagrams by enumerating policies was introduced in [1] and later applied in [2,10]. It presents two difficulties: 1) a number of policies which grows exponentially in the size of the system; 2) efficiently checking whether a given policy is strictly feasible. Previous methods for checking strict

---

[2] https://arxiv.org/abs/2505.02729.

feasibility already relied on polyhedral computations and lexicographic inequalities, but they reduced to an exponential number of (usual) linear inequality feasibility problems (see for instance [2,10]). We provide here the first feasibility check that returns in polynomial time. The challenge posed by the exponential number of policies seems irreducible. Indeed, stationary regimes are a generalization of fixed points of "positive" tropical polynomial systems. The existence of such a fixed point was shown to be NP-hard in [2]. Hence, we do not expect a general polynomial time algorithm for computing stationary regimes.

A classical theorem states that the dynamic programming equations of Markov decision processes admit an *invariant half-line* [22], another name for a stationary solution $z(t) = u + \rho t$. Moreover, the term ρ is uniquely determined. Theorem 3 partly extends this result to the case of "negative probabilities", showing that the throughput associated with a given policy is unique provided a semisimplicity condition is satisfied.

The modeling and staffing of emergency organizations, either emergency call centers, or medical emergency departments, has received much attention. These systems have been analyzed using probabilistic networks, particularly queuing networks, see e.g. [9,24]. The absence of monotonicity induced by the priority rules places these systems outside the scope of ordinary classes of exactly solvable models. In specific cases, "scaling limits" of queuing networks (limits of a family of discrete models with a scaling factor tending to infinity) have been obtained, leading to phase diagrams similar to those from our approach [9].

2 Petri Net Model of an Emergency Department

As a motivating example, we develop a Petri net model of an emergency department, illustrated in Fig. 1. This model captures the typical pathways followed by patients within the ED. It consists of the following steps: (i) administrative registration, (ii) triage, (iii) consultation, (iv) nursing care, (v) diagnostic tests, (vi) final exit consultation.

The model distinguishes the various types of human resources involved at each stage. Here, administrative registration is handled by medical secretaries, triage by a dedicated pool of nurses or physicians, consultations by junior or senior doctors, nursing care by nurses, and diagnostic tests by specialized technicians. We assume here that the exit consultation is ensured by senior doctors. Additionally, if a patient is initially consulted by a junior doctor, the pathway includes a step in which the junior doctor must validate the diagnosis or treatment with a senior doctor. Finally, certain steps may require material resources; for example, consultations and nursing care are typically conducted in dedicated treatment cubicles.

Petri nets provide a natural framework for modeling the synchronization between patients awaiting treatment and the availability of the necessary resources for each step of the care process. Patients, human resources, and material resources are represented by tokens that traverse the Petri net, passing through *places* (depicted by circular nodes) and *transitions* (solid rectangles).

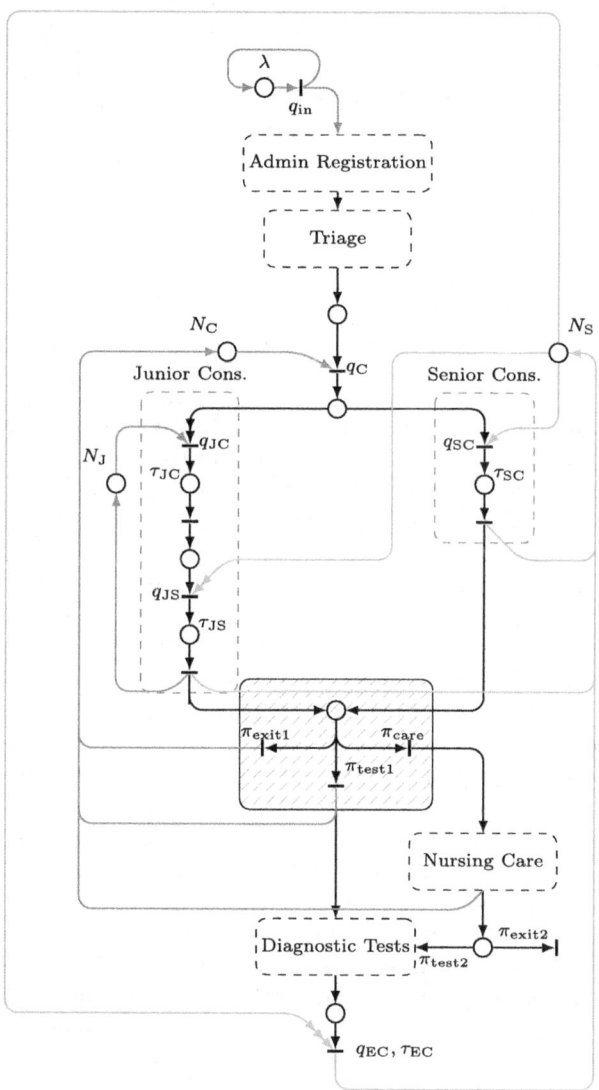

Fig. 1. Petri net model of an emergency department. Parameters prefixed by τ stand for holding times associated with places, those prefixed by π for routing proportions, and those prefixed by N for resources (initial marking).

Synchronization is modeled by the transitions. To illustrate this, consider the sub-Petri net shown in Fig. 2, which corresponds to a general task in the ED performed by a single, separate, category of resources. This pattern is replicated in the complete Petri net of Fig. 1 to represent administrative registration, triage, nursing care, and diagnostic tests. The place on the left in Fig. 2 represents the pool of available resources for the task, while the place at the top is a wait-

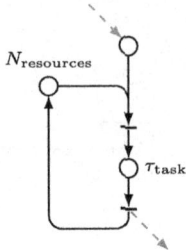

Fig. 2. Petri net of the administrative registration, triage, nursing, and exam procedures.

ing area for patients. As soon as each of the two places contains an available token, the top transition fires, consuming two upstream tokens and producing a single token in the downstream place. The latter corresponds to the task itself, and is associated with a holding time τ_{task}. Once this duration has elapsed, the token becomes available to fire the bottom transition, which, in turn, releases the resources back into the pool (via the left arc), and produces another token representing the patient (dashed arc), who proceeds to the next step in the pathway.

Another aspect of emergency departments that Petri nets naturally capture is the concurrent use of human and material resources. In the Petri net, this occurs when multiple transitions compete to consume tokens from a shared upstream place. Our model resolves this in two distinct ways, depending on the modeling needs: *proportion routing* or *priority routing*.

Proportion routing is illustrated by the gray-shaded box at the center of the Petri net in Fig. 1. The shared place collects patients exiting consultations with either junior or senior doctors who then follow one of three possible pathways depending on the severity of their condition: immediate exit, nursing care, or diagnostic tests. In the model, the token is routed to one of the three downstream transitions according to predefined proportions: π_{exit1} for immediate exit, π_{test1} for diagnostic tests, and π_{care} for nursing care. These three proportions sum up to 1.

Priority routing is illustrated by the circuit involving senior doctors, given by the arcs highlighted in orange in Fig. 1. Senior doctors are assigned to three distinct tasks, ordered by decreasing priority: exit consultation (transition q_{EC}), synchronization with junior doctors for diagnostic confirmation (q_{JS}), and initial consultation (q_{SC}). The level of priority is visually indicated by the number of arrowheads on the arcs leading to each corresponding transition: more arrowheads denote higher priority. This routing scheme specifies when the three downstream transitions are permitted to fire. For instance, transition q_{JS} can only fire if q_{EC} cannot (i.e., no patient is currently waiting for an exit consultation), regardless of whether patients are waiting for an initial consultation. We point out that priority rules can overlap. For example, patients waiting for initial consultation are assigned in priority to junior doctors (transition q_{JC}) over senior doctors (q_{SC}). This is consistent with the fact that senior doctors are primarily assigned to tasks requiring higher expertise and should only conduct initial consultations as a last resort, typically when the system becomes overcrowded.

The resources (human or material) of the emergency department are encoded in the *initial marking* of the Petri net, i.e., the number of tokens initially allocated in each place. When nonzero, this marking is indicated next to the corresponding place. In particular, N_C, N_J, and N_S represent the number of available cubicles, junior doctors, and senior doctors, respectively. The arrival of patients into the ED is modeled by the top component of the Petri net, which consists of a place with initial marking λ and a downstream transition that loops back to it, simulating a constant inflow of patients at rate λ.

Congestion Phase Diagram of the ED. Our goal is to assess the performance and congestion levels of an emergency department (ED) based on the available resources. As outlined in Sect. 1, we use the throughputs of transitions as a performance metric. Here, they correspond to the rates at which activities, such as consultations or diagnostic tests, are completed. We denote the throughput vector by ρ, where ρ_i represents the firing rate of transition q_i.

The congestion phase diagram provides a compact representation of how the performance of the ED evolves under varying resource allocations. It partitions the space of resources into polyhedral cells, each corresponding to a subset of resources that act as bottlenecks, limiting the flow of patients through the system. On each such cell, the throughput vector ρ is given by an explicit formula determined by the active bottlenecks.

As an illustration, Table 1 presents the congestion phase diagram generated by the algorithm developed in the following sections, applied to the Petri net model of the ED. Due to space constraints, the table restricts to the situation where only junior doctors, senior doctors, and cubicles are treated as potentially limiting resources (all other resources are assumed to be infinite, making the corresponding parts of the model fluid). Despite this simplification, these three resources capture the most intricate interactions in the system, particularly those arising from overlapping priority rules.

The first column of the table identifies the limiting resources and where they act as bottlenecks, the second exhibits the corresponding cell in the resource space, and the third provides the explicit expressions for the throughputs of the relevant transitions. Since the system dynamics are homogeneous (i.e., invariant under a simultaneous scaling of the arrival rate λ and the resources by the same factor), we represent every cell using the projective coordinates $(N_J/\lambda, N_S/\lambda, N_C/\lambda)$.

We refer to Sect. 6 for the analysis of the diagram as well as a short discussion of the ED model with all types of resources.

Table 1. The congestion phase diagram of the ED model of Fig. 1. We set $\pi_{\text{cont}} := \pi_{\text{exam1}} + \pi_{\text{care}}\pi_{\text{exam2}}$.

Bottleneck type	Phase region	Throughput expressions
Fluid phase		$\rho_C = \lambda$ $\rho_{JC} = \lambda$ $\rho_{SC} = 0$ $\rho_{JS} = \lambda$
Juniors limit junior cons.		$\rho_C = \lambda$ $\rho_{JC} = \frac{N_J}{\tau_{JC}+\tau_{JS}}$ $\rho_{SC} = \lambda - \frac{N_J}{\tau_{JC}+\tau_{JS}}$ $\rho_{JS} = \frac{N_J}{\tau_{JC}+\tau_{JS}}$
Cubicles limit senior cons. **Juniors** limit junior cons.		$\rho_C = \frac{(N_C-N_J)(\tau_{JC}+\tau_{JS})+N_J\tau_{SC}}{\Delta}$ $\rho_{JC} = \frac{N_J}{\tau_{JC}+\tau_{JS}}$ $\rho_{SC} = \frac{(N_C-N_J)(\tau_{JC}+\tau_{JS})-N_J\pi_{\text{care}}\tau_{\text{care}}}{\Delta}$ $\rho_{JS} = \frac{N_J}{\tau_{JC}+\tau_{JS}}$, where we set $\Delta := \pi_{\text{care}}\tau_{JC}\tau_{\text{care}} + \pi_{\text{care}}\tau_{JS}\tau_{\text{care}}$ $\qquad + \tau_{JC}\tau_{SC} + \tau_{SC}\tau_{JS}$
Juniors limit junior cons. **Seniors** limit senior cons.		$\rho_C = \frac{N_J(\tau_{SC}-\tau_{JS})+N_S(\tau_{JC}+\tau_{JS})}{\Delta'}$ $\rho_{JC} = \frac{N_J}{\tau_{JC}+\tau_{JS}}$ $\rho_{SC} = \frac{N_S(\tau_{JC}+\tau_{JS})-N_J(\pi_{\text{cont}}\tau_{EC}+\tau_{JS})}{\Delta'}$ $\rho_{JS} = \frac{N_J}{\tau_{JC}+\tau_{JS}}$, where we set $\Delta' := \pi_{\text{cont}}(\tau_{JC}\tau_{EC} + \tau_{EC}\tau_{JS})$ $\qquad + \tau_{JC}\tau_{SC} + \tau_{SC}\tau_{JS}$
Seniors limit junior-senior synch. (Juniors limit junior cons.).		$\rho_C = \frac{N_S}{\pi_{\text{cont}}\tau_{EC}+\tau_{JS}}$ $\rho_{JC} = \frac{N_S}{\pi_{\text{cont}}\tau_{EC}+\tau_{JS}}$ $\rho_{SC} = 0$ $\rho_{JS} = \frac{N_S}{\pi_{\text{cont}}\tau_{EC}+\tau_{JS}}$
Cubicles limit junior consultations.		$\rho_C = \frac{N_C}{\pi_{\text{care}}\tau_{\text{care}}+\tau_{JC}+\tau_{JS}}$ $\rho_{JC} = \frac{N_C}{\pi_{\text{care}}\tau_{\text{care}}+\tau_{JC}+\tau_{JS}}$ $\rho_{SC} = 0$ $\rho_{JS} = \frac{N_C}{\pi_{\text{care}}\tau_{\text{care}}+\tau_{JC}+\tau_{JS}}$
Seniors limit junior-senior synch. (Cubicles limit junior cons.).		$\rho_C = \frac{N_S}{\pi_{\text{cont}}\tau_{EC}+\tau_{JS}}$ $\rho_{JC} = \frac{N_S}{\pi_{\text{cont}}\tau_{EC}+\tau_{JS}}$ $\rho_{SC} = 0$ $\rho_{JS} = \frac{N_S}{\pi_{\text{cont}}\tau_{EC}+\tau_{JS}}$

3 Principle of the Algorithm

3.1 Piecewise Linear Dynamics of Petri Nets

We briefly recall how the piecewise linear dynamics of the form (D) can be explicitly constructed from a Petri net with priority rules. The counter functions are associated with transitions, i.e., $z_i(t)$ counts the number of times the transition q_i has fired up to and including time t. We denote by n the number of transitions in the Petri net. The set $\mathcal{A}_i$ then corresponds to the places located upstream the transition q_i. The parameter r_i^a is then set to the initial marking of the place a, i.e., it corresponds to the amount of resources allocated to this place. Holding times in the Petri net are gathered in the set $\mathcal{T}$. The way the parameters $(P_\tau^a)_{ij}$ are specified depends on the kind of routing used from the upstream place a. In the case of a proportion routing, $(P_\tau^a)_{ij}$ derives from the corresponding proportion. If the upstream place is subject to a priority rule, this parameter can also take ± 1 values. For a more detailed description, we refer to [1,2], where it is proved that an integer version of theÂădynamical system (D) (in which the counter functions are integer values and routing probabilities are replaced by preselection functions) is in correspondence with the semantics of the Petri net. The continuous setting we consider here, where counter functions collect fractions of tokens and can take real values, has been proved to be asymptotically tight under a scaling limit, for monotone systems [10, Chapter 2].

As an illustration, we provide in Fig. 3 a dynamical system that describes the part of the model of ED restricted to the interactions of patients with junior and senior doctors as well as cubicles (i.e., the dashed components in Fig. 1 are omitted). In this simplified model, we can reduce the dynamics to the five counter functions z_C, z_{JC}, z_{SC}, z_{JS} and z_{EC}, corresponding to the transitions q_C, q_{JC}, q_{SC}, q_{JS} and q_{EC} respectively. As shown in Fig. 3, the resources of the ED, namely N_J, N_S and N_C, appears in the "affine part" of the dynamics, i.e., in the parameters r_i^a, while the duration of tasks, the proportion of patients going to nursing care, exams, etc., are collected in the "linear part". The priority rules introduce subtractions of counter functions in some of the linear terms. As an example, the first term in the minimum specifying $z_{SC}(t)$ (third equation) represents the cumulative number of senior doctors that have been available from the senior doctor pool up to time t, i.e., $N_S + z_{JS}(t - \tau_{JS}) + z_{SC}(t - \tau_{SC}) + z_{EC}(t - \tau_{EC})$, to which we have subtracted the number $z_{JS}(t) + z_{EC}(t)$ of senior doctors that have been already assigned to the higher priority tasks, i.e., exit consultation and diagnosis validation with junior doctors.

3.2 Stationary Regimes and Lexicographic Constraints

We seek affine stationary regimes of the dynamics (D), i.e., solutions of form $z_i : t \mapsto u_i + \rho_i t$ for all $i \in [n]$. Substituting such solutions in (D) yields a system of lexicographic constraints over the pairs (ρ_i, u_i). Indeed, the terms achieving the minimum in (D) for all t large enough are fully determined by the lexicographic ordering: two affine functions $f(t) = w + \eta t$ and $g(t) = w' + \eta' t$

$$z_\text{C}(t) = \min\Big(\lambda t,\ N_\text{C} + (1 - \pi_\text{care})(z_\text{JS}(t - \tau_\text{JS}) + z_\text{SC}(t - \tau_\text{SC}))$$
$$+ \pi_\text{care}(z_\text{JS}(t - \tau_\text{JS} - \tau_\text{care}) + z_\text{SC}(t - \tau_\text{SC} - \tau_\text{care}))\Big)$$

$$z_\text{JC}(t) = \min\Big(N_\text{J} + z_\text{JS}(t - \tau_\text{JS}),\ z_\text{C}(t) - z_\text{SC}(t^-)\Big)$$

$$z_\text{SC}(t) = \min\Big(N_\text{S} + z_\text{JS}(t - \tau_\text{JS}) + z_\text{SC}(t - \tau_\text{SC}) + z_\text{EC}(t - \tau_\text{EC})$$
$$- z_\text{JS}(t) - z_\text{EC}(t),\ z_\text{C}(t) - z_\text{JC}(t)\Big)$$

$$z_\text{JS}(t) = \min\Big(N_\text{S} + z_\text{JS}(t - \tau_\text{JS}) + z_\text{SC}(t - \tau_\text{SC}) + z_\text{EC}(t - \tau_\text{EC})$$
$$- z_\text{SC}(t^-) - z_\text{EC}(t),\ z_\text{JC}(t - \tau_\text{JC})\Big)$$

$$z_\text{EC}(t) = \min\Big(N_\text{S} + z_\text{JS}(t - \tau_\text{JS}) + z_\text{SC}(t - \tau_\text{SC}) + z_\text{EC}(t - \tau_\text{EC})$$
$$- z_\text{SC}(t^-) - z_\text{JS}(t^-),\ \pi_{E1}(z_\text{JS}(t - \tau_\text{JS} - \tau_\text{test}) + z_\text{SC}(t - \tau_\text{SC} - \tau_\text{test}))$$
$$+ \pi_{E2}\pi_\text{care}(z_\text{JS}(t - \tau_\text{JS} - \tau_\text{care} - \tau_\text{test}) + z_\text{SC}(t - \tau_\text{SC} - \tau_\text{care} - \tau_\text{test}))\Big)$$

Fig. 3. The dynamical system over the counter functions of the ED Petri net.

satisfy $f(t) \leq g(t)$ for all t large enough if and only if $(\eta, w) \leq_\text{lex} (\eta', w')$, i.e., either $\eta < \eta'$, or $\eta = \eta'$ and $w \leq w'$. In this way, finding an affine stationary solution to (D) can be shown to be equivalent to solving the following system over the variables $(\rho, u) \in \mathbb{R}^n \times \mathbb{R}^n$

$$\forall i \in [n],\quad (\rho_i, u_i) = \operatorname*{lexmin}_{a \in \mathcal{A}_i} \Big([P^a]_i \rho,\ r_i^a + [P^a]_i u - [\bar{P}^a]_i \rho\Big) \tag{HL}$$

where lexmin stands for the minimum operator w.r.t. the lexicographic order $\leq_\text{lex}$, $P^a := \sum_{\tau \in \mathcal{T}} P_\tau^a$, $\bar{P}^a := \sum_{\tau \in \mathcal{T}} \tau P_\tau^a$, and $[M]_i$ stands for the ith row of the matrix M. We refer to [4, Lemma 7] for a proof.

As briefly explained in Sect. 1, our approach to solve the system (D) relies on policies. Formally, a *policy* $\sigma \colon [n] \to \cup_{i \in [n]} \mathcal{A}_i$ is a map such that $\sigma(i) \in \mathcal{A}_i$ for all $i \in [n]$. If $(\rho, u) \in \mathbb{R}^n \times \mathbb{R}^n$ is a solution of (HL), then there exists a policy σ such that, for all $i \in [n]$,

$$(\rho_i, u_i) = \Big([P^{\sigma(i)}]_i \rho,\ r_i^{\sigma(i)} + [P^{\sigma(i)}]_i u - [\bar{P}^{\sigma(i)}]_i \rho\Big),$$
$$(\rho_i, u_i) \leq_\text{lex} \Big([P^a]_i \rho,\ r_i^a + [P^a]_i u - [\bar{P}^a]_i \rho\Big) \quad \text{for all } a \in \mathcal{A}_i, a \neq \sigma(i). \tag{HL$_\sigma$}$$

Conversely, any solution to a system of the form (HL$_\sigma$) is also a solution to (HL), meaning that it corresponds to an affine stationary regime.

We point out that a policy associated with a stationary solution (ρ, u) (i.e., such that (ρ, u) is a solution of (HL$_\sigma$)) describes the set of the bottlenecks in the Petri net for this regime: for each $i \in [n]$, the transition q_i is bottlenecked by the upstream place associated with $a = \sigma(i) \in \mathcal{A}_i$.

As a consequence, determining the congestion phase diagram amounts to identifying, for every policy σ, the set C_σ of resources (r_i^a) for which the system (HL$_\sigma$) admits a solution (ρ, u) with $\rho \geq 0$. However, the trivial solution $(\rho_i, u_i) = (0, 0)$ for all $i \in [n]$, corresponding to a fully congested regime, is always feasible regardless of the policy (e.g., setting $r_i^{\sigma(i)} = 0$ for all i). More specifically, to facilitate the analysis of the performance of the system, we may be interested in the throughputs of a selection of transitions only. For instance, we may disregard some transitions like the input transition q_in in the ED model of Sect. 2 (whose throughput is supposedly nonzero except in a fully congested situation), or transitions corresponding to independent parts of the system. This motivates the introduction of a notion of *strictly feasible policies* w.r.t. a subset $I \subset [n]$ of transitions. These are the policies σ for which there exists a resource allocation (r_i^a) such that system (HL_σ) has a solution (ρ, u) with a throughput vector ρ that is nonnegative and such that at least one ρ_i for $i \in I$ is nonzero. The *congestion phase diagram* is then defined as the collection of cells C_σ where σ ranges over the strictly feasible policies. To compute this diagram, we propose to iterate over all policies σ, and for each one, determine whether it is strictly feasible and, if so, the corresponding cell C_σ and the set of stationary regimes defined by the solutions of system (HL_σ). This is the purpose of the techniques developed in Sects. 4 and 5.

4 Solving Lexicographic Inequalities in Polynomial Time

The systems (HL_σ) gives rise to the study of *lexicographic polyhedra* (or *lex-polyhedra* for short), which are defined as the set of solutions $x \in \mathbb{R}^N$ of finitely many *lexicographic linear inequalities* (*lex-inequalities* for short)

$$(\langle \alpha_i^1, x \rangle, \ldots, \langle \alpha_i^{d_i}, x \rangle) \leq_\text{lex} (b_i^1, \ldots, b_i^{d_i}) \;, \quad i \in [m] \tag{1}$$

where $\leq_\text{lex}$ stands for the lexicographic order over $\mathbb{R}^{d_i}$ ($d_i \geq 1$), the α_i^j are N-vectors ($i \in [m]$, $j \leq d_i$), and $\langle y, x \rangle := \sum_k y_k x_k$. The integer d_i is called the *depth* of the inequality. Note that depth-1 lex-inequalities precisely correspond to usual linear inequalities. Similarly, a usual linear equality can be simply encoded as a pair of opposed depth-1 lex-inequalities.

To the best of our knowledge, this way of extending polyhedra to a lexicographic setting has not been studied in the literature. Several works address lexicographic linear programming [19], in which only the objective function to be optimized is lexicographic, while the constraint set remains a standard polyhedron. Our lexicographic polyhedra can be seen as a generalization of this framework (sublevel sets of the objective function can be equivalently modeled as one lexicographic inequality), but the fact that multiple inequalities are lexicographic fundamentally changes the nature of the problem. Indeed, although lexicographic polyhedra are still convex sets, they can exhibit particular structures, even in the case of only two lexicographic inequalities, as illustrated in Fig. 4(a). Lex-polyhedra also encompass *non-necessarily closed polyhedra* [8] that have been widely applied in abstract interpretation to infer linear inequality program invariants, and that correspond to mixed systems of strict and non-strict linear inequalities. Indeed, a strict linear inequality $\langle \alpha, x \rangle < b$ can be equivalently encoded by a depth-2 lex-inequality $(\langle \alpha, x \rangle, 1) \leq_\text{lex} (b, 0)$.

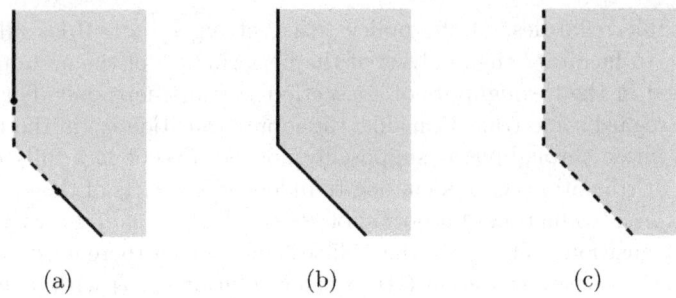

Fig. 4. (a) The lex-polyhedron defined by $(x_1, x_2) \geq_{\text{lex}} (0, 1)$, $(x_1, x_1) \geq_{\text{lex}} (-x_2, 1)$. In the boundary, solid lines and points are included, while dash lines are excluded. (b) The max-front polyhedron $\mathcal{P}^{f^*}$ and (c) its relative interior.

To make the exposition of the algorithm easier, we consider a lex-polyhedron $\mathcal{P}_{\text{lex}}$ defined by a system of lex-inequalities in *standard form*, i.e.,

$$Ax + s = b,$$
$$\forall i \in [m], (s_i^1, \ldots, s_i^{d_i}) \geq_{\text{lex}} 0 \qquad \text{(Lex-SF)}$$

where the variables are now $x = (x_1, \ldots, x_N)$ and $s = (s_i^j)_{i \in [m], j \in [d_i]}$, $A \in \mathbb{R}^{D \times N}$ and $b \in \mathbb{R}^D$, and $D := \sum_{i=1}^m d_i$ is the total depth. In consequence, $\mathcal{P}_{\text{lex}}$ is a subset of $\mathbb{R}^{N+D}$. We remark that reformulating systems of lex-inequalities of the form (1) to the system (Lex-SF) is simply achieved by introducing the *slack variables* s_i^j in every component of the lex-inequalities, and constraining these groups of variables to be nonnegative in the lexicographic sense. We refer to Example 1 at the end of the section for an illustration.

A *front* is a m-vector f such that $f_i \in [d_i + 1]$ for all $i \in [m]$. We denote $s^{<f}$ the vector consisting of the variables s_i^j for $i \in [m]$ and $j < f_i$, and s^f that of the variables $s_i^{f_i}$ for the $i \in [m]$ such that $f_i \leq d_i$. We set

$$\mathcal{P}^f := \left\{ (x, s) \in \mathbb{R}^{N+D} : Ax + s = b, \ s^{<f} = 0, \ s^f \geq 0 \right\}.$$

The sets $\mathcal{P}^f$ are (closed) convex polyhedra. Our aim is to approximate the lex-polyhedron $\mathcal{P}^{\text{lex}}$ by means of these polyhedra. Fronts can be ordered componentwise, i.e., $f \leq f'$ if $f_i \leq f'_i$ for all $i \in [m]$. Remark that $\mathcal{P}^{f'} \subset \mathcal{P}^f$ as soon as $f \leq f'$.

Lemma 1. *The front f^* defined by*

$$f_i^* := \max \left\{ j \in [d_i + 1] : \forall (x, s) \in \mathcal{P}^{\text{lex}}, \ s_i^1 = \cdots = s_i^{j-1} = 0 \right\} \text{ for all } i \in [m]$$

is the greatest front f such that $\mathcal{P}^{\text{lex}} \subset \mathcal{P}^f$.

We call f^* the *max-front*. Proposition 1 below states that the *max-front polyhedron* $\mathcal{P}^{f^*}$ is a tight approximation of $\mathcal{P}^{\text{lex}}$, in the sense that $\mathcal{P}^{\text{lex}}$ stands

```
 1: procedure COMPUTEMAXFRONT
 2:     f := (1, ..., 1)
 3:     repeat
 4:         I := ∅
 5:         for i ∈ [m] such that f_i ≤ d_i do
 6:             if sup {s_i^{f_i} : (x, s) ∈ P^f} ≤ 0 then I := I ∪ {i}
 7:         done
 8:         for i ∈ I do
 9:             f_i := f_i + 1
10:         done
11:     until I = ∅
12:     return f
13: end
```

Fig. 5. Computing the max-front

between $\mathcal{P}^{f^*}$ and its relative interior. An illustration is given in Fig. 4(b) and (c). We recall that the *affine hull* of a set S is the smallest (inclusionwise) affine subspace containing it, and that the *relative interior* relint S of S is its interior w.r.t. the subspace topology on its affine hull. For instance, the relative interior of a point is the point itself. The relative interior of a (closed) line segment between two points is the open line segment between the two points, etc. We note that the relative interior of a nonempty convex set S is always nonempty, and that its closure is equal to the closure of S. We refer to [27, Chapter 6] for a complete account on the topological properties of convex sets.

Proposition 1. *We have*

$$\text{relint } \mathcal{P}^{f^*} \subset \mathcal{P}^{\text{lex}} \subset \mathcal{P}^{f^*}.$$

Equivalently, $\mathcal{P}^{f^*}$ is the closure of $\mathcal{P}^{\text{lex}}$.

We now introduce the algorithm COMPUTEMAXFRONT that computes the front f^*; see Fig. 5. The principle of the algorithm is the following. It starts from the all-1 front, and then repeatedly increment the entries f_i of the current front f as soon as sup $\{s_i^{f_i} : (x, s) \in \mathcal{P}^f\} \leq 0$. The latter test is carried out by solving the linear program

$$\text{Maximize} \quad s_i^{f_i} \quad \text{subject to} \quad (x, s) \in \mathcal{P}^f,$$

and checking if the value of this linear program is nonpositive (by convention, this value is $-\infty$ is $\mathcal{P}^f$ if empty). This can be done in polynomial time [28, Chapter 13]. We obtain the following result:

Theorem 1. *The algorithm COMPUTEMAXFRONT computes the max-front f^* in polynomial time, by solving at most mD linear programs.*

Remark 1. As a corollary of Theorem 1, we can check in polynomial time if the lex-polyhedron $\mathcal{P}^{\text{lex}}$ is empty or not. Indeed, by Proposition 1, $\mathcal{P}^{\text{lex}}$ is nonempty if and only if the max-front polyhedron $\mathcal{P}^{f^*}$ is nonempty. The latter condition can be verified by first computing f^*, and then testing the emptiness of $\mathcal{P}^{f^*}$ in polynomial time using linear programming [28, Theorem 10.4].

Example 1. We consider a lex-polyhedron over the variables x, y, z and defined by the two lex-inequalities $(y, x) \leq_{\text{lex}} (x, z)$ and $(x, z) \leq_{\text{lex}} (y, y-1)$. It is easy to check that this lex-polyhedron is empty. In standard form, the system of constraints writes as the following set of equalities

$$y + s_1^1 = x, \quad x + s_1^2 = z, \quad x + s_2^1 = y, \quad z + s_2^2 = y - 1, \tag{2}$$

with the lex-inequalities $(s_1^1, s_1^2) \geq_{\text{lex}} 0$ and $(s_2^1, s_2^2) \geq_{\text{lex}} 0$.

We now give the details of the execution of the algorithm COMPUTEMAXFRONT. Initially, the front is given by $f = (1, 1)$, and the polyhedron $\mathcal{P}^f$ is defined by the equalities in (2) and the two inequalities $s_1^1 \geq 0$ and $s_2^1 \geq 0$. The first and third equalities entail that $s_1^1 = -s_2^1$ and so the supremum of s_1^1 as well as s_2^1 over $\mathcal{P}^f$ is 0 ($\mathcal{P}^f$ can be checked to be nonempty). Thus, the front is updated to $(2, 2)$ after the first iteration of the loop at Line 3, and the updated polyhedron $\mathcal{P}^f$ is now given by the equalities in (2) together with $s_1^1 = s_2^1 = 0$, and the inequalities $s_1^2 \geq 0$ and $s_2^2 \geq 0$. The equalities $s_1^1 = s_2^1 = 0$ implies that $x = y$, and by the second and fourth equalities of (2), we get that $z = x + s_1^2 \geq x = y$, and $y = z + s_2^2 + 1 \geq z + 1$. We deduce that $\mathcal{P}^f$ is empty. In this case, the supremum of s_1^2 as well as s_2^2 over $\mathcal{P}^f$ is $-\infty$, and so the front is updated to $(3, 3)$. This is indeed the max-front, since the original lex-polyhedron is empty.

Checking the Strict Feasibility of a Policy. We now come back to our original problem, i.e., determining if a policy σ is strictly feasible, and computing the associated cell C_σ of the congestion diagram. Recall that σ is strictly feasible if and only if, for some choice of resources (r_i^a), there is a solution $(\rho, u) \in \mathbb{R}^n \times \mathbb{R}^n$ to the lex-inequality system (HL$_\sigma$) such that $\rho \geq 0$ and $\rho \neq 0$. This leads to introducing the lex-polyhedron $\mathcal{P}_\sigma^{\text{lex}}$ over the stationary regime variables (ρ_i, u_i) ($i \in [n]$) and resource variables (r_i^a), defined by the equality and depth-2 lex-inequality constraints in the system (HL$_\sigma$) as well as n depth-1 inequalities $\rho_i \geq 0$ for $i \in [n]$. Once turned into standard form, every inequality corresponds to some entry of the max-front f^*. For convenience, we denote by $f^*_{\sigma_i}$ the entry of corresponding to the inequality $\sigma_i \geq 0$ ($i \in [n]$). By definition of the max-front, the policy σ is strictly feasible w.r.t. to a set $I \subset [n]$ of transitions if and only if $f^*_{\sigma_i} = 1$ for some $i \in I$. Moreover, denoting π the projection map onto the resource variables (r_i^a), the cell C_σ is precisely the image under π of the lex-polyhedron $\mathcal{P}_\sigma^{\text{lex}}$. We remark $\pi(\mathcal{P}_\sigma^{f^*})$ is a (closed) convex polyhedron, as a projection of a polyhedron. In consequence, we get the following theorem:

Theorem 2. *We can determine in polynomial time whether a policy σ is strictly feasible. Moreover, if σ is strictly feasible, we have*

$$\text{relint } \pi(\mathcal{P}_\sigma^{f^*}) \subset C_\sigma \subset \pi(\mathcal{P}_\sigma^{f^*}),$$

or, equivalently, the closure of C_σ is given by the convex polyhedron $\pi(\mathcal{P}_\sigma^{f^*})$.

5 Uniqueness of the Throughput Associated with a Policy

In this section, we consider a fixed policy σ, and we denote for brevity P_τ (resp. P and $\bar{P}$) the matrix with rows $[P_\tau^{\sigma(i)}]_i$ (resp. $[P^{\sigma(i)}]_i$ and $[\bar{P}^{\sigma(i)}]_i$). (We refer to Sect. 3.2 for the definition of the latter matrices.) Similarly, we denote by r the vector with entries $r_i^{\sigma(i)}$. We consider the affine dynamics determined by the policy σ, i.e.,

$$z_i(t) = r_i + \sum_{j,\tau} (P_\tau)_{ij} z_j(t-\tau) \;, \tag{3}$$

and look for stationary solutions $z(t) = u + \rho t$ $(u, \rho \in \mathbb{R}^n)$. Equivalently, this amounts to the system $u = r + Pu - \bar{P}$ and $\rho P = \rho$, which corresponds to the equality constraints from (HL_σ). We make the following assumption, which is among the sufficient conditions for the existence of stationary solutions identified in a previous work [4]:

Assumption 1. *The matrix P has 1 as an eigenvalue, and this eigenvalue is semisimple.*

Recall that an eigenvalue is *semisimple* if its algebraic and geometric multiplicities coincide. We next show that under Assumption 1, the throughput vector ρ associated with such a stationary solution is unique, and we provide an explicit formula in terms of the eigenvectors of the matrix P.

Denoting by q the multiplicity of the eigenvalue 1, and using the semisimplicity assumption, we can find a basis $v^1, \ldots, v^q$ of right eigenvectors of P, and a basis $m^1, \ldots, m^q$ of left eigenvectors of P, so that $Pv^k = v^k$ and $m^l P = m^l$, and such that $m^k v^l = \delta_{kl}$, for all $k, l \in [q]$, where δ_{kl} denotes Kronecker's delta function. See [20, §5,4], especially Eq. (5.36) on p. 41. Then, we define the *aggregated matrix* $\bar{M}(\tau) := (m^k \bar{P} v^l)_{k,l\in[q]} \in \mathbb{R}^{q\times q}$, and the vector $\bar{r} := (m^k r)_{k\in[q]} \in \mathbb{R}^q$.

Theorem 3. *Let (ρ, u) denote any stationary solution of the dynamics (3). Then, for generic values of the parameters $\tau \in \mathcal{T}$, the vector ρ is uniquely determined, being given by*

$$\rho = \sum_{k\in[q]} \lambda_k v^k \qquad \text{where} \qquad \lambda = (\lambda_k)_{k\in[q]} = (\bar{M}(\tau))^{-1} \bar{r} \;. \tag{4}$$

Theorem 3 actually extends a uniqueness result of the throughput that appeared in the setting of continuous timed Petri net [3, Theorem 7], by providing an explicit and more interpretable formula in terms of the spectral decomposition of the matrix P.

In the full version of the paper,[3] we show in Appendix B how Theorem 3 extends classical results from semi-Markov decision process theory to the current nonmonotone setting (i.e., where the matrix P may contain negative entries).

---

[3] https://arxiv.org/abs/2505.02729.

6 Application to Emergency Department Models

Implementation. The front-advancing algorithm COMPUTEMAXFRONT is implemented in Python. System equations of the form (D) are symbolically defined using SymPy, from which the corresponding lexicographic systems of the form (HL_σ) are constructed. Linear programs for checking the front are solved, after assigning numerical values to the time delay and routing proportion parameters, using the HiGHS solver, via the `linprog` function in the `scipy.optimize` package. Given a strictly feasible policy, we also use SymPy's `linsolve` function to symbolically solve the system of linear equations and obtain the associated throughput vector ρ.

The inequalities resulting from the max-front polyhedron are used to generate a congestion phase diagram with the Parma Polyhedra Library (PPL) interface for Python (pyppl). It is then projected onto the resource variables, from which we derive the vertices of the cells.

The Congestion Phase Diagram Revisited. For all the applications considered here, we consider strictly feasible policies w.r.t. the set I of transitions distinct from the input transition.

As described in Sect. 2, we applied the algorithm to the ED model, assuming only junior/senior doctors and cubicles as limiting resources, to produce the congestion phase diagram of Table 1. The matrices induced by the policies of this model satisfy the semisimplicity Assumption 1, implying that the throughputs for each policy are uniquely given as a function of the resources (Theorem 3), as reported in the third column. The values of the numerical parameters were fixed as follows: $\pi_{\text{care}} = 0.4$, $\pi_{\text{test1}} = 0.2$, $\pi_{\text{exit1}} = 0.4$, $\pi_{\text{test2}} = 0.7$, $\pi_{\text{exit2}} = 0.3$, $\tau_{\text{JC}} = 4$, $\tau_{\text{JS}} = 1$, $\tau_{\text{SC}} = 3$, $\tau_{\text{EC}} = 2$, $\tau_{\text{care}} = 5$, and $\tau_{\text{test}} = 6$.

Out of the 32 total policies, only 16 are strictly feasible w.r.t. I. Eight strictly feasible policies correspond to lower-dimensional faces, reflecting edge-case resource configurations that are unlikely in practice and not informative for decision-making. Two map to the same cell, a phenomenon further discussed below. In this way, we arrive at a diagram with 7 distinct full-dimensional cells, as reported in Table 1.

We now provide an interpretation of some cells in the diagram. The first two cells correspond to fluid regimes where all incoming patients λ are handled either by junior consultations (first cell) or, when the number of junior doctors decreases, by a mix of junior and senior consultations (second cell). Reducing the number of cubicles brings us to the third cell, the first congested regime, where there is a build-up of patients waiting to be allocated a cubicle. Here, the bottleneck in senior consultations shifts from doctor availability to the availability of patients awaiting consultation in a cubicle. This bottleneck affects senior consultations before the junior ones because the priority rule favors assigning patients to junior doctors. Thus, the logic of the priority rule is directly reflected in the structure of the congestion phase diagram. The remaining regimes reveal a non-trivial structure. For instance, we were surprised to find that there is no regime in the congestion phase diagram in which the exit consultations are bottlenecked by the senior doctors. This is explained by the fact that the flow of patients to the exit consultations is already capped by the sum of the lower-priority junior-senior synchronizations and senior consultations. Two more things remain to be noted. First, distinct policies may yield the same cell and stationary regime (ρ, u). This occurs in the last cell which corresponds to two policies. The reason for this is that the lack of senior doctors creates a bottleneck for the junior-senior synchronization which, due being higher priority than the senior consultations, renders the two different choices for bottlenecks on the senior consultations (upstream flow, or availability of senior doctors) equivalent. We ascribe this to the special structure of the model. Second, distinct cells may yield the same throughput vector ρ, such as the case of the fifth and last cells. In both, the lack of senior doctors is limiting the release of cubicles leading to the same dynamics for the system. However, we observe in the experiments that these policies have different vectors u (representing waiting rooms or storage levels associated with each place).

For the full model of ED, we obtain 352 strictly feasible policies w.r.t. I out of 512 total, yielding 76 full-dimensional cells where 64 of them are distinct. As an additional test, we have applied our algorithm to the model "EMS-B" of emergency call center of Samu de Paris (medical emergency aid) that was presented in [2, Figure 5]. The algorithm returns the same congestion diagram as the one that was previously computed by hand.

7 Concluding Remarks

We have developed a general algorithm to compute congestion diagrams of piecewise linear dynamical systems with priorities, such as the ones modeling emergency departments. Based on the allocation of human and material resources in the ED, the diagram allows for the identification of system congestion, the nature of bottlenecks, and provides a quantitative estimate of the additional resources required to achieve a fluid phase handling all patients. The heart of this algorithm is a method to check in polynomial time whether a given policy is (strictly) feasible, whereas earlier methods require an exponential time. A remaining open problem is to develop a polynomial time output sensitive algorithm to compute all realizable policies, i.e., an algorithm whose execution time is possibly exponential but that remains polynomially bounded in the number of strictly feasible policies. Another open problem is to reinforce Theorem 3: this result shows that every strictly feasible policy determines a unique throughput vector. However, one may not exclude that for a given allocation of resources, several policies may be strictly feasible and lead to different values of ρ. This pathology does not appear in the examples we analyzed; it remains to identify suitable assumptions implying that it is so. Finally, another open problem concerns the analysis of nonstationary behaviors: the conditions for the existence of the limit $\lim_t z(t)/t$ are still not understood beyond the monotone case.

Acknowledgement. The authors were partially supported by the URGE project of the Bernoulli Lab, joint between AP-HP & INRIA. They thank the members of the project, especially Youri Yordanov, Quentin Delannoy, Judith Leblanc, and Christine Fricker.

Data Availibility Statement. Notebooks containing the models, scripts, and tools to reproduce our experimental results are archived and publicly available at DOI 10.5281/zenodo.15724910.

References

1. Allamigeon, X., Bœuf, V., Gaubert, S.: Performance evaluation of an emergency call center: tropical polynomial systems applied to timed petri nets. In: Sankaranarayanan, S., Vicario, E. (eds.) Formal Modeling and Analysis of Timed Systems, pp. 10–26. Springer International Publishing, Cham (2015)
2. Allamigeon, X., Boyet, M., Gaubert, S.: Piecewise affine dynamical models of timed petri nets – application to emergency call centers. Fundamenta Informaticae **183**(3-4), 169–201 (2021). https://hal.archives-ouvertes.fr/hal-02550006
3. Allamigeon, X., Bœuf, V., Gaubert, S.: Stationary solutions of discrete and continuous petri nets with priorities. Perform. Eval. **113**, 1–12 (2017)
4. Allamigeon, X., Capetillo, P., Gaubert, S.: Stationary regimes of piecewise linear dynamical systems with priorities. In: Proceedings of the 28th ACM International Conference on Hybrid Systems: Computation and Control. HSCC 2025, Association for Computing Machinery, New York, NY, USA (2025)
5. Animobono, D., Scaradozzi, D., Zattoni, E., Perdon, A.M., Conte, G.: The model matching problem for max-plus linear systems: a geometric approach. IEEE Trans. Autom. Control **68**(6), 3581–3587 (2023)
6. Baccelli, F., Cohen, G., Olsder, G.J., Quadrat, J.P.: Synchronization and Linearity. Wiley, Hoboken (1992)
7. Baccelli, F., Foss, S.: Moments and tails in monotone-separable stochastic networks. Ann. Appl. Probab. **14**(2), 612–650 (2004). http://www.jstor.org/stable/4140422
8. Bagnara, R., Hill, P.M., Zaffanella, E.: Not necessarily closed convex polyhedra and the double description method. Formal Aspects Comput. **17**(2), 222–257 (2005)
9. Bœuf, V., Robert, P.: A stochastic analysis of a network with two levels of service. Queueing Syst. **92**(3–4), 30 (2019)
10. Boyet, M.: Piecewise Affine Dynamical Systems Applied to the Performance Evaluation of Emergency Call Centers. Ph.D. thesis, École polytechnique, Institut polytechnique de Paris (2022). http://www.theses.fr/2022IPPAX031/document
11. Cohen, G., Gaubert, S., Quadrat, J.: Max-plus algebra and system theory: where we are and where to go now. Annu. Rev. Control. **23**, 207–219 (1999). https://doi.org/10.1016/S1367-5788(99)90091-3
12. Cohen, G., Gaubert, S., Quadrat, J.P.: Asymptotic throughput of continuous timed Petri nets. In: Proceedings of the 34th Conference on Decision and Control. New Orleans (1995)
13. David, R., Alla, H.: Timed Continuous Petri Nets, pp. 159–229. Springer, Heidelberg (2010). https://doi.org/10.1007/978-3-642-10669-9_5

14. Di Loreto, M., Gaubert, S., Katz, R.D., Loiseau, J.J.: Duality between invariant spaces for max-plus linear discrete event systems. SIAM J. Control. Optim. **48**(8), 5606–5628 (2010). https://doi.org/10.1137/090747191
15. Farhi, N., Goursat, M., Quadrat, J.P.: Piecewise linear concave dynamical systems appearing in the microscopic traffic modeling. Linear Algebra Appl. **435**(7), 1711–1735 (2011). https://doi.org/10.1016/j.laa.2011.03.002
16. Gaujal, B., Giua, A.: Optimal stationary behavior for a class of timed continuous petri nets. Automatica **40**(9), 1505–1516 (2004)
17. Goltz, P., Schafaschek, G., Hardouin, L., Raisch, J.: Optimal output feedback control of timed event graphs including disturbances in a resource sharing environment. IFAC-PapersOnLine **55**(28), 188–195 (2022)
18. Heidergott, B., Olsder, G.J., van der Woude, J.: Max-plus at Work. Princeton Univ, Press (2005)
19. Isermann, H.: Linear lexicographic optimization. Oper. Res. Spektrum **4**(4), 223–228 (1982). https://doi.org/10.1007/BF01782758
20. Kato, T.: Perturbation Theory for Linear Operators. Springer, Heidelberg (1995)
21. Katz, R.D.: Max-plus (a, b)-invariant spaces and control of timed discrete-event systems. IEEE Trans. Autom. Control **52**(2), 229–241 (2007)
22. Kohlberg, E.: Invariant half-lines of nonexpansive piecewise-linear transformations. Math. Oper. Res. **5**(3), 366–372 (1980)
23. Komenda, J., Lahaye, S., Boimond, J.L., Boom, T.: Max-plus algebra and discrete event systems. IFAC-PapersOnLine **50**(1), 1784–1790 (2017)
24. L'Ecuyer, P., Gustavsson, K., Olsson, L.: Modeling bursts in the arrival process to an emergency call center. In: Rabe, M., Juan, A.A., Mustafee, N., Skoogh, A., Jain, S., Johansson, B. (eds.) Proceedings of the 2018 Winter Simulation Conference (2018)
25. Plus, M.: Max-plus-times linear systems. In: Blondel, V.D., Sontag, E.D., Vidyasagar, M., Willems, J.C. (eds.) Open Problems in Mathematical Systems and Control Theory, pp. 167–170. Springer (1999). max Plus is a collective name for M. Akian, G. Cohen, S. Gaubert, J.P. Quadrat and M. Viot
26. Recalde, L., Haddad, S., Silva, M.: Continuous petri nets: expressive power and decidability issues. In: Namjoshi, K.S., Yoneda, T., Higashino, T., Okamura, Y. (eds.) ATVA 2007. LNCS, vol. 4762, pp. 362–377. Springer, Heidelberg (2007). https://doi.org/10.1007/978-3-540-75596-8_26
27. Rockafellar, R.T.: Convex Analysis. Mathematical Series, Princeton University Press, Princeton (1970)
28. Schrijver, A.: Theory of Linear and Integer Programming. John Wiley & Sons Inc, New York (1986)

Author Index

A
Aldini, Alessandro 293
Allamigeon, Xavier 487
Ábrahám, Erika 1

B
Ballarini, Paolo 409
Barbot, Benoît 95, 429
Basset, Nicolas 95
Bernardo, Marco 293
Blohm, Pauline 389
Bouët-Willaumez, Olivier 429
Brand, Sebastiaan 371
Bruyère, Véronique 42
Budde, Carlos E. 21, 83

C
Capetillo, Pascal 487
Carnevali, Laura 21
Cry, Pierre 409

D
Dahlsen-Jensen, Mikael Bisgaard 314
Dang, Thao 95
Dang, Thi Kim Nhung 256
Dengler, Gabriel 21
Donzé, Alexandre 95

E
Esposito, Andrea 293
Esposito, Marco 95

F
Fievet, Baptiste 314
Finkel, Alain 237
Fitzsimmons, Maxwell 333
Fougea, Gaspard 237

G
Gall, Pascale Le 409
Garg, Abhinav 353
Garhewal, Bharat 42
Gaubert, Stéphane 487
Gerlach, Lina 1
Glesner, Sabine 176
Gros, Timo P. 134
Gürtler, Tobias 115

H
Haddad, Serge 237
Haesaert, Sofie 62
Hamscher, Anja 157
Hartmanns, Arnd 21, 83, 134
Herber, Paula 389
Hoese, Ivo 134
Horváth, András 409

J
Jain, Shreyas 237

K
Kaminski, Benjamin Lucien 115
Kogel, Paul 176
Kovács, Laura 275

L
Laarman, Alfons 371
Le Coënt Coënt, Adrien 429
Liu, Jun 333
Löding, Christof 1
Lopuhaä-Zwakenberg, Milan 256

M
Marin, Andrea 219
Meggendorfer, Tobias 83, 195
Menaschè, Daniel 219
Meyer, Joshua 134

Michel, Fabian 449
Mukund, Madhavan 353
Müller, Nicola J. 134

N
Nadali, Alireza 468
Nickovic, Dejan 95

O
Olliaro, Diletta 219

P
Pekergin, Nihal 429
Pérez, Guillermo A. 42
Peterseim, Benedikt 256
Petrucci, Laure 314

Q
Quist, Arend-Jan 371

R
Remke, Anne 389
Rossi, Sabina 219
Roy, Adwitee 353

S
Schmitt, Jens B. 157
Schön, Oliver 62
Schulz, Felix 389

Schwabe, Wolffhardt 176
Siegle, Markus 449
Soudjani, Sadegh 62
Srivathsan, B. 353
Staquet, Gaëtan 42
Stoelinga, Mariëlle 256

T
Trivedi, Ashutosh 468

V
Vaandrager, Frits W. 42
van de Pol, Jaco 314
van Dijk, Richard M. K. 371
Viswanathan, Gautham 353

W
Weininger, Maximilian 83, 195
Wienhöft, Patrick 83, 195
Wildberger, Lukas 157
Willemsen, Lisa 389
Winkler, Lorenz 275
Wolf, Verena 134

Y
Ye, Lina 237

Z
Zamani, Majid 468

Made in the USA
Monee, IL
03 May 2026